MW01518929

Suspensions: Fundamentals and Applications in the Petroleum Industry

ADVANCES IN CHEMISTRY SERIES 251

Suspensions: Fundamentals and Applications in the Petroleum Industry

Laurier L. Schramm, EDITOR
Petroleum Recovery Institute

American Chemical Society, Washington, DC

Library of Congress Cataloging-in-Publication Data

Suspensions: Fundamentals and applications in the petroleum
    industry / Laurier L. Schramm, editor.
        p.   cm.—(Advances in chemistry series, ISSN 0065–
    2393; 251)
        Includes bibliographical references and index.
        ISBN 0–8412–3136–2
        1. Particles.   2. Suspensions (Chemistry)   3. Petroleum
    engineering.
        I. Schramm, Laurier Lincoln.   II. Series.
    QD1.A355   no. 251
    [TP156.P3]
    540—dc20
    [665.5]   95–41569
                                                         CIP

The paper used in this publication meets the minimum requirements of American National Standard for Information Sciences—Permanence of Paper for Printed Library Materials, ANSI Z39.48–1984.   ∞

PRINTED IN THE UNITED STATES OF AMERICA

Advisory Board

Advances in Chemistry Series

FOREWORD

The ADVANCES IN CHEMISTRY SERIES was founded in 1949 by the American Chemical Society as an outlet for symposia and collections of data in special areas of topical interest that could not be accommodated in the Society's journals. It provides a medium for symposia that would otherwise be fragmented because their papers would be distributed among several journals or not published at all.

Papers are reviewed critically according to ACS editorial standards and receive the careful attention and processing characteristic of ACS publications. Volumes in the ADVANCES IN CHEMISTRY SERIES maintain the integrity of the symposia on which they are based; however, verbatim reproductions of previously published papers are not accepted. Papers may include reports of research as well as reviews, because symposia may embrace both types of presentation.

ABOUT THE EDITOR

LAURIER L. SCHRAMM is the manager for research and technology at the Petroleum Recovery Institute and adjunct associate professor of chemistry at the University of Calgary, where he lectures in applied colloid and interface chemistry. Dr. Schramm received his B.Sc. (Hons.) in chemistry from Carleton University in 1976 and Ph.D. in physical and colloid chemistry in 1980 from Dalhousie University, where he studied as a Killam and NRC Scholar. From 1980 to 1988 he held research positions with Syncrude Canada Ltd. in its Edmonton Research Centre.

His research interests have included many aspects of colloid and interface science applied to the petroleum industry, including research into mechanisms of processes for the improved recovery of light, heavy, or bituminous crude oils, such as in situ foam, polymer or surfactant flooding, and surface hot water flotation from oil sands. These mostly experimental investigations have involved the formation and stability of dispersions (foams, emulsions, and suspensions) and their flow properties, electrokinetic properties, interfacial properties, phase attachments, and the reactions and interactions of surfactants in solution.

Dr. Schramm is a Fellow of the Chemical Institute of Canada and served on the Local Section Executive. He is a member of the American Chemical Society and is a director of the Association of the Chemical Profession of Alberta. He has written more than 80 scientific publications and patents. This is his fourth ACS book, following *Emulsions: Fundamentals and Applications in the Petroleum Industry*, *Foams: Fundamentals and Applications in the Petroleum Industry*, and *The Language of Colloid and Interface Science*.

CONTENTS

GLOSSARY AND INDEXES

PREFACE

THIS BOOK PROVIDES AN INTRODUCTION to the nature, formation and occurrence, stability, propagation, and uses of suspensions in the petroleum industry. It is aimed at scientists and engineers who may encounter suspensions or apply suspensions, whether in process design, petroleum production, or research and development. The primary focus of the book is on the applications of the principles of particle dispersions, and it includes attention to practical processes and problems. Books available up to now are either principally theoretical (such as the colloid chemistry texts) or they focus on suspensions in general (like van Olphen's[1] classic book on clay colloid chemistry). The applications of, or problems caused by, suspensions in the petroleum industry area are quite diverse and have a great practical importance. The area contains a number of problems of more fundamental interest as well. Suspensions can be found, may require treatment, or may be applied to advantage in many parts of the petroleum production process: in reservoirs, in oilwells, and in surface processing operations. In these cases the suspensions may occur naturally or by design, and may be desirable or undesirable. In each case the presence or absence of dispersed particles can determine both the economic and technical successes of the industrial process concerned.

In this volume, a wide range of authors' expertise and experiences have been brought together to yield the first suspension book that focuses on the application and occurrence of suspensions in the petroleum industry. This broad range of authors' expertise has allowed for a variety of particle technology application areas to be highlighted. These areas serve to emphasize the different methodologies that have been successfully applied and also the wide variety of situations that occur in well-bores, porous media, flotation vessels, tailings ponds, and process plants. The material presented does not assume a knowledge of colloid chemistry, and thus the initial emphasis is placed on a review of the basic concepts important to understanding suspensions. As such, it is hoped that the book will be of interest to senior undergraduate and graduate students in science and engineering as well because topics such as this one are not normally part of university curricula.

Although the aim of the book is to provide an introduction to the field, it does so in a very applications-oriented manner. The first group

[1] van Olphen, H. *An Introduction to Clay Colloid Chemistry*, 2nd ed.; Wiley Interscience: New York, 1977.

of chapters sets out the fundamental principles in suspensions. In Chapters 1–5 the reader is introduced to suspension formation and stability, characterization, and flow properties. Chapters 6–9 introduce the reader to suspensions in porous media. Next, two groups of chapters address some of the more familiar petroleum industry occurrences of suspensions. Chapters 10–12 are aimed at the use of suspensions in near-well and oilwell processes such as the use of suspensions as drilling fluids. Chapters 13 and 14 address the occurrence and impact of suspensions in surface operations. The glossary is a comprehensive and fully cross-referenced dictionary of suspension terminology. A common theme in the chapters is the use of the fundamental concepts in combination with actual commercial and pilot-scale process experiences.

Overall, the book illustrates how to make and use desirable suspensions and how to approach destabilizing, or preventing the occurrence of undesirable suspensions. It also completes a natural trilogy, serving as a companion volume to my earlier books: *Emulsions: Fundamentals and Applications in the Petroleum Industry* and *Foams: Fundamentals and Applications in the Petroleum Industry*, both published by the American Chemical Society, Washington, DC.

Acknowledgments

I thank all the authors, who contributed considerable time and effort to their respective chapters. This book was made possible through the support of my family, Ann Marie, Katherine, and Victoria who gave me the time needed for the organization, research, and writing. I am also very grateful to Conrad Ayasse for his consistent encouragement and support. Throughout the preparation of this book many valuable suggestions were made by Gerhard Leopold, Hemanta Sarma, the reviewers of individual chapters, and by the editorial staff of ACS Books, particularly Barbara Pralle, Janet Dodd, Marc Fitzgerald, and Margaret Brown.

LAURIER L. SCHRAMM
Petroleum Recovery Institute
100,3512 33rd Street, N.W.
Calgary, Alberta T2L 2A6, Canada

FUNDAMENTALS

Suspensions: Basic Principles

Laurier L. Schramm

Petroleum Recovery Institute 100, 3512 33rd Street N.W., Calgary, Alberta T2L 2A6, Canada

This chapter introduces the occurrence, properties, and importance of suspensions in the petroleum industry. The principles of colloid science are central to an understanding of these suspensions. These principles may be applied to suspensions in different ways to achieve quite different results: a stable useful suspension in one application and the destabilization of an undesirable suspension in another.

Importance of Suspensions

Suspensions have long been of great practical interest because of their widespread occurrence in everyday life. Suspensions have important properties that may be desirable in a natural or formulated product or undesirable, such as an unwanted suspension in an industrial process. Some important kinds of familiar suspensions include those occurring in foods (batters, puddings, sauces), pharmaceuticals (cough syrups, laxatives), household products (inks, paints, "liquid" waxes), and the environment (suspended lake and river sediments, sewage). Suspensions are also quite important and widespread in the petroleum industry. In fact, suspensions may be encountered throughout each of the stages of petroleum recovery and processing (in reservoirs, drilling fluids, production fluids, process plant streams, and tailings ponds) as shown in the following list:

- migrating fines during secondary and enhanced oil recovery
- dispersions of asphaltenes in crude oils
- produced (well-head) solids in oil recovery
- drilling fluid (mud) suspensions
- well stimulation and fracturing suspensions
- well cementing slurries

0065–2393/96/0251–0003$17.50/0

- oil sand slurries in the hot water flotation process
- oil sands tailings ponds
- oilfield surface facility sludges

The various suspensions occurring in the petroleum industry may be desirable or undesirable. For example, the classic oilwell drilling fluids (drilling muds) are desirable suspensions. Here, a stable suspension is formulated and used to lubricate the cutting bit and carry cuttings up to the surface. Conversely, certain secondary and enhanced (tertiary) oil recovery processes, if not carefully designed, may cause in situ mobilization or swelling of clays in a reservoir, leading to drastic permeability reduction; in this case, the mobilized clays formed an undesirable suspension.

Suspensions may contain not just solid particles and water but also emulsified oil and even dispersed gas bubbles. In the large Canadian oil sands mining and processing operations, bitumen is disengaged from the sand matrix in suspensions created in large tumblers. The bitumen is then separated from the suspension by a flotation process in which the flotation medium is a suspension of fine particles that also contains emulsified oil (bitumen) and dispersed air bubbles (see Chapter 13).

The petroleum industry suspension applications and problems have in common the same basic principles of colloid science that govern the nature, stability, and properties of suspensions. The widespread importance of suspensions in general and scientific interest in their formation, stability, and properties have precipitated a wealth of published literature on the subject. This chapter provides an introduction intended to complement the other chapters on suspensions in this book. A good starting point for further basic information, although focused on clays, is van Olphen's classic book, *An Introduction to Clay Colloid Chemistry* (1). There are several other good books on suspensions (2, 3), and most good colloid chemistry texts contain introductions to suspensions and some of their properties (4–8).

Suspensions as Colloidal Systems

Definition and Classification of Suspensions. Colloidal particles (or droplets or bubbles) are usually defined as species having at least one dimension between ~1 and 1000 nm (a glossary of frequently encountered particle and suspension terms forms the final chapter of this book; additional information can be found in reference 9). A suspension is a special kind of colloidal dispersion: one in which a solid is dispersed in a continuous liquid phase. The dispersed solid phase is sometimes referred to as the internal (disperse) phase and the continuous phase as the external phase.

Two very different broad types of colloidal dispersions have been distinguished since Graham invented the term "colloid" in 1861 (*10*). Originally, colloids were subdivided into lyophobic and lyophilic colloids (if the dispersion medium is aqueous, then the terms hydrophobic and hydrophilic, respectively, are used). Lyophilic colloids are formed spontaneously when the two phases are brought together, because the dispersion is thermodynamically more stable than the original separated state. The term lyophilic is less frequently used in modern practice because many of the dispersions that were once thought of as lyophilic are now recognized as single-phase systems in which large molecules are dissolved. Lyophobic colloids, which include all petroleum suspensions, are not formed spontaneously on contact of the phases because they are thermodynamically unstable compared with the separated states. These dispersions can be formed with mechanical energy input via some form of agitation, such as that provided by a propeller-style mixer, a colloid mill, or an ultrasound generator. The resulting suspension may well have considerable stability as a metastable dispersion. One may also describe surface properties in terms such as hydrophilic and hydrophobic. For example, smectite clay particles, whose surfaces are strongly hydrophilic, can form quite stable suspensions in water—an example of "hydrophobic dispersion of hydrophilic particles."

Stability. Lyophobic suspensions are thermodynamically unstable but may be relatively stable in a kinetic sense, so it is crucial that stability be understood in terms of a clearly defined process. A consequence of the small particle size in many suspensions is that quite stable dispersions can be made. That is, the suspended particles may not settle out or float rapidly, and the particles may not aggregate quickly. Some use of the term stability has already been made without definition. In the definition of suspension stability, one considers stability against two different processes: sedimentation (negative creaming) and aggregation. Sedimentation results from a density difference between the two liquid phases. In aggregation two or more particles clump together, touching only at certain points and with virtually no change in total surface area. Aggregation is sometimes referred to as flocculation or coagulation, but for suspensions coagulation and flocculation are frequently taken to represent two different kinds of aggregation. In this case coagulation refers to the formation of compact aggregates in a primary potential energy of interaction minimum, whereas flocculation refers to the formation of a loose network of particles in a secondary potential energy of interaction minimum (*1, 9*). An example can be found in montmorillonite clay suspensions in which coagulation refers to dense aggregates produced by face–face oriented particle associations, and flocculation refers to loose aggregates produced by edge–face or edge–edge oriented particle associations (*1*). In aggregation the species retain their identity but lose their kinetic independence because the aggregate moves as a single unit. Kinetic stability can thus have different meanings. A system could be kinetically stable with respect to aggregation but unstable with respect to sedimentation.

Making Suspensions

Colloidal suspensions can be prepared by controlled precipitation of a salt from a supersaturated solution (the aggregation or condensation method), but in the petroleum industry it is more common for colloidal suspensions to be made by the degradation of large particles or aggregates into smaller particles, either dry with subsequent dispersion (size reduction to the order of a few micrometers) or directly in a slurry (size reduction to as small as a few tenths of a micrometer). Whether conducted wet or dry, this is the dispersion method. Factors that aid in the preparation of a suspension include favorable wetting conditions, the absence of strongly bound aggregates, fluid motions in the mixing cell that will submerge the particles, and shear to aid in disaggregation. The dispersion method can be implemented through the application of high shear forces, in a grinding mill or ultrasonic bath, to break up aggregates of particles. It can also be achieved through comminution, where a higher energy input is used to break apart more cohesive aggregates or to literally shatter particles into smaller sizes. To obtain a reasonably stable suspension, some stabilizing (peptizing) electrolyte or surfactant may have to be present as well. Laskowski and Pugh (*11*) describe and classify the major kinds of inorganic and polymeric dispersing agents used in mineral processing.

For clay minerals the natural processes of weathering and erosion tend to produce small particle sizes so that usually only mild dispersion in simple mixers, blenders, or ultrasonic baths are required. Also for clays, having inherent lattice charge means that when contacted with water, an electric double layer (EDL) is immediately created and no stabilizing (peptizing) electrolyte may be needed in this case. For other solids high energy and physical impact may be required to achieve small particles sizes. In this case grinding in ball or pebble mills may be required, using progressively finer grinding media to achieve finer particle sizes. Figure 1 shows examples of (a) a colloid mill and (b) a ball mill. The principles and operation of several kinds of dispersing equipment are discussed by Ross and Morrison (*7*). Fallenius (*12*) presents criteria for suspension for use in scaling turbine and propeller mixers and flotation cells.

Physical Characteristics of Suspensions

Particle Sizes. Not all suspensions exhibit the opaqueness with which they are usually associated. A suspension can be transparent if

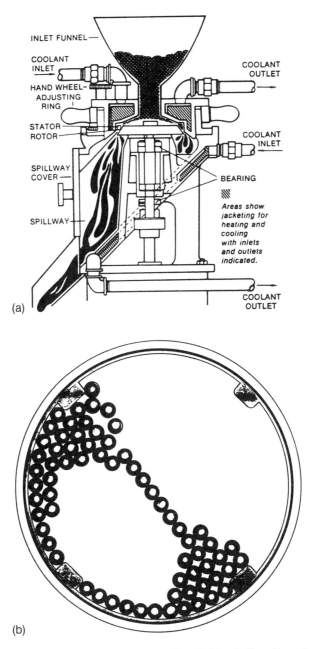

Figure 1.   Examples of (a) a colloid mill and (b) a ball mill used for the preparation of suspensions by the dispersion method. (Figure a courtesy of Premier Mill, Reading, PA. Figure b courtesy of Paul O. Abbé Inc., Little Falls, NJ.)

particles are sufficiently small compared with the wavelength of the illuminating light. If the particle sizes are 15 nm in diameter, even a 30% suspension will appear to be clear. If the particle sizes are about 1 μm in diameter, a dilute suspension will take on a somewhat milky blue cast. If the particles are very much larger, then the solid phase will become quite distinguishable. Aqueous suspensions may also exhibit different colors due to their scattering of light. A suspension of very fine gold particles (ca. 60 nm) appears red, whereas a suspension of only slightly larger particles (ca. 120 nm) appears blue.

It was stated previously that colloidal particles are between about 10^{-3} and 1 μm in diameter, although in practice, suspension particles are often larger. In fact, suspension particles usually have diameters greater than 0.2 μm and may be larger than 50 μm. Suspension stability is not necessarily a function of particle size, although there may be an optimum size for an individual suspension type. It is very common but generally inappropriate to characterize a suspension in terms of a given particle size because there is inevitably a size distribution. This is usually represented by a histogram of sizes or, if there are sufficient data, a distribution function.

In some suspensions, a particle size distribution that is heavily weighted toward the smaller sizes will represent the most stable suspension. In such cases changes in the size distribution curve with time yield a measure of the stability of the suspensions. The particle size distribution also has an important influence on the viscosity. For electrostatically or sterically interacting particles, suspension viscosity will be higher, for a given mass concentration, when particles are smaller. The viscosity will also tend to be higher when the particle sizes are relatively homogeneous, that is, when the particle size distribution is narrow rather than wide.

If the particle size is large enough and if the suspension is dilute enough, then optical microscopy can be used to determine the size and size distribution. Somewhat smaller particle-size suspensions can be characterized using cryogenic-stage scanning electron microscopy. If the suspension concentration is not too high and the particles are very small, light scattering can yield particle size information. When a beam of light enters a suspension, some light is absorbed, some is scattered, and some is transmitted. Many dilute fine suspensions show a noticeable turbidity given by

$$I_t/I_0 = \exp(-\tau l) \tag{1}$$

where I_t is the intensity of the transmitted beam, I_0 the intensity of the incident beam, τ is the turbidity, and l is the length of the path through the sample.

From Rayleigh theory, the intensity of light scattered from each particle depends largely on its size and shape and on the difference in refractive index between the particle and the medium. For a suspension, each spherical particle scatters light at an intensity I_d at a distance x from the particle, according to the following relationship:

$$I_d/I_0 \propto r^6/x^2\lambda^4 \tag{2}$$

where λ is the wavelength of the light and r is the particle radius. Because the scattering intensity is proportional to $1/\lambda^4$, blue light ($\lambda = 450$ nm) is scattered much more than red light ($\lambda = 650$ nm). With incident white light, a dilute suspension of $0.1-1$ μm-sized particles will, therefore, tend to appear blue when viewed at right angles to the incident light beam. If the particles are less than 50 nm or so, the suspension will appear to be transparent.

Concentrated suspensions are more difficult to study by optical methods. A detailed description of other approaches to characterizing particles is given in Chapter 2.

Ultramicroscopy.   When a testtube containing a dilute suspension of very small particles is held up to the light, it will appear to have a blue color due to Rayleigh scattering. Because this happens for particles so small they would be invisible under the light microscope (less than about a micrometer), this phenomenon suggests a way to observe particles that would otherwise be invisible. In the darkfield microscope, or ultramicroscope, the light scattered by small particles is viewed against a dark background. This method is used to observe the electrophoretic motions of colloidal particles, to be discussed later.

Rayleigh theory gives another relation from which particle sizes can be obtained. In the limit as the particle concentration goes to zero,

$$Kc/R_{90} = 1/M \tag{3}$$

where R_{90} is $R(\theta)$ for a scattering angle of $90°$, M is the molar mass of particles, c is the concentration of particles, and $K = (2\pi^2 n_0^2/N_A \lambda^4)(dn/dc)^2$; n_0 is refractive index of the solvent and N_A is Avogadro's number. The value dn/dc is measured with a differential refractometer. Once M is obtained, a knowledge of the particle density allows calculation of the average equivalent spherical diameter. This approach can be used to obtain the size of very small particles, ca. 20 nm diameter. For larger particles the theory is more involved.

Conductivity.   Of the numerous equations proposed (*12*) to describe the conductivity of dispersions (κ), one is cited here for illustration. The Bruggeman equation gives

$$(\kappa - \kappa_D)(\kappa_C/\kappa)^{1/3} = (1 - \phi)(\kappa_C - \kappa_D) \tag{4}$$

where ϕ is the dispersed phase volume fraction, κ_D is the conductivity of the dispersed phase, and κ_C is that of the continuous phase. If $\kappa_C \gg \kappa_D$, then $(\kappa/\kappa_C) = (1 - \phi)^{3/2}$. Further discussion of suspension conductivity and some practical examples for suspensions flowing in pipelines are given in Chapter 4.

The fact that the presence of solid particles can influence bulk conductivity forms the basis for a family of particle-size measuring techniques known as sensing-zone techniques. A well-known implementation of this is in the Coulter counter. A dilute suspension is allowed to flow through a small aperture (sensing zone) between two chambers. The conductivity (or resistance) between the chambers changes when a particle passes through the aperture. The degree of change is related to the particle volume, hence size.

The sensing-zone techniques are not limited to conductivity but may involve the measurement of capacitance. Figure 2 shows an example of the use of capacitance monitoring in a vertical sedimentation vessel. In this case (14) the effective suspension permittivity was measured and used to estimate the solids concentration (expressed in terms of volume fraction).

Rheology. The rheological properties of a suspension are very important. High viscosity may be the reason that a suspension is troublesome, a resistance to flow that must be dealt with, or a desirable property for which a suspension is formulated. The simplest description applies to Newtonian behavior in laminar flow. The coefficient of viscosity, η, is given in terms of the shear stress, τ, and shear rate, $\dot{\gamma}$, by

$$\tau = \eta\dot{\gamma} \tag{5}$$

where η has units of mPa·s. Many colloidal dispersions, including concentrated suspensions, do not obey the Newtonian equation. For non-Newtonian fluids the coefficient of viscosity is not a constant but is itself a function of the shear rate; thus

$$\tau = \eta(\dot{\gamma})\dot{\gamma} \tag{6}$$

It is common for industrial pumping and processing equipment to use shear rates that fall in the intermediate shear regime from about 10 to 1000 s^{-1} as illustrated in Table I. A convenient way to summarize the flow properties of fluids is by plotting flow curves of shear stress versus shear rate (τ vs $\dot{\gamma}$). These curves can be categorized into several rheological classifications (Figure 3). Suspensions are frequently pseudoplastic: as shear rate increases viscosity decreases. This is also termed

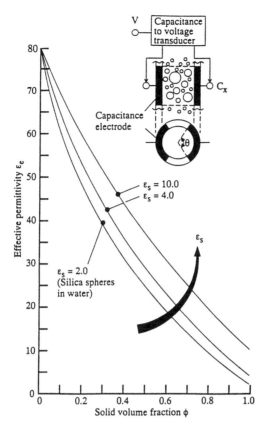

Figure 2. Capacitance monitoring in a sedimentation vessel (inset) and the relation between effective suspension permittivity ϵ_e and solids concentration in the suspension for spherical particles of permittivity ϵ_s in liquid of permittivity ϵ_L. (Reproduced from reference 14. Copyright 1992 Butterworth-Heinemann.)

Table I. Approximate Values of Shear Rate Appropriate to Various Processes

Process	Approximate Shear Rate (s^{-1})	Ref.
Very slow stirring	0.01–0.1	52
Reservoir flow in oil recovery	1–5	53
Mixing	10–100	52
Pumping	100–1000	52
Coating	10,000	52
Oilwell drilling fluid at the bit nozzle	10,000–100,000	54

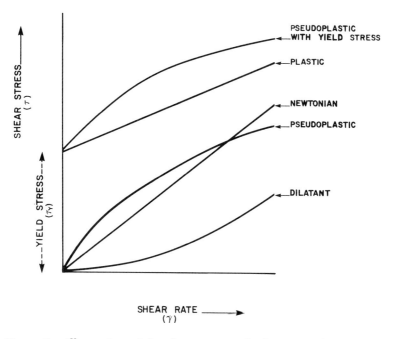

Figure 3. Illustration of the characteristic fluid types and their corresponding flow curves.

shear-thinning. A suspension may also exhibit a yield stress, that is, the shear rate (flow) remains zero until a threshold shear stress is reached, referred to as the yield stress (τ_Y), and then pseudoplastic or Newtonian flow begins. Some descriptions appropriate to different yield stresses are given in Table II. Pseudoplastic flow, which is time-dependent, is termed thixotropic, that is, at a constant applied shear rate viscosity

Table II.   Some Descriptions Appropriate to Different Yield Stresses

Yield Stress (Pa/s)	Description
<10	Easy to pour.
10–30	Thick, pours easily.
	Use conventional liquid designs.
30–40	Thick, hard to pour, forms peaks.
	Difficult to make flow under pump suction.
40–100	Flows poorly; will cleave to walls under gravity.
	Need to push into pump suction.
>100	Can build with it; will cleave to top of jar.
	Requires positive flow pump.

SOURCE: Adapted from reference 52.

decreases and in a flow curve hysteresis occurs. It is also not unusual for suspensions that are pseudoplastic at low to moderate solids concentrations to become dilatant (shear thickening) at high solids concentrations. In this case there is also a critical shear rate for the onset of shear thickening. Several other rheological classifications are covered in the glossary of this book.

Whorlow (15) and others (16, 17) described very useful experimental techniques. Very often, measurements are made with a suspension sample placed in the annulus between two concentric cylinders. The shear stress is calculated from the measured torque required to maintain a given rotational velocity of one cylinder with respect to the other. Knowing the geometry, the effective shear rate can be calculated from the rotational velocity. Less useful are the various kinds of simplified measuring devices found in many industrial plants and even in their technical support laboratories. Such devices may not be capable of determining shear stresses for known shear rates or may not be capable of operation at shear rates that are appropriate to the process under consideration. Instruments that are capable of absolute viscosity measurements provide much more useful information.

In an attempt to conduct rheological measurements on suspensions, a number of changes may occur in the sample chamber, making the measurements irreproducible and not representative of the original suspension. Prevalent among these changes is the sedimentation, or even centrifugal segregation, of solids, causing a nonuniform distribution within the measuring chamber. In the extreme sedimentation can cause complete removal of solids from the region in which measurements are made.

It is frequently desirable to be able to describe suspension viscosity in terms of the viscosity of the liquid continuous phase (η_0) and the amount of suspended material. A very large number of equations have been advanced for estimating suspension (or emulsion, etc.) viscosities. Most of these are empirical extensions of Einstein's equation for a dilute suspension of spheres:

$$\eta = \eta_0(1 + 2.5\phi) \tag{7}$$

where η_0 is the medium viscosity and ϕ is the dispersed phase volume fraction ($\phi < 1$). Examples of two empirical equations are the Oliver–Ward equation for spheres (9):

$$\eta = \eta_0(1 + a\phi + a^2\phi^2 + a^3\phi^3 + \cdots) \tag{8}$$

where a is an empirical constant, and the Thomas equation for suspensions:

$$\eta = \eta_0(1 + 2.5\phi + 10.5\phi^2 + 0.00273 \exp[16.6\phi]) \tag{9}$$

These equations apply to Newtonian behavior, or at least the Newtonian region of a flow curve, and they usually require that the particles not be too large and have no strong electrostatic interactions.

Other modifications have been made for application to suspensions of anisometric (unsymmetric) particles such as clays. In this case the intrinsic viscosity $[\eta]$, given by

$$[\eta] = \lim_{\phi \to 0} \lim_{\dot\gamma \to 0} (\eta/\eta_0 - 1)/\phi \qquad (10)$$

which is 2.5 for a dilute suspension of uncharged spheres, takes on a value that is different and more difficult to predict value. A useful such modification to Einstein's equation for dilute suspensions of anisometric particles is given by the Simha Equation, which is approximately

$$\eta = \eta_0(1 + a\phi/1.47b) \qquad (11)$$

where a is the major particle dimension and b the minor particle dimension.

A more detailed treatment of these kinds of relationships is given in Chapter 3.

Suspensions can show varying rheological, or viscosity, behaviors. Sometimes these properties are due to stabilizing agents in the suspension. However, typically particle–particle interactions are sufficient to cause the suspension viscosity to increase because of electrostatic interactions or simply particle "crowding."

Stability of Suspensions

Most suspensions are not thermodynamically stable. Rather, they possess some degree of kinetic stability, and it is important to distinguish the degree and the time scale of change. In this discussion of colloid stability, we explore the reasons why colloidal suspensions can have different degrees of kinetic stability and how these are influenced, and can therefore be modified, by solution and surface properties. Encounters between particles in a suspension can occur frequently due to Brownian motion, sedimentation, stirring or a combination of them. The stability of the dispersion depends on how the particles interact when this happens. The main cause of repulsive forces is the electrostatic repulsion between like charged objects. The main attractive forces are the van der Waals forces between objects.

Electrostatic Charges.   Most substances acquire a surface electric charge when brought into contact with a polar medium such as water. For suspensions, the origin of the charge can be due to

- ionization or surface hydrolysis, as when carboxyl and/or amino functionalities dissociate when proteins are put into water (pH dependent)

- ion adsorption, as when surfactant ions adsorb onto a solid surface

- ion dissolution, as when Ag^+ and I^- dissolve unequally from their crystal lattice when AgI is placed in water (Ag^+ and I^- are potential determining ions)

- ion diffusion, as when a dry clay particle is placed in water and the counterions diffuse out to form an EDL

The surface charge influences the distribution of nearby ions in the polar medium. Ions of opposite charge (counterions) are attracted to the surface, whereas those of like charge (coions) are repelled. An EDL, which is diffuse because of mixing caused by thermal motion, is thus formed.

Electric Double Layer. The EDL consists of the charged surface and a neutralizing excess of counterions over coions, distributed near the surface (Figure 4). The EDL can be viewed as being composed of two layers: an inner layer, which may include adsorbed ions, and a diffuse layer, in which ions are distributed according to the influence of electrical forces and thermal motion. Gouy and Chapman proposed a simple quantitative model for the diffuse double layer assuming, among other things, an infinite, flat, uniformly charged surface and point charge ions. Taking the surface potential to be ψ°, the potential ψ at a distance x from the surface is approximately

$$\psi = \psi^\circ \exp(-\kappa x) \tag{12}$$

The surface charge density is given as $\sigma^\circ = \epsilon\kappa\psi^\circ$, where ϵ is the permittivity; thus, ψ° depends on surface charge density and the solution ionic composition (through κ). $1/\kappa$ is called the double layer thickness and for water at 25 °C is given by

$$\kappa = 3.288 \sqrt{I} \tag{13}$$

where κ is given in nm^{-1}, I is the ionic strength, given by $I = (1/2) \Sigma_i c_i z_i^2$. For 1–1 electrolyte, $1/\kappa$ is about 1 nm in 10^{-1} M ionic strength solution and about 10 nm in 10^{-3} M ionic strength solution.

   In fact, an inner layer exists because ions are not really point charges, and an ion can only approach a surface to the extent allowed by its hydration sphere. The Stern model incorporates a layer of specifically adsorbed ions bounded by a plane, called the Stern plane (Figure 5). In

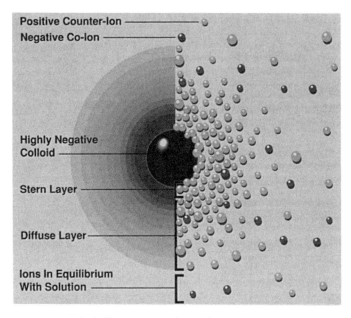

Figure 4. Simplified illustrations of the electrical double layer around a negatively charged colloidal suspension particle. The left view shows the change in charge density around the drop. The right view shows the distribution of ions around the charged drop. (Courtesy L. A. Ravina, Zeta-Meter, Inc., Staunton, VA.)

this case the potential changes from ψ° at the surface, to $\psi(\delta)$ at the Stern plane, to $\psi = 0$ in bulk solution.

Electrokinetic Phenomena. Electrokinetic motion occurs when the mobile part of the EDL is sheared away from the inner layer (charged surface). There are four types of electrokinetic measurements, electrophoresis, electroosmosis, streaming potential, and sedimentation potential, of which the first finds the most use in industrial practice. Good descriptions of practical experimental techniques in electrophoresis and their limitations can be found in references 18–20.

In electrophoresis an electric field is applied to a sample, causing charged particles or particles plus any attached material or liquid, to move toward the oppositely charged electrode. Thus, the results can only be interpreted in terms of charge density (σ) or potential (ψ) at the plane of shear. The latter is also known as the zeta potential. Because the exact location of the shear plane is generally not known, the zeta potential is usually taken to be approximately equal to the potential at the Stern plane (Figure 5):

$$\zeta \approx \psi(\delta). \tag{14}$$

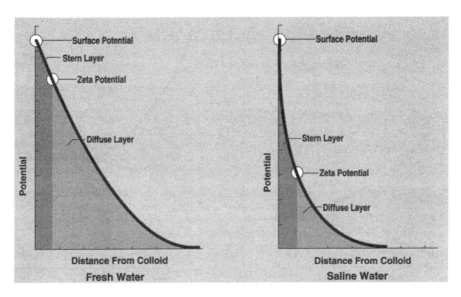

Figure 5.   Simplified illustration of the surface and zeta potentials for a charged suspension drop dispersed in high (saline water) and low (fresh water) electrolyte concentration aqueous solutions. (Courtesy L. A. Ravina, Zeta-Meter, Inc., Staunton, VA.)

In microelectrophoresis the dispersed particles are viewed under a microscope and their electrophoretic velocity is measured at a location in the sample cell where the electric field gradient is known. This must be done at carefully selected planes within the cell because the cell walls become charged as well, causing electroosmotic flow of the bulk liquid inside the cell.

The electrophoretic mobility, μ_E, is defined as the electrophoretic velocity divided by the electric field gradient at the location where the velocity was measured. It remains then to relate the electrophoretic mobility to the zeta potential (ζ). Two simple relations can be used to calculate zeta potentials in limiting cases:

Hückel Theory.   For particles of small radius, a, with "thick" EDLs, meaning that $\kappa a < 1$, it is assumed that Stokes law applies and the electrical force is equated to the frictional resistance of the particle, $\mu_E = \zeta\epsilon/(1.5\,\eta)$.

Smoluchowski Theory.   For large particles with "thin" EDLs, meaning particles for which $\kappa a > 100$, $\mu_E = \zeta\epsilon/\eta$. With these relations, zeta potentials can be calculated for many practical systems. Note that within each set of limiting conditions the electrophoretic mobility is independent of particle size and shape as long as the zeta potential is constant. For intermediate values of κa, the Henry equation and many other equations are available in the literature (18, 19, 21).

Repulsive Forces. In the simplest example of colloid stability, suspension particles would be stabilized entirely by the repulsive forces created when two charged surfaces approach each other and their EDLs overlap. The repulsive potential energy V_R for spherical particles is given approximately as

$$V_R = (B\epsilon k^2 T^2 a \gamma^2 / z^2) \exp[-\kappa x] \tag{15}$$

where the spheres have radius a and are separated by distance x. In Figures 6 and 7, H and d have the same meaning as x. B is a constant $(3.93 \times 10^{39} \text{ A}^{-2} \text{ s}^{-2})$, z is the counterion charge number, and

$$\gamma = (\exp[ze\psi(\delta)/2kT] - 1)/(\exp[ze\psi(\delta)/2kT] + 1) \tag{16}$$

In practice the situation may be more complicated. If particle surfaces are covered by long chain molecules (physically or chemically bonded to the surface), then steric repulsion between particles may be significant. This repulsion is due to an osmotic effect caused by the high concentration of chains that are forced to overlap when particles closely approach and also due to the volume restriction, or entropy decrease, that occurs when the chains lose possible conformations because of overlapping.

Dispersion Forces.   van der Waals postulated that neutral molecules exert forces of attraction on each other that are caused by electrical interactions between dipoles. The attraction results from the orientation of dipoles due to any of (1) Keesom forces between permanent dipoles, (2) Debye induction forces between dipoles and induced dipoles, or (3) London–van der Waals forces between induced dipoles and induced dipoles. Except for quite polar materials, the London–van der Waals dispersion forces are the most significant of the three. For molecules the force varies inversely with the sixth power of the intermolecular distance.

For dispersed particles (or droplets, etc.) the dispersion forces can be approximated by adding up the attractions between all interparticle pairs of molecules. When added this way, the dispersion force between two particles decays less rapidly as a function of separation distance than is the case for individual molecules. For two spheres of radius a in vacuum, separated by a small distance x, the attractive potential energy V_A can be approximated by

$$V_A = -Aa/12x \tag{17}$$

for $x < 10$–20 nm and $x \ll a$. The constant A is known as the Hamaker constant and depends on the density and polarizability of atoms in the particles. Typically, $10^{-20}\,J < A < 10^{-19}\,J$, or $2.5kT < A < 25kT$ at room temperature (*21*). When the particles are in a medium other than vacuum, the attraction is reduced. This can be accounted for by using an effective Hamaker constant

$$A = (\sqrt{A_2} - \sqrt{A_1})^2 \qquad (18)$$

where the subscripts denote the medium (1) and particles (2).

The effective Hamaker constant equation shows that the attraction between particles is weakest when the particles and medium are most chemically similar ($A_1 \approx A_2$). The Hamaker constants are usually not well known and must be approximated.

DLVO Theory.   Derjaguin and Landau (*23*) and, independently, Verwey and Overbeek (*24*) developed a quantitative theory for the stability of lyophobic colloids, now known as the DLVO theory. It was developed in an attempt to account for the observation that colloids coagulate quickly at high electrolyte concentrations, slowly at low concentrations, and with a very narrow electrolyte concentration range over which the transition from one to the other occurs. The latter defines the critical coagulation concentration (CCC). The DLVO theory accounts for the energy changes that take place when two particles approach each other and involves estimating the potential energy of attraction (dispersion forces) versus interparticle distance and the potential energy of repulsion (electrostatic forces) versus distance. These estimates, V_A and V_R, are then added together to yield the total potential interaction energy V. There are deviations from DLVO theory that appear at very small separation distances: the first is a short-range repulsive force that occurs in aqueous systems and may be due to an influence of the particle surfaces on hydrogen-bonding in nearby water molecules (*25*) and the second is a strong short-range repulsive force due to atomic electron cloud overlap, called Born repulsion. The DLVO theory has been developed for several special cases, including the interaction between two spheres, and refinements are constantly being made.

V_R decreases exponentially with increasing separation distance and has a range about equal to κ^{-1}, while V_A decreases inversely with increasing separation distance. Figure 6 shows a single attractive potential energy curve and two different repulsive potential energy curves, representing two very different levels of electrolyte concentration. The figure shows the total potential interaction energy curves that result in each case. It can be seen that either the attractive van der Waals forces or the repulsive electric double layer forces can predominate at different interparticle distances.

Where there is a positive potential energy maximum, a dispersion should be stable if $V \gg kT$, that is, if the energy is large compared to the thermal energy of the particles ($15kT$ is considered unsurmountable). In this case colliding particles should rebound without contact and the suspension should be stable to aggregation. If, on the other hand, the potential energy maximum is not very great, $V \approx kT$, then slow aggregation should occur. The height of the energy barrier depends on the surface potential, $\psi(\delta)$ and on the range of the repulsive forces, κ^{-1}. Figure 6 shows that a secondary potential energy minimum can occur at larger interparticle distances. If this is reasonably deep compared with kT, then a loose easily reversible aggregation should occur. Figure 7a illustrates the effects of altering the height of the potential energy maximum. Here computer simulations were conducted (26) for the interaction between two spherical particles based on DLVO theory. As particles approach each other and pass through the secondary minimum, they experience a small (case I) or a large (case II) energy barrier that has to be surmounted to reach the primary potential energy well. In this case the energy barrier was adjusted by reducing the concentration of indifferent electrolyte for case II. The effects on the proportions of primary particles versus aggregated particles are shown in Figure 7b. A smaller energy barrier to be surmounted results in a much higher proportion of aggregated versus primary particles.

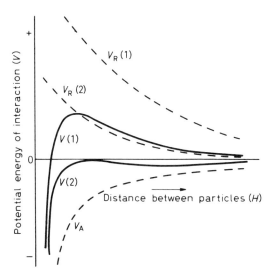

Figure 6.   The effect of different repulsive potential energy curves V_R (1) and V_R (2) on the total potential energy of interaction curves V (1) and V (2), for a given attractive potential energy curve. (Reproduced with permission from reference 4. Copyright 1981 Butterworth-Heinemann.)

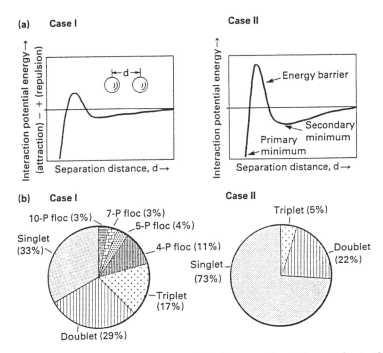

Figure 7. Interaction energy curves (a) for two spherical particles in the presence of two different concentrations of indifferent electrolyte. Also shown (b) are the aggregation statistics, in terms of the percentages of primary particles versus aggregates, obtained from computer simulations for each case. (Reproduced with permission from reference 26. Copyright 1992 Society of Chemical Industry.)

Practical Guidelines. It is apparent that the DLVO calculations can become quite involved, requiring considerable knowledge about the systems of interest. Also, there are some problems. For example, the prevailing assessment of the validity of the theory is changing as more becomes known about the influence of additional forces, such as those due to surface hydration. In addition, there is now considerable effort being devoted to the possibility of directly determining the forces between colloidal particles, which for the present can only be inferred. For a recent review on this topic and the use of the surface force apparatus, *see* reference 25. The DLVO theory nevertheless forms a very useful starting point in attempting to understand complex colloidal systems such as petroleum suspensions. There are empirical "rules of thumb" that can be used to give a first estimate of the degree of colloidal stability that a system is likely to have if the zeta potentials of the particles are known.

Many types of colloids tend to adopt a negative surface charge when dispersed in aqueous solutions that have ionic concentrations and pH typical of natural waters. For such systems one rule of thumb stems from observations that the colloidal particles are quite stable when the zeta potential is about -30 mV or more negative and quite unstable because of agglomeration when the zeta potential is between $+5$ and -5 mV. An expanded set of guidelines, developed for particle suspensions, is given by Riddick (20). Such criteria are frequently used to determine optimal dosages of polyvalent metal electrolytes, such as alum, used to effect coagulation in treatment plants.

Water treatment, whether for drinking water or for disposal of industrial wastes, inevitably involves removal of suspended solids (often referred to as turbidity) usually silt, clay, and organic matter. The charge on the solids is sufficiently negative to yield a stable dispersion that settles slowly and is difficult to filter. The solution to this problem is to reduce the zeta potential to values that permit rapid coagulation, increasing both sedimentation and filterability. A first step toward coagulating the suspension might be to add aluminum sulfate (alum), from which the trivalent aluminum ions will have a powerful effect on the zeta potential (remembering the Schulze–Hardy rule). Figure 8 shows an example of this effect. In practice, however, the alum required to reduce the zeta potential to below about -10 mV or so reduces the solution pH too much. Unreacted alum becomes carried to other parts of the plant and forms undesirable precipitates. As shown in Figure 9, a second step can be introduced then, in which a cationic polyelectrolyte is added to reduce the zeta potential to a near zero value (slightly positive in the example in Figure 9) but without changing the pH. As a final step, a high molecular weight anionic polymer may be added (molecular weight 500,000 to 1,000,000 or more) whose molecules can bridge between agglomerates, yielding very large rapid settling flocs.

Schulze–Hardy Rule. The transition from stable dispersion to aggregation usually occurs over a fairly small range of electrolyte concentration. This makes it possible to determine aggregation concentrations, often referred to as critical coagulation concentrations (CCC). The Schulze–Hardy rule summarizes the general tendency of the CCC to vary inversely with the sixth power of the counterion charge number (for indifferent electrolyte).

A prediction from DLVO theory can be made by deriving the conditions under which $V = 0$ and $dV/dx = 0$. The result is

$$CCC = (9.75B^2\epsilon^3k^5T^5\gamma^4)/(e^2N_AA^2z^6) \qquad (19)$$

showing that for high potentials ($\gamma \rightarrow 1$), the CCC varies inversely with z^6. As an illustration, suppose that for a hypothetical suspension equation 19 predicts a CCC of 1.18 M in solutions of sodium

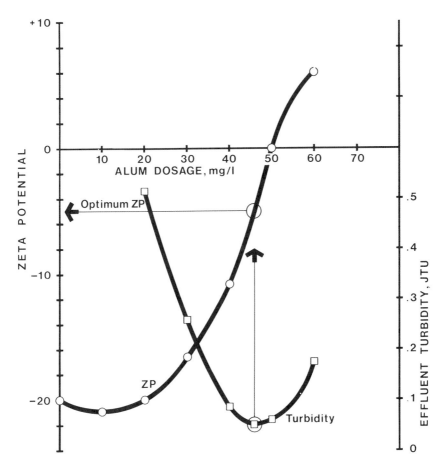

Figure 8. Illustration of the effects of alum treatment on particle Zeta potentials and turbidity levels in water treatment. (Courtesy L. A. Ravina, Zeta-Meter, Inc., Staunton, VA.)

chloride. The CCC in polyvalent metal chlorides would then decrease as follows:

Dissolved Salt	z	CCC (mol/L)
NaCl	1	1.18
$CaCl_2$	2	0.018
$AlCl_3$	3	0.0016

Protective Agents, Steric Stabilization, and Flocculation.

The transitions from stable dispersion to aggregation just described in terms of the CCC and the Schulze–Hardy rule apply best to suspensions

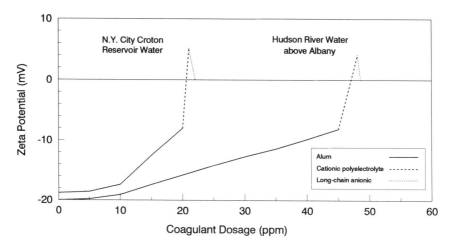

Figure 9.   Illustration of zeta potentials and coagulation of solids in New York city water treatment through sequential additions of aluminum sulfate (alum), cationic polyelectrolyte, and anionic polymer. (Reproduced with permission from reference 50. Copyright 1982 Zeta-Meter, Inc.)

in which the particles have only one kind of charge. But clay particles can carry positive and negative charges at the same time, on different parts of the particle. The clay minerals are composed (*1, 27*) of sheets of tetrahedrally coordinated silica and sheets of octahedrally coordinated alumina or magnesia. These sheets occur stacked upon one another, forming platelike layers or particles. When dispersed in aqueous solutions of near neutral to alkaline pH, the particles carry a net negative charge, largely due to isomorphic substitution of cations of lower charge for cations of higher charge within the lattice (e.g., Al^{+3} for Si^{+4} in tetrahedral sheets and Fe^{+2} or Mg^{+2} for Al^{+3} in the octahedral sheet; Figure 10). Meanwhile, the edges of clay particles may carry a positive charge in near neutral to acid pH solution because of protonation of various atoms exposed at the edges.

This heterogeneous charge distribution leads to a number of kinds of particle interaction orientations in clay suspensions as shown in Figure 11. Three basic modes are possible: face–face (FF), edge–face (EF), and edge–edge (EE). The different modes can be combined in different ways depending on the clay mineralogy (montmorillonite has higher negative charge density than kaolinite, for which the positive edge charging can be more significant), solution pH (directly determining the contributions from edge charging), solution ionic composition, and solution ionic strength (both of which affect aggregation overall and also the number of plates per tactoid in aggregates; *see* references 28 and 29).

The stability of a suspension can be enhanced (protection) or reduced (sensitization) by the addition of material that adsorbs onto particle sur-

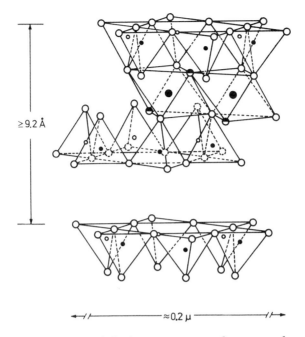

Figure 10. Illustration of the basic structure of smectite clay minerals. Lattice positions are assigned as silicon (• and ○); aluminum, magnesium, or iron (●); oxygen (○); and hydroxyl (◑). (Reproduced with permission from reference 27. Copyright 1968 McGraw Hill.)

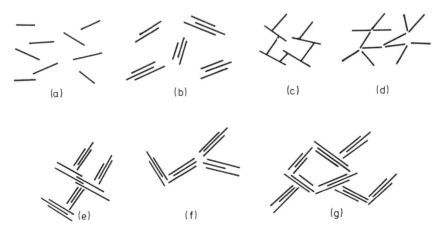

Figure 11. Some possible modes of clay particle association: (a) dispersed, (b) coagulated into tactoids, (c and d) flocculated primary particles, and (e–g) coagulated and flocculated particles. Face–face mode is shown as b, edge–face mode is c, edge–edge mode is d, and others are mixtures of these. (Reproduced with permission from reference 51. Copyright 1980 Laurier L. Schramm.)

faces (Figure 12). Protective agents can act in several ways. They can increase double layer repulsion if they have ionizable groups. The adsorbed layers can lower the effective Hamaker constant. An adsorbed film may necessitate desorption before particles can approach closely enough for van der Waals forces to cause attraction. If the adsorbed material extends out significantly from the particle surface, then an entropy decrease can accompany particle approach (steric stabilization). Oilfield suspensions of particles in oil may be stabilized by the presence of an adsorbed film formed from the asphaltene and resin fractions of the crude oil.

In steric stabilization adsorbed polymer molecules must extend outward from the particle surface yet be strongly enough attached to the surface that they remain adsorbed in the presence of applied shear. An example is a system of particles containing terminally anchored block copolymer chains having a hydrophobic portion of the molecule that is very strongly adsorbed on the particle surfaces and a hydrophilic part

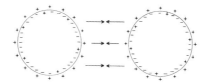

METHOD 1:   Mutual repulsion due to high Zeta Potential

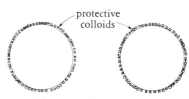

METHOD 2:   Adsorption of a small lyophilic colloid on a
larger electronegative colloid

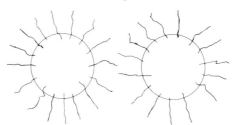

METHOD 3:   Steric hindrance due to adsorption of an
oriented nonionic polyelectrolyte

*Figure 12. Illustration of three means of promoting colloid stability.
(Courtesy of Zeta-Meter Inc., Staunton, VA.)*

of the molecule that extends outward from the particle surfaces. The size of the adsorbed molecule determines the extent of the long-range repulsive force between particles and also causes the primary minimum to disappear (Figure 13). Curve A shows an example for electrostatically stabilized particles. Curves B–D show examples for sterically stabilized particles in three different polymer stabilizer and solvent systems. As the figure shows, sterically stabilized particles are either stably dispersed or reversibly aggregated (flocculated) in a secondary minimum. By changing the nature of the medium, such a suspension can be adjusted back and forth between the stable and flocculated conditions. It is also possible to have particles stabilized by both electrostatic and steric stabilization; these are said to be electrosterically stabilized.

Flocculation of particles may be achieved through the addition to a suspension of high molecular mass (millions of g/mol) synthetic or natural polymers. These molecules act as bridging agents by adsorbing onto more than one particle at a time with a significant portion of the polymer chain still remaining in the aqueous phase. The bridging action causes the formation of porous flocs. The formation of such flocs has a dramatic effect on sedimentation rates, sediment volumes, and on the ease of filtration. Effective flocculation may occur over a narrow range of polymer concentration because too little polymer will not permit floc formation whereas too much polymer adsorption will eliminate the fraction of free particle surface needed for the bridging action (i.e., the polymer molecules will adsorb onto single particles in preference to bridging several particles). The nature of the flocculation is quite dependent on experimental conditions such as the nature and degree of agitation that may be present.

The process of stabilizing a suspension by reversing the processes of aggregation (coagulation and/or flocculation) is known as peptization, or deflocculation. This can be accomplished by attacking the mechanisms that cause, or caused, aggregation in the first place. For example, in the case of clay suspensions, some of the well-known peptizing agents act

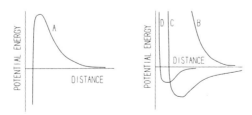

Figure 13. Potential energy diagrams for (A) electrostatically stabilized particles and (B–D) sterically stabilized particles with different degrees of steric stabilization. (Reproduced with permission from reference 7. Copyright 1988 Wiley.)

via the adsorption of anionic species at clay particle edges in sufficient quantity to reverse the charge. This weakens or eliminates the forces, causing EF and EE clay-plate associations. Given the small specific surface area of clay particle edges, it can be seen that this could be accomplished by very small additions of peptizing chemical. Figure 14 shows the effect of adding a peptizing agent to a clay suspension on the clay salt-flocculation value in a flocculation diagram. The point where all three curves intersect the vertical axis shows the flocculation value of the suspension to an indifferent salt (sodium chloride) alone. Curve 1 shows how addition of peptizing agent (sodium polymetaphosphate) increases the salt flocculation value. The curve shows that overdosing (overtreatment) is possible and that at very high additions the peptizer acts as a flocculating agent even without the original sodium chloride addition. Curve 2 shows the additive effect of using an agent that is similar to the original salt (in this case potassium chloride). Curve 3 illustrates the situation in which the agent added makes the suspension more susceptible to flocculation by the original salt (synergism).

Kinetics.   Thus far we have mostly been concerned with an understanding of the direction in which reactions will proceed. However, from an engineering point of view, it is just as important to know the rates at which such reactions will proceed. Two principal factors determine the rate of aggregation of particles in a suspension: the frequency of particle encounters and the probability that the thermal energy of the particles is sufficient to overcome the potential energy barrier to

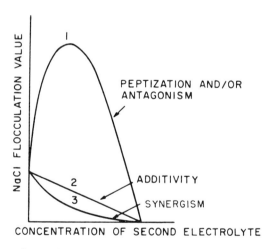

Figure 14.   Effects of a mixture of sodium chloride and different potential peptizing agents on the NaCl flocculation value (suspension stability). (Reproduced with permission from reference 1. Copyright 1977 Wiley.)

aggregation. The rate of aggregation can be given as $-(dn/dt) = k_2n^2$, where n is the number of particles per unit volume at time t and k_2 is the rate constant. For $n = n_0$ at $t = 0$, $1/n = k_2t + 1/n_0$. During the process of aggregation, k_2 may not remain constant.

If the energy barrier to aggregation is removed (e.g., by adding excess electrolyte), then aggregation is diffusion controlled—only Brownian motion of independent particles is present. For a monodisperse suspension of spheres, Smoluchowski developed an equation for this rapid coagulation:

$$n = n_0/(1 + 8\pi Dan_0t) \tag{20}$$

where a is the radius and the diffusion coefficient $D = kT/(6\pi\eta a)$. Now $k_2^0 = 4kT/(3\eta)$, where k_2^0 is the rate constant for diffusion controlled aggregation.

When there is an energy barrier to aggregation, only a fraction $1/W$ of encounters lead to attachment. W is the stability ratio $W = k_2^0/k_2$. Using W allows one to account for "slow coagulation" (hindered) times. In this case, the interaction energy and hydrodynamic viscous drag forces must be considered (4). For example, Alince and Van de Ven (30) studied the rate of destabilization of clay particles by pH adjustment and by cationic polyethylenimine addition. They found that the stability ratio could be correlated with the electrophoretic mobility of the clay particles and that the maximum rate of destabilization corresponded to conditions for which the electric charge on the clay particles or aggregates was nearly zero.

Finally, particles can also be brought into interaction distances by stirring or sedimentation where the relative motions of two adjacent regions of fluid, each carrying particles, can cause particle encounters. Coagulation due to such influence is called orthokinetic coagulation as distinguished from the Brownian-induced perikinetic coagulation. The theory for orthokinetic coagulation is much more complicated than that for perikinetic and is not dealt with here. It should be remembered, however, that shear can also cause dispersion if the energy introduced allows the interaction energy barrier to be overcome. (For more information on aggregation kinetics *see* references 31–33).

Electrostatic Properties in Nonaqueous Media.   Although suspensions most commonly comprise particles dispersed in aqueous media, the petroleum industry contains many examples of particles dispersed in nonaqueous media. Examples include precipitated asphaltenes in oil (*see* Chapter 8) and mineral solids dispersed in diluted froth in oil sands processing (*see* Chapter 13). Particles can be electrostatically stabilized in nonaqueous media, although the charging mechanism is different (7, 34). In a recent review Morrison (34) emphasized that many models are available for the behavior of electrical charges in nonaqueous media but few are universally accepted. The particles may acquire their charge by adsorption of ions or through the dissociation of ions from their surfaces. In one model of nonaqueous electrolyte solutions, the ions are considered to be held in large structures such as micelles that prevent their being neutralized. The electric field, its concentration dependence, and the electrical conductivity then are determined by the size, structure, and motions of these micelles.

In the petroleum industry flammable vapors are often present so both fire and explosion hazards exist. The conductivity of petroleum fluids is low, which allows the buildup of large potential gradients. The interaction of impurities in the petroleum fluids with pipe and tank walls apparently allows the generation of electrical charges (there is also a role for emulsified water droplets). The flow of the fluid in a tank or pipe allows the separation of charges and can cause sufficient charging for an electrostatic discharge, which in turn can cause an explosion. According to Morrison (34), appropriate safety precautions include keeping all containers and pipes electrically connected, increasing the electrical conductivity of the petroleum fluid by adding nonaqueous electrolytes, and keeping the environment free of oxygen.

The role of electrostatic repulsion in the stability of suspensions of particles in nonaqueous media is not yet clear. To apply the DLVO theory, one must know the electrical potential at the surface, the Hamaker constant, and the ionic strength to be used for the nonaqueous medium; these are difficult to estimate. The ionic strength will be low so the EDL will be thick, the electric potential will vary slowly with separation distance, and so will the net electric potential as the double layers overlap. For this reason the repulsion between particles can be expected to be weak. A summary of work on the applicability or lack of applicability of DLVO theory to nonaqueous media has been given by Morrison (34).

Sedimentation

Particles in a suspension will have some tendency to settle, possibly according to Stokes' law. In general, an uncharged spherical particle in a fluid will sediment out if its density is greater than that of the fluid. The driving force is that of gravity; the resisting force is viscous and is approximately proportional to the particle velocity. After a short period of time, the particle reaches terminal (constant) velocity dx/dt when the two forces are matched. Thus,

$$dx/dt = (2a^2(\rho_2 - \rho_1)g)/(9\eta) \tag{21}$$

where a is the particle radius, ρ_2 is the particle density, ρ_1 is the external fluid density, g is the gravitational constant, and η is the bulk viscosity. If the particle has a lower density than the external phase, then it rises instead (negative sedimentation, flotation, or creaming). In the cases of electrostatically interacting particles or particles with surfactant or polymeric stabilizing agents at the interface, the particles will interact, contrary to the assumption of this theory. Thus, Stokes' law will not strictly apply and may underestimate or even overestimate the real terminal velocity.

For dense particles in a suspension, the individual settling (kinetic) units may be primary particles, coagulated aggregates, or flocculated aggregates (flocs). If a given mass of solid is dispersed in liquid and placed in a sedimentation column (Figure 15), then the sedimentation volume, or specific sedimentation volume, can be determined as a function of time. A qualitative result is that the settling behaviors of the primary (deflocculated), coagulated, and flocculated particles are different and lead to different specific sedimentation volume–time curves, as shown in Figures 16 and 17. It can be seen that the packing densities in the particles translate into the same order of packing densities in the sediment, although given sufficient time these structures will change and become increasingly dense. A very simple settling test sometimes used in industrial plants is to establish a standard procedure for agitating a sample of suspension in a given size jar, allowing the suspension to stand, and measuring the time required for the particles to settle out. A time scale appropriate to the process under consideration is then set. For example, fast-settling suspensions might be those that settle out completely within 10 min. After the settling test, the jar can be agitated again to check the ease of resuspension.

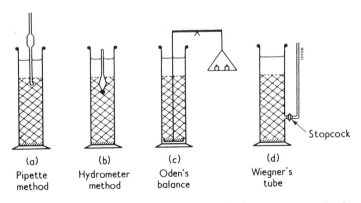

Figure 15. *Sedimentation columns used for the determination of sedimentation rates and particle size distributions in suspensions. (Reproduced with permission from reference 52. Copyright 1959 Karol J. Mysels.)*

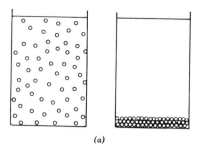

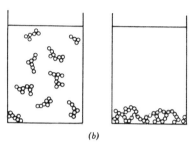

Figure 16.   Sedimentation of (a) primary particles leading to a close-packed sediment and (b) flocculated particles leading to a loose-packed voluminous sediment. (Reproduced with permission from reference 1. Copyright 1977 Wiley.)

A quantitative result is illustrated by the following example taken from Hiemenz (5). By using the Oden balance of Figure 15, one can measure the total mass W (in mass% units) of sedimenting clay particles that reach the balance pan at various times t. This is shown in Figure 18a. Here, W is made up of particles large enough to have fallen the full length of the column (w) and also a fraction of those that have are only able to fall part of the height of the column but that nevertheless

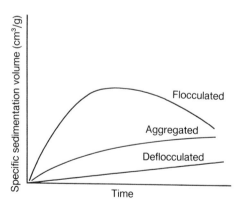

Figure 17.   Specific sedimentation volume versus settling time curves for deflocculated (primary) particles, coagulated aggregates, and flocculated aggregates of particles. (Reproduced with permission from reference 7. Copyright 1988 Wiley.)

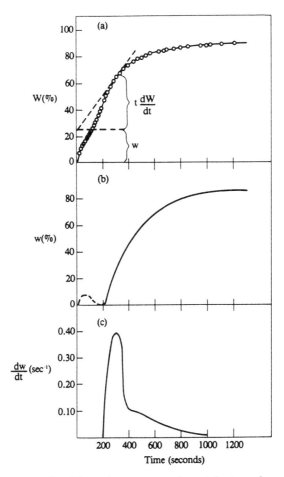

Figure 18.   Example of the sedimentation of particles in a clay suspension, monitored using an Oden balance, showing (a) the measured cumulative mass of particles collected versus time; (b) a graphical derivative of the first curve, with respect to time, showing the cumulative mass of oversize particles; and (c) a graphical second derivative of the first curve, with respect to time, showing the frequency by mass of particles. (Reproduced with permission from reference 5. Copyright 1986 Marcel Dekker.)

reach the balance pan $(t\,dW/dt)$. Here, dW/dt is the slope of the curve in Figure 18a at time t and measures the rate of accumulation of particles smaller than the cutoff size at that time. The intercepts of the tangents dW/dt drawn for various times t yield the values of w for each time. Figure 18b shows a plot of w versus t, the accumulated mass of particles greater than the cutoff size at any time. This is an integrated size distribution curve for particles greater than the cutoff size. Figure 18c shows a graphical differentiation of Figure 18b and is the particle size distribution curve. Although this is not a very accurate method, it shows that in this example the distribution has a significant peak corresponding to a clay particle size that settles in 300 s. By introducing an assumption about particle shape, one can apply a model and calculate particle sizes. Assuming that the particles are spheres and that Stokes' law applies the calculated peak, particle size in this example turns out to be about 24 μm diameter (using $\Delta\rho = 2.17$ g/mL and height 20 cm).

Enhanced Gravity Sedimentation.   It can be seen from Stokes' law that sedimentation will occur faster when there is a larger density difference and when the particles are larger. The rate of separation can be enhanced by replacing the gravitational driving force by a centrifugal field. Centrifugal force, like gravity, is proportional to the mass, but the proportionality constant is not g but $\omega^2 x$, where ω is the angular velocity (=$2\pi \times$ revolutions per s) and x is the distance of the particle from the axis of rotation. The driving force for sedimentation becomes $(\rho_2 - \rho_1) \omega^2 x$. Because $\omega^2 x$ is substituted for g, one speaks of multiples of g, or g's, in a centrifuge. The centrifugal acceleration in a centrifuge is not really a constant throughout the system but varies with x. Because the actual distance from top to bottom of a sedimenting column is usually small compared with the distance from the center of revolution, the average acceleration is used. The terminal velocity then becomes

$$dx/dt = (2a^2(\rho_2 - \rho_1)\omega^2 x)/(9\eta) \tag{22}$$

As an illustration, consider the problem of removing solids and water droplets from deaerated bituminous froth produced from the oil sands hot water flotation process (*see* Chapter 13 and reference 35). This is a nonaqueous suspension from which the particles and water droplets must be removed before upgrading and refining. At process temperature (80 °C) the suspension viscosity is similar to that of bitumen alone, but the density, because of the dispersed solids, is higher. Taking $\eta = 500$ mPa·s (35), $\rho_1 = 1.04$ g/mL and $\rho_2 = 2.50$ g/mL, the rate of settling of 10 μm diameter solid particles under gravitational force will be very slow:

$$dx/dt = (2a^2(\rho_2 - \rho_1)g)/(9\eta)$$
$$dx/dt = 1.59 \times 10^{-5} \text{ cm/s}$$
$$dx/dt = 1.37 \text{ cm/d}$$

In a commercial oil sands plant, a centrifuge process is used to speed up the separation. The continuous centrifuges can operate at 2500 g, the particles having to travel 9 cm to reach the product stream. With the centrifugal force added, the particle velocity would become

$$(dx/dt)' = (2a^2(\rho_2 - \rho_1)\omega^2 x)/(9\eta)$$

$$(dx/dt)' = 2500(dx/dt)$$

$$(dx/dt)' = 3.98 \times 10^{-2} \text{ cm/s}$$

This is 2500 times faster than with gravity alone, but the residence time in the centrifuge would have to be about 4 min. If to speed up the separation naphtha is added to the level of 25 vol%, it will lower the viscosity to about 4.5 mPa·s (36) and lower the density of the continuous phase to 0.88 g/mL. The particle velocity now becomes $(dx/dt)'' = 4.9$ cm/s, which yields a satisfactory residence time of about 2 s. In this particular example the centrifuges also have to separate out emulsified water droplets for which the calculations are analogous but the results are somewhat different (37).

In general, sedimentation will occur more slowly the greater the electrical charge on the particles and the higher the suspension viscosity. Although a distinct process, sedimentation does promote coagulation by increasing the particle crowding and hence the probability of particle–particle collisions.

Energetics of the Interfaces

In simple two-phase colloidal systems, a thin intermediate region or boundary, known as the interface, lies between the dispersed and dispersing phases. Interfacial properties are very important because dispersed colloidal particles have a large interfacial area. Even a modest interfacial energy per unit area can become a considerable total interfacial energy. There is more surface energy in a two-phase system when the dispersed phase is highly subdivided than there is when it is coarsely subdivided. This is why colloidal dispersions have unique properties and why they so often can only be prepared by applying high shearing forces to break down larger particles, droplets, or bubbles.

The ratio of surface or interfacial area to mass of material is termed the specific surface area of a substance, A_{sp}. Consider the specific surface area of two 1-g samples of silica spheres for which in sample 1 the spheres are 1 mm in diameter and in sample 2 they are 1 μm in diameter. The total mass of each is the same (density 2 g/mL), but they do not have the same amount of surface area. For n spheres of density ρ and radius R we have

$$A_{sp} = [(\text{\# particles})(\text{area/particle})]/[(\text{\# particles})(\text{mass/particle})]$$

$$A_{sp} = [n4\pi R^2]/[n(4\backslash 3)\pi R^3 \rho]$$

$$A_{sp} = 3/(\rho R) \tag{23}$$

Sample 1, containing the 1-mm diameter spheres, has a specific surface area of $A_{sp} = 0.0030$ m^2/g, whereas sample 2, containing the 1-μm diameter spheres, has a specific surface area of $A_{sp} = 3.0$ m^2/g. The sample of smaller particles has 1000 times more surface area. Schramm (37) provides an example of the emulsification of one barrel (159 L) of oil into water by repeatedly subdividing a large droplet into droplets until the initial radius of $r = 33.6$ cm becomes 0.64 μm. This represents an interfacial area increase of over five orders of magnitude.

Consider the molecules in a liquid. The attractive van der Waals forces between molecules are felt equally by all molecules except those in the interfacial region. This pulls the latter molecules toward the interior of the liquid. The interface thus has a tendency to contract spontaneously. This is the reason particles of liquid and bubbles of gas tend to adopt a spherical shape: it reduces the surface free energy, or surface tension. For the interface between a solid and a liquid, a similar situation applies, and an imbalance of intermolecular forces will still occur, except that the interface may not be able to curve. The surface free energy has units of mJ/m^2 (1 mJ/m^2 = 1 erg/cm^2) showing that area expansion requires energy. Surface free energies are usually described in terms of contracting forces acting parallel to the surface or interface. Surface tension (γ°), or interfacial tension (γ), is the force per unit length around a surface, or the free energy required to create new surface area. Thus, the units of surface and interfacial tension are mN/m (1 mN/m = 1 dyne/cm). These units for surface and interfacial tension are numerically equal to the surface free energy.

There are many methods available for the measurement of surface and interfacial tensions. Most of these techniques and their limitations are discussed in several good reviews (38–40). While numerous other properties have been investigated many times for aqueous colloidal suspensions, their surface and interfacial tensions have not received much attention. In fact, it is quite common in industrial practice, when dealing with such systems, to remove the solids and measure and report surface and interfacial tensions on a solids-free basis. This is despite the fact that it may be the properties of the suspensions proper that are important in many cases.

Surface tensions of aqueous solutions of inorganic electrolytes are well known to be larger than the surface tension of pure water at the same temperature. The surface tension of 2 M (10.5 mass%) sodium chloride solution is about 3.3 mN/m larger than the surface tension of water at the same (room) temperature. It is commonly believed that the increased surface tension is due to strong interactions between water molecules and the ionic solutes. However, many other solutes cause the surface tensions of their solutions to be smaller than the surface tension of water at the same temperature.

Salmang et al. (*41*) and Taylor and Howlett (*42*) investigated physical properties of ceramic casting slips, which typically consist of suspensions of kaolinite in water in the presence of deflocculating agents. They found that the commonly used casting slip suspensions can have surface tensions that are several mN/m greater than that of pure water. Brian and Chen (*43*) measured the surface tensions of suspensions of iron oxide and silicon dioxide particles in two liquids (water and isoparaffin). Their surface tensions for suspensions of these solids in water were larger than for pure water, whereas suspensions in isoparaffin had surface tensions that were smaller than for the pure liquid. Schramm and Hepler (*44*) used two different methods to measure the surface tensions of aqueous suspensions of sodium montmorillonite and observed that these surface tensions are larger than those of pure water at the same temperatures (Figure 19). Figure 19 shows that dispersed montmorillonite also increases the suspension–toluene interfacial tension compared with that of pure water–toluene. Menon et al. (*45–47*) found that the presence of fine solids at an oil–water interface could either raise or lower interfacial tensions and explained their results in terms of different degrees of repulsive interaction among particles present at the interface. High interfacial coverages with charged shale dust particles were found to raise interfacial tension, but when the negative charge of the clays was

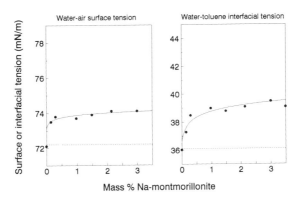

Figure 19. Suspension–air and suspension–toluene surface and interfacial tensions for aqueous suspensions of Na-montmorillonite, as measured with the du Nouy ring technique at 24.5 °C. The broken lines show the values measured for pure water. (Reproduced with permission from reference 44. Copyright 1994 National Research Council of Canada.)

almost completely countered by the adsorption of asphaltenes, interfacial tension lowering was observed. Another feature of the particle inter- action model is that dynamic surface or interfacial tensions might be expected because of the formation of a monolayer followed by particle rearrangements. Williams and Berg (48) support this finding and describe the phenomenon of adsorption and aggregation of colloidal particles at air–liquid interfaces.

It appears that aqueous suspensions of high charge density colloidal solids may exhibit raised surface and interfacial tensions compared with the corresponding solids-free systems and that the time dependence of suspension surface tensions can be appreciable. The effect of suspended particles of low charge density is much less clear. The effect of particle size on the surface or interfacial tensions of suspensions apparently has not been studied.

Contact Angles and Wetting.   Consider now what happens when a drop of oil in water comes into contact with a solid surface. The oil may form a bead on the surface or it may spread and form a film. A liquid having a strong affinity for the solid will seek to maximize its contact (interfacial area) and form a film. A liquid with much weaker affinity may form into a bead. The affinity is termed the wettability. Because there can be degrees of spreading, another quantity is needed. The contact angle, θ, in an oil–water–solid system is defined as the angle, measured through the aqueous phase, that is formed at the junction of the three phases. Whereas interfacial tension is defined for the boundary between two phases, the contact angle is defined for a three-phase junction.

If the interfacial forces acting along the perimeter of the drop are represented by the interfacial tensions, then an equilibrium force balance can be written as

$$\gamma_{W/O} \cos \theta = \gamma_{S/O} - \gamma_{S/W} \qquad (24)$$

where the subscripts refer to water, W, oil, O, and solid, S. This is Young's equation. The solid is completely water wetted if $\theta = 0$ and only partially wetted otherwise. This equation is frequently used to de- scribe wetting phenomena, so two practical points should be remem- bered. In theory, complete nonwetting by water would mean that θ = 180°, but this is not seen in practice. Also, values of $\theta < 90°$ are often considered to represent "water-wetting," whereas values of $\theta > 90°$ are considered to represent "non-water-wetting." This is a rather arbitrary assignment based on correlation with visual appearance of drops on surfaces.

These considerations come into play in petroleum emulsions. The so-called Pickering emulsions are emulsions stabilized by a film of fine particles. The most stable such emulsions occur when the contact angle is close to 90°, so that the particles will collect at the interface. One of the theories of emulsion type states that if an emulsifying agent is preferentially wetted by one of the phases, then more of the agent can be accommodated at the interface if that interface is convex toward that phase, that is, if that phase is the continuous phase. This works for both solids and surfactants as emulsifying agents (Bancroft's rule). Analogous to Bancroft's rule is that the liquid preferentially wetting the solid particles will tend to form the continuous phase. Thus, if there is a low contact angle ($\theta < 90°$ measured through the water phase), then an oil-in-water emulsion should form and a high contact angle ($\theta > 90°$) should produce a water-in-oil emulsion. Although there are many exceptions to such rules, they remain useful for making initial predictions.

In the earlier section on sedimentation, it was assumed that the particles were capable of sedimentation. But if dry particles are placed onto the surface of a body of liquid, they may not settle even though their density is greater than that of the liquid. The explanation provides an example of the actions of surface tension and contact angle. For a solid particle to float on a liquid surface, the upward pull of the meniscus around it (reflected in the surface tension and contact angle) must at least balance the weight of the particle. The surface tension and contact angle can be modified by adding species that alter the interfacial properties, for example, oils or surfactants. A needle can be made to float on water by coating it with oil so that the contact angle, θ, becomes $>90°$.

Consider a cube-shaped mineral particle having sides of length $L = 100$ μm and a density of 2.4 g/cm^3. If the particle is placed on the surface of a container of water of surface tension $\gamma = 72$ mN/m, it exhibits a contact angle of $\theta = 45°$ and a density difference $\Delta\rho = (2.4–1.0)$ g/cm^3. Ignoring the effects of immersion depth and taking upward force to be positive, the force due to surface tension is given by $-4L\gamma \cos\theta$, whereas the force due to gravity is given by $-L^3 \Delta\rho g$. The net force is then approximately $-2.03 \times 10^{-2} - 1.37 \times 10^{-5}$ mN $= -2.03 \times 10^{-2}$ mN, the particle sinks.

This particle that does not naturally float can be made to float. Suppose that a surfactant is added to the water so that the solution surface tension becomes $\gamma = 40$ mN/m and the contact angle becomes $\theta = 140°$. In this case the net force on the particle becomes $+1.22 \times 10^{-2} - 1.37 \times 10^{-5}$ mN $= +1.22 \times 10^{-2}$ mN, the particle now floats.

Density and contact angle are usually modified in a froth flotation process. Thus, "collector" oils or surfactants can be added to ores that adsorb on desirable ore particles increasing the contact angle (and promoting attachment to gas bubbles) but that do not adsorb much on undesirable particles (which do not then attach to the gas bubbles). This combined surface modification and gas bubble attachment allows the desired particles to be made lighter and selectively floated.

The simple example just presented forms the basis for a means of determining the hydrophobic index, an empirical measure of the relative wetting preference of very small solid particles. In one test method, solid particles of narrow size range are placed on the surfaces of a number of samples of water containing increasing concentrations of alcohol (thus providing a range of solvent surface tensions). The percentage alcohol solution at which the particles just begin to become hydrophilic and sink is taken as the hydrophobic index. The corresponding solvent surface-tension value is taken as the critical surface tension of wetting. The technique is also referred to as the film-flotation technique (49) or sink–float method.

List of Symbols

a	empirical viscosity constant (eq 8)
a	major particle dimension (eq 11)
a	particle radius (eqs 15, 17, 20–22)
A	Hamaker constant (eqs 17–19)
A_1	Hamaker constant for the medium (eq 18)
A_2	Hamaker constant for particles (eq 18)
A_{sp}	specific surface area (eq 23)
b	minor particle dimension (eq 11)
B	constant (eqs 15, 19)
c	concentration of particles (eq 3)
c_i	solution concentration of ions i
CCC	critical coagulation concentration (eq 19)
d	distance from a surface (same as x in eqs 12, 15, 17)
D	diffusion coefficient (eq 20)
e	charge on the electron (eqs 16, 19)
g	gravitational constant (eq 21)
H	distance from a surface (same as x in eqs 12, 15, 17)
I	ionic strength (eq 13)
I_d	intensity of light scattered by a particle in a suspension (eq 2)
I_0	intensity of incident light beam (eqs 1, 2)
I_t	intensity of transmitted light beam (eq 1)
k	Boltzmann's constant (eqs 15, 16, 19)
k_2	rate constant for aggregation
$k_2{}^0$	rate constant for diffusion controlled aggregation
K	Rayleigh constants (eq 3)
l	length of optical path through a sample (eq 1)
M	molar mass of particles (eq 3)
n	refractive index of a suspension (eq 3)

n	number of particles per unit volume at time t (eq 20)
n	number of spheres (eq 23)
n_0	refractive index of solvent (eq 3)
n_0	number of particles per unit volume at time $t = 0$ (eq 20)
N_A	Avogadro's number (eqs 3, 19)
r, R	particle radius (eqs 2, 23)
$R(\theta)$	Rayleigh ratio for a scattering angle of $\theta°$
R_{90}	Rayleigh ratio for a scattering angle of 90° (eq 3)
t	time (eqs 20–22)
T	absolute temperature (eqs 15, 16, 19)
V	total potential energy of interaction
V_A	attractive potential energy of interaction (eq 17)
V_R	repulsive potential energy of interaction (eq 15)
w	mass of particles falling the full length of a sedimentation column
W	in aggregation, stability ratio, $W = k_2^{0}/k_2$
W	in sedimentation, total mass of sedimenting particles
x	distance (eqs 2, 12, 15, 17, 21, 22)
z	counterion charge number (eqs 15, 16, 19)
z_i	charge number of ions i

Greek

γ	constant in repulsive potential energy equation (eqs 15, 16, 19)
γ	interfacial tension (eq 24)
$\dot{\gamma}$	shear rate (eqs 5, 6, 10)
$\gamma°$	surface tension
$\gamma_{S/O}$	solid–oil interfacial tension (eq 24)
$\gamma_{S/W}$	solid–water interfacial tension (eq 24)
$\gamma_{W/O}$	water–oil interfacial tension (eq 24)
ϵ	permittivity (eqs 15, 19)
ζ	zeta potential (eq 14)
η	viscosity (eqs 5–11, 21, 22)
η_0	viscosity of liquid continuous phase (eqs 7–11)
$[\eta]$	intrinsic viscosity (eq 10)
θ	contact angle (eq 24)
κ	conductivity of a dispersion (eq 4)
κ	$1/\kappa$ is the double layer thickness (eqs 12, 13, 15)
κ_D	conductivity of the dispersed phase (eq 4)
κ_C	conductivity of the continuous phase (eq 4)
λ	wavelength of light (eqs 2, 3)
μ_E	electrophoretic mobility

ρ	density (eq 23)
ρ_1	external fluid density (eqs 21, 22)
ρ_2	particle density (eqs 21, 22)
σ°	surface charge density
Σ_i	summation symbol
τ	turbidity (eq 1)
τ	shear stress (eq 5, 6)
τ_Y	yield stress (Figure 3)
ϕ	dispersed phase volume fraction (eqs 4, 7–11)
ψ°	surface electric potential (eq 12); ψ_o in Figure 5)
ψ	electric potential at distance x from a surface (eq 12)
$\psi(\delta)$	electric potential at the Stern plane (eqs 14, 16)
ω	angular velocity (eq 22)

References

1. van Olphen, H. *An Introduction to Clay Colloid Chemistry*, 2nd ed.; Wiley-Interscience: New York, 1977.
2. Yariv, S.; Cross, H. *Geochemistry of Colloid Systems for Earth Scientists;* Springer-Verlag: Berlin, Germany, 1979.
3. *Dispersions of Powders in Liquids*, 2nd ed.; Parfitt, G. D., Ed; Applied Science Pub. Ltd.: London, 1973.
4. Shaw, D. J. *Introduction to Colloid and Surface Chemistry*, 3rd ed.; Butterworth: London, 1981.
5. Hiemenz, P. C. *Principles of Colloid and Surface Chemistry*, 2nd ed.; Marcel Dekker: New York, 1986.
6. Adamson, A. W. *Physical Chemistry of Surfaces*, 5th ed.; Wiley-Interscience: New York, 1990.
7. Ross, S.; Morrison, I. D. *Colloidal Systems and Interfaces;* Wiley: New York, 1988.
8. *Colloid Science;* Kruyt, H. R., Ed.; Elsevier: Amsterdam, Netherlands, 1952; Vol. 1.
9. Schramm, L. L. *The Language of Colloid and Interface Science;* American Chemical Society: Washington, DC, 1993.
10. Graham, T. *Philos. Trans.* **1861**, *151*, 183.
11. Laskowski, J. S.; Pugh, R. J. In *Colloid Chemistry in Mineral Processing;* Laskowski, J. S.; Ralston, J., Eds.; Elsevier: Amsterdam, Netherlands, 1992; pp 115–171.
12. Fallenius, K. *Acta Polytech. Scand.* **1977**, *138*, 1–30.
13. Becher, P. *Emulsions: Theory and Practice*, 2nd ed.; Krieger: Malabar, FL, 1977.
14. Williams, R. A.; Simons, S. J. R. In *Colloid and Surface Engineering;* Williams, R. A., Ed.; Butterworth-Heinemann: Oxford, U.K., 1992; pp 55–111.
15. Whorlow, R. W. *Rheological Techniques;* Wiley: New York, 1980.
16. Van Wazer, J. R.; Lyons, J. W.; Kim, K. Y.; Colwell, R. E. *Viscosity and Flow Measurement;* Wiley: New York, 1963.
17. Fredrickson, A. G. *Principles and Applications of Rheology;* Prentice-Hall: Englewood Cliffs, NJ, 1964.
18. Hunter, R. J. *Zeta Potential in Colloid Science;* Academic: Orlando, FL, 1981.

19. James, A. M. In *Surface and Colloid Science;* Good, R. J.; Stromberg, R. R., Eds.; Plenum: New York, 1979; Vol. 11, pp 121–186.
20. Riddick, T. M. *Control of Stability through Zeta Potential;* Zeta Meter, Inc.: Staunton, VA, 1968.
21. O'Brien, R. W.; White, L. R. *J. Chem. Soc. Faraday Trans. 2* **1978,** *74,* 1607–1626.
22. Overbeek, J. Th. G. In *Colloidal Dispersions;* Goodwin, J. W., Ed.; Special Publication 43; Royal Society of Chemistry: London, 1982; pp 1–21.
23. Derjaguin, B. V.; Churaev, N. V.; Miller, V. M. *Surface Forces;* Consultants Bureau: New York, 1987.
24. Verwey, E. J. W.; Overbeek, J. Th. G. *Theory of the Stability of Lyophobic Colloids;* Elsevier: New York, 1948.
25. Luckham, P. F.; de Costello, B. A. *Adv. Colloid Interface Sci.* **1993,** *44,* 183–240.
26. Williams, R. A.; Jia, X. *Chem. Ind. (London)* **1992,** *11,* 409–412.
27. Grim, R. E. *Clay Mineralogy,* 2nd ed.; McGraw-Hill: New York, 1968.
28. Schramm, L. L.; Kwak, J. C. T. *Clays Clay Miner.* **1982,** *30,* 40–48.
29. Schramm, L. L.; Kwak, J. C. T. *Colloids Surf.* **1982,** *3,* 43–60.
30. Alince, B.; Van de Ven, T. G. M. *J. Colloid Interface Sci.* **1993,** *155,* 465–470.
31. Gregory, J. *Crit. Rev. Environ. Control* **1989,** *19,* 185–230.
32. Russel, W. B.; Saville, D. A.; Schowalter, W. R. *Colloidal Dispersions;* Cambridge University: Cambridge, U.K., 1991.
33. Overbeek, J. Th. G. In *Colloid Science;* Kruyt, H. R., Ed.; Elsevier: Amsterdam, Netherlands, 1952; Vol. 1, pp 194–244.
34. Morrison, I. D. *Colloids Surf.* A **1993,** *71,* 1–37.
35. Shaw, R. C.; Czarnecki, J.; Schramm, L. L.; Axelson, D. In *Foams: Fundamentals and Applications in the Petroleum Industry;* Schramm, L. L., Ed.; ACS Advances in Chemistry 242; American Chemical Society: Washington, DC, 1994; pp 423–459.
36. Schramm, L. L.; Kwak, J. C. T. *J. Can. Petrol. Technol.* **1988,** *27,* 26–35.
37. Schramm, L. L. In *Emulsions: Fundamentals and Applications in the Petroleum Industry;* Schramm, L. L., Ed.; ACS Advances in Chemistry 231; American Chemical Society: Washington, DC, 1992; pp 1–49.
38. Harkins, W. D.; Alexander, A. E. In *Physical Methods of Organic Chemistry;* Weissberger, A., Ed.; Interscience: New York, 1959; pp 757–814.
39. Padday, J. F. In *Surface and Colloid Science;* Matijevic, E., Ed.; Wiley-Interscience: New York, 1969; Vol. 1, pp 101–149.
40. Miller, C. A.; Neogi, P. *Interfacial Phenomena Equilibrium and Dynamic Effects;* Marcel Dekker: New York, 1985.
41. Salmang, H.; Deen, W.; Vroemen, A. *Ber. Dsch. Keram. Ges.* **1957,** *34,* 33–38.
42. Taylor, D.; Howlett, D. J. *Trans. J. Brit. Ceram. Soc.* **1974,** *73,* 19–22.
43. Brian, B. W.; Chen, J. C. *AIChE J.* **1987,** *33,* 316–318.
44. Schramm, L. L.; Hepler, L. G. *Can. J. Chem.* **1994,** *72,* 1915–1920.
45. Menon, V. B.; Nikolov, A. D.; Wasan, D. T. *J. Colloid Interface Sci.* **1988,** *124,* 317–327.
46. Menon, V. B.; Nikolov, A. D.; Wasan, D. T. *J. Dispersion Sci. Technol.* **1988/89,** *9,* 575–593.
47. Menon, V. B.; Wasan, D. T. *Colloids Surf.* **1988,** *29,* 7–27.
48. Williams, D. F.; Berg, J. C. *J. Colloid Interface Sci.* **1992,** *152,* 218–229.
49. Fuerstenau, D. W.; Williams, M. C. *Colloids Surf.* **1987,** *22,* 87–91.

50. *Zeta-Meter Applications Brochures;* Zeta-Meter, Inc.: Staunton, VA, 1982.
51. Schramm, L. L. Ph.D. Thesis, Dalhousie University, Halifax, Canada, 1980.
52. Mysels, K. J. *Introduction to Colloid Chemistry;* Interscience: New York, 1959.
53. Barnes, H. A.; Holbrook, S. A. In *Processing of Solid–Liquid Suspensions;* Ayazi Shamlou, P., Ed.; Butterworth-Heinemann: Oxford, U.K., 1993; pp 222–245.
54. Lake, L. W. *Enhanced Oil Recovery;* Prentice Hall: Englewood Cliffs, NJ, 1989; pp. 47–48.
55. Clark, R. K.; Nahm, J. J. In *Kirk–Othmer Encyclopedia of Chemical Technology,* 3rd ed.; Wiley: New York, 1982; Vol. 17, pp 143–167.

RECEIVED for review July 5, 1994. ACCEPTED revised manuscript January 6, 1995.

Particle and Suspension Characterization

Ahmed I.A. Salama and Randy J. Mikula

Western Research Centre, CANMET, Natural Resources Canada, P.O. Bag 1280, Devon, Alberta T0C 1E0, Canada

Particle and suspension characterization is essential for studying particle behavior in suspension and identifying interparticle interactions and association specifically for floc and aggregate and floc morphology. Microscopy remains the benchmark technique against which most others are compared, especially for particle sizing. It is particularly useful for identifying interparticle interactions and associations, specifically for floc or aggregate characterization and determination of floc morphology, which are difficult using other techniques. Therefore, microscopic methods are discussed in more detail than other methods. Aside from microscopic methods, scattering techniques offer the best alternative means to probe interparticle aggregation without disturbing the suspension.

PARTICLES AND SUSPENSIONS of many kinds play an important role in our interaction with the physical environment. They abound in earth, soil, water, and air. They are also present in chemicals and many other industrial products. If particles were spherical or cubical, characterization would be easy. Unfortunately, most of the particles present in the environment are of irregular size and shape. In many industrial processes, size characterization is a critical aspect of understanding particle behavior in suspension, the bulk properties of suspensions, and their bulk behavior. Therefore, it is important to understand the strengths and limitations of techniques used to characterize particles and suspensions.

Suspension characterization begins with characterization of the constituent particles and ions in solution. Bulk suspension behavior, settling, particle aggregation, and rheological properties are all dependent on the chemical properties of the particles and the nature of the suspension fluid. The wide array of analytical methods for materials char-

0065–2393/96/0251–0045$22.50/0

acterization (spectroscopic, wet chemical, X-ray diffraction, fluorescence, and so on) is not discussed here (1), although characterization methods pertinent to suspensions are discussed in some detail.

This chapter first outlines the principles, methods, and basic techniques of particle size characterization used in the petroleum and chemical industries. The most common techniques and methods used in particle size characterization are then briefly discussed. This is followed by a summary of applicable particle size ranges for different methods, including size ranges of the most common particles found in petroleum applications. Emphasis is given to microscopy and scattering techniques and their applications in the petroleum industry.

Microscopy remains the benchmark technique against which most others are compared, especially for particle sizing. It is particularly useful for identifying interparticle interactions and associations, specifically for floc or aggregate characterization and determination of floc morphology, which are difficult using other techniques. Therefore, microscopic methods are discussed in more detail than other methods.

Suspension behavior is determined largely by the particle interactions or degree of flocculation or aggregation. Separating the solids from the suspending fluid is necessary to determine mineralogy or particle shape but results in a loss of information regarding particle aggregation in suspension. Aside from microscopic methods, scattering techniques offer the best alternative means to probe interparticle aggregation without disturbing the suspension. The resolution of scattering methods depends on the radius of particles and wavelength of radiation used. As a result, resolution increases as one moves from light scattering to X-ray scattering to neutron scattering. Beyond differences in resolution, the opacity of the suspending media also determines which method might be suitable.

Solid particles can be viewed as a single phase (e.g., many sand samples or powders) or a mixed phase (e.g., coal, ores). However, dispersions can take different forms, depending on the dispersion medium and the particles being dispersed. Figure 1 illustrates the different types

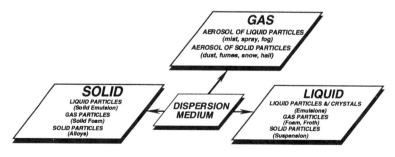

Figure 1. Different types of dispersions.

of dispersions (2). This chapter focuses on dispersions of solid particles in liquids (i.e., suspensions); however, some attention is given to aerosols of solid particles. Throughout the chapter, the term "particle" is used to denote both solid and liquid particles.

The interested reader can pursue this subject in detail by consulting the references listed and the excellent reviews (3–6). SI units are used throughout the chapter.

Particle Properties

Particle characterization is important in many petroleum and chemical applications. Therefore, it is essential to identify the main factors that control the behavior of particles in suspension. These factors include particle shape, size, size distribution, concentration, density, surface characteristics, and the dynamics of the suspension medium (7) (Figure 2).

Particle Shape. *Shape Factors.* The method of formation plays a major role in shaping the resultant particles. Particles generated by comminution, attrition, or disintegration resemble the parent material. However, if the method of formation is condensation from vapor or precipitation from solution, the smallest unitary particle may be spherical or cubical. Often condensation is followed immediately by solidification and the formation of chainlike aggregates (e.g., iron oxide fumes, carbon black). In liquid suspensions, similar particle aggregation or flocculation is important in determining suspension behavior.

Based on experimental data, it was found that for a collection of groups of particles with an average diameter D_{pi} for group i, the total particles surface area A_p can be expressed as (7, 8)

$$A_p = \alpha_s(\Sigma n_i D_{pi}^2) \tag{1}$$

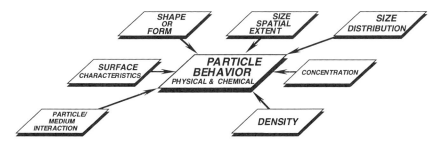

Figure 2. Main factors affecting particle behavior in a medium. (Adapted from reference 18. Copyright 1995 Academic.)

where α_s denotes the surface shape factor and n_i is the number of particles in group i. The total volume can be expressed as

$$V_p = \alpha_v(\Sigma n_i D_{pi}{}^3) \tag{2}$$

where α_v is the volume shape factor (7, 8). The surface and volume shape factors may be related to a particle shape factor k_p as

$$k_p = \frac{\alpha_s}{\alpha_v} \tag{3}$$

For spherical or cubical particles, it can be shown that the shape factor is 6. For irregular rugged particles, k_p values vary from 6 to 10. Other relationships can be obtained by using equations 1–3:

$$A_v = \frac{A_p}{V_p} = \frac{k_p}{D_p{}^*} \tag{4}$$

$$A_m = \frac{k_p}{\rho_p D_p{}^*} \tag{5}$$

$$D_p{}^* = \frac{\Sigma n_i D_{pi}{}^3}{\Sigma n_i D_{pi}{}^2} \tag{6}$$

where A_v, A_m, $D_p{}^*$, and ρ_p are specific surface area (surface area per unit volume), surface area per unit mass, specific diameter, and density of the particle material, respectively.

The surface and volume information allows an estimate of the equivalent spherical particle diameter used to study particle behavior in a suspension.

Fractal Geometry. Mandelbrot (9) introduced the basic concepts and theories of a new type of geometry, called fractal geometry, to describe rugged structures. The main idea behind fractal geometry is that the boundary of a rugged system can be described by a fractal dimension. Such a dimension can be considered as a measure of the space filling properties of the rugged structure boundary. In a fractal surface, the estimate of surface area ($A_p(\lambda_F)$) tends to increase without limit as the step size (resolution) λ_F of an elemental square decreases. This can be expressed as

$$A_p(\lambda_F) = k_a \lambda_F{}^{-(\delta-2)} \tag{7}$$

where k_a is a constant and the fractal dimension δ is >2. Therefore, a plot of $\ln(A_p(\lambda_F))$ versus $\ln(\lambda_F)$ will have a slope $-(\delta-2)$, where $\ln$ is the natural logarithm.

Fractal geometry has been applied to study carbon black agglomerates. It has also been used to study boundary and mass fractal dimensions of aerosol systems, fractal structures of fine particle systems, fragmentation, description of porous bodies, and gas adsorption (*10, 11*). Furthermore, particle aggregation and behavior in oil sands tailings suspensions have been described using fractal geometry (*12*). Using this technique, some tailings properties can be interpreted in terms of the fractal (or self-similar) nature of the particle aggregate.

Density.   Fragments or particles attrited from a solid have the same density as the parent material. However, if the material undergoes hydration or surface oxidization or if it agglomerates in clusters, its specific gravity will change. Particle density plays a major role in the separation of solids, as in centrifugation and gravitational sedimentation, for example.

Surface Characteristics and Interfacial Phenomena

The surface characteristics of small particles include surface area, rate of evaporation, adsorption or condensation of fluids, electrostatic charge, adsorption, adhesion, and scattering of radiation. In some cases, because of changes in the environment of a particle during sampling or particle size characterization, the particle may undergo changes in its size or state of aggregation and, therefore, its surface characteristics. Consequently, such changes must be considered in the selection of size characterization method or suitable sampling device, especially when the degree of aggregation or flocculation determines the properties of suspensions.

Surface Area.   An important characteristic of small particles is the rapid increase in exposed surface area per unit mass as size decreases, which leads to increased chemical reaction rate. Fine powders of organic and inorganic oxidizable materials (such as coal or iron) burn vigorously or explode violently when they are present as aerosols of solid particles. Furthermore, fine sand particles influence the performance of dewatering devices, clay behavior in suspension, and oil sands sludge characterization.

Evaporation and Condensation.   Evaporation and condensation are diffusion mass transfer processes that proceed at rates proportional to the surface area exposed. The temperature and partial pressure near the surface control the time required for small particles (e.g., water) to evaporate into still air. Assuming the diffusion process is perfectly static (i.e., no convection) and the droplet is sufficiently small and assuming first-order approximation, the evaporation time is given by (*7, 13*)

$$\tau = \frac{RT}{8M_{\mathrm{m}}} \frac{\rho_{\mathrm{p}} D_{\mathrm{p}}^{2}}{D \Delta_{\mathrm{p}}} \tag{8}$$

Finer particles can act as nuclei for condensation of moisture, leading to an increase in their size.

Electrostatic Charge.   Small particles naturally acquire charges as a result of electron transfer during contact or separation or during free ion diffusion. These charges reside on the particle surface in a moisture film or an adsorbed gas. Excess or deficiency of electrons on the particle surface represents the possessed electrostatic charge. The mechanisms responsible for producing natural charges on particles in aerosols of solid and liquid particles are shown in Figure 3 (7, 8, 14, 15): (a) contact-separation, electrification (triboelectrification) is produced during the contact and separation of two solid surfaces (surface work function); (b) spray electrification, separation of liquid by atomization results in charged liquid particles due to ion concentration in the particles; (c) electrolytic mechanism, in liquids containing mobile ionized solutes, charge transfer is effected by the motion of ions across an interface; (d) contact potential, arises from the differential work function of two dissimilar metals in contact, free electron transfer occurs across a potential barrier; and (e) ion diffusion in gas, ions may be produced by electrical discharges in air, natural radioactivity, or a flame.

Design of electrostatic classifiers and precipitators is based on the principle that the electrostatic charge generated on a particle is proportional to the particle surface area. Furthermore, the behavior of particles in an electric field is controlled by the presence and amount of the electrostatic charge residing on the particle surfaces.

Scattering Properties.   Scattering of radiation arises from inhomogeneities in a scattering medium, such as suspended dust or water droplets. Scattering is often accompanied by absorption, and both scattering and absorption remove energy from the incident beams. The change of intensity transmitted or scattered beams can be used to characterize the particle size.

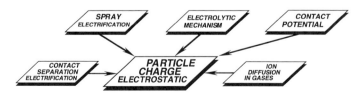

Figure 3.   Mechanisms responsible for producing natural charge on particles. (Adapted from reference 18. Copyright 1995 Academic.)

Most petroleum systems are opaque to conventional light-scattering techniques and require higher energy and shorter wavelength methods, such as X-ray or neutron scattering. Light-scattering techniques have been commercialized for some time and are widely used, mostly for suspensions in water. X-ray methods, although more expensive because of the cost of generating X-rays (versus light sources), are becoming more widely used. Neutron scattering, because of its cost, is still largely a research rather than routine characterization method.

Scattering techniques are used heavily in the petroleum industry for particle and suspension size characterization. They are discussed in detail later in the chapter.

Kinetic Behavior of Particles

The equivalent spherical particle diameter of an aggregate of irregularly shaped particles can be found by studying the inertial motion of particles in a medium. This inertial motion behavior is used in many applications, such as sedimentation vessels, electrostatic separators and precipitators, and particle collectors. The various forces that affect particle motion, shown in Figure 4, are briefly discussed below.

Field Forces.   In particle and suspension characterization, the encountered fields are gravitational, electric, and magnetic (*7, 8, 14, 15*).

Gravitational Field.   Forces are exerted on particles because of the earth's gravity.

Electric Field.   In an electric field, the particle comes under (1) *coulombic* force, due to the charged particle being in an external electric field; (2) *image effect*, due o the presence of an electrode near a ground where the ground acts as a mirror for the electrode and the image electrode has the same potential as the original, and the particle comes

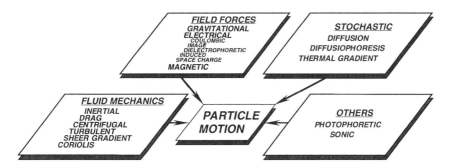

Figure 4.   Different forces affecting particle motion in a medium. (Adapted from reference 18. Copyright 1995 Academic.)

under the influence of both electrodes; (3) *induced charge*, where charge is acquired by a particle as a direct result of being near another charged particle; (4) *space charge*, which can be on electrodes, free ions, and particles; and (5) *dielectrophoretic*, when dielectric particles move as result of an applied external electric field.

Magnetic Field. Particles of magnetic materials are influenced by magnetic forces. Magnetic fields are rarely used in particle size characterization.

Fluid Mechanics and Particle Inertia. Particles in a medium experience (1) *drag*, the resistive force exerted on a particle as it moves in a medium; (2) *inertial forces*, due to the medium flow around the submerged particle or due to the particle path relative to the medium; (3) *centrifugal or vortex forces*, due to the rotational motion of the medium; (4) *Coriolis forces*, due to the linear and rotational motion of the medium; (5) *turbulent forces*, due to the convective transport of the medium; and (6) *shear gradient*, due to the relative movement of medium layers.

Stochastic Processes. Particle motion is a result of (1) *diffusion*, due to the concentration gradient (i.e., particles move from regions of higher density to regions of lower density); (2) *diffusiophoresis*, a specific process where concentration gradients of molecular species in gas result in particle motion (e.g., water vapor moves toward cold water surface carrying particles with it); and (3) *thermal gradient*, in which particles subject to temperature gradient tend to move from hotter to colder regions.

Other Processes. *Photophoresis.* Fine particles of solids in suspension in a medium produce unidirectional motion as a result of an intense beam of light (photons) directed at them. *Sonication.* Sonication is effected by rapidly alternating pressure gradient in the medium. Some applications of these processes are discussed later.

Medium Resistance (Drag). For a small spherical particle moving in a medium at low velocities (i.e., laminar flow), the drag (medium resistance) force acting on the particle is given by Stokes' law as (7, 8)

$$F_R = 3\pi\mu_m D_p V_{pm} = C_D \frac{A\rho_m V_{pm}^2}{2} \tag{9}$$

It is useful to relate the relative magnitudes of the inertial and viscous forces as a dimensionless Reynolds number

$$Re_p = \frac{\rho_m D_p V_{pm}}{\mu_m} \tag{10}$$

The relationship between drag coefficient C_D and Reynolds number can be found in any standard fluid mechanics textbook. However, for spheres with $Re_p < 1$, $C_D = 24/Re_p$.

In finite containers, the drag force acting on a particle increases because of two effects. Because the fluid is stationary within a finite distance of the particle, there is a distortion of the flow pattern that reacts back on the particle. The second effect occurs when the fluid streaming around the particle impinges on the container walls and is deflected back on the particle. Considering these two effects, the drag force may be modified as (8)

$$F_R = 3\pi \mu_m D_p V_{pm}\left(1 + \frac{k_R D_p}{L}\right) \tag{11}$$

where L represents the distance from the particle center to the container walls, $k_R = 0.563$ for a single wall or container bottom, and $k_R = 2.104$ for a cylindrical container (8).

Particle Motion.   *Motion in Gravitational or Drag Fields.*   The linear motion of a particle in a direction X relative to time t and under the influence of drag, gravitational, and buoyancy forces is governed by

$$\frac{\pi}{6}\rho_p D_p^3\left(\frac{d^2X}{dt^2}\right) = \frac{\pi}{6}(\rho_p - \rho_m)D_p^3 g - 3\pi\mu_m D_p \frac{dX}{dt} \tag{12}$$

where g denotes gravitational acceleration. If the particle starts from zero velocity, it will accelerate until it reaches a terminal velocity given by

$$\left(\frac{dX}{dt}\right)_{t\to\infty} = V_{gt} = \left[\frac{D_p^2(\rho_p - \rho_m)}{18\mu_m}\right]\cdot g \tag{13}$$

from which the particle diameter can be expressed as

$$D_p = \left[\frac{18\mu_m}{(\rho_p - \rho_m)g}\cdot V_{gt}\right]^{1/2} \tag{14}$$

Note that equations 16–18 are applicable for $Re_p < 1$.

For particles settling in air, it is usual to neglect the buoyancy correction, because ρ_p is in the order of unity, air density is about 10^{-3} g cm^{-3}, and viscosity of ambient air is 1.8×10^{-5} kg m^{-1} s^{-1}. Equation 13 then reduces to

$$V_{gt} = 3.03 \times 10^4 \rho_p D_p^2 \qquad (15)$$

where V_{gt}, ρ_p, and D_p are expressed in SI units.

Particles with diameters less than or close to the mean free path of the fluid molecules begin to slip between molecules and settle at a higher velocity than that predicted by equation 13. Considering the slip effect, the terminal settling velocity may be modified as (8)

$$V_{gtc} = V_{gt} C_c \qquad (16)$$

where

$$C_c = \left(1 + \frac{2\alpha\lambda}{D_p}\right) \qquad (17)$$

is the Cunningham slip correction factor, λ is the mean free path of medium molecules, and α is a constant of approximately 1 (8).

Particle–Particle Interaction and Hindered Settling. At solids concentrations of <0.5% by volume, the individual particles are on average so far apart they do not affect each other (i.e., no particle–particle interaction) as they move through the fluid (i.e., laminar flow). For practical purposes, there is no particle–particle interaction in suspensions where the ratio of particle diameter to interparticle distance is ≤0.1.

As the solids concentration in the suspension increases, interparticle distances become smaller and the particles start to interfere with each other. If the particles are not uniformly distributed, the effect may be a net increase in settling velocity, because the return flow due to volume displaced will predominate in particle-sparse regions (i.e., cluster formation). The cluster formation effect is significant only in suspensions that are nearly monodispersed. In practice, most suspensions are polydispersed and the clusters in such suspensions are short-lived. As a result, the settling rate steadily deteriorates with the increase in solids concentration because of the return flow being more uniform; this is known as hindered settling.

Hindered settling behavior may be described as Stokes' law correction (i.e., correcting the terminal settling velocity of a single particle) by introducing a multiplying factor, such as

$$V_{gth} = V_{gt} f(\varepsilon) \qquad (18)$$

$$c = 1 - \varepsilon \qquad (19)$$

where $f(\epsilon)$ is a porosity (concentration) function that may take different forms. The most popular and widely used is the Richardson and Zaki equation (16)

$$f(\varepsilon) = \varepsilon^{4.65} \qquad (20)$$

Another relationship was derived by considering the general motion of particles and taking into consideration acceleration of fluid displaced by the particles, wall effects, increase in area available for upward flow, increase in apparent viscosity (which is related to momentum transfer in the suspension), and decrease in gravitational force due to increase in buoyancy as the suspension becomes denser (*17*). The resulting relationship is

$$f(\varepsilon) = \frac{\varepsilon^2}{(1 + (1 - \varepsilon)^{1/3}) \exp \dfrac{5(1 - \varepsilon)}{3\varepsilon}} \qquad (21)$$

The difference between the relationships in equations 20 and 21 over the range of solids concentration ($0 < c < 40\%$) is small. Therefore, the much simpler equation of Richardson and Zaki is more suitable in practice.

Equations 20 and 21 apply only to free, particulate separation not affected by flocculation or aggregation and for all particles having the same density. Under gravity field, few shear forces exist and have little effect on flocs. Flocculation may be strong to change behavior of settling suspension to that of zone settling (i.e., all particles in slurry settle *en masse* at the same velocity irrespective of their size).

Motion in Rotational (Centrifugal or Vortex) Fields.   In a centrifugal or vortex field, the radial motion of a particle at a distance R from the center of rotation is governed by

$$\frac{\pi}{6} \rho_p D_p^{\,3} \left(\frac{d^2 R_r}{dt^2} \right) = \frac{\pi}{6} (\rho_p - \rho_m) D_p^{\,3} \omega^2 R_r - 3\pi \mu_m D_p \frac{dR_r}{dt} \qquad (22)$$

where ω is the angular velocity of the particle. If the particle movement outward is resisted by viscous drag and Stokes' law provides a reasonable approximation for the drag, the terminal radial velocity at equilibrium is given by

$$\left(\frac{dR_r}{dt} \right)_{t \to \infty} = V_{rt} = \left[\frac{D_p^{\,2} (\rho_p - \rho_m)}{18 \mu_m} \right] \cdot \omega^2 R_r \qquad (23)$$

To evaluate the performance of a centrifugal separation process, a dimensionless separation factor, SF, defined as the ratio of centrifugal acceleration to gravitational acceleration, is used

$$SF = \frac{\omega^2 R_r}{g} \qquad (24)$$

For dust-collecting cyclones (particles $>> 100 \; \mu m$), SF = 200, whereas conventional centrifuges used in precipitation of submicrometer particles and large molecules in liquid suspension have SF = 5000.

Motion in Electrostatic Fields.   When particles larger than 1 μm pass through a corona discharge, they acquire charges from electrons and adsorbed gas ions proportional to the surface area of the particle. The saturation charge acquired is given by (7, 14, 15)

$$Q_{pb} = ne = \pi \varepsilon_0 \varepsilon_1 \chi D_p^2 E \tag{25}$$

where

$$\chi = \frac{3\varepsilon_2}{\varepsilon_2 + 2\varepsilon_1} \tag{26}$$

For a particles $< 0.2 \; \mu m$, diffusion charging predominates and the charges acquired at time t are given approximately by (7, 8, 14, 15)

$$Q_{pd} = ne = \frac{D_p kT}{2e} \ln \left(1 + \frac{\pi D_p V_i N_0 e^2}{2kT} t \right) \tag{27}$$

Based on the acquired charge on particle, and if air resistance is approximated by Stokes' law, the particle terminal velocity is given by

$$V_{et} = C_c \left(\frac{EQ_p}{3\pi \mu_m D_p} \right) \tag{28}$$

where Q_p is Q_{pb} or Q_{pd}.

Motion in Thermal Gradient Fields.   It has long been observed that particles in a medium move from hotter to colder regions. In a thermal gradient field and at atmospheric pressure, the thermal force acting on a particle is given by (7, 8)

$$F_t = (-9\pi) \left(\frac{D_p}{2} \right) \left(\frac{\mu_m^2}{\rho_m T} \right) \left(2 + \frac{k_p}{k_m} \right)^{-1} \left(\frac{dT}{dX} \right) \tag{29}$$

The negative sign in the equation indicates that the force is in the direction of negative thermal gradient. By setting the thermal force equal to the resistive force of the medium, the terminal velocity of a particle is given as

$$V_{tt} = (-1.5) \left(\frac{C_c \mu_m}{\rho_m T} \right) \left(2 + \frac{k_p}{k_m} \right)^{-1} \left(\frac{dT}{dX} \right) \tag{30}$$

Particle velocity, flow velocity, and configuration are used to optimize the design geometry of thermal precipitators (*14, 15*).

Particle Size Characterization Techniques and Particle Size Ranges

To cover the scope of the different techniques used in particle size characterization, a survey was made of current literature. The results of this investigation are summarized in Figure 5 (*2, 7, 14, 18*). Also important is identification of the size ranges of most common particles found in the petroleum and chemical applications (*2, 7, 14, 18*) (Figure 6). Different types of industrially used particulate collecting equipment are summarized in Figure 7 (*2, 7, 14, 15, 18*).

Particle size characterization techniques can be broadly classified as direct and indirect techniques. In the direct techniques, the characterization is performed directly on the actual particle, whereas in the indirect techniques, the characterization relies on relationships between physical properties and particle size or shape.

Direct Techniques for Particle and Suspension Characterization

The direct techniques of particle size characterization include sieving and various microscopic methods, each discussed briefly. Details can be

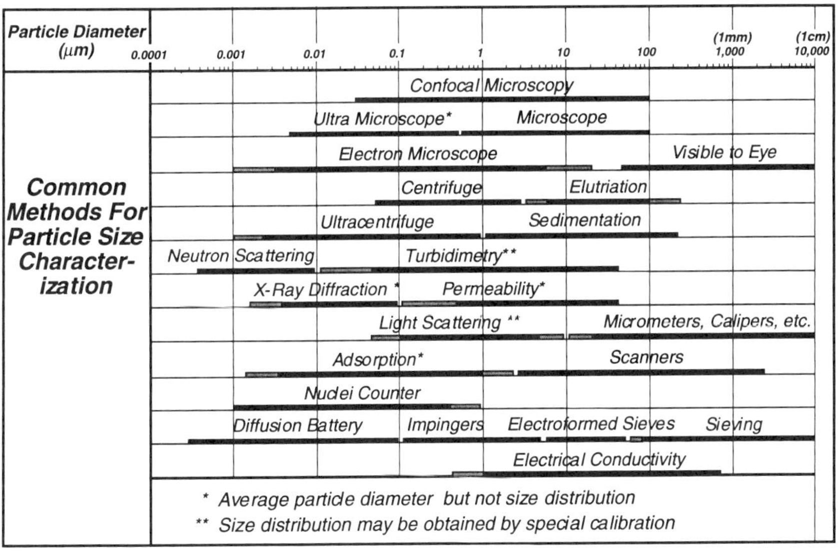

Particle Diameter (µm)	0.0001	0.001	0.01	0.1	1	10	100	(1mm) 1,000	(1cm) 10,000
Common Methods For Particle Size Characterization				Confocal Microscopy					
			Ultra Microscope*		Microscope				
		Electron Microscope					Visible to Eye		
				Centrifuge		Elutriation			
			Ultracentrifuge		Sedimentation				
	Neutron Scattering			Turbidimetry**					
		X-Ray Diffraction*		Permeability*					
			Light Scattering**			Micrometers, Calipers, etc.			
		Adsorption*			Scanners				
		Nuclei Counter							
	Diffusion Battery		Impingers	Electroformed Sieves		Sieving			
				Electrical Conductivity					

* Average particle diameter but not size distribution
** Size distribution may be obtained by special calibration

Figure 5. Common methods of particle size characterization. (Adapted from references 2 and 7.)

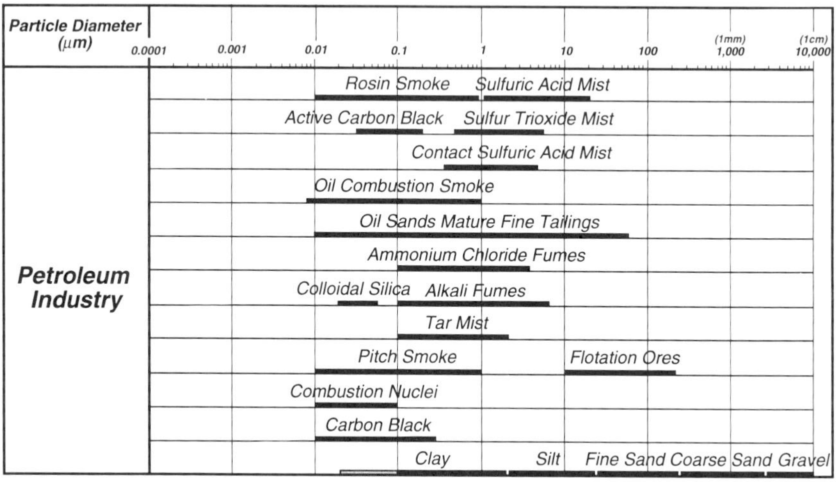

Figure 6. Typical particles and size ranges in the petroleum industry. (Adapted from references 2, 7, and 14.)

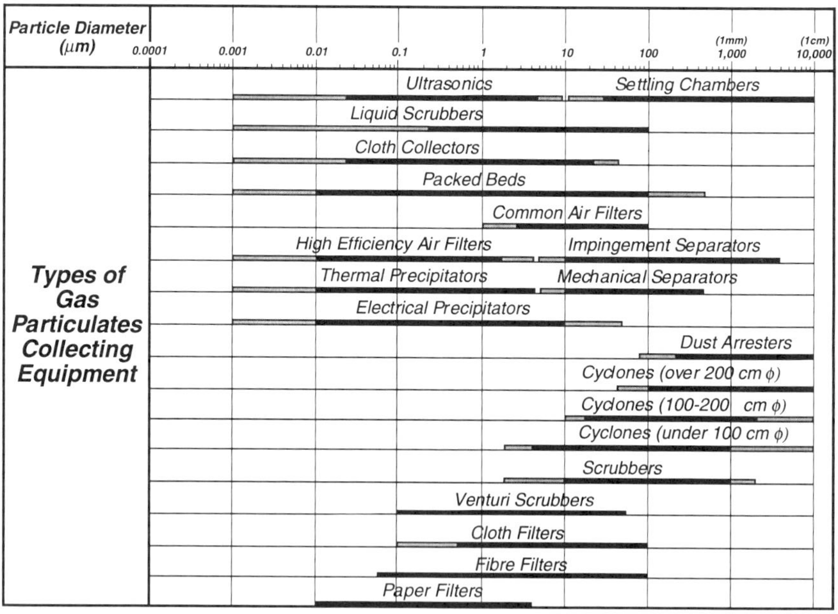

Figure 7. Different types of gas particulate collecting equipment. (Adapted from references 2, 7, 8, 14, and 18.)

found in the references listed. Suspension characterization is discussed in detail.

Sieving. Sieving is probably the most widely used technique for particle size characterization. The particles are classified based on size, regardless of any other particle characteristics such as density and surface properties. Particles of 5–20 μm are classified using micromesh sieves, whereas particles of 20–125 μm are classified in the standard woven wire sieves. Punched plate sieves are used to classify coarse particles (>125 μm). Punched plate sieves with circular and rectangular openings are commonly used in industrial applications.

The sieving test is conducted using up to 11 sieves stacked with progressively larger openings toward the top and placing the solids sample on the top sieve. A closed pan (receiver) is placed at the bottom of the stack. To facilitate the sieving process, the sieves are shaken using mechanical or ultrasonic means. Quantities of residues in each sieve are recorded and expressed as cumulative mass percentages against the nominal sieve aperture values.

Fine sieving can be conducted using machine, wet, hand, and air-jet sieving. Wet sieving is recommended for material originally suspended in a liquid and is required for solids samples that aggregate when dried. In wet sieving, the stack of sieves is filled with liquid and the sample fed is in slurry form at the top sieve. The sieving process can be accelerated by agitation, which can be accomplished by rinsing, vibration, reciprocating action, vacuum, ultrasonic agitation, or a combination of these.

Table I presents some of the different international sieve standards and the corresponding sieve types (8). Several sieve aperture progression ratios are commonly available, depending on the different international standards. In the United States, a progression ratio of $2^{1/2}$ is used. This ratio corresponds to successive particle groups of 2:1 in terms of particle

Table I. International Sieve Standards

Country	Standard	Sieve Type
United States	ASTM E11	Woven wire
	ASTM E161	Micromesh (Electroformed)
Germany	DIN 4188	Woven wire
	DIN 4187	Perforated plate
Great Britain	BS 410	Woven wire
France	AFNOR NFX 11-501	Woven wire
International	ISO R565 (1972)E	Woven wire Perforated plate

area. The progression ratio of $2^{1/3}$ $(10^{0.1})$, which has been adopted by France, corresponds to successive particle groups of 2:1 in terms of particle volume. The progression ratios of $10^{0.1}$ and $10^{0.05}$ are recommended for narrow size distributions.

The main factors that affect the particle passage through sieve aperture are the method of sieve shaking, the ratio of open area of sieve to total area, particle size distribution, the number of particles on the sieve (sieve loading), and the dimension and shape of the particle. Friability and cohesiveness of solid particles can also affect the sieving operation. Difficulty can also arise with high aspect ratio particles (i.e., needle-shaped or flaky particles).

The sieve openings and mesh numbers for U.S.A. Sieve Series and Tyler Equivalents (A.S.T.M. E-11-87) are shown in Table II.

Microscopy. Characterization of particulate from suspensions using microscopic methods is an effective method for establishing the size distribution of the particles and also their composition. Using properly prepared samples, automated image analysis techniques can be used that significantly reduce data collection times. Although the size distribution and composition of the suspended particles are important, the real strength of microscopic methods is the ability to observe particles in suspension and to determine how they interact and associate. There is little difference in microscopic methods applied to solids in suspension or to oil droplets in suspension (emulsions). As a result, the bulk of this discussion is borrowed from a similar chapter on emulsion characterization found in reference 19.

Direct observation of particles in suspension using microscopic techniques is especially useful because it makes it possible to characterize interparticle associations and flocculated structures. Suspension bulk behavior can often be better understood when related to microscopic observations of the interparticle structuring.

With conventional separated powder samples, size distributions obtained using various particle sizing techniques are often compared with values determined by microscopic methods in large part because of the ease of interpretation of the images (20-26). As with other characterization methods, using a representative sample is a prerequisite for microscopic characterization. It is somewhat more important with microscopy because of the small samples that are used, although many of the sample handling concerns discussed in this section apply equally to samples prepared for other characterization techniques. In the case of solid particulate systems prepared as dry samples, direct observation is justifiably the last word in characterization. In the case of suspensions where sedimentation and flocculation can change the nature of the sample, microscopic observation can have unique sample handling problems.

If these special sampling problems are addressed, microscopy can indeed provide the benchmark for the physical characterization of suspensions. Other methods can also be used to characterize and quantify interparticle associations and aggregations. These include NMR imaging, rheological properties, electrotonic amplification, among others that are discussed later.

The complex mathematical treatments used in light-scattering experiments and the experimental complexities of some of the other characterization techniques mean that, in general, great care is taken in the interpretation of the results and operators are aware of potential data reduction problems. In the case of microscopy, because "seeing is believing," there is a tendency to ignore sampling problems and to reach conclusions that are sometimes based on sampling artifacts or peculiarities of the microscopic observation technique.

As long as one is aware of the possible problems, microscopy is one of the most important suspension characterization tools. In the appropriate circumstances, it can give information about the relative amounts of the particulate that form the suspension and their interactions or associations. Various microscopic techniques can be used to define the physical nature of the sample and also the chemical composition, both mineral and organic.

Light Microscopy.   Light microscopy includes the use of transmitted light, reflected light, polarized light, fluorescence, and, more recently, techniques such as confocal microscopy. Each of these variations has particular strengths and applicability to suspensions.

Transmitted light microscopy requires a sample sufficiently thin to allow light to pass through the sample. This is only possible with particulate suspensions in water or light oil. With samples commonly encountered in the petroleum industry, opaque oil components often limit the utility of transmitted light microscopy. In addition, with samples thin enough to be transparent, interactions with the sample holder have to be understood.

Care must be taken to ensure the slide is properly prepared to accept the continuous phase. Water-wet solids suspended in an oil phase can interact differently when prepared on a hydrophilic glass surface compared with a hydrophobic sample holder. Careful observation and comparison of suspension morphology using both hydrophilic and oleophilic sample holders are sometimes required to confidently characterize a suspension. The transmitted light technique is limited by the opaque nature of most oil samples, and in cases where the sample cannot be made thin enough, an alternative technique using reflected light is available.

Table II. U.S.A. Sieve Series and Tyler Equivalents (A.S.T.M. E—11—87)

Sieve Designation		Sieve Opening		Nominal Wire Diameter		Tyler Screen Scale Equiv. Designation
Standard[a]	Alternative	mm	in (approx. equiv.)	mm	in (approx. equiv.)	
125 mm	5 in.	125	5	8	0.315	—
106 mm	4.24 in.	106	4.24	6.4	0.252	—
100 mm	4 in.[b]	100	4.0	6.3	0.248	—
90 mm	3½ in.	90	3.5	6.08	0.2394	—
75 mm	3 in.	75	3.0	5.8	0.2283	—
63 mm	2½ in.	63	2.5	5.5	0.2165	—
53 mm	2.12 in.	53	2.12	5.15	0.2028	—
50 mm	2 in.[b]	50	2.0	5.05	0.1988	—
45 mm	1¾ in.	45	1.75	4.85	0.1909	—
37.5 mm	1½ in.	37.5	1.5	4.59	0.1807	—
31.5 mm	1¼ in.	31.5	1.25	4.23	0.1665	—
26.5 mm	1.06 in.	26.5	1.06	3.9	0.1535	1.05 in.
25.0 mm	1 in.[b]	25.0	1.0	3.8	0.1496	—
22.4 mm	7/8 in.	22.4	0.875	3.5	0.1378	0.883 in.
19.0 mm	3/4 in.	19.0	0.75	3.3	0.1299	0.742 in.
16.0 mm	5/8 in.	16.0	0.625	3.0	0.1181	0.624 in.
13.2 mm	0.53 in.	13.2	0.53	2.75	0.1083	0.525 in.
12.5 mm	½ in.[b]	12.5	0.50	2.67	0.1051	—
11.2 mm	7/16 in.	11.2	0.438	2.45	0.0965	0.441 in.
9.5 mm	3/8 in.	9.5	0.375	2.27	0.0894	0.371 in.
8.0 mm	5/16 in.	8.0	0.312	2.07	0.0815	2½ mesh
6.7 mm	0.265 in.	6.7	0.265	1.87	0.0736	3 mesh
6.3 mm	¼ in.[b]	6.3	0.25	1.82	0.0717	—
5.6 mm	No. 3½[c]	5.6	0.223	1.68	0.0661	3½ mesh
4.75 mm	No. 4	4.75	0.187	1.54	0.0606	4 mesh
4.0 mm	No. 5	4.0	0.157	1.37	0.0539	5 mesh
3.35 mm	No. 6	3.35	0.132	1.23	0.0484	6 mesh
2.8 mm	No. 7	2.8	0.11	1.1	0.043	7 mesh

2.36 mm	No. 8	2.36	0.0937	1.0	0.0394	8 mesh
2.0 mm	No. 10	2.0	0.0787	0.9	0.0345	9 mesh
1.7 mm	No. 12	1.7	0.0661	0.81	0.0319	10 mesh
1.4 mm	No. 14	1.4	0.0555	0.725	0.0285	12 mesh
1.18 mm	No. 16	1.18	0.0469	0.65	0.0256	14 mesh
1.0 mm	No. 18	1.0	0.0394	0.58	0.0228	16 mesh
850 μm	No. 20	0.85	0.0331	0.51	0.0201	20 mesh
710 μm	No. 25	0.71	0.0278	0.45	0.0177	24 mesh
600 μm	No. 30	0.6	0.0234	0.39	0.0154	28 mesh
500 μm	No. 35	0.5	0.0197	0.34	0.0134	32 mesh
425 μm	No. 40	0.425	0.0165	0.29	0.0114	35 mesh
355 μm	No. 45	0.355	0.0139	0.247	0.0097	42 mesh
300 μm	No. 50	0.3	0.0117	0.215	0.0085	48 mesh
250 μm	No. 60	0.25	0.0098	0.18	0.0071	60 mesh
212 μm	No. 70	0.212	0.0083	0.152	0.006	65 mesh
180 μm	No. 80	0.18	0.007	0.131	0.0052	80 mesh
150 μm	No. 100	0.15	0.0059	0.11	0.0043	100 mesh
125 μm	No. 120	0.125	0.0049	0.091	0.0036	115 mesh
106 μm	No. 140	0.106	0.0041	0.076	0.003	150 mesh
90 μm	No. 170	0.09	0.0035	0.064	0.0025	170 mesh
75 μm	No. 200	0.075	0.0029	0.053	0.0021	200 mesh
63 μm	No. 230	0.063	0.0025	0.044	0.0017	250 mesh
53 μm	No. 270	0.053	0.0021	0.037	0.0015	270 mesh
45 μm	No. 325	0.045	0.0017	0.03	0.0012	325 mesh
38 μm	No. 400	0.038	0.0015	0.025	0.001	400 mesh
32 μm	No. 450		0.00126	0.0011		
25 μm	No. 500		0.00098	0.001		
20 μm	No. 635		0.00079	0.008		

*a*These standard designations corresponds to the values for test sieves apertures recommended by the International Standards Organization, Geneva, Switzerland.

*b*These sieves are not in the fourth root of 2 series, but they have been included because they are in common usuage.

*c*These numbers (3½ to 400) are the approximate number of openings per linear inch but it is preferred that the sieve be identified by the standard designation in millimeters or μm. 1000 μm = 1 mm.

Figure 8 is a schematic of the experimental setup for reflected light microscopy. Using reflected light, the sample can simply be put in a small container or well slide. A cover slip can be put over the sample and an oil immersion lens used. For suspensions in water, the objective lens can be immersed directly into the sample. Clays and other suspended solids are often transparent to white light, and in these cases, polarized light can be used to observe the clays. To enhance observation of the oil and to determine how the oil phase associates with the suspended solids, the fluorescence behavior of the organic phase is used. This involves incident light of violet or UV wavelength and observation of the fluorescent light in the visible region. The incident reflected beam is filtered out, and the returning light is due to the fluorescent behavior of the oil phase. Figure 9 illustrates the effect of using polarized light [for clays (right)] or fluorescence behavior [for the organic components (left)] to help characterize suspension components.

A potential problem with light microscopy, especially with high intensity mercury vapor lamps (for blue-violet incident light), is localized sample heating; however, for suspensions this phenomenon is generally not as important in creating image artifacts as interactions with the sample holder.

The availability of low-cost computing and image analysis capability has helped to reduce the time involved in quantification of microscopic analysis to determine size distribution. The comparison of images to rulers photographed under the same conditions and the use of split image microscopes (24–26) have largely been replaced by automated image analysis techniques. Image analysis (particle recognition) routines are coupled to computers that can control the electron microscope, allowing both particle size and composition to be determined automatically. Figures 10 and 11 illustrate the steps in this automated sizing and composition analysis. To be confident of an average size analysis within 10%, approximately 150 particles should be sized. To increase the confidence level to 5%, approximately 740 particles should be sized (23). Of course these numbers are only a rough guide, and the actual confidence levels will depend on the nature of the size distribution. Problems can occur when too much reliance is placed on automated image analysis, namely, inaccurate sizing of droplets or particle aggregates that are slightly out of focus. The fields of view to be analyzed, whether manually or using automated methods, have to be chosen carefully.

Because of the interactive nature of the microscopic technique, in other words, the human factor, there can be differences in size analyses by different operators and operator bias to either small or large particles. Missing large particles affects the mass distribution and neglecting small particles affects the number distribution. Size distributions may also be

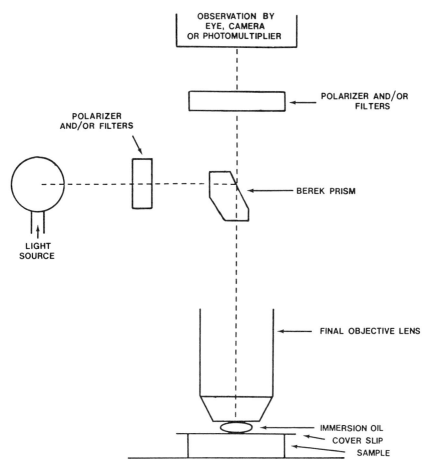

Figure 8. Schematic of the light microscope in reflectance mode. Air or oil immersion objectives may be used. Oil immersion objectives require a glass coverslip over the sample and an oil drop of appropriate refractive index to bridge the gap between the objective lens and the sample cover glass. This provides much higher resolution than air objectives. The light source can be plain or polarized white light (tungsten lamp) for observation of solids or appropriately filtered blue-violet light (high pressure mercury lamp) to excite fluorescence of certain organic components. Other techniques, such as dark field illumination, allow one to count particles without sizing them. The particles in this case are seen only as points of light on a dark background. (Adapted from reference 19. Copyright 1992.)

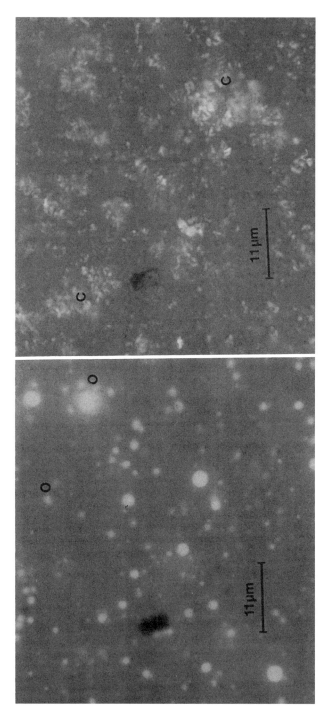

Figure 9.   White-light (polarized) photomicrograph in reflectance mode of a clay suspension with significant oil content. With polarized light, the clays (C) appear bright and the oil droplets cannot be seen at all. On the left is the same field of view but in fluorescence mode in which only the organic components can be seen. With conventional optical images, features above or below the focal plane will appear as points of light or have halos that prevent accurate sizing. (Adapted from reference 19. Copyright 1992.)

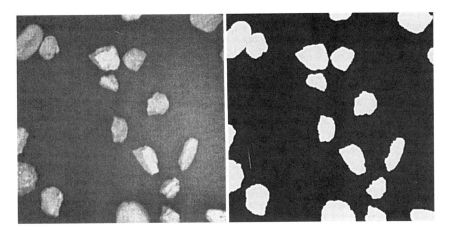

Figure 10. Scanning electron micrograph of particulates (left) and the computer-generated binary image (right). With computer control of the microscope stage and electron beam, the particles can be individually characterized as to composition and size distribution. With appropriate detector configurations, the computer can automatically identify the particles, size them, and control the SEM to determine the composition of the particulates. These methods are currently in common use only for separated dried particulates, although, in principle, similar techniques could be used for characterization of particulates in suspension.

biased by inappropriate sampling or data reduction. Of course these concerns are not limited to microscopic analysis.

With particles smaller than about 0.5 μm, one begins to approach the resolution limit of the light microscope, and often particles can be recognized but not sized properly because of limitations in both the resolution and the depth of field or focus in the optical system.

The confocal microscope solves some of these problems and adds a new dimension to light microscopic analysis. In conventional light microscopy, light from the entire illuminated area is returned via the lenses to form the image. As a result, areas of the sample that are out of the depth of field, and therefore out of focus, degrade image quality. With the confocal microscope, the laser light source is scanned across the sample and focused through a pinhole that accepts only the in-focus light from a defined focal depth. A schematic of the confocal microscope components is found in Figure 12. The confocal microscope digitizes the intensity information in a field of view and, by accurately moving the sample in the z-direction, makes it possible to reconstruct an image that is in focus over a significant depth in a sample (depending on how many in-focus z-slices of the image have been acquired). Through this reconstruction, it is possible to produce an optical image that provides

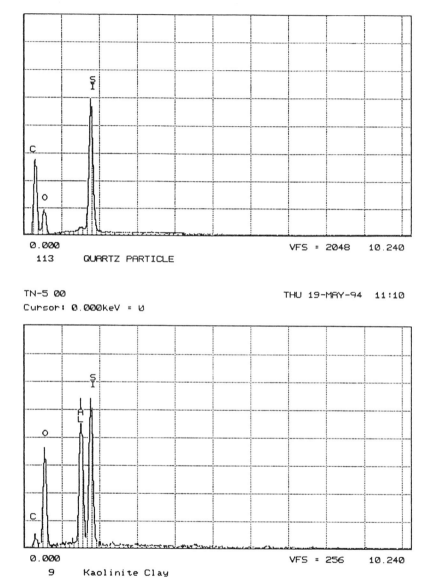

Figure 11.   X-ray spectra of selected particles from Figure 10. Particles can be characterized in terms of their X-ray spectra. For example, high Si is indicative of quartz, whereas Al and Si suggest clays. In certain cases, the Si:Al ratio can also distinguish between kaolinite and illite. Other components such as pyrite (Fe and S), rutile (Ti), ilmenite (Ti and Fe), and coke (C and S) can also be identified automatically by their characteristic X-rays.

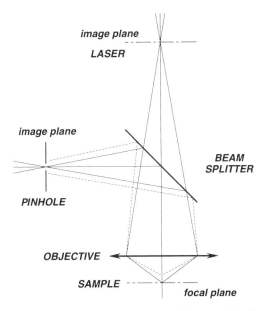

Figure 12. Schematic diagram of the scanning laser confocal microscope. The out-of-focus information that normally reaches the eyepiece (detector) and leads to difficulties in interpreting optical images is rejected because the optical path does not take it through the pinhole. By scanning the incident laser beam across the sample, a digitized image is constructed from the in-focus light rays that pass through the pinhole.

more information about associations between the water, oil, and solid phases and to observe interparticle associations in suspensions that would be ordinarily unseen because the floc structure could not be entirely in focus using conventional light microscopy. Figure 13 shows a scanning laser confocal microscope image of a concentrated suspension similar to that in Figure 9. This image represents a composition of over 50 in-focus slices. Using the summed images from the confocal microscope, resolution and depth of field are improved significantly.

Electron Microscopy. Both scanning and transmission electron microscopy have been used extensively to characterize suspensions (27–30). Transmission electron microscopy is somewhat less commonly used for suspensions and almost invariably involves the observation of replicas or metal reproductions of the sample. Scanning electron microscopy (SEM) is much more common and is analogous to conventional light microscopy, although the scanning electron microscope offers significant advantages over the light microscope in terms of depth of field and resolution. However, the vacuum environment and energy deposited by the electron beam means that sample handling and preparation are

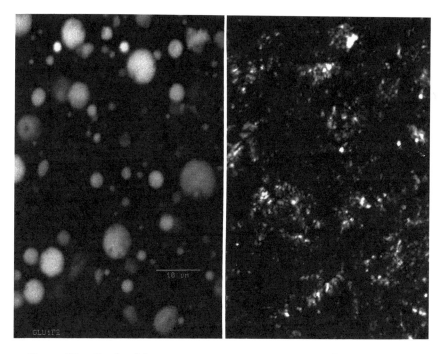

Figure 13.   Confocal laser scanning microscope images of a concentrated suspension using image acquisition at different wavelengths. On the right is a white-light image in which the oil phase is relatively transparent, and on the left is a fluorescent image in which the oil phase is observed and the mineral phase is essentially transparent. All of the features in this image appear in focus because it is made up of the sum of several in-focus planes.

much more difficult. The two main techniques commonly discussed in the literature are known as direct observation (or frozen hydrated observation) and the observation of replicas. Both techniques involve the fast freezing of the sample in a cryogen, such as liquid nitrogen, propane, or freon (among others). The frozen sample is then fractured to reveal the interior features. In the case of suspensions in water, a layer of water is usually sublimed away to observe the interparticle morphology more accurately. This fractured and sublimed surface can be coated with a metal film or observed directly. Often, the metal film is removed from the sample and observed as a replica. This type of procedure allows one to retain a permanent archive of the samples prepared, and the observation is the same as with any other electron microscope sample, with no concern about contamination of the microscope or beam damage to the sample. Transmission electron microscopy, with few exceptions, involves the creation of replicas because it depends on the electron beam passing through the sample, with regions of low or high density appearing as bright and dark areas. When replicas are used in transmission electron microscopy, the metal or carbon replica is shadowed with a second metal to accentuate the topography of the suspension on the replica. These methods are outlined in Figure 14.

SEM is relatively simple compared with transmission electron microscopy, and the images obtained are significantly easier to interpret. In addition, the direct observation option gives the potential to obtain much more information about the chemical composition of suspended solids.

With direct observation, the sample must be kept cold in the electron microscope, and care is required to prevent sample damage in the beam and to prevent microscope contamination. In addition, uncoated frozen samples are often difficult to image because of charging effects that distort the image. The benefit of using uncoated samples, however, is that electron beam interactions with the sample produce characteristic X-ray signals that allow one to identify components of the suspension. Furthermore, sublimation can be carried out in the electron microscope, enhancing features that may be important. With replicas, the sublimation step must be carried out before one has a chance to determine whether the sublimation was carried out for too long or too short a time.

The availability of high-speed computers and the ability to digitize electron signals at video rates mean that despite poor initial image quality in dealing with direct observation of frozen hydrated samples, it is possible to average several relatively noisy images to reduce the noise level and produce an image that is close in quality to that which can be obtained from replicas.

Figure 15 shows a quick-frozen suspension (oil sands mature fine tailings) of clays in a continuous water phase. After sublimation, the structuring of the suspended solids is clearly evident. Figures 16 and 17 show the same sample after vibration in which the structure has been destroyed.

Comparison of Light and Electron Microscopy.   Light and electron microscopic techniques are quite complimentary in terms of the information that they can provide. Light microscopy, in fluorescence mode or with polarized light, can provide information about the organic phases in a suspension sample. This is particularly useful in suspension samples that may be sterically stabilized by organic components or have a bi-wetted nature because of organic patches on an otherwise hydrophilic surface. Electron microscopy, through the X-rays excited in the sample, can also provide information about the inorganic or mineral phases present. This complimentary nature is shown in the series of photographs in Figures 18 and 19, where correlated confocal and cryo-SEM images illustrate the benefits in each technique.

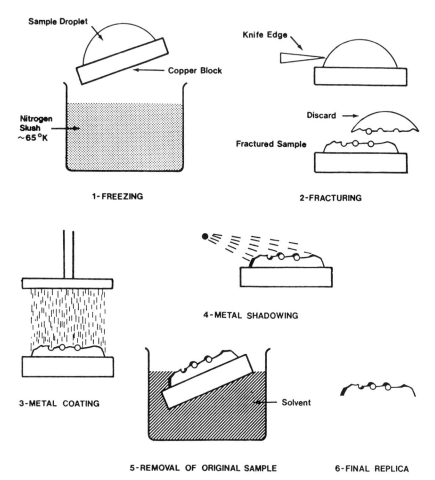

Figure 14.   The six main steps in freeze fracture sample preparation technique for electron microscopy. First, the sample is frozen rapidly to prevent ice crystal formation and sample distortion. After freezing, the sample is fractured to reveal the interior features and is then put into an electron microscope with appropriate cryogenic capability (frozen hydrated or direct observation). More commonly, a sample replica is created by coating the fractured surface with metal and dissolving the original sample; the replica is then observed in a conventional electron microscope. For transmission electron microscopy, where the image formation depends on differences in sample density, the replica must be shadowed or coated directionally with metal to provide a density contrast in the peaks and valleys of the replica. During the freezing and metal-coating steps, the sample must be kept in a vacuum to keep it frozen and to prevent frost deposition, which can sometimes be mistaken for the dispersed phase. With direct observation, it is possible to warm the sample carefully (to about 130–150 K) and sublime any frost layer away. This obviously cannot be done using a replica. Another advantage of direct observation is the X-ray information that can help identify the composition of the various suspension components. (Adapted from reference 19. Copyright 1992.)

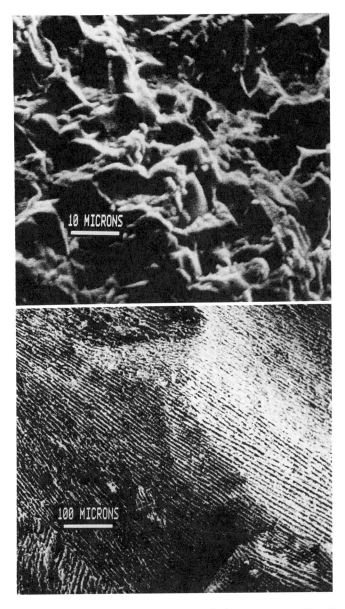

Figure 15.   Cryo-SEM image of an oil sands sludge suspension. Top: Sample frozen and fractured but before sublimation. The surface of the fracture is relatively featureless. Bottom: Sample after sublimation of some of the water matrix, revealing the "card house" nature of the clay particle association.

Figure 16. Cryo-SEM image of a similar oil sands sludge sample as in Figure 15. After mild vibration of the sample, the frozen, fractured, and sublimed sample shows disturbance of the structuring observed with a settled sample. The utility of microscopic methods in probing the important floc structure of suspensions rather than simply characterizing the component particles is illustrated.

The practical lower limit of suspension particle sizing using light microscopy is 0.5 μm. The limit is much lower using electron microscopy: 0.1 μm or less with direct observation of frozen samples in a scanning electron microscope and 0.01 μm or less using replicas and transmission electron microscopy. Sizes smaller than this can be recognized with each of these techniques, but quantification of the size distribution becomes difficult, as does characterization of interparticle interactions. Furthermore, at resolution levels of about 0.01 μm, it is extremely difficult to avoid artifacts and subsequent misinterpretations. As mentioned earlier, sample preparation is an extremely important consideration in both optical and electron microscopic techniques. With optical microscopy, interactions of the sample with the sample holder can affect interparticle interactions, especially with hydrophilic solids and glass (hydrophyilic) sample holders; with electron microscopy, artifacts due to the freezing process can affect interpretation of the results (20, 27).

Microscopic techniques are attractive for suspension characterization because they are capable of directly determining the size distribution of the suspended phase as well as chemical compositional infor-

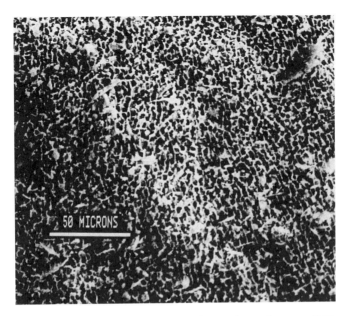

Figure 17.   Cryo-SEM of oil sands sludge after violent vibration (followed immediately by freezing) that destroyed the structure of the clay flocs.

mation about both organic and inorganic components. Most important is the ability to directly determine flocculation structures in the suspension and interactions between the suspended solids and the organic or oil phases. However, one must be aware of the limitations of this technique in terms of sample handling and preparation and of the danger of over interpreting images once they have been acquired.

NMR Imaging. Imaging using magnetic resonance properties has been in use for many years (*31*). Sometimes referred to as NMR microscopy (*31, 32*), the method involves no lenses or reflected or transmitted radiation. In fact, the image is created by changing the magnetic field gradient across and through a sample in an NMR instrument. The most common use of this technique is in medicine, although it has recently seen applications in the petroleum industry as well (*33*). Although resolution is limited by the ability to accurately define the magnetic field gradient and by sample properties that affect the nuclear spin resonance (such as the presence of paramagnetic material), it is possible to routinely achieve resolution on the order of millimeters and, in special cases, on the order of 10–100 μm. The advantage of the method is that it can monitor suspension behavior in otherwise completely opaque systems in real time. Figure 20 shows a series of magnetic resonance images

Figure 18.   Laser scanning confocal microscope image of the same structure as shown in Figure 15. In this case the sample was not frozen but shows the same ordering of the clays.

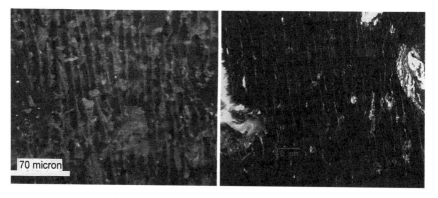

Figure 19.   Correlated images of the same sample observed using cryogenic SEM (left) and scanning laser confocal light microscopy (right). The confocal image in fluorescent mode (right) shows a concentration of fluorescing components that correlates with the clay structure.

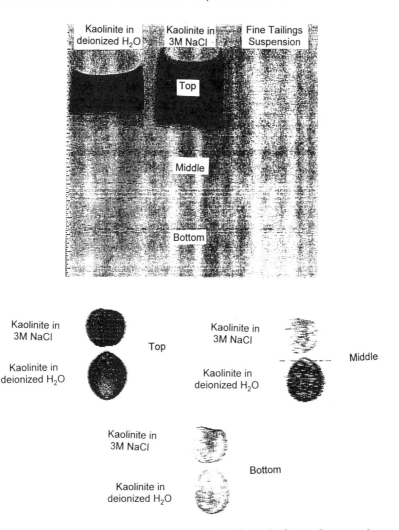

Figure 20. Magnetic resonance images of oil sands fine tailings settling behavior. This noninvasive technique allows observation of the sample segregation in a situation where the sample is otherwise completely opaque. The darker regions of the sample represent regions of higher water concentration or higher water mobility. A higher solids concentration is therefore represented by the lighter shading. The first image shows NMR images of three suspensions: kaolinite in deionized water (left), kaolinite in a 3 M salt solution (middle), and a suspension of fine tailings from an oil sands operation. The fine tailings showed no settling behavior at all, whereas the kaolinite in water showed some settling. The three slices taken at the top, middle, and bottom of the two kaolinite samples show that the technique can probe settling behavior without disturbing the sample. The kaolinite in 3 M salt shows a distinctly denser sediment via NMR imaging that could not be otherwise determined due to sample opacity. (Images provided by Dr. D. Axelson of the Petroleum Recovery Institute and by Dr. Peter Smith of the University of Alberta.)

of oil sands sludge and kaolinite settling in which the solids content gradient is clearly depicted by water imaging techniques using magnetic resonance. This settling behavior could be monitored as a function of time without disturbing the sample and therefore could become extremely important in characterizing flocculation and settling behavior in suspensions.

Flocculation Structure and the Degree of Dispersion. Determination of solids content and size distribution is only part of the characterization needed to understand the bulk behavior of suspensions. Suspension properties such as flocculation, sedimentation, and interfacial tension are discussed in Chapter 1. In an attempt to quantify the degree of dispersion of oil sands fine tailings, several empirical (as well as fundamental) tests were developed because direct observation of interparticle interactions (using microscopic techniques) is often tedious and time-consuming (34). One of these tests is simply the double hydrometer test in which a sample is evaluated "neat," followed by the standard method of dispersing the suspended solids. The difference in these two determinations can then be related to the degree of dispersion of the suspension. Of course, direct observation using microscopic methods is still the technique of choice. Many pertinent examples were already discussed in the section on microscopy.

Figure 21 shows an example of double hydrometer test data in which the suspension is evaluated with and without dispersants. The decrease in particle size when the particle floc aggregates are dispersed is a measure of the degree of flocculation. Figure 22 shows optical micrographs of two dispersed and undispersed suspensions of oil sands tailings, corresponding to the particle size distributions in Figure 21.

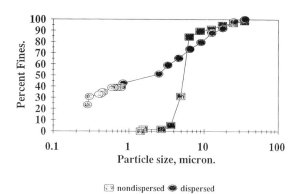

□ nondispersed ● dispersed

Figure 21. Hydrometer tests to determine particle size distribution in a suspension and the same suspension after dispersion of the particle flocs or aggregates. The difference in the two curves is an indication of the degree of dispersion or degree of flocculation or aggregation.

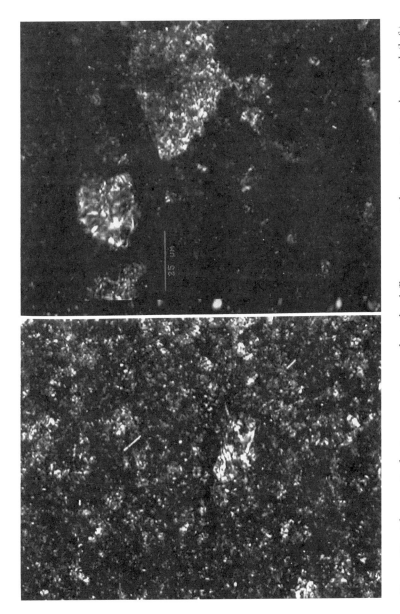

Figure 22. These optical microscope images show the difference in particle aggregation in dispersed (left) and nondispersed (right) oil sands tailings.

Viscosity (pumping behavior) and settling rate are the bulk properties of practical consequence in most petroleum applications, and these properties are clearly related to the fundamental way in which the particles in suspension interact with each other (the degree of dispersion or flocculation) and with the other components (oil and dissolved ions). The following section discusses some fundamental and empirical methods for characterizing the flocculation behavior of suspensions. Most of the examples are from oil sands fine tailings investigations, although the methods are applicable to many suspended solids systems.

Rheology.   Rheological properties are introduced in Chapter 1 and are discussed in detail in Chapter 3. The semiempirical approach of relating elastic modulus to the degree of dispersion of a suspension is all that is discussed here.

The elastic modulus (G′) of a system is a measure of how it reacts to an applied stress that is not severe enough to break any structure. This makes it a valuable method to quantify the degree of dispersion of particles in suspension. A lower G′ implies less interparticle structure or flocculation and therefore a greater degree of dispersion. Figure 23 shows the behavior of G′ as a function of pH for mature fine tailings from a commercial oil sands extraction process. The minimum corresponds to a maximum in the electrostatic repulsion of the particles as determined by electronic amplification (*34*).

Electrosonic Amplification.   This technique is a novel application of electrokinetics to characterize particulate interactions. Application

SUNCOR TAILINGS

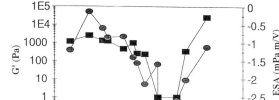

Figure 23.   The G′ or elastic modulus and the electrosonic amplitude (ESA) of an oil sands fine tailings as a function of pH. The G′ or response of the suspension structure to nondestructive shear is determined by the strength of the particle flocs. The particle interactions or floc formation is a function of the surface charges on the particles (determined by the electrosonic amplitude) and this, in turn, is sensitive to the pH of the system.

of an electric field to a particle in suspension causes distortion of the charged double layer, resulting in some particle motion. This particle motion is analogous to a sound wave in the sample cell. The electrosonic amplification method uses this behavior by either applying an electric field to a sample and monitoring the sound wave or by applying sound and monitoring the resulting electrical signal. In either case, the signal does not depend on actual observation of particle motion as with conventional microelectrophoresis. It is therefore extremely useful in characterizing opaque or high solids content suspensions. Figure 23 shows the electrosonic amplification behavior of oil sands mature fine tailings suspension as a function of pH. With conventional electrophoresis, the particles must be very dilute in suspensions. The mature fine tailings suspension has a solids concentration of about 30%. This particle charge (reflected in the magnitude of the ESA signal)–pH dependence is related to the determination of elastic modulus and confirms that the suspension is most disperse when there is a greater absolute value of charge on the particles.

Specific Resistance to Filtration. Specific resistance to filtration (SRF) is a standard method (*34*) for determination of the industrial amenability of a material to dewatering by filtration methods. Modifications of this method are useful empirical tools for evaluating interparticle interactions in concentrated suspensions. Figure 24 shows a schematic diagram of the apparatus in which the water released during filtration under standard conditions can be measured. Figure 25 shows data collected on oil sands mature fine tailings as a function of pH. As with the other methods, these data indicate that material is most difficult to handle at basic pHs.

Settling Tests. Settling tests are the most straightforward way to monitor the densification of suspensions and were discussed earlier. In the case of suspensions having high solids concentrations in which interparticle interaction and floc formation are important, settling tests performed under various field gradients can yield useful information. Low speed centrifugation is an easy way to enhance the gravity field under which settling generally occurs. Figure 26 shows settling results for low speed centrifugation of a fine tailings sample that has been allowed to settle slowly compared with one that was settled at a much higher force gradient (because of centrifugation). Slow settling allowed flocculation and interparticle associations that significantly strengthened the suspension structure compared with the particles that were settled under a higher gradient. The latter particles settled to a higher solids content, beyond the metastable structure set up with gravitational settling.

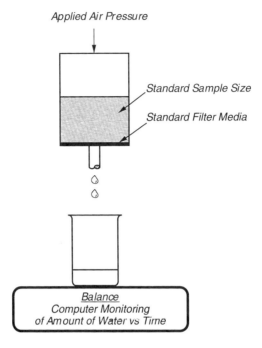

Figure 24.   A schematic of the apparatus used to measure the specific re-sistance to filtration. This is a useful empirical method for comparing sus-pensions. With a defined pressure for filtration, coupled with the quantifi-cation of the filtrate produced as a function of time, suspension characteristics can be compared.

Figure 27 shows the low speed centrifugation behavior of mature fine tailings from the oil sands extraction process as a function of pH. The greatest amount of water is released during centrifugation at pHs that correspond to less disperse suspensions. Figures 28 and 29 show shear stress versus shear rate for a suspension subjected to different pretreatments. In Figure 28, the sample was sheared prior to deter-mining shear stress versus shear rate. In Figure 29, the sample was allowed to set for 72 h, and this suspension shows a large yield point (indicative of gel strength) or shear stress at low shear rates. The time or treatment dependence of the rheological properties is related to the structures observed microscopically (Figures 15–17), with the sheared sample showing no yield stress or gel strength relative to a sample that was allowed to sit and form some interparticle structure.

Indirect Techniques for Particle and Suspension Characterization

The indirect techniques for particle size characterization include scat-tering techniques, sedimentation (gravitational or centrifugal), elutria-

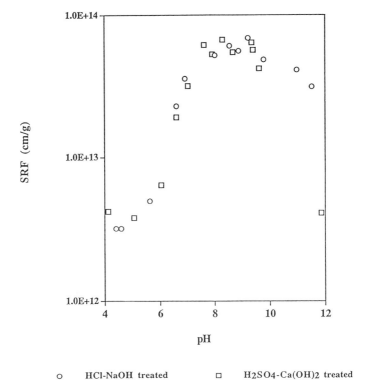

○   HCl-NaOH treated   □   H2SO4-Ca(OH)2 treated

Figure 25.   The specific resistance to filtration of oil sands sludge as a function of pH. This behavior correlates to the behavior of G' and ESA as a function of pH (Figure 23). The lowest resistance to filtration occurred at a pH where G' and ESA indicated the most dispersed suspension.

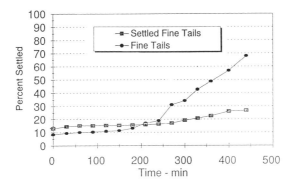

Figure 26.   Centrifugation behavior of fresh fine tailings compared to settled sludge. The fine tailings compacted more than the slowly settled sludge because no structure had been allowed to form that would resist settling and consolidation.

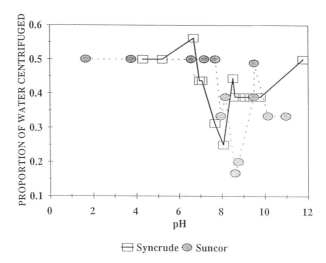

Figure 27.   Low speed centrifugation of oil sands fine tailings as a function of pH. This correlates with the behavior observed in specific resistance to filtration, G', and ESA in Figures 23 and 25.

tion, impaction, electrostatic precipitation, thermal precipitation, hydrodynamic chromatography, and electrical sensing zone techniques. Scattering techniques are heavily used in the petroleum industry for particle and suspension characterization and therefore are discussed in detail. Additional details on the other techniques can be found in the bibliographic references.

Sedimentation.   The fine solids particle size distribution can be determined by sedimentation of the fine solids in a suspension. Table III presents a classification of the methods used in sedimentation (gravitational or centrifugal).

Suspensions for sedimentation can be prepared using line start (two-layer) or the homogeneous suspension techniques. Figure 30 illustrates suspension sedimentation at the start of the test and at a subsequent time for each technique. In the two-layer technique, the solids sample is introduced at the top of a column of clear liquid. In the homogeneous suspension technique, the solids sample is uniformly suspended in the liquid. As the particles start to settle, the change in solids concentration at a particular fixed height with time or the sedimentation time rate is measured. The solids concentration or density measurement is used in the incremental methods, whereas the settling rate measurement is used in the cumulative methods. Incremental methods may be divided into fixed-time and fixed-depth methods, the latter being more popular, although combinations are sometimes used.

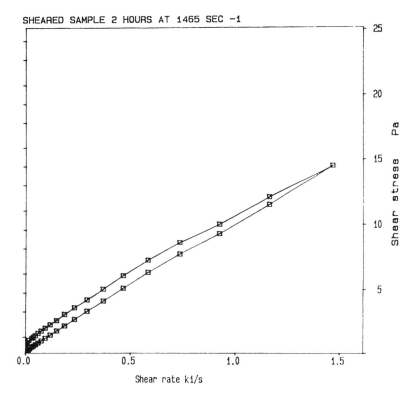

Figure 28.   Yield stress and gel strength as a function of shear rate. This oil sands fine tailings sample was sheared in rheometer to disrupt the floc structure before the run. The hysteresis in the curve is indicative of the thixotropy or shear thinning behavior of the suspension. This behavior is related to the structure changes observed using microscopy (Figures 16 and 17).

A fine solids sample is made up of three types of particles: primary particles, aggregates, and agglomerates. The primary particles are crystalline or organic structures bound together by molecular bonding, whereas the aggregates are primary particles tightly bound together at their point of contact by atomic or molecular bonding. The force required to break these bonds is considerable. In case of agglomerates, the primary particles are loosely bound together with weak van der Waals forces. It is often necessary to disperse the fine solids in a liquid before analysis. Dispersion is effected by the use of wetting agents that break down the agglomerates to their constituent parts. This process is facilitated by mechanical or ultrasonic agitation. Complications arise in the charac-

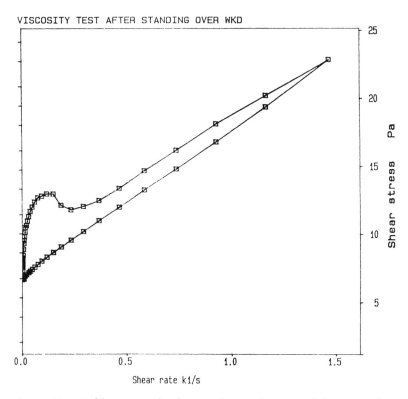

Figure 29. Yield stress and gel strength as a function of shear rate for a sample that has been allowed to sit for several hours (to allow floc structure to form). This behavior is related to the structure changes observed in Figure 15.

Table III.   Sedimentation Methods and Techniques

Incremental Methods	Cumulative Methods
Solids concentration variation	Sedimentation rate
Line start	Line start
Homogeneous suspension	Homogeneous suspension
Suspension density variation	
Line start	
Homogeneous suspension	

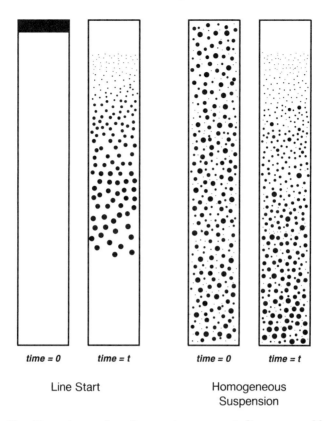

Figure 30. *Typical particle sedimentation pattern in line-start and homogenous gravitational sedimentation tests.*

terization of suspensions where the degree of aggregation, agglomeration, or flocculation is important. In these cases, careful sample handling along with comparisons of the dispersed and nondispersed size distributions are important.

Gravitational Sedimentation. There are four main gravitational sedimentation techniques: (1) sedimentation vessel wall pressure sensing, (2) manometry, (3) volume sample, and (4) mass sample, as shown schematically in Figure 31 (*8, 18, 35*).

Incremental Methods. Based on the results of motion of a particle in a gravitational or drag field reported earlier, it can be shown that the solids concentration at depth h can be related to the cumulative undersize mass distribution as (*8*)

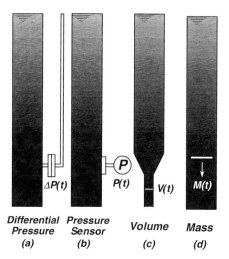

Figure 31. Main gravitational sedimentation techniques. (Adapted from reference 18. Copyright 1995 Academic.)

$$\frac{C(h,t)}{C(h,0)} = \frac{\int_{D_{p_{min}}}^{D_p(t)} f(D_p)dD_p}{\int_{D_{p_{min}}}^{D_{p_{max}}} f(D_p)dD_p} = \Psi(t) \qquad (31)$$

where $D_p(t)$ is given by

$$D_p(t) = \left[\frac{18\mu_m}{(\rho_p - \rho_m)g} \cdot \frac{h}{t}\right]^{1/2} \qquad (32)$$

and $dm(D_p)=f(D_p)dD_p$ represents the mass fraction having particle size between D_p and $D_p + dD_p$. In equations 31 and 32, it is assumed that the volume of particles in suspension is very small compared with the total volume of suspension. By plotting $100\Psi(t)$ against the free-falling particle diameter $D_p(t)$, the curve obtained shows the cumulative undersize percentage.

Similarly, it can be shown that (8)

$$\frac{C(h,t)}{C(h,0)} = \frac{\phi(h,t) - \rho_m}{\phi(h,0) - \rho_m} = \Phi(t) \qquad (33)$$

where $\phi(h,0)$ and $\phi(h,t)$ are the suspension density at a height h at time = 0 and time = t, respectively. Once again, by plotting $100\Phi(t)$ against $D_p(t)$, the cumulative undersize percentage curve is obtained.

One type of incremental method is the pipette method. Changes in concentration that occur within a settling suspension are determined by drawing off definite volumes at a set of discrete intervals of discrete time intervals using a pipette (7, 8). The solids concentration in the suspension is required to be between 0.2 and 1.0 vol%. If the concentration exceeds 1%, the hindered settling adversely affects the results of the analysis as discussed earlier. Initially, the fine solids sample is made into paste; this is followed by slow addition of the dispersing liquid, using a spatula plus mixing to form a slurry. Further dispersion may be carried out in an ultrasonic bath. The suspension is washed into a sedimentation vessel. The analysis starts with violent agitation of the vessel, avoiding the use of a stirrer. The container is continually inverted by hand for 1 minute. Because, initially, the suspended particles are not at rest, it is advisable to wait 1 minute before withdrawing samples. The use of a time scale progression of 2:1 that produces a $2^{1/2}$ particle size progression is common practice. Samples collected are prepared to determine the solids concentrations used in the determination of particle size cumulative mass distribution.

Hydrometers may be used to monitor the variation in density of a settling suspension. This method is used widely in the cement industry and has some applications in the petroleum industry. The method starts with a fully dispersed suspension and densities at known depths are recorded as the solid phase settles out (7, 8). The hydrometer technique is useful for quality control but not as an absolute method.

Divers are an extension of the hydrometer technique. They act as miniature hydrometers in which each diver is calibrated to a particular density. Several divers of different densities are added to the fully dispersed suspension and each will settle to a height where its density is equal to the suspension around it. A copper ring sealed in each diver interacts with an external search coil (energized by high frequency alternating current), and thus the location of the diver can be monitored.

The specific gravity balance may be used to monitor change within a settling suspension (7, 8). Such a balance comprises two bobs, one in clear fluid and the other in the suspension being studied. The bobs are connected to the two arms of a beam balance. The depth of immersion of the bobs is adjustable. The change in buoyancy is counterbalanced by means of solenoids that are connected to a pen recorder. From the settling behavior as monitored by the trace of pen recorder, the particle size distribution can be calculated.

Cumulative Methods. In cumulative methods the settling rate of suspensions or free-falling diameter are monitored. Cumulative methods have an advantage over incremental methods in that the amount of sample required is small (about 0.5 g), which reduces interaction between particles (i.e., hindered settling). This is an important advantage when only one small quantity of fine solids sample is available.

In the line start method, the size distribution may be directly determined by plotting the fractional weight settled against the free-falling diameter of particles. Special care must be exercised to eliminate the streaming problem, especially when the suspension at the top has higher density than the liquid.

In the homogeneous suspension method, a fine solids sample is considered with a mass distribution such that $dM = f(D_p)dD_p$, where dM represents the fractional mass of particles having a diameter between D_p and $D_p + dD_p$. Assume that the fine solids sample is completely dispersed in a liquid and consider a chamber of suspension of a height h. It can be reasoned that mass percent P that has settled out at time t is made up of two parts (8, 35):

1. All of the particles with a free-falling speed greater than that of $D_p(t)$ as given by Stokes' law or some related law, where $D_p(t)$ is the size of particle that has a velocity of fall h/t

2. Particles smaller than $D_p(t)$ that started at some intermediate position in the chamber. The falling velocity of one of these smaller particles is v, and the fraction of particles of this size that have fallen out at time t is (vt/h)

This mechanism can be represented mathematically as

$$P(t) = \int_{D_p(t)}^{D_{p_{max}}} f(D_p)dD_p + \int_{D_{p_{min}}}^{D_p(t)} \frac{vt}{h} f(D_p)dD_p \qquad (34)$$

which after some manipulation can be rewritten as

$$P(t) = M(t) + t\,\frac{dP(t)}{dt} \qquad (35)$$

Equation 35 may be written in a different form as

$$M(t) = P(t) - \frac{dP(t)}{d \ln (t)} \qquad (36)$$

The two terms in the right side of equation 35 are indicated on a typical sedimentation curve as shown in Figure 32. Both equations 35 and 36 can be used to find $M(t)$. The most obvious method is to tabulate t and $P(t)$ and thereby derive $dP(t)$, dt, and finally $M(t)$ (cumulative oversize percentage) versus $D_p(t)$ (equation 32). Equation 36 is recommended in suspensions of particles having a wide size distribution.

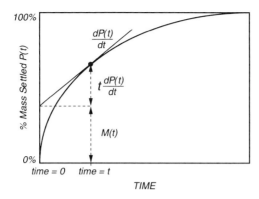

Figure 32.   Typical particle sedimentation curve.

Centrifugal Sedimentation.   Gravitational sedimentation for particle size characterization has limited flexibility. First, the only means of varying the particle velocity is by selecting a medium with different density or viscosity. Second, gravitational sedimentation cannot handle particles smaller than 5 μm. Third, most sedimentation devices suffer from the effects of convection, diffusion, and Brownian motion. These difficulties can be reduced by using centrifugation to speed up the settling process. In addition, the use of centrifugal field lowers the size limit and can reduce analysis time. As with gravitational methods, the data may be cumulative or incremental and the sample may be homogeneous or two-layer.

Calculations of size distribution from centrifugal data are more difficult than calculations from gravitational data, because particle velocities increase as they move away from the axis of rotation (i.e., the particle velocity depends on its radial position). One way to overcome this difficulty is to use a relatively small settling radial zone at a great distance from the center of rotation (i.e., the centrifugal force acting on all particles is approximately the same). Another solution is to use the line start technique. Figure 33 illustrates centrifugal sedimentation of suspensions using the line start and homogeneous suspension methods. Similar to gravitational sedimentation techniques, the most common techniques used in centrifugal sedimentation are schematically presented in Figure 34 (7, 8, 18).

Line Start Method.   Rewriting equation 23 as

$$V_{rt} = \frac{dR_{r2}}{dt} = \left[\frac{D_{pt}^2(\rho_p - \rho_m)}{18\mu_m}\right] \cdot \omega^2 R_{r2} \qquad (37)$$

and separation of variables and integration of equation 37 yields

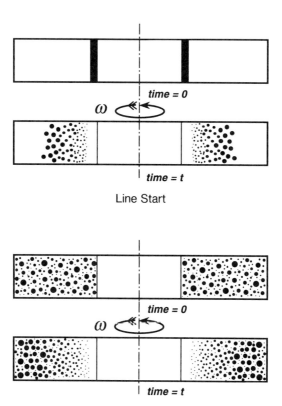

Figure 33. Typical particle sedimentation pattern in line-start and homogenous centrifugal sedimentation tests.

$$D_{\mathrm{p}}(t) = \left[\frac{18\mu_{\mathrm{m}}}{(\rho_{\mathrm{p}} - \rho_{\mathrm{m}})\omega^2 t} \cdot \ln\left(\frac{R_{r2}}{R_{r1}}\right) \right]^{1/2} \tag{38}$$

where t is the time for a particle of size $D_{\mathrm{p}}(t)$ to settle from the injection point (at distance R_{r1} from the center of rotation) to a radial distance R_{r2}. Therefore, at time t, all particles at R_{r2} will be of size $D_{\mathrm{p}}(t)$. Monitoring the percent solids or density of the suspension at specified intervals of time will produce the particle size distribution.

Homogeneous Suspension Method.   Equation 38 still applies; however, at time t, all particles of size greater than $D_{\mathrm{p}}(t)$ will have settled out radially to a distance R_{r2}. Conversion of the sedimentation curve into a cumulative curve is not as simple in this case as for gravitational sedimentation. Difficulties involved in evaluating the sedimentation curve in centrifugal fields may be overcome by assuming a constant

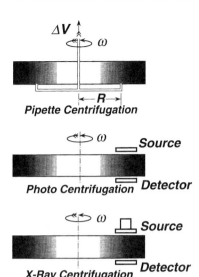

Figure 34. Main centrifugal sedimentation techniques. (Adapted from reference 18. Copyright 1995 Academic.)

centrifugal field, that is, the $(R_{r2} - R_{r1})$ interval is small enough to allow the approximation

$$\ln\left(\frac{R_{r2}}{R_{r1}}\right) \approx \frac{(R_{r2} - R_{r1})}{R_{r2}} \tag{39}$$

When the value $(R_{r2} - R_{r1})$ is $1/(20\ R_{r2})$, the cumulative curve can be obtained directly by the pipette technique with an error of 1%.

Elutriation. In fluid classification, various forces act to control the separation of dispersed particles. As discussed earlier, the kinetic behavior of particles is influenced by gravitation (as with elutriators) or centrifugal or Coriolis force (in classifiers). The medium is usually water or air. In general, all fluid classifiers can be divided into two classes: counterflow equilibrium (elutriation) and inverse flow separation. Elutriation is discussed briefly below (8).

In elutriation techniques, the field and the drag forces act in opposite directions, and particles leave the separation zone in one of two directions, depending on their size. Particles of a certain size stay in equilibrium in the separation zone. Grading is carried out in a series of vessels (cylindroconical form) of successively increasing diameter. Because fluid velocity decreases in each stage, coarse particles are retained in the smallest vessel and finer particles are retained in the successively larger vessels. For air elutriation, the analysis is considered complete if the rate of change of weight of residues is less than 0.2% of the initial weight in half an hour; for water elutriation the analysis ends when there is no sign of further classification.

Using Stokes' law, the particle size retained in an intermediate vessel can be predicted approximately as

$$D_p = \left[\frac{18\mu_m}{(\rho_p - \rho_m)g} \cdot V_{max} \right]^{1/2} \tag{40}$$

$$V_{max} = 2V = 2\frac{Q}{A} \tag{41}$$

In equation 41, it was assumed that the flow profile in the vessel is parabolic. Obviously, elutriation is only suitable for large particle suspensions. With small particles, sedimentation may be accelerated by using a centrifuge. This technique is used when classifying aerosols of solid particles using a stream of air that flows in the direction opposite to the centrifugal force.

Impaction. *Impactor.* Inertial impaction devices cause an air sample to be drawn into a round or rectangular nozzle where the gas velocity is increased. The jet from the nozzle is discharged against an adjacent flat surface, causing the air to diverge sharply. Particles in the air have more inertia than the air and continue forward as the air passes off to the sides, causing some particles to impact on the surface. To prevent the reentrainment of particles, a viscous material such as silicone fluid (or substrate) is used to coat the plate. The efficiency of impaction may be defined in terms of the dimensionless impaction factor I as (7, 8)

$$I^{1/2} = \left(\frac{C_c \rho_p D_{pa}^{\,2} V_j}{18\mu_m D_j} \right)^{1/2} \tag{42}$$

Classification of a particle cloud into discrete sizes using cascade impaction may be interpreted as measuring aerodynamic (equivalent spherical particle) diameter. Several impaction stages (cascade impactor) are used in the classification of a polydisperse cloud (*see* Figure 35).

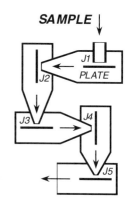

Figure 35. Cascade impactor. (Adapted from reference 18. Copyright 1995 Academic.)

The stages are arranged to allow jet velocity to increase with each suc-
ceeding stage (by successive reduction in jet diameter or width) and to
thereby cause particles of progressively smaller sizes to be impacted.
In effect, the cascade impactor classifies particles according to their
aerodynamic size. The aerodynamic diameter can be expressed in terms
of Stokes diameter as

$$D_{pa} = D_{Stokes}\rho_p^{1/2} \tag{43}$$

where D_{Stokes} is the measured diameter. The aerodynamic size is important because it controls the motion of a particle in an air stream.

Impinger.   Another sampling instrument, called an impinger, also
uses inertial impaction; however, deposition occurs at the bottom of a
fluid-containing vessel (7, 8). The downward-directed air jet displaces
the fluid and uncovers the bottom of the vessel. The particles that im-
pinge against the wet surface are subsequently washed off by the fluid.
Undeposited particles may be caught by air bubbles rising through the
fluid. The particles are usually examined in the liquid suspension. Water
is the most commonly used fluid.

Electrostatic Precipitation.   The electrostatic precipitator consists of an ionizing cathode at high potential surrounded by a collecting
anode; typically, these anodes consist of concentric cylinders, the inner
one often being a single wire. The gas suspension passes between the
cylinders, picks up charge, and travels to the anode where charge is
deposited. The transfer of electrons from one electrode to the other
produces an electric current. The magnitude of this current is proportional to the number of particles deposited. Particle sizes can be determined by varying the flow rate or applied potential (7, 8, 14, 15).
Classification in an electrostatic field by differences in charges is
related explicitly to particle size. The instrument consists of glass cylinder
with a central electrode. The inner surface of the cylindrical glass is
coated with a suitable material to act as the other electrode and to collect
samples. The principal advantages of this type of instrument are high
collection efficiency over a wide size range, low resistance, and high
flow rate capacity.

Thermal Precipitation.   Particles in a thermal gradient medium
move in the direction of negative gradient, that is, from hotter to colder
regions (7, 8). Based on this principle, the instrument typically consists
of two parallel round microscopic plates and a heated wire in between
as shown in Figure 36. The sample is drawn between the plates, and

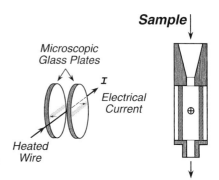

Figure 36. Thermal precipitation. (Adapted from reference 18. Copyright 1995 Academic.)

the particles are deposited on the glass plates and collected for further analysis.

Normally, a sample flow of $1-2$ cm^3 s^{-1} is recommended, and the collection efficiency is high for -5-μm particles. The collecting device may be modified so that the sample is collected directly on an electron microscope grid. Modifications of the basic design include means of centering the wire in position, substitution of wire by a ribbon to give more uniform deposits, and using inlet elutriators to exclude coarse particles. Application of thermal precipitation in gas cleaning plants has only rarely been attempted.

Hydrodynamic Chromatography.   Size information about colloidally suspended particles ($0.01-1$ μm) can be obtained by using the hydrodynamic chromatography (HDC) technique (36–38). A medium (aqueous solution) is pumped through a column packed with impermeable spheres. A pulse of colloidal suspension (0.2 cm^3) containing about 0.01 wt% polymer is injected into the flowing stream at the column entrance. The mobile phase from the column effluent is passed through a suitable detection system, such as a flow-through spectrophotometer of the type used in liquid chromatography, and the detector response of the colloid is detected as a function of elution time. An extra step is needed to determine the concentration of solids in the eluted solution.

It has been observed that the larger particles elute faster than the smaller ones. It has also been found that the smaller the packing diameter, the better the separation. Other factors that affect the rate of transportation through the bed are size of bed particles, ionic strength, flow velocities, and particle size of eluting particles. The method is applicable to size separation of particles between 0.02 and 1 μm, if they are rigid.

In general, HDC has been successfully applied to the size characterization of many polymer lattices. Moreover, it is expected that HDC has wide applicability to submicron particles, such as in lattices, carbon black, and colloidal silica.

Another extension of HDC is to replace the packed bed with a long capillary (39). Capillary particle chromatography requires 30 kPa pressure and has a separating range of 0.2–200 μm. In such techniques, the particle transit time is a logarithmic function of particle size.

Electrical Sensing Zone (Coulter Counter) As a particle passes through an orifice, the electrical resistance measured across the orifice changes as a function of the particle size (volume). This can be proved by considering Figure 37 and deriving the following relationships (7, 8):

$$\delta R_l = \frac{\Omega_l \delta l}{A_o} \tag{44}$$

$$\delta R_{lp} = \frac{\Omega_l \Omega_p \delta l}{\Omega_p (A_o - A) + \Omega_l A} \tag{45}$$

$$\delta(\Delta R_a) = (\delta R_l - \delta R_{lp})$$

$$\delta(\Delta R_a) = -\frac{\Omega_l}{A_o^2} A\delta l\left(1 - \frac{\Omega_l}{\Omega_p}\right) \frac{1}{\left[1 - \left(1 - \frac{\Omega_l}{\Omega_p}\right)\frac{A}{A_o}\right]} \tag{46}$$

Because of ionic inertia of the Helmholtz electrical double layer at the surface of the particles, the electrical resistivity becomes infinite; as a result, the term (Ω_l/Ω_p) can be neglected. Furthermore, if the particle cross-sectional area A_p is very small compared with the orifice area A_o, equation 46 reduces to

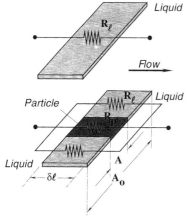

Figure 37. Electrical sensing zone technique.

$$\delta(\Delta R_a) = (\delta R_l - \delta R_{lp}) = -\frac{\Omega_l\, v_p}{A_o^2} \cdot A\delta l \qquad (47)$$

which indicates that the change in electrical resistance across the orifice is proportional to the particle volume, v_p.

Let A_o and A represent the orifice cross-sectional area and the particle cross-sectional area projected on the orifice face. Assuming spherical particles, several investigations have been conducted to integrate equation 47 giving

$$\Delta R_a = \frac{\Omega_l v_p}{A_o^2}\, F_i\!\left(\frac{D_p}{D_o}, L_o\right) \qquad (48)$$

where F_is are functions of the ratio (D_p/D_o) and L_o the orifice effective length (40–45). Equation 48 indicates that the change in electrical resistance (response) across the orifice is proportional to the particle volume v and is scaled by F_is. The functions F_is have closed form structures that are dependent on the mathematical and numerical approach adopted.

Practically, this method works well for spherical or near-spherical particles; however, it produces erroneous results for nonspherical particles. Moreover, this method is not suitable for porous material because the effective particle density is not known and the volume measured is the envelope volume. Special care is required to avoid crowding of the orifice; otherwise, special treatment is needed to analyze the instrument counts.

Size Distribution Using Scattering Properties.   The size range probed by various scattering techniques is a function of the wavelength; neutron scattering, X-ray scattering, and light scattering techniques are related in terms of the physical interaction between the radiation and the particles. As noted earlier in this chapter (Figure 5), these techniques can cover particle sizes from tens of nanometers to hundreds of microns. Because the resolution of the scattering techniques depends on the wavelength of the scattering source, neutron scattering has a greater resolution than X-ray scattering, which is in turn better than light scattering. Data reduction are similar for the three methods, but light scattering is limited by the sample opacity.

Light Scattering.   The most common commercially available sizing instruments depend on light scattering to obtain size information. The availability of inexpensive well-defined light from laser sources has resulted in a wide variety of scattering techniques using light and a number of related commercial instruments (46–52).

Light scattering can be broadly divided into time-averaged scattering, in which either spatial distribution or intensity is measured, and time-fluctuation scattering, which includes photon correlation spectrometry where scattering is correlated to the microscopic motion of individual scattering centers. These techniques have been discussed in detail in several reviews (52–57). Only a brief overview of the most common time-averaged methods are given here. These include Fraunhofer diffraction and light scattering at larger angles (Mei scattering), which are the basis of most commercially available sizing instruments.

Quasielastic light scattering or photon correlation spectrometry, Fraunhofer diffraction, and other techniques that depend on light have the same drawback—the opacity of most oil production samples makes them unsuitable for use. Typical problems with the theory and subsequent data reduction of the scattering information to a size distribution include an assumption of the nature of the size distribution (typically log-normal, although software is available with other options) and an inability to distinguish aggregates from large single particles. Characterizing high solids content suspensions is therefore not possible, nor is it possible to accurately determine flocculation behavior or interparticle interactions because the sample must be diluted and disturbed before flowing through the sample cell. It is also not possible to distinguish between mineral solids in a suspension and organic solids or emulsified oil, a disadvantage shared with the sensing zone techniques. In addition, the sample must be dilute enough to minimize multiple scattering.

Figures 38 and 39 illustrate the experimental setup for a light-scattering apparatus and show examples of the signal observed in Fraunhofer diffraction for two monodisperse and one polydisperse sample. The detection system in most commercial instruments is either an array of intensity sensors or a single detector with a moving mask that measures intensity differences of the overlapping concentric rings (not discernable in Figure 39). Although no calibration is necessary for monodisperse spherical systems, the data are output as an equivalent spherical diameter, and the range of applicability is generally for sizes exceeding 10 μm. For solid particulate systems, the refractive index is generally large, which means that the applicability of this technique can be extended to smaller sizes.

Most instrument manufacturers use scattering at larger angles, as well as diffraction, to probe the smaller sizes (less than about 10 μm). Scattering at larger angles involves a distinct dependence on refractive index, and various manufacturers use either the position or wavelength dependence of the scattered light at larger angles. Assumptions about an "average" refractive index or the nature of the size distribution (bimodal or log-normal, for instance) must be made to determine a size distribution from the light-scattering information at larger angles.

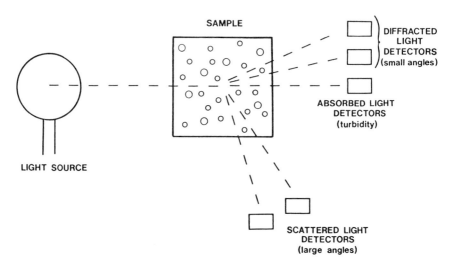

Figure 38.   Schematic of a light-scattering apparatus. Three techniques are illustrated here. (1) Attenuation of the incident light or turbidimetry can indicate the amount of suspended solids but offers no information about the size distribution. (2) Diffraction of the incident beam (Fraunhofer diffraction) provides size information for relatively large sizes where the particles are on the order of, or larger than, the wavelength of the incident light. (3) Scattering through large angles (Mie scattering) occurs with particles smaller than the wavelength of the incident light. This large-angle scattering can be affected by the refractive index of the scattering centers. Computer data handling reduces these signals to a size distribution. Variations of these basic techniques involve detection of scattered light as a function of angle, correlation of the scattered photons (photon correlation spectroscopy), and detection of scattering as a function of wavelength or polarization of the incident light. (Adapted from reference 19. Copyright 1992.)

Photon correlation spectroscopy involves monitoring the time dependence of light scattering from a single particle at a time. This time dependence is determined by Brownian motion of very small particles in suspension, which is in turn related to their size. This method extends the range of applicability for size characterization of suspended particles into the nanometre range and is available on many commercially available instruments.

X-ray and Neutron Scattering.   Small-angle X-ray and neutron scattering also can be used to probe size distributions, but at a much better resolution, down to the molecular level (about 4 Å with neutron scattering) (58–61). This level of detail in determining the size of suspended solids is often not applicable to samples commonly encountered in the petroleum industry, although the principles are the same as for light-scattering phenomena.

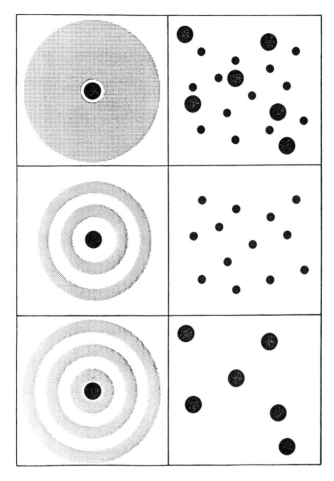

Figure 39.   Typical Fraunhofer diffraction patterns. In polydisperse systems the interpretation of the relationship between these patterns and the size distribution can be difficult and requires sensitive photomultipliers. The transmitted beam is blocked out and the detectors arranged outward from the center. In some cases a single detector has a moveable mask to measure diffracted light intensity as a function of position. As can be inferred from this pictorial, it is not possible to distinguish subtle differences in size distribution (i.e., log-normal vs. bimodal, and so on) and generally some assumption must be input to the data reduction programs. (Adapted from reference 19. Copyright 1992.)

All of the scattering techniques are beset with similar data reduction problems in that some assumptions must be made about the nature of the particles (refractive index and composition) and how they are associated. In most suspensions of interest in the petroleum industry, the samples are not dilute enough to rule out multiple scattering.

Acknowledgments

This work was supported in part by the Federal Panel on Energy Research and Development. Thanks to K.C. McAuley for preparation of the figures and photographs and to V.A. Munoz, who contributed all of the photographs related to the light and confocal microscopy. Thanks are also due to C.W. Angle, R. Zrobok, J. Kan, and Y. Xu, for contributing data (relating to oil sands fine tailings) on electrokinetics, rheology, low speed centrifugation, and specific resistance to filtration.

List of Symbols

A	an incremental (or total) projected area of particle normal to its motion (m^2)
A_o	an incremental (or total) cross-sectional area of orifice (aperture) (m^2)
A_{ez}	vessel cross-sectional area at equilibrium zone (m^2)
A_m	surface area per unit mass (m^2 kg^{-1})
A_p	particles surface area (m^2)
A_v	specific surface area (surface area per unit volume) (m^{-3})
c	solids concentration in a suspension by volume (fractional, per unit)
$C(h, t)$	solids concentration at depth h and time $= t$
$C(h, 0)$	solids concentration at depth h and time $= 0$
C_c	Cunningham slip correction factor
C_D	drag coefficient
D	diffusion coefficient of vapor from particle (m^2 s^{-1})
D_j	jet diameter or width (m)
D_o	orifice diameter (m)
D_p	particle diameter (m)
$D_p{}^*$	specific diameter (m)
D_{pa}	particle aerodynamic diameter (m)
D_{pi}	particle diameter in group i (m)
D_{pmax}	maximum particle diameter (m)
D_{pmin}	minimum particle diameter (m)
D_{Stokes}	diameter of a hypothetical sphere having the same terminal settling velocity as the particle (m)
$D_p(t)$	free-falling particle diameter (m)
e	electron charge (1.6022 $\times$ 10^{-19} C)
E	external electric field strength (V m^{-1}).

$f(D_p)$	mass distribution as a function of particle size D_p
F_i	function of the ratio (D_p/D_0) and L_0 (expression)
F_R	medium resistance (N)
F_t	force acting on a particle in the thermal field (N)
g	gravitational acceleration (ms^{-2})
G'	the elastic or storage modulus. With input of a small oscillatory shear at constant strain (in the linear viscoelastic region), G' is proportional to the average energy storage in the shear deformation cycle. With the small strains applied that do not break the structure, G' is a measure of the strength of the material (Pa)
h	depth (m)
I	dimensionless impaction factor
k	Boltzmann's constant (1.3807×10^{-23} J K^{-1})
$k1/s$	shear rate
k_a	constant in fractal area equation
k_p	shape factor
k_R	constant in equation for medium resistance
l	length (m)
L	distance from center of particle to the container walls (m)
L_o	orifice effective length (m)
$M(t)$	cumulative percentage oversize at size $D_p(t)$ (settled at time = t)
M_m	molecular weight of evaporating particulate material
n	number of electron charges acquired
n_i	number of particles in group i
N_o	ion density (ions m^{-3})
$P(t)$	mass percent settled out at time = t
Q	volumetric flow rate
Q_p	acquired (bombardment or diffusion) charge on particle (C)
Q_{pb}	saturation bombardment charge acquired (C)
Q_{pd}	diffusion charge acquired (C)
R	gas constant (8.3144 J mol^{-1} K^{-1})
R_a	electrical resistance across aperture (orifice) (ohms)
R_l	electrical resistance of liquid across aperture (orifice) (ohms)
R_{lp}	electrical resistance of liquid and particle across aperture (orifice) (ohms)
R_r	distance from the center of rotation (m)
Re_p	Reynolds number of particle
SF	cyclone separation factor
t	time (s)
T	absolute temperature (K)
dT/dX	temperature gradient in air (K m^{-1})
V	particle velocity (m s^{-1})
V_{et}	terminal velocity in electrostatic field (m s^{-1})

V_{gt}	terminal velocity in gravitational field (m s^{-1})
V_{gtc}	corrected terminal velocity in gravitational field (m s^{-1})
V_{gth}	terminal velocity in gravitational field having hindered settling (m s^{-1})
V_i	ion velocity (root mean square, m s$^{-1)}$
V_j	average air velocity at jet outlet (m s$^{-1)}$
V_{max}	maximum velocity = $2Q/A$ (m s^{-1})
v_p	particle volume (m^3)
V_p	particles volume (m^3)
V_{pm}	relative velocity between particle and medium (m s^{-1})
V_{rt}	terminal velocity in rotational field (m s^{-1})
V_{tt}	terminal velocity in thermal field (m s^{-1})
X	distance in linear direction (m)

Greek

α	constant (approximately one)
α_s	surface shape factor
α_v	volume shape factor
δ	fractal dimension > 2
δL	incremental length (m)
δR	incremental electrical resistance (ohms)
ε	porosity (fractional, per unit)
ε_0	permittivity of vacuum (8.8542×10^{-12} F m^{-1})
ε_1	relative permittivity of medium (gas) (fractional, per unit)
ε_2	relative permittivity of particle material (fractional, per unit)
κ_m	thermal conductivity of air (J m^{-1} s^{-1} K^{-1})
κ_p	thermal conductivity of particle material (J m^{-1} s^{-1} K^{-1})
λ	mean free path of medium molecules (m)
λ_F	resolution or side of an elemental square (m)
μ_m	medium viscosity (kg m^{-1} s^{-1})
π	constant, 3.14159
ρ_m	medium density (kg m^{-3})
ρ_p	density of particle material (kg m^{-3})
Ω_l	liquid electrical resistivity (ohm m)
Ω_p	particle electrical resistivity (ohm m)
τ	evaporation time (s)
$\phi(h, t)$	suspension density at height h and at time = t (kg^{-3})
$\phi(h, 0)$	suspension density at height h and time = 0 (kg^{-3})
ω	angular velocity of the particle (rad s^{-1})
Δ_p	pressure difference between pressure at particle surface and in surrounding fluid (N m^{-2})
$\Psi(t)$	relative suspension solids concentration as a function of time t

| $\Phi(t)$ | relative suspension density as a function of time t (fractional, per unit) |
| χ | ratio (fractional, per unit) |

References

1. Rao V. U. S. *Energy & Fuels* **1994**, *8*, 44.
2. Lapple, C. E. *Stanford Research Institute (SRI) Third Quarter Report;* Stanford Research Institute: Stanford, CA, 1961; pp 95–103.
3. *Modern Methods of Particle Size Analysis;* Barth H. G., Ed.; Chemical Analysis: A Series of Monographs on Analytical Chemistry and Its Applications; 1–75; John Wiley: New York, 1984; Vol. 73.
4. Barth Howard, G.; Sun, S. T. *Anal. Chem.* **1985**, *57*, 151R.
5. Barth Howard, G.; Sun, S. T.; Nikol, R. M. *Anal. Chem.* **1987**, *59*, 142R.
6. Miller, B. V.; Lines, R. *Crit. Rev. Anal. Chem.* **1988**, *20*, 75.
7. Silverman, L.; Bellings, C. E.; First, M. W. *Particle Size Analysis in Industrial Hygiene;* Academic: Orlando, FL, 1971.
8. Allen, T. *Particle Size Measurement*, 4th ed.; Chapman and Hall: New York, 1990.
9. Mandelbrot, B. B. *The Fractal Geometry of Nature;* W. Freeman Publishers: San Francisco, CA, 1983.
10. Kaye, B. H. *Part. Syst. Charac.*, in press.
11. Kaye, B. H. *A Random Walk Through Fractal Dimensions;* VCH Publishers: Weinheim, Germany, 1989.
12. Crickmore, P. F.; Schutte, R.; Cansgrove, J. *Proceedings of the 4th Unitar Conference on Heavy Crude and Tar Sands;* Alberta Oil Sands Technology and Research Authority: Edmonton, Canada, 1988; Vol. 1, pp 217–221.
13. Geankoplis, C. J. *Mass Transport Phenomena;* Holt, Rinehart, and Winston: New York, 1972.
14. Bohm, J. *Electrostatic Precipitators;* Chemical Engineering Monographs; Elsevier: New York, 1982; Vol 14.
15. Oglesby, S., Jr.; Nichols, G. B. *Electrostatic Precipitation;* Marcel Dekker: New York, 1978.
16. Richardson, J. F.; Zaki, W. N. *Chem. Eng. Sci.* **1954**, *3*, 65.
17. Barnea, E.; Mizrahi, J. *Chem. Eng. J.* **1973**, *5*, 171.
18. Salama, A. I. A. *Particle Size Analysis: Mechanical Techniques, Encyclopedia on Analytical Science;* Academic: London, 1995.
19. Schramm, L. L. *Emulsions: Fundamentals and Applications in the Petroleum Industry;* ACS Advances in Chemistry 231; American Chemical Society: Washington, DC, 1992; pp 1–49.
20. Mikula, R. J. *Emulsion Characterization;* CANMET Division Report 93.43; National Resources Canada: Devon, Canada, 1993.
21. Montgomery, D. W. *Rubber Age* **1964**, *45*, 759.
22. Ross, W. D. *Filtr. Sep.* **1973**, *10*, 587.
23. Dixon, W. J.; Massey F. J., Jr. *Introduction to Statistical Analysis*, 3rd ed.; McGraw-Hill: New York, 1969; p 550.
24. Loveland, R. P. *Photomicrography;* Wiley: New York, 1970.
25. Klein, M. V. *Optics;* Wiley: New York, 1970.
26. Schott, H.; Royce, A. E. *J. Pharm. Sci.* **1983**, *72*, 313.
27. Robards, A. W.; Sleytr, U. B. *Low Temperature Methods in Biological Electron Microscopy;* Elsevier: Amsterdam, Netherlands, 1985.
28. Menold, R.; Luttge, B.; Kaiser, W. *Adv. Colloid Int. Sci.* **1970**, *5*, 281.

29. Mikula, R. J. *Colloids Surf.* **1987**, *23*, 267.
30. Mikula, R. J. J. *Colloid Interface Sci.* **1988**, *121*, 273.
31. Kuhn, W. *Angew Chem. Int. Ed. Engl.* **1990**, 29.
32. Young, S. W. *Magnetic Resonance Imaging: Basic Principles*; Raven: New York, 1984.
33. Shaw, R. C.; Czarnecki, J.; Schramm, L. L.; Axelson, D. In *Foams: Fundamentals and Applications in the Petroleum Industry*; Schramm, L. L., Ed.; Advances in Chemistry 242; American Chemical Society: Washington, DC, 1984; pp 423–459.
34. Mikula, R. J.; Argk, C. W.; Zrobok, R.; Kan, J.; Xu, Y. *Factors That Determine Oil Sands Sludge Properties*; CANMET Division Report WRC 93–40; Natural Resources Canada: Devon, Canada.
35. Syvitski James, P. M. *Principles, Methods, and Applications of Particle Size Analysis*; Cambridge University: Cambridge, England, 1991.
36. Small, H. *Colloid Interface Sci.* **1974**, *48*, 147.
37. Small, H. *Chemtech* **1977**, *7*, 196.
38. Small, H.; Saunders, F. L.; Sole, J. *Adv. Colloid Interface Sci.* **1976**, *6*, 237.
39. Mullins, M. E.; Orr, C. *Int. J. Multiphase Flow* **1979**, *5*, 79.
40. De Blois, R. W.; Bean, C. P. *Rev. Sci. Instrum.* **1970**, *41*, 909.
41. Andersen, J. L.; Quinn, J. A. *Rev. Sci. Instrum.* **1971**, *42*, 1257.
42. Gregg, E. L.; Steidley, K. D. *Biophys. J.* **1965**, *5*, 393.
43. Grover, N. B.; Naaman, J.; Ben-Sasson, S.; Doljawski, F.; Nadav, E. *Biophys. J.* **1969**, *9*, 1398, 1415.
44. Smythe, W. R. *Phys. Fluids* **1961**, *4*, 756.
45. Smythe, W. R. *Phys. Fluids* **1964**, *7*, 633.
46. Goulden, J. D. S. *Trans. Faraday Soc.* **1958**, *54*, 941.
47. van der Waarden, M. *J. Colloid Sci.* **1954**, *9*, 215.
48. Mie, G. *Ann. Phys.* **1908**, *25*, 377.
49. Robillard, F.; Patitsas, A. J. *Can. J. Phys.* **1974**, *52*, 1571.
50. Robillard, F.; Patitsas, A. J.; Kaye, B. *Powder Technol.* **1974**, *9*, 307.
51. Lacharojana, S.; Caroline, D. *N.A.T.O. Adv. Study Inst. Ser., Ser. B* **1977**, *B23*, 499.
52. Nicholson, J. D.; Clarke, J. H. R. *Surfactants Solution* **1984**, *3*, 1663.
53. Sjoblom, E.; Friberg, S. *Colloid Interface Sci.* **1978**, *67*, 16.
54. Daniels, C. A.; McDonald, S. A.; Davidson, J. A. In *Emulsions, Lattices and Dispersions*; Becher, P.; Yudenfreund, M. N., Eds.; Marcel Dekker: New York, 1978; p 175.
55. Van der Hulst, H. C. *Light Scattering by Small Particles*; Chapman and Hall: London, 1957.
56. Kaler, E. W.; Davis H. T.; Scriven, L. E. *J. Chem. Phys.* **1983**, *79*, 5685.
57. Kaler, E. W.; Bennett, K. E.; Davis, H. T.; Scriven, L. E. *J. Chem. Phys.* **1983**, *79*, 5673.
58. Herbst, L.; Hoffmann, H.; Kalus, J.; Thurn, H.; Ibel, K. *Neutron Scattering in the Nineties*; International Atomic Energy Agency Bulletin; Atomic Energy Agency: Vienna, Austria, 1985.
59. Gunier A.; Fournet, G. *Small Angle Scattering of X-rays*; Wiley: New York, 1955.
60. Caldwell, K. D.; Li, J. *J. Colloid Interface Sci.* **1989**, *132*, 256.
61. Stewart, R. F.; Sutter, D. *Part. Sci. Technol.* **1986**, *44*, 251.

RECEIVED for review July 5, 1994. ACCEPTED revised manuscript March 16, 1995.

3

Rheology of Suspensions

Shijie Liu and Jacob H. Masliyah*

Department of Chemical Engineering, University of Alberta, Edmonton, Alberta T6G 2G6, Canada

This chapter is an in-depth review on rheology of suspensions. The area covered includes steady shear viscosity, apparent yield stress, viscoelastic behavior, and compression yield stress. The suspensions have been classified by groups: hard sphere, soft sphere, monodisperse, polydisperse, flocculated, and stable systems. The particle shape effects are also discussed. The steady shear rheological behaviors discussed include low- and high-shear limit viscosity, shear thinning, shear thickening, and discontinuity. The steady shear rheology of ternary systems (i.e., oil–water–solid) is also discussed.

RHEOLOGY IS THE STUDY OF THE RESPONSE OF MATERIALS to an applied shear stress or strain (*1*). In other words, rheology is a science of deformation (a typical response of solids to an applied strain, elasticity) and flow (a typical response of a fluid to an applied shear, viscosity). Sometimes, the methods that impose a strain are classified as the plasticity approach, whereas the methods that apply shear rates or shear stresses are termed the rheological approach (*2*). In this review, we focus on the rheological approach, whereas the plasticity approach is dealt with only briefly.

Definition of Suspension.   The rheology of suspensions deals with how suspensions respond to an applied stress or strain. The term suspension refers, in general, to dispersions of solids in fluids, although the term aerosol is conventionally used to refer to dilute suspensions of fine particles in a gas and the term emulsion is used to identify (concentrated) suspensions of particles in a gas or liquid in the field of fluidization. However, emulsion is conventionally defined as the dispersion of a liquid in another (immiscible) liquid. In a broader sense, emulsions are also considered as suspensions. In this chapter, we deal mainly with suspen-

* Corresponding author.

0065–2393/96/0251–0107$24.50/0

sions of solids in liquids, which are encountered often in the petroleum industry. For a gas dispersed in a liquid (i.e., foam), one should refer to Schramm and Wassmuth (3) and for a liquid dispersed in another liquid (i.e., emulsion), to Pal et al. (4).

Homogeneity of Suspensions.   Rheology is "the science of deformation and flow." Rheologists normally want to express the relation between deformation and flow on the one hand and stress or strain on the other hand by equations based on a homogeneous material, or a continuum. Obviously, the definition of a suspension entails that it is heterogeneous, that is, a dispersed system consisting of individual particles in a suspending fluid. Under what conditions can we treat a suspension as a continuous phase from a rheological view point?

The rheological behavior can for any mixture only be defined when average values of stress and strain are meaningful. Thus, if a rheological experiment involves a deformation over a length scale that is much greater than the dispersed particle size, then homogeneity or continuum can be observed. It may be useful to apply the representative elementary volume concept (*see* Chapter 5) for a suspension. One may not be interested in the response of any given individual particle but rather in the bulk response, that is, (volume) averaged properties.

Basic Rheological Concepts

The relationship between an applied stress or strain and the response of the material, shear rate, or deformation is the aim of the rheology of suspensions. Normally, both the stress and the strain are tensors with each having nine components. In simple shear, which is the most common way of determining the rheological behavior, the shear stress σ_{xy} (some literature also uses the symbol τ to stand for the shear stress) can be related to the shear rate $\dot{\gamma}$ by

$$\sigma_{xy} = \sigma = \mu\dot{\gamma} \tag{1}$$

where μ is the dynamic viscosity or simply called viscosity.

However, rheological measurements are also performed with other types of flow or stress fields. If a uniaxial extensional flow field is applied to a material, the stress distribution can be described by

$$\sigma_{xx} - \sigma_{yy} = \mu_E\dot{\epsilon} \tag{2}$$

where μ_E is the uniaxial (extensional) viscosity and $\dot{\epsilon}$ is the extensional strain rate.

In general, a material can be characterized based on the two types of rheological behavior, that is, viscous and elastic. A solid body is characterized by its elastic behavior when the deformation is fully recovered after removal of the applied stress below the rupture point value. A liquid is characterized by its viscous behavior when it flows under any stress. Commonly, many materials, like the suspensions, are characterized by both an elastic and a viscous response, so-called viscoelastic behavior. The type of the response depends on the time scale of the experiment. If a small strain or stress is applied very rapidly to a viscoelastic body, it will respond elastically. If the stress or strain is applied for a long time, the material will flow and hence show a viscous response.

It is possible to classify a particular system with the Deborah number as given by

$$De = \frac{t_E}{t_R} \quad (3)$$

where t_R is the relaxation time and t_E is the time scale of the experimental measurement. When De is large, the system behaves like a solid (elastic), and when De is small, it flows (viscous). When De is near unity, the response is viscoelastic. The relaxation time based on the Einstein–Smoluchowski relation for an isolated particle is given by

$$t_R = \frac{3\pi\mu_f d^3}{4kT} \quad (4)$$

where d is the particle diameter, μ_f is the viscosity of the continuous phase, k is Boltzmann's constant, $k = 1.3806 \times 10^{-23}$ J/K, and T is the absolute temperature.

The time scale of the experiment, t_E, for continuous shear flow is

$$t_E = \frac{1}{\dot{\gamma}} \quad (5)$$

and for a forced oscillation experiment, it is given by

$$t_E = \frac{1}{\omega} \quad (6)$$

where ω is the frequency of the oscillation.

For 10-μm particles suspended in water at room temperature, t_R is of the order of 10 min. Hence, it is only in dilute systems of small particles that complete relaxation of a suspension structure can take place within the shear rate range of most rotational viscometers on the market.

Time-Independent Rheology. In the simplest case, the shear stress σ is independent of time t and is proportional to the shear rate $\dot{\gamma}$ (equation 1). For this case, the fluid is called Newtonian (line 1 of

Figure 1). The viscosity μ is the slope of the line 1 in Figure 1a. The viscosity of a material is a function of temperature and pressure, but otherwise it is a material constant as shown in Figure 1b. Newtonian behavior is frequently met with homogenous fluids, but for suspensions it is restricted to dilute nonaggregating (nonflocculated) systems.

Frequently, however, the shear stress and shear rate relation is more complicated, as shown in Figure 1. Figure 1a depicts some common shear rate response to a shear stress, whereas Figure 1b shows the corresponding apparent viscosity variation versus the shear rate. Except for line 1, where the viscosity is independent of shear rate, the apparent viscosity is a function of shear rate (non-Newtonian behavior). For example, curve 2 shows an increasing apparent viscosity with shear rate. This system is called shear thickening or dilatant. Curve 3 shows a decrease in the apparent viscosity with shear rate and the system is said to be shear thinning. Curves 1 through 3 are normally generalized by a power law relation, where

$$\sigma = K\dot{\gamma}^n \tag{7}$$

Here n is the consistency (or power law) index and K is the consistency of the material. When $n > 1$, the system is shear thickening, whereas

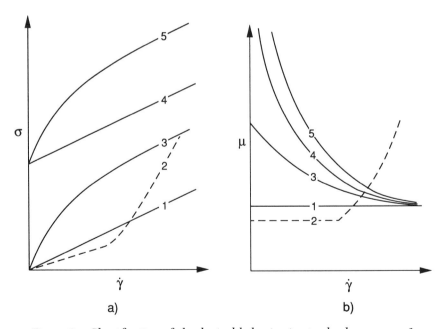

Figure 1.   *Classification of rheological behavior in steady shear: curve 1, Newtonian; curve 2, shear thickening; curve 3, shear thinning; curve 4, Bingham plastic; and curve 5, nonlinear plastic.*

for $n < 1$, the system is shear thinning. A Newtonian material has an index of unity, $n = 1$.

If the decrease in viscosity is very large at small shear rates, the system is sometimes called pseudoplastic (curves 4 and 5). Commonly, concentrated suspensions show a plastic behavior, that is, there is no response until a limiting yield stress σ_y has been exceeded. If the flow is linear above σ_y, the system is called Bingham plastic (curve 4) and can be expressed by the Bingham model (5):

$$\sigma = \sigma_y + \mu_p \dot{\gamma} \tag{8}$$

where the plastic viscosity μ_p is the slope of the flow curve when $\sigma > \sigma_y$. The yield stress σ_y is sometimes called the Bingham yield stress. The curve above the yield stress can also be nonlinear (curve 5). Such behavior can be described by a generalized power-law or the Herschel–Bulkley model:

$$\sigma = \sigma_y + K\dot{\gamma}^n \tag{9}$$

or by the Casson model (6):

$$\sigma^{1/2} = \sigma_y^{1/2} + K_C^{1/2}\dot{\gamma}^{1/2} \tag{10}$$

where K_C is Casson constant. It should be noted that fitting the same rheological data to different models can lead to different yield stresses. The accuracy of the evaluated yield stress is dependent on the applicability of the model used and should be treated strictly as a model parameter and not a material property. For a real suspension, the yield stress may not be a reality (7). The (apparent) yield stress, however, may be used just as a model for convenience (8). Several studies have been devoted for the determination of the yield stress, for example, Dzuy and Boger (9), Buscall et al. (10), Cheng (11), Yoshimura et al. (12), Astarita (13), Evans (14), Van der Aerschot and Mewis (15), Schurz (16), and De Kee and Chan Man Fong (17).

Often, one finds a linear relation at a very small shear rate, a shear thinning behavior at intermediate shear rate, and a linear relation at high shear rate values (*see*, e.g., Van Diemen and Stein [18] and Hunter [19]). In this respect, the Meter model may be of special interest if a system behaves like Newtonian fluid at both low and high shear rates:

$$\sigma = \dot{\gamma}\left[\mu_\infty + \frac{\mu_0 - \mu_\infty}{1 + (\sigma/\sigma_c)^m}\right] \tag{11}$$

where m is a power index, σ_c is the critical shear stress, and μ_0 and μ_∞ are the viscosities in the limits of very small shear rates and very high shear rates, respectively.

However, the Meter model may have complications in its application in computational flow simulations. A better curve-fitting model emerges by replacing the shear stress on the right hand side of equation 11 with a shear rate:

$$\sigma = \dot{\gamma}\left[\mu_\infty + \frac{\mu_0 - \mu_\infty}{1 + (\dot{\gamma}/\dot{\gamma}_c)^m}\right] \tag{12}$$

where $\dot{\gamma}_c$ is a critical shear rate. Equation 12 may be called Cross equation (20) or Van Wazer equation (21) and should have similar quality as compared with equation 11.

Another model of interest is given by Powell and Eyring (22),

$$\mu = \mu_\infty + (\mu_0 - \mu_\infty)\frac{\ln(t_{PE}\dot{\gamma} + 1)}{(t_{PE}\dot{\gamma})^n} \tag{13}$$

where t_{PE} is a characteristic time constant and n is an index constant, in most cases, $n \approx 1$.

Time-Dependent Rheology.   The rheological properties of suspensions are often time-dependent. If the apparent viscosity continuously decreases with time under shear with a subsequent recovery of the viscosity when the flow is ceased, the system is called thixotropic. The opposite behavior is called antithixopy or rheopexy. Figure 2 shows the time-dependent behaviors of suspensions. Curve 1 in Figure 2 illustrates a hysteresis produced by a thixotropic suspension, where con-

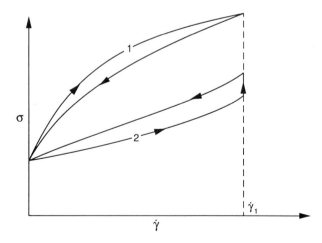

Figure 2.   Hysteresis of shear stress: curve 1, thixotropy with slow variation of shear rate; curve 2, rheopexy of slowly increasing, held steady at $\dot{\gamma}_1$, and decreasing $\dot{\gamma}$.

stant slow increase and decrease of the shear rates are applied. Sometimes, the loop formed by the increase and decrease of shear rates may not be singly connected, that is, the two lines may cross. Systems bearing this type of a behavior are, for example, pastes of clay in water, drilling mud, and crude oils. Modeling of the thixotropic behavior has not been well-established (23). For the time being, the Sestak model (24), which is a simplified version of the more general form of the Cheng's model based power-law concept (Herschel–Bulkley model), may be used

$$\sigma = \sigma_{y0} + K\dot{\gamma}^n + (\sigma_{y1} + K_1\dot{\gamma}^n)\lambda \qquad (14a)$$

$$\frac{d\lambda}{dt} = a(1 - \lambda) - b\dot{\gamma}^m\lambda \qquad (14b)$$

where λ is a memory parameter that has no physical significance; n is a form of power-law index; σ_{y0} denotes the permanent yield stress; σ_{y1} is the temporary yield stress component; K denotes consistencies; and a, b, and m are constants. All of the eight model parameters (constants) need to be determined experimentally for each system.

Curve 2 of Figure 2 shows a typical rheopexic hysteresis where the shear rate is increased slowly at $\dot{\gamma} < \dot{\gamma}_1$, held constant for some time at $\dot{\gamma} = \dot{\gamma}_1$ and is finally decreased gradually with a constant speed. This behavior is a reverse phenomenon of the thixotropy and is sometimes called antithixotropy. It occurs much less frequently than thixotropy and appears to be restricted to suspensions of very anistropic particles such as needles. In general, the models used for thixotropic systems can be adopted to simulate the behavior of the rheopexic systems.

Viscoelasticity. Viscoelastic materials are characterized by a combination of elastic and viscous properties. Thus, the shear stress is not only dependent on the rate of shearing but on the strain γ as well. In the simplest case, the viscoelastic behavior is governed by

$$\sigma = G\gamma + \mu\dot{\gamma} \qquad (15)$$

where G is the shear modulus of the system.

The four commonly used techniques to extract information on the viscoelastic behavior of suspensions are creep–compliance measurements, stress–relaxation measurement, shear–wave velocity measurements, and sinusoidal oscillatory testing (25–27). In general, transient measurements are aimed at two types of measurements, namely, stress relaxation, which is to measure the time dependence of the shear stress for a constant small strain, and creep measurement, which is to measure the time dependence of the strain for a constant stress.

The most widely used technique is the sinusoidal oscillatory or forced oscillation measurements. Oscillatory measurements consist of subjecting the medium under study to a continuously oscillating strain over a range of frequencies (Figure 3) and measuring the peak value of the stress σ_0 and the phase difference between the stress and the strain δ, that is,

$$\delta = \omega t_d \quad (16)$$

Here ω is the frequency of the applied stress. Normally, the amplitude of the applied strain, γ_0, is small enough to ensure that the system is in the linear viscoelastic region (i.e., the stress varies with strain linearly).

The applied strain or the strain resulting from an applied stress can be expressed in complex form by

$$\gamma = \gamma_0 e^{i(\omega t + \delta)} \quad (17)$$

The shear rate is given by

$$\dot{\gamma} = \frac{d\gamma}{dt} = i\omega\gamma_0 e^{i(\omega t + \delta)} = i\omega\gamma \quad (18)$$

In the linear viscoelastic region, the ratio of the stress to strain is given by

$$G^* = \frac{\sigma_0}{\gamma_0} = G' + iG'' \quad (19)$$

where G^* is the complex or dynamic modulus and is independent of the magnitude of the applied stress. G' is the storage modulus, which represents the in-phase stress-to-strain ratio, and G'' is the loss modulus, which represents the out-of-phase stress-to-strain ratio. We can write

$$G' = G^* \cos\delta \quad \text{and} \quad G'' = G^* \sin\delta \quad (20)$$

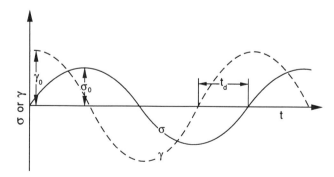

Figure 3.   Schematic of a viscoelastic response to an oscillating strain.

The complex notation may also be used for the viscosity as well:

$$\mu^* = \frac{G^*}{i\omega} = \mu' - i\mu'' \tag{21}$$

where

$$\mu' = \frac{G''}{\omega} \quad \text{and} \quad \mu'' = \frac{G''}{\omega} \tag{22}$$

By changing the frequency of the oscillation, the system under consideration behaves more like an elastic body when ω is small. However, the system behaves more like a viscous liquid at high frequencies.

We have seen that a simple apparent viscosity is used to characterize the stress–shear rate relationship for a non-Newtonian fluid. For a viscoelastic fluid, additional coefficients are required to determine the state of stress in any flow. For steady simple shear flow, the additional coefficients are given by the Criminale–Ericksen–Filbey equation

$$\sigma_{11} - \sigma_{22} = -\psi_1(\dot{\gamma})\dot{\gamma}_{21}^2 \tag{23}$$
$$\sigma_{22} - \sigma_{33} = -\psi_2(\dot{\gamma})\dot{\gamma}_{21}^2$$

where the functions of ψ_1 and ψ_2 are the primary and secondary normal stress coefficients. Subscripts 1, 2, and 3 for the shear stress and shear rate refer to the flow direction, shear axis, and neutral axis, respectively. The primary and secondary normal stress coefficients are strong functions of the shear rate and ψ_2 is normally about $-\psi_1/10$ with a positive value for ψ_2. Modeling of a viscoelastic fluid can be found in Bird et al. (28).

A general case of steady shear behavior often observed is that when the apparent viscosity is plotted against the shear rate, families of curves are found with different volume fraction of the dispersed solids. One typical curve of such a plot is shown in Figure 4. The features are that when the shear rate is less than $\dot{\gamma}_1$, a limiting (low shear) Newtonian behavior can be found as the three-dimensional structure of the suspension relaxes because $t_E < 1/\dot{\gamma}$, that is, elastic effect dominates. When $\dot{\gamma}_1 < \dot{\gamma} < \dot{\gamma}_2$, there is a marked shear thinning as the suspension structure changes are forced to occur because of the interparticle van der Waals attraction (29–31) as $t_E > 1/\dot{\gamma}$, the viscous effect becomes more pronounced. When $\dot{\gamma}_2 < \dot{\gamma} < \dot{\gamma}_3$, a two-dimensional arrangement parallel to the shear planes evolves, whereas shear thickening may be observed at high stresses owing to the rotational component of the shear field producing structural disorder (25) and the aggregates breakdown (18, 32–36).

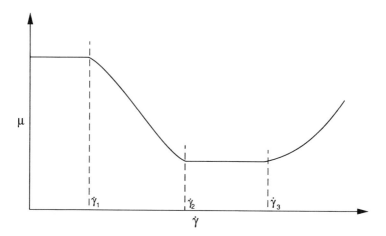

Figure 4.   Shear diagram for a suspension.

Suspension Structure and Forces Acting on the Suspended Particles

Some examples of suspension structures are illustrated in Figure 5. Figure 5a depicts a stable suspension with only short-range repulsive forces between the suspended fine particles. Hence, this system may settle as the particles move around each other into positions of lowest free energy, a consequence of the fact that the repulsive forces act between them. Figure 5b is a stable system for a more concentrated suspension. The

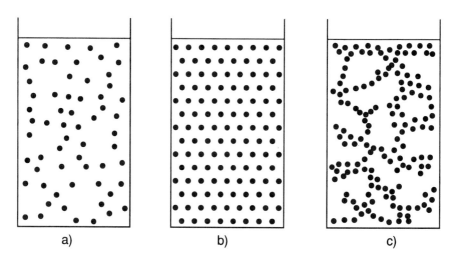

Figure 5.   Suspension structures: (a) disordered (stable); (b) ordered (stable); and (c) flocculated.

structure is very similar to a settled system of Figure 5a. This system may exist for a dilute system as well if long-range repulsive forces are strong enough. In Figure 5c, attractive forces are important, and the particles coagulate together on close approach to form aggregates or flocs. These flocs have often been found with an open disordered structure, which can be described using fractal concepts (37–39).

The forces acting between the particles will depend on the mechanism used to stabilize them, for example, electrostatic or steric, the size of the particles, the shape of the particles, the medium dielectric constant, the density of the particles, and the flow condition of the medium. The forces acting on particles and the interparticle potential energy (E) have been defined and discussed in Chapter 1 and hence only a short review is given here. Figure 6 depicts the range of the forces acting on particles that are exposed to a laminar flow field. Except for the long-range hydrodynamic forces, there are four major contributions to the forces, namely, Brownian forces, the London–Van der Waals forces (27, 40–44), the electrostatic double-layer potential (44–46), and polymeric interactions (47, 48).

Figure 7 illustrates the total interparticle potential, E, for colloidally stable systems and flocculated systems, where d is the particle diameter and r is the distance between the centers of two approaching particles. A colloidally stable suspension is characterized by a repulsive interaction (positive potential) when two particles approach each other (Figure 7a). Such a repulsion varies with distance, and hence it is termed "soft" repulsion. In the extreme, owing to the short range of the repulsive

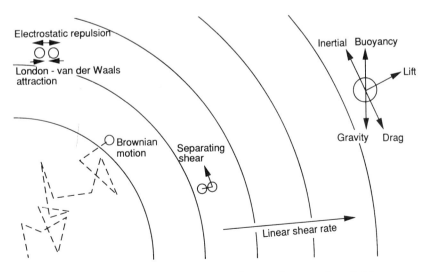

Figure 6.   Forces acting on particles in a laminar flow field.

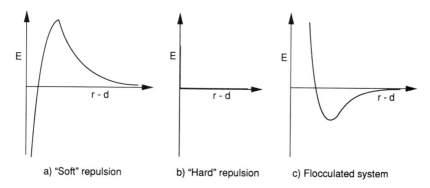

Figure 7. Total interparticle potentials for colloidally stable systems and flocculated systems.

forces in nature, a "hard" repulsion model may be sought, as shown in Figure 7b.

When attractive forces dominate, the suspension becomes flocculated because of the existence of a stable minimum potential, as shown in Figure 7c. The range (acting distance) of the attractive forces determines the degree of the flocculation. For weakly flocculated system, the attractive forces act in a longer range than those of a strongly flocculated system.

Rheological Measurements

The two main rheological properties of a suspension are the yield stress and the viscosity. Yield stress determines when the system becomes a fluid state and when is in a solid state, whereas viscosity determines the ability to flow. In this section, we start with the viscosity measurement. Although one can extract the yield stress from the complete viscosity–shear rate curve, it is helpful to measure the yield stress directly as well. The dynamic and transient measurements are also important for concentrated suspensions. However, because these two types of measurements can be blended into the measurements of the two main rheological properties with some modifications to the measuring instrument, we refer to their measurements only briefly when it is relevant to the discussion.

Viscosity Measurements.   Although it is not intended to give a complete review on the rheometry, some problems and related remarks on the pertinent experimental techniques are given for suspensions. The viscosity measurements are intended for relatively dilute suspensions where the shear rate can be defined. For a more complete listing of experimental techniques on rheometry, one is referred to Van Wazer et al. (21) and Whorlow (49). Whatever method is used, one should guard against deviations from homogeneity, separation of the suspension, that is, settling or creaming, into a region that is relatively crowded with solid particles and a region that is less concentrated. The former may be satisfied if the smallest dimension of the flow chamber is wider than, say 20 particle diameters, whereas the latter is perhaps best visually checked or estimated from hindered settling. Besides these two concerns, there is an analytical problem associated with the viscosity measurements, the wall effects on the flow field, especially for concentrated (50) and flocculated systems (51). This problem has not been resolved, although Pal et al. (4) suggested that a minimum of 20 particles should be ensured in any smallest dimension. For couette and parallel disc viscometers, Yoshimura and Prud'homme (52) studied the wall effects.

In practical applications, flow of the material through an orifice is perhaps the most frequently encountered rheological phenomenon. It is then natural to be used for the viscosity measurement of suspensions (53–55). However, the flow through an orifice is not precise in terms of shear measurement because the shear rate is not well defined under such circumstances. To meet this objection, the orifice is in most cases extended to a tube. This leads to the capillary flow type of viscometers, the simplest, and for Newtonian fluids, the most accurate type, comprising the familiar Ostwald und Ubbelohde viscometers. The fully developed axial velocity in the laminar regime is given by

$$v = \frac{\Delta p}{4L\mu} \left(r^2 - R^2 \right) \tag{24}$$

where v is the axial velocity, L is the length of the tube, r is the distance from the tube center, and R is the radius of the tube. The apparatus should be calibrated against the end effects. For example, the Marsh funnel viscometer, shown in Figure 8a, is frequently used in industry for the viscosity measurement of drilling mud.

In such a viscometer, the motion is rectilinear and everywhere parallel; thus, the shear rate is equal to the velocity gradient. In principle, this could be varied by applying various pressure differences over the tube. However, the most serious disadvantage for studying systems of any rheological complexity is that the shear rate varies from zero at the tube center to a maximum at the tube wall and is given by, for a Newtonian fluid,

$$\dot{\gamma} = \frac{dv}{dr} = \frac{\Delta p}{2L\mu} r \tag{25}$$

Another drawback to capillary viscometers is that time-dependent rheological behavior cannot be investigated because the flowing fluid has different shearing time history.

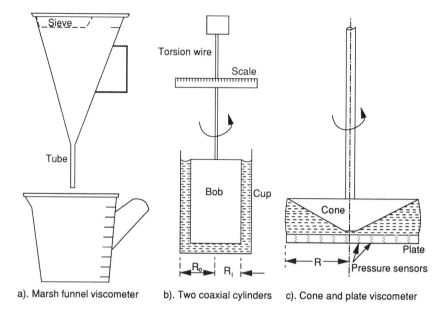

a). Marsh funnel viscometer   b). Two coaxial cylinders   c). Cone and plate viscometer

Figure 8.   Sketches of some commonly used viscometers.

Some of these problems are resolved by using the coaxial cylinder type of a viscometer, which consists of two coaxial cylinders (cup and bob), one of which rotates with an angular velocity of Ω (Figure 8b). The most common argument is to have the inner cylinder rotating with an angular velocity Ω_i and the outer cylinder being stationary. Thus, the angular velocity of the medium is a function of the distance r from the center of the axis:

$$\Omega = \frac{M}{4\pi\mu L}\left(\frac{1}{r^2} - \frac{1}{R_o^2}\right) \qquad (26)$$

where M is the torque exerted on the inner cylinder and R_o is the radius of the outer cylinder.

The shear rate for a fluid in the coaxial cylinder viscometer is given by

$$\dot{\gamma} = r\frac{d\Omega}{dr} = \frac{M}{2\pi L\mu r^2} \qquad (27)$$

which is also dependent on the position of the viscometer. However, because r is only allowed to change slightly in a typical cup-and-bob arrangement, $\dot{\gamma}$ is fairly constant as opposed to the case of a capillary viscometer. Moreover, all the material present in the viscometer is subjected to the shear for the same time during the experiment. The average shear rate is given by

$$\dot{\gamma}_{av} = \frac{M}{2\pi L \mu} \frac{\ln(R_o/R_i)}{R_o^2 - R_i^2} \tag{28}$$

In using such a viscometer, problems may arise from the curvilinear motion; a situation that is present even for a single fluid. If the motion is very strong, curvilinear laminar flow will change to Taylor vortex laminar flow, which will falsely exhibit a shear-thickening behavior. The centrifugal forces on the particles may change the local concentration of the particles and thus affect the suspension properties as well, especially for large particle and low viscosity suspending fluid systems.

The cone-and-plate viscometer shown in Figure 8c can give nearly a constant shear rate throughout the measurement volume chamber. This device is an improvement over the cup-and-bob rheometer, especially for small particle systems. For viscoelastic materials, this arrangement can measure the normal stresses with pressure sensors based on the flat surface. Although the wall effects can be eliminated to a certain extent, however, centrifugal clearing of the central region from the suspended particles can be a concern. It may not always be practical to obtain gaps greater than 20 particle diameters so that for suspensions having large particles, the coaxial cylinder geometry may be preferred because the gap between the cylinder is larger than that usually for the cone-and-plate arrangements. When the angular velocity is sinusoidal, both the cup-and-bob and cone-and-plate viscometers can be used to measure the viscoelastic response of a suspension.

A common problem for the viscosity measurements of suspensions is the presence of wall effects due to the presence of particles in the suspension. Some work has been done on the wall effects [*see*, e.g., Yoshimura and Prud'homme (52)]. Despite the efforts put into this subject, the words of Mewis (56) still stand: that there are no acceptable means in accounting for the wall effects or usable procedures to avoid wall effects available either experimentally or analytically. As shown by Wen et al. (57), the wall effects can lead to serious errors in viscosity measurements (Figure 9). The solid line in Figure 9 is the stress versus shear rate relation that they thought to be true for the system, whereas the symbols represent the experimental results from a rotational (cup-and-bob) viscometer with different gap size settings.

For concentrated suspensions, especially those near the random packing limit, the wall effect becomes so important that the viscosity may lose its meaning. Hence, other rheological properties, such as yield stress and wall boundary (slip) conditions, may be more meaningful for a concentrated system.

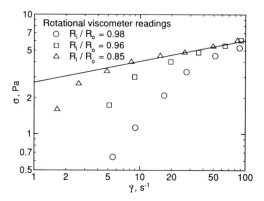

*Figure 9. Experimental results of concentric-cylinder viscometry for a 4%
bentonite slurry with different inner to outer cylinder radius ratios (57).*

Yield Stress Measurement.   The foundations of the rheological
treatment to fluids exhibiting a yield stress are due to Bingham (5).
Under steady flow conditions, it is common to neglect the contribution
from elastic deformation and to use the term Bingham fluid response.
Normally, the Herschel–Bulkley equation 9 is used to characterize the
flow.

When the slip condition is considered, flow of concentrated suspensions through capillary tubes may be characterized by the following
wall equation (58),

$$\sigma_w = \sigma_{wy} + \alpha_p p + \alpha_v v_{slip}{}^m \tag{29}$$

where σ_w is the wall shear stress, σ_{wy} is the wall shear yield stress, p is
the wall pressure, v_{slip} is the slip velocity, α_p and α_v are constants, and
m is the wall flow index. In practice, v_{slip}, as appeared in equation 29,
can be replaced by the mean velocity of the pipe (59). Most commonly,
the wall pressure effect is negligible and the wall flow index, m, is unity,
which reduces equation 29 to the Bingham wall boundary condition.
For dilute suspensions (60), the wall shear yield stress, σ_{wy}, is zero and
the Mooney wall boundary condition appears to prevail.

The slip effect can be so severe that one commonly observes a plug
flow of concentrated suspensions in capillary tubes. Hence, the method
does not form an intrinsically suitable basis for examining the bulk rheological characteristics. There are other problems associated with this
method as well, such as wall separation (a depletion zone exists in the
wall region), resulting in variation of the local particle concentration
profile.

The contraction (orifice) flows can be used to eliminate the particle
separation problem, although the flow can be very complex for single

fluid flow. Flow of concentrated suspensions through an orifice is unlikely to produce recirculation because of their low mobility. Benbow et al. (*61*) used this technique for catalyst pastes and applied an analytic solution based on a low bound plasticity method. For Herschel–Bulkley fluids, they introduced the following empirical extension to the plasticity solution for flow through orifice:

$$p_e = 2(\sigma_{Ey} + \lambda_0 u^\xi)\ln(d_t/d_o) \tag{30}$$

where p_e is the imposed extrusion pressure (difference), σ_{Ey} is the uniaxial yield stress, u is the mean velocity of the material in the orifice of diameter d_o, d_t is the tube diameter, and λ_0 and ξ are empirical constants. The major limitation to this solution is the assumption of homogeneous extension.

Another method of interest is the compression cell test, which is similar, to some degree, to the squeeze-film rheometer test and the cylindrical upsetting and compression ring test. A typical mechanism of the compression cell is shown in Figure 10. As is illustrated in Figure 10, there are also some problems associated with compression cell tests. The compression may not be uniform throughout the test region, where the shaded area is likely to be unyielding. Unlike the compression methods used for metals, the compression cell tests for suspensions need to be confined by a boundary (dashed lines in Figure 10). Depending on the confining boundary, the mechanism can also be slightly different. These methods have been described by Adams and co-workers (*2, 62, 63*), Kim and Luckham (*64*), and Homola and Robertson (*65*), among others. The compression cell test is essentially a plasticity approach toward characterizing the concentrated suspensions.

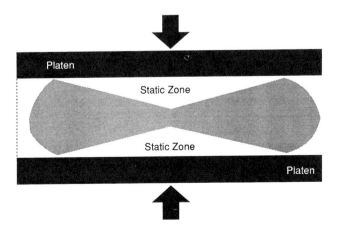

Figure 10. Schematic diagram of a compression test.

Steady Shear Rheology

Hard Sphere Systems.   A suspension made up of monosized (unimodal) hard spheres can be considered as a model suspension. This model suspension is the simplest case, where the flow is affected only by the hydrodynamic (viscous) interactions and Brownian motion. Although hard sphere systems are not frequently encountered in practice, they represent a very important idealization and starting point in evaluating the effect of interparticle potentials on the rheological behavior. The rheology of hard sphere systems under shear has been investigated by, among others, Walbridge and Waters (66), Krieger (67), Hoffman (68), Ackerson and Clark (69), Ackerson (70), Wagner and Russel (71), Jones et al. (72), and Woutersen and de Kruif (73).

The viscosity of a hard sphere suspension is a function of the volume fraction of particles. Generally, the viscosity can be expressed by a virial expansion:

$$\mu_r = \frac{\mu}{\mu_f} = 1 + k_E \phi + k_H \phi^2 + \cdots \tag{31}$$

where μ_r is the relative viscosity, which is the apparent viscosity of the suspension μ normalized by the suspending fluid viscosity μ_f and ϕ is the volume fraction of the particles. The term k_E is the Einstein constant and is sometimes called the intrinsic viscosity of the particles. Einstein constant reflects the single particle contribution to the viscosity at vanishing shear rate, $k_E = 2.5$ for hard spheres. The term k_H is the Huggins constant, which accounts for the particle pair interactions. For monodipersed (unimodal) hard sphere systems, Batchelor (74) obtained $k_H = 6.2$, and later Cichocki and Felderhof (75) refined his calculation and found $k_H = 6.0$. Equation 31 was found to be good for dilute systems when $\phi < 0.15$ (73, 76).

For concentrated systems, equation 31 loses its appeal because of the difficulties involved in evaluating the higher order coefficients. Instead, semiempirical models are more suitable. Such models are given by the Mooney equation (77), the Krieger–Dougherty equation (67), and the Quemada equation. In particular, the Quemada equation is the most used equation because of its simplicity and utility. The Mooney equation is given by

$$\mu_r = \exp\left(\frac{k_E \phi}{1 - \phi/\phi_{max}}\right) \tag{32}$$

where ϕ_{max} is the dense random packing limit volume fraction of the solids.

The Krieger–Dougherty equation is given by

$$\mu_r = \left(1 - \frac{\phi}{\phi_{max}}\right)^{-k_E \phi_{max}} \tag{33}$$

Heuristic derivations of equations 32 and 33 have been presented by Mooney (77), Krieger (67), Ball and Richmond (78), and Stein (79).

The Quemada expression is given by

$$\mu_r = \left(1 - \frac{\phi}{\phi_{max}}\right)^{-2} \tag{34}$$

Equation 34 is also referred to as the Maron–Pierce–Kitano model (55, 80–83).

Equation 32 is derived by starting with the assumption that μ_r is a function of ϕ. Let the total solid fraction be $\phi = \phi_1 + \phi_2$. If we treat the system as being made by adding additional amount ϕ_2 of solids into a system already containing ϕ_1 particle fraction, then it is reasonable to assume that

$$\mu(\phi) = \mu(\phi_1) \times \mu(\phi_2)/\mu(0)$$

which is consistent with

$$\mu_r = \exp(a_i \phi) \tag{35}$$

Additional refinements are that the viscosity becomes infinite at some volume fraction ϕ_{max}, where the motion of the suspended particles becomes inhibited. This hindering effect can be treated with an effective volume fraction through

$$\phi_{eff} = \frac{\phi}{1 - \dfrac{\phi}{\phi_{max}}} \tag{36}$$

And at the dilute limit, the Einstein equation, that is, equation 31 with second and higher order terms neglected, applies and it leads to $a_i = k_E$.

The derivation of equation 33 is achieved with a differential form of equation 31. Considering adding a very small amount of solids, $d\phi$, into the system that contains solid of volume faction ϕ, one would have

$$d\mu_r = k_E \mu_r d\phi \tag{37}$$

However, equation 36 applies strictly for adding a very small amount to a system of negligible solidsconcentration. Thus, an effective concentration should be used to give

$$d\mu_{\mathrm{r}} = k_{\mathrm{E}}\mu_{\mathrm{r}} \frac{d\phi}{1 - \dfrac{\phi}{\phi_{\mathrm{max}}}} \tag{38}$$

Integration of equation 38 leads to equation 33.

A more general model can be obtained by introducing a particle interaction parameter e (84) in modeling the hindering effect, and it is given by

$$d\mu_{\mathrm{r}} = k_{\mathrm{E}}\mu_{\mathrm{r}} \frac{d\phi}{\left(1 - \dfrac{\phi}{\phi_{\mathrm{max}}}\right)^{e}} \tag{39}$$

By a different choice of e value, one can derive many existing viscosity equations (84). For example, when $e = 2$, equation 39 can lead to the Mooney equation 32. When $e = 1$, equation 39 reduces to equation 38 and hence leads to the Krieger–Dougherty equation.

The same line of arguments can be made for the Quemada expression, for example, letting $e = 1$ and making appropriate assumptions when equation 39 is considered. However, one can notice that the Quemada expression is not very accurate for dilute systems where equation 31 may be useful. It is also possible to match equations 31 and 33 asymptotically to give a better equation that is suitable for the entire range of volume fraction,

$$\mu_{\mathrm{r}} = \left(1 - \frac{\phi}{\phi_{\mathrm{max}}}\right)^{-2} + \left(k_{\mathrm{E}} - \frac{2}{\phi_{\mathrm{max}}}\right)\phi + \left(k_{\mathrm{H}} - \frac{6}{\phi_{\mathrm{max}}^{2}}\right)\phi^{2} \tag{40}$$

Equation 40 may be called the modified Quemada equation.

Figure 11 shows the low shear limit relative viscosity variation with volume fraction for hard sphere systems. The experimental data used are those of Jones et al. (72), Wan der Werff and de Kruif (85), Papir and Krieger (86), and Mewis et al. (87). All of the models are based on the maximum random dense packing limit of $\phi_{\mathrm{max}} = 0.64$. We observe that the modified Quemada equation is the best among the models. The least accurate model is the Mooney equation. A better agreement can be found for the Mooney equation by modifying the volume fraction correction term, equation 36, in the following manner:

$$\mu_{\mathrm{r}} = \exp\left(\frac{k_{\mathrm{E}}\phi}{\sqrt{1 - \phi/\phi_{\mathrm{max}}}}\right) \tag{41}$$

However, the prediction by equation 41 is not as good as equation 40. Hence, equation 40 should be used to estimate the relative viscosity of the hard sphere systems.

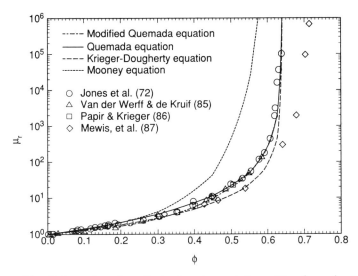

Figure 11.   Relative low shear limit viscosity variation with volume fraction.

In deriving equation 32 or equation 33, it is assumed that ϕ_{max} is the solid volume fraction at which the suspended particles cease moving. Thus, the forces, such as shearing, that can disturb the suspension structure and hence improve the mobility of particles will have an effect on the value of ϕ_{max}. This is confirmed by the fact that a value of $k_H = 6.0$ is observed at low shear limit, that is, $\dot{\gamma} \rightarrow 0$ and at high shear limit, $\dot{\gamma} \rightarrow \infty$, $k_H = 7.1$ is found. Typical values of ϕ_{max} have been found with the use of Quemada's equation as $\phi_{max} = 0.63 \pm 0.02$ in the low shear limit and $\phi_{max} = 0.71 \pm 0.02$ in the high shear limit for submicrometer-sized sterically stabilized silica spheres in cyclohexane (*72, 85, 88*).

Figure 12 shows the variation of the high shear limit relative viscosity variation with particle volume fraction. One can observe that large discrepancies are present in the experimental data among different studies. This indicates the difficulty in measuring the viscosity of suspensions. Many factors can affect the experimental measurements. For instance, the uniformity of the particles, properties of the suspending medium, the wall effects of the viscometer, and even the time of the experiment (*92*).

Jones and co-workers (*72, 88*) found that the suspension viscosity variation with shear rate can be fitted fairly well by the Cross equation, equation 12, with $m = 0.5 - 0.84$. Both the low and high shear limit relative viscosities, $\mu_{r0}, \mu_{r\infty}$, can be expressed by the Quemada's equation with $\phi_{max} = 0.63$ and 0.71, respectively.

The viscosity change with shear rate can be scaled by the Peclet number, which is defined by

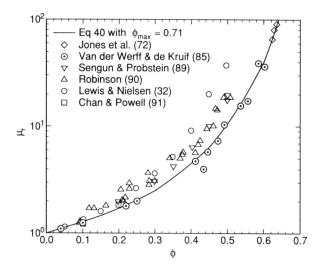

Figure 12. High shear limit relative viscosity variation with volume fraction.

$$Pe = \mu_f \dot{\gamma} \frac{d^3}{8kT} \qquad (42)$$

where k is Boltzmann's constant, $k = 1.3806 \times 10^{-23}$ J/K, and d is the particle diameter. Hence, the size effect can be included into the viscosity correlation quite elegantly. It should be noted that the hard sphere scaling is, by no means, unique. Reduced shear stress is also commonly used in the literature, and one may find physical importance in using the reduced shear stress (93). However, the Peclet number is the better representation in terms of utility because it does not depend on the term of interest, the viscosity of the suspension.

Figure 13 shows a typical plot of the steady shear relative viscosity versus the Peclet number for polystyrene spheres of various sizes suspended in various fluids. The success of the Peclet number scaling is well observed. One can also observe that the viscosity is higher when the shear rate is small, and at both high and low shear limits, the viscosity curve shows a plateau, corresponding to the high and low shear limit Newtonian behavior. The explanation for this behavior has been, in part, discussed earlier for the random packing limit of the particles.

The characterization of suspension microstructure by Wildemuth and Williams (34) resulted in a shear-dependent maximum packing limit fraction $\phi_{max}(\sigma)$. This more fundamental approach may find usefulness in the future. If we understand the microstructure change with shear and other factors, the viscosity of a suspension may be related to a single parameter, $\phi_{max}(\sigma)$.

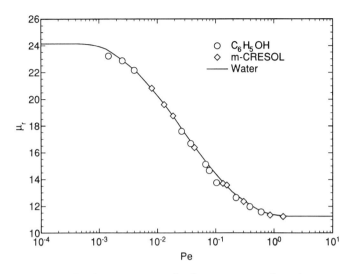

Figure 13. Steady shear viscosity of polystyrene particles of various sizes suspended in different fluid media with a volume fraction of 0.5. (67).

Soft Sphere Systems.   Suspensions are normally colloidally stabilized by the electric double-layer interaction (electrostatic) and attached polymer chains (steric). The rheological behavior is then strongly dependent on the separation distance between the particles. In electrostatically stabilized (electrostatic) systems, the range of the electric double-layer interaction potential, expressed by Debye length scale (44), is strongly dependent on the ionic strength of the suspending medium. In sterically stabilized (steric) systems, the thickness of the stabilizing polymer chain attached to the particles can change with the state of the suspending medium due to the degree of solvation of the polymer.

It is natural to replace the particle diameter with an effective particle diameter to include the effect of the polymer chain and the electric double layer to the hard sphere. Let

$$d_{\text{eff}} = d + 2\Delta = d\left(1 + \frac{2\Delta}{d}\right) \tag{43}$$

where Δ is the thickness of the stabilizing polymer layer or the effective electrostatic repulsive layer thickness. For electrostatically stabilized systems, Buscall (93) gave a semiempirical model for the effective soft sphere diameter as

$$d_{\text{eff}} = d + \int_d^\infty \left\{1 - \exp\left(\frac{8E(r)}{8kT + \mu\dot{\gamma}r^3/K_B}\right]\right\}dr \tag{44}$$

where K_B is a shear effect constant, which is dependent on the volume fraction of the particles. $K_B \approx 0.1$ may be used (93). $E(r)$ is the pairwise interaction potential. Normally, the surface charge is small, hence the term in the exponential can be treated as small. Equation 44 may be reduced to

$$\frac{\mu \dot{\gamma} d_{eff}^3}{E(d_{eff}) - kT} = K_B \tag{45}$$

The error introduced by this approximation is insignificant, at least when the Peclet number is large.

The volume fraction of the particles for a soft system is to be replaced by an effective volume fraction. The effective volume fraction is given by

$$\phi_{eff} = \phi \frac{d_{eff}^3}{d^3} \tag{46}$$

Figure 14 shows the variation of the steady shear relative viscosity at the high shear limit with the effective volume fraction as defined by equation 46 for poly(methyl methacrylate) (PMMA) suspensions of different sizes in decalin sterically stabilized by means of grafted poly(12-hydroxystearic acid) chains with a degree of polymerization of 5. The stabilizing polymer layer thickness is 9 ± 1 nm, in particular, $\Delta = 9$ nm

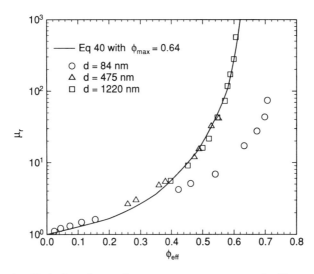

Figure 14.   High shear limit relative viscosity variation with effective volume fraction for various particle sizes (87).

for d = 84 nm, Δ = 8 nm for d = 475 nm, and Δ = 10 nm for d = 1220 nm. Hard sphere scaling principles and the use of ϕ_{eff} apply only when the ratio of the thickness of the stabilizing polymer layer to the particle radius $2\Delta/d$ is small. When the thickness of the polymer layer is increased, particularly at high volume fractions, the relative viscosity decreases drastically from that with the same hard sphere scale suspensions (87, 88, 94, 95). Prestige and Tadros (96), Kim and Luckham (97), and Liang et al. (95) used a different approach to estimate the thickness of the stabilizing polymer layer by fitting the viscosity with ϕ_{eff} to the Krieger–Dougherty equation while leaving Δ as a free parameter to be determined in the process. Their studies showed that Δ decreases with increasing volume fraction, indicating a substantial compression of the attached polymer layer.

Krieger and Equiluz (98) studied the influence of the suspending fluid ionic strength on the rheology of electrostatically stabilized monodisperse polystyrene latex particles. Figure 15 shows some results of Krieger and Equiluz (98) for latex particles suspended in aqueous solutions of HCl at various concentration levels (different ionic strength). The theoretical model of Buscall (93) is presented as lines in Figure 15. It can be observed that the relative suspension viscosity decreases dras-

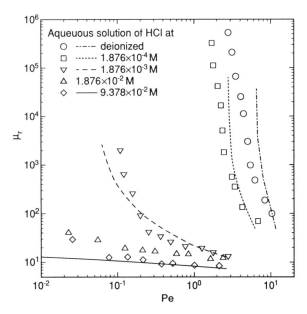

Figure 15.   Variation of steady shear μ_r with Pe for polystyrene spheres of d = 220nm suspended in aqueous solutions of HCl at ϕ = 0.4 (98).

tically with increasing ionic strength (increasing concentration of HCl). The decrease is much more substantial at lower shear (smaller Peclet number) region, owing to the fact that the mobility of the particles is strongly restricted by the electrostatic charge at low concentration of HCl. Hence, these suspensions are expected to behave solid-like at rest. When the Peclet number is large, the steady shear behavior is controled by the hydrodynamic interactions and hence the relative viscosity varies less with ionic strength. The prediction by the soft sphere scaling model of Buscall (93) shows fairly good agreement with the experimental data.

For sterically stabilized suspensions, the polymer thickness Δ changes with the medium temperature and shear rate. It can be said that the effective polymer thickness decreases with compression pressure exerted on the polymer chain. In other words, for the same suspension, when the volume fraction is low, the polymer layer thickness is likely to have more effect on the steady shear viscosity than when the volume fraction is high. However, when the ratio of the polymer thickness to the particle diameter is large, the suspending fluid inside the polymer layer can reduce the effect of the attached polymer.

Figure 16 shows the steady shear relative viscosity variation with the effective Peclet number, Pe_{eff}, based on the effective particle diameter at each temperature level, and the temperature for a PMMA suspension. The particles of 0.8 μm are sterically stabilized by a thick layer of terminally anchored poly(dimethylsiloxane) and suspended in n-hexadecane at the volume fraction of $\phi = 0.282$. The data points are

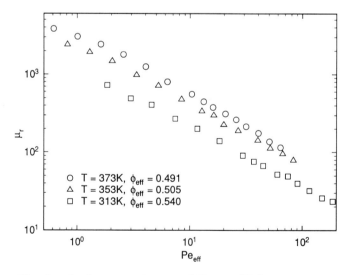

Figure 16.   Steady shear μ_r variation with Pe_{eff} and T for a steric suspension of d = 0.8 μm *and* ϕ = 0.282 (99).

from Croucher and Milkie (99). Because the actual effective polymer layer thickness is dependent on the density (volume fraction) of the polymer fibers stemming out of the particles as well as the length of the fibers, the effective particle diameter as defined by equation 43 does not scale the shear rate well as is shown in Figure 16. The relative viscosity increases with temperature, whereas the attached polymer layer is thinner when the temperature is raised. The noncollapsing of the data points for different temperatures may be attributed to the difference in ϕ_{eff}. However, an examination of ϕ_{eff} values reveals that a surprising behavior of higher viscosity for lower ϕ_{eff} is displayed. This "unexpected" behavior can be explained by the rigidity of the thick polymer layer. When temperature is higher, the particles become more crowded resulting in a decrease in Δ and an increase in the volume fraction of the fibers in the polymer layer. Hence, the rigidity of the overall polymer layer after raising the temperature is stronger compared with that at a lower temperature. The stronger polymer layer results in a more effective barrier and gives a higher viscosity.

Particle Shape Effect.   To this point, we have been dealing only with spherical particle suspensions. When the particles have irregular shapes, the rheological properties are expected to be very different from those of the spherical particle suspensions. Consider, for example, a simple system of cylindrical fibre suspensions. Because the particles are expected to align in the direction of the flow or shear, the viscosity needs to be treated as a second-order tensor, that is, the values of the viscosity under the same condition are different when different directions are referred. Only at the low (zero) shear limit may the particles be randomly distributed and have an isotropic rheological behavior.

Let the relative viscosity (normalized by the suspending fluid viscosity) as measured in the direction of the cylinder axis (longitudinal direction) as μ_{rL} and the relative transverse viscosity be μ_{rT}. At the low shear limit and in a dilute system, the viscosity is expected to be isotropic. Eshelby (*100*) obtained

$$\mu_{rL} = \mu_{rT} = 1 + 2\phi + \cdots \tag{47}$$

When the fiber suspension is concentrated, Christensen (*101*) used the lubrication theory to derive the limiting behavior

$$\mu_{rT} \rightarrow \frac{3^{3/2}\pi}{8(1 - \phi/\phi_{max})^{3/2}} \quad \text{as} \quad \frac{\phi}{\phi_{max}} \rightarrow 1 \tag{48}$$

$$\mu_{rL} \rightarrow \frac{3^{1/2}\pi}{(1 - \phi/\phi_{max})^{1/2}} \quad \text{as} \quad \frac{\phi}{\phi_{max}} \rightarrow 1 \tag{49}$$

It should be noted that the relative viscosity dependence on the volume fraction for suspensions of spherical particles using the same treatment as above gives

$$\mu_r \rightarrow \frac{C}{(1 - \phi/\phi_{max})} \quad \text{as} \quad \frac{\phi}{\phi_{max}} \rightarrow 1 \tag{50}$$

Equation 50 does not agree with the experimentally observed Quemada equation. However, by comparing equations 48 and 49 with 50, we may expect that the longitudinal viscosity of the cylindrical fiber suspension is smaller than the viscosity of spherical particle suspensions at a concentrated state, whereas the transverse viscosity is higher than the viscosity of spherical suspensions.

For a suspension of rigid spheroids, Leal and Hinch (102) determined the Einstein constant (intrinsic viscosity) for the limiting case of large axis (aspect) ratio, r_e = long axis/short axis → ∞, as follows:

$$k_{ET} = 2 + \frac{0.312 r_e}{\ln(2r_e) - 1.5} \tag{51}$$

$$k_{EL} = 3.183 - \frac{1.792}{r_e} \tag{52}$$

where k_{ET} is the Einstein constant in the transverse directions and k_{EL} is the Einstein constant in the longitudinal direction.

At low shear limit, it is possible to maintain a random structure for suspensions of rodlike particles. The low shear viscosity can be measured by, for example, falling ball rheometry (103, 104) without disturbing the random structure significantly and traditional rheometry at incipient shearing (105, 106). The hydrodynamic behavior of isotropic suspensions is also modeled by Claeys and Brady recently in a three-paper series (107–109). However, one may find little use of the isotropic viscosity because the particles will align with the flow and rotate in real situations to reduce the shear loss. Hence, the values of viscosity obtained from the falling ball rheometry will be always much higher than those obtained from a steady shear rheometry. The longitudinal and transverse viscosity approaches are not problem-free either because the particles do not necessarily all align well in the direction of the flow or shear, whereas a distribution of alignment is expected.

A more fundamental approach is to consider the rheological properties with dynamical properties. For a given rate-of-strain tensor $\mathbf{E}$ and moments of the orientation vector $\mathbf{p}$, Batchelor (110) derived an expression for the bulk average deviatoric stress σ for a suspension of non-Brownian fibers of large aspect ratio given by

$$\sigma = \mu(\langle\mathbf{pppp}\rangle - \tfrac{1}{3}\delta\langle\mathbf{pp}\rangle): \mathbf{E} + 2\mu_f\mathbf{E} \tag{53}$$

where μ_f is the suspending fluid viscosity and μ is now assigned as the viscosity of the fiber suspension. Equation 53 indicates that the shear stress is a strong function of the rate-of-strain as well as the orientation of the fibers.

For $r_e > 20$ and freely rotating particles, Simha (*111*) found that the Einstein constant for spheroids and rigid rods is given by

$$k_E = \frac{14}{15} + \frac{r_e^2/15}{\ln(2r_e) - c_a} + \frac{r_e^2/5}{\ln(2r_e) - c_a + 1} \tag{54}$$

where c_a is a constant. For spheroids, $c_a = 1.5$ and for rigid rods, $c_a = 1.8$. Goldsmith and Mason (*112*) obtained the Einstein constant based on a time-averaged optimum orientation given by

$$k_E = \frac{r_e^3}{3[\ln(2r_e) - c_a](r_e + 1)^2} \tag{55}$$

Haber and Brenner (*113*) obtained a relation for ellipsoids.

The Einstein constant for a fiber suspension is also a function of shear rate. The ratio of low to high shear limit Einstein constant is $r_e/1.17$ for rods and $r_e/31$ for discs (*114*). The low shear limit Einstein constant is given by

$$k_E = a_0 + a_1 r_e, \qquad \text{for } r_e < 50 \tag{56}$$

where a is a constant. For discs, $a_0 = 2.51$ and $a_1 = 0.1127$, and for rods, $a_0 = 2.34$ and $a_1 = 0.1636$ (*115*). Shaqfeh and Fredrickson (*116*) and Shaqfeh and Koch (*117*) gave the following formulae:

$$k_E = \frac{\pi l^3/(3V_p)}{\ln(1/\phi) + \ln\ln(1/\phi) + A(\phi)} - 1 \tag{57}$$

where l is the longest dimension of the fiber, V_p is the volume of one particle, $A(\phi)$ is a weak function of ϕ and is dependent on the shape of the fiber.

Kitano et al. (*82*) found that the relative viscosity of a suspension of cylindrical rods can be estimated by the Quemada equation with the maximum packing fraction given by

$$\phi_{max} = 0.53 - 0.013r_e, \qquad \text{for } 5 < r_e < 30 \tag{58}$$

By an asymptotic matching of the dilute limit homogeneous viscosity with the Quemada equation, Phan-Thien and Graham (*118*) obtained the following equation valid for a wide range of solid concentration:

$$\mu_r = 1 + k_E \frac{1 - 0.5\ \phi/\phi_{max}}{(1 - \phi/\phi_{max})^2}\ \phi \tag{59}$$

where k_E is the intrinsic viscosity of the suspension and may be estimated by equation 56.

Figure 17 shows the experimental results on the low shear relative viscosity for suspensions of cylinders and spheroids of Phan-Thien and Graham (*118*) for a good range of particle volume fractions. These data agree with the experimental results of Ganani and Powell (*119*), although the latter deal only with dilute suspensions. From Figure 17, we observe the agreement between the predictions based on equation 59 and the experimental data is fairly good. Equation 40 may also be used, but the value of the Huggins constant must be provided in the concentration range shown in Figure 17. We find that the following formulae may be used with equation 40 based on the data of Kitano et al. (*82*) and Phan-Thien and Graham (*118*):

$$k_H = 3k_E/\phi_{max} \tag{60}$$

In a uniaxial elongational flow, the Trouton, or elongational viscosity of an anisometric particle suspension can be estimated by equation 59 as well with an intrinsic viscosity given by

$$k_E = \frac{2r_e^2}{3[\ln(2r_e^2) - 1.5]} \tag{61}$$

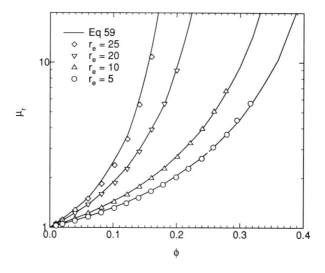

Figure 17.   Low shear limit viscosity variation with solid volume fraction and aspect ratio for suspensions of spheroids and cylinders (118).

Polydispersity Effect. In the previous sections, we discussed the rheology of monodispersed particles, that is, all of the solids have the same particle size and shape. In this section, we consider the case where the suspension is made up of solids having different sizes.

Polydispersity can have a significant influence on the rheological behavior of suspensions. For the same solid volume fraction, it has been known that particle size distribution can greatly reduce the hydrodynamic resistance of random packs of spherical particles (*120*) because of the reduction of channel sizes in the packs as compared with the unimodal systems. Apart from the channel (gap between particles) size effect, a multimodal system can have significantly lower void ratio as well. The Quemada equation suggests that lowering of the random packing void ratio alone would lead to a significant reduction of the suspension viscosity at a given solid volume fraction. Several studies have shown that the use of a bimodal or a continuous broad particle size distribution can lower the viscosity and increase the maximum attainable solid volume fraction (*36, 53–55, 121–136*). The most complete studies to date for concentrated suspensions are the studies of Rodriguez et al. (*130*) and Chang and Powell (*55*). The latter used an orifice viscometer instead of the traditional rotational viscometers.

For bimodal sterically stabilized silica suspensions in cyclohexane, Figure 18 shows the variation of the Einstein and Huggins constants with the relative volume fraction of the larger particles in the total solids,

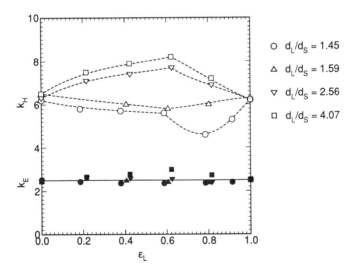

Figure 18. Variation of the Einstein and Huggins constants with the ratio of the large to small particle size and volume fraction content of the large particles (73). The filled symbols mean the same as the open symbols.

$\epsilon_L = \phi_L/\phi$, for different size ratios of large to small particle, d_L/d_S. The experimental conditions of the data (73) correspond to the low shear limit. It can be seen that the bidispersity has a relatively small effect on the Einstein constant, whereas a more pronounced effect on the Huggins constant can be observed. When the particle size ratio is small, the Huggins constant k_H experiences a minimum. However, when the particle size ratio is greater than 1.6, a maximum value in k_H can be observed. The low shear limit viscosity at vanishing solid volume fraction for bidispersed and polydispersed hard sphere systems is also modeled numerically by Wagner and Woutersen (137) and Jones (138). These two theoretical studies revealed that the Einstein constant remains constant, whereas the Huggins constant decreases slightly with increasing degree of polydispersivity.

Figure 19 shows the steady shear relative viscosity variation with the relative volume fraction of the large particles, ϵ_L, for various large to small particle size ratios, d_L/d_S, and different total solid volume fraction, ϕ, for bimodal concentrated dispersions of submicron resin particles at a shear stress level of 0.0155 Pa. The experimental data are taken from Hoffman (129). We can observe that the relative viscosity exhibits a minimum near $\epsilon_L = 0.8$. The minimum viscosity behavior is more-pronounced at large particle diameter ratios, d_L/d_S. The bimodal system viscosity can be several order of magnitudes lower than the corresponding monodispersed systems when the larger particles composed of about

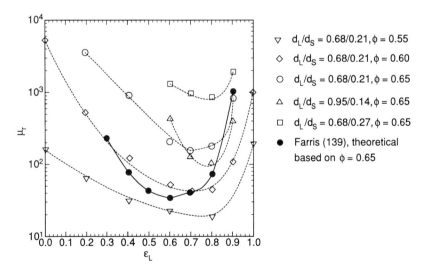

Figure 19.   Effect of composition of bimodal suspensions of colloidally stable polymer particles on the viscosity measured at a steady shear level of 0.0155 Pa (129).

80% of the solids in volume. As would be expected, the relative viscosity increases with increasing total solid volume fraction for a given ϵ_L value as well. The predictions based on Farris (*139*) are also shown in the figure. Farris (*139*) hypothesized that multimodal systems can be treated as suspending large particles (the large size component) into a homogeneous suspension made up of the small particles. This treatment is believed to hold for the case where the particle size ratio is very large, so that the large size particles "see" the suspension of the small particles as a continuous phase. It holds only for the case when one of the components is dominant in quantity.

Kim and Luckham (*140*) suggested that the relative dynamic viscosity of a bidispersed suspension may be estimated by using the product of the two component relative viscosities each computed from the Krieger–Dougherty equation as if they were alone in the suspension. This treatment has been commonly used since Farris (*139*). However, it is valid only when concentration of either component is large, that is, $\epsilon_L \rightarrow 1$ or $\epsilon_L \rightarrow 0$, and the particle sizes are very different from each other, that is, $d_S \ll d_L$. When the two concentrations are similar, that is, $\epsilon_L \approx 0.5$, the estimation of the apparent viscosity using this approach gives a much lower value than the experimental values. This behavior is same for dynamic viscosity as the steady shear viscosity.

Rodriguez et al. (*130*) and Chang and Powell (55) showed that the traditional Quemada equation and other models can be used to correlate the steady shear viscosity of bidispersed systems provided that the correct value of the maximum packing fraction, ϕ_{max}, is used. Figure 20 shows the universal scaling for both the low and the high shear limit relative viscosities as a function of the reduced solid volume fraction for monodisperse and bidisperse latex particle suspensions. Figure 21 shows the low shear limit maximum packing fraction, ϕ_{max}, variation with the relative volume fraction of the large particles, ϵ_L, for various large to small particle size ratios, d_L/d_S. We can observe that the maximum packing fraction ϕ_{max} reaches a maximum between $\epsilon_L = 0.6$ and 0.8 and the maximum value increases with the particle diameter ratio, d_L/d_S. The phenomenon of higher maximum packing limit, ϕ_{max}, for bimodal systems than unimodal systems is a reversed relation of the relative shear viscosity versus the relative volume fraction of the large particles, ϵ_L, and particle diameter ratio, d_L/d_S.

Rodriguez et al. (*130*) showed that the behavior of steady shear viscosity versus the shear stress for a binary system is no different from that of a monodisperse suspension. A universal equation may be obtained for both the monodisperse and bidisperse suspensions. Hence, we may further infer from the literature results available that the steady shear viscosity for a polydispersed (multimodal) suspension can be treated the

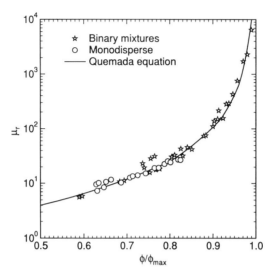

Figure 20.   Universal low and high shear limit viscosity vs. solid volume fraction curve for monodisperse and bidisperse systems (130).

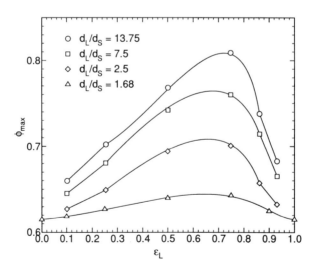

Figure 21.   Maximum packing fraction variation with composition of particles and large to small size ratio for binary mixtures (55, 130).

same way as one would for a monodisperse (unimodal) suspension by using ϕ_{max} to normalize ϕ.

Shear Thickening and Discontinuity Behaviors. Although continuous shear thinning behavior is commonly observed, shear thickening behavior can also be observed for concentrated suspensions of nonaggregating solid particles when subjected to the appropriate conditions (*141*).

Metzner and Whitlock (*142*) were probably the first to observe the shear thickening behavior of suspensions. The systematic studies on this subject have been conducted by Hoffman (*68, 143, 144*), Strivens (*145*), Wagstaff and Chaffey (*124*), Woodcock (*146*), Laun (*147*), Boersma et al. (*148, 149*), Yilmazer and Kalyon (*150*), and Laun et al. (*151*). However, a better understanding of the shear thickening behavior is needed to predict the conditions for which shear thickening would occur.

Figure 22 shows shear thickening behavior of colloidally stable suspensions for both monodisperse and bidisperse poly(vinyl chloride) particles in dioctyl phthalate. The monodisperse system was studied by Hoffman (*68*) and the bidisperse system was taken from Boersma et al. (*149*). The concentrated suspensions of Figure 22 show either continuous or discontinuous shear thickening behavior depending on the solid volume fraction and the composition of the solids. The severity of the shear thickening increases with particle concentration, possibly ϕ/ϕ_{max}, whereas the critical Peclet number (based on the average particle di-

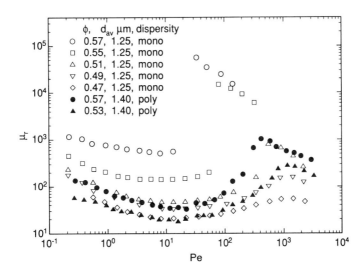

Figure 22.   Shear thickening behavior of suspensions of monodisperse and bidisperse polyvinyl chloride particles in dioctyl phthalate (68, 149).

ameter) for the onset of shear thickening decreases with increasing particle concentration. For a more concentrated monodisperse system, a large jump in the steady shear viscosity is observed at the critical Peclet number and followed by a shear thinning behavior, whereas the transition to the next shear thinning behavior is smooth for less concentrated or bidisperse systems. Owing to the increase in maximum packing limit volume fraction or reduction in the effective solid concentration for multimodal systems, the shear thickening behavior is delayed for the same particle concentration as compared to monodipersed systems (152).

Hoffman (68) attributed the shear thickening behavior to some order–disorder transition of particle microstructure, which was also observed by Ackerson and Pusey (153) and Wagner and Russel (71). This model was continued by Boersma et al. (148). They assumed that the shear thickening occurs when the shear forces overcome the interparticle forces. Hence, a dimensionless group, N_d, the ratio of the viscous to the repulsive forces, may be used to describe the transition:

$$N_d = \frac{(3/2)\pi\mu_f d^2 \dot{\gamma}}{2\pi\epsilon_0\epsilon_r E_0{}^2} \tag{62}$$

where ϵ_0 is the permittivity of vacuum, ϵ_r is the dielectric constant of the suspending fluid medium, and E_0 is the surface charge (zeta) potential of the particles. This group relates to the case of constant surface potential of electrostatically stabilized systems. Shear thickening is predicted to occur when $N_d > 1$. In addition, the shear thickening can occur in sterically stabilized concentrated systems as well (154, 155).

Shear thickening behavior appears almost like a jump in the steady shear viscosity versus shear rate, μ versus $\dot{\gamma}$, or μ_r versus Pe plots, especially for more concentrated suspensions. Time dependence is thus expected to occur. Boersma et al. (149) demonstrated that strong time dependence can be observed when a steady shear is applied near the critical point for which the shear thickening occurs.

Laun et al. (151) investigated different suspension systems near the transition to shear thickening. Their studies show that the order–disorder assumption may not hold, although there exist some structural changes near the transition. It is certain, however, that shear thickening behavior is associated with the loss of long-range order of the particles. However, the loss of long-range order does not necessarily produce shear thickening (85, 156, 157).

Ackerson and Clark (158), Chen and Zukoski (159), Stevens et al. (160), and Chen et al. (161) observed a different kind of discontinuous viscosity response to an order–disorder transition for electrostatically stabilized suspensions at a relatively low solid concentrations, where the viscosity displays a discontinuous drop at certain shear rate. Apart from the obvious changes in the viscosity versus shear rate plot, Chen and co-workers (*156*, *162*) observed a rather peculiar discontinuity in the viscosity–shear stress or shear stress versus shear rate plot for a solid concentration between the shear melting and shear thickening values. As shown in Figure 23, there is no visible change in the viscosity versus shear rate behavior, although the suspension order changed from a strained-crystal microstructure to a shearing layer microstructure. The system consisted of a polystyrene latex particles of 0.146 nm diameter suspended in an aqueous solution of 0.001 M NaCl. The solid fraction is 0.33. In the transition region, the long-range orientational order is lost and the suspension takes a polycrystal-line appearance. Despite the changes in the suspension structure, the steady shear viscosity shows little effect on the μ versus $\dot{\gamma}$ plane (Figure 23). However, when the shear stress is plotted against the shear rate, as shown in Figure 24, discontinuity is observed.

It can be concluded that the shear melting, thickening, and the shear stress discontinuity are the results of the shear-induced suspension structure change. The irregularities occur when several suspension structures coexist in the system in separate zones. In the region where the irregularity of melting shear or shear stress discontinuity occurs, the shear stress may decrease with shear rate or remain constant for a particular region of shear rate (*163*). On the shear stress versus shear rate plane, hysteresis exists. The irregularities (shear melting, thickening, and the shear stress discontinuity) are more severe upon increasing shear

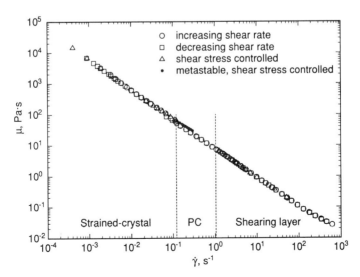

Figure 23. Steady shear viscosity vs. shear rate diagram for 0.146 μm polystyrene latex suspension in $10^{-3}M$ NaCl (162).

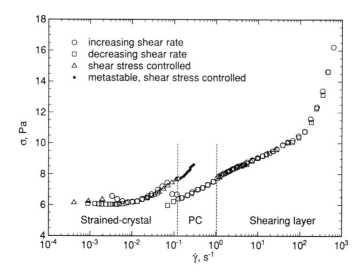

Figure 24.   Steady shear stress vs. shear rate diagram for 0.146 μm poly-styrene latex suspension in $10^{-3}M$ NaCl (162).

rates. Imhof et al. (163) also found that the critical shear stress over the high frequency storage modulus is a constant, 0.033 ± 0.003, independent of the solid volume fraction and ionic strength. The critical shear stress values are believed to the same as those of Chen and Zukoski (159). Hence, depending on the concentration of the solids, composition of the solids, and the stabilizing methods, these irregular behaviors can all be attributed to a certain force balance. A universal criterion may be awaiting to be found in this regard. As is suggested by Chen et al. (156), the same model of Hoffman (68) and Boersma et al. (148) may be worth pursuing.

Flocculated Systems.   Flocculation is commonly caused by a dominant attractive interparticle potential. Flocs are disordered meta-stable structures with very long relaxation times. Flocculated systems are normally associated with a history-dependent rheological behavior. Because the relaxation time is strongly dependent on the strength of the interparticle attraction, it is convenient to distinguish between two types of flocculated systems: weakly flocculated and strongly flocculated (164). Weakly flocculated systems with an interparticle potential of 1 $< -E/kT < 20$ are able to attain a reproducible rest state after shear in a reasonably short time. In strongly flocculated systems in which the interparticle potential is large, $-E/kT > 20$, the diffusion of particles is absent and no reproducible rest state can be attained after shearing. As a result, the rheological behaviors of these two types of flocculated systems are distinguishably different.

Weakly flocculated systems have been studied by Buscall et al. (*165*), Heath and Tadros (*166*), Goodwin et al. (*167*), Patel and Russel (*168*, *169*), Otsubo (*170–172*), Buscall et al. (*51, 173*), Woutersen and de Kruif (*174*), and Nakai et al. (*175*). The weak flocculation can be obtained by several means, including secondary-minimum flocculation, depletion flocculation, polymer-bridging flocculation, and incipient flocculation. Details of the various mechanisms of interparticle attraction can be found in Russel et al. (*27*) and Somasundaran and Yu (*176*). Normally, flocculated systems have a solid volume fraction of no less than 0.2. When $\phi < 0.2$, an attractive system will settle down quickly. When $\phi > 0.2$, a flocculated suspension can be maintained easily for a period of time for a steady shear measurement to be completed.

The system studied by Woutersen and de Kruif (*174*) is sterically stabilized suspension of silica particles in a marginal solvent, benzene. When the temperature was lowered, the magnitude of the interparticle attraction increased. The steady shear behavior changes drastically by lowering the temperature. However, in the same temperature range and conditions for slightly smaller particles, small angle neutron scattering study revealed no permanent flocculation (*177*). The rheological measurements also showed no time history dependence. These effects are due to the fact that the two opposing potential forces (positive Brownian motion and attractive forces) are of the same order of magnitude in the weak attractive region. Wouterson and de Kruif (*174*) suggested that temporary aggregates are formed in the process. At a high temperature, the viscosity of the system is relatively invariant with shear rate. At a low temperature, however, a severe shear thinning behavior is displayed.

Although most flocculated systems display a shear thinning behavior, Otsubo (*171*) observed dilatant (shear thickening) behavior in a weakly flocculated suspension of styrene–methyl acrylate copolymer (SMAC) particles of 80 nm suspended in an aqueous polymer solution. The particle volume fraction was in the range of 0.15–0.3. Shear thickening behavior was found to occur within a particular range of bridging polymer, poly(acrylic acid) (PAA), concentrations. When the bridging polymer concentration was either low or high, the system displayed shear thinning only, with some difference in the degree of severity. The shear thickening became more pronounced when the temperature was lowered (Figure 25), corresponding to a more attractive system. The shear thickening may be irreversible so that if preshearing is applied, the shear thickening is no longer attainable (*178*), which is similar to the behavior of the irregularities in stable systems (*163*).

Nakai et al. (*175*) studied the sterically stabilized silica suspension in aqueous hydroxypropyl(methyl)cellulose solution (HPMC). They confirmed, in part, the results of Otsubo (*171*): that both shear thinning

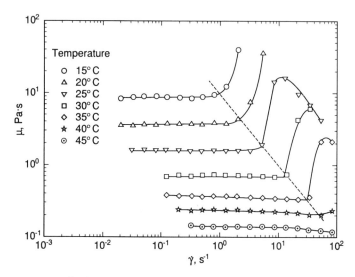

*Figure 25.   Steady shear viscosity variation with shear rate for 15% styrene–
methyl acrylate copolymer suspension in a 2 wt% PAA solution (171).*

and shear thickening can occur for a weakly flocculated system. In addition, rheopexy was observed for 5% silica suspensions and stress overshooting behavior was observed for 7.5% silica suspensions.

Strongly flocculated systems are characterized by an irreversible nonrecoverable rheological behavior when shear rate is cycled up or down due to the fact that the Brownian motion is far too weak compared with the interparticle attraction to result in a reproducible state. Because the rheological behavior is not reversible for a strongly flocculated system, the time-dependent steady shear behavior is termed shear degradation and should be distinguished from thixotropy, where a reversible behavior is assumed. Firth and Hunter (19, 179, 180) extensively studied the steady shear behavior of mainly aqueous TiO_2 and flocculated PMMA suspensions. They developed an elastic floc model to link the floc structure and the strength of the interparticle attraction to the rheological behavior. To obtain reproducible flow curves, the flocculated suspensions were first subjected to a very strong shear. After this treatment, relatively good agreement between the model predictions and the experimental results was found. They determined the Bingham yield stress and obtained

$$\sigma_B \sim \phi^2 \qquad (63)$$

The stronger the flocculation is, the more severely shear thinning behavior is observed, up to a sudden (catastrophic) drop in viscosity at a critical shear rate (*15, 51, 181, 182*). Figure 26 shows a catastrophic viscosity drop on the viscosity versus shear rate plane. This behavior is a result of large loose flocs breakup and the formation of smaller denser flocs. With increasing shear rate, the viscous forces reduce the size of the flocs while increasing the density of the flocs (*183–186*). Increasing shear rate also results in the release of suspending liquid trapped inside the flocs, which results in a decrease in the effective volume fraction of the aggregates.

   The qualitative difference between weakly and strongly flocculated systems may also be manifested in the occurrence of a yield stress. Weakly flocculated systems commonly show an apparent low shear viscosity (Figure 27), whereas a strongly flocculated system typically shows a yield stress. The low shear limit viscosity for weakly flocculated systems is shown in Figure 27 from various sources. Owing to the formation of flocs immobilized certain fraction of suspending liquid as well, the steady shear viscosity for a flocculated system is higher than that for a hard sphere system at a given solid concentration. The relative viscosity may be approximated by the following equation:

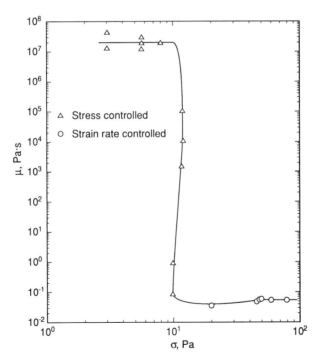

Figure 26.   Steady shear viscosity for a strongly flocculated suspension of 2.5% fumed silica in methyl laurate measured with different rheometries (15).

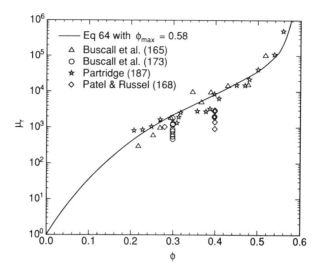

Figure 27. Low shear viscosity variation with solid volume fraction for weakly flocculated systems.

$$\mu_r = \frac{a_v \phi^3}{\phi_{max} - \phi} + 1 \qquad (64)$$

where a_v is a constant depending on the structure of the flocs. The line shown in Figure 27 is for $a_v = 2.3 \times 10^4$. Owing to the high viscosity values at low shear limit, a yield stress may also be assumed as a convenience for weakly flocculated suspensions. Especially for a concentrated suspension, a long time is needed for the flow to occur after the application of a low shear rate (stress). Figure 28 shows the Bingham yield stress versus the solid volume fraction diagram for several weakly flocculated systems (*173*). It can be observed that the same scaling for the low shear viscosity may also be applied to the Bingham yield stress:

$$\sigma_B = \frac{a_y \phi^3}{\phi_{max} - \phi} + \sigma_{B0} \qquad (65)$$

where a_y and σ_{B0} are constants that depend on the structure of the suspension. The curve shown in Figure 28 is for $a_y = 0.24$ Pa and $\sigma_{B0} = 0.05$ Pa. From Figure 28, one observes that the maximum packing limit ϕ_{max} is also a function of the particle size. The particle size dependence of Bingham yield stress is understood by the fact that different degree of flocculation is obtained for different particle sizes for the same system.

    Because of the size of a floc, the wall effects become significant in the viscosity and other rheological measurements for a flocculated sus-

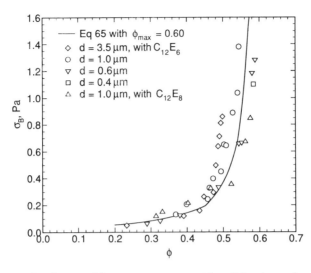

Figure 28. Bingham yield stress variation with solid volume fraction for weakly flocculated systems of different particle sizes and different layer of thickness (173).

pension. Because the measurement devices are limited (normally intentionally) in size, the assumption of a continuum cannot be justified. Apart from the continuum assumption, the wall slip problems can be more severe than for stable systems (51). More studies are needed to elucidate the wall effects of flocculated and stable systems.

Buscall et al. (51) pointed out that one common way of avoiding the wall effects is to roughen the surfaces of the measuring vessel. This technique may be appropriate for stable systems provided that the walls are roughened using the same particles and with the same concentration as that in the testing suspension. In addition, the suspension has to behave as a single phase, that is, no apparent relative particle-fluid movement. However, for flocculated systems, it is rather difficult to simulate a suspension structure on the wall. Hence, measurements with a roughened wall are at best as "good" as those with smooth surfaces (Figure 29). It is interesting to note that Buscall et al. (51) observed no wall effect contribution from the outer cylinder of the cup-and-bob rheometer. In a manner similar to the wall effects for flow through a fixed bed (188), strong wall effects are present in the low shear regime.

Solids and Droplets Codispersed Systems: Solids–Oil–Water Ternary Mixtures.   Mixtures of oil and solids dispersed in water and mixtures of water and solids dispersed in oil are commonly encountered in the petroleum industry (189). It is of great interest in studying the rheological behavior of ternary systems, where two distinct materials

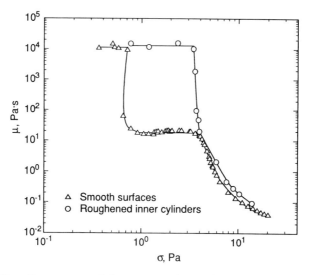

Figure 29.   Comparison of flow curves obtained for a nonaqueous acrylic copolymer suspension in low-aromatic white spirit containing poly(iso-butylene) (PIB) by a coaxial cylindrical rheometer (51).

act as the dispersed phases. The ternary systems were also reviewed earlier by Pal et al. (*4*). The rheological behaviors for ternary systems with low solid content, as are commonly encountered in the petroleum industry, are much simpler to describe than when both of the dispersed materials have high concentration.

For fluid-in-fluid (liquid-in-liquid and gas-in-liquid) dispersions, Taylor extended Einstein's treatment with interface effects neglected (*190, 191*) and obtained

$$k_{\mathrm{E}} = \frac{5\mu_{\mathrm{d}} + 2\mu_{\mathrm{f}}}{2(\mu_{\mathrm{d}} + \mu_{\mathrm{f}})} \tag{66}$$

where μ_{d} is the viscosity of the droplet phase fluid. Oldroyd (*192, 193*) obtained the Einstein constant by including the interface effects:

$$k_{\mathrm{E}} = \frac{5\mu_{\mathrm{d}} + 2\mu_{\mathrm{f}} + \mu_{\mathrm{i}}/d}{2(\mu_{\mathrm{d}} + \mu_{\mathrm{f}} + \mu_{\mathrm{i}}/5d)} \tag{67}$$

where μ_{i} is the interface viscosity and is expressed as

$$\mu_{\mathrm{i}} = 2\mu_{\mathrm{si}} + 3\mu_{\mathrm{ei}} \tag{68}$$

with μ_{si} representing the shear viscosity and μ_{ei} the extensional viscosity of the interface.

Equations 66 and 68 indicate that the droplet behaves like a solid particle only when the viscosity ratio of the dispersed phase to the continuous phase is large. For liquid-in-liquid dispersions, the modified Quemada equation, Krieger–Dougherty equation, and Mooney equation are still applicable provided that the maximum packing limit and the Einstein constant are left as adjustable parameters for a given system.

One of the unique rheological features of emulsions is that the apparent viscosity of the emulsion can drop below the viscosity of the continuous phase when the concentration of the dispersed phase is low, normally below 0.1 in volume fraction (*194*). When solids are added to the emulsion, the apparent viscosity can decrease even further and the volume fraction of the dispersed phase at which minimum viscosity occurs increases with increasing solids content. Figure 30 shows the apparent viscosity of water-and-sand-in-bitumen, μ_{wsb}, variation with the solid-free water volume fraction, β_w, for two shear rate values. The experimental data were provided by Yan (private communication), where the system consists of 52 μm sand particles treated with hexadecyltrimethylammonium bromide (**HAB**) and water droplets of a Sauter mean diameter of 9 μm dispersed in bitumen at 60 °C. The sand particle volume fraction on water-free basis is $\beta_s = 0.193$. The range of the water droplet volume fraction, on a solid-free basis, β_w is between 0 and 0.4. It can be observed that a minimum viscosity is present at a solid-free water droplet volume fraction of about 0.1. For a lower solid concentration, $\beta_s = 0.113$, the minimum apparent viscosity is found at $\beta_w = 0.05$

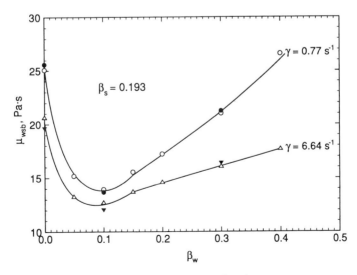

Figure 30.   Effect of water addition to a sand-in-bitumen suspension at 60 °C. (Data from Y. Yan, private communication, 1990.)

(*194*). The viscosity reduction by water addition is not due to the presence of the surfactant (HAB). For the sand-in-bitumen suspension, the viscosity variation is shown in Figure 31. It can be observed from Figure 31 that the sand-in-bitumen suspensions are slightly shear thinning type. A low shear limit viscosity is observed, although a yield stress may be assumed if Casson's model is used (*194*). The shear thinning behavior is more severe when the solid volume fraction is increased.

Shear thinning behavior is commonly observed in emulsions. When solids are added to emulsions, both shear thinning and shear thickening behaviors are observed. Figure 32 shows the rheograms of clay-and-oil in water measured by Yan (private communication), where the clay-free oil volume fraction, β_O = 0.2. Part of Figure 32, $\dot{\gamma} > 10$ s^{-1}, was also shown by Yan et al. (*195*). The systems are shear thinning, which is similar to the pure suspension of clay in water and oil in water emulsion (*195*). In the low shear regime, the systems display some discontinuities on the shear stress versus shear rate plane, and the discontinuity is more severe when the clay volume fraction, ϕ_c, is about 0.12. For pure clay in water suspensions, similar behavior is also observed, whereas the discontinuity is much reduced as the oil concentration is increased.

Figure 33 shows the apparent viscosity variation with shear rate for 44 μm glass beads suspended in a 50% bitumen-in-water emulsion at 25 °C. The viscosity ratio of the dispersed liquid phase (bitumen) to the

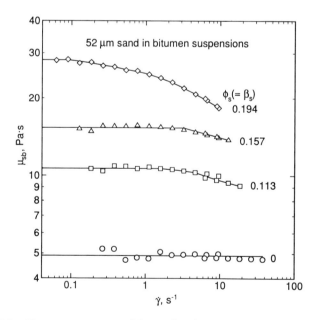

Figure 31.   Viscosity variation of the sand-in-bitumen suspensions with shear rate at different sand volume fractions at 60 °C (194).

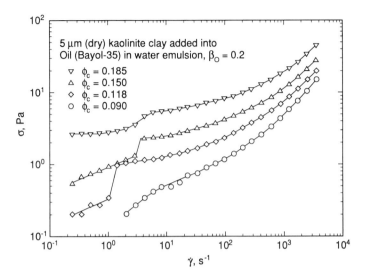

Figure 32.   Rheograms of kaolinite clay and Bayol-35 oil dispersed in water at 25 °C. (Data from Y. Yan, private communication, 1990.)

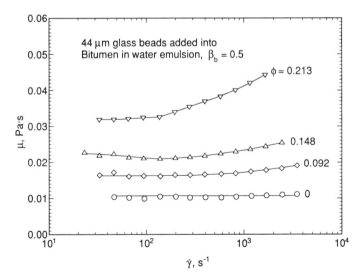

Figure 33.   Viscosity variation with shear rate for 44 μm glass beads suspended in 50% bitumen-in-water emulsion at 25 °C (196).

continuous phase (water), μ_d/μ_f, is 306,000. The bitumen droplets have a Sauter mean diameter of 9.1 μm. This system displays shear thickening behavior. The shear thickening appeared to be more pronounced for the case where the viscosity ratio of the dispersed phase fluid to the continuous phase fluid is high (196). The shear thickening behavior as observed by Yan et al. (196) is monotonous, that is, the viscosity does not decrease with increasing shear rate at high shear rates. They found that when μ_d/μ_f is small, for the cases of μ_d/μ_f = 2.4 (Bayol oil droplets of diameter 10.6 μm in water) and μ_d/μ_f = 150 (paraffin oil droplets of diameter 10.2 μm in water), both the emulsions and the ternary systems (all with 44 μm polystyrene beads added) display shear thinning behavior. When μ_d/μ_f is large, as for the case of corena oil droplets of diameter 9.6 μm in water, μ_d/μ_f = 1040, the emulsion is Newtonian and the ternary system (with 44 μm polystyrene beads added) displays shear thickening behavior. When the solid particle size is large, all the ternary systems are nearly Newtonian, irrespective of the values of the viscosity ratio, μ_d/μ_f.

Liquid droplets cannot be treated the same as solid particles in their codispersed systems. This behavior has been indicated by equation 66 or 68, in which the Einstein constant increases with increasing viscosity ratio of the dispersed phase to the continuous phase. As is shown by Yan et al. (195, 197, 198), liquid droplet and solid particle effects are additive only when the solid concentration is low, say ϕ_s < 0.05, and when both solid particles and liquid droplets have comparable sizes. However, when the particle-to-droplet size ratio is large, the particles and the droplets become additive (192) for a wider solid concentration range (Figures 34 and 35). The apparent viscosity of the system may be "added" in terms of the two distinct model systems: pure emulsion characterized by solid-free dispersed phase volume fraction and pure suspension characterized by the volume fraction of the solids. The additive rule for the ternary systems is similar to the rule for bimodal solid particle suspensions due to Farris (139):

$$\frac{\mu_{sdf}}{\mu_f} = \frac{\mu_{df}(\beta_d)}{\mu_f} \times \frac{\mu_{sf}(\phi_s)}{\mu_f} \qquad (69)$$

where the subscripts f, d, and s stand for continuous fluid phase, dispersed liquid droplet phase and solid particle phase, respectively. Equation 69 holds at least for a low volume fraction of solid particles, say ϕ_s < 0.25, and the solid particle-to-droplet size ratio is large, d_s/d_d > 3 if the viscosity of the fluid forming the droplet phase is not too far from the continuous phase, where no concentration limit is posed on the dispersed droplet phase. The critical size ratio increases with the viscosity ratio of the fluid forming the dispersed phase to the continuous

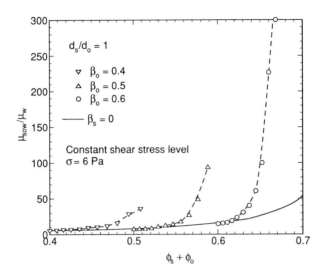

Figure 34.   Variation of the relative viscosity with the total volume fraction of the dispersed phases for 9 μm silica sands and light oil in water at 25 °C (197).

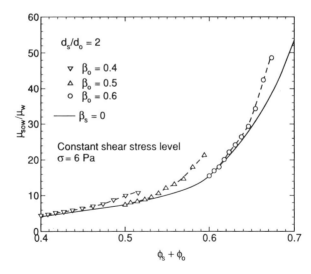

Figure 35.   Variation of the relative viscosity with the total volume fraction of the dispersed phases for 19 μm glass beads and light oil in water at 25 °C (197).

phase, μ_d/μ_f (*196*). Furthermore, each component of equation 69 can be estimated by the Quemada equation 34 or the modified Quemada equation 40. When the particle size is smaller than the droplet size, $d_s/d_d < 1/3$, the role of the droplet phase and the solid particle phase is simply reversed in equation 69 (*see* reference 199). However, one should be aware that the addition of small particles may be acting as a stabilizing agent for the emulsion, where the small particles tend to be adsorbed in the interfaces between the droplet and the continuous phase (*200–202*).

Viscoelasticity

Concentrated suspensions commonly display viscoelatic behavior. The viscoelastic properties can be measured by oscillatory tests (*26*). Comparing with steady shear measurements, oscillatory measurements are made under small deformations, at which the suspension structure is only slightly perturbed. Hence, oscillatory measurements are suitable for correlating rheological behavior with structural data and interparticle potentials, even for strongly flocculated systems that show irreversible changes when subjected to large deformations.

When oscillatory measurements are made, the interpretation is simplified if the measurements are carried out in the linear viscoelastic regime. In this region, the shear modulus of the system, G, is independent of strain, γ, and it can be assumed that no irreversible changes of the suspension structure occur. The linear viscoelastic region is usually ensured in the test by performing a pretest at some fixed frequency of, say 0.1 or 1 Hz, in which the amplitude of the strain is increased and the shear modulus is recorded. Figure 36 depicts such a measurement on a sterically stabilized suspension: PMMA particles suspended in decalin with chemically grafted chains of poly(12-hydroxystearic acid) on the particle surface (*203*). The limit of the linear viscoelastic region can be defined as the deformation at which the storage modulus G' starts to decrease. It can be observed that the linearity limit decreases with increasing particle size and increasing solid concentration. The nonlinearity region can often be observed earlier by an increase in the loss modulus G''.

Stable Systems.   The viscoelastic response of a concentrated noncoagulating suspension is strong when the average distance between the suspended particles is of the same order as the distance at which the interparticle repulsive forces become important. Hence, the viscoelastic behavior originates from the interparticle repulsive potential. Several studies have been carried out on hard sphere systems (*72, 204, 205*), steric systems (*88, 94, 203, 206*), and electrostatic systems (*163,*

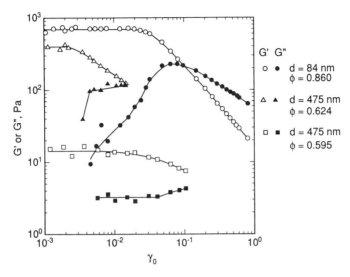

Figure 36. Storage and loss moduli variations with strain for a steric PMMA suspension in decalin at ω = 1 Hz or 2π s⁻¹ (203).

207, 208). Prestidge and Tadros (96), Liang et al. (95), and Tadros et al. (209) studied the viscoelastic responses of both electrostatically and sterically stabilized systems.

Figure 37 shows the viscoelastic behavior of a sterically stabilized suspension of PMMA particles in decalin. The storage modulus G' increases with increasing frequency and volume fraction of the particles. With increasing volume fraction, the average distance between particles decreases, causing a progressively stronger overlap of the polymer layers. This strong overlap results in an increase in the elasticity of the suspension, reflected in an increase in the storage modulus.

The increase in the storage modulus with increasing frequency can be understood through the relaxation time t_R of the suspension. At low frequencies, the characteristic experimental time scale t_E, based on equation 6, is longer than t_R, and the perturbed structure is able to relax during the oscillation. This results in a mainly viscous response (small storage modulus), because the suspension is able to dissipate most of the energy. With increasing frequency, t_E approaches t_R, resulting in a viscoelastic behavior where the suspension has both a viscous and an elastic response. When the frequency is increased further, t_E becomes smaller than t_R, the perturbed suspension structure is unable to relax during oscillation and the suspension displays an elastic response. In this region, the storage modulus is larger than the loss modulus and the shear moduli tends to be constant,

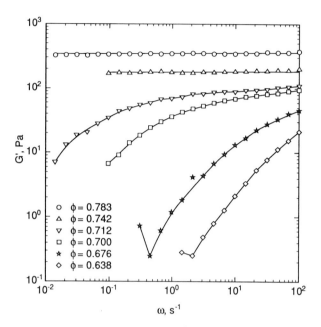

Figure 37. Storage moduli variation with frequency at different volume fractions for sterically stabilized suspensions of 84 nm PMMA in decalin at 293 K (203).

$$G^* \approx G' = G_\infty \qquad (70)$$

which is demonstrated by high solid content cases of Figure 37. The high frequency limit shear modulus, G_∞, is only a property of the suspension because no structural change is expected under high frequency oscillations. Hence, G_∞ can be related to the structure and interparticle potential of the suspension. The high frequency limit shear modulus G_∞ has been modeled by Wagner (210).

Figure 38 shows the shear moduli variation with solid volume fraction for sterically stabilized polystyrene latex suspensions at an oscillating frequency of 1 Hz. One can observe that the particle size does not have a significant effect on the shear moduli when the particles are relatively large, say $d > 0.5$ μm. For small particles, the size effect becomes noticeable. In all these cases, the shear moduli depict a gradual change from a more viscous ($G'' > G'$) behavior to a more elastic ($G'' < G'$) behavior with increasing solid volume fraction.

Figure 39 shows the shear moduli variation with solid volume fraction for the electrostatically stabilized suspension of 1.4 μm polystyrene latices in aqueous solutions of NaCl. At the lower NaCl concentration (10^{-5} M), the double-layer thickness, $1/\kappa$, is 100 nm, and therefore the suspension show "soft" type interaction due to the extended double

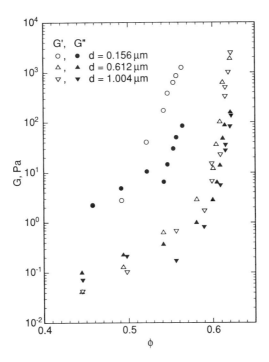

Figure 38.   Shear moduli variation with volume fraction at a frequency of 1 Hz for steric polystyrene latex suspensions of various sizes. (209.)

layer. For the higher NaCl concentration case, the double layer is only 10 nm and is much smaller than the particle size; thus, the suspension is close to a hard-sphere system.

Flocculated Systems. The viscoelastic responses of flocculated systems are strongly dependent on the suspension structure. The suspension starts to show an elastic response at a critical solid volume fraction of $\phi_{ct} = 0.05 - 0.07$, at which the particles form a continuous three-dimensional network (*211–213*). The magnitude of the elastic response for flocculated suspensions above ϕ_{ct} depends on several parameters, such as the suspension structure, interparticle attraction forces and particle size, and shape and volume fraction. Buscall et al. (*10*) found that the volume fraction dependence of the storage modulus follows a power-law behavior.

$$G' \sim \phi^m \tag{71}$$

with a power of $m = 2 - 5$ (*10, 212–216*).

Figure 40 shows the experimental results of Chen and Russel (*215*), which demonstrate that an increase in the magnitude of the interparticle

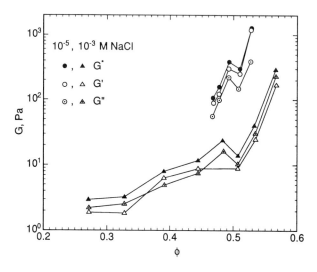

Figure 39. Shear moduli variation with volume fraction for electrostatic suspension of polystyrene latex in aqueous solution of NaCl at ω = 1 Hz (206).

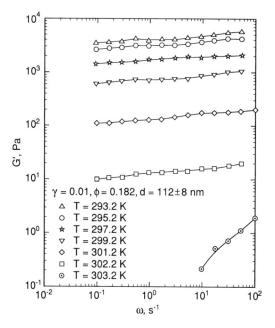

Figure 40. G' variation with ω at temperatures between 20 and 30 °C for a coated silica suspension with d = *112 nm in hexadecane (215).*

attraction results in an increase in the storage modulus, provided that the suspension structure remains minimally disturbed. Lowering the temperature below 303.2 K causes the suspension to form a particle network with stronger elastic responses. The change in G' depicts the sharp change in the suspension properties from a liquidlike (viscous) at 303.2 K to a solidlike (elastic) response at 302.2 K. Further reducing the temperature results in an increase in the magnitude of the interparticle attraction, which leads to an increase in the storage modulus. The system becomes more elastic. Similar behavior showing the importance of the interparticle forces is also reported by Otsubo (*217*) with different particle sizes instead of varying the temperature.

Kawaguchi and co-workers systematically studied the steady shear, transient shear, and dynamic rheological properties of a weakly flocculated silica suspension in aqueous solutions of HPMC. To help us understand the viscoelastic behavior of the suspensions, the results of Kawaguchi and Ryo (*218*) and Nakai et al. (*176*) are summarized here. The systems shown in Figures 41–43 are for steric silica suspensions in a 2.0 wt% aqueous solution of HPMC having a molecular weight of 321,000. Figure 41 shows the shear moduli responses with changing frequency for three systems of silica content: 7.5, 5.0, and 2.5 wt%. We observe, for example, that the loss modulus crosses over the storage modulus at $\omega \approx 3$ s^{-1} for the 5.0 wt% silica suspension. This indicates that the time scale at which the 5.0 wt% silica suspension switches from a more elastic to a more viscous behavior is about $t_E \approx 1/3$ s. From the

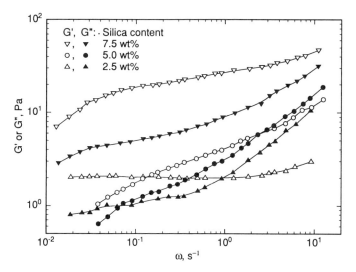

Figure 41.   Shear moduli dependence on frequency for silica suspensions in 2.0% aqueous HPMC solutions (218).

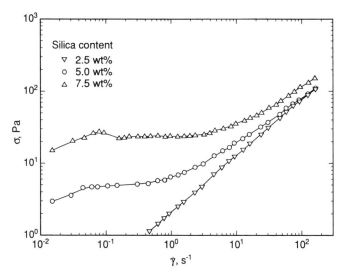

Figure 42. Steady shear stress dependence on shear rate for silica suspensions in 2.0% aqueous HPMC solutions (218).

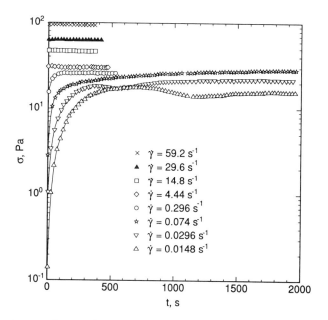

Figure 43. Evolution of transient shear stress for 7.5 wt% silica suspensions in 2.0% aqueous HPMC solution (176).

steady shear stress versus shear rate diagram (Figure 42), one can observe
that when the shear rate is below 3 s^{-1}, the suspension exhibits a relatively constant shear stress with an increase of shear stress when the
shear rate is above 3 s^{-1}. The time scale is well described by equations
5 and 6. For the most concentrated suspension, 7.5 wt% silica, the critical
time scale as given by $\dot{\gamma}^{-1}$ is less than 0.1 s as shown in Figures 41 and
42. From Figure 43, one can observe that the shear stress is a function
of time when the shear rate is below 10 s^{-1}. As the shear rate is lowered,
the time dependence of the shear stress becomes more pronounced and
a longer time is needed for the system to reach a steady state. Both
rheopexy (i.e., shear stress increases with time at a constant applied
shear rate) and stress overshooting (i.e., the shear stress is higher than
the steady state value) can be observed in Figure 43 at shear rates below
0.07 s^{-1}.

Compression Rheology

Callagham and Ottewill (*219*) were among the first to study the compressive and decompressive yield stress of suspensions. Their system
consisted of fine aqueous montmorillonite clay suspensions. When the
tests were performed at different ionic strengths (10^{-4}–10^{-1} M NaCl),
they found systematic variations in the compressive yield stress due to
the long-range electrostatic double-layer interaction. The suspension
was compressed at a low ionic strength and it showed a compressive
stress response over a much larger clay concentration range, starting at
10 wt% clay. In addition, a hysteresis was found upon decompression.

Cairns et al. (*220*) studied monodispersed PMMA latex with a particle
diameter of about 156 nm suspended in dodecane and stabilized by
covalently attached poly(12-hydroxystearic acid) with a thickness of
$\Delta = 9$ nm. They found that the resistance to compression starts to increase drastically around $\phi = 0.55$, and at $\phi = 0.566$ the compression
resistance was very strong, indicating that the suspension would not
yield to compression when the maximum packing limit is reached.

Most colloidal stable suspensions show more or less reversible response to compression and decompression. However, in the case of flocculated suspensions, the compressive properties are irreversible. In
concentrated flocculated suspensions, a continuous particle network
forms. The particle network can support some stress up to a critical
value. Once this critical stress, also called the compressive yield stress
P_y, is exceeded, the network consolidates to a higher volume fraction
with a higher critical stress.

Buscall et al. (*212*) found that the compressive yield stress could be
fitted to a power law form with respect to the solid volume fraction

$$P_y \sim \phi^m \tag{72}$$

with $m \approx 4$. Equation 72 is in agreement with the percolation theory (213) when the solid volume fraction is well above the percolation threshold, ϕ_c. They also studied the particle size effect on P_y and fitted P_y to d by a power law form

$$P_y \sim d^n \tag{73}$$

with a best fit of $n = -2$, although they would have preferred $n = -2.3$. This can be understood by considering the fact that the compressive yield stress is inversely proportional to the particle surface area. When the surface area is large, it is easier to make the particles to contact with each other and break up the floc structure. For a spherical particle, the surface area is proportional to the square of its diameter.

The compressive yield properties are strongly dependent on the suspension structure. Mills et al. (186) showed that extensive preshearing of a flocculated suspension can result in a much denser floc morphology (in fractal terms), which was reflected by a decrease in the compressive yield stress by one order of magnitude.

The compressive yield stress can also be measured with the centrifuge method, where the sedimentation height as a function of centrifugal acceleration is measured (221–223). Because the centrifugal acceleration can disturb the suspension structure as well, it is expected that the measured compressive yield stress depends on the applied forces especially at a lower solid volume fraction. At a higher solid volume fraction, the centrifuge method can still measure the compressive yield stress that is more or less independent of the acceleration used. However, a strong acceleration is needed to make the suspension consolidate. Figure 44 shows the compressive yield stress variation with volume fraction of solids under various centrifugal accelerations (based on the study of Auzerais et al., i.e., reference 23). The system is a strongly flocculated silica suspension in hexadecane with an initial solid volume fraction of 0.096. The lines are plotted based on the usual scaling of the flocculated systems:

$$P_y = \frac{P_0 \phi^n}{\phi_{max} - \phi} \tag{74}$$

For the high centrifugal speed cases, the (dashed) curve of $P_0 = 160$ kPa and $n = 6$ follows the experimental data fairly closely. For lower speeds, the maximum packing limit is much reduced. However, the same scaling is indicated in the figure. The solid line is for $P_0 = 30$ kPa and $n = 3$. The maximum packing limit used for the two lines corresponds to the high shear limit. The power of 3 is the same as for the low shear viscosity and Bingham yield stress dependence on the solid volume fraction. The higher power for a fixed speed is due to the flocculation of

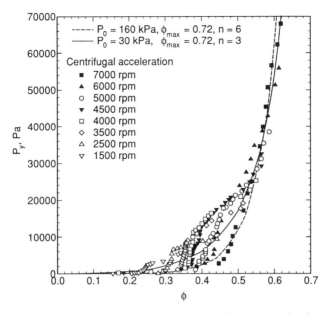

Figure 44. Compressive yield stress variation for a strongly flocculated suspension in hexadecane (d = 0.08 μm) with solid volume fraction and centrifugal acceleration (223).

the suspension. High energy input is necessary to break up the flocs, and low speeds simply would not make the suspension consolidate completely.

Bergstrom et al. (224) studied the compressive yield stress of alumina suspensions in decalin with different degrees of flocculation by adsorbing different fatty acids at the alumina–decalin interface. Figure 45 shows their results (symbols) as compared with a set of correlations (curves), all based on equation 74, where a power of 3 is used. The properties corresponding to those of the figure are tabulated in Table I. It can be observed that both P_0 and ϕ_{max} are strongly dependent on the magnitude of the attractive forces. As the attractive force increases, the degree of flocculation increases. For highly flocculated systems, the maximum packing fraction of the solids is reduced and the force required to break up the flocs is greatly increased. Because the compression experiments do not disturb the suspension structure to a significant extent, the maximum packing limit is, by and large, retained for the flocculated systems.

The compression yield stress is directly comparable with the osmotic pressure for a given suspension system. The osmotic pressure has been measured by Barclay et al. (225), Rohrsetzer et al. (226), Goodwin et al. (227), and Bonnet-Gonnet et al. (228). The osmotic pressure displays the same trend as the compressive yield stress.

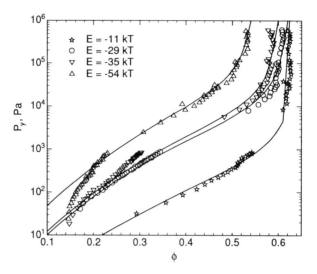

Figure 45. Compressive yield stress variation with solid volume fraction for flocculated systems (224).

Summary

This chapter has been devoted to an in-depth review of the rheological behavior of suspensions. A most significant problem still exists in today's studies in that the wall effects are not quite well understood. The difficulty in isolating wall effects has contributed to the scatter of experimental data from one rheologist to another. Because the rheological studies are directly related to the hydrodynamic behavior of suspensions, the understanding of the rheological behavior cannot be isolated from the hydrodynamic studies. It is for this reason that the rheological studies to date are very important even though the wall effects are present in all the experimental studies.

One unique scaling behavior has been uncovered in this chapter for the low shear limit viscosity, equation 64, Bingham yield stress, equation 65, and the compressive yield stress–osmotic pressure, equation 74, for

Table I.   Parameters of the Flocculated Systems
Shown in Figure 45

Added Fatty Acid	$-E_0$ (kT)	P_0 (kPa)	ϕ_{max}
Propionic	54	210	0.54
Pentanoic	35	6.0	0.595
Heptanoic	29	5.2	0.615
Oleic	11	0.40	0.63

flocculated systems. Although they are different in nature, they all show the same scaling behavior for the solid concentration.

List of Symbols

$A(\phi)$	weak function of ϕ, dimensionless
a	empirical constant, dimensionless
a_i	intrinsic viscosity, dimensionless
a_v	constant in viscosity scaling equation, dimensionless
a_y	constant in yield stress scaling equation, dimensional
a_0, a_1	particle shape factors, dimensionless
b	empirical constant, dimensional
C	constant, dimensionless
c_a	constant, dimensionless
De	Deborah number, dimensionless
d	diameter of particle or dispersed droplet, dimensional
d_{eff}	effective soft sphere diameter, dimensional
d_L	diameter of the large particle, dimensional
d_o	diameter of the orifice, dimensional
d_S	diameter of the small particle, dimensional
d_t	tube diameter, dimensional
E	electrostatic potential, dimensional
$\mathbf{E}$	rate-of-strain tensor, dimensional
$E(r)$	pairwise interaction potential, dimensional
E_0	surface charge (zeta) potential, dimensional
e	particle interaction parameter, dimensionless
G	shear modulus, dimensional
G^*	dynamic modulus, dimensional
G'	storage modulus, dimensional
G''	loss modulus, dimensional
K	kelvin
K	consistency coefficient, dimensional
K_B	shear effect constant, dimensionless
K_C	Casson constant, dimensional
K_1	consistency, temporary part, dimensional
k	Boltzmann's constant, $k = 1.3806 \times 10^{-23}$ J/K
k_E	Einstein constant or intrinsic viscosity, dimensionless
k_{EL}	longitudinal Einstein constant, dimensionless
k_{ET}	transverse Einstein constant, dimensionless
k_H	Huggins constant, dimensionless
l	longest dimension of the particle, dimensional
L	tube length, dimensional
M	torque, dimensional
m	empirical constant, dimensionless

N_d	the ratio of the viscous to the repulsive forces, dimensionless
Pe	Peclet number, dimensionless
p_e	extrusion pressure, dimensional
P_y	compressive yield stress, dimensional
P_0	scaling factor for compressive yield stress, dimensional
p	pressure, dimensional
$\mathbf{p}$	particle orientational vector
R	radius, dimensional
R_i	inner radius, dimensional
R_o	outer radius, dimensional
r	radial coordinate, dimensional
r_e	ratio of the long axis and the short axis, dimensionless
T	absolute temperature, dimensional
t	time, dimensional
t_d	time shift between the stress and strain, dimensional
t_E	experimental time scale, dimensional
t_{PE}	Powell-Eyring characteristic time, dimensional
t_R	relaxation time, dimensional
u	axial velocity, dimensional
V_p	volume of one particle, dimensional
v	velocity, dimensional

Greek

α_p, α_v	parameters, dimensional
β	$\beta_i = (V_i/V_i + V_f)$, the ratio of the ith (a given dispersed) phase volume, V_i, to the summation of the volume of the ith dispersed phase, V_i, and the continuous phase, V_f, dimensionless
Δ	polymer chain length, dimensional
δ	phase difference between stress and strain
γ	strain, dimensionless
γ_0	amplitude of the oscillating strain, dimensionless
$\dot{\gamma}$	shear rate, dimensional
$\dot{\gamma}_c$	critical shear rate, dimensional
$\dot{\gamma}_1$	a given shear rate, Figure 2, or the first transitional shear rate, Figure 4, dimensional
$\dot{\gamma}_2, \dot{\gamma}_3$	second and third transitional shear rates, Figure 4, dimensional
$\dot{\epsilon}$	extensional strain rate, dimensional
ϵ_L	volume fraction of the large particles in the total volume of the dispersed phases, dimensionless
ϵ_r	dielectric constant of the suspending fluid, dimensionless
ϵ_0	permittivity of vacuum, $\epsilon_0 = 8.85 \times 10^{-12}$ C/Vm
λ	memory parameter, dimensionless

λ_0	empirical constant, dimensional
μ	viscosity, dimensional
μ^*	dynamic (complex) viscosity, dimensional
μ'	storage viscosity, dimensional
μ''	loss viscosity, dimensional
μ_d	droplet phase viscosity dimensional
μ_E	uniaxial (extensional) viscosity, dimensional
μ_{ei}	extensional viscosity of the interface, dimensional
μ_i	interface viscosity, dimensional
μ_r	relative viscosity, dimensionless
μ_{rL}	relative longitudinal viscosity
μ_{rT}	relative transverse viscosity
μ_{si}	shear viscosity of the interface, dimensional
ξ	empirical constant, dimensionless
ρ	density, dimensional
σ	stress, dimensional
σ_B	Bingham yield stress, dimensional
σ_{B0}	Bingham yield stress of the suspending fluid, dimensional
σ_c	critical stress, dimensional
σ_{Ey}	uniaxial yield stress, dimensional
σ_w	wall shear stress, dimensional
σ_{wy}	wall yield stress, dimensional
σ_y	yield stress, dimensional
σ_{y0}	permanent yield stress, dimensional
σ_{y1}	temporary yield stress, dimensional
σ_0	amplitude of the oscillating stress, dimensional
ϕ	volume fraction of a given dispersed phase in the total volume of the dispersion, dimensionless
ϕ_c	percolation threshold
ϕ_{ct}	critical volume fraction: lower bound value at which a flocculated suspension shows elastic response, dimensionless
ϕ_L	volume fraction of the large particle, dimensionless
ψ_1	primary normal stress coefficient, dimensional
ψ_2	secondary normal stress coefficient, dimensional
Ω	angular velocity or rotational speed, dimensional
Ω_i	rotational speed on the inner cylinder, dimensional
ω	frequency, dimensional

Superscripts

m	power, empirical constant
n	power, consistency index

Subscripts

av	average
b	bitumen
c	clay particles
d	dispersed liquid droplets
E	experimental
eff	effective
ei	interfacial due to extension
f	suspending fluid
i	interface
L	longitudinal
max	maximum limit
o	oil droplets
p	plastic
s	solids
sb	solid suspended in bitumen
s, d, f	solids and liquid droplets dispersed in a suspending fluid
si	interfacial due to shear
slip	slip
sow	solid-and-oil-in-water system
T	transverse
w	water or water droplets
wsb	water-and-solid-in-bitumen system
xx	principal component of the x-direction
xy	x-directional component on the xz plane
yy	principal component of the y-direction
y	yield value
∞	at the high shear limit
0	at the low shear limit
11	diagonal (principal) component of the flow direction
21	shear-directional component on the shear-neutral (flow) plane
22	diagonal (principal) component of the shear direction
33	diagonal (principal) component of the neutral direction

References

1. Barnes, H. A.; Hutton, J. F.; Walters, K. *An Introduction to Rheology*; Elsevier: Amsterdam, Netherlands, 1989.
2. Adams, M. J.; Briscoe, B. J.; Kamjab, M. *Adv. Colloid Interface Sci.* **1993**, *44*, 143–182.
3. Schramm, L. L.; Wassmuth, F. In *Foams: Fundamentals and Applications in the Petroleum Industry*; Schramm, L. L., Ed.; Advances in Chemistry 242; American Chemical Society: Washington, DC, 1994; pp 3–45.

4. Pal, R.; Yan, Y.; Masliyah, J. In *Emulsions: Fundamentals and Applications in the Petroleum Industry;* Schramm, L. L., Ed.; ACS Advances in Chemistry 231; American Chemical Society: Washington, DC, 1992; pp 131–170.
5. Bingham, E. C. *Fluidity and Plasticity;* McGraw-Hill: New York, 1922.
6. Casson, N. In *Rheology of Disperse Systems;* Mill, C. C. Ed.; Pergamon: New York, pp 84–104.
7. Barnes, H. A.; Walters, K. *Rheol. Acta* 1985, *24,* 323–326.
8. Hartnett, J. P.; Hu, R. Y. Z. *J. Rheol.* 1989, *33,* 671–679.
9. Dzuy, N. Q.; Boger, D. V. *J. Rheol.* 1985, *29,* 335–347.
10. Buscall, H. A.; Mill, P. D. A.; Yates, G. E. *Colloids Surf.* 1986, *18,* 341–358.
11. Cheng, D. C. H. *Rheol. Acta.* 1986, *25,* 542–554.
12. Yoshimura, A.; Prud'homme, R. K.; Princen, H. M.; Kiss, A. D. *J. Rheol.* 1987, *31,* 699–710.
13. Astarita, G. *J. Rheol.* 1992, *36,* 1317.
14. Evans, I. D. *J. Rheol.* 1992, *36,* 1313–1316.
15. Van der Aeschot, E.; Mewis, J. *Colloid Surf.* 1992, *69,* 15–22.
16. Schurz, J. *J. Rheol.* 1992, *36,* 1319–1312.
17. De Kee, D.; Chan Man Fong, C. F. *J. Rheol.* 1993, *37,* 775–776.
18. Van Diemen, A. J. G.; Stein, H. N. *J. Colloid Interface Sci.* 1982, *71,* 316–336.
19. Hunter, R. J. *Adv. Colloid Interface Sci.* 1982, *17,* 197–211.
20. Cross, M. M. *J. Colloid Sci.* 1965, *20,* 417–437.
21. Van Wazer, J. R.; Lyons, J. W.; Kim, J. W.; Cowell, R. E. *Viscosity and Flow Measurements;* Interscience: New York, 1963.
22. Powell, R. E.; Eyring, H. *Nature (London)* 1944, *154,* 427–428.
23. Mewis, J. *J. Non-Newtonian Fluid Mech.* 1979, *6,* 1–20.
24. Sestak, J. In *Progress and Trends in Rheology II: Proceedings of the Second Conference of European Rheologists;* Steinkopff Verlag: Darmstadt, Germany, 1988; pp 18–24.
25. Goodwin, J. W. *Solid/Liquid Dispersion;* Academic: Toronto, Canada, 1987.
26. Marin, G. In *Rheological Measurements;* Collyer, A. A.; Clegg, D. W., Ed.; Elsevier: New York, 1988; pp 297–343.
27. Russel, W. B.; Saville, D. A.; Schowalter, W. R. *Colloidal Dispersions;* Cambridge University: Oxford, England, 1989.
28. Bird, R.; Armstrong, R.; Hassager, D. *Dynamics of Polymeric Liquids;* Wiley: New York, 1987.
29. Tsai, S. C.; Ghazimorad, K. *J. Rheol.* 1990, *34,* 1327–1332.
30. Tsai, S. C.; Viers, B. *J. Rheol.* 1987, *31,* 483–494.
31. Tsai, S. C.; Zammouri, K. *J. Rheol.* 1988, *32,* 737–750.
32. Lewis, T. B.; Nielsen, L. E. *Trans. Soc. Rheol.* 1968, *12,* 421–443.
33. Castillo, C.; Williams, M. C. *Chem. Eng. Commun.* 1979, *3,* 529–547.
34. Wildemuth, C. R.; Williams, M. C. *Rheol. Acta* 1984, *23,* 627–635.
35. Tsenoglou, C. *J. Rheol.* 1990, *34,* 15–24.
36. Tsai, S. C.; Botts, D.; Plouff, J. *J. Rheol.* 1992, *36,* 1291–1305.
37. *Kinetics of Aggregation and Gelation;* Family, F.; Landau, D. P., Eds.; North Holland: Amsterdam, Netherlands, 1984.
38. Meakins, P. *Ann. Rev. Phys. Chem.* 1988, *39,* 237–267.
39. Jullien, R.; Meakin, P. *J. Colloid Interface Sci.* 1989, *127,* 265–272.
40. Hamaker, H. C. *Physica* 1937, *4,* 1058–1072.
41. Lifshitz, E. M. *Sov. Phys. JETP* 1956, *2,* 73–83.
42. Tabor, D.; Winterton, R. H. S. *Proc. Roy. Soc. London* 1969, *A312,* 435–450.

43. Horn, R. G.; Israelachvili, J. N. *J. Chem. Phys.* **1981**, *75*, 1400–1411.
44. Masliyah, J. H. *Electrokinetic Transport Phenomena*; Technical Publication Series #12; Alberta Oil Sands Information Services, Alberta Energy: Edmonton, Canada, 1994.
45. Hiemenz, P. C. *Principles of Colloid and Surface Chemistry*; Marcel Dekker: New York, 1977.
46. Carnie, S. L.; Chan, D. C. Y.; Stankovich, J. *J. Colloid Interface Sci.* **1994**, *165*, 116–128.
47. Napper, D. H. *Polymeric Stabilization of Colloidal Dispersions*; Academic: London, 1983.
48. *The Effect of Polymers on Dispersion Properties*; Tadros, Th. F., Ed.; Academic: London, 1982.
49. Whorlow, R. E. *Rheological Techniques*; Ellis Horwood, Ltd.: Chichester, England, 1980.
50. Kalyon, D. M.; Yaras, P.; Aral, B.; Yilmazer, U. *J. Rheol.* **1993**, *37*, 35–53.
51. Buscall, R.; McGowan, I. J.; Morton-Jones, A. J. *J. Rheol.* **1993**, *37*, 621–641.
52. Yoshimura, A.; Prud'homme, R. K. *J. Rheol.* **1988**, *32*, 53–67.
53. Chong, J. S.; Christiansen, E. B.; Baer, A. D. *J. Appl. Polym. Sci.* **1971**, *15*, 2007–2021.
54. Storms, R. F.; Ramarao, B. V.; Weiland, R. H. *Powder Tech.* **1990**, *63*, 247–259.
55. Chang, C.; Powell, R. L. *J. Rheol.* **1994**, *38*, 85–98.
56. Mewis, J. *Int. Chem. Eng. Symp. Ser. No. 91 Powtech '85, Part Technol.*; Pergamon: New York, 1985; pp 83–93.
57. Wen, O.; Mitschka, P.; Tovchigrechko, V. V.; Kovalevskaya, N. D.; Yushkina, T. V.; Pokryvalo, N. A. *Chem. Eng. Commun.* **1985**, *32*, 153–170.
58. Benbow, J. J.; Bridgwater, J. In *Tribology in Particulate Technology*; Briscoe, B. J.; Adams, M. J., Eds.; Adam Hilger: London, 1987; p 80.
59. Benbow, J. J.; Jazayeri, S. H.; Bridgwater, J. *Powder Tech.* **1991**, *65*, 391–401.
60. Jastrzebski, Z. D. *I&EC Fundam.* **1967**, *6*, 445–454.
61. Benbow, J. J.; Oxley, E. W.; Bridgwater, J. *Chem. Eng. Sci.* **1987**, *42*, 2151–2162.
62. Adams, M. J.; Mullier, M. A.; Seville, J. P. K. *Powder Tech.* **1994**, *78*, 5–13.
63. Adams, M. J.; Biswas, S. K.; Briscoe, B. J.; Kamyab, M. *Powder Tech.* **1991**, *65*, 381–392.
64. Kim, I. T.; Luckham, P. F. *Colloids Surf.* **1992**, *68*, 243–259.
65. Homola, A.; Robertson, A. A. *J. Colloid Interface Sci.* **1976**, *54*, 286–297.
66. Walbridge, D. J.; Waters, J. A. *Disc. Faraday Soc.* **1966**, *42*, 294–300.
67. Krieger, I. M. *Adv. Colloid Interface Sci.* **1972**, *3*, 111–136.
68. Hoffman, R. L. *Trans. Soc. Rheol.* **1972**, *16*, 155–173.
69. Ackerson, B. J.; Clark, N. A. *Physica A* **1982**, *118*, 221–249.
70. Ackerson, B. J. *J. Rheol.* **1990**, *34*, 553–590.
71. Wagner, N. J.; Russel, W. B. *Phys. Fluids A* **1990**, *2*, 491–502.
72. Jones, D. A. R.; Leary, B.; Boger, D. V. *J. Colloid Interface Sci.* **1991**, *147*, 479–495.
73. Woutersen, A. T. J. M.; de Kruif, C. G. *J. Rheol.* **1993**, *37*, 681–693.
74. Batchelor, G. K. *J. Fluid Mech.* **1977**, *83*, 97–117.
75. Cinchocki, B.; Felderhof, B. U. *J. Chem. Phys.* **1988**, *89*, 3705–3709.
76. Selim, M. S.; Al-Naafa, M. A.; Jones, M. C. *AIChE J.* **1993**, *39*, 3–16.
77. Mooney, M. *J. Colloid Sci.* **1951**, *6*, 162–170.
78. Ball, R. C.; Richmond, P. *Phys. Chem. Liq.* **1980**, *9*, 99–116.

79. Stein, H. N. In *Encyclopedia of Fluid Mechanics Vol. 5: Slurry Flow Technology;* Cheremisinoff, N. P., Ed.; Gulf Publishing Co.: Houston, TX, 1986.
80. Roscoe, R. *Br. J. Appl. Phys.* **1952**, *3*, 267–269.
81. Maron, S. H.; Pierce, P. E. *J. Colloid Sci.* **1956**, *11*, 80–95.
82. Kitano, T.; Kataoka, T.; Shirota, T. *Rheol. Acta* **1981**, *20*, 207–209.
83. Metzner, A. B. *J. Rheol.* **1985**, *29*, 739–775.
84. Sudduth, R. D. *J. Appl. Polym. Sci.* **1993**, *48*, 25–36.
85. Van der Werff, J. C.; de Kruif, C. G. *J. Rheol.* **1989**, *33*, 421–454.
86. Papir, Y. S.; Krieger, I. M. *J. Colloid Interface Sci.* **1970**, *34*, 126–130.
87. Mewis, J.; Frith, W. J.; Strivens, T. A.; Russel, W. B. *AIChE J.* **1989**, *35*, 415–422.
88. Jones, D. A. R.; Leary, B.; Boger, D. V. *J. Colloid Interface Sci.* **1992**, *150*, 84–96.
89. Sengun, M. Z.; Probstein, R. F. *PhysicoChem. Hydrodyn.* **1989**, *11*, 229–241.
90. Robinson, J. V. *Phys. Colloid Chem.* **1949**, *53*, 1042.
91. Chan, D.; Powell, R. L. *J. Non-Newtonian Fluid Mech.* **1984**, *15*, 165.
92. Jomha, A. I.; Merrington, A.; Woodcock, L. V.; Barnes, H. A.; Lips, A. *Powder Tech.* **1991**, *65*, 343–370.
93. Buscall, R. *J. Chem. Soc.; Faraday Trans.* **1991**, *87*, 1365–1370.
94. Ploen, H. J.; Goodwin, J. W. *Faraday Discuss. Chem. Soc.* **1990**, *90*, 77–90.
95. Liang, W.; Tadros, Th. F.; Luckham, P. F. *J. Colloid Interface Sci.* **1992**, *153*, 131–139.
96. Prestidge, C.; Tadros, Th. F. *J. Colloid Interface Sci.* **1988**, *124*, 660–665.
97. Kim, I. T.; Luckham, P. F. *J. Colloid Interface Sci.* **1991**, *144*, 174–184.
98. Krieger, I. M.; Eguiluz, M. *Trans. Soc. Rheol.* **1976**, *20*, 29–45.
99. Croucher, M. D.; Milkie, T. H. *Faraday Discuss. Chem. Soc.* **1983**, *76*, 261–276.
100. Eshelby, J. D. *Proc. Roy. Soc. London* **1957**, *A241*, 376–396.
101. Christensen, R. M. *J. Rheol.* **1993**, *37*, 103–121.
102. Leal, L. G.; Hinch, E. J. *J. Fluid Mech.* **1971**, *46*, 685–703.
103. Milliken, W. J.; Gottlib, M.; Graham, A. L.; Mondy, L. A.; Powell, R. L. *J. Fluid Mech.* **1989**, *202*, 217–232.
104. Powell, R. L.; Mondy, L. A.; Stoker, G. G.; Milliken, W. J.; Graham, A. L. *J. Rheol.* **1989**, *33*, 1173–1188.
105. Bibbo, M. A.; Dinh, S. M.; Armstrong, R. C. *J. Rheol.* **1985**, *29*, 905–929.
106. Bibbo, M. A. Ph.D. Dissertation, Massachusetts Institute of Technology, 1987.
107. Claeys, I. L.; Brady, J. F. *J. Fluid Mech.* **1993**, *251*, 411–442.
108. Claeys, I. L.; Brady, J. F. *J. Fluid Mech.* **1993**, *251*, 443–477.
109. Claeys, I. L.; Brady, J. F. *J. Fluid Mech.* **1993**, *251*, 479–500.
110. Batchelor, G. K. *J. Fluid Mech.* **1971**, *46*, 813–829.
111. Simha, R. *J. Phys. Chem.* **1940**, *44*, 25–34.
112. Goldsmith, H. L.; Mason, S. G. In *Rheology;* Erich, F. R. Ed.; Academic: Orlando, FL, 1967; Vol. 4, pp 86–250.
113. Haber, S.; Brenner, H. *J. Colloid Interface Sci.* **1984**, *97*, 496–514.
114. Brenner, H.; Condiff, D. W. *J. Colloid Interface Sci.* **1974**, *47*, 199–264.
115. Utracki, L. A. In *Rheological Measurement;* Collyer, A. A.; Clegg, D. W., Eds.; Elsevier Applied Science: New York, 1988; pp 479–594.
116. Shaqfeh, E. G.; Fredrickson, G. H. *Phys. Fluids A.* **1190**, *2*, 7–24.
117. Shaqfeh, E. G.; Koch, D. L. *Phys. Fluids A.* **1990**, *2*, 1077–1093.

118. Phan-Thien, N.; Graham, A. L. *Rheol. Acta* **1991**, *30*, 44–57.
119. Ganani, E.; Powell, R. L. *J. Rheol.* **1986**, *30*, 995–1013.
120. Carman, P. C. *Trans. Inst. Chem. Eng.* **1937**, *15*, 150–166.
121. Lee, D. I. *J. Paint Technol.* **1970**, *42*, 579–587.
122. Parkinson, C. S.; Matsumoto, S.; Sherman, P. *J. Colloid Interface Sci.* **1970**, *33*, 150–160.
123. Goodwin, J. W. *Colloid Sci.* **1975**, *2*, 246–293.
124. Jeffrey, D. J.; Acrivos, A. *AIChE J.* **1976**, *22*, 417–432.
125. Wagstaff, I.; Chaffey, C. E. *J. Colloid Interface Sci.* **1977**, *59*, 53–62 and 63–75.
126. Goto, H.; Kuno, H. *J. Rheol.* **1984**, *28*, 197–205.
127. Poslinski, A. J.; Ryan, M. E.; Gupta, P. K.; Seshadri, S. G.; Frechette, F. J. *J. Rheol.* **1988**, *32*, 751–771.
128. Smith, P. A.; Haber, R. A. *J. Am. Ceram. Soc.* **1992**, *75*, 290–294.
129. Hoffman, R. L. *J. Rheol.* **1992**, *36*, 947–965.
130. Rodriguez, B. E.; Kaler, E. W.; Wolfe, M. S. *Langmuir* **1992**, *8*, 2382–2389.
131. Shapiro, A. P.; Probstein, R. F. *Phys. Rev. Lett.* **1992**, *68*, 1422–1425.
132. Sudduth, R. D. *J. Appl. Polym. Sci.* **1993**, *48*, 37–55.
133. Sudduth, R. D. *J. Appl. Polym. Sci.* **1993**, *50*, 123–147.
134. Sudduth, R. D. *J. Appl. Polym. Sci.* **1994**, *52*, 985–996.
135. Probstein, R. F.; Sengun, M. Z.; Tseng, T.-C. *J. Rheol.* **1994**, *38*, 811–829.
136. Sweeny, K.; Geckler, R. *J. Appl. Phys.* **1954**, *25*, 1135–1144.
137. Wagner, N. J.; Woutersen, J. T. M. *J. Fluid Mech.* **1994**, *278*, 267–287.
138. Jones, R. B. *Physica A*, **1994**, *212*, 43–60.
139. Farris, K. J. *Trans. Soc. Rheol.* **1968**, *12*, 281–301.
140. Kim, I. T.; Luckham, P. F. *Powder Tech.* **1993**, *77*, 31–37.
141. Barnes, H. A. *J. Rheol.* **1989**, *33*, 329–366.
142. Metzner, A. B.; Whitlock, M. *Trans. Soc. Rheol.* **1958**, *2*, 239–254.
143. Hoffman, R. L. *J. Colloid Interface Sci.* **1974**, *46*, 491–506.
144. Hoffman, R. L. *Mater. Res. Soc. Bull.* **1991**, *16*, 32–37.
145. Strivens, T. A. *J. Colloid Interface Sci.* **1976**, *57*, 476–487.
146. Woodcock, L. V. *Chem. Phys. Lett.* **1984**, *111*, 455–461.
147. Laun, H. M. In *Progress and Trends in Rheology II: Proceedings of the Second Conference of European Rheologists*; Steinkopff Verlag: Darmstadt, Germany, 1988; pp 287–290.
148. Boersma, W. H.; Laven, J.; Stein, H. N. *AIChE J.* **1990**, *36*, 321–332.
149. Boersma, W. H.; Baets, P. J. M.; Laven, J.; Stein, H. N. *J. Rheol.* **1991**, *35*, 1093–1120.
150. Yilmazer, U.; Kalyon, D. M. *Polym. Comp.* **1991**, *12*, 226–232.
151. Laun, H. M.; Bung, R.; Schmidt, F. *J. Rheol.* **1991**, *35*, 999–1034.
152. D'Haene, P.; Mewis, J. *Rheol. Acta* **1994**, *33*, 165–174.
153. Ackerson, B. J.; Pusey, P. N. *Phys. Rev. Lett.* **1988**, *61*, 1033–1036.
154. Choi, G. N.; Krieger, I. M. *J. Colloid Interface Sci.* **1986**, *113*, 101–113.
155. Mashall, L.; Zukosi, C. F. *J. Phys. Chem.* **1990**, *94*, 1164–1171.
156. Chen, L. B.; Chow, M. K.; Ackerson, B. J.; Zukoski, C. F. *Langmuir* **1994**, *10*, 2817–2829.
157. Bradbury, A.; Goodwin, J. W.; Hughes, R. W. *Langmuir* **1992**, *8*, 2863–2872.
158. Ackerson, B. J.; Clark, N. A. *Phys. Rev. Lett.* **1981**, *46*, 123–126.
159. Chen, L. B.; Zukoski, C. F. *J. Chem. Soc. Faraday Trans. I.* **1990**, *86*, 2629–2639.

160. Stevens, M. J.; Robbins, M. O.; Belak, J. F. *Phys. Rev. Lett.* **1991**, *66*, 3004–3007.
161. Chen, L. B.; Zukoski, C. F.; Ackerson, B. J.; Hanley, H. J. M.; Straty, G. C.; Barker, J.; Glinka, C. J. *Phys. Rev. Lett.* **1992**, *69*, 688–691.
162. Chen, L. B.; Ackerson, B. J.; Zukoski, C. F. *J. Rheol.* **1994**, *38*, 193–216.
163. Imhof, A.; van Blaaderen, A.; Dhont, J. K. G. *Langmuir* **1994**, *10*, 3477–3484.
164. Bergstrom, L. In *Surface and Colloid Chemistry in Advanced Ceramics Processing*; Pugh, R. J.; Bergstrom, L., Eds.; Marcel Dekker: New York, 1994; pp 193–244.
165. Buscall, R.; McGowan, I. J.; Mumme-Young, C. A. *Faraday Discuss. Chem. Soc.* **1983**, *76*, 277–290.
166. Heath, D.; Tadros, Th. F. *Faraday Discuss. Chem. Soc.* **1983**, *76*, 203–218.
167. Goodwin, J. W.; Hughes, R. W.; Partridge, S. J.; Zukoski, C. F. *J. Chem. Phys.* **1986**, *85*, 559.
168. Patel, P. D.; Russel, W. B. *J. Rheol.* **1987**, *31*, 599–618.
169. Patel, P. D.; Russel, W. B. *J. Colloid Interface Sci.* **1989**, *131*, 201–210.
170. Otsubo, Y. *Langmuir* **1990**, *6*, 114–118.
171. Otsubo, Y. *Langmuir* **1992**, *8*, 2336–2340.
172. Otsubo, Y. *Langmuir* **1994**, *10*, 1018–1022.
173. Buscall, R.; McGowan, I. J.; Mumme-Young, C. A. *Faraday Discuss. Chem. Soc.* **1990**, *90*, 115–127.
174. Woutersen, A. T. J. M.; de Kruif, C. G. *J. Chem. Phys.* **1991**, *94*, 5739–5750.
175. Nakai, Y.; Ryo, Y.; Kawaguchi, M. *J. Chem. Soc. Faraday Trans.* **1993**, *89*, 2467–2472.
176. Somasundaran, P.; Yu, X. *Adv. Colloid Interface Sci.* **1994**, *53*, 33–49.
177. Duits, M. H. G.; May, R. P.; Vrij, A.; de Kruif, C. G. *Langmuir* **1991**, *7*, 62–68.
178. Otsubo, Y. *Adv. Colloid Interface Sci.* **1994**, *53*, 1–32.
179. Firth, B. A. *J. Colloid Interface Sci.* **1974**, *57*, 257–265.
180. Firth, B. A.; Hunter, R. J. *J. Colloid Interface Sci.* **1976**, *57*, 248–256.
181. Willey, S. J.; Macosko, C. W. *J. Rheol.* **1978**, *22*, 525–545.
182. Velamakanni, B. V.; Chang, J. C.; Lange, J. C.; Pearson, D. S. *Langmuir* **1990**, *6*, 1323–1325.
183. Sonntag, R. C.; Russel, W. B. *J. Colloid Interface Sci.* **1986**, *113*, 399–413.
184. Sonntag, R. C.; Russel, W. B. *J. Colloid Interface Sci.* **1987**, *115*, 378–395.
185. Potanin, A. A. *J. Colloid Interface Sci.* **1991**, *145*, 140–157.
186. Mills, P. D. A.; Goodwin, J. W.; Grover, B. W. *Colloid Polym. Sci.* **1991**, *269*, 949–963.
187. Partridge, S. J. Ph.D. Thesis, University of Bristol, England, 1985.
188. Liu, S.; Afacan, A.; Masliyah, J. H. *Chem. Eng. Sci.* **1994**, *49*, 3565–3586.
189. Marjerrison, D. M.; Sayre, J. A. *J. Can. Pet. Tech.* **1988**, *27*, 68–72.
190. Taylor, G. I. *Proc. Roy. Soc.* **1932**, *A138*, 41–48.
191. Taylor, G. I. *Proc. Roy. Soc.* **1934**, *A146*, 501–523.
192. Oldroyd, J. C. *Proc. Roy. Soc.* **1953**, *A218*, 122–132.
193. Oldroyd, J. C. *Proc. Roy. Soc.* **1955**, *A232*, 567–577.
194. Yan, Y.; Masliyah, J. H.; Pook, G. *AOSTRA J. Res.* **1992**, *8*, 149–157.
195. Yan, Y.; Pal, R.; Masliyah, J. *Ind. Eng. Chem. Res.* **1991**, *30*, 1931–1936.
196. Yan, Y.; Masliyah, J. *Can. J. Chem. Eng.* **1993**, *71*, 852–857.
197. Yan, Y.; Pal, R.; Masliyah, J. *Chem. Eng. Sci.* **1991**, *46*, 985–994.

198. Yan, Y.; Pal, R.; Masliyah, J. *Chem. Eng. Sci.* **1991,** *46,* 1823–1828.
199. Pal, R.; Yan, Y.; Masliyah, J. *Chem. Eng. Sci.* **1992,** *47,* 967–970.
200. Yan, N.; Masliyah, J. H. *J. Colloid Interface Sci.* **1994,** *168,* 386–392.
201. Yan, N.; Masliyah, J. H. *Colloid Surf. A* **1995,** *96,* 229–242.
202. Yan, N.; Masliyah, J. H. *Colloid Surf. A* **1995,** *96,* 243–252.
203. Frith, W.; Strivens, T. A.; Mewis, J. *J. Colloid Interface Sci.* **1990,** *139,* 55–62.
204. Van der Werff, J. C.; de Kruif, C. G.; Blom, C.; Mellema, J. *Phys. Rev. A* **1989,** *39,* 795–807.
205. Tadros, Th. F. *Langmuir* **1990,** *6,* 28–35.
206. D'Haene, P. Ph.D. Dissertation, K. V. Leuven, Belgium, 1992.
207. Goodwin, J. W.; Gregory, T.; Stiles, J. A. *Adv. Colloid Interface Sci.* **1982,** *17,* 185–195.
208. Buscall, R.; Goodwin, J. W.; Hawkins, M. W.; Ottewill, R. H. *J. Chem. Soc. Faraday Trans. I.* **1982,** *78,* 2873–2887.
209. Tadros, Th. F.; Liang, W.; Costello, B.; Luckham, P. F. *Colloid Surf. A* **1993,** *79,* 105–114.
210. Wagner, N. J. *J. Colloid Interface Sci.* **1993,** *161,* 169–181.
211. Seaton, A. A.; Glandt, E. D. *J. Chem. Phys.* **1987,** *86,* 4668–4677.
212. Buscall, R.; Mills, P. D. A.; Goodwin, J. W.; Lawson, D. W. *J. Chem. Soc. Faraday Trans. I.* **1988,** *84,* 4249–4260.
213. Kanai, H.; Navarrete, R. C.; Macosko, C. W.; Scriven, L. E. *Rheol. Acta* **1992,** *31,* 333–344.
214. Sonntag, R. C.; Russel, W. B. *J. Colloid Interface Sci.* **1987,** *116,* 485–489.
215. Chen, M.; Russel, W. B. *J. Colloid Interface Sci.* **1991,** *141,* 564–577.
216. Khan, S. A.; Zoeller, N. J. *J. Rheol.* **1993,** *37,* 1225–1235.
217. Otsubo, Y. *J. Colloid Interface Sci.* **1992,** *153,* 584–586.
218. Kawaguchi, M.; Ryo, Y. *Chem. Eng. Sci.* **1993,** *48,* 393–400.
219. Callagham, I. C.; Ottewill, R. H. *Discuss. Faraday Chem. Soc.* **1974,** *57,* 110–118.
220. Cains, R. J. R.; Ottewill, R. H.; Osmond, D. W. J.; Wagstaff, I. *J. Colloid Interface Sci.* **1976,** *54,* 45–51.
221. Buscall, R. *Colloid Surf.* **1982,** *5,* 269–283.
222. Buscall, R.; White, L. R. *J. Chem. Soc. Faraday Trans. I.* **1987,** *83,* 873–891.
223. Auzerais, F. M.; Jackson, R.; Russel, W. B.; Murphy, W. F. *J. Fluid Mech.* **1990,** *221,* 613–639.
224. Bergstrom, L.; Schilling, C. H.; Aksay, I. A. *J. Am. Ceram. Soc.* **1992,** *75,* 3305–3314.
225. Barclay, L.; Harrington, A.; Ottewill, R. H. *Kolloid–Z. Z. Polym.* **1972,** *250,* 655–666.
226. Rohrsetzer, S.; Kovacs, P.; Nagy, M. *Colloid Polym. Sci.* **1986,** *264,* 812–816.
227. Goodwin, J. W.; Ottewill, R. H.; Parentich, A. *Colloid Polym. Sci.* **1990,** *268,* 1131–1140.
228. Bonnet-Gonnet, C.; Belloni, L.; Cabane, B. *Langmuir* **1994,** *10,* 4012–4021.

RECEIVED for review July 5, 1994. ACCEPTED revised manuscript January 19, 1995.

Flow of Suspensions in Pipelines

H. A. Nasr-El-Din

Laboratory R&D Center, Saudi Aramco, P.O. Box 62, Dhahran 31311, Saudi Arabia

Slurry pipelines are used in many industrial applications. Several parameters are often needed by the operator, including critical deposit velocity, solids concentration, and particle velocity profiles. This chapter first reviews important formulas used to predict critical deposit velocity both in Newtonian and non-Newtonian (power-law) carrier fluids. Various methods to measure local velocity and solids concentration profiles in slurry pipelines are discussed. Local solids concentration can be measured by sample withdrawal technique. However, the sample should be withdrawn at isokinetic conditions. Sampling downstream of tees and elbows can result in significant errors in measuring solids concentration. Gamma-ray absorption methods can be used; however, two scans are needed to obtain local solids concentration. Bulk velocity of conductive slurries can be obtained using magnetic flowmeters mounted on a vertical section of the pipe. Local particle velocity can be obtained using conductivity probes. NMR methods can be used to measure concentration and particle velocity profiles but are limited to small-diameter pipes. Vertical solids concentration of coarse slurries flowing in a horizontal pipeline exhibits a positive gradient near the bottom of the pipe. Traditional models to predict these profiles are given, and new mathematical models and computer software to determine these profiles are introduced.

SOLID–LIQUID FLOWS are encountered in a variety of applications ranging from food to mining industries (*1*). Unlike single fluid flow in pipes, slurry flow in pipelines is complex. The complexity of these flows has necessitated the use of empirical equations in the design of slurry handling equipment, often leading to expensive systems. This complexity depends on the physical properties of the solid particles, for example, particle density, shape, and mean diameter. It also depends on the viscosity and density of the carrier fluid and, finally, on the operating con-

0065–2393/96/0251–0177$19.25/0

ditions, for example, bulk velocity, solids concentration, and pipe diameter. It is extremely important to understand how the above parameters affect the type of flow in slurry pipelines. Such understanding is important to properly design a slurry pipeline. Many aspects of this design, including corrosion–erosion, pressure drop, and pumping requirements, depend on such understanding.

Velocity and concentration profiles are two important parameters often needed by the operator of slurry handling equipment. Several experimental techniques and mathematical models have been developed to predict these profiles. The aim of this chapter is to give the reader an overall picture of various experimental techniques and models used to measure and predict particle velocity and concentration distributions in slurry pipelines. I begin with a brief discussion of flow behavior in horizontal slurry pipelines, followed by a revision of the important correlations used to predict the critical deposit velocity. In the second part, I discuss various methods for measuring solids concentration in slurry pipelines. In the third part, I summarize methods for measuring bulk and local particle velocity. Finally, I review models for predicting solids concentration profiles in horizontal slurry pipelines.

Classification of Flow Behavior

Flow behavior of slurry flow in horizontal pipelines depends on many factors, including flow rate; pipe shape, characteristic length, and orientation; physical properties of the carrier fluid; and particle size, density, and concentration. Two extremes of slurry behavior, homogeneous and heterogeneous flow, occur as a result of particular interactions of these factors as reported by Satchwell et al. (2) and Wasp et al. (3).

Homogeneous flow in a horizontal pipe is characterized by uniform solid concentration profiles. A homogeneous slurry results when particle addition simply alters the carrier fluid density and rheological characteristics. The effects of various operating parameters on the rheological properties of solid–liquid systems are discussed in more detail in Chapters 1 and 2. A uniform concentration profile is usually obtained when particles having low settling velocities are used. Such low settling velocities are encountered when the density of the solid particles approaches that of fluid, for example, polystyrene-water slurry systems having a density ratio, ρ_s/ρ_f, of 1.05, where ρ_s and ρ_f are the densities of the solid particles and carrier fluid, respectively.

The other extreme of slurry behavior in a horizontal pipe, heterogeneous flow, is characterized by a pronounced variation in the local solids concentration with position in the pipe. The particle settling velocity in this case is high. This implies that the density of the solid particles is higher than the working fluid, for example, sand–water slurry systems ($\rho_s/\rho_f = 2.65$). Also, the particle mean diameter for these slurries is large. Because of the gravitational force, the solid particles of heterogeneous slurries are concentrated in the lower half of the pipe. The vertical concentration profiles of strongly heterogeneous slurries may exhibit a positive concentration gradient near the bottom of the pipe (4–6). Typical solids concentration profiles of heterogeneous slurries are discussed later.

The pressure drop and pumping requirements for slurry transportation are functions of slurry flow type. The flow curve (shear stress versus shear rate) is also a strong function of the slurry type (7). Figure 1 shows the variation of the wall shear stress as a function of the nominal shear rate for homogeneous and heterogeneous slurry flow in a horizontal pipe. At high shear rates, heterogeneous fluid response (curve A) tends to parallel that of the carrier fluid. As the slurry velocity is decreased, that is, at low shear rates, the vertical solids concentration gradient increases until either a stationary or slowly moving bed of deposited particles appears along the pipe bottom. The slurry velocity at which a bed of particles forms is defined as the critical deposit velocity, V_D. A further decrease in the slurry velocity below V_D leads to increased friction loss and may also result in pipe plugging.

If shutdown occurs while pumping a heterogeneous slurry, solids will deposit in a stationary bed along the pipe bottom. To resume the operation of the slurry pipeline, it becomes necessary to resuspend these solids to remove them from the pipe. If the fluid flow rate over the settled solids is gradually increased, a response similar to curve A of Figure 1 is obtained. With increasing shear rate, the wall shear stress decreases until a minimum is reached. The fluid velocity that corresponds to this minimum shear stress is the critical resuspension velocity, V_S (7).

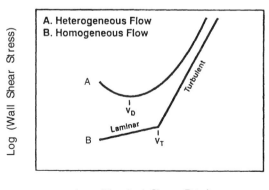

Figure 1. Flow curves for homogeneous and heterogeneous slurries. (Reproduced with permission from reference 7. Copyright 1991.)

Curve B of Figure 1 displays the flow curve of a homogeneous slurry. At high shear rates, a steep linear turbulent flow regime occurs. As the slurry velocity is decreased in the turbulent regime, a sudden transition to laminar flow regime occurs. The transition velocity, V_T, corresponds to this change in flow regime and remains the same whether approached from turbulent or laminar flow directions (7).

Critical Deposit Velocity

The critical deposit velocity for settling slurries is an important factor in the design of slurry pipelines. Below this velocity, a stationary bed of particles forms on the bottom of the pipe. Long-distance pipelines are generally designed to operate above this velocity. Because of its importance, several studies have been conducted to examine the effects of various operating parameters on the critical deposit velocity. Reviews of these studies are given by Wiedenroth and Kirchner (8), Carleton and Cheng (9), Thomas (10), and Hanks (11). In this chapter, however, a brief discussion on the most important correlations used to predict the critical deposit velocity is given.

One of the best known correlations for the deposit velocity of sand and gravel–water slurries is that of Durand and Condolios (12):

$$V_D = F_1 \left[2 \ g \ D(S - 1) \right]^{1/2} \tag{1}$$

where $S = \rho_s/\rho_f$, g is the gravitational acceleration, D is the pipe inside diameter, and F_1 is an empirical function. F_1 equals 1.34 for particles sizes $\geq$ 2 mm (13).

Oroskar and Turian (14) evaluated critical deposit velocity correlations obtained by Durand (15), Kao and Wood (16), Newitt et al. (17), and Zandi and Govatos (18). Oroskar and Turian (14) developed a critical deposit velocity correlation based on balancing the energy required to suspend particles with that dissipated by a fraction, χ, of turbulent eddies present in the flow. They estimated χ based on the assumption that only those eddies possessing instantaneous velocities equal to or greater than the hindered settling velocity of the particles are effective in maintaining particles in suspension. Their equation is

$$\frac{V_D}{\sqrt{gd_{50}} \ (S-1)} = 1.85 \ C_v^{0.1536} \ (1 - C_v)^{0.3564} \ (d_{50}/D)^{-0.378} \ \tilde{N}_{\mathrm{Re}}^{0.09} \chi^{0.30} \tag{2}$$

where d_{50} is the mean particle diameter, C_v is the discharge solids concentration, and $\tilde{N}_{\mathrm{Re}}$ is a modified Reynolds number defined as

$$\tilde{N}_{\mathrm{Re}} = D\rho_f [2 \ gd_{50} \ (S - 1)]^{1/2}/\mu_f \tag{3}$$

The coefficient and exponent values of equation 2 were obtained from regression analysis of 357 data points.

The Oroskar–Turian's correlation and previous ones were developed to determine the critical deposit velocity of Newtonian carrier fluids with various particle sizes and concentrations. Shah and Lord (7) generalized equation 2 to extend its capability to correlate the critical deposit velocity for non-Newtonian carrier fluids (power law). The parameter χ was eliminated from equation 2 because of its insignificant contribution to the correlation results and because it would be undefined for the laminar flow regime of non-Newtonian fluids. The generalized form of equation 2, which can be applied to either critical deposit (V_D) or resuspension velocity (V_S), is as follows:

$$\frac{[V_D] \text{ or } [V_S]}{\sqrt{gd_{50} (S-1)}} = Y \, C_v^{0.1536} \, (1 - C_v)^{0.3564} \, (d_{50}/D)^{-w} \tilde{N}_{Re}^{\ z} \qquad (4)$$

where $\tilde{N}_{Re}$ is defined in equation 3, with the apparent viscosity used instead of the fluid viscosity. The coefficient Y and the exponents w and z are empirical constants.

Other correlations to determine the critical deposit velocity are given by Hanks (11), Sommerville (19), Roco and Shook (20), and Gillies and Shook (21).

Solids Concentration Measurement in Slurry Pipelines

A number of methods have been used to measure solids concentration in slurry pipelines. Reviews of these methods are given by Kao and Kazanskij (22), Baker and Hemp (23), Heywood and Mehta (24), and recently by Shook and Roco (25). In general, the principle of any of these methods is to find a specific property that is significantly different for the two phases, for example, relative permittivity (dielectric constant), electrical conductivity, density, refractive index, or absorption of electromagnetic radiation. Solids concentration can be determined by measuring this property for the mixture and then using a calibration curve. Any of these methods will give inaccurate measurement if the values of the specific property of the two phases approach one another or if the solids concentration is very low. In the following section, various methods for measuring local concentration, including sampling, conductivity, capacitance gamma-ray absorption, and NMR, are discussed.

Sample Withdrawal Methods.   Sample withdrawal is widely used in industry to measure in situ solids concentration, composition, and size distribution in solid–liquid systems (26). It is probably the only reliable method for use at low solids concentration. It is also used to calibrate and evaluate newly developed methods of measuring solids concentration (27). A number of methods of sampling differ primarily in the geometry of the sampling device. Figure 2 shows schematic diagrams of the most commonly used sampling devices.

Serious errors in measuring solids concentration and particle size distribution arise as a result of improper sampling. The effectiveness of a sampling device is usually expressed as the ratio of the measured solids concentration to the upstream solids concentration (28). Three main factors can cause the sampling efficiency of a sampling device to deviate from unity (i.e., ideal sampling): particle inertia, particle bouncing, and flow structure ahead of the sampler. In the following sections, the effect of these factors on the performance of various sampling devices is discussed.

Particle Inertia.   Particle inertia is a major source of sampling errors when the densities of the two phases are significantly different. Because particle inertia is different from that of an equivalent volume of fluid, particle motion does not follow the distorted fluid. Consequently, sample solids concentration and composition will be significantly different from those in the slurry handling equipment, for example, pipes, mixing tanks, and so on. Sampling errors due to inertia depend on how the sampling device disturbs the flow field upstream of the sampling port and how the particles respond to this disturbance.

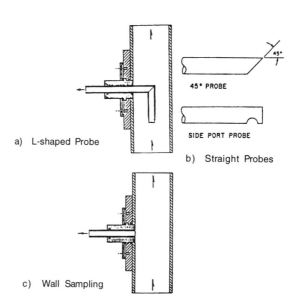

Figure 2.   Sampling devices. (Reproduced with permission from reference 26. Copyright 1989 Gulf Publishing Company.)

Sample Withdrawal Using L-Shaped Probes. Thin L-shaped probes are commonly used to measure local solids concentration in slurry pipelines (29–34). However, serious sampling errors may arise as a result of particle inertia. To illustrate the effect of particle inertia on the performance of L-shaped probes, consider the flow field ahead (upstream) of a sampling probe located at the center of a pipe, as shown in Figure 3. The probe has zero thickness, and its axis coincides with that of the pipe. The fluid ahead of the sampler contains particles of different sizes and densities. Figure 3a shows the fluid flow when sampling with a velocity equal to the upstream local velocity (isokinetic sampling). Of course, the probe does not disturb the flow field ahead of the sampler, and, consequently, sample solids concentration and composition to equal those in the pipe.

Sampling with a velocity different from the upstream local velocity (anisokinetic sampling) distorts the fluid flow ahead of the sampler. This distortion depends on the ratio of the sampling velocity, U, to the upstream local velocity, U_o. If the sampling velocity ratio ($U:U_o$) is less than unity, the fluid diverges away from the probe as shown in Figure 3b. Particles of low inertia will follow the fluid flow and will miss the probe, whereas those of high inertia will move in straight lines like bullets. As a result, the sample obtained has a higher solids concentration, is more coarse, and has more dense particles than those in the pipe.

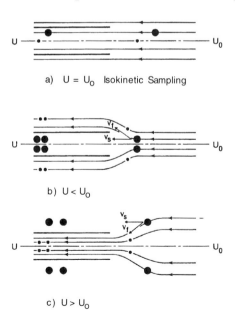

Figure 3. *Isokinetic and anisokinetic sampling. (Reproduced with permission from reference 26. Copyright 1989 Gulf Publishing Company.)*

An opposite trend occurs if the sampling velocity ratio is greater than unity. In this case, the fluid converges into the probe (Figure 3c), but the particles will respond according to their inertia; particles of low inertia follow the fluid into the probe, whereas those of high inertia will miss the probe. The sample, in this case, has a lower solids concentration and more fine and light particles than those upstream of the probe.

The preceding discussion indicates that the sampling efficiency for thin L-shaped probes is a function of two parameters: the deviation from the isokinetic conditions and the response of the particles to the deflection of the fluid upstream of the sampler. The deviation from the isokinetic conditions is a function of the sampling velocity ratio $(U:U_o)$, whereas the particle response is a function of the ratio of particle inertia to fluid drag. This ratio, in a dimensionless form, is known as the particle inertia parameter, the Stokes number, or the Barth number (K), defined as

$$ K = \frac{\rho_s \, d_{50}{}^2 \, U_o}{18\mu_f D_{sm}} \tag{5} $$

where μ_f is the fluid viscosity and D_{sm} is the sampler diameter. The effect of particle inertia on the sampling efficiency for thin L-shaped probes has been studied extensively in fluid–solid systems of low solid concentrations. Reviews on the performance of thin L-shaped probes to sample from gas–solid systems were given by Fuchs (35) and Stevens (36).

Unlike gas–solid systems, few investigations have been conducted on sampling from solid–liquid systems. Rushton and Hillestad (29) measured solids concentration profiles in vertical and horizontal slurry pipelines by using different sampling techniques. Nasr-El-Din et al. (34) examined both theoretically and experimentally the performance of thin L-shaped probes when used to sample from slurry pipelines. Figure 4 compares the predicted solids line and measured sampling efficiency $(C:C_o)$ obtained at different sampling velocity ratios $(U:U_o)$. C and U are the sample solids concentration and velocity, respectively. C_o and U_o are the local solids concentration and velocity upstream of the probe. Particle inertia parameter (k) for sampling using L-shaped probes is $(\rho_s \, d_{50}{}^2 U_o / 18\mu_f R_{sm})$ and particle Reynolds number (Re_o) is $\rho_f \, d_{50} \, U_o/\mu_f$. Figure 4 shows good agreement between the model prediction and the experimental measurements for sand particles having a mean particle size of 0.45 mm and an upstream local solids concentration of 17.3 vol%. The dashed line in Figure 4 represents model prediction obtained by neglecting the effect of solids concentration (infinite dilution) on the sampling efficiency. Obviously, the effect of the solids concentration on the predicted sampling efficiency is significant and should be considered when sampling from concentrated slurries is considered.

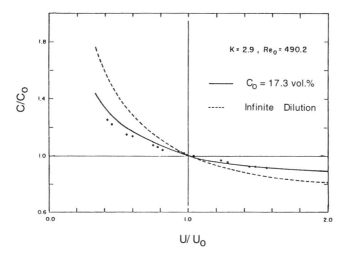

Figure 4. Predicted and observed sampling efficiencies for the 0.45-mm sand particles. (Reproduced with permission from reference 34. Copyright 1984.)

Wall Sampling.   Another way to collect a sample from a slurry pipeline is by withdrawing a sample from an opening in the wall (Figure 2c). This method of sampling, known as wall sampling, is widely used in industry, not only for slurry pipelines (29, 37–39) but also for mixing vessels (40–43) and slurry heat exchangers (44). The advantage of this technique is its simplicity of operation, because it uses a small aperture in the wall of the pipe and does not disturb the flow with a probe. However, the main disadvantage is that the sampling efficiency is a strong function of particle inertia and the solids distribution upstream of the sampler.

Wall Sampling from Vertical Slurry Pipelines.   Moujaes (44) used wall sampling to measure solids concentration in upward vertical slurry flows. He found the sample concentration to be consistently lower than the true values in the pipe. Torrest and Savage (45) studied collection of particles in small branches. The sampling transport efficiency, E, defined as the ratio of the solids flow rate in the branch to that in the main pipe, was found to be a function of the single particle settling velocity (W_0) and the upstream bulk velocity (U_b) as follows:

$$E = 158.7 \times Q \left[\frac{40\ (W_0 + U_b) - 58.4}{1 - 125(W_0 + U_b)} \right] \tag{6}$$

where Q is the branch flow rate (m³/s) and ($W_0 + U_b$) is in meters per second. Equation 6 is valid for the range of $0.04 \leq (W_0 + U_b) \leq 0.4$ m/s.

Nasr-El-Din and co-workers (37, 46) studied wall sampling from an upward vertical slurry flow. They found that this type of sampling caused serious errors in measuring solids concentration and particle size distribution. Figure 5 shows that the sampling efficiency, C:C_v, for sidewall sampling from a vertical pipeline is always less than unity. The sampling efficiency is a strong function of particle size and sampling velocity ratio. Figure 6 depicts that the sample mean particle diameter using wall sampling is smaller than that in the pipe, especially at low sampling velocity ratios.

In a later study, Nasr-El-Din et al. (47) examined wall sampling from a vertical slurry pipeline. The effects of particle size, fluid viscosity, solids concentration, and sampling velocity ratio on the sampling efficiency were studied in detail. They developed the following semi-empirical correlation to predict the sampling efficiency at sampling velocity ratios > 3:

$$C/C_v = \exp[-A(1 - \alpha)^a\, k^b] \qquad (7)$$

where α is the solids volume fraction and k is related to the particle inertia parameter, defined in equation 5, by

$$k = K(\rho_s - \rho_f)/\rho_s \qquad (8)$$

The best fit of 37 data points gave $A = 0.46$, $a = 1$, and $b = 0.5$. Figure 7 shows a plot of the sampling efficiency as a function of $(1 - \alpha)\sqrt{k}$. It can be seen that equation 7 predicts the experimental data fairly well.

Wall Sampling from Horizontal Slurry Pipelines.   Unlike wall sampling from a vertical slurry flow, the sampling efficiency of wall sampling from a horizontal slurry pipeline may exceed unity in some cases. Nasr-El-Din et al. (38, 39) showed that the sampling efficiency for wall sampling from a horizontal slurry pipeline is a strong function of the sampler orientation (upward, sideways, and downward). Figure 8 shows the effect of the sampling velocity ratio, U:U_b, on the sampling efficiency, C:C_v, for sand particles of $d_{50} = 0.39$ mm with the sampler orientation as a parameter. Slurries having a solids concentration of 12.5 vol% and a bulk velocity of 2.65 m/s were used with a sampler of 8 mm ID. The effect of the sampling velocity ratio on the sampling efficiency depends on the sampler orientation. For sampling from the top of the pipe (top wall sampling), the sampling efficiency increases as the sampling velocity ratio is increased. For sampling from the side of the pipe (side wall sampling), the C:C_v versus U:U_b relationship is similar to that of the top wall sampling. For sampling from the bottom of the pipe (bottom wall sampling), the sampling efficiency exhibits a maximum at a sampling

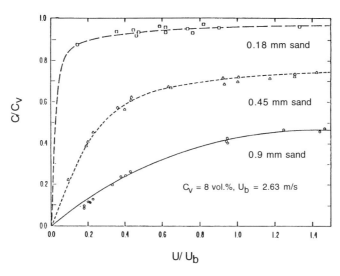

Figure 5.   Effect of particle size on the sampling efficiency of side wall sampling from vertical slurry flow. (Reproduced with permission from reference 37. Copyright 1985.)

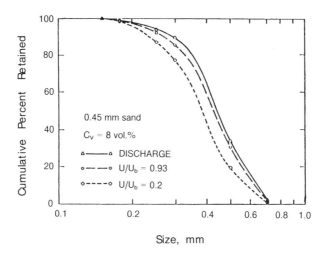

Figure 6.   Effect of sampling velocity ratio on the sample particle size distribution. (Reproduced with permission from reference 37. Copyright 1985.)

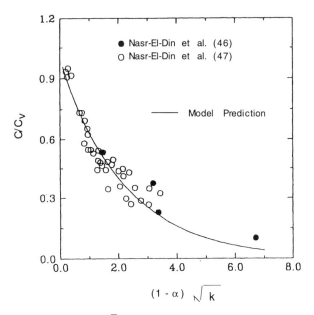

Figure 7.   Effect of $(1 - \alpha)\sqrt{k}$ on the sampling efficiency. (Reproduced with permission from reference 47. Copyright 1991.)

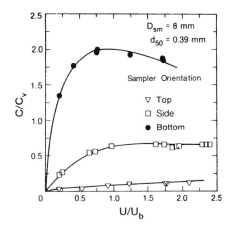

Figure 8.   Effect of sampling velocity ratio on the sampling efficiency for sampling from horizontal slurry pipeline ($d_{50} = 0.39$ mm). (Reproduced with permission from reference 38. Copyright 1989.)

velocity ratio of nearly 0.8. The sampling efficiency for top or side wall sampling is always less than unity. This trend is similar to that observed for wall sampling from vertical slurry pipelines (37). The sampling efficiency for the bottom wall sampling is higher than unity except at velocity ratios <0.12. At a given velocity ratio, the sampling efficiency strongly depends on the sampler orientation. This trend is reasonable and is due to the nonuniform vertical concentration profile upstream of the sampler for the sand fraction used.

Figure 9 shows the effect of the sampling velocity ratio on the sampling efficiency for sand particle of $d_{50} = 0.08$ mm, at an upstream solids concentration of 11.6 vol%, and a bulk velocity of 2.55 m/s using a sampler of 4 mm ID. For the top and side wall sampling, the effect of the sampling velocity on the sampling efficiency is not significant. Similar to the results obtained with the 0.39-mm sand fraction, the sampling efficiency of the bottom wall sampling exhibits a maximum. However, the value of this maximum for the 0.08-mm sand is significantly less than that obtained with the 0.39-mm sand. The effect of the sampler orientation on the sampling efficiency of the 0.08-mm sand is much less than that noted for the coarser sand. This result is reasonable because the vertical concentration profile for the 0.08-mm sand upstream of the sampler is uniform. The results shown in Figure 9 indicate that a representative sample for the 0.08-mm sand with an error of less than 3% of the value can be obtained using side wall sampling at velocity ratios greater than 0.25.

Similar to sampling from vertical slurry pipelines, the sampling velocity ratio affects the sample particle size distribution. However, in the

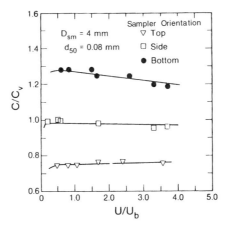

Figure 9.   Effect of sampling velocity ratio on the sampling efficiency for sampling from horizontal slurry pipeline ($d_{50} = 0.08$ mm). (Reproduced with permission from reference 38. Copyright 1989.)

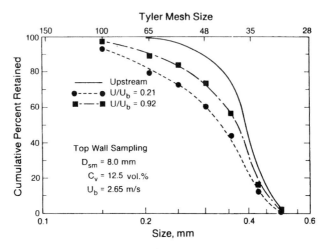

Figure 10. Effect of sampling velocity ratio on the sample particle size distribution for d_{50} = 0.39 mm: top wall sampling. (Reproduced with permission from reference 38. Copyright 1989.)

case of sampling from a horizontal pipe, this effect depends on the sampler orientation. Figures 10 and 11 display the measured particle size distribution for the 0.39-mm sand at an upstream solids concentration of 12.5 vol% and a bulk velocity of 2.65 m/s using a sampler of 8 mm ID. Figure 10 shows the sample particle size distribution obtained at different sampling velocity ratios using top wall sampling as compared

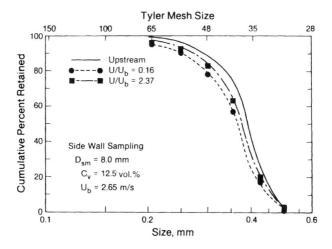

Figure 11. Effect of sampling velocity ratio on sample particle size distribution for d_{50} = 0.39 mm: side wall sampling. (Reproduced with permission from reference 38. Copyright 1989.)

with that in the main pipe. At a velocity ratio of ≈ 0.2, the mean particle diameter of the sample is significantly smaller than that in the main pipe upstream of the sampler. As the sampling velocity ratio is increased, the sample mean diameter approaches that in the pipe. A similar trend is obtained for the side wall sampling as shown in Figure 11.

A comparison of the sample size distribution obtained at a velocity ratio of ≈ 0.2 using the top wall sampling (Figure 10) and the side wall sampling (Figure 11) indicates that the sample mean diameter for the former is significantly smaller than that obtained in the latter case. Unlike the sample size distribution for the top and side wall sampling, the effect of the sampling velocity on the sample particle size distribution for the bottom wall sampling is not significant.

Particle Bouncing.   A second source of sampling errors occurs as a result of particle bouncing effects. A typical example of this effect is sampling particles of high inertia using thick (blunt) L-shaped probes. In this case, particles may hit the probe wall, lose some of their inertia, and enter the probe. Consequently, the sample solids concentration is higher than the upstream local concentration, even when $U{:}U_o = 1$.

The effect of particle bouncing on the sampling efficiency of thick L-shaped probes was first noted in gas–solid systems by Whitely and Reed (*48*). They found that the sampling efficiency for thick L-shaped probes was higher than unity at $U{:}U_o = 1$. To account for the effect of particle bouncing on the sampling efficiency at the isokinetic velocity, Belaev and Levin (*49*) and Yoshida et al. (*50*) proposed an empirical equation. This equation can be written in a slightly different form as

$$C/C_o = 1 + B(2\,T + T^2) \tag{9}$$

where T is the probe relative wall thickness (wall thickness/probe radius) and B is the fraction of particles that hit the nozzle edge and enter the probe.

To establish the performance of blunt probes when used to sample from liquid–solid systems, a set of L-shaped probes of different thicknesses was tested by Nasr-El-Din and Shook (*51*). Figure 12 shows the effect of the probe relative wall thickness on the sampling efficiency for sand particles of $d_{50} = 0.45$ mm at a local solids concentration of 10 vol%. The sampling efficiency at $U{:}U_o = 1$ is greater than unity, an observation found for sampling sand particles using thick probes. As the relative wall thickness is increased, $C{:}C_o$ at $U{:}U_o = 1$ increases. Also, to obtain the correct concentration using these probes, samples should be taken at a critical velocity ratio greater than the isokinetic one. It is worth noting that the increase in the sample solids concentration at isokinetic conditions is much less than the corresponding values obtained from equation 9 with $B = 0.5$.

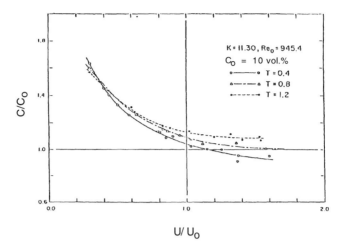

Figure 12.   Effect of the probe relative wall thickness on the sampling efficiency using thick L-shaped probes. (Reproduced with permission from reference 51. Copyright 1985.)

Figure 13 shows the sampling efficiency versus the velocity ratio for L-shaped probes having a tip angle θ of 18° and probe relative wall thicknesses of 0.4, 0.8, and 1.2. The 0.08-mm sand at 6.3 vol% discharge solids concentration, and 2.63 m/s bulk velocity was used in these experiments. At this tip angle, the increase of $C:C_o$ at $U:U_o = 1$ is eliminated.

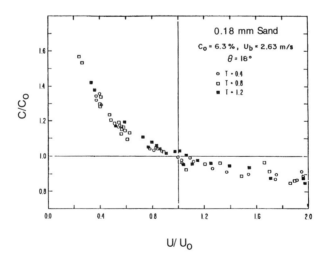

Figure 13.   Sampling efficiencies for L-shaped probes having a tip angle of 18° and various relative wall thicknesses. (Reproduced with permission from reference 51. Copyright 1985.)

This result confirms the explanation given previously about the bouncing effect and agrees with the trends obtained by Whitely and Reed (*48*) in gas–solid systems.

Figure 14, from Nasr-El-Din et al. (*52*), depicts the effect of the probe relative wall thickness on the sampling efficiency for polystyrene particles of 0.3 mm mean diameter and a density of 1050 kg/m³. Samples were taken from the center of the pipe at a mean solids concentration of 37 vol% and a bulk velocity of 3.4 m/s, with L-shaped probes of relative wall thicknesses of 0.05, 0.5, 0.8, and 1.2. Unlike the results obtained with the sand particles, shown in Figure 12, the effect of the sampling velocity on the sampling efficiency is not significant. This result is reasonable because the density of the polystyrene particles is very close to water. This finding implies that these particles follow the fluid into the probe, and, consequently, the sampling efficiency is unity, no matter what the sampling velocity.

To account for particle rebound and inertia effects simultaneously, a modification was introduced by Nasr-El-Din and Shook (*51*). Figure 15 compares the calculated sampling efficiency for a thick L-shaped probe having a relative wall thickness of 0.8, considering the inertial effect alone and with the particle bouncing effect, with the experimental measurements. Clearly, the agreement is much better when particle inertia and bouncing effects are considered.

A second example of the particle bouncing effect is sampling using straight probes. Although thick L-shaped probes are more practical than thin probes, they require a relatively large aperture in the wall of the pipe. Straight probes are robust, simple to construct, require a minimum

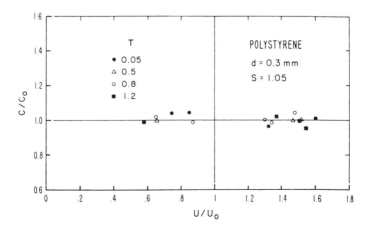

Figure 14. Effects of sampling velocity ratio and probe relative thickness on the sampling efficiency of the polystyrene particles. (Reproduced with permission from reference 52. Copyright 1986.)

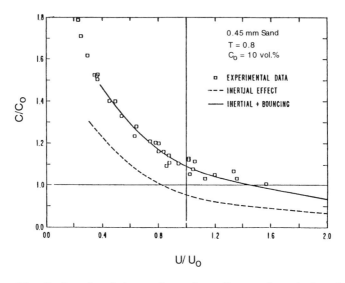

Figure 15.   Predicted and observed sampling efficiency for a thick L-shaped probe have a relative wall thickness of 0.8. (Reproduced with permission from reference 51. Copyright 1985.)

size of aperture in the wall of the pipe, and can be withdrawn after sampling. The performance of two different straight probes, a side-port probe, and a 45° probe (Figure 2b) for measuring solids concentration of liquid–solid systems was examined (51). Figure 16 shows $C:C_o$ versus $U:U_o$ for the thin-walled L-shaped and the circular-port probes. For the

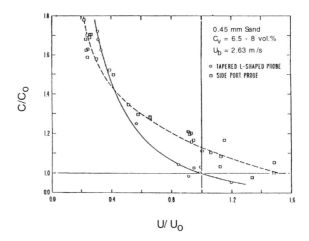

Figure 16.   Sampling efficiency for circular port and thin-walled L-shaped probes. (Reproduced with permission from reference 51. Copyright 1985.)

circular-port probe, the sampling efficiency is greater than unity at the isokinetic velocity. Thus, to obtain the correct concentration, the sampling velocity ratio would have to be greater than unity. Nasr-El-Din and Shook (51) showed that the performance of the 45° probe was also different from that of thin L-shaped probes.

Flow Structure Ahead of the Sampler.   A third source of sampling errors is not directly related to the geometry of the sampling device but to the flow structure ahead of the sampler. Obviously, if the flow field ahead of the sampler is strongly three dimensional, it will be very difficult to obtain a representative sample. To illustrate this point, consider the flow field downstream of a 90° elbow. Whenever a fluid flows along a curved pipe, a pressure gradient must occur across the pipe to balance the centrifugal force. The pressure is greatest at the wall farther from the center of curvature (pressure wall) and lowest at the nearer wall (suction wall). Because of inertia, the fluid in the core moves across the pipe from the suction wall toward the pressure wall and returns to the inner edge along the wall, as shown in Figure 17. A pair of symmetrical counterrotating vortices is formed as a result of the fluid inertia. This secondary flow is superimposed on the main stream, so the resultant flow consists of helical motion on each side of the plane of the bend passing through the axis of the pipe. The strength of secondary flow depends, among other factors, on the flow Reynolds number and the curvature of the elbow (53).

The helical flow downstream bends, and tees, affects sampling in two ways. First, because of the helical motion, it is very difficult to align the probe axis with the fluid velocity vector. Consequently, and because of the inertial effect, sample concentration will be always less than the upstream concentration (54). Second, the inertial effects on the elbow

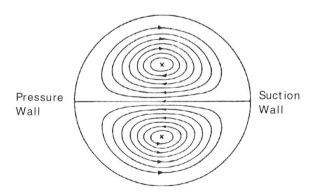

Figure 17.   Secondary flow downstream of a 90° elbow. (Reproduced with permission from reference 58. Copyright 1987.)

plane and the centrifugal force on a plane perpendicular to that of the elbow will produce a nonuniform solids distribution downstream of the elbow. Therefore, it is important to measure solids concentration across the pipe cross-section and not to rely on a single measurement.

A few studies considered the solids distribution downstream of elbows. Ayukawa (55) and Toda et al. (56) observed an accumulation of coarse particles at the outer wall of vertical bends. Toda et al. (57) noted some changes in the solids distribution downstream of 90° bends. However, no concentration measurements were taken.

Nasr-El-Din and Shook (58) studied solids distribution in a vertical pipe downstream of a 90° elbow. They tested sand–water slurries of various solid concentrations and particle sizes. The slurry flows were turbulent, and the particle Stokes number (inertia parameter) based on the pipe diameter and bulk velocity varied from 0.5 to 3. The solids distribution downstream of the elbow was found to be a function of the radius of curvature of the elbow, solids concentration, and particle size.

Figure 18 depicts the solids concentration profile in a vertical pipe 22 pipe diameters downstream of a short-radius elbow. The concentration profile is symmetrical, and a minimum solids concentration appears at the center of the pipe. This variation in solid concentration across the pipe is evidently a consequence of the centrifuging action of the secondary flow that is generated by the bend upstream. Figure 18 also shows that the concentration profiles are concentration dependent, and

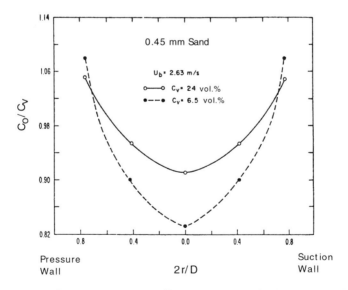

Figure 18.   Solid concentration profiles downstream of a short radius elbow. (Reproduced with permission from reference 58. Copyright 1987.)

as the solids concentration is increased, the profiles become flatter. Other results (58) indicate that these profiles depend on the particle size and the radius of curvature of the elbow.

To establish a uniform concentration profile downstream of a 90° elbow, Nasr-El-Din and Shook (58) inserted straightening vanes 10 cm long just downstream of a short-radius elbow. Figure 19 shows the effect of these vanes on the concentration profile of 0.45-mm sand particles. Although the concentration becomes flatter in the presence of the vanes, a distinct minimum at the center of the pipe still exists. These results imply that the solids are already distributed at the exit of the elbow, and the vanes merely increase the rate of diffusion of the particles.

The concentration profiles discussed so far were obtained in a vertical pipeline downstream of a 90° elbow with a horizontal approach. Colwell and Shook (59) examined concentration profiles in a horizontal slurry pipeline downstream of a 90° elbow. According to their results, a length of at least 50 pipe diameters downstream of the elbow is needed to obtain fully developed concentration profiles.

Conductivity Methods.   The electrical conductivity of a mixture of two or more phases is an important property of the mixture. Many details regarding the mixture's structure can be inferred from its electrical conductivity.

According to the nature of the dispersed phase in the mixture, uses of electrical conductivity can be divided into two major groups. In the first group, the dispersed phase (the solid particles in slurry systems or the oil droplets in oil-in-water emulsions) consists of loose particles dispersed in a continuous phase (matrix). The particles have a defined shape

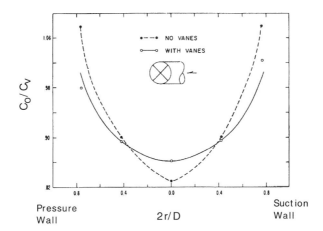

Figure 19.   Effect of straightening vanes on solids concentration profiles. (Reproduced with permission from reference 58. Copyright 1987.)

and size distribution, but the concentration of the dispersed phase is less than the corresponding maximum packing concentration. In this group, the electrical conductivity of the mixture can be used to measure the concentration and the size distribution of the dispersed phase within the system. Typical examples of such systems are measuring concentration of the dispersed phase in mixing tanks (60–62), in slurry pipelines (27, 63), in oil-in-water emulsion pipelines (64, 65), in three-phase fluidized beds (66), sedimentation vessels (67), and three-phase reactors (68).

In the second group, the solid-phase concentration is high, and solids particles are either loose but in contact or consolidated. In this case, the solid phase is the matrix, whereas the liquid phase is the dispersed phase. In this group, the mixture electrical conductivity can be used to measure the effective porosity of the porous medium (69, 70). Also, if two immiscible fluids, for example, oil and water, are present in a porous medium, the electrical conductivity can be used to measure the relative saturations of the two fluids and to give an indication of the wettability of the porous medium (71, 72).

The electric conductivity methods are widely used in both categories because they are simple to operate and give quick response, accurate results, and a continuous reading, that is, they can be used as a measuring element in any control loop.

Besides these examples, knowing the relationship between the mixture conductivity and the porosity or the concentration of the dispersed phase is important. Such a relation can be used to predict other transport coefficients such as the diffusion coefficient, dielectric constant, and thermal conductivity. Of course, such coefficients are useful in many practical applications.

Although estimating the electrical conductivity of a mixture of two or more phases looks simple and straightforward, it is a very complicated problem, both theoretically and experimentally. This complexity explains the huge volume of work devoted to solving this problem since the pioneering work of Maxwell (73) and Lord Rayleigh (74).

Definition of the Mixture Conductivity.   If a liquid–solid mixture is placed between two electrodes of different potential, the resulting potential difference will cause a current to flow from the electrode of higher potential to the one of lower potential. The current and potential gradient are related by the following diffusion-type equation:

$$I = \lambda_m \nabla \phi \qquad (10)$$

where I and ϕ are the volume-average values of the current and the potential, respectively. The proportionality constant λ_m is the effective conductivity of the mixture. For a homogeneous and isotropic mixture,

λ_m is a scalar quantity, whereas for a homogeneous and anisotropic mixture, λ_m is a second-order symmetrical tensor.

In the following sections, various expressions to determine the conductivity of homogeneous and isotropic solids–liquid mixtures are discussed.

Mathematical Description of Mixture Conductivity. In very general terms, the electrical conductivity of a mixture is a function of the electrical conductivity of its constituents, their relative amounts, and their distribution within the system. Models and expressions to predict the mixture effective conductivity can be divided according to the degree of complexity of the mixture into two major categories. In the first category, the mixture contains particles of definite shape (e.g., spheres, spheroids, and ellipsoids) at low solid concentrations. For these mixtures, describing the boundary conditions is straightforward. Also, the effect of the surrounding particles can be neglected. For this category, rigorous solutions are available for particles of simple geometrical shapes. A rigorous solution in this case means solving Laplace's equation for the potential using appropriate boundary conditions.

The second category includes mixtures of high solid concentrations. Unlike dilute systems, particle–particle interactions cannot be neglected. Also, it is very difficult to describe the boundary conditions. Because solution of equation 10 is basically a boundary value problem, no rigorous solution is available for concentrated mixtures, except for ordered arrays. To overcome these problems, various approaches have been considered. In this chapter, approximate solutions based on Maxwell's theory and empirical formulas are discussed.

Effective Conductivity of Dilute Mixtures. The simplest best-defined case is a cluster of spherical particles dispersed in a liquid and located in a uniform electrical field. If the particles have the same conductivity as the liquid, the potential around the particles will not be distorted, and the mixture conductivity is equal to that of the liquid. If the particles have a lower conductivity, the streamlines will diverge away from the particles, and the mixture conductivity will be lower than that of the liquid. If the particles have a higher conductivity, the streamlines will converge into the particle, and the mixture conductivity will be higher than that of the liquid.

Maxwell (73) calculated the potential distribution for a single spherical particle immersed in a conducting medium and subjected to a uniform electrical field. He solved Laplace's equation within the two phases subject to continuity of potential and continuity of the normal component of the current density at the surface of the particle. Maxwell then extended his single-sphere solution to dilute mixtures and obtained the following expression for λ_m:

$$\lambda_m = \lambda_f \left[\frac{2\lambda_f + \lambda_s - 2(\lambda_f - \lambda_s)C}{2\lambda_f + \lambda_s + (\lambda_f - \lambda_s)C} \right] \quad (11)$$

where λ_f and λ_s are the electrical conductivities of the liquid and solid phases, respectively, and C is the volumetric concentration of the dispersed phase. The assumptions used to derive equation 11 are very important:

- The particles are spherical, of uniform size, and have the same electrical conductivity.
- The electrical field around a particle is not affected by the presence of other particles, that is, the particle diameter is much smaller than the distance between the particles. Obviously, this condition can be met only for very dilute mixtures.
- The effect of surface conductance is negligible.
- The solid–liquid mixture is homogeneous and isotropic.

Equation 11 indicates that the mixture conductivity does not follow the additivity rule, which is sometimes used as a simplifying assumption. The mixture conductivity is independent of particle size for monosized spheres. This condition is observed to be true in practice provided that the particle size is much smaller than the spacing between the two sensor electrodes. Equation 11 satisfies the following three limiting conditions:

1. As $C \to 0$, $\lambda_m \to \lambda_f$.
2. As $C \to 1$, $\lambda_m \to \lambda_s$.
3. As $\lambda_s \to \lambda_f$, $\lambda_m \to \lambda_f$, for all solid concentrations.

The second condition can be obtained only with mixtures having an infinitely wide distribution. For monosized particle, C cannot be greater than the maximum packing concentration (C_M). Furthermore, the third condition can be used to measure the conductivity of the dispersed phase by using solutions of known conductivities.

For a mixture of nonconducting spheres ($\lambda_s = 0$) in a conducting liquid, equation 11 reduces to

$$\lambda_m = \lambda_f \left[\frac{2(1 - C)}{2 + C} \right] \quad (12)$$

Effective Conductivity of Concentrated Mixtures. So far, the conductivity of dilute mixtures of random spheres has been considered. This case has defined boundaries and, consequently, equation 10 has a rigorous solution. Unfortunately, a rigorous solution is not possible for random concentrated suspensions where it is difficult to describe the boundaries. Because of this difficulty, it was necessary to introduce more simplifying assumptions. In this section, the most important approaches to predict λ_m are reviewed.

The first approach to estimate the conductivity of concentrated suspensions was introduced by Bruggeman (75). Basically, his derivation is an extension of Maxwell's theory. According to Bruggeman, a suspension of high solids concentration is formed by continuously adding particles (dispersed phase) to the liquid (matrix). The addition process starts with the smallest particles; then, in each step, larger particles are added. At any step, the suspension of smaller particles is treated as a continuum with a conductivity that can be calculated from Maxwell's relationship. The conductivity of the suspension (after adding larger particles) can be determined by applying Maxwell's equation once more. This process is repeated up to the desired solids concentration.

Regarding Bruggeman's assumptions, it is worth noting two points. First, at each step, the suspension of smaller particles is not a continuum. Second, the suspension must have an infinite range of particle sizes. This situation is seldom encountered in practice.

Using these assumptions and applying Maxwell's equation, Bruggeman derived the following implicit equation for λ_m:

$$(\lambda_m - \lambda_s)((\lambda_m/\lambda_f)^{-0.33} = (1 - C)(\lambda_f - \lambda_s) \tag{13}$$

For nonconducting solids in a conducting liquid, equation 13 gives

$$\lambda_m = \lambda_f(1 - C)^{1.5} \tag{14}$$

De La Rue and Tobias (76) measured the conductivities of random suspensions of spheres, cylinders, and sand particles in aqueous solutions of zinc bromide of approximately the same densities as the particles. They found the suspension conductivity could be calculated from the following expression:

$$\lambda_m = \lambda_f(1 - C)^x \tag{15}$$

where $x = 1.5$ for solids concentrations in the range 0.45–0.75. Equation 15 is similar to that of Bruggeman (75).

Begovich and Watson (66) found experimentally that the mixture conductivity in a liquid–solid fluidized bed is proportional to the liquid holdup, that is,

$$\lambda_m = \lambda_f(1 - C) \tag{16}$$

Another linear expression for λ_m was given by Machon et al. (61):

$$\lambda_m = \lambda_f(1 - a_1 C) \tag{17}$$

where a_1 is a constant to be determined experimentally. Machon et al. found this constant by measuring the conductivity of a bed of nonmoving particles. The bed solids concentration was in the range 0.6–0.65. According to equation 17, $\lambda_m = \lambda_f$ for $C = 0$; this result is similar to that obtained using equation 11. However, at $C = 1$, equation 17 does not agree with Maxwell's prediction unless $a_1 = 1$. This observation and the fact that it has no theoretical justification suggest that equation 17 should be used with caution. Other empirical correlations for the mixture conductivity are given by Holdich and Sinclair (67).

A comparison of Maxwell, Burggeman, and Begovich and Watson's expressions to predict the conductivity of solid–liquid mixtures with $\lambda_s = 0$ is given in Table I. This table shows the normalized mixture resistance, $(R_m - R_f)/R_f$, as a function of the dispersed-phase concentration. R_m and R_f are the mixture and fluid resistances, respectively. Maxwell's (73) and Bruggeman's (75) relationships give very similar results at low solid concentrations. However, at higher solid concentrations, Bruggeman's equation gives higher values. Begovich and Watson's (66) expression predicts lower values at all solid concentrations, and the deviation from the other two relationships increases as the concentration is increased. This comparison indicates that the relationship between λ_m and C is not linear, except at very low solid concentrations.

Conductivity methods have been successfully used to measure solid concentration in multiphase systems, where the conductivity of continuous phase is greater than zero, for example, sand–water slurries and oil-in-water emulsions (65). These methods have been used to measure bulk and local solids concentration in the slurry pipelines. Nasr-El-Din et al. (27) developed a four-electrode conductivity probe for measuring local solids concentration in slurry pipelines (Figure 20). The probe is

Table I.   Comparison Between Various
Expressions for $(R_m - R_l)/R_l$

C (%)	Ref. 66	Ref. 73	Ref. 75
10	0.111	0.167	0.171
20	0.25	0.375	0.398
30	0.429	0.643	0.708
40	0.667	1.0	1.52
50	1.0	1.5	1.829
60	1.5	2.05	2.953
70	2.33	3.5	5.086
80	4.0	6.0	10.18
90	9.0	13.5	30.623
100	Infinite	Infinite	Infinite

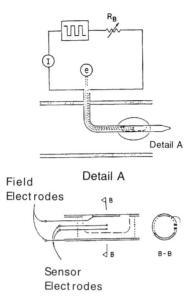

Figure 20.   A conductivity probe for local solids concentration measurement in slurry pipelines. (Reproduced with permission from reference 27. Copyright 1987.)

capable of measuring solid concentration over a small volume of space. The probe was tested against gamma-ray absorption techniques and found to perform fairly well.

Figure 21, from Foster et al. (77), shows the vertical solids concentration profile for sand particles of 0.18 mm mean diameter, a bulk velocity of 2 m/s, and discharge solids concentration of 5 vol%. Good

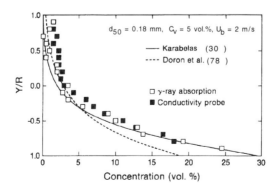

Figure 21.   Concentration profiles obtained with gamma-ray and the conductivity probe. (Reproduced with permission from reference 77. Copyright 1992.)

agreement between the gamma-ray method and conductivity probe measurements was obtained. The figure displays some scatter at the top of the pipe where the solids concentration is very low and both methods are subject to error. The solid and dashed lines of Figure 21 represent predictions based on Karabeles (30) and Doron et al. (78) models, respectively. Details about these models are given later.

It should be mentioned that the accuracy of conductivity methods depends on maintaining a constant conductivity for the working fluid throughout the measurements. The electrical conductivity of electrolytes (e.g., tap water) is a strong function of temperature and ionic strength. Nasr-El-Din et al. (27) showed that increasing temperature or ionic strength of the carrier fluid can cause significant changes in the output of the conductivity probe that they developed. These changes, if not compensated for, can introduce significant errors in solids concentration measurements.

Capacitance Methods. Capacitance methods have been used to measure solids concentration in slurry pipelines (79). This method requires the dielectric constant of the solids and the carrying fluid to be significantly different. Sand–water slurry is a good example to use the capacitance method. In this case, the dielectric constant for water is 80, whereas that of the sand particles is 5. The method relies on the variation of the dielectric constant of the mixture, E_m, with the solids concentration, C. For homogeneous slurries of spherical particles at low solids concentration, Maxwell's correlation can be used to predict the dielectric constant of the mixture. However, several investigators assumed that the relationship of the dielectric constant of the mixture and solids concentration was linear, as follows:

$$E_m = E_f(1 - C) + E_s C \qquad (18)$$

where E_f and E_s are the dielectric constants for the fluid and solid phases, respectively. It should be clearly mentioned that equation 18 can be used only for slurries having solids concentration $<<1\%$. Several correlations to predict the permitivitty of concentrated systems are reviewed by Pal (65) and Louge and Opie (80).

Capacitance methods have been used to measure the concentration of dispersed phase in slurry (79) and emulsion (65) pipelines in which the continuous phases have low conductivity, for example, water-in-oil emulsions. The advantage of the capacitance methods over the conductivity methods is that the effect of temperature on the dielectric constant of water is not significant. According to Pal (65), capacitance methods are not effective to measure the concentration of dispersed phase, when the continuous phase has high conductivity.

Gamma-Ray Absorption Method.   The gamma-ray is an electromagnetic radiation pulse of very short wave. The gamma-ray emission energy will penetrate most materials placed in its path of radiation but is partially absorbed in proportion to the mass of material it passes through. By using this property, the gamma-ray absorption technique was developed for measuring the mass density of an unknown substance.

In the solid–liquid flow applications, medium and low energy gamma-ray sources, such as ^{137}Cs, were most commonly used. When a lead-shielded housing is used as the holder of the source isotope, the small radiation energy is well contained and will not cause any health hazard. Besides the energy source, a gamma-ray density measurement device requires another essential part: gamma-ray detector. The two most common substances used in the detector are NaI (TI) (thallium-activated sodium iodide) and Ge(Li) (lithium drifted germanium).

For solid–liquid systems, the established correlation between the density of matter and absorption of gamma energy can be applied. Mathematically, this correlation is (22)

$$I(E) = I_R(E) \, \exp\{-D_t(\rho_m\bar{\mu}_m E)\} \tag{19}$$

where $I(E)$ and $I_R(E)$ are gamma-ray intensities after penetrating the mixture and the known reference, respectively, D_t is the transmission length, ρ_m is the mean mass density of the mixture, and $\bar{\mu}_m(E)$ is the mass attenuation coefficient of gamma radiation.

When water is used as the carrier fluid, it is more convenient to express the gamma-ray intensity, I, in relation to the intensity recorded for clear water, I_w, in the same system. This relationship is

$$r(E) = \frac{I(E)}{I_w(E)} = \exp\{-D_t[\rho_m\bar{\mu}_m(E) - \mu_w(E)]\} \tag{20}$$

where the mass density of water, ρ_w, is considered as unity in SI units. A solution to equation 20 for the mass density of the mixture, ρ_m, can be expressed as

$$\rho_m = \frac{D_t\mu_w(E) - \ln(r(E))}{D_t\mu_m(E)} \tag{21}$$

Gamma-ray methods have been used to measure solids concentration in slurry pipelines. Usually, the gamma-ray source and detector are mounted in the opposite sides of the pipe as shown in Figure 22. It should be mentioned that the gamma-ray absorption method in this case will measure solids concentration averaged over a chord in the pipe. It is worth noting that the chord-average concentration can be different from local solids concentration in some applications. Examples of such

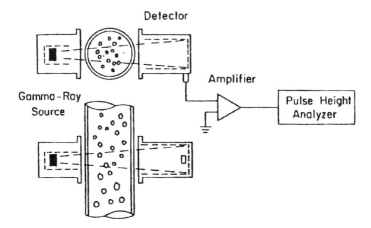

Figure 22.   Gamma-ray assembly for measuring solids concentration in slurry pipelines. (Reproduced with permission from reference 22. Copyright 1979.)

cases are flow of coarse solids in pipelines, where solids are concentrated in the pipe core, and flow of slurry in elbows, where solids are concentrated near the pipe wall (52). To measure local solid concentration in a slurry pipeline, two gamma-ray scans in the x and y directions will be needed. Details regarding this procedure are given by McKibben and Shook (81).

NMR Methods.   Altobelli et al. (82) and Sinton and Chow (83) studied solid velocity and concentration profiles of flowing slurries using NMR flow imaging techniques. These profiles were obtained from the displacement of a tagged slice oriented perpendicular to the flow direction using fast Fourier reconstruction algorithms (83).

NMR imaging methods are noninvasive, can be used for nonconducting fluids, can give 3-dimensional distribution of solids, and can produce local velocity and concentration simultaneously (84). However, more research is needed to improve the spatial and temporal resolutions. Also, these methods are limited to pipes of small diameter.

Velocity Measurement in Slurry Pipelines

Bulk Velocity.   *Electromagnetic Flowmeters.* An electromagnetic flowmeter is based on the principle of electromagnetic induction as expressed by Faraday's law. An electric current is induced in a conductor when it moves across a magnetic field. The flowmeter consists of a pair of electrodes placed diametrically opposite one another on the same section of pipe (Figure 23). A conductive slurry flowing through the

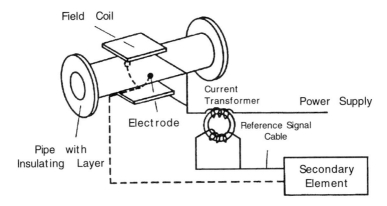

Figure 23.   A conductivity probe for measuring local particle velocity. (Reproduced with permission from reference 88. Copyright 1983.)

meter passes through the magnetic field normally generated by a pair of electric coils of opposing polarities, and a voltage is induced. As the magnetic flux and the length of the conductor have fixed values, the signal voltage can be measured and is directly proportional to the mean flow velocity. Meters can be designed to cope with fluids having as low conductivity as 0.1 μmho cm^{-1}. Thus, most water-based slurries and oil-in-water emulsions give no problem when this meter is used. However, the flow velocity of slurries whose suspending liquid is a hydrocarbon liquid cannot be measured by this meter.

For slurry systems, electromagnetic flowmeters are often installed in a vertical pipe to reduce the risk of solids deposition on one of the electrodes and to avoid inaccuracies caused by asymmetric velocity profiles encountered in horizontal pipes. However, if horizontal mounting is unavoidable, then the meter should never be installed with the electrodes at the ends of a vertical diameter, because the electrode at the top of the pipe would be affected by the occasional air bubble passing (*24*).

Magnetic flowmeters have been used to measure bulk velocity in vertical slurry flow, where the flow is homogeneous. In horizontal slurry flow of highly settling solids, the flow is heterogeneous. Even in this case, Shook and Roco (*25*) indicated that magnetic flowmeters mounted on horizontal slurry pipelines produced accurate velocity measurements.

Other methods to measure bulk velocity in slurry pipelines include venturi meter (*22*) and elbow meter (*85*).

Particle Velocity.   *Cross-Correlation Methods.* The cross-correlation methods involve at least two sensors a known distance apart. These sensors may detect changes in electrical conductivity, capacitance, and transmitted or reflected ultrasound (24). The time of passage of individual particles between the two sensors enables the average particle velocity to be computed. Beck and co-workers (86, 87) developed a cross-correlation technique to measure average particle velocity in slurry pipelines. When nonconducting particles cross an electric field, the resistance of the fluid increases, which creates an electrical signal. If two sensors are placed in this field, two signals, $x(t)$ and $y(t)$, will be generated. The cross-correlation function between the two signals is

$$R_{xy}(\tau) = \operatorname*{LIM}_{T \to \infty} \frac{I}{T} \int_0^T x(t)y(t + \tau)dt \qquad (22)$$

The average transit time, τ_{av}, can be obtained from equation 22. The average particle velocity, v_p, can be calculated from

$$v_p = \frac{L}{\tau_{av}} \qquad (23)$$

where L is the spacing between the two sensors.

Beck and co-workers (86, 87) placed the sensors on the pipe wall. As a result, their measurements give particle velocity averaged over the pipe cross-section. Mounting the sensors on the pipe wall is useful for vertical slurry flow where the velocity profile is uniform. This technique, however, is not so useful for the flow of settling slurries in horizontal pipelines. In this case, the solids are not uniformly distributed over the pipe cross-section, and, as a result, particle velocity is a strong function of position in the pipe.

To measure local particle velocity in slurry pipelines, Brown et al. (88) modified Beck et al.'s (86, 87) conductivity method. They developed a new conductivity probe where four electrodes are mounted on an L-shaped probe. The probe has two field electrodes and two sets of sensor electrodes separated by a known distance (Figure 24). The probe is capable of measuring particle velocity in vertical and horizontal slurry

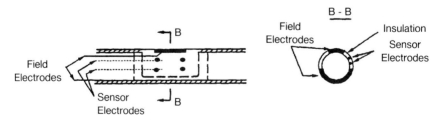

Figure 24.   Doppler ultrasonic flowmeter. (Reproduced with permission from reference 22. Copyright 1979.)

pipelines. It was successfully tested in 50-mm and 0.5-m slurry pipelines (25, 88).

Ultrasonic Flowmeters. Ultrasonic methods have been used to measure flow velocity and concentration in slurry pipelines (22) and emulsion pipelines (65). There are three methods of ultrasonic flow meter applications: transmission of ultrasonic wave, beam deflection, and frequency shift method (22). The frequency shift method (the ultrasonic Doppler flowmeter) consists of a transducer and an electronic control box. The transducer is either clamped on the outside of the pipe or inserted into the pipe so that it is flush with the inside of the pipe wall. The transducer comprises the sensors to transmit and receive the Doppler signal. These sensors are either in a single transducer or in two separate transducers. The control box processes transmitted and received signals (Figure 25).

When the ultrasonic beam is projected into a flowing solid–liquid mixture at an angle θ_i, some acoustic energy is scattered by the suspended particles. The frequency of the reflected acoustic wave can be expressed as

$$f_r = \frac{a - v_p \cos \theta_i}{a + v_p \cos \theta_i} f_t \tag{24}$$

where f_r and f_t are the received and transmitted frequencies, respectively, v_p is the particle velocity, and a is the speed of the acoustic wave. The frequency shift is

$$\Delta f = f_t - f_r = \frac{2 f_t v_p \cos \theta_i}{a + v_p \cos \theta_i} \tag{25}$$

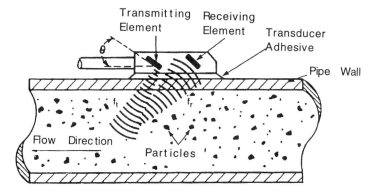

Figure 25. Magnetic flowmeter for measuring bulk velocity in slurry pipelines. (Reproduced with permission from reference 24. Copyright 1988.)

When a is $>>v_p$, the $(v_p \cos \theta_i)$ term can be neglected. Hence, the particle velocity is given by (22)

$$v_p = \frac{a \, \Delta f}{2 \, f_t \cos \theta_i} \tag{26}$$

Modeling of Slurry Flow in Pipelines

Vertical Solids Concentration Profiles in Horizontal Pipes. For slurry flow in a horizontal pipe, solids concentration and particle velocity profiles are not uniform, especially for highly settling slurries. These profiles are functions of particle settling velocity, bulk velocity, pipe diameter, and shape (89). A brief review of various models to predict vertical solids concentration in a horizontal pipe is given later.

The original model used to explain the vertical solids concentration profile for steady fully developed slurry flow is that of Schmidt (90) and Rouse (91):

$$\varepsilon_s \, dc/dy + W_0 c = 0 \tag{27}$$

where c is the local solids concentration, y is the vertical distance measured from the bottom of the pipe, ε_s is a particle diffusion coefficient, and W_0 is the terminal setting velocity of a single particle. For a spherical particle, W_0 is

$$W_0 = \{[4 \, gd_{50}(\rho_s - \rho_f)]/[3 \, \rho_f C_D]\}^{0.5} \tag{28}$$

C_D is the particle drag coefficient for a spherical particle and is expressed as (92)

$$C_D = (24/Re_p)(1 + 0.15 \, Re_p^{0.687}) \text{ for } Re_p \leq 1000$$
$$C_D = 0.44, \text{ for } Re_p > 1000 \tag{29}$$

Re_p is the particle Reynolds number defined as

$$Re = \rho_f W_0 \, D/\mu_f \tag{30}$$

Equation 27 simply states that the solid flux $(W_0 c)$ due to gravity is balanced by the upward flux due to turbulent diffusion $(\varepsilon_s \, dc/dy)$. Equation 27 does not take the volume of the particles into account. Therefore, its validity is limited to suspensions having very low solids concentrations. Hunt, as mentioned by Karabelas (30), extended the diffusion equation by accounting for the presence of the particles and obtained

$$\varepsilon_s \, dc/dy + W_0(1 - c)c = 0 \tag{31}$$

In the limit of $c \ll 1$, equation 31 becomes equation 27. Karabelas (30) integrated equation 31 assuming that $\varepsilon_s = \varepsilon_f$, where ε_f was the fluid momentum eddy diffusivity, and that the solids concentration was a function of the vertical coordinate only. He predicted the vertical solids concentration profile across a horizontal pipe to be

$$c(Y') = [1 + [2I_1(J)/(V J)] \exp(J Y')]^{-1} \tag{32}$$

Y' is the vertical distance measured from the center of the pipe divided by the pipe radius. V is a function of the average solids concentration, C, and is defined as

$$V = C/(1 - C) \tag{33}$$

$I_1 (J)$ is the modified Bessel function of the first order. J is a dimensionless parameter defined as

$$J = W_0/\zeta u_* \tag{34}$$

ζ is a dimensionless particle diffusion coefficient (ε_s/R_{u*}) and u_* is the friction velocity, defined as

$$u_* = U_b(f/2)^{0.5} \tag{35}$$

where U_b is the bulk velocity and f is the friction factor for the flow of a single fluid in a pipe. The friction factor can be calculated from the Colebrook formula (93):

$$1/f^{0.5} = -2.0 \log [(\varepsilon/3.7 D) + 2.51/(f^{0.5} Re)] \tag{36}$$

where ε/D is the pipe roughness and Re is the pipe Reynolds number, defined as

$$Re = \rho_f U_b D/\mu_f \tag{37}$$

Prediction of the vertical concentration profile of concentrated slurries is much more complicated than that given by equation 32. First, the particle settling velocity is a function of solids concentration, which in return is a function of position. Second, ε_s is different from ε_f and is a function, among other factors, of the position in the pipe. To overcome these difficulties, many modifications have been introduced to equation 27. Shook et al. (4) used the hindered settling multiplier $(1 - c)^n$ on the second term of equation 27 to account for the effect of solids concentration on the particle settling velocity. Other researchers (94) integrated equation 27 by assuming mean values for the particle diffusion coefficient and the particle settling velocity. Based on these assumptions, Doron et al. (78) obtained the following expression for the solids vertical concentration profile in a horizontal pipe:

$$c(y) = c(\text{bottom}) \exp(-Wy/\zeta Ru_*) \tag{38}$$

Where W is the particle hindered settling velocity and ζ is a dimensionless particle diffusion coefficient, both are calculated at solids concentration averaged over the pipe cross-sectional area. c (bottom) is the solids concentration at the bottom of the pipe that can be calculated from (78):

$$c(\text{bottom}) = (\pi/2)C \int_{-\pi/2}^{\pi/2} \exp[-(W/\zeta u_*) \sin (y)] \cos^2 (y) \, dy \tag{39}$$

The modifications introduced into the diffusion equation extended the range of its applicability. However, for higher solids concentrations, especially with coarse particles, the solids concentration profile exhibits a positive concentration gradient near the bottom of the pipe (4–6). Prediction of the concentration profile for these cases requires the inclusion of more forces, for example, Bagnold dispersive stresses, to balance the gravity force in equation 27. The effect of these forces on solids concentration profile is discussed in the next section.

Lateral Concentration Profiles in Horizontal Pipes. Previous theoretical studies for horizontal pipe and channel flow have concentrated on the variation of solids concentration in the vertical direction. In this case, gravity (including buoyancy), fluid–particle drag, turbulent diffusion, and particle–particle interaction effects must occur. In this section, the variation of solids concentration in a horizontal plane through the pipe axis is examined.

According to Prandtl (95), the particle diffusion coefficient is related to the liquid momentum eddy diffusivity ε_f defined in terms of the turbulent shear stress (τ_{ij}) and the time-average strain rate (γ_{ij})

$$\gamma_{ij} = -(\nabla V + \nabla V^T); \qquad \rho_f \varepsilon_f = \frac{\tau_{ij}}{\gamma_{ij}} \tag{40}$$

ε_f is a function of position, being given approximately by the expression of Roco and Frasineanu (96):

$$(\varepsilon_f/Du_*) = 0.073(0.54 + r_1^2)(1 - r_1^2) = 0.073 \, g(r_1) \tag{41}$$

where $r_1 = 2r/D$ is the dimensionless radial position.

In dilute open-channel flows, solids concentration profile measurements showed ϵ_s to be greater than ε_f by an amount that varied with particle diameter (97). Pipeline flow measurements at low concentrations (25, 98) also showed differences of this type.

Slurry pipelines usually use concentrations that are much higher than those for which equation 27 was originally proposed and volumetric concentrations in excess of 30 vol% are common. Using a closed rectangular channel which approximated a one-dimensional flow, Shook et al. (4) measured the vertical solids distribution by the gamma-ray absorption method. At higher solid concentrations, values of dC/dy from equation 27 were found to be much too low. These could be corrected approximately with a hindered settling multiplier $(1 - C)^n$ on the particle flux term $(W_0 c)$ in equation 27. The exponent n was the value from the correlation of Richardson and Zaki given in Wallis (92). Although this heuristic correction did allow the diffusion model to be extended, quantitative predictions were still difficult because of doubt concerning the appropriate value of ε_f to be used in a slurry flow, where the mixture density significantly varies with position in the pipe.

A systematic deviation from this modified diffusion equation was observed with coarse particles. This led to a positive concentration gradient near the bottom of the pipe in some circumstances (4–6). This deviation was attributed to the effect of the dispersive stress discovered by Bagnold (99).

To incorporate the dispersive stress effect, the diffusion model must first be written as a form of force balance. Including the hindered settling correction, the appropriate equation 27 would be

$$\left[\frac{3C_D \rho_f \varepsilon_s^2}{4d_{50}C^2(1 - C)^{2n}}\right]\left(\frac{\partial C}{\partial y}\right)\left|\frac{\partial C}{\partial y}\right| = -(\rho_s - \rho_f)\,g \tag{42}$$

The left side of equation 42 represents turbulent diffusion and mixing effects. This balances the immersed weight of the particles in situations where interparticle contacts are not important.

The dispersive stress evidently contributes an additional term to the force balance. This stress is strongly dependent on solids concentration. According to Hanes and Inman (100), this stress requires a finite interparticle shear strain rate and would not exist in a sliding bed of solids. In the latter case, the immersed weight of the particles would be transmitted to the pipe wall by interparticle Coulombic friction. The stress resulting from this type of contact was denoted the "supported load" (94).

Because of these complications, the vertical concentration distributions are difficult to interpret quantitatively. Evidence of a lateral (x-direction) variation has been reported occasionally (5) in conjunction with methods for determining local solids concentration. In this case, gravity and supported load complications would not arise, and the differential equation that describes the lateral solid concentration distribution would be

$$\left[\frac{3C_D \rho_f \varepsilon_s^2}{4d_{50}C^2(1 - C)^{2n)}}\right]\left(\frac{\partial C}{\partial x}\right)\left|\frac{\partial C}{\partial x}\right| = \frac{-1}{C}\left(\frac{\partial \tau_{sxx}}{\partial x}\right) \tag{43}$$

where τ_{sxx} is the dispersive stress.

Previous experimental studies of the dispersive stress used simple shear devices rather than turbulent pipe flows. Bagnold identified two regions of behavior: the macroviscous and the inertial regimes, distinguished by a dimensionless group B, defined as

$$B = \frac{\rho_s d_{50}^2 \lambda_c^{1/2} \gamma}{\mu_f} \tag{44}$$

where λ_c is the linear concentration: the ratio of the particle diameter to the mean separation distance between surfaces. In terms of the maximum solids packing concentration, C_M,

$$\lambda_c = \left[\left(\frac{C_M}{C} \right)^{1/3} - 1 \right]^{-1} \tag{45}$$

For $B < 40$, stresses were considered to be transmitted by fluid friction and

$$\tau_{sxx} = \mu_f f(C) \gamma \tag{46}$$

For $B > 450$, intergranular contact is the transfer mechanism and, according to the results summarized by Hanes and Inman (100),

$$\tau_{sxx} = \beta \rho_s d_{50}^2 F(C) \, (2 + \alpha_c) \gamma \, | \gamma | \tag{47}$$

where

$$\beta = \frac{(1 + e)}{[30(1 - e)]} \tag{48}$$

α_c is a function of solids concentration that specifies the contact process, e is the coefficient of restitution of the particles, and

$$F(C) = \frac{C^2(2 - C)}{(1 - C)^3} \tag{49}$$

From an experimental standpoint, the most convenient plane to examine the lateral concentration variation is the one through the pipe axis. In this case, x can be replaced by radial distances.

Figure 26 shows solids concentration as a function of distance from the pipe axis for slurries of spherical polystyrene beads of a mean diameter of 0.3 mm at a bulk velocity of 3.4 m/s. The concentration reported as C_v is the in situ mean values for the whole pipe (discharge solids concentration). This concentration will differ somewhat from the mean values for the lateral planes through the pipe axis.

At the lower mean solid concentrations, the profiles are flat. However, at the comparatively high concentration of 45 vol%, evidence exists

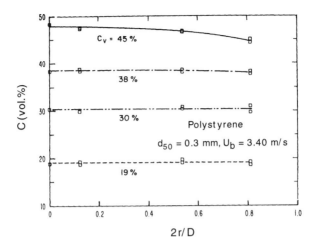

Figure 26. Lateral concentration distribution for 0.3-mm polystyrene particles at 3.4 m/s. (Reproduced with permission from reference 6. Copyright 1987.)

of a concentration variation of the type that could be attributed to the dispersive stress. Measurements could not be taken closer to the pipe wall because of the limitations of the probe's physical size.

Figure 27 displays the lateral concentration variation for slurries of much 1.4-mm polystyrene particles at 3.4 m/s. Although not spherical, these particles were approximately isometric with shapes resembling

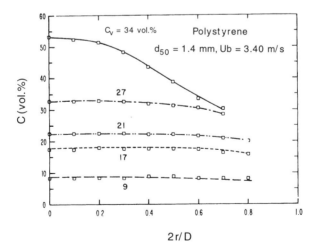

Figure 27. Lateral concentration distribution for 1.4-mm polystyrene particles at 3.4 m/s. (Reproduced with permission from reference 6. Copyright 1987.)

cubes with rounded corners and edges. Figure 27 shows more pronounced migration from the wall than Figure 26.

These results show that for nearly neutrally buoyant particles of 0.3–1.4 mm, comparatively high mean concentrations are required to produce conditions in which lateral migration can be detected in the central core of the pipe.

The sand slurry experiments were important because of the profound difference in the quantity $(S - 1)$: 1.65 compared with 0.05 or 0.06 for polystyrene. Figure 28 illustrates the lateral solids concentration for sand particles of 0.9 mm diameter. Here, the variation of in situ concentration with position was significant even at a solids concentration of 25 vol%.

The effect of the dispersive stress can be detected in vertical concentration profiles, where it occurs in combination with gravity. Figure 29 illustrates the insensitivity of concentration to elevation, resulting from the very low settling velocity of the 0.3-mm polystyrene particles. Even at the highest concentration, the concentration variation is only slightly greater than the scatter in the measurements.

For the 1.4-mm polystyrene particles, the results shown in Figure 30 are more interesting. A significant difference occurs in concentration between the bottom and the top of the pipe at the lowest mean concentration. At this concentration, the lateral variation was shown to be minimal in Figure 26 at a mean concentration of 21 vol%; a steeper concentration gradient occurs near the top of the pipe where the migration tendency acts in combination with gravity. Near the bottom of

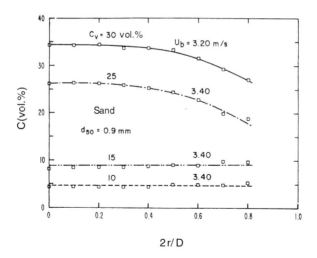

Figure 28. Lateral concentration distribution for 0.9-mm sand at 3.2–3.4 m/s. (Reproduced with permission from reference 6. Copyright 1987.)

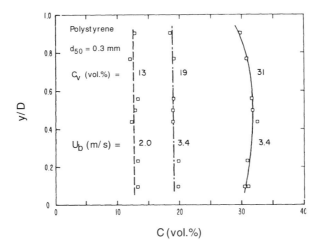

*Figure 29.   Vertical concentration profiles for 0.3-mm polystyrene particles.
(Reproduced with permission from reference 6. Copyright 1987.)*

the pipe, gravity is evidently insufficient to maintain a negative concentration gradient and the dispersive stress causes a concentration reversal. At the highest concentration, the profile is close to symmetrical when the dispersive stress becomes dominant compared with other effects.

Considering these results, from the velocity independence of the profiles and the assumption that ε_s varies as u^*, it can be concluded that the shearing process must have been inertial in these experiments. The time average value of γ for these flows in the region $r_1 > 0.5$ can

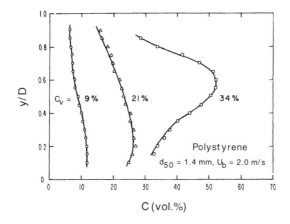

*Figure 30.   Vertical concentration profiles for 1.4-mm polystyrene particles.
(Reproduced with permission from reference 6. Copyright 1987.)*

be estimated from the law of the wall in terms of the von-Karman coefficient κ:

$$\gamma = \frac{2u_*}{\kappa D(1 - r_1)} \tag{50}$$

Velocity distribution measurements in the polystyrene slurries showed that κ in the slurry is somewhat higher than the clear fluid value of 0.4. Using estimated γ values, the Bagnold numbers were <450 in all cases and <40 for the fine particles. It appears Bagnold's criterion does not apply to these turbulent flows.

If the shearing process is inertial, then for $r_1 > 0.5$

$$\frac{\partial C}{\partial r_1}\left|\frac{\partial C}{\partial r_1}\right| = -\beta_1\left(\frac{d_{50}{}^3 S}{D^3 C_D}\right)C(1 - C)^{2n} \cdot g^{-2}(r_1)\left(\frac{\varepsilon_f}{\varepsilon_s}\right)^2\left|\frac{\partial}{\partial r_1}\left[\frac{F(C)(2 + \alpha_c)}{(1 - r_1)^2}\right]\right| \tag{51}$$

where

$$\beta_1 = 16.7\frac{(1 + e)}{\kappa^2(1 - e)} \tag{52}$$

Equation 51 indicates that the profiles change rapidly at high values for r_1. This trend agrees qualitatively with the experiments.

The experimental measurements in Figure 26 indicate little lateral variation at low solids concentrations. This implies that the right hand side of equation 51 should be a strong function of C. However, if one uses previous estimates of ε_s (94) and α_c (100), equation 51 does not predict this strong concentration dependence of $\partial C/\partial r_1$. The most likely source of the discrepancy is in the value of n used for the correction term. Although a multiplier of this type is necessary to explain vertical concentration profiles, it seems likely that the n values from the Richardson–Zaki equation are too high.

Equation 51 is rather cumbersome, especially when there are so many unknown quantities. The simpler approximation below may be useful

$$\frac{-\partial C}{\partial r_1} = \frac{\phi C^m}{(1 - r_1)^p} \tag{53}$$

with m and $p > 1$ and

$$\phi = \phi\left(\frac{d_{50}}{D}, S, C_D, \ldots\right) \tag{54}$$

Only the measurements for the 1.4-mm polystyrene particles showed sufficient variation to define m. Using the C values at $r_1 = 0$ and $r_1 = 0.7$, m was found to be approximately 3.

Because the lateral variation is important for all planes passing through the cross-section, it was desirable to demonstrate that the experimental conclusions were valid at other locations. Chord-average concentrations in 0.18-mm sand slurries were measured with a gamma-ray absorption apparatus, using a horizontal beam of 1 mm diameter. These are compared with the local values, measured at the pipe midplane with the conductivity probe in Figure 31. Below about 40 vol%, little difference occurs between the mean (gamma ray) and the midpoint ($x = 0$) values, implying flat profiles. Above 40 vol%, a significant difference exists, and this difference is similar at considerably different y/D values.

Figure 31 illustrates the strong effect of gravity on the concentration profiles for sand, in comparison with the nearly neutrally buoyant polystyrene. The combined effects of the dispersive stress and hindered settling begin to become significant as solids concentration exceeds 40 vol%. The characteristic reversal occurs near 50 vol%, where the dispersive stress is dominant for these fine particles.

Because it is desirable to have mean values of the correlating coefficients in equation 53 for the whole pipe, measurements of the type shown in Figure 31 are probably more convenient to use in further experimental studies of these effects.

The earlier discussion shed some light into traditional methods for predicting solids concentration profiles in slurry pipeline. These models significantly depend on empirical equations and assumptions that need to be verified. Another way to predict velocity and concentration profiles is to start with the continuity and momentum equations for each phase.

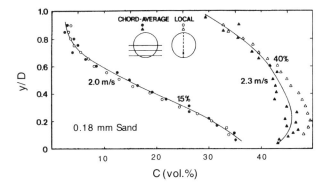

Figure 31. Local and chord-average concentrations for the 0.18-mm sand. (Reproduced with permission from reference 6. Copyright 1987.)

Other terms that describe particle-liquid, particle–particle, and wall effects will be also needed. Conservation of mass and momentum, in addition to interaction terms, result in a set of nonlinear partial differential equations. Such a set of equations can be solved numerically using finite-difference or finite-element techniques.

Computer software is now available to solve such complicated sets of differential equations. However, these packages still need the interaction terms and boundary conditions for both phases. More details about these methods and software are given by Shook and Roco (25), Ding et al. (84), Lyczkowski and Wang (101), Grbarcic et al. (102), Bouillard (103), and Binder and Hanratty (104).

Conclusions

Slurry flow in pipelines is governed by many factors. Depending on the physical properties of the solid particles, the carrying fluid, and operating conditions, the slurry flow can be homogeneous or heterogeneous.

Various techniques are available to measure velocity and solids concentration profiles in slurry pipeline. Sample withdrawal using an L-shaped probe can give a representative sample at isokinetic conditions. Other sample devices will produce significant errors that must be corrected. Conductivity probes can be used to measure local velocity and concentration profiles simultaneously. However, the carrier fluid should be conductive. NMR imaging methods do not disturb the flow with a probe; however, they are limited to pipes of small diameter.

Many experimental techniques and models to predict solids distribution in slurry pipelines are discussed in this chapter. Still more research work, both theoretically and experimentally, is needed to understand such complex flow.

List of Symbols

c local solids concentration (vol%)
C sample solids concentration (vol%)
C_o local upstream solids concentration (vol%)
C_D drag coefficient (dimensionless)
C_M maximum packing solids concentration (vol%)
C_v discharge solids concentration (vol%)
D pipe inside diameter (m)
D_{sm} sampler diameter (m)
d_{50} particle mean diameter (m)
E_f fluid relative permittivity
E_m mixture relative permittivity
E_s solids relative permittivity

f	friction factor (dimensionless)
g	gravitational acceleration (m/s^2)
$I_1(J)$	Bessel function of the first order
J	dimensionless parameter, $W_0/\zeta u_*$
K	particle inertia parameter, $\rho_s d_{50}{}^2\, U_b/18\mu_f\, D_{SM}$
n	Richardson and Zaki's exponent (dimensionless)
$\tilde{N}_{Re}$	modified Reynolds number, defined in equation 3
P	fluid pressure (Pa)
Q	volumetric flow rate (m^3/s)
R	pipe radius (m)
Re	pipe Reynolds number, $\rho_f\, DU_b/\mu_f$
Re$_p$	particle Reynolds number, $\rho_f d_{50} W_0/\mu_f$
R_f	fluid resistance, ohm
R_m	mixture (slurry) resistance, ohm
S	density ratio, ρ_s/ρ_f
t	time (s)
T	probe relative wall thickness (wall thickness/probe radius), dimensionless
U	sampling velocity (m/s)
U_b	bulk velocity in the pipe (m/s)
U_o	local upstream velocity (m/s)
u_*	friction velocity, $U_b(f/2)^{0.5}$ (m/s)
v_p	local particle velocity (m/s)
V_D	deposition velocity (m/s)
V_S	resuspension velocity (m/s)
V_T	transition velocity (m/s)
W_0	terminal setting velocity of a single particle (m/s)
W	hindered settling velocity (m/s)
y	vertical distance measured from the bottom of the pipe (m)
Y	vertical distance measured from the center of the pipe (m)
Y'	vertical distance measured from the center of the pipe, y/R (dimensionless)

Greek

α	solids volume fraction
γ	shear rate (s^{-1})
ε_f	fluid momentum eddy diffusivity (m^2/s)
ε_s	particle diffusion coefficient (m^2/s)
λ	conductivity (S)
μ	viscosity (Pa·s)
$\bar{\mu}_m$	mass attenuation for gamma ray
ρ	density (kg/m^3)
τ	shear stress (Pa)

Subscripts

f fluid
m mixture
p particle
s solids

Acknowledgments

I thank the management of Saudi Aramco Oil Company for permission to publish this paper and Mr. L. Monteiro and Ms. E. B. Bernardo for typing this manuscript.

References

1. Crowe, C. T. *J. Fluids Eng.* **1993**, *115*, 341.
2. Satchwell, R. M.; Sharma, M. P.; Miller, R. L. *J. Energy Res. Tech.*, **1988**, *September*, 141.
3. Wasp, E. J.; Kenny, J. P.; Gandhi, R. L. *Solid-Liquid Flow-Slurry Pipeline Transportation;* Series on Bulk Materials Handling, Trans Tech Publications: Rockport, MA, 1977.
4. Shook, C. A.; Daniel, S. M.; Scott, J. A.; Holgate, J. P. *Can. J. Chem. Eng.* **1968**, *46*, 238–244.
5. Scarlett, B.; Grimley, A. In *Proceedings of the 3rd Conference on the Hydraulic Transport of Solids in Pipes;* BHRA Fluid Engineers: Cranfield, U.K., 1974; paper D3, pp 24–37.
6. Nasr-El-Din, H. A.; Shook, C. A.; Colwell, J. *Int. J. Multiphase Flow* **1987**, *13*, 661.
7. Shah, S. N.; Lord, D. L. *AIChE J.* **1991**, *37*, 863.
8. Wiedenroth, W.; Kirchner, M. *Proceedings of the 2nd International Conference on Hydraulic Transport of Solids in Pipes;* BHRA Fluid Engineers: Cranfield, U.K., 1972; paper E1, pp 1–22.
9. Carleton, A. J.; Cheng, D. C-H. *Proceedings of the 3rd International Conference on Hydraulic Transport of Solids in Pipes;* BHRA Fluid Engineers: Cranfield, U.K., 1974; paper E5, pp 57–74.
10. Thomas, A. D. *Int. J. Multiphase Flow* **1979**, *5*, 113.
11. Hanks, R. W. In *Slurry Flow Technology, II;* Cheremisinoff, N. P., Ed.; Gulf Publishing Company: Houston, TX, 1986.
12. Durand, R.; Condolios, E. In *Proceedings of the Colloquium on the Hydraulic Transport of Coal;* National Coal Board: London, 1952; paper IV, pp 39–52.
13. Wani, G. A. In *Slurry Flow Technology, II;* Cheremisinoff, N. P., Ed.; Gulf Publishing Company: Houston, TX, 1986.
14. Oroskar, A. R.; Turian, R. M. *AIChE J.* **1980**, *26*, 550.
15. Durand, R. *Proceedings of the Colloquium on the Hydraulic Transport of Coal and Solid Materials in Pipes;* National Coal Board: London, 1972.
16. Kao, T. Y.; Wood, D. J. *Trans. Soc. Min. Eng. of AIME* **1974**, *255*, 39.
17. Newitt, D. M.; Richardson, M. A.; Turtle, R. B. *Trans. Inst. Chem. Eng.* **1955**, *33*, 79.
18. Zandi, I.; Govatos, G. *J. Am. Soc. Civ. Eng., J. Hydr. Div.* **1967**, *93*, HY3, 145.

19. Sommerville, D. R. *AIChE J.* **1991,** 37, 274.
20. Roco, M. C.; Shook, C. A. *AIChE J.* **1985,** 31, 1401.
21. Gillies, R. G.; Shook, C. A. *Can. J. Chem. Eng.* **1991,** 69, 1225.
22. Kao, D. T.; Kazanskij, I. *Proceedings of the 4th International Technical Conference on Slurry Transportation;* Slurry Transport Association (now Coal and Slurry Technology Association): Washington, DC, 1979; p 102.
23. Baker, R. C.; Hemp, J. *Fluid Eng. Ser. (BHRA)* **1981,** 8, 3.
24. Heywood, N. I.; Mehta, K. In *Proceedings of the 11th International Conference on the Hydraulic Transport of Solids in Pipes;* BHRA Fluid Engineering Centre: Cranfield, U.K., 1988; paper C2.
25. Shook, C. A.; Roco, M. C. *Slurry Flow Principles and Practice;* Butterworth-Heinemann: Stoneham, MA, 1991.
26. Nasr-El-Din, H. A. In *Encyclopedia of Environmental Control Technology;* Cheremisinoff, N. P., Ed.; Gulf Publishing Company: Houston, TX, 1989; Vol. 3, pp 389–422.
27. Nasr-El-Din, H. A.; Shook, C. A.; Colwell, J. *Int. J. Multiphase Flow* **1987,** 13, 365.
28. Nasr-El-Din, H. A. In *Emulsions: Fundamental and Applications in the Petroleum Industry;* Schramm, L. L., Ed.; Advances in Chemistry 231; American Chemical Society: Washington, DC, 1992; pp 171–217.
29. Rushton, J. H.; Hillestad, J. Y. American Petroleum Institute: Washington, DC, 1964; paper 52–64, p 517.
30. Karabelas, A. J. *AIChE J.* **1977,** 23, 426.
31. Hayashi, H. A.; Sampei, T.; Oda, S.; Ohtomo, S. *Proceedings of the 7th International Conference on Hydraulic Transport of Solids in Pipes;* British Hydrodynamics Research Association: Cranfield, UK, 1980; paper D2, p 149.
32. Iinoya, K. *Kagaku Kogaku Ronbunshu* **1970,** 34, 69.
33. Akers, R. J.; Stenhouse, J. I. T. *Proc. Inst. Mech. Eng.* **1976,** 45.
34. Nasr-El-Din, H. A.; Shook, C. A.; Esmail, M. N. *Can. J. Chem. Eng.* **1984,** 62, 179.
35. Fuchs, N. A. *J. Atmos. Environ.* **1975,** 9, 698.
36. Stevens, D. C. *J. Aerosol Sci.* **1986,** 17, 729.
37. Nasr-El-Din, H. A.; Shook, C. A.; Esmail, M. N. *Can. J. Chem. Eng.* **1985,** 63, 746.
38. Nasr-El-Din, H. A.; Afacan, A.; Masliyah, J. H. *Chem. Eng. Commun.* **1989,** 82, 203.
39. Nasr-El-Din, H. A.; Afacan, A.; Masliyah, J. H. *Int. J. Multiphase Flow* **1989,** 15, 659.
40. Rushton, J. H. *AIChE Symp. Ser.* **1965,** 10, 3.
41. Barresi, A.; Baldi, G. *Chem. Eng. Sci.* **1987,** 42, 2949.
42. Sharma, R. N.; Das, H. C. L. *Coll. Czech. Chem. Commun.* **1980,** 45, 3293.
43. MacTaggart, R. S.; Nasr-El-Din, H. A.; Masliyah, J. H. *Chem. Eng. Sci.* **1993,** 48, 921.
44. Moujaes, S. F. *Can. J. Chem. Eng.* **1984,** 62, 62.
45. Torrest, R. S.; Savage, R. W. *Can. J. Chem. Eng.* **1975,** 53, 699.
46. Nasr-El-Din, H. A; Shook, C. A. *Int. J. Multiphase Flow* **1986,** 12, 427.
47. Nasr-El-Din, H. A.; Afacan, A.; Masliyah, J. H. *Trans. Inst. Chem. Eng. Part A, Chem. Eng. Res. Des.* **1991,** 69, 374.
48. Whitely, A. B.; Reed, L. E. *J. Inst. Fuel* **1959,** 32, 316.
49. Belyaev, S. P.; Levin, L. M. *J. Aerosol Sci.* **1972,** 3, 127.

50. Yoshida, H.; Yamashita, K.; Masuda, H.; Iinoya, K. *J. Chem. Eng. Jpn.* **1978**, *11*, 48.
51. Nasr-El-Din, H. A.; Shook, C. A. *J. Pipeline Div. Am. Soc. Civ. Eng.* **1985**, 5, 113.
52. Nasr-El-Din, H. A.; Shook, C. A.; Colwell, J. *Hydrotransport* **1986**, *10*, 191.
53. Ito, H. *JSME Int. J.* **1987**, *30*, 543.
54. Lundgren, D. A.; Durham, M. D.; Mason, K. W. *Am. Ind. Hyg. Assoc. Q.* **1978**, *39*, 640.
55. Ayukawa, K. *Bull. JSME* **1969**, *12*, 1388.
56. Toda, M.; Ishkawa, T.; Sait, S.; Maeda, S. *J. Chem. Eng. Jpn.* **1973**, 6, 140.
57. Toda, M.; Komori, N.; Sait, S.; Maeda, S. *J. Chem. Eng. Jpn.* **1972**, 5, 4.
58. Nasr-El-Din, H. A.; Shook, C. A. *J. Pipeline Div. Am. Soc. Civ. Eng.* **1987**, 6, 239.
59. Colwell, J. M.; Shook, C. A. *Can. J. Chem. Eng.* **1988**, *66*, 714.
60. Lee, K. T.; Beck, M. S.; McKeown, K. M. *Meas. Control* **1974**, 7, 341.
61. Machon, V.; Fort, I.; Skrivanek, J. *Proceedings of the 4th European Conference on Mixing*; BHRA Fluid Engineering Centre: Cranfield, U.K., 1982; p 289.
62. MacTaggart, R. S.; Nasr-El-Din, H. A.; Masliyah, J. H. *Sep. Technol.* **1993**, 3, 151.
63. Ong, K. H.; Beck, M. S. *Meas. Control* **1975**, 8, 453.
64 Pal, R.; Rhodes, E. *Proceedings of the 3rd Multi-Phase Flow and Hear Transfer Symposium*; Clean Energy Research Institute: Coral Gables, FL, 1983.
65. Pal, R. *Colloids Surf.* **1994**, *84*, 141.
66. Begovich, J. M.; Watson, J. S. *AIChE J.* **1978**, *24*, 351.
67. Holdich, R. G., Sinclair, R. G. *Powder Technol.* **1992**, *72*, 77.
68. Uribe-Salas, A; Gomez, C. O., Finch, J. A. *Chem. Eng. Sci.* **1994**, *49*, 1.
69. Perez-Rosales, C. *J. Pet. Technol.* **1976**, *28*, 819.
70. Perez-Rosales, C. *J. Soc. Pet. Eng. J.* **1982**, *22*, 531.
71. Sweeney, S. A.; Jennings, H. Y. Jr. *J. Phys. Chem.* **1960**, *64*, 551.
72. Keller, G. V. *Oil Gas J.* **1953**, 62.
73. *A Treatise on Electricity and Magnetism*, 3rd ed.; Maxwell, J. C., Ed.; Dover: New York, 1954; Vol. 1, Article 314.
74. Lord Rayleigh *Phil. Mag.* **1892**, *34*, 481.
75. Bruggeman, D. A. G. *Ann. Physik* **1935**, *24*, 636.
76. De La Rue, R. M.; Tobias, C. W. *J. Electrochem. Soc.* **1959**, *106*, 827.
77. Foster, J; Nasr-El-Din, H. A.; Masliyah, J. H. *Can. J. Chem. Eng. J.* **1992**, *70*, 3.
78. Doron, P.; Granica, D.; Barnea, D *Int. J. Multiphase Flow* **1987**, *13*, 535–549.
79. Keska, J. In *Proceedings of the 5th International Conference on the Hydraulic Transport of Solids in Pipes*; BHRA Fluid Engineering Centre: Canfield, U.K., 1978; paper G-5.
80. Louge, M.; Opie, M. *Powder Technol.* **1990**, *62*, 85.
81. McKibben, M.; Shook, C. A. *J. Pipeline Div. Am. Soc. Civ. Eng.* **1987**, 6, 291.
82. Altobelli, S. A., Givler, R. C.; Fukushima, E. *J. Rheol. (N. Y.)* **1991**, *35*, 721.
83. Sinton, S. W.; Chow, A. W. *J. Rheol.* **1991**, *35*, 735.
84. Ding, J; Lyczkowski R. W.; Sha, W. T., Altobelli, S.; Fukushima, E. *Powder Technol.* **1993**, *77*, 301.

85. Colwell, J. M.; Shook, C. A. *J. Pipeline Div. Am. Soc. Civ. Eng.* **1989**, *7*, 243.
86. Beck, M. S.; Calvert, G.; Hobson; J. H.; Lee, K. T.; Mendies, P. J. *Meas. Control* **1971**, *4*, 133–138.
87. Beck, M. S.; Green, R. G.; Hammer, E. A.; Thorn, R. *Inst. Meas. Control* **1983**, *16*, 1–12.
88. Brown, N. P.; Shook, C. A.; Peters, J.; Eyre, D. *Can. J. Chem. Eng.* **1983**, *61*, 597.
89. Rasteiro, M. G.; Rebola, M. M.; Scarlett, B. *Hydrotransport* **1988**, *11*, 49–61.
90. Schmidt, W. *Probl. Kosm. Phys.* **1925**, 7.
91. Rouse, H. *Trans. Am. Soc. Civ. Eng.* **1937**, *102*, 463–505.
92. Wallis, G. B. *One-Dimensional Two-Phase Flow*; McGraw-Hill: New York, 1969.
93. Gerhart, P. M.;. Gross, R. J. *Fundamentals of Fluid Mechanics*; Addison-Wesley: Reading, MA, 1985.
94. Roco, M. C.; Shook, C. A. *Can. J. Chem. Eng.* **1983**, *61*, 494–503.
95. Prandtl, L. *Essentials of Fluid Dynamic*; Blackie: London, 1952.
96. Roco, M. C.; Frasineanu, G. *Stud. Cercet. Mec. Apl.* **1977**, *36*, 311.
97. Raudkivi, A.J. *Loose Boundary Hydraulic*; Pergamon: Oxford, England 1967.
98. Sharp, B. B.; O'Neill, I. C. *J. Fluid Mech.* **1971**, *45*, 575–584.
99. Bagnold, R. A. *Proc. R. Soc. Ser. A* **1954**, *225*, 49–64.
100. Hanes, D. M.; Inman, D. L. *J. Fluid Mech.* **1985**, *150*, 351–380.
101. Lyczkowski, R. W.; Wang, C. S. *Powder Technol.* **1992**, *69*, 285.
102. Grbarcic, Z. B.; Garic, R. V.; Vukovic, D. R.; Hadzismajlovic, Dz. E.; Littman, H.; Morgan, M. H.; Jovanovic, S. Dj. *Powder Technol.* **1992**, *72*, 183.
103. Bouillard, J. X. *Powder Technol.* **1994**, *78*, 99.
104. Binder, J. L.; Hanratty, T. T. *AIChE J.* **1993**, *39*, 1581.

RECEIVED for review July 28, 1994. ACCEPTED revised manuscript June 7, 1995.

Principles of Single-Phase Flow Through Porous Media

Shijie Liu and Jacob H. Masliyah*

Department of Chemical Engineering, University of Alberta,
Edmonton T6G 2G6, Canada

Porous media are both permeable and dispersive to a traversing fluid. Flow of a single-phase fluid in porous media is not only of practical interest but also of fundamental significance in characterizing the porous media. In this chapter, the characteristics of porous media are introduced from both fundamental and application points of view. A continuum approach is used. The volume-averaged equations are used to describe the flow, where the momentum dispersion has been neglected. The relations between Darcy's law–Brinkman's equation and the volume-averaged Navier–Stokes equation are described. The Forchheimer hypothesis, Ergun equation, and Liu–Afacan–Masliyah equation are briefly described in terms of coupling of the viscous and inertial effects on the single-phase flow in porous media. Discussions are provided on the concept and modeling of areal porosity, tortuosity, permeability, and shear factor. A curved passage model is discussed in terms of the shear factor and pressure-drop modeling for flow through porous media. Bounding wall effects are discussed through a simple approach. Examples of flow simulations in porous media (i.e., slightly compressible flow in oil reservoirs and incompressible flow in fixed beds) are provided.

Definitions

Porous medium is a material consisting of a solid matrix with interconnected pores. The interconnected pores are responsible for allowing a fluid to traverse through the material. For the simplest situation, the medium is saturated with a single fluid ("single fluid flow"). In "multiphase fluid flow," several fluids (liquids and/or gas) share the open pores. Porous media are classified as unconsolidated and consolidated.

* Corresponding author.

0065–2393/96/0251–0227$22.00/0

Beach sand, packed or fixed beds, soil, and gravel are unconsolidated porous media. Cloth, most naturally occurring rocks such as sandstone and limestone, concrete, bricks, paper, and wood are consolidated porous media.

The most common way of deriving the laws governing the average or macroscopic variables is to begin with the standard continuity and momentum equations and to average them over volumes or areas containing many pores. A macroscopic variable is defined by an appropriate average over a sufficiently large representative elementary volume (REV). REV is a conceptual space unit that bears the same meaning as the physical "point" for fluid continuum. An REV is the minimum space unit at which the porous medium of concern can be treated as a continuous medium. At continuum, the porous medium grain or pore structure is invisible. The length scale of the REV is larger than the grain or pore scale but smaller than the length scale of the entire flow domain.

The porosity, ϵ, of a porous medium is the fraction of the total volume of the medium that is occupied by the open pores. The porosity is an average quantity, or a bulk property. In practice, only the effective porosity, that is, only the interconnected pores, are useful for traversing fluids. Dead ends and isolated pores have a negligible effect on a single fluid flow. Media that have the same average value of porosity may be very different in their pore structure and flow capacity. A common method for measuring the porosity of a porous medium is the liquid displacement test. The basic assumption for the success of the liquid displacement test is that the open pores can be filled or replaced by a working liquid. By measuring the weight difference between the porous medium itself and with a liquid saturated in the pores, the void space can be calculated.

For natural porous media, the porosity does not normally exceed 0.6. For petroleum reservoirs, the porosity is typically between 0.1 and 0.4. For beds of solid spheres of uniform diameter, the porosity can vary between the rhombohedral packing of 0.2595 and the cubic packing of 0.4764. Nonuniform grain size tends to yield smaller porosities than uniform grains. Industrial random packs and porous foams, on the other hand, can offer a porosity of as high as 0.99. Table I shows the porosities of common porous materials.

The specific surface or surface per unit volume, a_P, of a porous medium is defined as the ratio of the total open pore surface area to the volume of the solids. The equivalent spherical diameter, d_s, is the diameter of an equivalent sphere that has the same surface area per unit volume of the solid material forming the porous medium.

The flow properties that describe the matrix from the view point of a flowing fluid are sometimes called pseudotransport properties, such as the permeability, dispersion coefficient, and tortuosity. The perme-

Table I. Properties of Common Porous Materials

Material	Porosity ϵ	Permeability k (m^2)	Surface Area per Unit Volume a_p (m^2/m^3)
Agar-agar	0.57 ~ 0.66	2×10^{-14} ~ 4.4×10^{-13}	7×10^5 ~ 8.9×10^5
Black slate powder	0.12 ~ 0.34	4.9×10^{-14} ~ 1.2×10^{-13}	
Brick		4.8×10^{-15} ~ 2.2×10^{-13}	
Catalyst (Fischer-Tropsch, granules)	0.45		5.6×10^7
Coal	0.02 ~ 0.12		
Concrete (bituminous)		1×10^{-13} ~ 2.3×10^{-11}	
Concrete (ordinary mixes)	0.02 ~ 0.07		
Copper powder (hot compacted)	0.09 ~ 0.34	3.3×10^{-10} ~ 1.5×10^{-9}	
Cork board		2.4×10^{-11} ~ 5.1×10^{-11}	
Crushed mica	0.86 ~ 0.94		
Crushed halite	0.43 ~ 0.52		
Crushed quartz	0.41 ~ 0.54		
Crushed calcite	0.40 ~ 0.55		
Fibreglass	0.88 ~ 0.93		5.6×10^4 ~ 7.7×10^4
Granular crushed rock	0.45		
Limestone	0.015 ~ 0.2	1×10^{-18} ~ 4.5×10^{-10}	
Random packing; Berl saddles	0.60 ~ 0.83	1×10^{-7} ~ 3.9×10^{-7}	105 ~ 899
Random packing; Intalox saddles	0.75 ~ 0.94		89 ~ 894
Random packing; Pall rings	0.87 ~ 0.97		85 ~ 345
Random packing; Raschig rings	0.69 ~ 0.95		62 ~ 787
Random packing; Spheres	0.36 ~ 0.42		
Random packing; Super Intalox	0.79 ~ 0.94		89 ~ 253
Sand	0.37 ~ 0.50	2×10^{-11} ~ 1.8×10^{-10}	1.5×10^4 ~ 2.2×10^4
Sandstone	0.08 ~ 0.38	5×10^{-16} ~ 3×10^{-12}	
Silica grains	0.65		
Silica powder	0.37 ~ 0.49	1.3×10^{-14} ~ 5.1×10^{-14}	6.8×10^5 ~ 8.9×10^5
Soil	0.43 ~ 0.54	2.9×10^{-13} ~ 1.4×10^{-11}	
Wire crimps	0.68 ~ 0.76	3.8×10^{-9} ~ 1×10^{-8}	2.9×10^3 ~ 4×10^3

SOURCE: Data taken from reference 1.

ability of a porous medium is its conductance to flow. The dispersion coefficient describes the mixing caused by a fluid flowing through an interconnected network of pores. The tortuosity is the ratio of the distance between two fixed points to the free pathway length in a porous medium sample. The tortuosity measures the influence of the pore structure on the conductivity of the sample. The capillary pressure represents the interfacial forces resulting from constrictions. These properties are bulk properties and apply to media of a finite size that contain enough pores to yield an average value.

Darcy's Law

Henri Darcy's (2) investigations into the hydrology of the fountains of Dijon and his experimental studies on steady unidirectional flow in a uniform medium revealed a proportionality between flow discharge rate and the applied pressure drop. This linear relationship is referred to as Darcy's law.

Darcy's law is expressed as

$$\frac{Q}{A_t} = q = -\frac{k}{\mu}\frac{\Delta p}{L} \tag{1}$$

where Q is the flow rate, A_t is the total cross-sectional area of the porous medium, μ is the dynamic viscosity of the fluid, Δp is the pressure drop across the porous medium of thickness L, and k is the intrinsic permeability of the porous medium. The permeability can be measured by forcing a known volumetric flow rate of fluid through a porous medium of known length and cross-sectional area.

Values of k for natural porous materials vary widely. Typical permeability values for soils are clean gravel $10^{-9}\sim10^{-7}$ m^2, clean sand $10^{-11}\sim10^{-10}$ m^2, peat $10^{-13}\sim10^{-11}$ m^2, stratified clay $10^{-16}\sim10^{-13}$ m^2, and unweathered clay $10^{-20}\sim10^{-16}$ m^2. Table I shows typical values of permeability. In some literature, especially in the early studies, the unit of Darcy (cm^{2}cP/atms $= 0.987 \times 10^{-12}$ m^2) is also used.

Although the permeability k is supposed to be a property of the porous medium, experimental studies often indicate a larger value for gases than for liquids, especially for media having small pores. This phenomenon appears to be a difference in the physics of the fluid–solid interface at the microscopic level. Liquids usually obey no-slip boundary conditions at the solid wall, whereas gases may slip along the solid wall because of the small scale of the pore cross-section (1, 3). The resulting discrepancy in the macroscopic permeabilities is known as the Klinkenberg effect, and, in general, the apparent intrinsic permeability k_{app} can be expressed as

$$k_{app} = k + \frac{k^*}{\rho} \qquad (2)$$

where ρ is the density of the fluid and k^* is a parameter dependent on both the fluid and the porous medium properties. The values of k^* are determined experimentally. For liquids, the values of k^* are very small, and one can treat $k^* = 0$.

Equation 1 has been generalized to, and has been widely used for, multidimensional flows. In a multidimensional situation, the permeability k is replaced by a second-order tensorial permeability $\underline{\underline{k}}$ that is dependent on the directional properties of the pore structure. The quantity q is replaced by a superficial velocity vector $\mathbf{v}$. Darcy's law in a multidimensional form is given by

$$\mathbf{v} = -\mu^{-1}\underline{\underline{k}} \cdot \nabla p \qquad (3)$$

Darcy's law was empirically derived from a unidirectional flow and lacks the flow viscous diffusion effects, that is, $\nabla^2 \mathbf{v}$ term, and hence it is not valid at the interface of a porous medium–solid and porous medium–free flow.

Brinkman's Equation

Brinkman (4) added a diffusion term to the Darcy's law, leading to

$$\nabla p = -\mu\underline{\underline{k}}^{-1} \cdot \mathbf{v} + \breve{\mu}\nabla^2 \mathbf{v} \qquad (4)$$

where $\breve{\mu}$ is an effective viscosity. He also found that the effective viscosity $\breve{\mu}$ should be the same as the fluid viscosity μ to accommodate for the available experimental data. Brinkman's equation becomes

$$\nabla p = -\mu\underline{\underline{k}}^{-1} \cdot \mathbf{v} + \mu\nabla^2 \mathbf{v} \qquad (5)$$

Equation 5 has been used for multidimensional flows by many investigators as the governing flow equation for flow through porous media, including Neale et al. (5) and Nandakumar and Masliyah (6). Brinkman's equation is, like Darcy's law, inertial-free and hence valid only for very weak flows. Few theoretical studies gave some heuristic accountability for Brinkman's equation by using volume averaging of the Navier–Stokes equation for flow in porous media (7–9). However, these studies gave different viscosities to the second term on the right side of Brinkman's equation 5. The traditional volume averaging technique assumed that the pressure (9) or the pressure gradient (7, 8) inside the solid material was zero and would influence the flow in the pores.

Forchheimer Hypothesis

To account for the nonlinear behavior of the flow in porous media, Forchheimer (10) hypothesized that the pressure drop for flow in a packed bed is a direct result from the viscous (linear in origin) and the inertial (quadratic) effects. Forchheimer's hypothesis in a multidimensional form can be expressed as

$$-\nabla p = \mu \underline{\alpha} \mathbf{v} + \rho |\mathbf{v}| \underline{\beta} \cdot \mathbf{v} \qquad (6)$$

where $\underline{\alpha}$ and $\underline{\beta}$ are constant tensors of second order, and like the permeability $\underline{k}$, they depend on the structure of the porous medium. The Forchheimer hypothesis has been generally accepted as an extension to Darcy's law for high flow rates. The Forchheimer relation attracted much attention in the literature. The qualitative relation was heuristically obtained by many authors through various means. Among others, Blick (11) used a model of a bundle of parallel capillary tubes with orifice plates spaced along the tubes. A balance of static forces was applied to obtain a Forchheimer-type relationship. Coulaud et al. (12) obtained the Forchheimer relation by correlating their own numerical results on flow across bundles of cylinders arranged in a regular pattern. These authors obtained a deterministic equation in which the coefficients can be determined without the need for direct experiments. Other investigators used the continuum approach to arrive at the Forchheimer relation, for example, Dullien and Azzam (13), Barak and Bear (14), and Cvetkovic (15). In general, the continuum approach was used as a means to explain the physics behind the Forchheimer relation and not to quantify the pressure drop for flow through porous media. The form of Forchheimer equation is not uniquely defined through volume averaging based on different hypotheses (16). To rigorously quantify the pressure drop for flow-through porous media, the process involved with the continuum approach would be formidable.

It should be noted that Ward (17) attempted to cast equation 6 into a more usable form by relating $\underline{\alpha}$ and $\underline{\beta}$ to the known porous medium properties for isotropic porous media. In a one-dimensional situation, Ward's model gave

$$\alpha = \frac{1}{k}, \qquad \beta = \frac{c_F}{k^{1/2}} \qquad (7)$$

where c_F is the form-drag coefficient of the solid material. Ward (17) assumed that c_F was a universal constant and equal to 0.55. Joseph et al. (18) considered Ward's model to be the appropriate modification to Darcy's law. Equations 6 and 7 have been regarded more useful than all other models by Nield and Bejan (19). However, the fact is that c_F

changes significantly when the porous medium is changed (*19*). Hence, Ward's model does not improve the utility of the Forchheimer hypothesis, where two different unknown parameters, k and c_F, are used.

Ergun Equation

Equations 1, 5, and 6 contain unknown constants k, α, and β that are properties of the porous media. Hence, the previously mentioned equations are strictly speaking qualitative expressions and have to a large extent a limited scope of utility. Kozeny (*20*) introduced a model that represents the porous media as bundles of straight passages. The Kozeny's theory was consequently revised by Carman (*21*). With the Kozeny–Carman theory, the porous media properties can be accounted for in the pressure drop–flow rate relationship. One-dimensional pressure drop of flow through porous media has been correlated semiempirically by various investigators based on the Kozeny's theory and Forchheimer's hypothesis. The most notable study is by Ergun (*22*). The correlation given by Ergun (*22*) received most attention and has been termed the Ergun equation, which can be written as

$$-\frac{\Delta p}{L}\frac{d_s^2\epsilon^3}{\mu q(1-\epsilon)^2} = 150 + 1.75\frac{d_s q\rho}{(1-\epsilon)\mu} \tag{8}$$

where d_s is the equivalent spherical diameter of a representative unit solid element (particle). For a bed of uniform particles, d_s is given by

$$d_s = \frac{6}{a_P} \tag{9}$$

The Ergun equation was empirically obtained based on a more general form of the dimensionless pressure drop equation, namely, the Blake–Kozeny equation, which was also theoretically derived by Irmay (*23*). Irmay's model yields

$$-\frac{\Delta p}{L}\frac{d_s^2\epsilon^3}{\mu q(1-\epsilon)^2} = a_E + b_E\frac{d_s q\rho}{(1-\epsilon)\mu} \tag{10}$$

where a and b are shape factors of the porous medium.

The Ergun equation has been revised by various investigators, to name a few, Wentz and Thodos (*24*), Handley and Heggs (*25*), Tallmadge (*26*), Hicks (*27*), and Jones and Krier (*28*). A recent and a more accepted correlation (modified Ergun equation) is given by MacDonald et al. (*29*). The Ergun equation has also been used in estimating the permeability of unconsolidated porous media (*30*). However, MacDonald et al. (*29*) noted that the Ergun equation is valid over a certain porosity range and is not valid for the entire porosity range. This may be attributed to the oversimplification in the Kozeny–Carman theory in which parallel straight passages were assumed as the internal structure of a porous medium.

Liu–Afacan–Masliyah (LAM) Equation

Although Ergun equation is widely accepted in predicting the pressure drop for flow-through porous media, it is a known fact that the Ergun equation or its modified forms overpredict the pressure drop by as much as 100% at high porosity and underpredict the pressure drop by as much as 300% for low porosity medium such as sandstones (31). A more accurate equation has been developed by Liu et al. (32) based on a revised Kozeny–Carman theory.

Let us define a normalized pressure drop factor, f_v, by

$$f_v = -\frac{\Delta p}{L} \frac{d_s^2 \epsilon^{11/3}}{\mu q (1 - \epsilon)^2} \tag{11}$$

For a wide range of porosity and different type of porous media, Liu et al. (32) obtained the following relation for a unidirectional flow (one-dimensional):

$$f_v = 85.2 + 0.69\, Re_m \frac{Re_m^2}{16^2 + Re_m^2} \tag{12}$$

where Re_m is the modified Reynolds number and is defined by

$$Re_m = \frac{1 + (1 - \epsilon^{1/2})^{1/2}}{(1 - \epsilon)\epsilon^{1/6}} \frac{d_s \rho q}{\mu} \tag{13}$$

Equation 12 is valid for both Darcy and Forchheimer flow regimes. Figure 1 shows a general sketch of f_v variation with Re_m for a porous medium.

Since equation 12 is valid when Darcy's law is applicable, the equivalent spherical particle diameter, d_s, and the intrinsic permeability, k, can be related by

$$d_s = 9.23(1 - \epsilon)\epsilon^{-11/6} k^{1/2} \tag{14}$$

It is helpful to write LAM equation 12 in the absence of the grain size. In terms of the porosity and the intrinsic permeability, Liu et al.'s model gives

$$-\frac{k}{\mu} \frac{\Delta p}{L} = \left(1 + 0.0081 Re_m \frac{Re_m^2}{16^2 + Re_m^2}\right) q \tag{15}$$

and

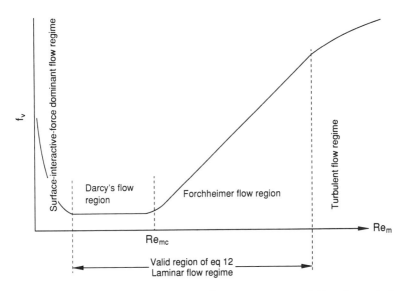

Figure 1.   Normalized pressure drop factor variation with Re_m *for a single fluid flow in a porous medium.*

$$Re_m = 9.23 \frac{1 + (1 - \epsilon^{1/2})^{1/2}}{\epsilon^2} \frac{k^{1/2} \rho q}{\mu} \qquad (16)$$

This model has been tested for packed beds of monosized spheres for porosities as low as $\epsilon = 0.36$ and for porous fiber mats having a porosity as high as $\epsilon = 0.94$. We modify this model to give an improved accuracy for flow through consolidated sandstones or rocks.

Volume-Averaged Navier–Stokes Equation

The governing equations for a single Newtonian fluid flow in an infinitely permeable medium are given by

$$\frac{\partial \rho}{\partial t} + \nabla \cdot (\rho \mathbf{v}^*) = 0 \qquad (17)$$

and

$$\frac{\partial (\rho \mathbf{v}^*)}{\partial t} + \nabla \cdot (\rho \mathbf{v}^* \mathbf{v}^*) + \nabla p^* - \mu \nabla^2 \mathbf{v}^* = 0 \qquad (18)$$

where the superscript * is used to denote microscopic values. Equation 17 is usually referred to as the continuity equation, and equation 18 is the momentum equation.

The volume-averaging technique has been developed by Whitaker (7), Slattery (8), Bear (33), and Lundgren (9). Assuming that the solid matrix of the porous medium is incompressible, immobile, and not supported by the fluid, equations 17 and 18 can be subjected to a volume-averaging procedure in an REV and reduce to the following equations (32):

$$\epsilon \frac{\partial \rho}{\partial t} + \nabla \cdot (\rho \mathbf{v}) = 0 \tag{19}$$

and

$$\frac{\partial(\rho \mathbf{v})}{\partial t} + \nabla \cdot (\rho \mathbf{v}\mathbf{v}/\epsilon) + \nabla p - \mu \nabla^2 \mathbf{v} + \mu F \mathbf{v} = 0 \tag{20}$$

where F is the shear factor that needs to be defined. The shear factor can be related to the one-dimensional pressure drop of the porous medium locally through the following relationship:

$$F = \frac{\Delta p}{\mu L |\mathbf{v}|} \tag{21}$$

When the model of Liu et al. (32) is invoked, equation 21 takes the following form:

$$F = f_v \frac{(1-\epsilon)^2}{d_s^2 \epsilon^{11/3}} = \frac{f_v}{85.2k} \tag{22}$$

Unlike conventionally used volume-averaged equations, equation 20 does reduce to the Brinkman equation 5 when the inertial effects are negligible and to Darcy's law, equation 3, when both inertial and multidimensional effects are negligible.

Equation 20 shows that a porous medium is permeative, that is, a shear factor exists to account for the microscopic momentum loss. Our preliminary study recently reveals that, however, a porous medium is not only permeative but dispersive as well. The dispersivity of a porous medium has been traditionally characterized through heat transfer (in a single- or multifluid flow) and mass transfer (in a multifluid flow) studies. For an isothermal single-fluid flow, the dispersivity of a porous medium is characterized by a flow strength and a porous medium property-dependent apparent viscosity. For simplicity, we discuss the single-fluid flow behavior in this chapter without considering the dispersivity of the porous medium.

Scope

Flow in porous media can be classified as a single- or multiphase flow. In a limiting sense, single-phase flow need not be considered alone because multiphase flows may be degenerated to a single-phase flow by invoking a mixture or homogeneous model. However, modeling of a single-phase flow is simpler when considered by itself other than as a limiting case. For isothermal homogeneous multiphase or single-phase flow in the absence of inertial effects and bounding wall effects, there is one single pseudotransport coefficient, the permeability k that characterizes the porous medium. The flow in a porous medium is also governed by its strength or the modified Reynolds number, Re_m. In cases where the temperature does vary, a thermal dispersion coefficient and a thermal shear factor are also required to characterize the flow. Normally, the thermal F-factor (shear factor) is neglected.

Multiphase flow can be further divided into two categories: miscible and immiscible flows. For a miscible flow, the mixing is described by an additional pseudotransport coefficient, the dispersion coefficient K. For an immiscible flow, the capillary pressure p_c or relative permeability k_r is another additional pseudotransport coefficient that is needed to characterize the flow.

Porous Media Description

Characterization of porous media based on the pore (microscopic) level is carried out for the purpose of understanding, modeling, and sometimes controling the macroscopic behavior and properties of the medium. The macroscopic (bulk) properties needed to relate to the pore description are porosity, permeability, tortuosity, and connectivity. When one examines a sample of a porous medium, for example, sandstone, it is obvious that the number of pore sizes, shapes, orientations, and interconnections is enormous. Furthermore, even the identification of a pore is not unique. Because of this complexity, pore structure is often characterized based on an idealized model. A true description is not realistic for a natural porous medium.

Pore Size Distribution.   The pore structure is sometimes interpreted as a characteristic "pore size," which is sometimes ambiguously called "porosity." More generally, "pore structure" is characterized by a "pore size distribution," characteristic of the sample of the porous medium. The pore size distribution $f(\delta)$ is usually defined as the probability density function of the pore volume distribution with a corresponding characteristic pore size δ. More specifically, the pore size distribution function at δ is the fraction of the total pore volume that has a characteristic pore size in the range of δ and $\delta + d\delta$. Mathematically, the pore size distribution function can be expressed as

$$f(\delta) = \frac{dF(\delta)}{d\delta} \tag{23}$$

where $F(\delta)$ is the probability distribution function. Conversely, one has

$$\int_0^\infty f(\delta)d\delta = 1, \qquad F(\delta) = \int_0^\delta f(\delta)d\delta \qquad (24)$$

Although the volume parameter can be measured directly, the characteristic pore size is always calculated from some measured physical quantities in terms of some pore structure model. Hence, the pore size distribution is not a well-defined property because, by and large, it depends very much on the particular method used in its determination.

The determination of a pore size distribution consists of measuring a given physical quantity in relation to another physical quantity under the control of an operator and varied in the experiment. Although the technology has been advancing toward direct measurements, the complexity normally involved with a porous medium is preventing an accurate detailed characterization. If one does obtain a complete characterization, it is difficult to justify its cost over its utility. Hence, the experiments have been focused on indirect methods. For instance, in liquid porosimetry, the volume of liquid penetrating the porous medium sample is measured as a function of the pressure imposed on the liquid (34). In the vapor sorption experiment, the volume of gas absorbed on the pore surface is measured as a function of the gas pressure. The volume of liquid displaced miscibly is measured as a function of the volume of displacing liquid injected into the sample in a miscible displacement experiment. The list keeps going on; however, no concrete relation can be made between the measured physical quantity and the pore size distribution. Hence, often a pore size distribution should always be associated with a pore structure model. A different method or different pore structure model can give a very different characterization of the pore size distribution. For a more complete review of the techniques used in determining the pore size distributions and their related problems, the reader is to refer to Dullien (31). Although the inaccuracies relating to an operator or a particular working fluid can be resolved given the proper technique (34), the problems relating to the "irregularities" of the porous medium sample is not easy to eradicate. As an example, Figure 2 exemplifies the error that can be introduced by the characterization of mercury intrusion porosimetry data. Figure 2a shows a possible pore structure with a throat diameter δ and a bulk diameter δ_b. Figure 2b shows the "measured pore structure" based on a 1-dimensional uniform size model. Now that it becomes clear, one can interpret the pores very differently based on different pore structure models. Figure 3 shows an example given by Klinkenberg (35). Two different methods are used to measure the same samples that have a permeability of $k = 9.75 \times 10^{-13}$ m^2 and a porosity of $\epsilon = 0.289$. Figure

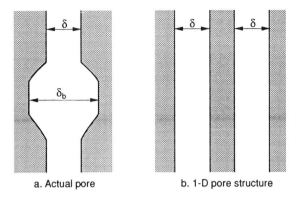

Figure 2.   Illustration of possible errors introduced by the characterization of mercury intrusion porosimetry data. δ is the pore throat diameter and δ_b is the pore body diameter.

3 illustrates that very different results are obtained for the same porous medium sample when different methods are used. One can appreciate the reality of errors that can be introduced by any measuring technique.

One-Dimensional Pore Structure Models and Pore Size.   Experimental data invariably have been characterized in terms of an arbitrary model of pore structure. The most common method consists of a bundle of parallel capillary tubes of equal length and a distributed size. The pore size would be unique only if the pores were tubes of uniform size and cross-section or spherical bodies. As neither is the

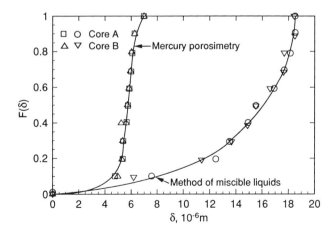

Figure 3.   Cumulative pore size distribution for Bentheim sandstone, with respect to different methods of measurements (35).

case, the pore size needs to be defined, that is, a characteristic pore size should be used. A simple definition is to identify the characteristic pore size with the hydraulic or equivalent pore diameter, d_e, which is very close to twice the mean radius of curvature of the interface separating two immiscible fluids in mechanical equilibrium in the pore for the special case of zero contact angle, except for curved polygon cross-sections (Table II) where the diameter of the inscribed circle is much smaller than the hydraulic diameter. The pore size is given by

$$\delta = \frac{2r_m}{\cos\theta} = \frac{4}{\cos\theta}\left(\frac{1}{r_1} + \frac{1}{r_2}\right)^{-1} \qquad (25)$$

where θ is the contact angle.

The pore size distribution can be obtained from capillary pressure measurements or mercury porosimetry. The capillary pressure is related to the specific free energies of the interface between fluids and between the fluid and the capillary wall. At mechanical equilibrium, the surface free energy between the fluids is a minimum. The equilibrium condition is expressed by the Laplace equation:

$$p_c = \frac{2\sigma}{r_m} = \frac{4\sigma\cos\theta}{\delta} \qquad (26)$$

where p_c is the capillary pressure and σ is the interfacial tension.

Table II.   List of Comparative Values of the Reciprocal Hydraulic Diameter and Twice the Reciprocal Mean Radius of Curvature in Capillary

Cross-Section	$2/r_m$	$4/d_e$
Circle	$2/a$	$2/a$
Ellipse		
$a/b = 11/10$	$20/11\ b$	$1.910/b$
$a/b = 12/10$	$5/3\ b$	$1.837/b$
$a/b = 15/10$	$4/3\ b$	$1.683/b$
$a/b = 2/1$	$3/2\ b$	$1.542/b$
$a/b = 5/1$	$6/5\ b$	$1.338/b$
$a/b = 10/1$	$11/10\ b$	$1.294/b$
$a/b = 20/1$	$21/20\ b$	$1.280/b$
Parallel plates	$1/b$	$1/b$
Rectangle	$1/a + 1/b$	$1/a + 1/b$
Equilateral concave circular arc quadrilateral*	$2(\sqrt{2} + 1)/a$	$2\pi(4 - \pi)^{-1}/a$
Equilateral concave circular arc triangle*	$\sqrt{3}(2 + \sqrt{3})/a$	$2\pi(2\sqrt{3} - \pi)^{-1}/a$
Equilateral triangle	$4\sqrt{3}/a$	$4\sqrt{3}/a$

*a is the radius of the circular arcs forming the polygon.
SOURCE: Data taken from reference 36.

The definition of hydraulic (equivalent pore) diameter d_e for a capillary tube of uniform cross section is

$$d_e = 4 \; \frac{\text{area of cross section}}{\text{perimeter of wetted cross section}} \tag{27}$$

In general, for a capillary of irregular shape, the minimum value of the ratio given by the right side of equation 27 must be found at a given location by varying the orientation of the sectioning plane. The minimum value is then, by definition, the hydraulic diameter of the capillary at the given point. Both definitions of the hydraulic diameter d_e and the mean radius of curvature r_m are best suited for the case of pore throats (neck) only, which control both the capillary penetration by a nonwetting fluid into the medium and the flow rate of fluids through the medium. The size of a pore body is best dealt with by using photomicrographs of sections made through the porous medium where the pore body is made visible. However, in most studies, the pore size is simply modeled through the pore throat value and that is where the errors arise (*see* Figure 2).

The pore size distribution is usually measured based on straight capillaries having a uniform cross-section. Under this condition, the pore neck diameter can be assigned to serve as the diameter of the capillary. If a given pressure p_c is applied to a fluid-filled porous medium, the saturation of the medium will be a function of the applied pressure. The relation between the saturation S and the capillary pressure p_c can be found for a nonwetting fluid as

$$S_n = \int_{4\sigma \cos \theta / p_c}^{\infty} f(\delta) d\delta \tag{28}$$

where $f(\delta)$ is the pore size distribution function. For a wetting fluid, the saturation is given by

$$S_w = \int_0^{4\sigma \cos \theta / p_c} f(\delta) d\delta \tag{29}$$

Figure 4 shows typical capillary pressure curves for a porous medium sample. Notice that the figure indicates strong hysteresis. For a porous medium initially saturated with a wetting fluid, $S_w = 1$, there is a threshold or entry pressure that must be applied before a nonwetting fluid can penetrate into the sample. Normally, the pore diameter corresponding to this threshold pressure, evaluated from equation 26, is called the "breakthrough" diameter or entry pore diameter, δ_e. At first, only the largest pores are open for the nonwetting fluid. Afterward, as S_w decreases, smaller pores starts to open for the nonwetting fluid and p_c

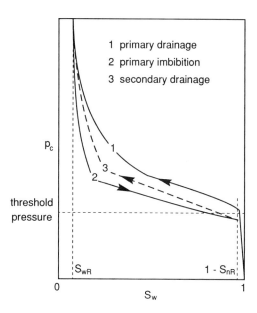

Figure 4.   Typical capillary pressure curves for a porous medium.

increases along the primary drainage curve. An irreducible saturation level occurs for the wetting fluid, S_{wR}, that no matter how high the pressure is increased, the wetting fluid cannot be displaced. During resaturation with a wetting fluid, the wetting fluid enters the smaller pores first, and p_c decreases along the primary imbibition curve. An irreducible saturation level exists also for the nonwetting fluid, S_{nR}. Subsequently, on redraining the wetting fluid, the capillary pressure follows the secondary drainage curve. Any attempt to drain or to resaturate between S_{wR} and $1 - S_{nR}$ would result in curves called scanning curves. The fact that the capillary pressure is lower when saturation changes in the imbibition direction than when it changes in the drainage direction indicates that imbibition requires less energy than drainage owing to the curvature direction of the front.

More sophisticated one-dimensional models have included dead-end pores, ink-bottle pores, pockets or turner structures, and also, periodically constricted tubes. Two-dimensional and three-dimensional network models have also been developed. The first proposal for a two-dimensional network model was given by Fatt (37, 38). Whereas Fatt primarily dealt with immiscible displacement, Simon and Kelsey (39) used a two-dimensional network model for the simulation of miscible displacement. The first three-dimensional network model is due to Irmay (40). Subsequently, the three-dimensional network model or percolation theory for porous medium has received much attention [*see*, e.g., Chatzis and Dullien (*41*) and Dullien (*31*)]. For a review of these and other models, the reader is referred to Dullien (*31*). Although the network models and the percolation theories are introduced with intention to characterize the porous medium structure, the actual situation is very different. The network models and percolation theories are often applied to model some porous medium properties but they generally failed to represent the real structure of the porous medium, even though the models are often as complicated as the real structures (*42*, *43*).

Field Level Characterizations.   Figure 5 depicts a reservoir and an exploring well. The flow simulation in such an oil field is normally done by using Darcy's law because of the large domain of interest and slow-moving nature of the fluids in the reservoir. It is assumed that the oil in the reservoir is homogeneous or single phase and slightly compressible. Furthermore, the flow in a reservoir can be approximated by a two-dimensional model by ignoring the effects of the top and the bottom of the reservoir. The continuity equation can be written as

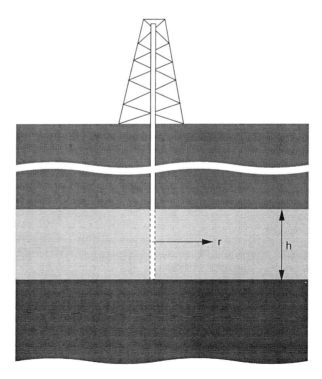

Figure 5.   A schematic diagram of a reservoir and its exploring well.

$$\epsilon \frac{\partial \rho}{\partial t} + \nabla \cdot (\rho \mathbf{v}) = 0 \tag{30}$$

Darcy's law for an isotropic porous medium is given by

$$\mathbf{v} = -\frac{k}{\mu} \nabla p \tag{31}$$

Substituting equation 31 into equation 30, we obtain

$$\epsilon \mu \frac{\partial \rho}{\partial t} = \nabla \cdot (\rho k \nabla p) \tag{32}$$

For a compressible liquid of constant compressibility c, one can write

$$\rho = \rho_0 e^{c(p - p_0)} \tag{33}$$

Substituting equation 33 into equation 32 gives the governing equation for a slightly compressible liquid:

$$\epsilon \mu c \frac{\partial p}{\partial t} = \nabla \cdot (k \nabla p) \tag{34}$$

Let

$$\alpha_k = \frac{k}{\epsilon \mu c} \tag{35}$$

be the hydraulic diffusivity, and equation 34 can be rewritten as

$$\frac{\partial p}{\partial t} = \alpha_k \nabla^2 p \tag{36}$$

The solution to equation 36 can be found in numerous forms in the literature. Earlogher (44) writes a general solution to equation 36 in terms of the initial pressure, p_i, for a reservoir of porous rock with a single well with source Q as

$$p_i - p = \frac{Q B_v \mu}{4 \pi k h} [p^*(t, r, r_w, C_D, \text{geometry}, \ldots) + S_{\text{kin}}] \tag{37}$$

where B_v is the formation volume factor (the ratio of the fluid density measured in stock tanks to the fluid density in the reservoir), C_D is the wellbore storage constant (relative to a circular cylindrical wellbore), $p^*(. \ . \ .)$ denotes the dimensionless pressure solution of equation 36 for the given geometry, S_{kin} denotes the skin effect, and r_w is the wellbore radius given by

$$r_{we} = r_w e^{skin/2} \qquad (38)$$

Here, r_{we} is the effective wellbore radius. The skin effect is defined as any factor at or around the wellbore that causes a change in the pressure drop, as compared with the pressure drop that occurs when the stratum is homogeneous and the well fully penetrates the sand (45–48). It is also implied that the inertial effects are attributed to the skin effects because Darcy's law is used to model the flow and hence the skin effects can be different for higher production rates.

For simplicity, we assume that the reservoir domain is a cylinder with a thickness of h and a radius l. For a constant rate withdraw from (production) or injection into the reservoir, the governing flow equation 36 may be further reduced to

$$\frac{1}{\alpha_k}\frac{\partial p}{\partial t} = \frac{1}{r}\frac{\partial}{\partial r}\left(r\frac{\partial p}{\partial r}\right) \qquad (39)$$

For the moment, let us consider the case of an infinite radial system. Assume that at a point the pressure increases or decreases (flow into or flow out of the reservoir) over a formation thickness h. The initial and boundary conditions are

$$\begin{cases} p(r, 0) = p_i, & \text{for all } r \\ 2\pi r_w hv(r_w, t) = Q = \text{constant}, & t > 0 \\ p(\infty, t) = p_i, & t > 0 \end{cases} \qquad (40)$$

Similarity solution can be sought for equation 39. Let

$$t_D = \frac{4\alpha_k t}{r^2} \qquad (41)$$

and assuming that the dimensionless time t_D is the only independent variable, we have

$$\begin{cases} \dfrac{\partial p}{\partial t} = \dfrac{dp}{dt_D}\dfrac{\partial t_D}{\partial t} = \dfrac{4\alpha_k}{r^2}\dfrac{dp}{dt_D} \\ \dfrac{1}{r}\dfrac{\partial}{\partial r}\left(r\dfrac{\partial p}{\partial r}\right) = \dfrac{1}{r}\dfrac{\partial t_D}{\partial r}\dfrac{d}{dt_D}\left(r\dfrac{\partial t_D}{\partial r}\dfrac{dp}{dt_D}\right) = \dfrac{2t_D}{r^2}\dfrac{d}{dt_D}\left(2t_D\dfrac{dp}{dt_D}\right) \end{cases} \qquad (42)$$

Hence, the partial differential equation 39 can now be reduced to the following ordinary differential equation:

$$t_D\frac{d}{dt_D}\left(t_D\frac{dp}{dt_D}\right) - \frac{dp}{dt_D} = 0 \qquad (43)$$

The initial and boundary conditions can be reduced to

$$
\begin{cases}
p = p_i, & t_D = 0 \\
\dfrac{dp}{dt_D} = \dfrac{Q\mu}{4\pi hk t_D}, & t_D = \infty
\end{cases}
\tag{44}
$$

To have the similarity procedure valid, it is assumed that the wellbore radius is very small and a point sink is used to simulate the well.

The solution to equation 43 with the boundary conditions given by equation 44 can be obtained as

$$
p = p_i - \frac{Q\mu}{4\pi kh} Eif(t_D)
\tag{45}
$$

where $Eif(x)$ is an exponential integral function and is defined by

$$
Eif(x) = -Ei(-x^{-1}) = \int_{x^{-1}}^{\infty} \frac{du}{ue^u} = \ln x - \gamma - \sum_{n=1}^{\infty} \frac{(-1)^n}{n!n} x^{-n}
\tag{46}
$$

Here γ is Euler's constant, $\gamma = 0.57721566490\ldots$. The exponential integral function has a useful asymptotic form for small time scale and is given by

$$
Eif(x) = xe^{-1/x} \sum_{n=0}^{n=N} (-1)^n n! x^n + R_N(x), \quad \text{for } N \le \frac{1}{x}
\tag{47}
$$

where the truncation error term is given by

$$
|R_N(x)| = \left| (-1)^{N+1}(N+1)! \int_{x^{-1}}^{\infty} \frac{du}{u^{N+1}e^u} \right| \le (N+1)! x^{N+2} e^{-1/x}
\tag{48}
$$

The asymptote given by equation 47 is not a normal convergent series, that is, one cannot take infinite terms. It is given here for the purpose of quick estimation of the exponential integral function at short times, say, $t_D < 0.2$. Only a few terms are needed to reach an accurate result for $Eif(t_D)$, whereas many more terms are needed when equation 46 is used at short times.

The solution expressed by equation 45 is satisfactory for use with finite sources and sinks of finite radius as long as they are far away from the well:

$$
t_D = 4\alpha_k r^{-2} t < 0.01
\tag{49}
$$

Hence, this solution is good for large systems or for short times.

The exponential integral function, $Eif(x)$, has an asymptotic form for small systems or long times, given by

$$Eif(x) \approx \ln(x) - 0.57722 = \ln(x/4) + 0.80908 \qquad (50)$$

Since equation 39 is linear, superposition of component solutions is permitted. It becomes straightforward to construct solutions for a finite domain reservoir; for multiple production and injection wells as long as they are far away from each other; for pressure buildup, that is, turning off a production well; and for pressure drawdown.

The reservoir boundaries are normally determined by the constant rate pressure drawdown test. A plane boundary at a distance l from the well would have the same effect on the pressure as the presence of a second well producing at the same rate and at a distance $2l$ from the first well in a reservoir of infinite size. An analytical solution can be found through superposition for the bottom-hole pressure p_w, $p(r = r_w, t)$, in which the effect of the boundary is represented by the effect of the second (image) well (49):

$$p_w = p_i - \frac{Q\mu}{4\pi kh}(\ln t_{D\text{real}} - 0.5772) - \Delta p_{\text{skin}} - \frac{Q\mu}{4\pi kh}Eif(t_{D\text{image}}) \qquad (51)$$

where $t_{D\text{real}}$ and $t_{D\text{image}}$ are the dimensionless times for the real well (first) and the second (image) well, respectively, and are defined by

$$t_{D\text{real}} = \frac{4\alpha_k t}{r_w^2} = \frac{4kt}{\epsilon\mu c r_w^2}, \qquad t_{D\text{image}} = \frac{4\alpha_k t}{(2l)^2} = \frac{kt}{\epsilon\mu c l^2} \qquad (52)$$

In deriving equation 51, it is assumed that $t_{D\text{real}}$ is large; in other words, equation 51 is not valid at initial time. From equation 51, one can observe that p_w is a linear function of $\ln(t_{D\text{real}})$ without the image well, or $t_{D\text{image}} = 0$, after production has started, or $t_{D\text{real}}$ is large. Hence, the difference between the measured value, p_w, at a later time and the straight line (not the tangential line) extrapolated value, p_w', to the same time (from the initial time on the bottom-hole pressure drawdown curve) is

$$p_w' - p_w = \frac{Q\mu}{4\pi kh}Eif(t_{D\text{image}}) \qquad (53)$$

From equation 53, one can estimate the reservoir boundary with the known production rate, formation thickness h, and the permeability k.

Similarly, the bottom-hole pressure buildup (shut-in) curves can be constructed to measure the product of the formation thickness and the permeability.

Porosity and Permeability Distributions. The porosity and permeability are bulk properties of a porous medium. They are normally expected to be constant for a given porous medium. However, because of the anistropic or heterogenous behaviors of natural porous materials, these properties may change when different portion or sample of the same porous medium are used. For example, the rock materials in an oil reservoir formation are often found to have a vast distribution of porosities and permeabilities. However, characterization of the permeability distribution at a field level is very difficult. Figure 6 shows the permeability distribution for the layers of the formation of an oil reservoir in which no order of location is recorded. Normally, only an average value of the permeability for an entire reservoir is obtained from the pressure shut-in curve.

The porosity and permeability distributions have been measured in the laboratory for natural porous medium samples, such as sandstone [*see* Freeze (*51*), King (*52*), and Drummond and Horgan (*53*)]. For example, Law (*54*) found that the porosity is close to be normally distributed. Law (*54*) and Henriette et al. (*55*) showed that the permeability distributions are skewed and close to a log-normal distribution curve.

Areal Porosity and Tortuosity.   "Areal porosity" or "areosity," A_p, is defined as the effective areal ratio of the open pore cross-section to the bulk space. A more strict definition of areosity was introduced by Ruth and Suman (*56*). However, the areosity as defined by them is not a property of the porous medium only but a property of both the porous medium and the transport strength of the fluid such as the flow strength and electric current strength. The areal porosity is undoubtedly a very useful quantity for a bundled or ensemble passage model because it represents the ratio of the total passage cross-sectional area to the total cross-sectional area of the porous medium at a given planar section.

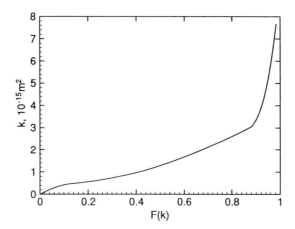

Figure 6. *Cumulative permeability distribution for layers of an oil reservoir* (*50*).

The areal porosity has been invariably considered to be equivalent to the porosity, ϵ, of the medium (*1, 31, 33*). This is by and large due to the success and popularity of the straight bundle passage models and nonconnecting constricted passage models, in which the ratio of the total open pore area to the total cross-sectional area for the direction under consideration has to be equivalent to the (volume) porosity of the porous medium. However, as it is expected from the difference in the definitions, it is no surprise that Dullien and Mahta (*57*) found their measured (areal) porosity is significantly different from the measured (volume) porosity with a different technique for the same porous medium sample.

The areal porosity can be measured using optical methods. Before measurements, the sample sections must be polished. It is often necessary to impregnate the pores with some material such as wax, plastic, or Wood's metal to make the pores more visible or to distinguish between interconnected and blocked pores. When impregnating the sample from outside, evidently only the interconnected pores will be penetrated (*31*).

The areal porosity has been modeled by Liu et al. (*32*) for an isotropic porous medium. They argued that the void space ratio in all directions should be the same. The porosity is the product of the void space ratios in the orthogonal directions. This argument leads to

$$A_p = \epsilon^{2/3} \tag{54}$$

Until accurate experimental measurements are available, equation 54 is to be used to estimate the areal porosity.

The tortuosity, τ, of a porous medium is defined as the ratio of the distance between two fixed points and the tortuous passage followed by a fluid element of a single fluid saturated in the porous medium when traversing the two points. It may be viewed as a "line porosity" because, by definition, tortuosity is a one-dimensional property of the porous medium. It can be related to the formation factor, or formation conductivity factor, F_ϵ by

$$\tau = \frac{1}{F_\epsilon \epsilon} \tag{55}$$

Based on a nonconnecting passage model, as we pointed out earlier, $A_p = \epsilon$, Suman and Ruth (*58*) concluded that ϵ as in equation 55 should be replaced by A_p.

It is generally believed that the pores are interconnected and the formation factor can be related with porosity by Archie's law (*59*):

$$F_\epsilon \equiv \frac{\sigma_w}{\sigma_o} = \frac{a_c}{\epsilon^m} \tag{56}$$

where σ_o and σ_w are the electric conductivity of a nonshaly formation sample completely saturated with an ionic solution and the electric conductivity of the pure ionic solution itself, respectively, a_c is an empirical constant, and m is the "cementation factor" and is supposed to be between 1.3 and 2.5 for various rocks. For random suspensions of spheres, cylinders, and sand in aqueous solutions of zinc bromide of approximately the same densities as the particles, De La Rue and Tobias (60) found $a_c = 1$ and $m = 1.5$, which very accurately correlates for porosity ranging from 0.6 to 0.95. The expression given by De La Rue and Tobias (60) is also found true for packed beds and fused glass beads of porosity as low as $\epsilon = 0.2$ (61–63).

The averaging theory of Suman and Ruth (58) for a nonconducting solid matrix porous medium leads to

$$F_\epsilon = \frac{1}{A_p IL} \int_{V_o} J dV + \frac{\sigma_w}{A_p IL} \int_{S_i} E \cdot n dS_i \qquad (57)$$

where I is the electric current, J is the electric current density, V_0 is the free space volume in an REV, E is the electrical potential, S_i is the solid-free space interface, and n is the out normal of S_i. When the detailed pore structure is known, the Suman–Ruth equation 57 can be used to evaluate the formation factor directly.

Maxwell (64) theoretically derived an expression for dilute uniform nonconducting spheres imbedded in a fluid continuum given by

$$F_\epsilon = \frac{3 - \epsilon}{2\epsilon} \qquad (58)$$

Maxwell's equation does not agree with the experimental results of De La Rue and Tobias (60), except for $\epsilon \to 1$.

The so-called "Humble" formula has been widely used for oil reservoir formations and consolidated sandstones. The Humble formula is given by

$$F_\epsilon = \frac{0.62}{\epsilon^{2.15}} \qquad (59)$$

When considering an isotropic porous medium, Liu et al. (32) argued that the tortuous passage bears a two-dimensional property. With the argument that either direction should have the same void space ratio for an isotropic porous medium and the porosity is a product of the two directions, the tortuosity can be determined as

$$\tau = \epsilon^{1/2} \qquad (60)$$

Equation 60 is in agreement with the formation factor experiments of De La Rue and Tobias (*60*), Johnson et al. (*62*), and Wong et al. (*63*). The tortuosity is also discussed by Epstein (*65*) and Puncochar and Drahos (*66*). We conclude that for random packs and suspensions of $\epsilon > 0.2$, equation 60) should be used as the tortuosity of the system.

For consolidated porous medium of $\epsilon < 0.45$, the Humble equation

$$\tau = 1.61\epsilon^{1.15} \tag{61}$$

can be used if the value of τ is not available. It should be noted that the above equation does leave a large margin in the accuracy of the tortuosity value. For some sandstones, the porosity can be about the same, whereas the tortuosity can differ significantly (Table III).

Table III shows some tortuosity values for various sandstones, where τ_c is the tortuosity value computed from equation 61. We observe that the prediction by equation 61 is poor because of the heterogeneity of the sandstones. Hence, whenever possible, the measured value should be used. The tortuosity value can be estimated rather well from the pore size distribution results with a proper pore structure model (*31*).

Single-Phase Fluid Flow

Single-phase fluid flow in porous media is a well-studied case in the literature. It is important not only for its application, but the characterization of the porous medium itself is also dependent on the study of a single-phase flow. The parameters normally needed are porosity, areal porosity, tortuosity, and permeability. For flow of a constant viscosity Newtonian fluid in a rigid isotropic porous medium, the volume averaged equations can be reduced to the following: the continuity equation,

$$\epsilon \frac{\partial \rho}{\partial t} + \nabla \cdot (\rho \mathbf{v}) = 0 \tag{62}$$

and the momentum equation

$$\frac{\partial (\rho \mathbf{v})}{\partial t} + \nabla \cdot \left(\frac{\rho \mathbf{v} \mathbf{v}}{\epsilon} \right) + \nabla p - \mu \nabla^2 \mathbf{v} + \mu F \mathbf{v} = 0 \tag{63}$$

For the simple case of creeping flow, that is, no noticeable inertial effects, the shear factor is simply the inverse of the intrinsic permeability,

$$F = \frac{1}{k} \tag{64}$$

Table III. Porosity, Tortuosity, Entry Pore Diameter, and Permeability of Sandstones

Sandstone	ϵ	τ	τ_c	$\delta_e \ (m)$	$k \ (m^2)$
Boise	0.232	0.5187	0.301	1.04×10^{-4}	5.30×10^{-12}
Bartlesville	0.270	0.3696	0.358	3.2×10^{-5}	1.2905×10^{-12}
Tuscaloosa	0.28	0.476	0.373		8.59×10^{-13}
Berea 108	0.178	0.0897	0.222	2.9×10^{-5}	6.05×10^{-13}
Berea BE1	0.220	0.2936	0.283	2.6×10^{-5}	3.95×10^{-13}
Noxie 47	0.250	0.4449	0.328	2.8×10^{-5}	2.94×10^{-13}
St. Meinrad	0.247	0.3457	0.323	1.8×10^{-5}	2.07×10^{-13}
Torpedo	0.246	0.4174	0.322	1.2×10^{-5}	1.82×10^{-13}
Big Clifty	0.192	0.2568	0.242	1.8×10^{-5}	1.62×10^{-13}
Noxie 129	0.238	0.4460	0.310	1.2×10^{-5}	1.456×10^{-13}
Cottage Grove	0.216	0.4347	0.277	1.3×10^{-5}	1.14×10^{-13}
Clear Creek	0.191	0.3963	0.240	1.5×10^{-5}	9.33×10^{-14}
Pennsylvanian	0.17	0.290	0.210		5.82×10^{-14}
Bandera	0.191	0.2125	0.240	5.4×10^{-6}	8.49×10^{-15}
Whetstone	0.134	0.0805	0.160	2.9×10^{-6}	7.40×10^{-16}
Belt series	0.083	0.0307	0.092	1.4×10^{-6}	3.55×10^{-16}

SOURCE: Data taken from reference 31.

As we mentioned earlier, in writing equation 63, we neglected the momentum dispersion effects. Normally, the dispersion is not important, especially when the flow domain is large.

Permeability. Permeability is the hydraulic conductance of a medium defined with direct reference to Darcy's law. In a somewhat more general sense, the shear factor is the hydraulic resistivity of the medium. When the term permeability is used, one normally refers to linear flow systems (no inertial effects).

Darcy's law is valid for creeping flow (negligible inertial effects) in a large-scale porous medium in which the possible disturbances that may be introduced into the system at the boundaries have negligible effects on the flow. In this case, both the time evolution term and the viscous diffusion term can be dropped from the momentum equation, and we obtain

$$\nabla p + \frac{\mu}{k} \mathbf{v} = 0 \tag{65}$$

In the simplest situation, where the flow domain is confined to a straight channel, integration of equation 65 leads to

$$q = \frac{k}{\mu} \frac{p_1 - p_2}{L} \tag{66}$$

where q is the superficial flow velocity and L is the flowing passage length. Equation 66 states that the superficial flow velocity through a porous bed is proportional to the pressure drop across the bed. The intrinsic permeability is normally measured through a pressure drop measurement for a given flow rate with equation 66, which can be rearranged to give

$$k = \frac{Q}{A_t} \frac{\mu L}{p_1 - p_2} \tag{67}$$

where Q is the fluid discharge rate and A_t is the cross-sectional area of the porous medium normal to the main flow direction.

Normally, the permeability of a porous medium can be related to other porous medium properties through a model of the porous medium structure. Pragmatically, the complexity of the flow rules out any rigorous analytical attempts to resolve the problem. Ideally, one would like to use heuristic arguments to derive an expression for k or F in terms of universal constants and easily measurable properties of the porous material and the flowing fluid. Such attempts have been made by many researchers in the past. A large collection of these studies has been documented by Dullien (31). Of note are the straight parallel passage model (20, 21), cell model (67, 68), bundles of periodically constricted parallel tubes (69–77), network models (78–82), cut-and-random-rejoin-type models (83, 84), and the randomly oriented straight cylindrical tube model of Haring and Greenkorn (85). Many empirical correlations exist as well. Some of these and other deterministic relations are collected by Panigrahi and Murty (86). For flow in granular beds, the most successful one is that due to the Kozeny–Carman theory, which contains universal constants and easily measurable porous medium properties, d_s and ϵ, and fluid properties, ρ and μ.

Following the Kozeny–Carman theory, we consider a curved flow passage of uniform cross-section and an arbitrary shape to model the pore structure of a porous medium. A sketch of such a flow passage is depicted by Figure 7. In the limit of creeping flow, Liu and Masliyah (87) found that the pressure drop in a curved or helical pipe is similar to that in a straight pipe. In analogy to the Poiseuille equation, the pressure gradient in a curved passage for a fully developed flow at zero flow strength can be written as

$$\frac{\Delta p}{L} = \frac{16 k_1 \mu}{d_e^2 \tau} u_e \tag{68}$$

where Δp is the pressure drop, L is the apparent length parallel to the main flow, d_e is the equivalent diameter of the passage, and u_e is the average (characteristic) velocity in the passage or the characteristic interstitial velocity. The term k_1 is expected to be a universal constant, that is, not a property of porous media. For a straight tube (passage), k_1 is a constant that weakly depends on the shape of the passage. For a constant τ and by using the Dupuit's assumption of $u_e = q/\epsilon$, equation 68 becomes the Blake–Kozeny equation. Here, q is the superficial velocity. Equation 68 was also given by Carman (21) as a modified Kozeny's equation; however, Carman (21) and other investigators considered τ to be a universal constant, that is, independent of the porous structure. For example, Haring and Greenkorn (85) gave τ^{-1} as 2.25.

Figure 8 shows a REV that depicts flow passages in a porous medium. As is shown in Figure 8, the pores (passages) are interconnected so that

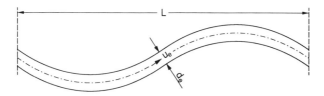

Figure 7.   A sketch of the model curved passage.

——— Solid-Free volume interface
– · – · – Boundary of a representative elementary volume

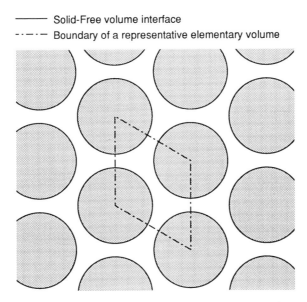

Figure 8.   Flow passages in a REV in a porous medium.

fluid mixing among the passages is inevitable and a higher pressure drop than otherwise in a single passage is expected. The curvature and cross-sectional area variation of the passage also result in a higher pressure drop than otherwise for flow through a straight uniform cross-sectional passage. These effects are not all inertial in origin; however, they are more pronounced when the flow rate is increased. If one considers that the passages are parallel and straight, k_1 will have a fixed value. The shape factors for a few types of cross-sectional passages have been given by Carman (21), and more values can be found in Ward-Smith (88). The most probable passage cross-section at the pore neck in a packed bed of spheres is shown in Figure 9. Some relevant values are $k_1 = 2$ for circular tubes, $k_1 = 1.7784$ for square ducts, $k_1 = 1.6662$ for equilateral triangular ducts, and $k_1 = 0.812$ for equilateral concave circular arc quadrilateral ducts (Figure 9b). It is clear that the value of k_1 does not change dramatically with the passage shape and the difference in k_1 does not normally exceed 100% for different shapes. In reality, we would expect the value of k_1 to be different from the values given above, by and large, because of the fact that not all the passages are parallel to the direction of the flow. Liu et al. (32) gave

$$k_1 = 2.37 \qquad (69)$$

based on the experimental data on granular beds.

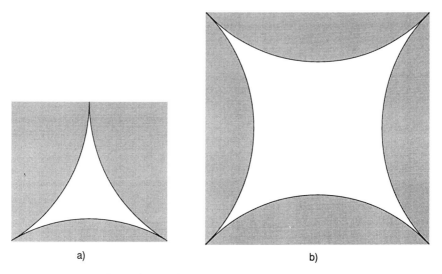

a)   b)

*Figure 9.   Equilateral triangle (a) and equilateral quadrilateral (b) formed
by tangent circular arcs.*

The characteristic interstitial velocity u_e can be related to the superficial velocity by equating the time required for a fluid element to travel with the superficial velocity q for a fixed apparent length L within the porous medium and the time for a fluid element to travel with the characteristic interstitial velocity u_e in the pore for a fixed passage length of L/τ. Hence, we obtain

$$u_e = \frac{q}{\tau A_p} \qquad (70)$$

where A_p is the areal porosity of the porous medium. It should be noted that u_e does not necessarily have to be the average velocity in the parallel passages of a uniform cross section.

From equations 68–70 and equation 66, we obtain

$$k = \frac{d_e^2 \tau^2 A_p}{16 k_1} = 0.0264 d_e^2 \tau^2 A_p \qquad (71)$$

Equation 71 is the basic equation that relates permeability of a porous medium to its other properties. However, equation 71 contains the hydraulic diameter of the passage (pore), tortuosity, and areal porosity of the medium, which may not be easily accessible. For example, sandstones or rock formations have irregular pore structure and often have inconsistent pore size measurement values (*see* previous section). It is rather difficult to measure the average hydraulic pore diameter. On the other hand, for random suspensions and random packs of spheres, cylindrical rods, fibers, and sands, it is possible to define the hydraulic diameter given the size of the particles, although the measurement of the pore size is just as difficult as for sandstones. Hence, we discuss equation 71 with respect to consolidated porous media (sandstone, and so on) and random packs, including fixed solid matrix of more orderly systems where the diameter of the grain is defined.

The equivalent passage diameter d_e is given by

$$d_e = 4 \frac{\text{cross sectional area normal to flow}}{\text{wetted perimeter}}$$

$$d_e = 4 \frac{\text{volume of free space in medium}}{\text{wetted surface area}} = \frac{2\epsilon d_s}{3(1 - \epsilon)} \qquad (72)$$

where d_s is the equivalent spherical diameter of the particles or fibers forming the porous structure. It is defined as

$$d_s = 6 \frac{V_P}{S_P} = \Phi_s \left(\frac{6V_P}{\pi}\right)^{1/3} \qquad (73)$$

Here V_P, S_P, and Φ_s are the volume, the effective surface area, and the sphericity of the particle, respectively.

The characteristic interstitial velocity can be obtained by substituting equations 54 and 60 into equation 70 to yield

$$u_e = q\epsilon^{-7/6} \qquad (74)$$

Hence, this model for the characteristic interstitial velocity as given by equation 74 gives a higher value than the intrinsic phase average interstitial velocity or the interstitial velocity based on the Dupuit's assumption, which is given by

$$u_\epsilon = q\epsilon^{-1} \qquad (75)$$

Because of the parallel (not interconnected) and uniform cross-sectional area models used by Du Plessis and Masliyah (74, 75), their pore velocity is also given by equation 75.

The final equation for the pressure drop in porous media when the flow is very weak can be obtained by substituting equations 54, 72, and 60 into equation 71, leading to

$$k = 0.0117 \frac{\epsilon^{11/3} d_s^{\,2}}{(1 - \epsilon)^2} \qquad (76)$$

For known specific surface per unit volume, a_P, equation 76 takes the following form:

$$k = 0.423 \frac{\epsilon^{11/3}}{(1 - \epsilon)^2 a_P^2} \tag{77}$$

As a comparison, the equation due to the Kozeny–Carman's theory is given by

$$k = \frac{1}{36k_0} \frac{\epsilon^3 d_s^2}{(1 - \epsilon)^2} \tag{78}$$

where k_0 is the Kozeny–Carman constant and $k_0 = 5$ according to Carman (21), whereas $k_0 = 4.1667$ given by the Blake–Kozeny equation. Comparing equation 78 with equation 76, we note a difference in the porosity dependence.

The correction on the tortuosity term in Kozeny–Carman equation was initiated by Foscolo et al. (89); however, they repeated the error of Kozeny's work, which was pointed out later by Epstein (65). The tortuosity model of Foscolo et al. (89) was not in agreement with experiments, although the same line of investigation was continued by Puncochar and Drahos (66).

It is interesting to note that in Kozeny–Carman theory, τ was considered as a universal constant and was embedded into the Kozeny–Carman constant. Although it was considered separately by some investigators, the tortuosity was treated as a constant (invariant with porosity). Equation 76 also indicates that the correlations in the literature based on the Kozeny's analysis or Blake–Kozeny equation are valid only for a narrow porosity range and cannot be extrapolated to different range of porosity without loss of accuracy. This behavior was noted by MacDonald et al. (29), who suggested that a better correlation might be obtained by replacing ϵ^3 in the Ergun equation with $\epsilon^{3.6}$.

Permeability for a Rock Formation.   For natural consolidated porous medium, however, the definitions of the equivalent spherical diameter and the specific surface area per unit volume are not widely used because of its difficulty in determination and relation to other measurable quantities. Just to serve as a comparison, we give the permeability equation based on the previous passage model with the tortuosity given by equation 61 and assuming that the areal porosity equation 54 still holds. The permeability can then be given by

$$k = 1.09 \frac{\epsilon^{14/3}}{(1 - \epsilon)^2 a_P^2} \tag{79}$$

Equation 79 is difficult to make use of because of the lack of experimental data on the specific surface area. However, as a quick check, if one is to assume that the Kozeny–Carman equation is valid at $\epsilon = 0.4$, using the porosity dependence difference between equations 79 and 78, we obtain

$$\frac{k_{eq\,79}}{k_{eq\,78}} = \left(\frac{\epsilon}{0.4}\right)^{5/3} \tag{80}$$

where $k_{eq\,78}$ and $k_{eq\,79}$ denote the permeability values as estimated by equations 78 and 79, respectively. Using equation 80, we found that equation 78 overpredicts the permeability by 217% at $\epsilon = 0.2$ and 360% at $\epsilon = 0.15$. In other words, the pressure drop as predicted by Kozeny–Carman equation is around 300% lower than that based on the model of Liu et al. (32). It is not totally surprising that the Kozeny–Carman equation underpredicts the pressure drop for sandstones by 300% (31).

Hence, at this point we may conclude that equation 71 is correct for both granular beds and sandstones and, subsequently, equation 77 or 76 should be used for granular beds and equation 79 should be used for consolidated porous media of low porosity, $\epsilon < 0.4$.

To use the model for k as given by equation 79, we must relate the effective passage diameter to the commonly measured properties of the sandstones, for example, the breakthrough diameter or entry pore diameter, δ_e. It is possible to relate the characteristic passage diameter to the entry pore diameter through pore volume balance and hence the following expression may be sought

$$\delta_e^{\,2} = c_e A_p \tau d_e^{\,2} \tag{81}$$

where c_e is a constant. Substituting equation 81 into equation 71, we obtain

$$k = \frac{0.0264}{c_e}\,\tau \delta_e^{\,2} \tag{82}$$

When equation 61 is used, the permeability equation becomes

$$k = \frac{0.0425}{c_e}\,\epsilon^{1.15}\delta_e^{\,2} \tag{83}$$

Using the data shown in Table III, we determine that

$$c_e = 9.1 \tag{84}$$

Figure 10 shows the quality of equation 83, where k is the experimental value as shown in Table III and k_c is the computed value from equation 83 using also the entry pore diameter δ_e as given in Table III. From Figure 10, we observe that equation 83 predicts the permeability of sandstones fairly well in comparison with the experimental data.

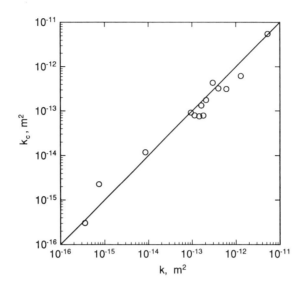

Figure 10. Permeability predicted by equation 83, k_c, *in comparison with the permeability,* k, *shown in Table III.*

However, it should be noted that the form of equation 83 is similar to those of Dullien (*31*), Katz and Thompson (*90*), Haring and Greenkorn (*85*), and Macmullin and Muccini (*91*). Because of the difficulty in obtaining a consistent value for the characteristic pore diameter for consolidated media, the permeability correlation equation can be very different based on the experimental data of different investigators.

Shear Factor, *F.*   The shear factor *F* is a generalized hydrodynamic resistivity of porous media. It appears in the momentum equation 63 and is needed to solve the problem of single-phase flow in porous media. The shear factor *F* can be related to the pressure drop of a unidirectional flow without bounding wall effects, that is, in a one-dimensional medium through equation 21. In this section, we give a detailed account for the derivation of the expressions for f_v and *F.*

The laminar flow, outside the surface interactive force dominant region, in porous media may be subdivided into two flow regimes, Darcy's flow and Forchheimer flow. Darcy's flow refers to the creeping flow, or flow with negligible inertial effect, whereas Forchheimer flow refers to flow with noticeable inertial effect. These two regimes have been observed by pressure drop measurements (*92*) and laser-Doppler velocimetry (LDV) measurements (*93*), and the two regimes do not have a clear transition. Up to this point, we have been dealing with creeping flow only. For simplicity, we focus on granular media from here on.

Toward the end of this chapter, the shear factor for the consolidated media is given.

To study the inertial effect on the fluid flow in porous media, we define a pore or particle Reynolds number based on the curved passage model as described earlier,

$$Re_p = \frac{3d_e \rho u_e}{2\mu} = \frac{d_s \rho q}{(1 - \epsilon)\epsilon^{1/6}\mu} = \frac{6\rho q}{(1 - \epsilon)\epsilon^{1/6}\mu a_P} \qquad (85)$$

where $3/2$ is introduced for convenience. When Re_p is small, equation 76 can be used to evaluate the shear factor, and it can be rearranged to give

$$F = \frac{\Delta p}{\mu L |v|} = 85.2 \frac{(1 - \epsilon)^2}{\epsilon^{11/3} d_s^2} \qquad (86)$$

To simplify equation 86, a modified friction factor is defined

$$f_v = f_m Re_p = -\frac{\Delta p}{L} \frac{d_s^2 \epsilon^{11/3}}{\mu q (1 - \epsilon)^2} \quad \text{or} \quad f_m = -\frac{\Delta p}{L} \frac{d_s \epsilon^{23/6}}{q^2 \rho (1 - \epsilon)} \qquad (87)$$

where f_v is the normalized pressure drop factor and f_m is a modified Fanning friction factor. For a unidirectional flow in a one-dimensional medium, $q = |v|$, and

$$F = f_v \frac{(1 - \epsilon)^2}{d_s^2 \epsilon^{11/3}} \qquad (88)$$

Equation 76 becomes

$$f_v|_{Re_p \to 0} = 85.2 \qquad (89)$$

When Re_p is large, the pressure drop will have a noticeable inertial effect. The inertial effects can be subdivided into two factors: the mixing effect and curvature effect. The mixing effect refers to the effect of passage cross-sectional area variation and interpassage branching (or networking), whereas the curvature effect refers to the curvature effect of the passage. If one is only concerned about the mixing effect, Re_p would be satisfactory in relating the inertial effects. The excess pressure drop due to mixing is proportional to the Reynolds number. For curvature effect to be included, a characteristic curvature ratio, λ_m, should also be defined. To determine the characteristic curvature ratio in a porous medium, Liu et al. (32) argued that

$$\lambda_m = r_0 + r_1 \epsilon^n \qquad (90)$$

where r_0, r_1, and n are treated as universal constants.

The power n can be thought of as an index for the flow multidimensionality. Because the flow passages inside the porous media may be regarded as two dimensional, the value of n should be expected to be $= \frac{1}{2}$. If the porosity were zero, the passage would be extremely curved. That is, $\lambda_m = 1$ for $\epsilon = 0$, which leads to $r_0 = 1$. When porosity is unity, that is, free space, the passage becomes straight. In other words, $\lambda_m = 0$ when $\epsilon = 1$, which gives $r_1 = -r_0 = -1$. Hence, we obtain

$$\lambda_m = 1 - \epsilon^{1/2} \qquad (91)$$

Because a fully developed helical flow is not expected to be established in porous media, the excess pressure drop due to the curvature effect should be expected to be proportional to the Dean number (curvature ratio times the Reynolds number). Hence, the total inertial effects can be better defined by a modified Reynolds number that includes both the mixing and the curvature components. Liu et al. (32) found that

$$f_{v\infty} = BRe_m \qquad (92)$$

where $f_{v\infty}$ is the total inertial contributions to the normalized pressure drop, B is a universal constant that is equal to 0.69, and Re_m is a modified Reynolds number, given by

$$Re_m = (1 + \lambda_m^{1/2})Re_p = \frac{1 + (1 - \epsilon^{1/2})^{1/2}}{(1 - \epsilon)\epsilon^{1/6}} \frac{d_s\rho q}{\mu} \qquad (93)$$

The unified normalized pressure drop factor or the shear factor can be built based on the derived Reynolds number and the total inertial effects at high flow rates. The Forchheimer hypothesis normally becomes involved at this step. However, for very small Re_m flows, it is evident from the experimental data that f_v changes very little with varying Re_m, and when Re_m is high, f_v changes linearly with Re_m. This type of a friction loss dependence on the flow rate is due to the influence of the secondary flow in the flow passage and is similar to that found for helical pipe flows. The secondary flow in a porous medium can be produced by the mixing action from joining passages and changing passage cross-sectional area and by the curvature of the passage. When the flow in the passage is weak, the secondary flow originated from either mixing or curvature does not give rise to a significantly higher pressure drop than that for a straight uniform cross-sectional passage. Only when the flow is strong, the secondary flow can significantly increase the pressure drop in comparison to a straight uniform cross sectional passage. Liu (94) and Liu and Masliyah (87) used the following fRe correlation for laminar flow in a helical pipe:

$$fRe = (fRe)_s + \frac{Dn^{2n}}{C_{Dn}^{2n} + Dn^{2n}} B_{Dn} Dn^n \qquad (94)$$

where f is the Fanning friction factor, Re is the pipe Reynolds number, Dn is the flow Dean number, and $(fRe)_s$ is the value of fRe at $Dn = 0$ or for a straight pipe. C_{Dn} is a constant that signals the importance of the inertial effects, and it is proportional to the critical value of Dn at which the inertial effect starts to influence fRe. The term n is the asymptotic order of dependence. It should be noted that fRe is proportional to $Dn^{1/2}$ when Dn is very large, that is, $n = \frac{1}{2}$. By using the same form of a correlation for porous media and replacing Dn with Re_m, Liu et al. (32) obtained the variation of f_v ($= f_m Re_p$) with Re_m. Noting that the asymptotic order of dependence on Re_m is $n = 1$, we obtain

$$f_v = f_m Re_p = -\frac{\Delta p}{L} \frac{d_s^2 \epsilon^{11/3}}{\mu q (1 - \epsilon)^2} = A + \frac{Re_m^2}{C^2 + Re_m^2} BRe_m \qquad (95)$$

where $A = 85.2$ is the value of f_v at $Re_m = 0$, $B = 0.69$ is the inertial term constant at large Re_m, and $C = 16$ can be deduced from experiments signaling the transition from Darcy's region to Forchheimer's flow region. The relation between Darcy's flow to Forchheimer's flow in porous media is similar to that between Poiseuille flow and centrifugal force dominated flows in a curved channel. The inertial effect does not appear at low Reynolds number values and hence the Forchheimer hypothesis does not apply at small Reynolds numbers.

The shear factor can now be defined based on the above results as

$$F = \frac{(1 - \epsilon)^2}{\epsilon^{11/3} d_s^2} \left(85.2 + 0.69 \frac{Re_v^2}{16^2 + Re_v^2} Re_v \right) \qquad (96)$$

where Re_v is the local modified Reynolds number and is defined by

$$Re_v = \frac{1 + (1 - \epsilon^{1/2})^{1/2}}{(1 - \epsilon)\epsilon^{1/6}} \frac{d_s \rho |\mathbf{v}|}{\mu} \qquad (97)$$

In terms of the permeability, k, rather than the particle diameter, we have

$$F = k^{-1} + 0.0081 k^{-1} \frac{Re_k^2}{16^2 + Re_k^2} Re_k \qquad (98)$$

where Re_k is the local modified Reynolds number based on the permeability k

$$Re_k = 9.23 \frac{1 + (1 - \epsilon^{1/2})^{1/2}}{\epsilon^2} \frac{k^{1/2} \rho |\mathbf{v}|}{\mu} \qquad (99)$$

or in a more general form,

$$Re_k = 9.23 \frac{1 + (1 - \epsilon^{1/2})^{1/2}}{A_p^{3/2}\tau^2} \frac{k^{1/2}\rho\,|\mathbf{v}|}{\mu} \tag{100}$$

For a rock formation, or when the Humble formula applies, the local modified Reynolds number may take the following form to use with equation 99:

$$Re_k = 3.56 \frac{1 + (1 - \epsilon^{1/2})^{1/2}}{\epsilon^3} \frac{k^{1/2}\rho\,|\mathbf{v}|}{\mu} \tag{101}$$

Hence, we have formed the closure for the single-phase flow in porous media using a model of the shear factor for both consolidated and granular media. For the term closure, we mean that the shear factor F of equation 63, which was introduced through averaging procedures, is now defined.

Wall Effect.   Wall effects on fluid flow in porous media manifest themselves in three ways. First, the bed porosity is nearly unity very close to the wall (smaller than the particle size scale), and it decreases to its mean value as one moves away from the bounding wall toward the center of the bed. Near the bounding wall, different degree of freedom in the flow path is imposed as compared with the bed center. There is a significant effect of the momentum dispersion near the wall. Second, the bounding wall itself provides extra surface area, in addition to that of the solid matrix of the bed, which comes in contact with the traversing fluid. Third, the bounding wall provides a barrier to the fluid so that no fluid element can cross the bounding wall. The last effect is also termed the multidimensional effect and can be treated separately with the volume-averaged governing equations (32).

Three simple distinct approaches have been developed for incorporating the wall effects into the pressure drop versus flow rate relations. The first approach used to correct for the bounding wall effects hinges on the fact that the additional surface area provided by the bounding wall must be taken into account while evaluating the equivalent hydraulic diameter in the context of the capillary bundle approach (21). Based on the vast amount of experimental data available at the time, Carman (21), Sullivan and Hertel (95), and Coulson (96) observed that the bounding wall effect on the inertial term is generally much smaller than that on the viscous (Darcy) term. Carman (21) also concluded that the bounding wall surface area is not equivalent to the particle surface area of the granular beds when treating the viscous term. Metha and Hawley (97), however, treated the bounding wall surface area exactly the same as one would treat the particle surface area. The different degree of influence of the bounding wall on the viscous term and the inertial term may be explained by a combination of the momentum dispersion and the porosity variation near the wall.

To the degree of uncertainties and variety of the beds studied, these investigators were able to incorporate the wall effects into the Ergun equation by replacing the particle diameter d_s with a wall modified particle diameter d_p,

$$d_p = d_s/M \tag{102}$$

where M is the wall effect factor, given by

$$M = 1 + \frac{2d_s}{3(1 - \epsilon)D} \tag{103}$$

Here, D is the diameter of the column. For a more general empirical approach, the coefficient in equation 103 needs to be changed or the particle diameter change needs to be corrected partially to correlate the wall effects on the viscous term (*21, 96, 98*).

This approach was also found to correlate power law fluid flow through granular beds of column to particle diameter ratio $5.8 < D/d_s < 20$ in the creeping (Darcy) flow regime (*99*).

The second approach involves dividing the bed into three regions: the wall, transition, and central zones (*100*). By incorporating the detailed porosity variation in the radial direction such as that of Roblee et al. (*101*), Benenati and Brosilow (*102*), Haughey and Beveridge (*103*), Ridgway and Tarbuck (*104*), or Pillai (*105*), Cohen and Metzner (*100*) were able to derive a pressure drop–flow rate relationship for Newtonian and power law fluids. Subsequently, Nield (*106*) demonstrated that a two-region model can yield a pressure drop prediction just as good. More realistically, one may make the modeling more cumbersome than the three-region model by including the angular porosity variation of Muller (*107*). However, these treatments would all jeopardize the use of volume averaging technique by inclusion of an unnecessarily small scale pointwise porosity variation and hence limit their utility.

The third approach is evaluating the constants of the Ergun equation, or Irmay equation, 10, for each column to particle diameter ratio and then correlating these constants with the diameter ratio (*91, 108, 109*). This approach is feasible for a wider diameter ratio range than the first two approaches. However, this approach loses its appeal due to the lack of fundamental background (*99*).

Still based on the first approach, Liu et al. (*32*) studied the wall effects while excluding the multidimensional effects. By heuristic arguments, they derived wall effect correction factors for both viscous and inertial terms. For the viscous term, A of equation 95, is to be replaced by

$$A_w = AC_w^2 = A\left[1 + \frac{\pi d_s}{6D(1 - \epsilon)}\right]^2 \tag{104}$$

where A_w denotes the constant A as corrected due to the presence of a wall. The wall effect corrected inertial term is given by replacing B of equation 95 with

$$B_w = BC_{wi} = B\left[1 - \frac{\pi^2 d_s}{24D}\left(1 - \frac{d_s}{2D}\right)\right] \tag{105}$$

where B_w denotes the constant B as corrected due to the presence of a wall.

The wall-effect-corrected one-dimensional pressure drop is given by

$$f_v = f_m Re_p = 85.2\left[1 + \frac{\pi d_s}{6D(1 - \epsilon)}\right]^2$$
$$+ 0.69\left[1 - \frac{\pi^2 d_s}{24D}(1 - 0.5d_s/D)\right]Re_m \frac{Re_m^2}{16^2 + Re_m^2} \tag{106}$$

The previous LAM equation is the equivalent to the Ergun equation while accounting for wall, viscous, and inertial effects. Essentially, equation 106 is a one-dimensional pressure drop equation.

The shear factor corrected for wall effects based on the model of Liu et al. (32) is given by

$$F = \frac{(1 - \epsilon)^2}{\epsilon^{11/3} d_s^2}\left\{85.2\left[1 + \frac{\pi d_s}{6D(1 - \epsilon)}\right]^2\right.$$
$$+ 0.69\left[1 - \frac{\pi^2 d_s}{24D}(1 - 0.5d_s/D)\right]Re_v \frac{Re_v^2}{16^2 + Re_v^2}\right\} \tag{107}$$

Equation 107 can be referred to as a multidimensional model for the flow in porous media.

The Ergun equation, when corrected for the wall effects, can be written as

$$f_v = \epsilon^{2/3}\left\{150\left[1 + \frac{\pi d_s}{6D(1 - \epsilon)}\right]^2\right.$$
$$+ 1.75\left[1 - \frac{\pi^2 d_s}{24D}(1 - 0.5d_s/D)\right]\frac{\epsilon^{1/6}Re_m}{1 + (1 - \epsilon^{1/2})^{1/2}}\right\} \tag{108}$$

Correspondingly, the shear factor based on the Ergun equation is

$$F = \frac{(1 - \epsilon)^2}{\epsilon^3 d_s^2} \left\{ 150\left[1 + \frac{\pi d_s}{6D(1 - \epsilon)} \right]^2 \right.$$

$$\left. + 1.75\left[1 - \frac{\pi^2 d_s}{24D} (1 - 0.5 d_s/D) \right] \frac{\epsilon^{1/6} Re_v}{1 + (1 - \epsilon^{1/2})^{1/2}} \right\} \qquad (109)$$

Equation 108 is a wall modified one-dimensional Ergun equation, and equation 109 can be considered as the wall modified Ergun equation in a multidimensional format.

Multidimensional Effects.   In the previous section, we studied the wall effect on the shear factor. To give a full account of the wall effects, we now look at the no-slip flow effect posed by the containing wall (multidimensional effect) on the total pressure drop. For simplicity, let us rewrite the normalized pressure drop factor, f_v, based on the permeability of the medium rather than the particle diameter,

$$f_v = -85.2 \frac{k}{\mu q} \frac{\Delta p}{L} \qquad (110)$$

Equation 110 may be treated as a generalized form of f_v.

A porous medium bed confined in a circular tube is commonly used. Figure 11 is a sketch of the cylindrical system. Under such conditions, the only averaged velocity component is the axial velocity u. Hence,

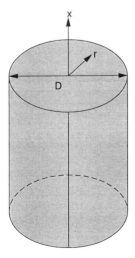

Figure 11.   A sketch of a cylindrical system.

the volume-averaged governing equations may be simplified to one single ordinary differential equation:

$$Fu - \frac{1}{r}\frac{d}{dr}\left(r\frac{du}{dr}\right) = -\frac{1}{\mu}\frac{dp}{dx} = \frac{q}{85.2k}f_v \tag{111}$$

where q is the average superficial flow velocity through the bed,

$$\int_0^{D/2} rudr = \frac{D^2q}{8} \tag{112}$$

Let

$$Y = \frac{u}{q} - \frac{f_v}{85.2kF} \tag{113}$$

then

$$r^2\frac{d^2Y}{dr^2} + r\frac{dY}{dr} - Fr^2Y = 0 \tag{114}$$

For simplicity, we assumed that F does not vary with u, that is, the flow is in the Darcy's flow regime and the second term of the shear factor is zero. Because the variable Y is expected to have a finite value at the center of the pipe, the solution of equation 114 gives

$$Y = \frac{u}{q} - \frac{f_v}{85.2kF} = HI_0(F^{1/2}r) \tag{115}$$

where H is a constant and $I_0(x)$ is the zeroth order modified Bessel function of first kind.

$$I_0(x) = \sum_{n=0}^{\infty} \frac{x^{2n}}{4^n(n!)^2} \tag{116}$$

Using the no-slip boundary condition at the wall, we obtain

$$H = -\frac{f_v}{85.2kF}\frac{1}{I_0(F^{1/2}D/2)} \tag{117}$$

Hence, the final solution can be written as

$$u = q\frac{f_v}{85.2kF}\left[1 - \frac{I_0(F^{1/2}r)}{I_0(F^{1/2}D/2)}\right] \tag{118}$$

In the limiting case of $F = 0$, it can be shown that equation 118 reduces to the Hagen–Poiseuille flow equation, that is,

$$\lim_{F \to 0} u = q \lim_{F \to 0} \frac{f_v}{85.2kF}\left[1 - \frac{I_0(F^{1/2}r)}{I_0(F^{1/2}D/2)}\right]$$

$$\lim_{F \to 0} u = q \lim_{F \to 0} \frac{f_v}{85.2kF}\frac{1 + (F^{1/2}D/2)^2 - 1 - (F^{1/2}r)^2}{1 + (F^{1/2}D/2)^2}$$

$$\lim_{F \to 0} u = q\left[1 - \left(\frac{r}{D/2}\right)^2\right]\lim_{F \to 0}\frac{f_v(D/2)^2}{85.2k} = 2q\left[1 - \left(\frac{r}{D/2}\right)^2\right] \quad (119)$$

We now return to the general solution. Substituting equation 118 into equation 112, we can obtain

$$f_v = \frac{85.2kF}{1 - \frac{4I_1(F^{1/2}D/2)}{F^{1/2}DI_0(F^{1/2}D/2)}} = 85.2kF\frac{1 + \sum\limits_{n=1}^{\infty}\frac{F^n D^{2n}}{16^n(n!)^2}}{\sum\limits_{n=1}^{\infty}\frac{F^n D^{2n}}{16^n(n-1)!(n+1)!}} \quad (120)$$

Let C_{md} be a coefficient for multidimensional effect on the normalized pressure factor,

$$C_{md} = \frac{1}{1 - \frac{4I_1(F^{1/2}D/2)}{F^{1/2}DI_0(F^{1/2}D/2)}} = \frac{1 + \sum\limits_{n=1}^{\infty}\frac{F^n D^{2n}}{16^n(n!)^2}}{\sum\limits_{n=1}^{\infty}\frac{F^n D^{2n}}{16^n(n-1)!(n+1)!}} \quad (121)$$

and

$$\frac{f_v}{85.2kF} = C_{md} \quad (122)$$

Equation 121 may be approximated by

$$C_{md} = 1 + 0.00043 \times 10^{-4}[e^{9.9(F^{1/2}D/2)^{-0.2}} - 1] \quad \text{for } \frac{F^{1/2}D}{2} > 1 \quad (123)$$

Figure 12 shows the multidimensional effect on the normalized pressure drop factor from both the exact solution, equation 121, and its approximation, equation 123. It can be observed that equation 123 gives a fairly good approximation to the exact solution. The multidimensional effect is significant when the normalized bed radius is small, say, $F^{1/2}D/2 < 100$. When the normalized bed radius is large, $C_{md} \to 1$.

The final solution to the velocity field can be obtained by substituting equation 121 into equation 118 to yield

$$u = qC_{md}\left[1 - \frac{I_0(F^{1/2}r)}{I_0(F^{1/2}D/2)}\right] \quad (124)$$

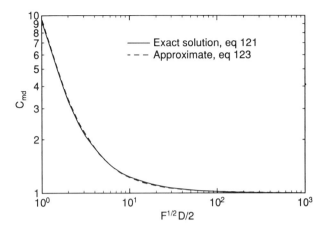

Figure 12.   No slip effect on normalized pressure drop factor.

Figure 13 shows the axial velocity profiles computed from equation 124. One can observe the variation of the axial velocity with the normalized bed radius, $DF^{1/2}/2$. When the normalized bed radius is zero, the axial velocity displays a parabolic profile that corresponds to the Hagen–Poiseuille solution. As the normalized bed radius increases, the axial velocity profile flattens. When the normalized bed radius is infinite, the axial velocity corresponds to a unidirectional flow (flat) profile.

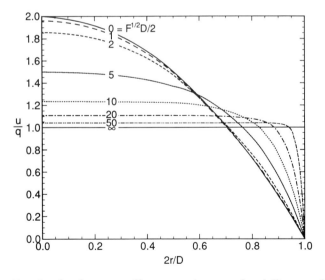

Figure 13.   Axial velocity profile across the pipe for different values of normalized pipe diameter.

From equation 124, we observe that the coefficient of multidimensional effect is also related to the ratio of the superficial velocity near the center line to the average superficial velocity, that is,

$$\frac{f_v}{f_{v1D}} = C_{md} = \frac{u\,(r=0)}{q}\,\frac{I_0(F^{1/2}D/2)}{I_0(F^{1/2}D/2) - 1}, \quad \text{for } F^{1/2}D/2 > 0 \quad (125)$$

Hence, we expect that the larger the multidimensional effect is, the larger the deviation of the center line velocity from the average velocity. In the extreme case where $F = 0$, or the medium is infinitely permeable, the velocity ratio reaches 2, whereas the coefficient of the multidimensional effect is infinite.

Steady Flow in Packed Beds of Monosized Spherical Particles. Steady incompressible fully developed flow in porous media confined in a circular pipe can be treated with a single differential equation as given by equation 111. The inertial effects are only reflected in the shear factor term. Two purposes are served in this section: to verify the integrity of the models presented earlier, including the passage model on shear factor and wall effects on the flow, and to show the flow behavior itself. The flow problem is solved numerically with a central difference method. An abundance of experimental data are available in the literature. However, we confine ourselves to the laminar flow regime for a packed bed of spherical particles. We make use of the latest available data presented by Fand et al. (*110*) for a packed bed with weak wall effects and the experimental data of Liu et al. (*32*).

Figures 14 and 15 show the normalized pressure drop factor for a densely packed bed of monosized spherical particles. For $Re_m < 7$, f_v is fairly independent of Re_m, and at high Re_m values, it increases fairly linearly with Re_m. The data points are the experimental results taken from Fand et al. (*110*), where the bed diameter is $D = 86.6$ mm and the particle diameter is $d_s = 3.072$ mm. One can observe that the 2-dimensional model of Liu et al. (*32*), referred to as equation 107, agrees with the experimental data fairly well in the whole range of the modified Reynolds number. From Figure 14, one observes a smooth transition from the Darcy's flow to Forchheimer flow regime. The one-dimensional model of Liu et al. (*32*) (i.e., equation 106) showed only slightly smaller f_v value. Hence, the no-slip effect or two-dimensional effect for this bed is small. As shown in Figures 14 and 15, the Ergun equation consistently underpredicts the pressure drop. The deviation becomes larger when flow rate is increased.

Figures 16 and 17 show the experimental results of Liu et al. (*32*) as well as the theoretical predictions for the packed bed of large glass beads where $D = 4.47$ mm, $d_s = 3.184$ mm, and $\epsilon = 0.6007$. One observes

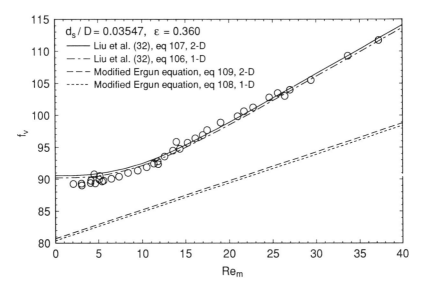

Figure 14. Variation of pressure drop factor with modified Reynolds number for a densely packed bed of monosized spherical particles at low flow rates. The symbols are experimental data taken from reference 110.

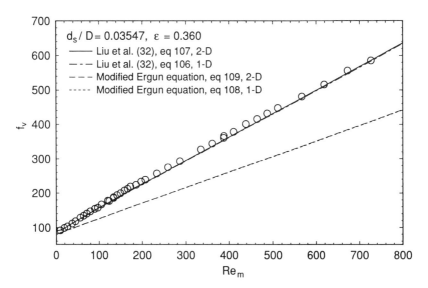

Figure 15. Variation of pressure drop factor with modified Reynolds number for a densely packed bed of monosized spherical particles at high flow rates. The symbols are experimental data taken from reference 110.

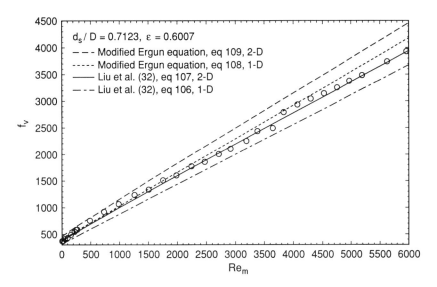

Figure 16. Pressure drop factor variation with modified Reynolds number for a packed bed having large glass beads in the high Re_m range (32).

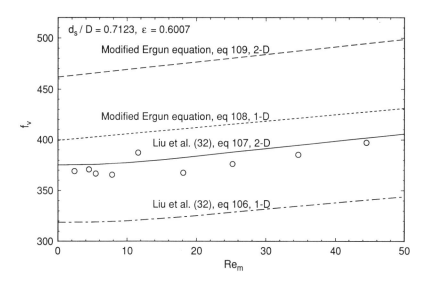

Figure 17. Pressure drop factor variation with modified Reynolds number for a packed bed of large glass beads in the low Re_m range (32).

that the two-dimensional model of Liu et al. (32) predicts the experimental results fairly well for the entire range studied, $Re_m < 6000$. The deviation between the two-dimensional model predictions and the experimental data is within 10%. The wall effect modified Ergun equation overpredicts the experimental data in the entire range studied.

Figures 18 and 19 show the experimental data of Liu et al. (32) for the packed bed of small glass beads as compared with the theoretical predictions. Here, D = 4.47 mm, d_s = 1.917 mm, and ϵ = 0.4529. We observe that the 2-dimensional model of Liu et al. (32) agrees fairly well with the experimental results. The modified Ergun equation predicts the experimental data well in the low Re_m range as is shown in Figure 16 and underpredicts the experimental data as Re_m is increased (Figure 18).

Table IV gives a summary of the packed beds that we made use of in this section. The term C_{md} reflects the two-dimensional effects for Re_m = 0. A value of zero, for $100(C_{md} - 1)$, would indicate no two-dimensional effects. We can observe that the wall effects on the viscous term, C_w^2, range from about 6% for the experimental data of Fand et al. (110) to 274% for the packed bed of Liu et al. (32) with a d_s/D = 0.7123 as used here. The wall effects on the inertial term, C_{wi}, range from around 1 to 19%. The two-dimensional effects are also significant for the packed beds of large particle to tube diameter ratios. The 2-dimensional model of Liu et al. (32) predicts quite well over a wide range of wall effects. In contrast, the wall modified Ergun equation significantly underpredicts at low porosity (Figures 15 and 19) and overpredicts at high porosity (Figures 16 and 17) the experimental data.

Steady Flow Through a Cylindrical Bed of Fibrous Mat. The model of Liu et al. (32) was also tested against high porosity cases for the steady incompressible flow through a fibrous mat of D = 25.5 mm, d_s = 0.167 mm, and ϵ = 0.9301.

Figure 20 shows the calculated pressure drop factor and the experimental values. We observe that the model of Liu et al. (32) predicts the experimental pressure drop both in the Darcy's flow regime, the transition, and the Forchheimer regimes. The two-dimensional model gives a much better prediction than that using the one-dimensional model. The Ergun equation significantly overpredicts the experimental data.

The coefficient for multidimensional effect, equation 121 or 123, is strictly speaking valid for Darcy's flow only. However, it is possible to estimate the multidimensional effect. Because the multidimensional effect is only significant near the wall, the shear factor used for evaluating the multidimensional coefficient may take an average value in the boundary layer. Let

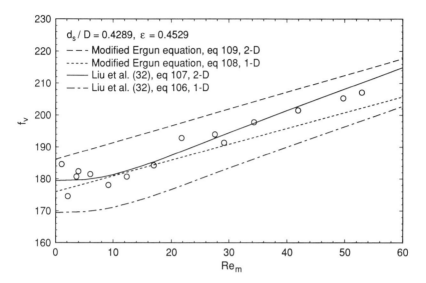

Figure 18.   Pressure drop factor variation with modified Reynolds number for a packed bed of small glass beads in the low Re_m range (32).

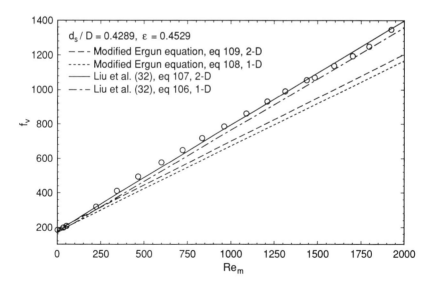

Figure 19.   Pressure drop factor variation with modified Reynolds number for a packed bed of small glass beads in the high Re_m range (32).

Table IV.   Summary of the Characteristics of the Packed Beds
Used in This Study

D, mm	d_s, mm	d_s/D	ε	Ref	C_w^2	C_{wi}	$100(C_{md} - 1)$
86.6	3.072	0.03547	0.360	110	1.0589	0.9857	0.38
4.47	3.184	0.7123	0.6007	32	3.7405	0.8114	15.08
4.47	1.917	0.4289	0.4529	32	1.9894	0.8614	5.56

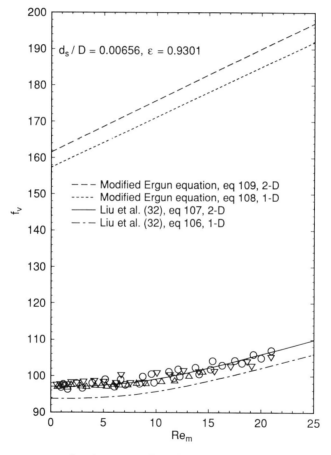

$d_s / D = 0.00656$, $\varepsilon = 0.9301$

– – – Modified Ergun equation, eq 109, 2-D
– – – – Modified Ergun equation, eq 108, 1-D
——— Liu et al. (32), eq 107, 2-D
–·– Liu et al. (32), eq 106, 1-D

Figure 20.   Normalized pressure drop factor variation with modified Rey-nolds number for the fine foam (32).

$$F_D = \frac{3f_{v1D0} + f_{v1D}}{4 \times 85.2k} \frac{D^2}{4} \qquad (126)$$

where f_{v1D} is the normalized pressure drop factor, f_v, evaluated by a one-dimensional model, for example, equation 126 for a granular bed, and f_{v1D0} is the value of f_{v1D} at $Re_m = 0$. The coefficient for multidimensional effect can then be given by

$$C_{md} = 1 + 0.00043 \times 10^{-4}(e^{9.9F_D^{-0.1}} - 1) \qquad (127)$$

and the normalized pressure drop factor becomes

$$f_v = f_{v1D}C_{md} = f_{v1D}[1 + 0.00043(e^{9.9F_D^{-0.1}} - 1)] \qquad (128)$$

Figures 21 and 22 show the normalized pressured drop estimated by equation 128 for a packed $D = 5.588$ mm, $d_s = 3.040$ mm, and $\epsilon = 0.5916$. The experimental data are taken from Fand and Thinakaran (92). We can observe that the approximate solution, equation 128, predicts fairly well the experimental results and is very good in representing the exact numerical solution of the governing equations. For clarity, Figure 21 is an expanded region for the small Re_m values. Even with this scale, we observe that the approximate solution is very close to the exact numerical solution.

It was found that equation 128 is a good prediction for the normalized pressure drop factor when used for data in Figures 14–22. The use

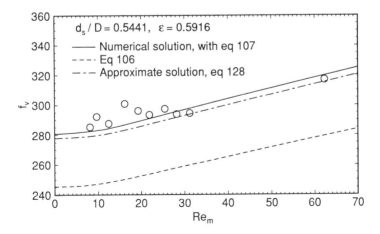

Figure 21. f_v *variation with* Re_m *for a packed bed of monosized spherical particles at low flow rate. The symbols represent the experimental data taken from reference 92.*

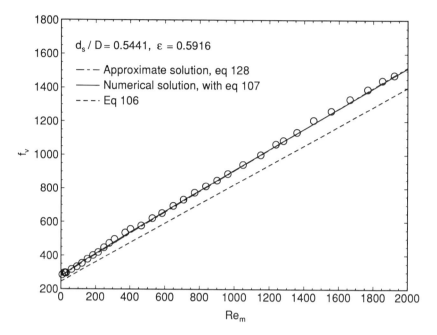

Figure 22.   Normalized pressure drop factor variation with Re_m *for a packed bed of monosized spherical particles at high flow rate. The symbols represent the experimental data taken from reference 92.*

of equation 128 eliminates the need to solve for the two-dimensional flow in a cylindrical porous medium bed, and its use is of the same complexity as that for a one-dimensional problem. However, the predictive accuracy has been vastly improved.

Summary

In this chapter, we introduced the characteristics of porous media from both fundamental and application points of view. We listed some relevant definitions and properties and useful models dealing with flow though porous media and reviewed the description of porous media to introduce to the reader the techniques used in modeling the porous media and the properties of the porous media and their relationship to the models. Field level characterization is actually a direct utilization of the models for transport in porous media, and our section on this may be considered as a study on the transient compressible flows. Because of the importance of the tortuosity and areal porosity, we devoted a section to discuss them. These two properties are not newly defined, but they do generate some confusion in the literature. For instance, the areal porosity was thought to be identical to the porosity for all the porous media, although experimental observations indicate otherwise. This is by and large due to the view point that the pores in a porous medium may be treated as bundles of straight nonintersecting channels. The tortuosity is a classical concept; however, when modeling a transport phenomenon in porous media, different groups of investigators used different values. Even for the same porous medium when dealing with different transport phenomena, different values were assigned.

We then discussed the modeling for single-fluid phase flow in porous media. In particular, the shear factor and permeability model of Liu et al. (32) is discussed in detail. The bounding wall effects are presented. This section completed the modeling requirements for single-phase incompressible flow in porous media. We showed how to solve the governing equations for flow in porous media and an approximate solution of the pressure drop for an incompressible flow through a cylindrically bounded porous bed was constructed.

The usage of the flow equations can be summarized as follows. For the case of a one-dimensional single fluid flow, either equation 106 or 108 can be used to predict the normalized pressure drop factor in a porous medium. The determined normalized pressure drop factor is related to the pressure drop by equation 11. For the simple case of packed spherical beads, d_s and ϵ are known a priori. The Reynolds number is evaluated using equation 93. For random packs of nonspherical particles, the particle's sphericity needs to be known. Equation 73 can be used to estimate d_s. For the case of consolidated porous medium, one can estimate d_s from the knowledge of the intrinsic permeability using equation 14.

For the case of multidimensional or composite flows, the complete averaged Navier–Stokes equation needs to be solved. They are given by equation 19 and 20. The shear factor F is given by either equation 107 or 109. The evaluation of d_s follows that for the case of one-dimensional flow above.

It should be noted that the porosity ϵ of the porous bed used in equation 20 is not assumed to be constant except for short range (particle diameter scale) variations. Therefore, when solving a problem having an interface of free space and porous medium, there is no need to specify a boundary condition for the interface. The variation of ϵ would suffice to distinguish between the porous medium and the free flow. Moreover, in the free flow region, F becomes zero.

When the porous medium is contained in a cylindrical bed, the multidimensional effect is given by equation 127. When the pressure drop is needed, there is no need to solve the governing equation numerically. Equations 106 and 128 can be used for the estimation of pressure drop as well as scaling-up of a packed bed.

Comparison between the models given by the modified Ergun equation 108 or 109 and by LAM equation 106 or 107 with available experimental data would indicate a preference to using LAM-type equations.

List of Symbols

A	first (viscous) constant of LAM equation, dimensionless, $A = 85.2$
A_p	areal porosity, dimensionless
A_t	total cross-sectional area of the porous medium perpendicular to the main flow direction, dimensional
A_w	wall effect corrected first LAM equation constant, dimensionless
a	radius of a circle, long axis of an ellipse, long side width of a rectangle, radius of circular arcs, or side width of an equilateral triangle, all are dimensional
a_c	constant, Archie's law, dimensionless
a_E	passage shape factor: the first (viscous) constant of the Ergun equation, dimensionless.
a_p	specific surface or surface per unit volume, dimensional
B	second (inertial) constant of LAM equation, dimensionless, $B = 0.69$
B_{Dn}	inertial term constant for a curved duct flow, dimensionless
B_v	formation volume factor, dimensionless
B_w	wall effect corrected second LAM equation constant, dimensionless
b	short axis of an ellipse, width of a parallel plate or short side width of a rectangle, all are dimensional
b_E	second (inertial) constant of Ergun equation, dimensionless
C	third (viscous dominant to inertial dominant flow transition) constant of LAM equation, dimensionless
C_D	wellbore storage constant, dimensionless
C_{Dn}	viscous to inertial flow transitional parameter for curved duct flow, dimensionless
C_{md}	multidimensional effect coefficient, dimensionless
C_w	wall effect coefficient (viscous) constant for LAM or Ergun equation, dimensionless
C_{wi}	wall effect coefficient (inertial) constant for LAM or Ergun equation, dimensionless
c	compressibility, dimensional
c_F	form-drag coefficient, Ward's model, dimensionless
D	diameter of containing cylinder for a porous medium bed, dimensional

Dn	dean number, dimensionless
d_e	hydraulic diameter of pores or equivalent passage diameter, dimensional
d_p	modified particle diameter, dimensional
d_s	equivalent particle diameter, dimensional
E	electrical potential, dimensional
$Eif(.)$	modified exponential integral function
$Ei(.)$	exponential integral function
F	shear factor, dimensional
F_D	average shear factor, dimensional
F_ϵ	formation factor, dimensionless
$F(.)$	probability distribution function
$f(.)$	distribution function
f	fanning friction factor, dimensionless
f_m	modified Fanning friction factor, dimensionless
f_v	pressure drop factor, dimensionless
f_{v1D}	value of f_v evaluated by one-dimensional model
f_{v1D0}	value of f_{v1D} at $Re_m = 0$
h	formation height of a reservoir, dimensional
H	constant in equation 115
I	electrical current, dimensional
$I_0(x)$	modified Bessel function of zeroth order
$I_1(x)$	modified Bessel function of first order
J	electrical current density, dimensional
k	intrinsic permeability, dimensional
$\underline{\underline{k}}$	second-order tensorial k
$\overline{k}_{app}$	apparent intrinsic permeability, dimensional
k_c	calculated value of k, dimensional
k_0	Kozeny–Carman constant, dimensionless
k_1	flow passage shape factor, dimensionless
L	thickness of the porous medium parallel to the main flow direction, dimensional
l	radius of the reservoir, dimensional
LAM	Liu–Afacan–Masliyah
LDV	laser-Doppler velocimetry
m	cementation factor
M	wall effect factor, dimensionless
n	index
n	out normal of surface S_i
p	pressure, dimensional
p^*	microscopic pressure, dimensional
$p^*(...)$	dimensionless pressure solution of equation 36
p_0	reference pressure, dimensional
p_i	initial pressure ($t = 0$), dimensional

p_c	capillary pressure, dimensional
p_w	bottom-hole pressure, dimensional
p_w'	extrapolated bottom-hole pressure, dimensional
Q	flow rate, dimensional
q	superficial unidirectional (1-dimensional) or average axial flow velocity, dimensional
Re_m	modified Reynolds number, dimensionless
Re_v	local modified Reynolds number, dimensionless
$R_N(.)$	truncation error of a function at nth term
Re_p	pore Reynolds number, dimensionless
REV	representative elementary volume
r	radial coordinate, dimensional
r_m	mean radius of the capillary pore, dimensional
r_w	wellbore radius, dimensional
r_{we}	effective wellbore radius, dimensional
r_1, r_2	meniscus radii of the immicible fluids in the porous medium pore, dimensional
S	saturation ratio, dimensionless
S_i	solid-free space interface
S_{kin}	skin effect factor, dimensionless
S_n	saturation ratio of the nonwetting phase, dimensionless
S_{nR}	irreducible saturation ratio of the nonwetting phase, dimensionless
S_P	surface area of particle, dimensional
S_w	saturation ratio of the wetting phase, dimensionless
S_{wR}	irreducible saturation ratio of the wetting phase, dimensionless
t	time, dimensional
t_D	dimensionless time, equation 41
t_{Dimage}	dimensionless time based on the image well, equation 52
t_{Dreal}	dimensionless time based on the real well, equation 52
u	superficial axial velocity, dimensional
u_e	characteristic pore velocity, dimensional
V	volume, dimensional
V_P	particle volume, dimensional
V_0	volume of the free space in an REV
v	superficial radial velocity, dimensional
$\mathbf{v}$	superficial velocity vector, dimensional
$\mathbf{v}^*$	microscopic velocity vector, dimensional
x	variable or axial coordinate
Y	variable defined by equation 113

Greek

α	first Forchheimer constant, dimensional
$\underline{\underline{\alpha}}$	second-order tensorial α

α_k	hydraulic diffusivity, dimensional
β	second Forchheimer constant, dimensional
$\underline{\beta}$	second-order tensorial β
Δ	difference
δ	pore size, dimensional
δ_b	pore body diameter, dimensional
δ_e	entry pore diameter, dimensional
ϵ	porosity, dimensionless
θ	contact angle
λ_m	characteristic curvature ratio, dimensionless
μ	viscosity, dimensional
$\breve{\mu}$	effective viscosity, dimensional
ρ	density, dimensional
Σ	summation
σ_w	electric conductivity of a fluid
σ_0	electric conductivity of a porous medium saturated with a conducting fluid
τ	tortuosity, dimensionless
τ_c	calculated value of tortuosity, dimensionless
Φ_s	particle sphericity, dimensionless

Acknowledgments

We thank the Natural Sciences and Engineering Research Council of Canada and NCE for financial support.

References

1. Scheidegger, A. E. *The Physics of Flow Through Porous Media;* University of Toronto: Toronto, Canada, 1974.
2. Darcy, H. P. G. *Les Fontaines Publique de la Ville de Dijon;* Victor Dalmont: Paris, 1856; English translation: Muskat, M. *Flow of Homogenous Fluids Through Porous Media;* McGraw-Hill: New York, 1937.
3. Klinkenberg, L. J. *Drill. Prod. Pract.* **1941,** 200–213.
4. Brinkman, H. C. *Appl. Sci. Res.* **1949,** *A1,* 27–34.
5. Neale, G., Epstein, N.; Nader, W. *Chem. Eng. Sci.* **1973,** *28,* 1865–1874.
6. Nandakumar, K.; Masliyah, J. H. *Can. J. Chem. Eng.* **1982,** *60,* 202–211.
7. Whitaker, S. *Chem. Eng. Sci.* **1966,** *21,* 291–300.
8. Slattery, J. C. *AIChE J.* **1969,** *15,* 866–872.
9. Lundgren, T. S. *J. Fluid Mech.* **1972,** *51,* 273–299.
10. Forchheimer, P. *Z. Ver. Dtsch. Ing.* **1901,** *45,* 125–127.
11. Blick, E. F. *I&EC Process Des. Dev.* **1966,** *1,* 90–94.
12. Coulaud, O., Morel, P.; Caltagirone, J. P. *J. Fluid Mech.* **1988,** *190,* 393–407.
13. Dullien, F. A. L.; Azzam, M. I. S. *AIChE J.* **1973,** *19,* 222–229.
14. Barak, A. Z.; Bear, J. *Adv. Water Res.* **1981,** *4,* 54–66.
15. Cvetkovic, V. D. *Transp. Porous Media* **1986,** *1,* 63–97.

16. Ruth, D. W.; Ma, H. *Transp. Porous Media* **1992**, *7*, 255–264.
17. Ward, J. C. *ASCE J. Hydraul. Div.* **1964**, *90*, 1–12.
18. Joseph, D. D., Nield, D. A.; Papanicolaou, G. *Water Resourc. Res.* **1982**, *18*, 1049–1052 and *19*, 591.
19. Nield, D. A.; Bejan, A. *Convection in Porous Media;* Springer-Verlag: New York, 1992.
20. Kozeny, J. *Ober Kapillare Leitung das Wassers in Boden;* S. Ber. Wierner Akad. Abt., *IIa*, 1927; pp 136–271.
21. Carman, P. C. *Trans. Inst. Chem. Eng.* **1937**, *15*, 150–166.
22. Ergun, S. *Chem. Eng. Progr.* **1952**, *48*, 89–94.
23. Irmay, S. *J. Geophys. Res.* **1958**, *39*, 702–707.
24. Wentz, C. A.; Thodos, G. *AIChE J.* **1963**, *9*, 81–84.
25. Handley, D.; Heggs, P. J. *Trans. Inst. Chem. Eng.* **1968**, *46*, T251–T264.
26. Tallmadge, J. A. *AIChE J.* **1970**, *16*, 1092–1093.
27. Hicks, R. E. *Ind. Eng. Chem. Fundam.* **1970**, *9*, 500–502.
28. Jones, D. P.; Krier, H. *J. Fluid Eng.* **1983**, *105*, 168–172.
29. MacDonald, I. F.; El-Sayed, M. S.; Mow, K.; Dullien F. A. L. *Ind. Eng. Chem. Fundam.* **1979**, *18*, 199–208.
30. Dullien, F. A. L. *Porous Media, Fluid Transport and Pore Structure;* Academic: Orlando, FL, 1979.
31. Dullien, F. A. L. *Porous Media, Fluid Transport and Pore Structure,* 2nd ed.; Academic: Toronto, Canada, 1992.
32. Liu, S.; Afacan, A.; Masliyah, J. H. *Chem. Eng. Sci.* **1994**, *49*, 3565–3586.
33. Bear, J. *Dynamics of Fluids in Porous Media;* Dover: New York, 1972.
34. Miller, B.; Tyomkin, I. *J. Colloid Interface Sci.* **1994**, *162*, 163–170.
35. Klinkenberg, L. J. *Pet. Trans. AIME* **1957**, *210*, 366.
36. Carman, P. C. *Soil Sci.* **1941**, *52*, 1–14.
37. Fatt, I. *Pet. Trans. AIME* **1956**, *207*, 144–160.
38. Fatt, I. *Pet. Trans. AIME* **1956**, *207*, 160–164.
39. Simon, R.; Kesley, F. J. *SPE J.* **1972**, *12*, 345–351.
40. Irmay, S. *Bull. Res. Coun. Isr.* **1955**, *5A(1)*, 84.
41. Chatzis, I.; Dullien, F. A. L. *J. Can. Pet. Technol.* **1977**, *16*, 97–108.
42. Garboczi, E. J. *Powder Technol.* **1991**, *67*, 121–125.
43. Matthews, G. P.; Moss, A. K.; Spearing, M. C.; Voland, F. *Powder Technol.* **1993**, *76*, 95–107.
44. Earlougher, C. R., Jr. *Advances in Well Test Analysis;* Monograph Vol. 5; Society of Petroleum Engineers of AIME: New York, 1977.
45. Hurst, W. *Pet. Eng. J.* **1953**, *25(11)*, B6–B16.
46. Van Everdingen, A. F. *Trans. AIME* **1953**, *198*, 171–176.
47. Hawkins, M. F., Jr. *Trans. AIME* **1956**, *207*, 356–357.
48. Odeh, A. S. *Trans. AIME* **1980**, *269*, 964–965.
49. Slider, H.C. *"Slip," Worldwide Practical Petroleum Reservoir Engineering Methods;* PennWell: Tulsa, OK, 1983.
50. Stiles, W. M. E. In *Petroleum Transactions 2: Water Flooding;* American Institute of Mining, Metallurgical, and Petroleum Engineers: New York, 1948; pp 101–105.
51. Freeze, R. A. *Water Resour. Res.* **1975**, *11*, 725–741.
52. King, P. R. *J. Phys. A: Math. Gen.* **1987**, *20*, 3935–3947.
53. Drummond, I. T.; Horgan, R. R. *J. Phys. A: Math. Gen.* **1987**, *20*, 4661–4672.

54. Law, J. *Pet. Trans. AIME* **1944**, *155*, 202–222.
55. Henriette, A.; Jacquin, C. G.; Adler, P. M. *Physicochem. Hydrodyn.* **1989**, *11*, 63–80.
56. Ruth, D. W.; Suman, R. *Transp. Porous Media* **1992**, *7*, 103–125.
57. Dullien, F. A. L.; Mehta, P. N. *Powder Technol.* **1971**, *5*, 179–193.
58. Suman, R.; Ruth, D. W. *Transp. Porous Media* **1993**, *12*, 185–206.
59. Archie, G. E. *Trans. AIME* **1942**, *146*, 54–61.
60. De La Rue, R. E.; Tobias, C. W. *J. Electrochem. Soc.* **1959**, *106*, 827–833.
61. Sen, P. N.; Scala, C.; Cohen, M. H. *Geophysics* **1981**, *46*, 781–795.
62. Johnson, D. L.; Plona, T. J.; Scala, C.; Pasierb, F.; Kojima, H. *Phys. Rev. Lett.* **1982**, *49*, 1840–1844.
63. Wong, P.-Z.; Koplik, J.; Tomanic, J. P. *Phys. Rev.* **1984**, *B30*, 6606–6614.
64. Maxwell, J. C. *A Treatise on Electricity and Magnetism*, 2nd ed.; Oxford University: London, 1881.
65. Epstein, N. *Chem. Eng. Sci.* **1989**, *44*, 777–779.
66. Puncochar, M.; Drahos, J. *Chem. Eng. Sci.* **1993**, *48*, 2173–2175.
67. Happel, J. *AIChE J.* **1958**, *4*, 197–201.
68. Philipse, A. P.; Pathmamanoharan, C. *J. Colloid Interface Sci.* **1993**, *159*, 96–107.
69. Turner, G. A. *Chem. Eng. Sci.* **1958**, *7*, 156–165.
70. Turner, G. A. *Chem. Eng. Sci.* **1959**, *10*, 14–21.
71. Payatakes, A. C.; Tien C.; Turian, R. *AIChE J.* **1973**, *19(I)*, 58–67; *19(II)*, 67–76.
72. Dullien, F. A. L. *AIChE J.* **1975**, *21*, 299–307.
73. Azzam, M. I. S.; Dullien, F. A. L. *Chem. Eng. Sci.* **1977**, *32*, 1445–1455.
74. Du Plessis, J. P.; Masliyah, J. H. *Transp. Porous Media* **1988**, *3*, 145–161.
75. Du Plessis, J. P.; Masliyah, J. H. *Transp. Porous Media* **1991**, *6*, 207–221.
76. Du Plessis, J. P. In *Computational Methods in Water Resources IX, Vol. 2: Mathematical Modelling in Water Resources*; Russel, T. F.; Ewing, R. E.; Brebbia, C. A.; Gray, W. G.; Pinder, G. F., Eds.; Computational Mechanics: Boston, MA, 1992.
77. Ma, H.; Ruth, D. W. *Transp. Porous Media* **1993**, *13*, 139–160.
78. Lin, C.; Cohen, M. H. *J. Appl. Phys.* **1982**, *53*, 4152–4165.
79. Koplik, J. *J. Fluid Mech.* **1982**, *119*, 219–247.
80. Koplik, M. J.; Lin, C.; Vermette, M. *J. Appl. Phys.* **1984**, *56*, 3127–3131.
81. Ioannidis, M. A.; Chatzis, I. *Chem. Eng. Sci.* **1993**, *48*, 951–972.
82. Sotirchos, S. V.; Zarkanitis, S. *Chem. Eng. Sci.* **1993**, *48*, 1487–1502.
83. Childs, E. C.; Collis-George, N. *Proc. Roy. Soc. London A* **1950**, *201*, 392–405.
84. Brutsaert, W. *Water Resour. Res.* **1968**, *4*, 425–434.
85. Haring, R. E.; Greenkorn, R. A. *AIChE J.* **1970**, *16*, 477–483.
86. Panigrahi, M. R.; Murty, J. S. *Chem. Eng. Sci.* **1991**, *46*, 1863–1868.
87. Liu, S.; Masliyah, J. H. *J. Fluid Mech.* **1993**, *251*, 315–353.
88. Ward-Smith, A. J. *Internal Fluid Flow: The Fluid Dynamics of Flow in Pipes and Ducts*; Oxford University: New York, 1980.
89. Foscolo, P. U.; Gibilaro, L. G.; Waldram, S. P. *Chem. Eng. Sci.* **1983**, *38*, 1251–1260.
90. Katz, A. J.; Thompson, A. H. *Phys. Rev.* **1986**, *B34*, 8179–8181.
91. Macmullin, R. B.; Muccini, G. A. *AIChE J.* **1956**, *2*, 393–403.
92. Fand, R. M.; Thinakaran, R. *Trans. ASME: J. Fluids Eng.* **1990**, *112*, 84–88.

93. Dybbs, A.; Edwards, R. V. In *Fundamentals of Transport Phenomena in Porous Media*; Bear, J.; Corapcioglu, M. Y., Eds.; Martinus Nijihoff: Hingham, MA, 1984.
94. Liu, S. Ph.D. Dissertation, University of Alberta, Edmonton, Canada, 1992.
95. Sullivan, R. R.; Hertel, K. L. *J. Appl. Phys.* **1940**, *11*, 761–765.
96. Coulson, J. M. *Trans. Inst. Chem. Eng.* **1949**, *27*, 237–257.
97. Metha, D.; Hawley, M. C. *I&EC Process Des. Dev.* **1969**, *8*, 280–282.
98. Dolejs, V. *Int. Chem. Eng.* **1978**, *18*, 718–723.
99. Rao, P. T.; Chhabra, R. P. *Powder Technol.* **1993**, *77*, 171–176.
100. Cohen, Y.; Metzner, A. B. *AIChE J.* **1981**, *27*, 705–715.
101. Roblee, L. H. S.; Baird, R. M.; Tierney, J. W. *AIChE J.* **1958**, *4*, 460–464.
102. Benenati, R. F.; Brosilow, C. W. *AIChE J.* **1962**, *8*, 359–361.
103. Haughey, D. P.; Beveridge, G. S. G. *Chem. Eng. Sci.* **1966**, *21*, 905–916.
104. Ridgway, K.; Tarbuck, K. J. *Chem. Eng. Sci.* **1968**, *23*, 1147–1155.
105. Pillai, K. K. *Chem. Eng. Sci.* **1977**, *32*, 59–61.
106. Nield, D. A. *AIChE J.* **1983**, *29*, 688–689.
107. Muller, G. E. *Powder Technol.* **1993**, *77*, 313–319.
108. Reichelt, W. *Chem. Ing. Tech.* **1972**, *44*, 1068–1071.
109. Srinivas, B. K.; Chhabra, R. P. *Int. J. Eng. Fluid Mech.* **1992**, *5*, 309.
110. Fand, R. M.; Kim, B. Y. K.; Lam, A. C. C.; Phan, R. T. *Trans. ASME: J. Fluids Eng.* **1987**, *109*, 268–274.

RECEIVED for review July 28, 1994. ACCEPTED revised manuscript June 6, 1995.

SUSPENSIONS IN POROUS MEDIA

Permeability Decline Due to Flow of Dilute Suspensions Through Porous Media

H. A. Nasr-El-Din

Lab R&D Center, Saudi Aramco, P.O. Box 62, Dhahran 31311, Saudi Arabia

Suspension flow in porous media is encountered in many industrial applications. In the oil industry, suspended solids present in injected waters can cause significant damage around the wellbore or deep in the formation. Depending on the physical properties of the solid particles, the porous medium, and operating conditions, solids can form external or internal filter cake, or just flow through the media without causing any damage. External filter cake formation causes a fast and sharp drop in permeability or injectivity of the formation. Reversing the flow direction can recover some of the damaged permeability. Internal filter cake formation cases a gradual or steady drop in permeability. Reversing the flow direction will not recover the damaged permeability. Increasing solids concentration or particle size will cause more damage to formation. Injection of low-salinity water into sandstone reservoirs can trigger fines migration and clay swelling. Both factors can damage the formation. Injection of water that is incompatible with the formation brine may cause precipitation of insoluble sulfates that can plug the formation. Stimulation (or acidizing) the formation can also produce solid particles that can damage the formation. Corrosion by-products (e.g., iron sulfide) can block the flow paths and reduce the permeability of the formation. Many experimental and modeling studies to predict formation damage due to flow of suspensions in porous media are discussed in this chapter. Solids can be present in injected waters or be generated in the formation. More research is needed to predict flow of suspensions in porous media when solid particles invade and are generated in the formation simultaneously.

THE FLOW OF SUSPENSIONS through porous media has many practical applications. The process of filtration is a very important separation process in which a porous medium (filter) is used to remove suspended solids. Another practical example of suspensions flow in porous media is encountered in the oil industry. Water injection is used to recover crude oil on a large scale throughout the world (1). The presence of suspended solids in water can cause major problems (2–23). The suspended particles can plug the formation and, as a result, the amount of water that can be injected to the formation will diminish with time. A third example, also encountered in the oil industry, is disposing of large amounts of water that are produced with the crude oil. This water is usually disposed of by injection into the formation, using disposal wells. Again, the presence of solids or oil droplets in the disposed water will reduce the injection capacity of these wells.

Injection waters used in the oil industry may be taken from various sources: seawater, fresh water, subterranean water, and production water, which was initially brought to the surface with the crude oil. Injection waters may contain different kinds of particulate materials (formation particles, corrosion products, insoluble carbonates, or sulfates, iron compounds, oil-in-water emulsions, and bacteria) that may be deposited in the rock pores. Because this deposition may lead to well impairment, injection water may have to be treated before it is injected. Oil-producing wells can also be damaged during drilling or workover. Solids present in drilling muds and workover fluids can invade the formation and damage the wellbore area (3, 10, 24, 25).

Depending on the size of the particles and the pore size distribution of the porous medium, solid particles can be removed on the surface of the porous medium (filter cake) or inside the medium (deep filtration). If the particle size is larger than the pore throat diameter of the porous media, then particles will be separated on the face of the porous medium (i.e., form a skin) and will not deeply penetrate the porous medium (Figure 1). If the particle diameter is very small in comparison with pore

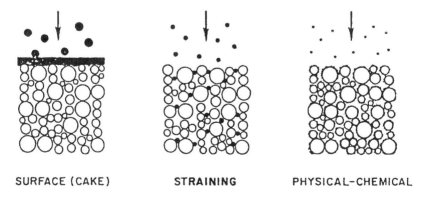

SURFACE (CAKE)   STRAINING   PHYSICAL–CHEMICAL

Figure 1.   Types of formation plugging.

throat diameters, then the particle will flow in the porous medium without causing any change to the fluid paths (i.e., they will not damage the medium).

According to Barkman and Davidson (2), loss of injectivity due to suspended solids occurs by one of the following mechanisms:

1. Particles larger than one-third of the pore diameter bridge pore entrances at the formation face to form an external filter cake.

2. Particles smaller than one-third, but larger than one-seventh of the pore diameter invade the formation and are trapped, forming an internal filter cake.

3. Particles smaller than one-seventh of the pore diameter cause no formation impairment, because they are carried through the formation.

Other mechanisms for particle capture include particle settling, inertia, and hydrodynamics (1). When a particle is captured in a pore, the pore throat diameter is reduced by a factor dependent on particle size and pore throat diameter. The reduced diameter of the pore throat will lead to a lower permeability. Permeability damage due to flow of suspensions increases with particle size and solids concentration.

Another source of solids is the oil reservoir itself. Formation fines can be mobilized and flow in the formation as a suspension. A typical example of this case is the injection of a low-salinity brine into a sandstone reservoir. Sandstone reservoirs contain varying amounts of clay particles (e.g., smectite, kaolinite, and illite). Injection of water having a salinity less than that of the oil reservoir can cause fines migration (kaolinite) and clay swelling (smectite). These clay particles move with the fluid, deposit at the pore throats, and cause plugging of the formation. This will lead to loss of injectivity of injection and disposal wells. Many studies have considered permeability decline due to fines migration and clay swelling (26–36). A brief review of fines migration will be given later.

Solids can form upon mixing two incompatible waters. In this case, calcium or stronsium sulfate precipitates that may affect the permeability of the porous medium. Solid particles can also form as a result of microbial corrosion. Sulfate-reducing bacteria (SRB) form iron sulfide and biomass as corrosion by-products that may plug the formation.

Damage in injection and producing wells can be removed in most cases by stimulating the well. In this case, an acid is injected to remove the damage. The type of the acid and its volume depend on the lithology of the formation (carbonate or sandstone) and the nature of the plugging material. Hydrochloric acid at 15 or 28 wt% is used to stimulate carbonate reservoirs. Various combinations of hydrochloric and hydrofluoric acids are used in sandstone reservoirs (37).

There is some similarity between flow of oil-in-water emulsions in porous media with that of suspensions. The former has been studied extensively (38–47). Some of the results obtained is related to suspension flow in porous media and will be also discussed.

In this review, however, the influence of solid particles and oil droplets on fluid flow in porous media will be examined. This chapter starts by defining the permeability and other physical properties of porous media. Various mechanisms of formation plugging due to flow of suspensions through porous media will be discussed. Models to determine permeability decline as a function of time will be reviewed.

Properties of Porous Media

A porous medium is simply a solid containing void spaces. Several properties of porous media that affect the flow of suspensions in these media are as follows: porosity, permeability, and pore size distribution of the rock. In the following sections, a brief description of these properties is given.

Porosity.   The fraction of total (bulk) volume occupied by the voids is defined as the porosity of the porous medium. A porous medium can be classified according to the type of porosity involved. In sandstone and unconsolidated sand, the voids are between sand grains, and this type of porosity is known as intergranular. Carbonate rocks are more complex and may contain more than one type of porosity. The small voids between the crystals of calcite or dolomite constitute intercrystalline porosity (47). Often carbonate rocks are naturally fractured. The void volume formed by fractures constitutes the fracture porosity. Carbonate rocks sometimes contain vugs, and these carbonate rocks constitute the vugular porosity. Still some carbonate formations may contain very large channels and cavities, which constitute the cavernous porosity.

Absolute and Effective Permeability.   The permeability of a porous medium is a measure of the fluid conductivity of the medium that determines the flow rate at a given pressure gradient. The permeability of a porous medium can be determined using Darcy's equation as follows. The porous medium, usually a core, is saturated under vacuum with a fluid of a known viscosity. The fluid is then injected into the core in a linear mode at a constant flow rate under isothermal flow conditions. If the porous medium is homogeneous, isotropic, the core is placed in horizontal direction, and Reynolds number of the injected fluid is less than unity, then Darcy's law can be used to determine permeability. Darcy's law is

$$q = \frac{KA}{\mu} \frac{\Delta P}{L} \tag{1}$$

where q is the flow rate (mL/s), K is the permeability (darcies), A is the cross-sectional area of the medium (cm^2), μ is viscosity of the fluid (mPa·s) and $\Delta P/L$ is the pressure gradient (atm/cm).

For the case of multiphase flow in a porous medium, the flow rate of a phase j can be calculated from

$$q_j = \frac{K_j A}{\mu_j} \frac{\Delta P_j}{L} \tag{2}$$

where K_j is the effective permeability of fluid j, μ_j is its viscosity and q_i is its flow rate.

Pore Size Distribution. The pore size distribution of a porous medium is a measure of the average size of the pores and the variability of pore sizes. It is usually determined by mercury porosimetry. In this technique the pressure required to force mercury into a pore can be related to the radius of the pore by

$$P_c = \frac{2\sigma_{ca} \cos(\theta)}{r} \tag{3}$$

where P_c is the capillary pressure, r is the radius of the pore, σ is the surface tension of mercury, and θ is the contact angle. By measuring the volume of mercury entering into a sample as a function of the applied pressure, the pore size distribution of the porous medium can be determined from the following equation (48):

$$d(r) = \frac{P}{r} \frac{d(V_0 - V)}{dP} \tag{4}$$

where $d(r)$ is the differential pore radius (m), P is the pressure (Pa), V_0 is the pore volume of the sample (m^3), and V is the volume of mercury injected (m^3).

The permeability of a porous medium can be calculated from the pore size distribution obtained by the mercury injection method (1) using

$$K = 10.24 \, (\sigma \cos \theta)^2 \phi \lambda \int_{x=0}^{x=1} \frac{dx}{(P_c)^2} \tag{5}$$

where K is the permeability (mD), σ is the surface tension (dyn/cm), λ is the lithology factor, x is the fraction of pore space filled, ϕ is the porosity, and P_c is the capillary pressure (psia).

Permeability Decline Due to Flow of Suspensions

An important aspect of the flow of suspensions through porous media is that the solid particles can change the morphology of the pores. This change can simply occur by plugging some of the pore throats. As a result, the flow resistance of the medium will increase or its permeability will diminish. Permeability impairment is usually assessed by examining the variation of the permeability ratio, K/K_i, with respect to time, where K is the permeability at any time and K_i is the initial permeability.

There are several chemical and mechanical factors that can result in the generation or release of fine solid particles. These particles can plug the formation by several mechanisms that will be discussed in the following sections.

Chemical Factors.   There are several mechanisms to damage a formation that involve chemical factors. The most important and well defined mechanisms are the following: (1) fines migration and clay swelling, (2) mixing of incompatible waters, and (3) corrosion products. A brief discussion of each mechanism follows.

Fines Migration and Clay Swelling.   This mechanism is mainly related to sandstone formations. These formations contain nearly 10 wt% clay particles (3, 49). According to Almon and Davies (50) there are four major types of clay minerals that exist in petroleum reservoirs: kaolinite, smectite, illite, and chlorite. These clay particles are present in the formation in the form of small particles (typically <2 μm) and they are plately in shape. These particles are attached to the formation by London–van der Waal attraction forces. Injection of a low-salinity water (low ionic strength or different pH) can mobilize kaolinite particles and cause formation damage via pore throat plugging. Smectite can swell when it comes in contact with low-salinity waters. As smectite swells, it reduces the size of the flow paths and, as a result, the permeability decreases. Chlorite, an iron-rich clay, can cause formation damage during stimulation. Hydrochloric acid reacts with the iron present in chlorite. When the acid is spent and the pH is greater than 2–3, ferric hydroxide, a gelatinous material, will precipitate and block the face of the formation (3, 50). Other minerals present in the reservoir (e.g., pyrite, hematite, magnetite, and siderite) can react with hydrochloric acid and precipitate iron once the acid is spent (17).

The question now is why a change in salinity or pH can trigger fines migration? This point has been extensively studied over the last few decades and is discussed in Chapter 7 of this book. However, a brief explanation of this phenomenon will be given here. A fine particle attached to another larger particle (Figure 2) is subjected to several colloidal forces, including London–van der Waals attraction force, Born

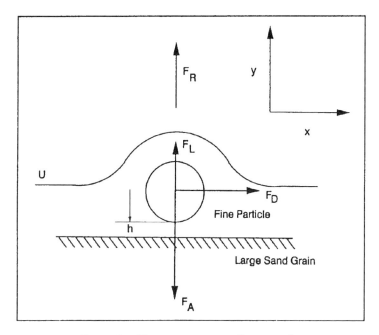

Figure 2. Forces acting on a fine particle.

repulsion force, electric double layer repulsion, and hydration forces (51). The London–van der Waals and the electric double layer repulsion forces are the dominant colloidal forces (52).

The attraction force between a small particle and a large sand grain depends on the distance between the two particles, h, the Hamaker constant, A_H, and the radius of the fine particle, a. For a small spherical particle attached to a larger sand grain, the attraction force, F_A, is (53)

$$F_A = - \frac{aA_H}{6h^2} \tag{6}$$

The electric double layer repulsion force, F_R, for the same system is (53)

$$F_R = \frac{4\pi\epsilon\kappa a\phi_1\phi_2 \exp(-\kappa h)}{[1 + \exp(-\kappa h)]} \tag{7}$$

where ϕ_1 and ϕ_2 are the surface potentials of the two particles, ϵ is the dielectric constant of the solution, and κ is the Debye–Hückel reciprocal length. It should be mentioned that equation 7 applies for ϕ_1 and ϕ_2 $\approx$ 50–60 mV and $\kappa a \geq 10$.

The electric double layer thickness is a function of the fluid ionic strength. For aqueous solutions having a symmetric electrolyte (e.g., NaCl), the electric double layer thickness, $1/\kappa$, at 25 °C is (54)

$$1/\kappa = \frac{0.304 \times 10^{-9}}{I^{0.5}} \qquad (8)$$

where I is the molar ionic strength defined as

$$I = 0.5 \sum C_i Z_i^2 \qquad (9)$$

where C_i is the concentration of an ion i and Z_i is its valency.

The electric double layer repulsion force is a strong function of the ionic strength of the surrounding fluid. Injection of a low-salinity brine will increase the thickness of the double layer, the repulsion force increases, and its effect on fines migration becomes very significant (Figure 3). Fogler and co-workers (28–30, 34) examined the colloidal forces acting on a fine particle in detail. According to their results, clay particles will detach and damage the formation below a critical salt concentration, C_{sc}. The critical salt concentration is a function of cation type in the injected fluid, type, and amount of clays present in the porous medium.

Permeability Reduction in Presence of Oil.   Muecke (55) and Sarkar and Sharma (56) examined permeability reduction due to chemical fines migration in Berea core samples saturated with brine and at residual oil saturation. Liu and Civan (49) modeled permeability impairment due to salinity shock with and without the presence of residual oil. In the first experiment, fresh water was injected into a Berea sandstone core saturated with 3 wt% sodium chloride. In the second experiment, the fresh water was injected into a similar core at residual oil saturation. Figure 4 shows that their model predicted the experimental data fairly

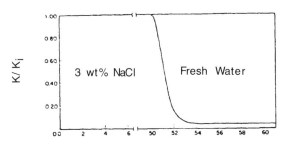

Figure 3.   Permeability decline due to chemical fines migration. (Reproduced with permission from reference 29. Copyright 1983.)

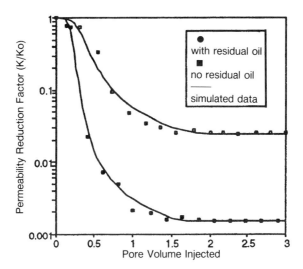

Figure 4. Permeability reduction due to chemical fines migration for core with and without residual oil. (Reproduced with permission from reference 49. Copyright 1993.)

well. Both experimental data and model predictions indicate that formation damage in the presence of oil is less pronounced.

Mixing Incompatible Waters. The second mechanism of formation damage due to chemical factors occurs when the injection water is incompatible with the formation brine. Mixing of incompatible brines can produce solid precipitates if supersaturation is reached. Blending a water that contains high sulfate ion concentration with a formation water that is rich in calcium ion is a good example for mixing two incompatible fluids. In this case, calcium sulfate will precipitate once its solubility product is exceeded. Calcium sulfate can precipitate in the following forms (57): gypsum or dihydrate ($CaSO_4 \cdot 2H_2O$), metastable hemihydrate ($CaSO_4 \cdot \frac{1}{2} H_2O$), and anhydrite ($CaSO_4$). The crystalline form of calcium sulfate depends on the operating conditions, for example, temperature and total dissolved solids (58, 59).

Compatibility tests should be conducted using reservoir brines at reservoir temperature. Formation of calcium sulfate scale is a serious problem because such a scale cannot be removed with hydrochloric or hydrofluoric acid, but can be removed, however, by first injecting sodium carbonate and then acidizing using hydrochloric acid. EDTA can be also used to remove calcium sulfate scale (37).

Corrosion Products. The third mechanism of formation plugging is related to corrosion products. There are several types of corrosion (e.g., electrochemical and microbial) that can produce solid particles that can affect fluid flow in porous media. Sulfate-reducing bacteria (SRB) cause microbial corrosion and produce solid particles and biomass. These bacteria live in water under aneorobic conditions, causing major corrosion to water pipelines, pumps, and other handling equipment (60, 61). According to von Wolzogen Kuhr and van der Vlugt (62), SRB attack iron as follows:

$$4Fe \rightarrow 4Fe^{2+} + 8e \qquad \text{Anodic reaction}$$
$$8H_2O \rightarrow 8H^+ + 8OH^- \qquad \text{Dissociation of water}$$
$$8H^+ + 8e \rightarrow 8H \qquad \text{Cathodic reaction}$$
$$SO_4^{-2} + 8H \rightarrow S^{-2} + 4H_2O \qquad \text{Cathodic depolarization by SRB}$$
$$Fe^{+2} + S^{-2} \rightarrow FeS \qquad \text{Corrosion product}$$
$$3Fe^{+2} + 6OH^- \rightarrow 3Fe(OH)_2 \qquad \text{Corrosion product}$$

Overall reaction

$$4Fe + SO_4^{-2} + 4H_2O \rightarrow FeS + 3Fe(OH)_2 + 2OH$$

Figure 5 illustrates microbial corrosion by SRB (63). The corrosion products FeS and $Fe(OH)_2$ flow with the injection water as suspended solids. They deposit in the formation and cause significant reduction in the injectivity of water injection wells (64). The presence of SRB is characterized by the formation of tiny black iron sulfide particles and hydrogen sulfide smell is interesting to note. Sulfate-reducing bacteria

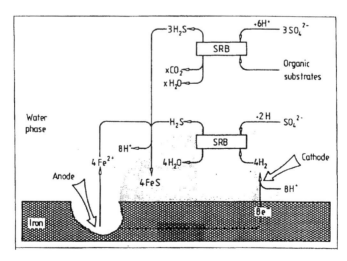

Figure 5.   SRB microbial corrosion. (Reproduced with permission from reference 63. Copyright 1987.)

can also cause formation plugging by themselves. They grow and form colonies that restrict fluid flow in the formation. These bacteria require degradable organic compounds as nutrients (e.g., lactate and acetate). They can live in saline waters up to 80 °C and pH from 5.5 to 8.5 (63).

Mechanical Factors. The most important mechanisms of formation damage by mechanical factors are (1) mechanical fines migration and (2) invasion of foreign particles.

Mechanical Fines Migration. A fine particle attached to the surface of the sand particle is subjected to several hydrodynamic forces (51, 65). The two major hydrodynamic forces that act on a fine particle are the lift force and the viscous-drag force (Figure 2). The lift force, F_L, acts in a direction normal to the flow direction (y-direction) and its magnitude is (66)

$$F_L = 81.2a^2\mu\nu^{-0.5}(du/dy)^{0.5}U \qquad (10)$$

where ν is the kinematic viscosity, a is the radius of the fine particle, μ is the fluid viscosity, du/dy is the velocity gradient, and U is the fluid velocity at the center of the particle, $y = a$, where y is the vertical distance measured from the surface of the sand grain.

The fluid viscous drag force, F_D, acts in the flow direction (x-direction) and is given by (66)

$$F_D = 1.7(6\pi)\mu aU \qquad (11)$$

Equations 10 and 11 indicate that the relative importance of the hydrodynamic forces increases with increasing fluid velocity, viscosity, or the radius of the fine particle. Both hydrodynamic forces tend to dislodge the fine particle from the sand grain. According to Herzig et al. (12), drag and hydrodynamic forces are dominant for particle diameters >30 μm. It should be mentioned that there is a critical velocity (32, 33) above which mechanical fines migration will occur. Exceeding this velocity either around injection or producing wells will cause fines migration. Figure 6, from Miranda and Underdown (67), shows a typical decline in permeability ratio as the flow rate was increased. More details on particle release mechanism due to hydrodynamic forces are given by Gotoh et al. (68), Hubbe (69), and Chamoun et al. (70).

Invasion of Foreign Particles. Invasion of solid particles can cause severe damage around the wellbore. In this case, the solids plug some of the pores and, as a result, the formation permeability diminishes. According to Wojtanowicz et al. (71), the decline of permeability with respect to time can give an indication of the plugging mechanism. At a constant flow rate experiments, the pressure response that occurs due

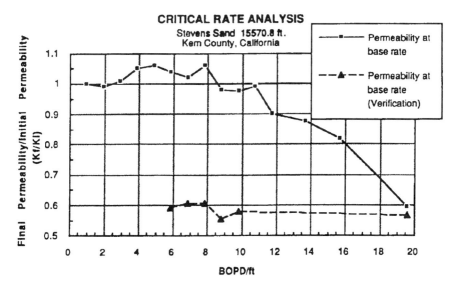

Figure 6.   Permeability decline due to mechanical fines migration. (Reproduced with permission from reference 67. Copyright 1993.)

to foreign solids invasion can be a linear, hyperbolic, or quadratic function of time. In the following section, the response of various plugging mechanisms is briefly discussed.

Three basic mechanisms of blocking formation pores were analyzed by Wojtanowicz et al. (71): gradual pore blocking, single pore blocking (screening), and cake forming (straining). Sodium chloride solutions containing various drilling mud concentrations were injected into a core having a permeability of 670 md. Injection rate was 10 mL/min. Contaminant, drilling mud, concentration was varied from 0.2 to 3%. Figure 7 shows the permeability ratio as a function of the cumulative pore volume for the three mechanisms. It is important to note that the straining mechanism causes the fastest decline in the permeability ratio. Screening mechanism results in a drop in permeability at a rate that increases with time. Gradual blocking reduces the permeability ratio over a long time period. It is worth noting that the blocking mechanism is a function of concentration of contaminant. Increasing the contaminant concentration results in shifting the blocking mechanism from gradual blocking to straining.

Gradual Pore Blocking.   This mechanism includes continuous capture of fine particles at the rock walls due to retention forces. For

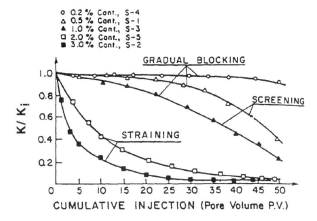

Figure 7. Permeability response for three blocking mechanisms. (Reproduced with permission from reference 71. Copyright 1988.)

linear flow through a core at constant flow rate, the permeability response to gradual blocking is

$$\sqrt{K/K_i} = 1 - \alpha t \tag{12}$$

where K is the permeability at a time, t, and K_i is the initial permeability. Figure 8 shows a plot of $\sqrt{K/K_i}$ as a function of time. At a total suspended solids (TSS) of 405 mg/L the relationship is linear as indicated by equation 12. This result indicates that the gradual blocking mechanism is dominant under these conditions. However, at higher TSS values, the gradual blocking mechanism is dominant at the initial stages of injection only. At later injection stages the experimental data deviate from the initial linear relationship, indicating that another mechanism, screening, is dominant.

Single Pore Blocking (Screening). This mechanism occurs when the particle size is close to the pore throat diameter. In this case foreign particles will block the pores, thus eliminating them from the flow system. The permeability response is

$$K/K_i = 1 - \alpha_1 t \tag{13}$$

Figure 9 illustrates the same experimental data shown in Figure 8; however, K/K_i is plotted as a function of time. At the lowest TSS ex-

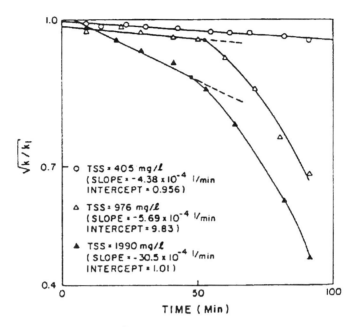

Figure 8.   Foreign particle invasion, gradual pore blocking mechanism. (Reproduced with permission from reference 71. Copyright 1988.)

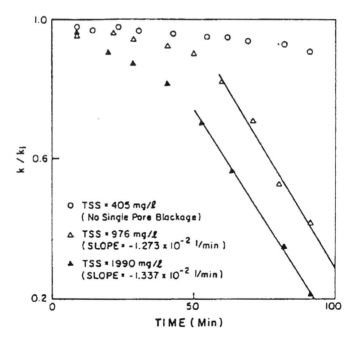

Figure 9.   Foreign particle invasion, single pore blocking. (Reproduced with permission from reference 71. Copyright 1988.)

amined, the screening or single pore blocking mechanism is not observed. Only at the higher TSS values the screening mechanism becomes dominant at injection times >50 min.

Cake Forming (Straining). In this case, particle size is greater than the pore size. Filter cake builds up at or close to the formation face. The permeability response is

$$K/K_i = \frac{1}{1 + \alpha_2 t} \tag{14}$$

or

$$K_i/K = 1 + \alpha_2 t \tag{15}$$

Figure 10 illustrates the variation of K_i/K with time. TSS in the injected water were higher than those examined in Figures 8 and 9. The relationship is linear during the whole injection time. This result indicates that the cake-forming mechanism is dominant.

The results shown in Figures 8–10 clearly indicate that the presence of solids in injection waters can damage the formation. The rate of per-

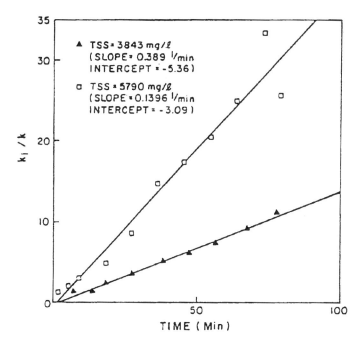

Figure 10. Foreign particle invasion, filter cake forming mechanism. (Reproduced with permission from reference 71. Copyright 1988.)

meability decline for a specific porous medium strongly depends on solids concentration.

Effect of Particle Size and Solids Concentration on Permeability Reduction.   Two important parameters affect flow of suspensions in porous media: particle size and solids concentration. Figure 11 displays the effect of particle size on the permeability ratio. Very small particles ≈ 2 μm do not cause any damage to the core and K/K_i is unity. Increasing particle size results in more damage to the core through screening mechanism. Figure 12 shows that increasing solids concentration causes formation damage in a shorter period of time.

The most important aspect, from a practical point of view, that can be inferred from Figures 11 and 12 is that solids can cause serious injectivity problem that can affect oil production. The cost of removing this damage, through stimulation, also adds to the problem. Therefore, it is extremely important to maintain very low levels of suspended solids or oil droplets in injection waters. Removal of solids from injection waters can be achieved by using proper filters, hydrocyclones, and so on. It is also recommended to use filters during workover operations. This will significantly reduce potential formation damage.

Mobilization and Capture of Formation Solids.   Permeability damage occurs also as a result of solids mobilized or precipitated in the formation, as explained earlier. Figure 13 shows the decline in per-

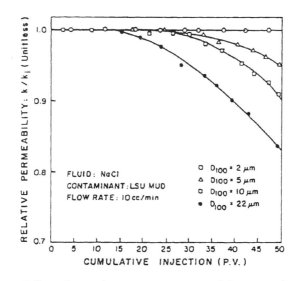

Figure 11.   Effect of particle size on core permeability ratio. (Reproduced with permission from reference 71. Copyright 1988.)

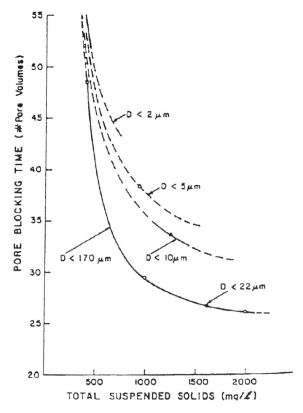

Figure 12.   Effect of solids concentration on core permeability ratio. (Reproduced with permission from reference 71. Copyright 1988.)

meability due to in situ mobilized solids. Similar to the damage observed with foreign solids invasion, filter cake mechanism causes the most damage to the formation. Wojtanowicz et al. (71) assumed an exponential behavior for the fines mobilization and, as a result, the blocking mechanisms can be modeled as follows.

Gradual Blocking and Sweeping.   This mechanism occurs when the size of mobilized particles is significantly smaller than the pore size. Permeability response is

$$\sqrt{K/K_i} = 1 - \alpha_3 t \exp(-f_r t) \qquad (16)$$

Where f_r is the particle release coefficient. According to equation 16, a plot of $\ln [(1 - K/K_i)/t)]$ versus t should yield a straight line.

Single Pore Blocking.   This mechanism occurs when the size of mobilized particles is close to that of the pore throat. The permeability response is

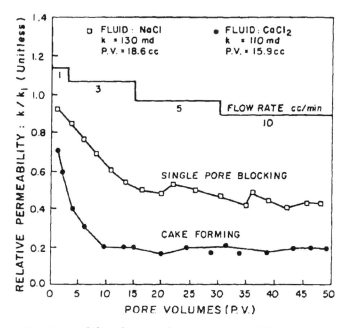

Figure 13. Permeability damage due to in-situ mobilized particles. (Reproduced with permission from reference 71. Copyright 1988.)

$$K/K_i = 1 - \alpha_4[1 - \exp(-f_r t)] \tag{17}$$

A straight line will be obtained when $\ln\left[(C - 1 + K/K_i)/C\right]$ is plotted versus time. The constant C is

$$C = \frac{Y_1 Y_2 - Y_3^2}{Y_1 + Y_2 - 2Y_3} \tag{18}$$

where Y_1, Y_2, and Y_3 are the values of the function $Y(t)$ at times t_1, t_2, and t_3, respectively. The function $Y(t)$ is defined as

$$Y(t) = 1 - K/K_i \tag{19a}$$

and

$$t_3 = 0.5(t_1 + t_2) \tag{19b}$$

Cake Forming.   In this case the size of mobilized particles is larger than the pore throat size. The permeability response is

$$K_i/K = 1 + \alpha_5[1 - \exp(-f_r t)] \tag{20}$$

$Y(t)$ is defined as

$$Y(t) = (K_i/K) - 1 \qquad (21)$$

Equations 12–21 indicate that the damage that results from in situ mobilization is more difficult to characterize. Also, identifying the blocking mechanism will be very difficult task if the suspension invading the formation is also incompatible with the formation brine.

Models to Predict Injectivity Decline Due to Solids Invasion

Two important parameters are usually needed by production engineers dealing with water injection or disposal wells. These parameters are (1) well injectivity and (2) invasion depth.

Well Injectivity. Barkman and Davidson (2) were among the first to model injectivity decline due to foreign solids invasion. They considered that after a specific period of time, known as the bridging time, the decline in injectivity is due to filter cake formation only. According to the filtration theory, the cumulative volume is proportional to square root of filtration time, provided that the pressure drop across the filter remains constant. Figure 14 shows the variation of the cumulative vol-

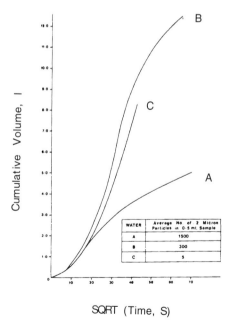

Figure 14. Barkman–Davidson plot. SQRT is square root. (Reproduced with permission from reference 2. Copyright 1972.)

ume with the square root of time. This plot is known as the Barkman–Davidson's plot.

Barkman and Davidson (2) used the slope (S) of the straight line portion of the plot to define a water quality ratio, W/K_c, such that

$$W/K_c = \left(\frac{A}{S}\right)^2\left(\frac{2\rho_c}{\rho_w}\frac{\Delta P}{\mu}\right) \qquad (22)$$

where W is the concentration of suspended solids, K_c is the permeability of the filter cake, ρ_w and ρ_c are the densities of water and filter cake, respectively. ΔP is the pressure drop across the filter, A is the surface area, and μ is the viscosity of water.

Barkman and Davidson considered plugging due to external filter cake formation only. Other plugging mechanisms can have an important impact on the rate of permeability decline, as was observed by Mitchell and Finch (15), who found that the bridging time was a function of the plugging mechanism.

Eylander (16) modified Barkman and Davidson's (2) equation for internal formation plugging and introduced the filter cake porosity parameter (0.2). His equation is based on internal filter cake formation that grows towards the wellbore.

Eylander's equation for the time, t_α, required for the injectivity to decrease to a fraction, α, where $\alpha = q/q_i$, is

$$t_\alpha = \left[\frac{2\pi r_w^2 H(1 - \phi_c)}{q_i C_s}\right]\left(\frac{K_c}{K_m}\right) \ln\left(\frac{r_e}{r_w}\right)\left[\left(\frac{1}{\alpha}\right)^2 - \left(\frac{1}{\alpha}\right)\right] \qquad (23)$$

Where q is the injection rate at time t, K_c and K_m are the permeability of the filter cake and porous medium, respectively. H is the height of injected zone, r_e and r_w are the radii of the external drainage and wellbore, respectively. C_s is the fraction solids concentration in the injected water, and ϕ_c is the porosity of the filter cake.

In the case of internal filter cake formation, Eylander's equation is

$$t_\alpha = \left[\frac{2\pi D^2 H \phi_m(1 - \phi_c)}{q_i C_s}\right]\left[\frac{\phi_m K_c}{K_m - (\phi_m K_c)}\right] \ln\left(\frac{r_e}{r_w}\right)\left[\left(\frac{1}{\alpha}\right)^2 - \frac{1}{\alpha}\right] \qquad (24)$$

Where D is the invasion depth and ϕ_m is the porosity of porous medium.

Eylander used core flood tests to predict filter cake properties. Equations 23 and 24 can be used to determine the injections well's half time ($\alpha = 0.5$). Eylander, however, suggested further research is needed to better determine the invasion depth, D.

Pautz and Crocker (20) adapted Barkman and Davidson's internal cake permeability equation at a constant injection flow rate and derived the following expression to calculate the permeability of the internal cake:

$$\frac{\rho_c}{W}\left(\frac{K_i}{K_d} - 1\right) = \frac{K_i}{K_c}(PV - PV_B) \tag{25}$$

Where PV is the number of pore volumes injected and PV_B is the number of pore volumes injected to the onset of particles bridging. K_c is the permeability of internal cake, and K_d is the permeability of the damaged core.

Rege and Fogler (*17*) modeled formation damage due to solids invasion in radial geometry. A radial network model was developed to simulate formation damage due to deep filtration. According to Rege and Fogler, the initial capture probability is

$$\text{Initial Capture Probability} = 4\left[\left(\frac{\theta a}{R_0}\right)^2 - \left(\frac{\theta a}{R_0}\right)^3\right] + \left(\frac{\theta a}{R_0}\right)^4 \tag{26}$$

where a is the radius of the fine particle, R_0 is the radius of the bond, θ is a parameter. When a particle is captured in a pore, the increase in the pressure drop, $\Delta P_{particle}$, can be determined from

$$\Delta P_{particle} = \frac{12\mu a U_0}{R_0^2}\ [1 - (1 - a/R_0)^2]^2 K_1 \tag{27}$$

where K_1 is a function of (a/R_0) and U_0 is the center line velocity.

The total pressure drop across the bond is

$$\Delta P_{total} = \Delta P_{clean\ tube} + \Delta P_{particle} \tag{28}$$

The effective radius after N particles have deposited is

$$\frac{1}{R_{New}^4} = \frac{1}{R_0^4} + 0.1875 \sum_{i=1}^{N} \frac{a_i}{L}\ [1 - (1 - a_i/R_0)^2]^2 K_1 \tag{29}$$

Rege and Fogler model predictions indicated that decreasing fluid velocity leads to higher value of the capture probability. Figure 15 shows Rege and Fogler model prediction for linear and radial geometries. More damage can be seen in radial flow. This result is important because it shows more damage around the wellbore area where the flow is radial and less damage deep in the reservoir where the flow is nearly linear.

van Velzen and Leerlooijer (*18*) developed a model to predict internal filter cake formation. According to their model, the cumulative injected volume to reduce injectivity to α is

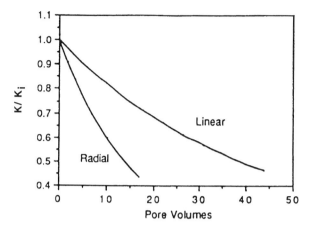

Figure 15.   Permeability damage due to flow of suspensions in linear and radial cores. (Reproduced with permission from reference 17. Copyright 1991.)

$$V_\alpha = \frac{(1 - \alpha)}{\alpha} \frac{2\pi r_w^2 H\phi_m \ln [r_e/r_w]}{C_s N_R} \tag{30}$$

Where α is the injectivity index, defined as

$$\alpha = \frac{[q/(-\Delta P)]t = t}{[(q/-\Delta P)]t = 0} \tag{31}$$

N_R is a dimensionless number for the rate of permeability impairment

$$N_R = \delta_t I_t \tag{32}$$

δ_t is a damage factor, given by

$$\delta_t = \left[\frac{K_m}{K_i} - 1\right] \frac{1}{1 - \phi_c} \tag{33}$$

and I_t is an exponential integral given in table format by van Velzen and Leerlooijer (18). The injectivity reduction time, t_α, for a constant injection pressure is

$$t_\alpha|_{(-\Delta P)_c} = \frac{(1 + \alpha)}{\alpha} \frac{V_\alpha}{2q_i} \tag{34}$$

and for a constant injection rate

$$t_\alpha|_{q=c} = \frac{V_\alpha}{q_i} \tag{35}$$

van Velzen and Leerlooijer also determined the invasion depth, $i_{i,\alpha}$. They defined the invasion depth as the depth where the local fractional permeability is 0.98. Their equation is

$$D = r_\mathrm{w} \sqrt{\left\{ 1 - \frac{1}{\lambda} \ln \left[0.02 \, \frac{\alpha}{1 - \alpha} \cdot \frac{1}{2\lambda \ln (r_\mathrm{e}/r_\mathrm{w})} \right] \right\}} - r_\mathrm{w} \qquad (36)$$

where λ is $((\pi r_\mathrm{w}^2 H \, \lambda_\mathrm{v}) q_\mathrm{i})$, and λ_v is the volumetric filter coefficient. To validate their model, van Velzen and Leerlooijer examined flow of dilute suspensions containing 2 ppm suspended solids in linear and radial cores. Figure 16 shows injectivity as a function of cumulative volume for both core geometry. The particle size of the suspension was 0.19 of the mean pore size of the core. Initial injection velocity was 3 cm/min. The injectivity of the core decreased in both cases, with the radial core showing significant damage at higher injection volumes. These results illustrate the significant damage that might occur in the field if the injection waters contain as a low TSS as 2 ppm of small particles. The damage in this case is deep in the formation. The solids slowly deposited on the pore surfaces, reduced the flow paths and damaged the core.

Increasing the initial injection velocity keeps the solid suspended and, as a result, less damage occurs as shown in Figure 17. It is worth noting that increasing the velocity from 3 to 8.7 cm/min increase the half-life by more than an order of magnitude.

The solid and dashed lines in Figures 16 and 17 are the model prediction. The model predicted the experimental data fairly well. Figure 18 depicts the effect of flow velocity on the permeability decline. It can

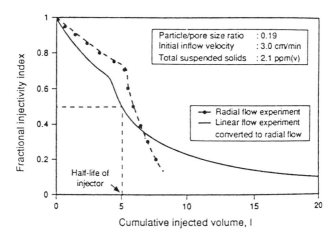

Figure 16. Injection decline at a low flow rate. (Reproduced with permission from reference 18. Copyright 1992.)

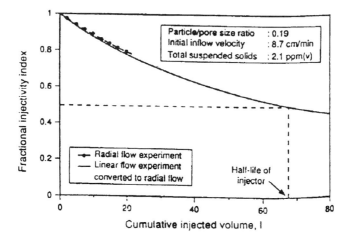

Figure 17.   Injection decline at a high flow rate. (Reproduced with permission from reference 18. Copyright 1992.)

be seen that the rate of permeability impairment strongly depends on the injection velocity and particle size.

Flow of Oil-in-Water Emulsions in Porous Media

Flow of oil-in-water emulsions in porous media is very similar to that of suspensions. Emulsion flow in porous media has been studied extensively in the literature (38–47) and has been recently reviewed by Kokal et al. (47). Unlike solid particles, oil droplets can be removed from a pore throat by increasing the pressure drop across the pore or simply by increasing the flow rate. Consider a single droplet of oil entering a pore throat smaller than itself, as shown in Figure 19. The radius of curvature of the leading edge is smaller than the radius of curvature of the trailing edge of the droplet in the pore throat, and consequently the capillary pressure is greater at the front of the droplet than at its back. A certain pressure is then required to force the droplet through the constriction. This Laplacian differential pressure required to move the droplet through the pore throat is given by

$$dP = 2\sigma\left[\frac{1}{r_1} - \frac{1}{r_2}\right] \tag{37}$$

where r_1 and r_2 are the radii of curvature at the leading and trailing edges of the drop, respectively. If the actual pressure differential across the pore is less than that predicted by equation 37, the oil droplet will plug the pore throat.

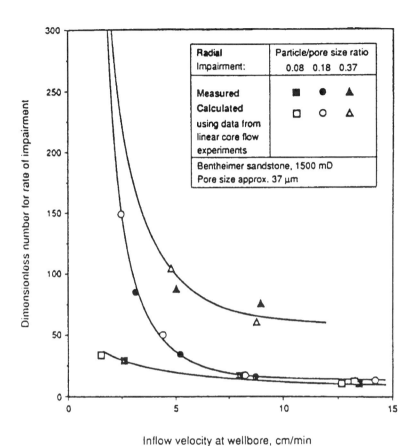

Figure 18. Effect of flow rate of the rate on permeability impairment. (Reproduced with permission from reference 18. Copyright 1992.)

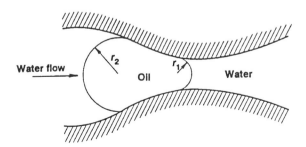

Figure 19. Emulsion blockage mechanism.

Concluding Remarks

Suspension flow in porous media is governed by many factors. Depending on the physical properties of the solid particles, the porous media, and operating conditions, solids form external cake, internal filter cake, or just flow through the media without causing any damage.

External filter cake formation causes a fast and sharp drop in permeability or injectivity of the formation. Internal filter cake formation cases a steady drop in permeability. Increasing solids concentration or particle size will cause more damage to formation.

Injection of low-salinity waters into sandstone reservoirs can trigger fines migration and clay swelling. Injection of water that is incompatible with the formation brine may cause precipitation of calcium, barium, and strontium sulfates, which can plug the formation.

Many experimental and modeling studies to predict formation damage due to flow of suspension through porous media are discussed in this chapter. Still, there is a room for more research to have deep understanding for such complex problem.

Acknowledgments

The author thanks Saudi Aramco for giving permission to publish this paper and Miss E. B. Bernardo for typing this manuscript.

List of Symbols

a	particle radius, m
A	cross-sectional area of medium, m^2
A_H	Hamaker constant, J
C_{cs}	critical salt concentration
C_i	concentration of an ion i, mol/L
C_s	solids volume fraction at the injection surface
D	invasion depth, m
f_r	coefficient of particle release, 1/s
F_A	London–van der Waals attraction force, N
F_D	viscous drag force, N
F_L	lift force, N
F_R	electric double layer repulsion force, N
h	distance between particles, m
H	height of injected zone, m
$i_{i,\alpha}$	invasion depth
I	ionic strength, mol/L
K	permeability, d
K_1	a function of (a/R_o)
K_c	permeability of filter cake, d

K_d	permeability of damaged core, d
K_i	initial (undamaged) permeability, d
K_m	permeability of porous medium, d
K_{pc}	permeability
L	length, m
N_R	permeability impairment
P	pressure, Pa
ΔP	pressure drop, Pa
$\Delta P_{particle}$	pressure drop due to captured particle, Pa
P_c	capillary pressure, Pa
PV	pore volumes injected
PV_B	pore volumes injected at bridging
q	flow rate, m^3/s
q_j	flow rate of phase j
r	radius of pore, m
r_e	drainage radius, m
r_w	wellbore radius, m
r_1	radius of curvature at the leading edge of the drop, m
r_2	radius of curvature at the trailing edge of the drop, m
R_{new}	effective bond radius after N particles deposited
R_o	radius of bond between particles of the porous medium
S	slope of Barkman–Davidson plot
t_α	injectivity reduction time
t_α	time for injectivity to decrease by α fraction
U_o	center line velocity, m/s
v	kinematic viscosity
V	volume of mercury injected, m^3
V_α	cumulative injected volume to reduce the injectivity to a fraction α
V_o	sample pore volume, m^3
W	weight of particles per unit volume of water
x, y, z	Cartesian coordinates, m
Z_i	valence of an ion i

Greek

α	injectivity index
δ_t	damage factor
Θ	a parameter
θ_{ca}	contact angle, rad
κ	Debye–Hückel reciprocal length
λ	dimensionless filtration constant

λ_v	volumetric filter coefficient, l/s
μ	fluid viscosity, Pa·s
ν	kinematic viscosity, m²/s
ρ	density, kg/m³
ρ_c	density of filter cake, kg/m³
ρ_w	density of water, kg/m³
σ	surface tension, dyn/cm (eq 5)
σ	surface tension of mercury, mN/m (eq 3)
ϕ	porosity
ϕ_1	surface potential of particle 1
ϕ_2	surface potential of particle 2
ϕ_c	porosity of filter cake, volume fraction
ϕ_m	porosity of porous medium, volume fraction
ϵ	dielectric constant, F/m
$1/\kappa$	electric double layer thickness, m

Acronyms

EDTA	(ethylenediaminetetraacetic acid)
SRB	sulfate-reducing bacteria
TSS	total suspended solids

Subscripts

c	cake
i	initial
j	phase
1	leading edge index
2	trailing edge index
x	x-direction
y	y-direction
z	z-direction

References

1. Todd, A. C.; Somerville, J. E.; Scott, G. Presented at Formation Damage Control Symposium, Bakersfield, CA, February 13–14, 1984; paper SPE 12498.
2. Barkman, J. H.; Davidson, D. H. *J. Pet. Technol.* **1972**, *24*, 865.
3. Liu, X.; Civan, F., Presented at the International Symposium on Oil Field Chemistry, San Antonio, TX, February 14–17, 1995; Paper SPE 28980.
4. Davidson, D. H., Presented at the 54th Annual Fall Meeting of the SPE, Las Vegas, NV, September 23–26, 1979; Paper SPE 8210.
5. Patton, C. C. *J. Pet. Technol.* **1990**, *42*, 1238.

6. Todd, A. C.; Brown, J.; Noorkami, M.; Tweedie, J. A. Presented at the International Symposium on Oilfield and Geothermal Chemistry, Dallas, TX, January 22–24, 1979; paper SPE 7883.
7. Rickford, R. L.; Finney, T. P., Presented at SPE Production Operations Symposium, Oklahoma City, OK, March 13–14, 1989; paper SPE 18886.
8. Khatib, Z. I.; Vittal, S. *SPE Prod. Eng.* **1991**, *6*, 233.
9. Pang, S.; Sharma, M. M. Presented at the SPE 69th Annual Technical Conference and Exhibition, New Orleans, LA, September 25–28, 1994; paper SPE 28489.
10. Krueger, R. F. *J. Pet. Technol.* **1986**, *38*, 131.
11. Baghdikian, S. Y.; Sharma, M. M.; Handy, L. L. *SPE Res. Eng.* **1989,** *May*, 213.
12. Herzig, J. P.; Leclerc, D. M., Le Goff, P. *Ind. Eng. Chem.* **1970**, *65*, 8.
13. Imdakm, A. O.; Sahimi, M. *Chem. Eng. Sci.* **1991**, *46*, 1977.
14. Arcia, E. M.; Civan, F. *J. Can. Pet. Technol.* **1992**, *21*, 27.
15. Mitchell, R. W.; Finch, E. M. *J. Pet. Techol.* **1981**, *271*, 1141.
16. Eylander, J. G. R. *SPE Res. Eng.* **1988**, *3*, 1287.
17. Rege, S. D.; Fogler, H. S. *Chem. Eng. Commun.* **1991**, *108*, 67–83.
18. van Velzen, J. F. G.; Leerlooijer, K. Presented at the SPE International Symposium on Formation Damage Control, Lafayette, LA, February 26–27, 1992; paper SPE 23822.
19. Mehdizadeh, A. M.; Hashemi, R.; Caothien, S. Presented at the SPE International Symposium on Formation Damage Control, Lafayette, LA, February 26–27, 1992; paper SPE 23826.
20. Pautz, J. F.; Crocker, M. E. Presented at the SPE Production Operation Symposium, Oklahoma City, OK, March 13–14, 1989; paper SPE 18888.
21. Giorgi, D. P. Presented at the SPE Latin American Petroleum Engineering Conference, Rio De Jenerio, Brazil, October 14–19, 1990; paper SPE 21143.
22. Lancey, T. W.; Kammula, K. Presented at the Pacific Coast Oil Show and Conference, Bakersfield, CA, 1986.
23. McDowell-Boyer, L. M.; Hunt, J. R.; Sitar, N. *Water Resour. Res.* **1986**, *22*, 1901.
24. Porter, K. E. *J. Pet. Technol.* **1989**, *41*, 780.
25. Abrams, A. *J. Pet. Technol.* **1977**, *29*, 586.
26. Gray, D. H.; Rex, R. W. *Proceeding of the 14th National Conference on Clays and Clay Minerals;* Pergamon: London, 1966; p. 113.
27. Mungan, N. *J. Can. Pet. Technol.* **1968**, *7*, 113.
28. *Surface Phenomena in Enhanced Oil Recovery;* Khilar, K. C.; Fogler, H. S.; Shah, D. O., Eds; Plenum: New York, 1981.
29. Khilar, K. C.; Fogler, H. S. *Soc. Pet. Eng. J.* **1983**, *23*, 55.
30. Khilar, K. C.; Fogler, H. S. *J. Colloid Interface Sci.* **1984**, *101*, 214.
31. Lever, A.; Dawe, R. A. *J. Pet. Geol.* **1984**, *7*, 97.
32. Gruesbeck, C.; Collins, R. E. *Soc. Pet. Eng. J.* **1982**, *22*, 847.
33. Gabriel, G. A.; Inamdar, G. R. Presented at the 58th Annual Technical Conference and Exhibition, San Francisco, CA, October 5–8, 1983; paper SPE 12168.
34. Kia, S. F.; Fogler, H. S.; Reed, M. G.; Vaidya, R. N. *SPE Prod. Eng.* **1987**, *2*, 277.
35. Nasr-El-Din, H. A.; Maini, B. B.; Stanislav, P. *AOSTRA J. Res.* **1991**, *7*, 1.
36. Sharma, M. M.; Chamoun, H.; Sita Rama Sarma, D. S. H.; Schechter, R. S. *J. Colloid Interface Sci.* **1992**, *149*, 121.

37. Economides, M. J.; Nolte, K. G. *Reservoir Stimulation;* Prentice Hall: Englewood Cliffs, NJ, 1989.
38. McAuliffe, C. D. *J. Pet. Technol.* **1973,** *25,* 727.
39. Hofman, J. A. M. H.; Stein, H. N. *Colloids Surf.* **1991,** *61,* 316.
40. Islam, M. R.; Farouq Ali, S. M. *J. Can. Pet. Technol.* **1994,** *33,* 59.
41. Soo, H.; Radke, C. J. *Ind. Eng. Chem. Fundam.* **1984,** *23,* 242–347.
42. Soo, H.; Radke, C. J. *J. Colloid Interface Sci.* **1984,** *102(2),* 462–476.
43. Soo, H.; Radke, C. J. *AIChE J.* **1985,** *31,* 1926.
44. Soo, H.; Radke, C. J. *Chem. Eng. Sci.* **1986,** *41(2),* 263–272.
45. Soo, H.; Williams, M. C.; Radke, C. J. *Chem. Eng. Sci.* **1986,** *41(2),* 273–281.
46. Coskuner, G. *Oilfield Emulsions;* Report No. 1988–18; Petroleum Recovery Institute: Calgary, Canada, 1988.
47. Kokal, S. L.; Maini, B. B.; Hoo, R. In *Emulsions: Fundamentals and Applications in the Petroleum Industry;* Schramm, L. L., Ed.; Advances in Chemistry 231; American Chemical Society: Washington, DC, 1992; pp 219–262.
48. Updegraff, D. M. In *Microbial Enhanced Oil Recovery;* Zajic, J. E.; Cooper, D. G.; Jack, T. R.; Kosaric, N., Eds.; PennWell: Tulsa, OK, 1983; Chapter 6.
49. Liu, X.; Civan, F. Presented at the Production Operations Symposium, Oklahoma City, OK, March 21–23, 1993; paper SPE 25429.
50. Almon, W. R.; Davies, D. K. In *Clays and the Resource Geologist;* Longstaffe, F. J., Ed.; Mineralogical Association of Canada: City, Canada, 1981; Vol. 7, Chapter 5.
51. Vitthal, S.; Sharma, M. M.; Sepehrooni, S. Presented at the Formation Damage Control Symposium, Bakersfield, CA, February 8–9, 1988; paper SPE 17146.
52. Cerda, C. M. *Clays Clay Miner.* **1988,** *36,* 491.
53. Okada, K.; Akagi, Y.; Kogure, M.; Yoshioka, N. *Can. J. Chem. Eng.* **1990,** *68,* 343.
54. Hunter, R. J. *Zeta Potential in Colloid Science: Principles and Applications;* Academic: London, 1981.
55. Mueke, T. W. *J. Pet. Technol.* **1979,** *31,* 144.
56. Sarkar, A. K.; Sharma, M. M. *J. Pet. Technol.* **1990,** *May,* 646–652.
57. El-Hattab, M. I. Presented at the Middle East Oil Technical Conference, Manama, Bahrain, March 14–17, 1983; paper SPE 11449.
58. Vetter, O. J.; Phillips, R. C. *J. Pet. Technol.* **1970,** *October,* 1299.
59. Dunn, T. L.; MacGowan, D. B.; Surdam, R. C. Presented at the Rock Mountain Regional Meeting and Low-Permeability Research Symposium, Denver, CO, April 15–17, 1991; paper SPE 21839.
60. Lapointe, D. A.; Muhsin, M. A. Presented at the SPE Middle East Oil Technical Conference, Manama, Bahrain, November 16–19, 1991; paper SPE 21367.
61. Lee, W.; Lewandowski; Nielsen, P. H.; Hamilton, W. A. *Biofouling* **1995,** *8,* 165.
62. von Wolzogen Kühr, C. A. H.; van der Vlugt, L. W. *Water (Berchem, Belg.)* **1934,** *18,* 147.
63. Cord-Ruwisch; Kleinitz, W.; Widdel, F. *J. Pet. Technol.* **1987,** *January,* 97.
64. Rosnes, J. T.; Graue, A.; Lieu, T. Presented at the SPE Formation Damage Control Symposium, Lafayette, LA, February 22–23, 1990; paper SPE 19429.

65. Cerda, C. M. *Colloids Surf.* **1987,** *27,* 219.
66. Hubbe, M. A. *Colloids Surf.* **1984,** *12,* 151.
67. Miranda, R. N.; Underdown, D. R. Presented at the Production Operations Symposium, Oklahoma City, OK, March 21–23, 1993; paper SPE 25432.
68. Gotoh, K.; Iriya, M.; Tagawa, M. *Colloid Polym. Sci.* **1983,** *261,* 805.
69. Hubbe, M. A. *Colloids Surf.* **1985,** *16,* 249.
70. Chamoun, H.; Schechter, R. S. Sharma, M. M. *Oil-Field Chemistry: Enhanced Recovery and Production Stimulation;* Borchard, J. K.; Yen, T. F., Eds.; ACS Symposium Series 396; American Chemical Society: Washington, DC, 1989; pp 548–559.
71. Wojtanowicz, A. K.; Krilov, Z.; Langlinais, J. P. *Trans. ASME* **1988,** *110,* 34.

RECEIVED for review August 9, 1994. ACCEPTED revised manuscript July 7, 1995.

Fines Migration in Petroleum Reservoirs

Brij Maini, Fred Wassmuth, and Laurier L. Schramm

Petroleum Recovery Institute, 100, 3512 33rd Street N.W., Calgary, Alberta, T2L 2A6, Canada

Production of petroleum is often hampered by damage to the permeability of reservoir rocks resulting from interaction of injected fluids with the porous rock formation. Fine particles of clays and other minerals are often found attached to the pore walls of reservoir rocks. The interaction between injected fluids and the rock can cause their movement by a combination of mechanical shear forces, colloid–chemical reactions and geochemical transformations. This chapter reviews several different aspects of the fines migration process.

The nature and properties of common clay minerals found in petroleum reservoirs are briefly discussed to set the stage for a review of the colloidal and hydrodynamic forces acting on the fine particles. This is followed by a review of reported experimental studies of permeability damage by fines movement under purely hydrodynamic forces.

Migration of fines triggered by colloidal interactions is discussed by reviewing the roles of various process variables, including salinity, ionic strength, pH, ion exchange capacity, and temperature. An in depth review of the available phenomenological and theoretical models of fines migration is then presented. Finally, ways of minimizing the permeability damage are briefly discussed.

OIL AND NATURAL GAS RESERVOIRS are found in sedimentary rocks which are either porous sandstone or porous limestone. The oil is held within these permeable rocks by structural or stratigraphic traps and by capillary forces (1). The oil is produced from the reservoir by making it flow into a production well, initially under its own pressure (primary production) and later by injection of water (secondary production) or other displacing fluids (tertiary production) into the reservoir via injec-

0065–2393/96/0251–0321$20.75/0

tion wells placed some distance away from the production well. Thus the production of oil involves flow of fluids (oil, gas, and water) through permeable porous rock formations. How easily the oil can flow into the production well depends on the permeability of the porous rock. Therefore, any mechanism which causes a reduction in the rock permeability would be detrimental to sustained oil production.

Almost all reservoir rocks contain fine particles and clay minerals within the rock matrix. These fine particles can be quartz fragments, amorphous silica, feldspar, carbonate fragments, or clay minerals. Although the bulk of the rock matrix is held together tightly, either by mineral cements that bind individual grains or by the confining stress of the overburden rock, these fine particles are only loosely attached to the pore walls. Therefore, they are susceptible to mobilization due to mechanical forces and colloid–chemical reactions during flow of fluids through the porous rock. Such mobilization of fine particles often results in severe loss of the rock permeability due to subsequent trapping of the mobilized particles by a process analogous to deep-bed filtration. Permeability damage due to fines migration is a major concern in reservoir processes such as water flooding and acidization.

Although all fines can become mobilized, irrespective of their mineral composition, the most severe problem is caused by fines belonging to the "clay fraction." In petroleum literature, the term "clay fraction" is used to denote either the particle size or the type of mineral involved. As a particle size term, clay refers to particles smaller than 4 μm. As a mineral type term, clay refers to silicate minerals with a crystal structure similar to that of mica. As the size of clay minerals is small and the structure platy, their surface area is large. Consequently, clays play a disproportionately large role in physicochemical processes involving surface phenomena occurring within porous rock formations. In this chapter we examine the mechanisms causing migration of clays and other fines during flow through porous formations and the problems caused by such migration. Depending on their origin, the clays present in reservoir rocks are classified into two groups: allogenic and authigenic clays (2, 3).

Allogenic (or Detrital) Clays

The term allogenic is used interchangeably with detrital. Allogenic clays are formed before deposition of the sediments and are mixed with the sand during the deposition or shortly after the deposition. Allogenic clays can occur in many forms. Sand-sized clay particles are found as part of the rock matrix in sandstones. Clays also occur as thin laminating layers. Another frequently encountered form is called "rip up" clasts, which are fragments eroded from mud layers. Fragments of older shales and mudstones can also be deposited with the sand. Another interesting form is biogenic pellets, which are produced by ingestion and excretion of mud by living organisms.

Clays introduced shortly after deposition occur as burrows lining or filling, grain coatings, and randomly distributed flocculated aggregates. Figure 1 depicts various forms of allogenic clays in sedimentary rocks.

Authigenic Clays

The term authigenic refers to minerals that have formed in their sedimentary depositional setting, either as precipitates from solution (neo-formed clays) or by chemical transformation of a pre-existing clay or any other precursor mineral (transformed clays). Authigenic clays occur as pore linings, pore fillings, grain replacements, or fracture and vug fillings. Figure 2 shows some of the more commonly encountered forms of authigenic clays. Because these clays form within the porous rock, they exhibit superior crystalline habits and relatively larger size of the individual crystals.

Structure and Properties of Clay Minerals

The most frequently encountered clay minerals found in sedimentary rocks are kaolinite, montmorillonite, and illite. These are all phyllosilicates with a crystal structure similar to that of micas: sheet-layer structures with strong covalent bonding within each sheet and among the two- or three-sheet layers belonging to the unit structure and only weak bonding (van der Waals attraction or hydrogen bonding) between the adjacent layered structures (4).

The basic building block of silicate minerals, including clays, is the SiO_4^{4-} silica tetrahedron, containing a silicon atom surrounded by four oxygen atoms in a tetrahedral configuration. These silica tetrahedra can link together by sharing oxygens in various ways. Phyllosilicates are characterized by silica tetrahedra linked to form sheets. Tetrahedral sheets consist of six-member rings of tetrahedra each of which share three of the four oxygens. The unshared oxygens in one sheet all point in the same direction. The chemical formula for the tetrahedral sheet can be written as $Si_4O_{10}^{4-}$. Because it has a negative charge, it can exist only in combination with cations. The tetrahedral structure can only accommodate smaller cations such as silicon and aluminum. Clays contain two types of sheets. The tetrahedral sheet in clays is always associated with an octahedral sheet that contains cations surrounded by six nearest neighbors. This configuration can accommodate larger cations, such as Al^{3+}, Mg^{2+}, and Fe^{3+}. The simplest clay structure contains one tetrahedral and one octahedral sheet to form a two-layer structure. The unshared oxygens in the tetrahedral sheet become part of the octahedral

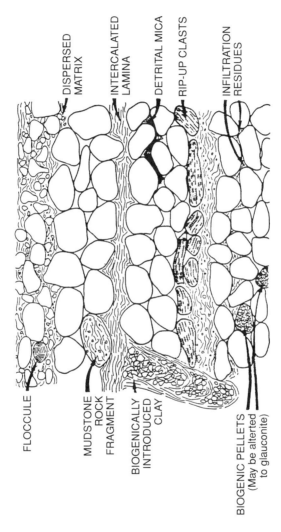

Figure 1.   Modes of occurrence of allogenic clay in sandstones.

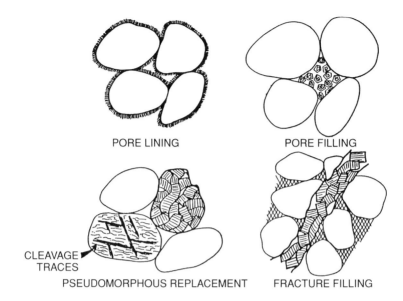

PORE LINING   PORE FILLING

CLEAVAGE
TRACES
PSEUDOMORPHOUS REPLACEMENT   FRACTURE FILLING

Figure 2. Modes of occurrence of authigenic clay in sandstones.

sheet. The two-layer structure consists of three planes of OH^- or O^{2-}, which sandwich a plane of octahedral cations and a plane of silicons. A single flake of kaolinite consists of one Si–O tetrahedral sheet and one Al–OH octahedral sheet.

The other major type of layer contains two tetrahedral sheets, which sandwich one octahedral sheet. The unshared oxygens of both tetrahedral sheets become part of the octahedral sheet. The whole layer contains four planes of oxygens (OH^- or O^{2-}) and three planes of cations.

Cation substitutions frequently occur within the sheet structure. For example, silicon in the tetrahedral sheet can be replaced with aluminum and the aluminum in the octahedral sheet can be replaced with magnesium or ferrous ion. These isomorphic substitutions cause a charge imbalance within the sheet that is balanced by adsorption of cations between the layers. This is the origin of most charge density in three-layer clays.

Characteristics of Common Forms of Clay Minerals. Although many different forms of clays are found in reservoir rocks, the more frequently encountered clays from fines migration perspective are kaolinite, illite, and montmorillonite (5). A brief description of their characteristics is included in the following.

Kaolinite. Particles of kaolinite are easily recognized by their characteristic pseudohexagonal plate form. Several plates are usually found stacked together in booklet or card-pack structure. Individual plates generally range from 3 to 20 μm in diameter (3). The chemical formula for kaolinite may be written as $Al_4Si_4O_{10}(OH)_8$. It is a two-layer clay containing one tetrahedral sheet and one octahedral sheet. Kaolinite is not known to show significant ionic substitution and no expansion of its lattice occurs when it comes in contact with water.

Illite. Illite generally occurs as irregular flakes with lath-like projections. This growth form varies considerably. Authigenic illite is often found attached to sand grains as sheets that curl away from the point of attachment. Primary particles of illite are generally smaller than 2 μm. It is a three-layered clay mineral that usually contains potassium ion between the unit layers to balance the charge deficiency of the unit structure. Like kaolinite, no swelling of its lattice occurs in contact with water.

Montmorillonite. Montmorillonite belongs to the smectite group of clays. It occurs as a crinkly coating on sand grains. This mineral is a three-layered clay that is very susceptible to swelling in contact with water. Montmorillonite is also known to readily exchange cations and has a high cation exchange capacity.

Chemical Properties. An important chemical property of clays, which directly affects fines migration is the cation exchange capacity (CEC) (6–9). CEC is a measure of the capacity of a clay to exchange cations. It is usually reported in units of milliequivalents per 100 g of clay (meq/100 g). The CEC depends on the concentration of exchangeable cations in the diffuse Gouy–Chapman layer (*see* later). This concentration depends on the total particle charge, which may vary with pH. Unless stated otherwise, the reported values of CEC are measured at neutral pH. CEC values (meq/100g) of common clay minerals are as follows: smectites, 80–150; vermiculites, 120–200; illites, 10–40; kaolinite, 1–10; and chlorite, <10 (*10*).

Electrical Properties. The particles of clay minerals usually carry two types of charge: structural (or permanent) and surface charge, which is pH dependent (*4*). The permanent charge is due to ion substitution or structural imperfections. The surface charge is caused by adsorption of charged ions or the chemical reactions that occur at the surface of clays. An important cause of the surface charge is the hydrolysis of broken Si–O and Al–OH bonds along the edges. The net charge can be either positive or negative, depending on the clay structure and the pH and salinity of the solution.

It has already been pointed out that fines movement results from the interaction between injected fluids and the reservoir rock and is thought to be due to some combination of mechanical shear forces, colloid–chemical reactions, and geochemical transformation effects (*11–20*). Of these, increasing attention is being paid to the mechanism of particle dispersion involving the increased electrostatic repulsion between particles that can be brought about by changes to the electrolyte solution environment. This can create situations in which the colloidal forces and hydrodynamic forces combine to cause fines migration. The main colloidal forces are electrostatic repulsion forces between like-charged objects and van der Waals attractive forces between objects.

The clay particles present in reservoir rocks normally carry a net negative electric surface charge. The extent of electric double-layer interaction between particles strongly depends on the electrolyte composition and concentration. As described in Chapter 1, Derjaguin and Landau, and independently Verwey and Overbeek, developed a quantitative theory for the stability of lyophobic colloids, now known as the DLVO theory. The DLVO theory accounts for the energy changes that take place when two particles approach each other and involves estimating the energies of attraction and repulsion versus particle separation distance. Thus, the DLVO theory, which describes the stability of charged colloidal particles, together with a consideration of ion exchange reactions, can potentially be used to describe effects such as (*11–13, 15, 16*):

- the decrease in permeability that accompanies flooding with solutions of decreasing NaCl concentration
- the less severe decrease in permeability that accompanies flooding with $CaCl_2$ solutions
- the sensitization to fines migration–induced permeability damage brought on by exposure to NaCl solutions

The permeability reduction may be due to the dislodging of fine particles and subsequent trapping in pore restrictions, or the swelling of clay particles, or both.

Because injected fluids can alter the local pH and electrolyte environment, a number of mineral or alkali reactions may result (*17*). These can change the surface properties of the minerals. Accordingly the surface charge properties under conditions of varying solution pH are also of interest. Electrokinetic measurements are a sensitive measure of changes in fine particle stability to be expected from small changes in solution properties and have been shown to be a good indicator of permeability reduction potential (*12, 16, 18*).

Forces Acting on Fines

Repulsive Forces. Most substances, when brought into contact with a polar liquid, develop an electrical charge at their surface by one of several different mechanisms. For sandstones (quartz and clay) and carbonates (limestones and dolomite) the principal mechanisms are as follows:

- ionization of surface groups, for example, the pH-dependent dissociation of surface hydroxyls

- differential dissolution of ions from a crystal lattice, for example, the unequal dissolution of calcium and carbonate ions from calcite, when placed in water

- isomorphous replacements in the crystal lattice, for example, the replacement of Si^{4+} by Al^{3+} in the crystal structure of an aluminosilicate, or clay mineral, causing the lattice to become negatively charged

- ion adsorption, for example, the specific adsorption of hydroxy–metal complexes

The surface charge influences the distribution of nearby ions in the polar medium. Ions of opposite charge (counterions) are attracted to the surface while those of like charge (coions) are repelled. An electric double layer (EDL), which is diffuse because of mixing caused by thermal motion, is thus formed. As discussed in Chapter 1, the EDL consists of the charged surface and a neutralizing excess of counterions over coions, distributed near the surface. The EDL can be viewed as being composed of two layers:

- an inner Stern layer that may include adsorbed ions

- a diffuse Gouy–Chapman layer where ions are distributed according to the influence of electrical forces and thermal motion

For a charged surface in aqueous solution the surface potential decays with distance away from the surface to a reduced potential at the Stern plane boundary, to zero in bulk solution. The Stern plane potential is readily estimated from electrokinetic measurements.

Electrokinetic motion occurs when the mobile part of the electric double layer is sheared away from the inner layer (charged surface). Of the four types of electrokinetic measurements, electrophoresis finds the most use in industrial practice. Streaming potential measurements give an indication of the average surface potential in a porous rock but are strongly influenced by the pore size distribution and may not be sensitive to the contributions of individual mineral constituents (19). Electrophoresis measurements require crushing of the rock but have the advantage of being convenient for establishing equilibrium with different solutions and provide information about individual rock components (e.g., clay vs. silica components of sandstones). In microelectrophoresis, suspended particles are viewed under a microscope, an electric field is applied, causing charged particles, and any attached material or liquid, to move toward the oppositely charged electrode. The particle electro-phoretic velocity is measured at carefully selected planes within the sample cell where the electric field gradient and electroosmotic flow of bulk liquid inside the cell are known. The electrophoretic mobility, electrophoretic velocity divided by the electric field gradient at the location where the velocity was measured, can be interpreted in terms of potential at the plane of shear; known as the zeta potential. The zeta potential is usually taken to be approximately equal to the potential at the Stern plane. Methods for estimating zeta potentials on the basis of electrophoretic mobility measurements are discussed in Chapter 1. Other descriptions of practical experimental techniques in electropho-resis and their limitations can be found in references 20 and 21. Typi-cally, electrokinetic measurements are made on particles suspended in electrolyte solutions having ionic strengths in the range 10^{-4} to 10^{-2} M. However, the electrolyte solutions found in petroleum reservoirs may have ionic strengths ranging from a few tenths to several molar and very little work involving reservoir rock surface charge measurements in practical brines has been conducted (22).

In the simplest example of fines migration due to colloidal forces particles would be repelled from a rock surface by the repulsive forces created when two charged surfaces are near to each other and their electric double layers overlap. Various models have been used including the plate–plate, sphere–sphere, cylinder–plate, or sphere–plate ar-rangements (16, 18, 23). Assuming for simplicity a plate–plate model, the repulsive energy V_R can be calculated using zeta potentials to ap-proximate the potential at the shear (Stern) plane, $\psi(\delta)$, in the following equation:

$$V_R = (64 n_0 k_B T \Upsilon^2/\kappa) \exp[-\kappa h] \tag{1}$$

where n_0 is the number of ions per unit volume away from the surfaces, the plates are separated by distance h, and

$$\Upsilon = (\exp[ze\psi(\delta)/2k_B T] - 1)/(\exp[ze\psi(\delta)/2k_B T] + 1) \tag{2}$$

where z is the counterion charge number.

There is another repulsive force at very small separation distances (< 0.5 nm) where hydration layers or the particles' atomic electron clouds overlap, causing a strong repulsion, called Born repulsion.

Attractive Forces. Molecules exert forces of attraction on each other, called dispersion forces, which are caused by the orientation of dipoles due to any of (1) Keesom forces between permanent dipoles, (2) Debye induction forces between dipoles and induced dipoles, or (3) London–van der Waals forces between induced dipoles and induced dipoles. Except for quite polar materials the London–van der Waals dispersion forces are the most significant of the three. By adding up the attractions between all interparticle pairs of molecules the dispersion force between two (plate) particles separated by a small distance h, the attractive energy V_A can be approximated by

$$V_A = -H/(12\pi h^2) \tag{3}$$

Different assumptions lead to different expressions for V_A (18, 24). The constant h is known as the Hamaker constant and depends on the density and polarizability of atoms in the particles. Typically 10^{-20} J $< H < 10^{-19}$ J, or $0.25\ kT < H < 25\ kT$ at room temperature (24, 25). When the particles are in a medium other than vacuum, the attraction is reduced. In addition, this can be accounted for by using an effective Hamaker constant (26). The Hamaker constants are usually not well known and must be approximated.

The DLVO theory accounts for the energy changes that take place when two particles approach each other, and involves estimating the energies of attraction and repulsion versus separation distance. These, V_A and V_R, are then added together to yield the total interaction energy V_T. V_R decreases exponentially with increasing separation distance and has a range about equal to κ^{-1}, whereas V_A decreases inversely with increasing separation distance. Chapter 1 gives examples of attractive and repulsive energy curves and the total interaction energy curves that result showing that either the attractive forces or repulsive electric double layer forces can predominate at different interparticle distances. The total colloidal force between particles F_T is then obtained as

$$F_T = \partial V_T/\partial h = \partial(V_A + V_R)/\partial h \tag{4}$$

Hydrodynamic Forces. The hydrodynamic forces acting on a fine particle near a larger rock grain have been estimated in several ways (23, 27–30). By assuming that the liquid flow field past the rock grains is generally not affected by the fine particles and that for separations of several particle diameters or more the fine particle moves along a streamline of undisturbed flow, the work of Goren and O'Neill (28) and Spielman and Fitzpatrick (29) has led to equations for the tangential force. An example is

$$F_H = -\mu a_P^3 A_S u \cos \theta_P F(h')/a_S^2 \tag{5}$$

where μ is the liquid viscosity, a_P is the fine particle radius, a_S is the rock grain radius, A_S is a dimensionless parameter describing the flow model used, u is the superficial velocity, θ_P is the angle a line between the centres of the fine particle and rock grain makes with the direction of flow, and $F(h)$ contains several constants and functions of h', $h' = h/a_p$, where h is the separation distance between fine particle and rock grain.

To determine the conditions for fine particle detachment the magnitude of the hydrodynamic forces near a rock grain surface have been estimated by Cerda (27) assuming spherical particles and rock porosities in the range 0.2 to 0.4 as

a_P (μm)	a_S (μm)	μ (cm/h)	F_H (pN)
1	50–200	1	0–2
5	100	20	40–80

These forces are very small compared with typical colloidal forces, which can be of the order of 1000–2000 pN in dilute electrolyte solutions. Figure 3 shows an example of colloidal forces calculated between a fine quartz particle and a sandstone grain. To be of the same order of magnitude as the colloidal forces, a superficial velocity of the order of 200 cm/h would be required in some reservoirs (23, 27). This is not to say that mechanically induced fines migration does not occur in some reservoirs at lower velocities. Critical flow velocities below which fines migration does not occur have been reported to span the range from about 14 to 900 cm/h (31). Recognizing that there will be exceptions, as a first approximation it may be appropriate to consider fine particle mobilization to be independent of fluid velocity and consider simply the influence of solution and interfacial chemistry on the colloidal forces.

Influence of Solution Chemistry.   For many solids H^+ and OH^- are potential determining ions, so the surface charge is pH-dependent and can be either positive or negative. Therefore the pH at which the surface is electrically neutral, the point of zero charge (pzc), is a characteristic property of the solid. The pH of zero zeta potential is referred to as the isoelectric point (iep). Table I shows some ranges for isoelectric points for a number of minerals of interest in various dilute aqueous solutions. The ranges are broad, probably due to variations in mineral compositions, pretreatments, and aqueous compositions. Although the effect of adding an indifferent electrolyte is predictable from EDL theory and does not change the iep, the specific adsorption of certain multivalent ions can be more complex. Such adsorption does not require high concentrations and can cause a shift in the iep, and even additional iep's.

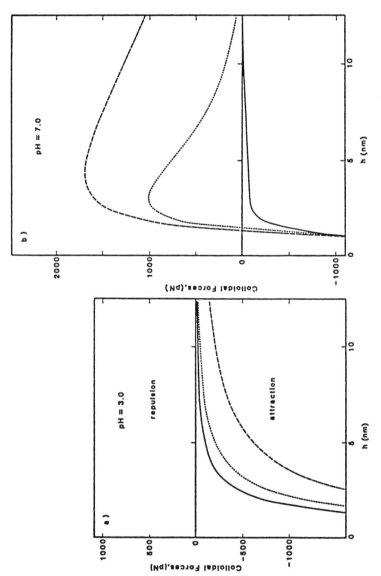

Figure 3. Interaction colloidal forces between a fine-grained particle and a sand grain surface. Computations made for (1) ——— 10^{-1} M, (2) ······ 10^{-2} M, and (3) — — — 10^{-3} M NaCl.

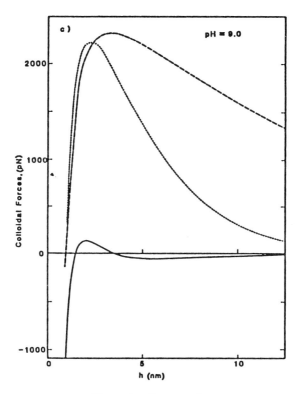

Figure 3.–Continued.

When quartz or other silicates are immersed in water, dissociation of surface hydroxyls (silanols) makes the surface positive at low pH and negative at high pH. Sandstones often contain a significant amount of clays, such as kaolinite, illite, and smectite. These clays are layered aluminosilicate minerals whose layers are disrupted at particle edges, leaving surface hydroxyl groups that can dissociate as already described for quartz. The crystal lattice, on the other hand, may carry a pH-independent negative charge caused by isomorphic substitutions in the crystal lattice (32, 33). The isoelectric points for kaolinite and smectite given in Table I vary considerably with electrolyte concentration because of the interplay of the potentially differently charged edges and faces (34). For carbonates limestone or dolomite in water, surface charge can be generated by unequal dissolution of lattice ions. These minerals are more soluble than quartz or clays, and various surface hydrolysis products can be formed with each of Ca^{2+}, Mg^{2+}, and CO_3^{2-}. Accordingly measurement and interpretation of the surface charges of these minerals are more difficult (35) and the Table I iep's are more diverse. Generally, high pH will favor an excess concentration of negative species (HCO_3^-

Table I.   Rock and Mineral Isoelectric Points
from the Literature

Rock–Mineral	Isoelectric Point	Refs.
Berea sandstone	3–5	28, 37
Quartz	1.5–3.7	28, 38–41
Kaolinite	<1–5	28, 41–45
Smectite	<2.5	43, 45
Calcite	<5–10.8	26, 28, 40, 41, 46–51
Dolomite	<6 or 7	28, 40, 52

and $CO_3{}^{2-}$) at the surface, whereas low pH will favor an excess of Ca^{2+}, $CaHCO_3{}^+$, and $CaOH^+$. Additional data for sandstone, quartz, kaolinite, and carbonate particles under reservoir electrolyte solution conditions are given in reference 22.

Besides potential-determining ions, other electrolytes can have a strong influence on surface charge. The effect of adding trivalent or divalent ions versus monovalent ions is accounted for by the ionic strenth term in EDL theory. Within a given valence state the lyotropic series indicates, in decreasing order, the effectiveness of different ions in terms of coagulating power:

cations: $Cs^+ > Rb^+ > K^+ > Na^+ > Li^+$

anions: $CNS^- > I^- > Br^- > Cl^- > F^- > NO_3{}^- > ClO_4{}^-$

In suspensions the transition between stable dispersion of particles and aggregation of particles usually occurs over a fairly small range of electrolyte concentration, making it possible to determine aggregation concentrations, often referred to as critical coagulation concentrations (CCC). The Schulze–Hardy rule summarizes the general tendency of the CCC to vary inversely with the sixth power of the counterion charge number (for indifferent electrolyte). An illustration is given in Chapter 1. In fines migration an analogous concentration may be defined (18, 36, 37) for the threshold onset of colloidally induced fines migration that occurs when the salt concentration of flowing solution falls below a certain value, the critical salt concentration (CSC).

Experimental Studies of Permeability Damage by Mechanical Fines Migration

As discussed in the preceding section, the hydrodynamic force exerted by a fluid flowing through a pore on a fine particle attached to the wall of the pore is directly proportional to the velocity of the fluid and its viscosity (*see* eq 5). Although it was stated that the hydrodynamic forces are small compared with typical colloidal forces under normal flow conditions encountered in the interior regions of petroleum reservoirs, their magnitude can be large enough to mobilize fines in the region close to injection and production wells where fluid velocities are high. Therefore a possibility exists for permeability damage by mechanically induced fines migration in the immediate vicinity of wells. This type of fines migration has been investigated in several experimental studies (*38–44*) and the effects of several important parameters have been established. Most of the available information pertains to flow of a single fluid (generally brine) through the rock. The reported information on fines migration under multiphase flow conditions is relatively sparse. Therefore we will focus on what has been reported on mechanical fines migration under single-phase flow conditions by examining the effects of various process parameters.

Effects of Flow Velocity. The role of flow velocity was first elucidated by Gruesbeck and Collins (*39*). They observed that permeability damage by mechanical fines migration occurred only when the fluid velocity or flow rate was higher than a critical value. Several subsequent studies have confirmed this finding and reported the experimentally determined values of critical velocity in different systems, such as Berea sandstone, core plugs from several reservoirs, and unconsolidated sandpacks (*11, 40–44*). The existence of a critical velocity can be explained by considering the influence of velocity on the likelihood of mobilized particles bridging at pore constrictions. At low velocities, only a small number of fines are mobilized, and these dispersed fines can align themselves to work their way one by one through the pore constrictions. At high velocity the fines are in rapid motion and interfere with each other and bridge in a "brush heap" manner. It is apparent that the conditions leading to mobilization of fines and their subsequent bridging would depend on the type of fines involved and other characteristics of the rock fluid system. Therefore the value of the critical velocity is system-specific and usually needs to be measured experimentally. Figure 4 shows a typical critical velocity measurement in a reservoir core.

Gruesbeck and Collins (*39*) also observed that with continued flow at a constant velocity, higher than the critical velocity, the permeability continues to decline for some time but eventually stabilizes. The extent of permeability reduction depends on the margin by which the critical velocity is exceeded, being more severe at higher flow velocities.

A partial recovery of permeability in damaged cores following a reversal of the flow direction has been noted in many experimental studies (*11, 37, 39–45*). This type of behavior has been referred to as "check valve" plugging and is an indicator of pore throat blockage by fines.

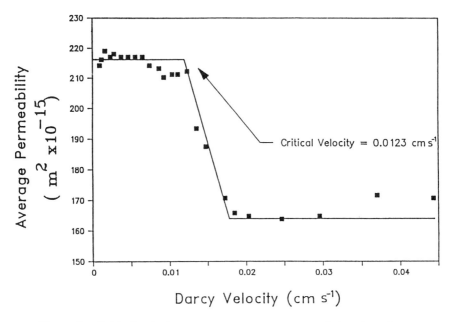

Figure 4.   Critical velocity measurement in an unconsolidated sand core.

The permeability recovery is only temporary and the permeability rapidly declines again with continued flow in the reversed direction as pore throats in the opposite direction become clogged by particle bridges.

By means of micromodel tests, Muecke (38) examined the role of particle wettability and surface–interfacial forces in determining particle mobility. He concluded that particles will move only when the fluid that wets them was mobile. The water-wet fines remained immobile during the flow of oil at irreducible water saturation. He also observed that the particle bridges that formed at pore throats could be easily disrupted by pressure disturbances and flow reversals. Gabriel and Inamdar (41) reported that no critical velocity existed for the flow of a chemically compatible, nonwetting oil at connate water saturation. In contrast to Muecke's visual observations and the findings of Gabriel and Inamdar, other studies have reported significant permeability damage during high rate flow of oil at connate water saturation. Gruesbeck and Collins (39) found that the permeability of Berea sandstone was damaged by flow of a white oil at connate brine saturation when the flow velocity exceeded 0.14 cm/s. Similar damage was observed in a field core also, for which the critical velocity was found to be 0.24 cm/s. It should be noted that the critical velocity for damage by flow of the white oil at connate water saturation was much higher than the critical velocity for damage by flow of brine. Miranda and Underdown (43) reported that well-defined and distinct critical rates existed for damage by flow of crude oil in two different zones of Miocene Stevens Formation in Kern County, California. They found that a critical rate can also exist for flow of natural gas at connate water saturation. These studies suggest that, although fines are mobilized more readily by flow of a wetting fluid, even the flow of a nonwetting fluid at sufficiently high rates can cause permeability damage.

Effect of Fluid Viscosity. The flow-induced fines migration occurs when the hydrodynamic forces acting on fines become larger than the binding forces that hold the fines on pore surfaces. Other factors being equal, the viscous drag exerted by the flowing fluid would be proportional to the fluid viscosity. Therefore, permeability damage could occur at a lower velocity when higher viscosity fluids, such as polymer solutions, are injected into a reservoir.

Gruesbeck and Collins (39) evaluated the effect of increasing brine viscosity (by addition of a polymer) on the critical velocity for permeability damage. They observed a roughly proportional decrease in the critical velocity when the viscosity was increased by a factor of 10. However, they were unable to draw definitive conclusions regarding the quantitative effect of fluid viscosity because of a very limited amount of data being available.

To our knowledge, no systematic study of the effect of viscosity on critical velocity or the extent of permeability damage has been reported.

Effect of Rock Characteristics. Rock characteristics that affect mechanically induced fines migration include the following: absolute permeability, pore geometry and pore size distribution, size distribution of fines, amount and type of fines present in the rock, and the location of fines with respect to the pores.

The absolute undamaged permeability appears to affect the susceptibility of porous rocks for permeability damage. The more permeable rocks tend to have larger pores and pore throats. Particle bridges are more difficult to form across large pore throats and would be less stable when they do get formed. Experiments carried out in Berea sandstone cores, which are known to contain nonswelling clays, revealed that when distilled water follows brine, more permeable cores did not plug completely, but expelled a large amount of migratory fines in the effluent (46). Nasr-El-Din et al. (44) reported that the permeability damage in a high-permeability unconsolidated sand core from a heavy oil reservoir by mechanical fines migration was insignificant even at the high flow velocity of 250 m/day. The core contained 3 wt% kaolinite and 1 wt% illite. Migratory fines were detected in the core effluent, but no permeability damage was observed even when a 56 mPa·s glycerin solution was injected at 68 m/day velocity.

The relationship between the pore size distribution and the size distribution of fine particles has been noted to play an important role in pore throat bridging (47, 48). Studies of deep bed filtration have shown that the ratio of pore size to particle size affects the removal efficiency (49). Studies of the flow of suspended particles through porous media indicate that certain specific particle size distributions can be transported through porous media with given pore size distribution with little damage to porosity and permeability (50, 51). In deep bed filtration, particles are captured by two mechanisms: straining capture at pore throats and surface capture at pore walls. The surface capture occurs when the Brownian motion brings a particle close to a wall. The capture efficiency of a given deep bed filter would be minimum at a specific particle size. It increases by more efficient straining capture as the particles become larger and by more efficient surface capture as the particles become smaller. A similar situation exists in permeability damage by mechanical fines migration. Once the fines are mobilized, the plugging of pore constrictions will be controlled by the ratio of particle diameter to pore throat diameter. At a certain ratio of average particle to pore throat size, a large number of particles will be transported through the rock and eluted without any permeability damage.

The amount and the type of fines present will certainly affect the extent of permeability damage. Relatively clean formations, that is, rocks containing only a small amount of clays and other fines are less susceptible to formation damage by fines migration (44). Conflicting results have been reported on the influence of clay types involved. Egbogah (40) reported that sandstones containing the highest content of kaolinite were more prone to permeability damage and had lower critical velocity. Leone and Scott (11) found that no significant fines migration problem occurred in the cores from a zone containing a high fraction of kaolinite, whereas pronounced fines migration damage was observed in core from the other zone in which kaolinite was present only in minor amounts. It is apparent that other factors are involved in controlling the susceptibility of a given rock to fines migration damage.

A very important factor is the location and distribution of clay particles within the rock. The total clay content or clay type are not dependable indicators of the susceptibility of the rock to damage. The location of clays and the clay growth form tend to control the degree of damage susceptibility. A rock in which most of the clay is confined to shale streaks or mud "rip up" clasts is likely to be less susceptible to damage compared with the rock in which clay is present in the pore lining form.

In unconsolidated sand formations, the net overburden pressure also appears to affect fines migration. Coskuner and Maini (42) reported that the critical velocity at which permeability damage is initiated decreases with increasing net confining pressure. They suggested that the decrease in critical velocity is due to the effect of net confining pressure on permeability, pore sizes, and pore throat sizes.

Fines Migration in Multiphase Flow.

Most of the preceding discussion was concerned with flow of brine at 100% brine saturation. From a petroleum engineering perspective this brine staturation represents a condition that is rarely encountered in the field. A more relevant situation would be flow of brine at residual oil saturation and commingled flow of oil and brine. When fines migration occurs in the presence of two immiscible fluids, additional factors such as the wettability of the medium and that of the fines and the relative permeability characteristics become important. Therefore, it is important to consider the effect of the presence of a second immiscible fluid on fines migration and permeability damage.

Mungan (52, 53) suggested that the problem of fines migration and water sensitivity may be less severe in the presence of oil in the core. Extraction of residual oil tends to remove any organic coating present on the pore walls in the native state and makes the rock more susceptible to damage. Clementz (54) reported that the adsorption of petroleum heavy ends on pore surfaces can stabilize clays and prevent fines migration damage.

Muecke (38) investigated the behaviour of fines under two-phase flow conditions using a micromodel. He reported that fines migrate only when the phase that wets them is mobile. Therefore if the fines are water-wet, as would be expected in most sandstone formations, the flow of oil at connate water saturation will not cause any permeability damage. As long as only oil is flowing, the fines will be restricted to the immobile connate water layer. Muecke (38) suggested that in simultaneous two-phase flow, the local pressure disturbances due to capillary effects keep the fines mobilized and reduce the stability of particle bridges at pore constrictions.

Sarkar and Sharma (55) reported the results of an experimental investigation of fines migration in two-phase flow. Although the fines migration in this study was triggered by chemical means, some of their findings are relevant to this discussion. The damage ratio (ratio of damaged permeability to initial permeability) at 100% water saturation was 1/1000. In the presence of residual saturation of a nonpolar mineral oil, the damage ratio was 1/50. Thus it would appear that the presence of residual oil saturation reduces the extent of damage even when the oil is a nonpolar refined mineral oil. Because the wettability of fines in Berea sandstone is likely to remain unchanged in the presence of this oil, and the nonpolar oil is not expected to provide a protective surface coating, the mechanism involved in reducing the damage appears to be unrelated to surface or interfacial effects. Because the endpoint relative permeability to water in the Berea core is typically less than 10%, the effective permeability to water after damage was nearly same in both cases. At 100% water saturation the final permeability of the damaged core represents the flow conductance of particle bridges across pore throats. It is conceivable that the same type of bridges are formed in the presence of residual oil in the core and their conductance is not affected by the oil because oil is unable to enter the microporosity of the bridges.

Another interesting result reported by Sarkar and Sharma (55) was that after saturating the core initially with a polar crude oil, the damage ratio was only 1/1.5 and the permeability decline was very slow. Thus, allowing the rock surfaces to come in direct contact with the polar oil did make the rock less susceptible to damage. This effect is apparently related to adsorption of polar components on the rock surfaces and altered wettability of the fines.

Experimental Studies of Permeability Damage by Chemical Fines Migration

Chemical fines migration refers to the situation in which the changes in the chemical environment within the porous rock initiate dispersion of fines and permeability damage. This change in chemical environment is brought about primarily by a change in the composition of the fluid flowing through the rock. However a change in temperature by conductive heating may also be a contributing factor. As noted earlier, the colloidal forces acting on fine particles, and keeping them attached to pore walls in the undamaged rock, depend strongly on the composition of the fluid present. When this fluid is water, the electric double-layer repulsion between the pore wall and the fine particle depends on the concentration and type of ions present. Therefore, a change in the concentration or type of ions present in water can alter the balance between forces acting on the fine particle from a condition favoring attachment to the wall to a condition favoring detachment. When this happens, the fines detach from the walls, migrate with the flowing fluid, and eventually become trapped in downstream pore constrictions. Consequently, the permeability of the rock is reduced. Such permeability damage has been a big concern in secondary oil recovery operations involving injection of water to displace oil. Several experimental studies aimed at evaluating the role of different factors involved have been reported (12, 18, 36, 55–65). Because of the diverse nature of rocks, clays and other fines, and ionic species involved, the quantitative results obtained with a particular rock–fluid system may not apply to other systems. However, a reasonably good qualitative understanding of the chemical fines migration process has evolved. The effect of various factors that play a role in this process will be reviewed next.

Effect of Changes in Water Salinity (Ionic Strength). It was noted in several earlier studies that the permeability of some sandstones was severely damaged when they were brought in contact with fresh water. Such reduction in the permeability was initially attributed to swelling of water sensitive clays, such as montmorillonite, and the accompanying reduction in the pore size due to a larger volume being occupied by swelled clays. For some time, the phrase "clay swelling" became synonymous with permeability reduction due to fresh water contact. However, it soon became apparent that clay swelling alone could not account for some of the observed behavior. Land and Baptist (56) noted that there was no correlation between the amount of swelling clay present and the extent of permeability damage by fresh water contact. They concluded that the water sensitivity of the sand is not necessarily a result of pore blockage due to the increased volume occupied by the swollen clays and may be a result of dispersion and subsequent transportation of clay minerals to pore constrictions. They also noted that permeability reduction could occur in formations that do not contain expandable clay minerals.

Mungan (52, 53) investigated the water sensitivity of Berea sandstone and other formations containing only nonexpandable clays. His experiments showed that the primary cause of permeability reduction was blocking of pore passages by dispersed particles and occurred regardless of the type of clay involved. He noted that fresh water or 30,000 ppm NaCl brine could be injected into a new sample of Berea sandstone without any permeability loss. However, after injection of the brine, the permeability was readily damaged by fresh water. The dispersion and permeability damage by fresh water after a brine injection depends on the type of rock and the clay distribution in the rock as well as the type and concentration of brine (13, 37). A very significant finding of Mungan's study was that the severe permeability reduction occurred only when the core experienced a step change in salinity. A slow and gradual change in salinity caused no permeability reduction. This finding lead to the development of the "water shock" test for evaluating the water sensitivity of reservoir rocks. The results of a typical water shock experiment are shown in Figure 5. The core permeability decreases sharply soon after the flow is switched from brine to fresh water and reaches a value more than two orders of magnitude smaller than the undamaged permeability after only two pore volumes of fresh water injection.

Critical Salt Concentration (CSC). It was noted in several studies (13, 64–67) that the chemical fines migration occurs only when the

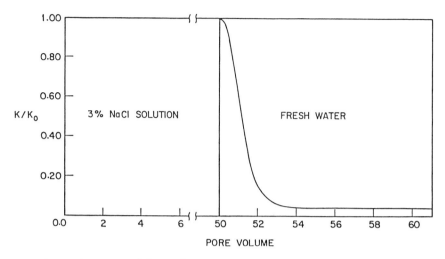

Figure 5. Permeability reduction in a typical water shock experiment.

salt concentration decreases below a critical value, which depends on the characteristics of the rock and the salt involved. If equilibrium is established between a reservoir rock and a salt solution at concentration higher than the CSC, and then the salt concentration in the flowing fluid is reduced below the CSC, a sharp reduction in permeability occurs. Changing salt concentrations at levels higher than the CSC or lower than the CSC causes no damage.

Figure 6 shows how the critical salt concentration is determined. A core sample is saturated with a suitable salt solution and mounted in a standard core holder and coreflood apparatus. Salt solution is then injected at a specified superficial velocity. Subsequently the salt concentration is reduced in small steps until a decrease in permeability is observed (36).

In agreement with the statements made regarding the influence of hydrodynamic forces in the previous section, Khilar et al. (18, 36) have found at best a weak dependence of CSC on the superficial velocity (for 3 to 568 cm/h). Because in practise one deals with solutions of mixed salts, Khilar et al. (18) have introduced the critical total ionic strength (CTIS) to improve the predictions for solutions containing multivalent ions such as calcium. A number of studies have shown that where freshwater flooding of sandstones may drastically decrease permeability due to fines migration, suitable adjustment of the flooding solution composition to above the CSC or CTIS can decrease or eliminate the permeability reduction (12, 18, 36). In these cases the solution compositions are adjusted so as to reduce the Zeta potential at the particle surfaces, which reduces the repulsive colloidal forces. Thus the same factors that

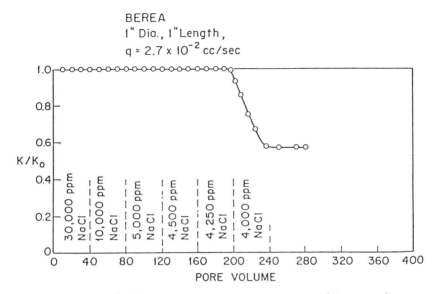

Figure 6. Critical salt concentration in water sensitivity of Berea sandstone.

are used in colloid science to reduce the stability of particle suspensions may be used in formulating injection solution compositions for the prevention of fines migration:

- diffuse layer compression with simple electrolyte
- Stern layer adsorption with multivalent ions
- surface charge reduction by adjusting (usually decreasing) the pH

Figure 7 gives an example of diffuse layer compression and Stern layer adsorption in which the normalized permeability (k/k_i where k_i is the initial permeability) of a Berea sandstone is plotted versus number of pore volumes of liquid injected. In separate experiments either sodium chloride solution or calcium chloride solution was injected, followed by fresh water injection. The calcium treatment made the rock relatively insensitive to subsequent fresh waterflooding because adsorbed calcium ions reduced the surface potential and were not quickly ion exchanged off the surfaces. In the sodium case only diffuse layer compression was occurring so that fresh waterflooding quickly reduced the electrolyte concentration below the CSC level and the permeability decreased to less than 1% of its original value. Figure 8 shows the results from similar experiments in which cores were saturated with different 0.51 M salt solutions and then "shocked" by injecting fresh water.

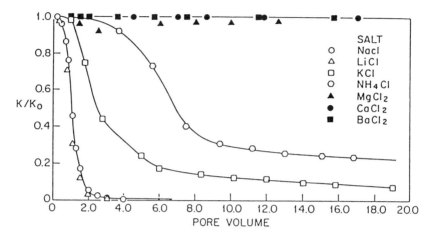

BEREA SANDSTONE
1" Dia., 2" Length
Flow Rate ≈ 100 cc/hr

Figure 7. Comparison of permeability reduction for various salt solutions.

Critical Rate of Salinity Decline (CRSD).

It was noted previously that no permeability damage occurred when the salinity was changed very slowly, whereas severe damage occurred when the salinity was changed abruptly. It is therefore expected that as the rate of salinity change is varied from the abrupt change to very slow change, at some point the damage will become small. Khilar (69) noted that a system-specific "critical rate of salinity decrease" exists below which the damage due to salinity change is minimized or totally eliminated. He explained the existence of CRSD by suggesting that when the salinity is decreased gradually, the fines are released at a slower rate and, therefore, the concentration of fines in the flowing suspension remains low. This low concentration of fines does not lead to the "log-jam" of particles at the pore throats, allowing the particles to be transported through the core without being trapped.

Effect of pH.   The pH of the flowing fluid is an important factor in the fines migration process (5, 12, 52, 58). Mungan (52) noted that injection of strong acids or bases could cause permeability damage. Under very high or very low pH conditions, the permeability damage is caused by dissolution of the matrix material, which produces fine particles of varied mineral composition. Somerton et al. (58) found that the water sensitivity of reservoir sands was related to the pH response exhibited by the rock after the contact with fresh water. Most sandstone cores showed an increase in the effluent pH after the switch was made

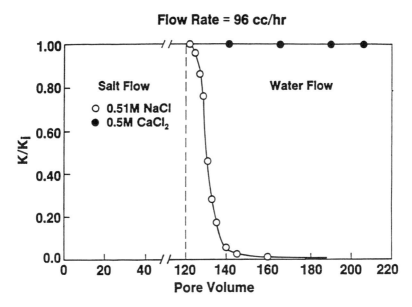

Figure 8.   Effect of calcium chloride on water sensitivity of Berea sandstone.

to flow of fresh water. However, some cores showed a decline in the effluent pH, and these cores were found to suffer no permeability loss from the salinity shock.

Kia et al. (*12*) conducted a systematic study of the effect of pH on colloidally induced fines migration in Berea sandstone by determining the effect of salinity shock at different values of pH. They found that the release of fines and subsequent formation damage can be prevented by adjusting the pH of the flowing solution. A drastic reduction in permeability was found to occur when the salinity shock was given at a pH above 6. As the pH of the brine and the fresh water was decreased below 6, the extent of damage also decreased. No permeability damage occurred below a pH value of 4.8.

The effect of pH on fines migration can be explained by considering the variation in surface charge of clay particles with pH. Under neutral pH conditions clay particles are generally negatively charged. Low pH conditions lead to binding of potential-determining H^+ ions at the edges. This imparts a small net positive charge to some clay particles at low pH. The pore walls also acquire their surface charge by adsorption of potential-determining ions from the fluid present in the pore. This charge is also positive at low pH and negative at high pH. Clay surfaces are attracted to positively charged pore walls at low values of pH. Furthermore, the attraction between the positively charged edges and negatively charged surfaces leads to considerable surface to edge type flocculation.

Consequently, clay particles do not disperse at low values of pH. At high pH, both the clay particles and pore walls become negatively charged. This produces a significant increase in the repulsive force and leads to the detachment of particles from the wall.

Role of Ion Exchange.   The role of ion exchange between the flowing fluid and the rock minerals becomes apparent when one considers that the chemical environment experienced by the clay particles downstream of the injection point can be considerably different from the injected fluid because of exchange of ions between the rock and the fluid. Both the ionic composition and the pH of the fluid can change by ion exchange. Although it was noted in several earlier studies that the composition and pH of core effluent differs significantly from the injected fluids, the vital role of ion exchange in formation damage was elucidated by Vaidya and Fogler (63, 64). During fresh water injection into a brine (NaCl)-saturated core, there is an exchange between the Na^+ adsorbed on clays and H^+ ions in the injected water. This exchange causes the pH of the flowing fluid to increase. The increased pH alters the surface potential of clay particles and the pore walls, as discussed in the preceding section. The zeta potential of kaolinite, which is a measure of the total charge of kaolinite particles, is shown as a function of pH in Figure 9. As the pH increases the zeta potential becomes increasingly negative in the presence of monovalent cations. Additional data on the dependence of zeta potential on pH for reservoir rock particles has been reported by Schramm et al. (22). For Ca^{2+} the zeta potential remains relatively low even at high pH. The fines release in salinity shock experiment is caused by the increase in pH that comes about by the ion exchange process. Conditions that prevent the increase in pH also prevent fines migration. Thus injection of a low pH fluid, containing an

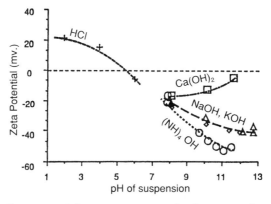

Figure 9.   Zeta potential measurements on kaolinite as a function of pH.

excess of H^+ ions allows the Na^+ to be replaced with H^+ without the solution becoming alkaline. Consequently, fines release and permeability damage does not occur in salinity shock experiments at low pH.

The ion exchange also explains why no permeability damage occurs when the salinity is changed slowly. When the change in salinity is slow, only a limited exchange between the adsorbed Na^+ and H^+ occurs during the injection of one pore volume of fluid. This exchange causes only a small increase in the pH of this fluid. Although by the time salinity decreases to a low value, most of the Na^+ is replaced by H^+, this exchange takes place with a large volume of injected fluid and the total change in H^+ concentration of this large volume of fluid remains small. Consequently, the in situ pH value never becomes high enough to trigger fines release.

Effect of Temperature.　Temperature can affect the fines migration and permeability reduction process in several ways. It has a direct effect on the electric double layer repulsion that increases with the absolute temperature (*see* eq 1). It also affects the adsorption and ion exchange behavior. Allen et al. (*70*) reported that formation damage decreased with increasing temperature during flow of brine in the presence of oil. However, Udell and Lofty (*69*) reported that permeability damage due to stress-induced silica dissolution at grain contacts increased with increasing temperature. Khilar and Fogler (*36*) found that the critical salt concentration increased with increasing temperature.

Although only a limited amount of experimental data is available on the effect of temperature, it is apparent that a modest change in temperature has only a minor effect on the process. When a large change in temperature is involved, as may be encountered in thermal oil recovery processes, the temperature effects can be more pronounced. However, at temperatures involved in thermal recovery operations, other formation damage mechanisms, such as mineral dissolution and reprecipitation and colloidal iron plugging, may also be involved.

Mathematical Modeling of Fines Migration in Reservoirs.
Modeling of fines migration aims at predicting the nature and conditions leading to the retention and release of particles in porous media. Historically, the modeling of fines migration was attempted through phenomenological modeling. However, this approach did not give further insight into the deposition and retention mechanisms. On this accord, more theoretical approaches were undertaken.

Phenomenological modeling uses a set of partial differential equations that characterize the fines migration process by means of model parameters. The values of these phenomenological parameters are attained through experiments. Phenomenological modeling can also be broadly classified as macroscopic modeling, since the parameters are determined from lab scale experiments.

Theoretical modeling tries to give a quantitative explanation of the fines migration mechanism by means of derived mathematical formulations. The mathematical formulations take into account the dynamic interactions of the fines with other suspended particles, the solid substrate of the porous medium, and the liquid phases surrounding the particles. These microscopic theories of fines migration are intended to give insight into the release, transport, and deposition of fines.

The modeling efforts are concentrated on variables that can be easily measured through experiments. Two of the more accessible quantities are the effluent concentration of fines and the pressure difference across a section of porous material through which the transport fluid is being injected.

The effluent concentration can be predicted by solving the mass conservation equation. The conservation equations of particulate matter consider the change in concentration of particulate and change of porosity with time. The amount of fines retained in the porous medium is represented by σ, while u signifies the superficial velocity of the incompressible transport fluid. For constant volumetric, incompressible flow, neglecting dispersion and gravitational effects, the one dimensional conservation equation follows.

$$\frac{\partial(\sigma + \phi C)}{\partial t} + u\frac{\partial C}{\partial x} = 0 \tag{6}$$

An auxiliary relationship has to be set up to describe the process of fines deposition and release. A general rate equation can be set up such that G represents a generic function and γ represents a vector of parameters. Note, that the deposition rate depends explicitly on the suspended particle concentration and the concentration of retained fines.

$$\frac{\partial\sigma}{\partial t} = G(\gamma, C, \sigma) \tag{7}$$

Solving the coupled, partial differential equations (eqs 6 and 7), allows for the prediction of the effluent concentration profile.

The deposition and release of particles within the core causes a local change in pore geometry, resulting in a local variation of the pressure gradient. A generic expression for the pressure variation is given in equation 8.

$$\frac{\partial p}{\partial x} \bigg/ \left(\frac{\partial p}{\partial x}\right)_0 = F(\beta, \sigma) \tag{8}$$

Integration of equation 8 over the length of the porous medium generates an expression for the pressure difference across the core.

In the following sections the phenomenological and theoretical modeling approaches are reviewed. Some of these models are quite complex, yet they follow the basic principles introduced previously.

Phenomenological Modeling. *Filtration Theory.* Modeling of fines or particulate migration was first considered in deep bed filtration problems. The introduction of a more convenient time variable

$$\theta = t - \int_0^x \frac{\phi}{u} \, dx \tag{9}$$

allows the filtration equation to be written in its original form (*see* Appendix A):

$$\frac{\partial \sigma}{\partial \theta} = -u \frac{\partial C}{\partial x} \tag{10}$$

In practice the difference between θ and t is very small.

On the basis of experimental data, various expressions were proposed for the rate of filtration in a filter bed. Iwasaki (71) originally proposed a linear rate expression, which also implies a logarithmic distribution of fines throughout the filter bed.

$$\frac{\partial \sigma}{\partial \theta} = \lambda u C \tag{11}$$

$$\frac{\partial \sigma}{\partial z} = -\lambda C \tag{12}$$

The filter coefficient, λ, varies as deposited material changes the morphology of the porous medium and as conditions surrounding the collection sites change. It has been noted that the filtration coefficient increases as fines migrate through a clean filter bed; the retained fines increase the specific surface area. This increase in λ is short-lived, and the magnitude of the filter coefficient decreases as additional fines are retained. Since Iwasaki published his notes on filtration in 1937, numerous variations of the rate expression have been recorded (72).

A capillary tube model can be used to estimate the permeability of the medium before fines deposition or release has occurred. The Carman–Kozeny equation uses the diameter of the substrate particles, d_g, and the tortuosity of the medium, τ, to evaluate the effective permeability of the porous medium.

$$-\left(\frac{\partial p}{\partial x}\right)_0 = \frac{72\tau(1 - \phi_0)^2}{\phi_0{}^3 d_g{}^2} \mu u \tag{13}$$

To model the permeability reduction, due to fines deposition, the generic function $F(\beta, \sigma)$ from equation 8 can be expressed through an empirical correlation (73).

$$F(\beta_1, \beta_2, \sigma) = \frac{1}{(1 - \beta_2 \sigma)^{\beta_1}} \qquad (14)$$

The empirical parameters β_1 and β_2 are usually adjusted to fit experimental pressure responses.

Formation Damage Model.   The early attempts of modeling formation damage due to fines migration in petroleum reservoirs made use of the equations obtained from deep bed filtration studies. However, in these early studies basic experimental evidence leading to permeability impairment was neglected. Ohen and Civan (74) published a versatile, phenomenological model that includes the effects of clay swelling, external particles invasion, and fines generation, migration, and retention.

For a compressible transport liquid that contributes to swelling of the clays, a more complex continuity equation has to be specified.

$$\frac{\partial(\phi \rho_l^f + \sigma_l^s)}{\partial t} + \frac{\partial(\rho_l^f u)}{\partial x} = 0 \qquad (15)$$

where ρ_l^f is the mass concentration of the liquid, σ_l^s is the amount of liquid adsorbed by the clays. Civan et al. (75) assumed that liquid diffuses into the pore walls according to Ficks law, after initial contact. The adsorption rate is proportional to the mass concentration difference of the transport fluid and the initial mass concentration of liquid in the solid matrix, $\rho_{l,i}^f$; B is an adsorption rate coefficient.

$$\frac{\partial \sigma_l^s}{\partial t} = B(\rho_l^f - \rho_{l,i}^f)/\sqrt{t} \qquad (16)$$

The conservation equation for the fines is similar to equation 6. However now the distinction is made between external particle deposition, σ_p^s, and re-entrainment of indigenous particles, σ_p^{s*}.

$$\frac{\partial(\phi \rho_p^f + \sigma_p^s - \sigma_p^{s*})}{\partial t} + \frac{\partial(f \rho_p^f u)}{\partial x} = 0 \qquad (17)$$

The fraction f is referred to as the particle transport efficiency and represents the fraction of the pore cross-sectional area through which the particles may pass. Evaluation of f is based on a size-exclusion principle.

$$f = 1.0 - E_p \int_{d_p}^{d_{th}} f_y d_y \qquad (18)$$

where E_p is the plugging efficiency, f_y is the pore throat size distribution function, d_{th} is the lower pore size, and d_p represents the particle diameter. This size exclusion principle, generates a straining or plugging action that causes a fraction of the fines to be retained along each section of the porous medium. (*see* Figure 10).

The net rate of external fines accumulation is equal to the rate of fines captured minus the rate of fines re-entrained ($\partial\sigma_p^s/\partial t = r_r - r_e$). Civan et al. (75) assumed that the retention rate, r_r, of the fines is proportional to the flux of the flowing phase in the pore volume available to the flow of fines. Also, the rate of re-entrainment, r_e is proportional to an excess pressure gradient once the limiting pressure gradient, $((\partial p/\partial x)_{cr})$ for the onset of fines re-entrainment, has been surpassed.

$$\frac{\partial\sigma_p^s}{\partial t} = k_1 u \rho_p^f \phi - k_2 \sigma_p^s \left(\left(\frac{-\partial p}{\partial x} \right) - \left(\frac{-\partial p}{\partial x} \right)_{cr} \right) \tag{19}$$

The loosely anchored indigenous particles can be categorized into two major groups: expansive authigenic clay minerals and nonexpansive clay minerals. A simple kinetic expression is set up for the generation of these indigenous fines. The variable r' represents the fraction of expansive clay minerals with respect to the total releasable fines, σ_p^{s*}.

$$\frac{\partial\sigma_p^{s*}}{\partial t} = [k_1^* r' \sigma_p^{s*} + k_2^*(1 - r')\sigma_p^{s*}] \left(\left(\frac{-\partial p}{\partial x} \right) - \left(\frac{-\partial p}{\partial x} \right)_{cr} \right) \tag{20}$$

To refine the kinetic rate expression, Civan and Ohen coupled the expansion rate of the indigenous clays with the diffusive flux of water into the porous matrix (*see* eq 16); as the particles expand, they experience an increase in viscous drag forces exerted by the moving fluid, such that the particles become more easily entrained by the flow. Furthermore, it is assumed that the release rate is proportional to the exposed pore surface area, a_u, after the deposition of external particles. A simple ex-

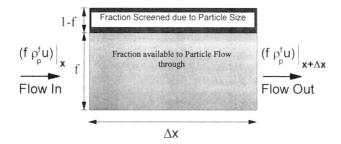

Figure 10. Fraction of pore volume available to particle flow.

ponential function, dependent on the deposit concentration, $\sigma_p{}^s$, is effective in calculating the exposed surface area.

$$a_u = a_s e^{-k_5\sigma_p{}^s} \qquad (21)$$

After some rearrangement, the refined kinetic expression takes on a more complex form:

$$\frac{\partial \sigma_p{}^{s*}}{\partial t} = k_4 e^{-k_5\sigma_p{}^s}\sigma_p{}^{s*}$$

$$\times [\alpha - r' \exp(-k_3 B(\rho_l{}^f - \rho_{l,i}{}^s)\overline{V}t)]\left(\left(\frac{-\partial p}{\partial x}\right) - \left(\frac{-\partial p}{\partial x}\right)_{cr}\right) \quad (22)$$

The effective porosity, ϕ_{eff}, available for the transport of fines depends on the factor f, calculated from the size exclusion principle, and the amount of porosity lost to fines deposition and clay swelling.

$$\phi_{\text{eff}} = f\left(\phi_i - \frac{(\sigma_p{}^s + \sigma_{sw}{}^s)}{\rho_p}\right) \qquad (23)$$

It remains to establish, the pressure trend throughout the porous medium as the particles filter out of solution. As long as no filtration has taken place, Darcy's law can be applied. However, the pressure profile will change according to deposition of filtrate on the matrix of the porous medium. The permeability reduction can be approximated with the Carman–Kozeny relation:

$$\frac{k}{k_0} = \left(\frac{\phi_{\text{eff}}}{\phi_0}\right)^3 \qquad (24)$$

The Forcheimer equation incorporates the pressure response due to fines deposition.

$$\frac{\partial p}{\partial x} = -\frac{\mu_f}{N_{nD}k}u \qquad (25)$$

where N_{nD} represents the non-Darcy flow number. A correlation for N_{nD} was established by Geertsma (76).

Equations 15 to 25 demonstrate a versatile approach to the phenomenological modeling. Ohen and Civan (74) present, additional, auxiliary relationships that make this model deterministic. In all there are 12 parameters $(k_1, k_2, \ldots, (\partial p/\partial x)_{cr})$ to be determined. Some of the parameters can be directly measured by experiment, some are evaluated through sensibility tests, and the rest are determined by history matching experimental results with simulations.

A number of factors lead to the mobilization of fines: ionic strength and composition of the fluid, surface potential of the particle and porous substrate, concentration of the releasable fines, fluid shear forces, swelling of the clay minerals, and temperature. The phenomenological coefficients optimized for one set of conditions will have to be re-evaluated for another set of conditions. As such, phenomenological modeling is not ideal for demonstrating the fundamental mechanisms and a more theoretical approach has to be taken.

Theoretical Modeling. The phenomenological modeling is based on matching the experimental results by adjusting the phenomenological coefficients. This approach does not shed light on the retention and deposition mechanisms of fines. Civan (77) compared modeling aspects of various theoretical and phenomenological formation damage models. Simplified, structural models of porous media were proposed to investigate, in detail, the dynamics of filtration by mathematical means. Two of these models are particularly noteworthy: the unit bed element (UBE) model and the network model. In the UBE model, the porous medium is divided into series of UBEs of length l_e (*see* Figure 11). Each of the unit elements consists of a number of collectors that attract the fines. The network model represents the porous medium as an array of pores interconnected by throats (*see* Figure 12). The average number of throats connected to a pore specifies the coordination number of the network model. Both methods have given valuable insight into the filtration process and warrant further discussion.

Even though the two models are based on a different porous medium representation, the fundamental approach is to evaluate the net force that is acting on the fines and to determine the likelihood of the particles

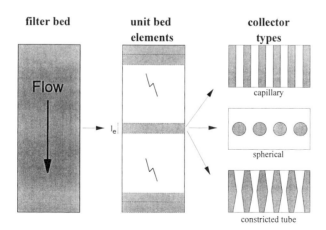

Figure 11. UBE model.

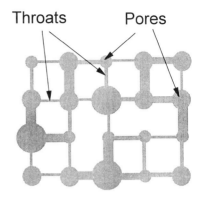

Figure 12. Two-dimensional network model.

to be retained by substrate under the influence of this force. Some of the individual forces acting on a particle are listed here:

1. electrostatic forces
2. gravitational forces
3. dispersion forces
4. hydrodynamic forces
5. mechanical stress

Unit Bed Element Modeling.   The filter bed is divided into many cross-sectional elements along the axial direction of flow. If the number of elements is large, then it may be assumed that the filtration coefficient λ is constant throughout each UBE, at an instant in time. The removal efficiency, η, of each element, m, is based on the difference of the filtrate concentration entering and leaving the element.

$$\frac{C_{(m-1)} - C_{(m)}}{C_{(m-1)}} = \eta \tag{26}$$

The removal efficiency can be related to the filtration coefficient by integrating equation 11.

$$\lambda = \frac{1}{l_e} \ln \left(\frac{1}{1 - \eta} \right) \tag{27}$$

The selection of the type of particle collector is somewhat arbitrary, usually limited by practical considerations. The stream lines of fluid flow around the collectors should be similar to those in a porous medium and yet simple to calculate. As a particle carried by the fluid comes into the proximity of the collector and experiences the net interaction force between itself and the collector, the trajectory of the particle deviates from that of the fluid streamlines. The actual path of the particle is determined through the equation of motion on the basis of the force balances. Not all the particle trajectories have to be calculated, only the limiting particle stream line. The limiting stream line is defined as the trajectory separating the ones that collide with the collector and those that do not. Any particle trajectory between the collector and a limiting trajectory will be deposited on the collector. Particles travelling external to this limiting trajectory will not come in contact with the collector. The calculation of the stream lines and trajectories is beyond the scope of this chapter; only the basic analytical steps are indicated here [Tien and Payatakes (72)]:

1. specification of the collector type, geometry, size, and size distribution
2. determination of streamlines around the collector
3. establishing the forces acting on the particle

The volumetric flow rate between two calculated stream lines, Ψ_1' and Ψ_2', is equal to the difference between their respective values, $q_{12} = \Psi_1' - \Psi_2'$. Assuming that all the particles transported within the limiting stream function and the collector get deposited, then a relationship can be established between the stream lines and the removal efficiency.

$$\eta = \frac{\Psi_{collector}' - \Psi_{lim}'}{\Psi_{collector}' - \Psi_{reference}'} \tag{28}$$

During the initial stages of fines deposition the simplified stream functions of the spherical model are well suited for modeling the deposition process. However, in the later stages the pore blocking mechanism becomes more important and the constricted tube model is more representative of this situation. The calculations here are carried out for relatively large particles (diameter greater than 1 μm).

Even though the theoretical removal efficiencies, η, vary according to which type of particle collector was chosen, some general observations hold true for most of the collectors. Some of the observations made by Payatakes et al. (78) for a constricted tube model by systematically varying pertinent dimensionless groups are listed here.

Gravitational Force.

$$N_g = \frac{2(\rho_p - \rho_l)a_p^2 g}{9\mu v} \tag{29}$$

Sensitivity studies showed that η is a strong increasing function of the gravitational group N_g, when the magnitude of N_g is small. This dependence on N_g declines as the magnitude of N_g gets larger.

Particle–Collector Interactions.   Dispersion forces and double-layer forces were the two interaction forces considered between the particle and collector. The London dispersion forces can be expressed with the Hamaker constant H, the distance between the two particles δ, and a retardation factor α_{sp}.

$$F_{\mathrm{L}} = -\frac{2Ha_{\mathrm{p}}^{3}}{3\delta^{2}(2a_{\mathrm{p}} + \delta)^{2}} \alpha_{sp} \tag{30}$$

The London dispersion forces are attractive forces that increase as the separation distance δ decreases between the particle and collector. The dimensionless group, N_{L}, is associated with the dispersion forces.

$$N_{\mathrm{L}} = \frac{H}{9\pi\mu a_{\mathrm{p}}^{2}v} \tag{31}$$

The double layer force, experienced by two approaching bodies, is determined by the potential of each body Ψ_{01} and Ψ_{02}, the distance between them δ, the reciprocal thickness of the double layer κ, and the dielectric constant of the liquid ϵ.

$$F_{\mathrm{E}} = \frac{\epsilon a_{\mathrm{p}}\kappa(\Psi_{01}^{2} + \Psi_{02}^{2})}{2} \left(\frac{2\Psi_{01}\Psi_{02}}{(\Psi_{01}^{2} + \Psi_{02}^{2})} - e^{-\kappa\delta} \right) \frac{e^{-\kappa\delta}}{1 - e^{-2\kappa\delta}} \tag{32}$$

The double layer force may either be attractive, Ψ_{01} and Ψ_{02} have opposing signs, or repulsive, Ψ_{01} and Ψ_{02} have the same sign. The dimensionless groups that Payatakes et al. (*74*) investigated for the double layer are

$$N_{\mathrm{DL}} = \kappa a_{\mathrm{p}}; \quad N_{\mathrm{E1}} = \frac{\epsilon\kappa(\Psi_{01}^{2} + \Psi_{02}^{2})}{12\pi\mu u} ; \quad N_{\mathrm{E2}} = \frac{2(\Psi_{01}^{2}\Psi_{02}^{2})}{(\Psi_{01}^{2} + \Psi_{02}^{2})} \tag{33}$$

Even though the dimensionless groups for the double layer are independent of each other, they are still tied together through the force equation.

The sensitivity studies were set up such that the double-layer forces were always repulsive. In this instance deposition occurred only when the dispersion forces were able to overcome the repulsive force. In Figure 13 the sensitivities of the removal efficiency to each group are sketched. As the London dispersion forces, N_{L} increase to the point where they counteract the repulsive forces, there is a dramatic increase in removal efficiency, up to three orders in magnitude. Conversely, increasing the magnitude of the repulsive force, N_{E1}, causes a dramatic decrease in the removal efficiency. The effect of the reciprocal double layer thickness, N_{DL}, is also presented. As N_{DL} is increased, the effective

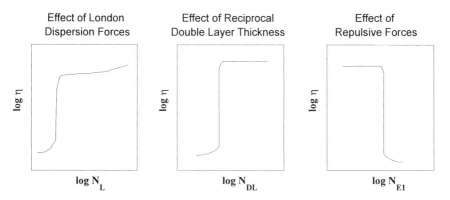

Figure 13. Particle–collector interactions.

repulsive forces decrease; once a critical point is surpassed, the removal efficiency again increases tremendously. These results are extremely important because they show exactly how fines can be captured on the collectors by varying the chemistry of the surroundings fluid. For instance, changing the pH, composition, or the ionic strength of the carrying fluid will influence all of the dimensionless numbers: N_L, N_{DL}, and N_{E1}.

Most of the unit bed element modeling takes place during the initial onset of fines deposition in a clean filter bed. As more and more fines get deposited, the morphology of the collector changes. As such, the streamlines around the collector and the attractive and repulsive force balances around the collector will also change. Mackie et al. (79) had some success in modeling the dendritic effect of fines deposition. Unfortunately this model works well only at the initial stages of deposition since the simplified stream functions of a spherical collector have been used.

Through the UBE models the basic deposition mechanisms of fines can be probed. However these models center their analysis at one or a series of unit cells. Furthermore, the predictions focus on the deposition and entrainment of particles. Often fines are released from substrate of the porous medium, and it becomes necessary to understand the releasing mechanism to avoid potential plugging problems.

Network Modeling. Network models have been proposed such that the interconnectivity and the stochastic nature of the porous medium can be investigated [Sharma (80), Sharma and Yortsos (81), and Rege and Fogler (82)]. As mentioned previously, a network representation consists of a number of pore sites interconnected by bonds. In the model by Sharma and Yortsos (81), the bonds have negligible contribution toward the pore volume. Here, population balances are proposed for a number of particle types; each of the particles types is characterized by a population density and size distribution function

1. suspended particles (N_s, f_s)
2. attached particles (N_A, f_A)
3. trapped particles (N_T, f_T)

These particle classifications originated from Khilar and Fogler (60) (*see* Figure 14). The conservation equation for a given size of suspended particle takes into account the rate of deposition due to surface forces, r_A and the rate of mechanical trapping, r_T. Dispersion is neglected, the porosity is assumed to be constant, and the transport fluid is incompressible, moving at a superficial flow velocity, u. The deposition rates, r_A, and the trapping rates, r_T, contribute further to the attached particle density, N_A, and trapped particle density, N_T, respectively. The populations of the particles of interest are pursued by three coupled, partial differential equations.

$$\phi \frac{\partial(N_s f_s)}{\partial t} + u \frac{\partial(N_s f_s)}{\partial x} = -(r_A + r_T) \qquad (34a)$$

$$\frac{\partial(N_A f_A)}{\partial t} = r_A \qquad (34b)$$

$$\frac{\partial(N_T f_T)}{\partial t} = r_T \qquad (34c)$$

Sharma and Yortsos considered the porous substrate to be either a perfect sink or a perfect source in the absence of an energy barrier; thus the net forces are either completely attractive or repulsive. For these situations the deposition rate expression is set up with the variable ξ; where ξ is set to 0 for conditions of particle release, and set to 1 for particle deposition.

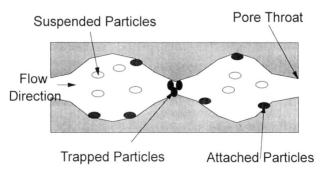

Figure 14.   Particle classifications.

$$r_A = -(1 - \xi)k_r N_A f_A + \xi k_d N_s f_s \qquad (35)$$

Under the assumption of the substrate being a perfect sink or source, the evaluation of the rate constants k_r or k_d is simplified, since the rate limiting step is now diffusion controlled.

$$k_{r/d} = 2\pi(Dl_p)^{2/3}N_p \int_o^\infty (2r^2 u_r)^{1/3} dr \qquad (36)$$

where D represent the diffusion coefficient of the particles, u_r signifies the mean velocity through a throat of radius r, and l_p represents the effective pore length. The rate at which particles are trapped is proportional to the number density of the particular size of suspended particles and the flow rate, u. The function $P(r_s)$ represents the fraction of particles of radius r_s that are trapped in pore throats of the medium due to size exclusion.

$$r_T = P(r_s)N_s f_s \frac{u}{l_p} \qquad (37)$$

Thus, Sharma and Yortsos managed to express the deposition mechanism and trapping mechanism in terms of physical principals. The effect of deposited and trapped material on the pressure response is established next.

The permeability reduction occurs via two mechanisms: (1) trapping (particles larger than the pore throat size) and (2) deposition (particles attached to the pore walls). The same two mechanisms also contribute to pore throat plugging. The porous medium is characterized by the number density of unplugged pore throats, N_p, and the size distribution of open throats f_p.

$$\frac{\partial(N_p f_p)}{\partial t} = - \left(r_{pd}^1 + r_{pd}^2\right) \qquad (38)$$

where r_{pd}^1 represents the rate of pore plugging due to gradual deposition, and r_{pd}^2 equals the rate of pore plugging due mechanical trapping of fines larger than the pore throat.

Sharma and Yortsos make use of the effective medium theory (83) to evaluate the network flow distribution. The effective medium theory is based on the premise that the pressure difference across a particular pore is equal to the mean pressure difference across the porous medium plus a local fluctuation in pressure.

$$\Delta p_p = \Delta p_m + \Delta p_f \qquad (39)$$

In addition, conductivities, g_r, are defined for each pore of radius r_p. It should be noted that for cylindrical pores the conductivity is proportional to the pore radius raised to some power ($g_r \propto r_p{}^n$, $2 < n < 4$). Based on the pore size distribution, a conductance distribution can be set up, $G(g_r)$, for the network. The effective medium conductivity, g_m, is evaluated through the following integral.

$$\int_0^\infty \frac{g_m - g_r}{g_r - (1 - z'/2)g_m} G(g_r)dg_r = 0 \qquad (40)$$

Extensions of the effective medium theory allows for the evaluation of the mean fluid velocity (84) inside a pore, u_r. Thus the connection is made between the local velocity in unplugged pores and the local pore radius.

$$u_r = \frac{ur_p{}^2}{8k}\left(1 + \frac{r_m{}^4 - r_p{}^4}{r_p{}^4 + \left(\frac{z}{2} - 1\right)r_m{}^4}\right) \qquad n = 4 \qquad (41)$$

Here, the coordination number, z', defines the average number of throats connected to a pore. As fines get deposited on the pore wall the effective radius of the pore decreases, causing a change in the local velocity. The rate of pore-radius change can be related to the deposition rate of particles on the pore wall. Furthermore, as the number of unplugged pores changes with time due to particle trapping, so will the conductance distribution G. Thus, the combination of particle trapping and particle deposition influence the effective medium conductivity. The ratio of current mean conductivity to the original mean conductivity is equal to the ratio of current permeability to original permeability.

$$\frac{k}{k_0} = \frac{g_m}{g_{m0}} \qquad (42)$$

The network model, as developed by Sharma and Yortsos, is designed for micrometer-sized particles (< 2 μm) that exhibit Brownian motion.

Sharma and Yortsos define the characteristics of the migrating fines and medium through distribution functions and average quantities such as the coordination number, but Rege and Fogler (85) follow the path of individual particles through a predefined maze of pores and throats. The network model consists of interconnected bonds that represent the pore throats in the medium. Bond lengths, bond radii, and particles sizes are assigned, either randomly or in some related manner. The particles are introduced at the entrance of the network and then are transported through the bonds, where they may be captured or pass through. At the nodes of the network a particle can be selected to flow through one of several available bonds. A flow biased principle is applied here, given that a bond with the greater flow exhibits the highest probability of particle transport. Rege and Fogler also consider two mechanism for particle capture, straining and deposition. Pores are plugged by particles with size greater than the bond diameter. Particles of size smaller than the bond diameter can be deposited on the walls of the bond. The capture probability shown below was first introduced by Stein (86).

$$\text{Capture Probability} = 4*[(\omega a_p/a_b)^2 - (\omega a_p/a_b)^3] + (\omega a_p/a_b)^4 \quad (43)$$

where a_b equals the bond radius and a_p refers to the particle radius. The variable ω is a parameter that encompasses the particle deposition and release forces. If the ratio $(\omega a_p/a_b)$ is greater than or equal to 1, then the capture probability is set to 1, and the particle is captured (*see* Figure 15). Rege and Fogler postulate, on the basis of experimental observations, an exponential relationship between the local bond velocity v and the lumped parameter ω.

$$\omega = \omega_0 \exp\left(\frac{-v}{v^*}\right) \quad (44)$$

where ω_0 is a constant and v^* is a critical velocity. Usually both these variables need to be fit to experimental results.

As particles are deposited on the bond walls, the average bond radius decreases. The average bond radius depends on the number of particles deposited and the particle radius. In addition, the pressure drop across the bond, due to drag on the deposited particles has to be reevaluated. Rege and Fogler assume that particles are deposited on dendrites and experience the drag from the average flow velocity at the center of the bond.

$$\Delta p_{total} = \Delta p_{clean \ tube} + \Delta p_{particle} \quad (45)$$

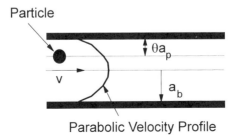

Figure 15. Particle capture probability.

The flow diagram in Figure 16 depicts the algorithm that is used to advance the individual fines through the network. The statistical nature of the problem requires that several runs need to be made to arrive at an averaged solution.

Rege and Fogler point out that their model can predict the evolution of the filter coefficient. When particles get deposited in a bond, the average bond radius, r_b, decreases, thereby increasing the capture

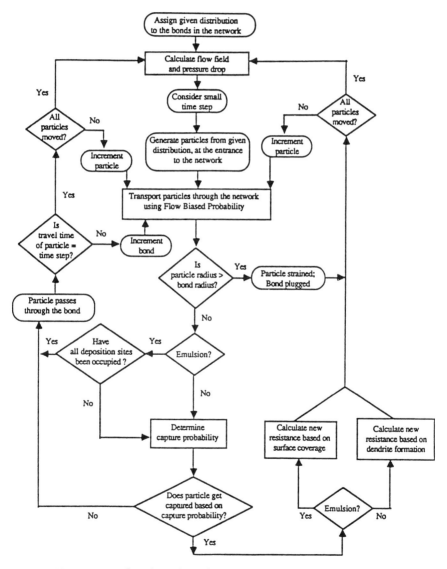

Figure 16.   Flowsheet describing essential features of the model.

probability (*see* eq 43). Constricting the bond radius causes the bond velocity v to increase. This in turn decreases the parameter ω (eq 44), and accordingly the capture probability decreases as well. Therefore, two competing effects have been isolated, the decrease in bond radius and the increase in bond velocity. If the velocity effect is dominant right from the onset of filtration, then the filter coefficient monotonically decreases throughout the experiment. However, if the decrease in bond radius is the initial governing effect, than a temporary increase in the filter coefficient is experienced until the velocity effect dominates again and the filter coefficient begins to decrease. By adjusting the magnitude of the critical velocity v^*, it is possible to demonstrate both of these scenarios.

Fines Migration in Two-Phase Systems.   Up to this point only the fines migration in single-phase flow systems has been discussed. However, most of the fines migration problems in reservoirs deal with two-phase systems. Liu and Civan (87) present a model that characterizes fines migration in two-phase flow. In addition to solving the particle transport equations, the solution of the aqueous phase and oleic phase transport equations is nontrivial. An excellent review on the mass conservation equations for multiphase phase flow in porous media is given in the book by Lake (88). Aziz and Settari (89) present a number of recipes to solve the phase transport equations. The following discussion focusses on the particle transport equations, assuming that the solution to the phase transport is readily available.

The mass transport equation of particles for a flowing phase in a multiphase system (eq 46) has a similar appearance as the transport equation for particles in a single-phase system (eq 6).

$$\phi S_1 \frac{\partial C_{i,1}}{\partial t} + u_1 \frac{\partial C_{i,1}}{\partial x} + \frac{\partial \sigma_{i,1}{}^T}{\partial t} = 0 \qquad (46)$$

The phase saturation, S_1, and the superficial phase velocity, u_1, are used to characterize the volume fraction and flow velocity of the aqueous phase $(1 = w)$ and of the oleic phase $(1 = o)$. The net rate particle loss of species i in phase 1 is accounted for by the term $(\partial \sigma_{i,1}{}^T/\partial t)$. The particle loss can be classified further: particle loss due to deposition or re-entrainment $(\sigma_{i,1}{}^d)$, particle loss due to pore throat blocking $(\sigma_{i,1}{}^*)$, and particle loss due to transfer from one fluid phase to the next $(\sigma_{i,1}{}')$.

$$\frac{\partial \sigma_{i,1}{}^T}{\partial t} = \frac{\partial \sigma_{i,1}{}^d}{\partial t} + \frac{\partial \sigma_{i,1}{}^*}{\partial t} + \frac{\partial \sigma_{i,1}{}'}{\partial t} \qquad (47)$$

Liu and Civan propose phenomenological relationships for the individual mechanisms of particle loss. First-order rate equations are chosen to approximate these mechanisms.

$$\frac{\partial \sigma_{i,1}{}^{d}}{\partial t} = k_{i,1}{}^{d1} \sigma_{i,1}{}^{d}(C_{sc,w} - C_{s,w}) - k_{i,1}{}^{d2} \sigma_{i,1}{}^{d}(u_1 - u_{1c}) \qquad (48a)$$

$$\frac{\partial \sigma_{i,1}{}^{*}}{\partial t} = k_{i,1}{}^{*} u_1 C_{i,1} \qquad (48b)$$

$$\frac{\partial \sigma_{i,1}{}'}{\partial t} = k_{i,1}{}' C_{i,1} \qquad (48c)$$

Fines are migrated in the medium if the salinity of the aqueous phase, $C_{s,w}$, is below a critical value or if a critical flow velocity is surpassed $(u_1 > u_{1c})$. The critical salinity concept does not apply to fines migration in the oleic phase, therefore $k_{i0}{}^{d1} = 0$. It is assumed that the rate at which fines are trapped at pore throats is proportional to the mass flux of the particles, $(u_1 C_{i,1})$. Furthermore, for fines that transfer from one mobile phase to the next, the transfer rate is set proportional to the fines concentration. These rate equations can be simplified depending on the wettability of the particles: oil-wet, water-wet, or intermediately wet.

Liu and Civan propose that absolute permeability reduction can be attained through the modification of the Carmen–Kozeny equation.

$$\frac{K}{K_0} = \left((1 - f) \, k_f + f \frac{\phi}{\phi_0} \right)^3 \qquad (49)$$

where f represents the fraction of pore volume available to flow and k_f allows for fluid seepage through the plugged sections of the core.

Chemical Treatments for Minimizing Permeability Damage.

An obvious method of avoiding permeability damage by clay defloccula-tion is to maintain a sufficiently high salt concentration in the injected water. On the basis of the kinds of considerations discussed in the pre-ceding sections, Scheuerman and Bergersen (90) have formulated field-selection guidelines for the injection of brines into reservoir formations. However, it is often not practical to maintain the ionic strength of in-jected water above the required critical total ionic strength. Under such conditions, it becomes desirable to reduce the fresh water sensitivity of the formation by pretreating with clay-stabilizing chemicals. Because of the high cost, such treatments are usually applied only to the near wellbore region. Moderate permeability damage in the regions far away from the wellbore can usually be tolerated.

The resistance of fine particles to mobilization can be enhanced not only by the kinds of electrolyte treatments discussed earlier but also by the addition of other, nonsimple electrolyte, materials that adsorb onto particle surfaces and reduce repulsive colloidal forces (91–95). Successful treatments are referred to as clay stabilization or fines stabilization in

the context of fines migration, but it should be noted that this terminology is exactly opposite to the meaning of fines stabilization in colloid
science where the stabilization is of dispersed particles. The chemicals
for clay stabilization can be divided into four categories (96): (1) simple
inorganic cations, (2) inorganic-cationic polymers, (3) cationic surfactants, and (4) organic-cationic polymers.

Simple inorganic cations can readily exchange with the counterions
on clay surfaces and may make the clays at least temporarily resistant
to deflocculation. The treatment efficacy depends on the type of cations
used. In general the flocculating power of cations is a function primarily
of the valence and secondarily of the specific cation used. The divalent
cations are 50 to 100 times more effective in flocculating clays than
monovalent cations. Within the same valence group, the flocculating
power increases with decreasing hydrated ionic radius. Thus the flocculating power decreases in the order $Cs^+ > Rb^+ > NH_4^+ > K^+ > Na^+$
$> Li^+$ for monovalent cations and $Ba^{2+} > Sr^{2+} > Ca^{2+} > Mg^{2+}$ for divalent
cations (90). The resistance to deflocculation imparted by adsorption of
such simple cations is of a temporary nature because these cations can
be stripped from the clay surfaces by ion exchange when they come in
contact with the original brine.

Inorganic-cationic polymers formed by hydrolyzable metal ions are
much more effective in flocculating clays (96–98). The attraction between a negatively charged clay surface and its exchangeable counterion
increases exponentially with the charge on the cation. Therefore polynuclear cations, such as hydroxyaluminum or zirconium oxychloride,
which carry net charge of 6+ or 8+ (or more), are held much more
tenaciously by the clay surfaces. The polyvalent cations also greatly
reduce the expandable atmosphere of exchangeable counterions around
the clays. Large polynuclear ions also sorb to more than one clay particle
and thereby crosslink the particles together. Such crosslinking makes
deflocculation more difficult. Unfortunately, this type of treatment requires either high or low pH condition (depending on the specific metal
ion used), and for this reason is not always applicable.

Cationic surfactants adsorb strongly on clay surfaces by cation exchange. The fatty tails of these adsorbed surfactants impart oil-wetness
to the clay surfaces and shield the clays from direct contact with water.
This shielding has an obvious stabilizing effect; however, this change in
wettability often results in undesirable side effects, such as a decrease
in oil relative permeability. Moreover, because of the possibility of mul-
tilayered adsorption (formation of surface micelles), a high surfactant
concentration is required to satisfy the cation exchange capacity of the
clays, which can make such treatments rather expensive.

Currently the most popular clay stabilizers are organic-cationic
polymers, such as quaternary amine polymers. A monomolecular layer

of these polymers is adsorbed on clay surfaces by cation exchange. Because of the large size of polymer molecules, there is a large probability that the molecule will be adsorbed on more than one clay particle. Thus the polymer crosslinks many clay particles. Deflocculation of such crosslinked particles is unlikely because of the necessity of simultaneous release of all cationic sites on the particle. There is little or no alteration of wettability associated with such treatments.

The concentration of treating chemical required for clay stabilization depends on the type and amount of clays involved. Since these chemicals adsorb by cation exchange, a logical approach would be to determine the cation exchange capacity of the formation. The required concentration of treating chemical in the injected solution can then be calculated on basis of the need to completely satisfy the cation exchange capacity in the contacted zone.

Summary

In this chapter we presented a brief overview of the information available in the open literature on fines migration in petroleum reservoirs. Almost all information included in this review pertains to sandstone formations. We started with a discussion of the type of fines and clays found in petroleum reservoirs and important characteristics of clays. This was followed by a discussion of hydrodynamic and colloidal forces acting on fines. Experimentally measured results on the effects of various parameters, such as flow velocity, salinity, ionic strength, pH, ion exchange capacity, and temperature on the fines migration process were reviewed. The available phenomenological and theoretical models of fines migration were then discussed. Finally ways of minimizing permeability damage by injecting clay-stabilizing chemicals were briefly reviewed.

Although carbonate reservoirs account for more than half of the current worldwide oil production, fines migration in carbonate reservoirs has not received much attention in the literature. Carbonate formations generally contain less clays and have much higher concentrations of divalent ions in their native brines. Therefore, such formations are generally not damaged by exposure to fresh water. Nonetheless there are other types of fines in carbonate rocks that can migrate and cause damage.

List of Symbols

a	particle radius (μm)
a'	specific surface area (length^2)
a_b	bond radius (length)
a_c	collector radius (length)
a_p	particle radius (length)

a_s	rock grain diameter (length)
a_u	uncovered specific surface area (length^2)
A	cross-sectional area (cm^2)
A_s	dimensionless parameter
B	adsorption rate coefficient (time$^{-1/2}$)
C	volume concentration of particles in a single phase (length^3/length^3)
$C_{i,1}$	mass concentration of particle i in flowing phase 1 (mass · length^{-3})
$C_{s,w}$	salinity mass concentration in the aqueous phase 1 (mass · length^{-3})
$C_{(m)}$	volume concentration in unit bed element m (length^3/length^3)
CCC	critical coagulation concentration (concentration units)
CSC	critical salt concentration (concentration units)
CTIS	critical total ionic strength (concentration units)
d_p	particle diameter (length)
d_{th}	lower limit of pore size (length)
d_g	grain size diameter (length)
D	particle diffusion coefficient (length^2 · time^{-2})
DLVO	Derjaguin, Landau, Verwey, Overbeek theory
e	elementary electronic charge (C)
E_p	plugging efficiency (dimensionless)
EDL	electrical double layer
f	fraction of pore cross-sectional area available for particle flow (dimensionless)
f_A	attached particle size distribution (dimensionless)
f_p	open pore throats size distribution (dimensionless)
f_s	suspended particle size distribution (dimensionless)
f_T	trapped particle size distribution (dimensionless)
f_y	pore throat size distribution function (length^{-1})
F	generic function (dimensionless)
F_H	tangential force, (mass · length · time^{-2})
F_L	London dispersion force (dimensionless)
F_T	colloidal force, (mass · length · time^{-2})
g	gravitational acceleration (length · time^{-2})
g_m	effective medium conductivity (length^3)
g_r	conductivity for a pore of radius r (length^3)
G	conductance distribution
G	generic function (time^{-1})
h	separation distance (length)
H	Hamaker constant, (mass · length^2 · time^{-2})
I	ionic strength (mass)
iep	isoelectric point (pH units)

k	permeability (length^2)
$k_1 \ldots k_5$	rate constants
k_B	Boltzmann constant (mass $\cdot$ length^2 $\cdot$ time^{-2} $\cdot$ K^{-1})
k_d	deposition rate constant (time^{-1})
k_f	fluid seepage (dimensionless)
k_i	initial permeability (length^2)
$k_{i,1}{}^*$	rate constant for pore throat blocking of species i in phase 1 (time^{-1})
$k_{i,1}$	rate constant of particle i transfer between fluid phases (time^{-1})
k_0	original permeability (length^2)
k_r	release rate constant (time^{-1})
l	phase identifier
l_e	length of unit bed element (length)
l_p	effective pore length (length)
n	exponent relating pore radius to conductivity
n_o	number of ions per unit volume, (length^{-3})
N_A	number density of attached particles (number $\cdot$ length^{-3})
N_{DL}, N_{E1}, N_{E2}	dimensionless numbers for the double layer (dimensionless)
N_g	gravitational group (dimensionless)
N_L	London dispersion group (dimensionless)
N_{nD}	non-Darcy flow number (dimensionless)
N_0	number of unit bed elements (number)
N_p	number of unplugged pores per unit volume (number $\cdot$ length^{-3})
N_s	number density of suspended particles (number $\cdot$ length^{-3})
N_T	number density of trapped particles (number $\cdot$ length^{-3})
p	pressure (mass $\cdot$ length^{-1} $\cdot$ time^{-2})
p_f	local pressure fluctuation (mass $\cdot$ length^{-1} $\cdot$ time^{-2})
p_m	uniform external pressure (mass $\cdot$ length^{-1} $\cdot$ time^{-2})
p_p	pore pressure (mass $\cdot$ length^{-1} $\cdot$ time^{-2})
$P(r_s)$	trapped fraction of particles with radius r_s (dimensionless)
pzc	point of zero charge (pH units)
pzr	point of zeta potential reversal (pH units)
q	volumetric flow rate (length^3 $\cdot$ time^{-1})
Q	surface charge (C)
Q_E	electrokinetic charge (C)
r	radial distance variable (length)
r'	fraction of expansive clays (dimensionless)
r_A	deposition rate of attached particles (number $\cdot$ length^{-3} $\cdot$ time^{-1})

r_e	re-entrainment rate (mass $\cdot$ length^{-3} $\cdot$ time^{-1})
r_m	mean pore radius (length)
r_p	pore radius (length)
$r_{pd}{}^1$	plugging rate due to deposition (number $\cdot$ length^{-3} $\cdot$ time^{-1})
$r_{pd}{}^2$	plugging rate due to mechanical trapping (number $\cdot$ length^{-3} $\cdot$ time^{-1})
r_r	retention rate (mass $\cdot$ length^{-3} $\cdot$ time^{-1})
r_T	particle trapping rate (number $\cdot$ length^{-3} $\cdot$ time^{-1})
S_1	saturation of phase 1 (length^3/length^3)
t	independent time variable (time)
T	temperature (Kelvin)
u	superficial velocity (length $\cdot$ time^{-1})
u_1	superficial velocity of phase 1 (length $\cdot$ time^{-1})
v	interstitial velocity (length $\cdot$ time^{-1})
v^*	critical interstitial velocity (length $\cdot$ time^{-1})
v_E	electrophoretic velocity (cm/s)
V	applied electric field potential (V)
V_A	attractive energy (mass $\cdot$ length^2 $\cdot$ time^{-2})
V_R	repulsive energy (mass $\cdot$ length^2 $\cdot$ time^{-2})
V_T	total interaction energy (mass $\cdot$ length^2 $\cdot$ time^{-2})
x	independent distance variable (length)
z	counterion charge number
z'	coordination number (dimensionless)

Greek Letters

α	empirical constant (dimensionless)
α_{sp}	separation factor (dimensionless)
β	vector of parameters
β_1, β_2	empirical parameters
γ	vector of parameters
δ	separation between particle and collector (length)
ϵ	dielectric constant of the liquid (mass $\cdot$ length $\cdot$ time^{-2})
η	removal efficiency (dimensionless)
κ	reciprocal Stern layer thickness, (length^{-1})
μ	viscosity of fluid, (mass $\cdot$ length^{-1} $\cdot$ time^{-1})
μ_f	viscosity of fluid and fines (mass $\cdot$ length^{-1} $\cdot$ time^{-1})
ϕ	porosity (length^3/length^3)
ϕ_d	porosity of deposits, (length^3/length^3)
ϕ_{eff}	effective porosity (length^3/length^3)
ϕ_0	original porosity (length^3/length^3)
λ	filtration coefficient (length^{-1})

Ψ_{01} potential of body 1, (dimensionless)

Ψ_1 stream function (length$^3 \cdot$ time^{-1})

θ corrected time variable (time)

θ_P angle between direction of flow and line from fine particle to rock grain (radians)

σ volume of deposited matter per unit volume of porous medium (length^3/length^3)

$\sigma_{i,1}{}^T$ net particle loss (source/sink) of species i in phase 1, (mass $\cdot$ length^{-3})

$\sigma_{i,1}{}^d$ particle loss due to deposition of species i in phase 1, (mass $\cdot$ length^{-3})

$\sigma_{i,1}{}^*$ particle loss due to pore throat blocking of species i in phase 1, (mass $\cdot$ length^{-3})

$\sigma_{i,1}{}'$ particle loss due to transfer between fluid phases, (mass $\cdot$ length^{-3})

σ_l^s mass of liquid adsorbed by the clays per unit volume (mass $\cdot$ length^{-3})

σ_p^s mass of particles deposited from external sources (mass $\cdot$ length^{-3})

σ_p^{s*} mass of particles re-entrained from indigenous sources (mass $\cdot$ length^{-3})

ρ_p density of particle (mass $\cdot$ length^{-3})

ρ_l density of liquid (mass $\cdot$ length^{-3})

ρ_l^f mass concentration of the transport liquid (mass $\cdot$ length^{-3})

$\rho_{l,i}^s$ initial mass concentration of liquid in the rock matrix (mass $\cdot$ length^{-3})

τ tortuosity (length/length)

ω lumped parameter (dimensionless)

$\psi(\delta)$ electric field potential at the outer boundary of the Stern layer

References

1. Skirvin, R. T.; Ausburn, B. E. In *Petroleum Engineering Handbook*; Bradley H. W., Ed.; Society of Petroleum Engineers: Richardson, TX, 1987; Chapter 29.
2. Carrigy, M. A.; Mellon, G. B. *J. Sediment. Petrol.* **1964**, *32(3)*, 461.
3. Wilson, M. D.; Pittman, E. D. *J. Sediment. Petrol.* **1977**, *47(1)*, 3.
4. Eslinger, E.; Pevear, D. *Clay Minerals for Petroleum Geologists and Engineers*; SEPM Short Course Notes No. 22; Society of Economic Paleontologists and Mineralogists: Tulsa, OK, 1988.
5. Bakhsh, S. Ph.D. Thesis, Loughborough University of Technology, 1991.
6. Hill, D. G. Presented at the SPE Formation Damage Symposium, Lafayette, LA, March 24–25, 1982; paper SPE 10656.
7. Somerton, W. H.; Radke, C. J. Presented at the 1st SPE/DOE Symposium on EOR, Tulsa, OK, April 20–23, 1980; paper SPE 8845.

8. Swartzen-Allen, S. L.; Matijevic, E. *Chem. Rev.* **1974**, *74(3)*, 385.
9. Almon, W. R.; Davies, D. K. *Trans. Gulf Coast Assoc. Geol. Soc.* **1978**, *28*, 1–6.
10. Drever, J. I. *The Geochemistry of Natural Waters* Prentice Hall, Englewood Cliffs, NJ, 1982; p 329.
11. Leone, J. A.; Scott, E. M. *SPE Reservoir Eng.* **1988**, *3*, 1279–1286.
12. Kia, S. F.; Fogler, H. S.; Reed, M. G.; Vaidya, R. N. *SPE Prod. Eng.* **1987**, *2*, 277–283.
13. Azari, M. and Leimkuhler, J., Formation Permeability Damage Induced by Completion Brines, *J. Pet. Technol.* **1990**, *42*, 486–492.
14. Dahab, A. S.; Omar, A. E.; Gassier, M. M.; Awad-El-Kariem, H. *Technol. Rev. L'Inst. Fr. Pet.* **1989**, *44*, 583–593.
15. Jones, F. O. *J. Pet. Technol.* **1964**, *April*, 441–446.
16. Cerda, C. M.; Non-Chhom, K. *Proceedings of the 4th UNITAR/UNDP Conference on Heavy Crude and Tar Sands;* Alberta Oil Sands Technology and Research Authority: Edmonton, Canada, 1988; paper 20.
17. Thornton, S. D. *SPE Reservoir Eng.*, **1988**, *3*, 1153–1160.
18. Khilar, K. C.; Vaidya, R. N.; Fogler, H. S. *J. Pet. Sci. Eng.* **1990**, *4*, 213–221.
19. Kuo, J-F.; Sharma, M. M.; Yen, T. F. *J. Colloid Interface Sci.* **1988**, *126(2)*, 537–546.
20. Hunter, R. J. *Zeta Potential in Colloid Science: Principles and Applications;* Academic: Orlando, FL, 1981.
21. James, A. M. In *Surface and Colloid Science;* Good, R. J.; Stromberg, R. R., Eds.; Plenum: New York, 1979; Vol. 11, pp 121–186.
22. Schramm, L. L.; Mannhardt, K.; Novosad, J. J. *Colloids Surf.* **1991**, *55*, 309–331.
23. Cerda, C. M. *Colloids Surf.* **1987**, *27*, 219–241.
24. Hiemenz, P. C. *Principles of Colloid and Surface Chemistry;* 2nd ed.; Dekker: New York, 1986.
25. Overbeek, J. Th. G. In *Colloidal Dispersions;* Goodwin, J. W., Ed.; Special Publication 43; Royal Society of Chemistry: London, 1982; pp 1–21.
26. Gregory, J. *J. Colloid Interface Sci.* **1981**, *83*, 138–145.
27. Cerda, C. M. *Clays Clay Miner.* **1988**, *36*, 491–497.
28. Goren, S. L.; O'Neill, M. E. *Chem. Eng. Sci.* **1971**, *26*, 325–338.
29. Spielman, L. A.; Fitzpatrick, J. A. *J. Colloid Interface Sci.* **1973**, *42*, 607–623.
30. Goldman, A. J.; Cox, R. G.; Brenner, H. *Chem. Eng. Sci.* **1967**, *22*, 653–660.
31. Buller, D. C.; Harper, T. R. *Proceedings of the Europe Oil & Gas Conference;* Comm. European Communities: 1991; pp 182–191.
32. Grim, R. E. *Clay Mineralogy,* 2nd ed.; Mc-Graw-Hill, Toronto, Canada, 1968.
33. van Olphen, H. *An Introduction to Clay Colloid Chemistry,* 2nd ed.; Wiley-Interscience: New York, 1977.
34. Williams, D. J. A.; Williams, K. P. *J. Colloid Interface Sci.* **1978**, *65*, 79–87.
35. Thompson, D. W.; Pownall, P. G. *J. Colloid Interface Sci.* **1989**, *131*, 74.
36. Khilar, K. C.; Fogler, H. S. *J. Colloid Interface Sci.* **1984**, *101*, 214–224.
37. Azari, M.; Leimkuhler, J. *Proceedings of SPE Formation Damage Control Symposium;* Society of Petroleum Engineers: Richardson, TX, 1990; paper SPE 19431, pp 237–244.
38. Muecke, T. W. *J. Pet. Technol.* **1979**, *February*, 144.

39. Gruesbeck, C.; Collins, R. E. *Soc. Pet. Eng. J.* **1982,** *December,* 847.
40. Egbogah, E. O. Presented at the 35th Annual Technical Meeting of the Petroleum Society of CIM, Calgary, Alberta, June 10–13, 1984; CIM paper 84–35–16.
41. Gabriel, G. A.; Inamdar, G. R. Presented at the 58th Annual Technical Conference and Exhibition, San Francisco, CA, October 5–8, 1983; paper SPE 12168.
42. Coskuner, G.; Maini, B. *J. Pet. Sci. Eng.* **1990,** *4,* 105–117
43. Miranda, R. M.; Underdown, D. R.; In *Proceedings of Production Operations Symposium;* Society of Petroleum Engineers: Richardson, TX, 1993; paper SPE 25432.
44. Nasr-El-Din, H. A.; Maini, B. B.; Stanislav, P. *AOSTRA J. Res.* **1991,** 7, 1–15.
45. Amaefule, J. O.; Padilla, P. C.; McCaffery, F. G.; Teal, S. L. Presented at the Formation Damage Control Symposium, Bakersfield, CA, February 13–14, 1984; paper SPE 12499.
46. Thomas, R. C.; Crowe, C. W.; Simpson, B. E. Presented at the 51st Annual Fall Technical Conference and Exhibition, New Orleans, LA, October 3–6, 1976; paper SPE 6007.
47. Todd, A. C.; Brown, J.; Noorkami, M.; Tweedie, J. A. Presented at the SPE International Symposium on Oilfield and Geothermal Chemistry, Houston, TX, January 22–24, 1979; paper SPE 7883.
48. Todd, A. C.; Somerville, J. E.; Scott, G. Presented at the Formation Damage Control Symposium, Bakersfield, CA, February 13–14, 1984; paper SPE 12498.
49. Tien, C.; Payatakes, A. C. *AIChE J.* **1979,** *25(5),* 737.
50. Donaldson, E. C.; Baker, B. A. Presented at the Annual Fall Technical Conference and Exhibition, Denver, CO, October 9–12, 1977; paper SPE 6905.
51. O'Melia, C. R.; Ali, W. *Prog. Water Technol.* **1978,** *10(5/6),* 167.
52. Mungan, N. *J. Pet. Technol.* **1965,** *17,* 1449.
53. Mungan, N. *J. Can. Pet. Technol.* **1968,** *7,* 113.
54. Clementz, D. M. *J. Pet. Technol.* **1977,** *29,* 1061.
55. Sarkar, A. K.; Sharma, M. M. In *Proceedings of The SPE California Regional Meeting;* Society of Petroleum Engineers: Richardson, TX, 1988; paper SPE 17437.
56. Land, C. S.; Baptist, O. C. *J. Pet. Technol.* **1965,** *17,* 1213–1218.
57. Monaghan, P. H.; Sallathiel, R. A.; Morgan, B. E.; Kaiser, A. D. *Trans. AIME* **1959,** *216,* 209.
58. Somerton, W. H.; Chen, S. P.; Schuh, M. J.; Yen, J. P. Presented at the Formation Damage Control Symposium, Bakersfield, CA, February 13–14, 1984; paper SPE 12500.
59. Lever, A.; Dawe, R. A. *Mar. Pet. Geol.* **1987,** *4,* 112.
60. Khilar, K. C.; Fogler, H. S. *Soc. Pet. Eng. J.* **1983,** *February,* 55.
61. Khilar, K. C.; Fogler, H. S.; Ahluwalia, J. S. *Chem. Eng. Sci.* **1983,** *38(5),* 789.
62. Kia, S. F.; Fogler, H. S.; Reed, M. G. *J. Colloid Interface Sci.* **1987,** *118(1),* 158.
63. Vaidya, R. N.; Fogler, H. S. *Proceedings of the SPE Formation Damage Control Symposium;* Society of Petroleum Engineers: Richardson, TX, 1990; pp 125–132; paper SPE 19413.
64. Vaidya, R. N.; Fogler, H. S. *Colloids Surf.* **1990,** *50,* 215–229.

65. Omar, A. E. *J. Pet. Sci. Eng.* **1990,** *4,* 245–256.
66. Kolakowski, J. E.; Matijevic, E. *J. Chem. Soc. Faraday Trans.* **1979,** *75,* 65.
67. Rowel, D. L.; Payne, D.; Ahmed, N. *J. Soil. Sci.* **1969,** *20,* 176.
68. Udell, K. S.; Lofty, J. D. Presented at the SPE California Regional Meeting, Bakersfield, CA, March 27–29, 1985; paper SPE 13655.
69. Khilar, K. C. Ph.D. Thesis, The University of Michigan, 1981.
70. Allen, F. L.; Riley, S. M.; Strassner, J. E. Presented at the Formation Damage Control Symposium, Bakersfield, CA, February 13–14, 1984; paper SPE 12488.
71. Iwasaki, T. *J. Am. Water Works Assoc.* **1937,** *29,* 1591.
72. Herzig, J. P.; Lelerc, D. M.; LeGoff, P. *Ind. Eng. Chem.* **1970,** *6,* 8.
73. Ives, K. J.; Pienvichitr, V. *Chem. Eng. Sci.* **1965,** *20,* 965.
74. Ohen, H. A.; Civan, F. SPE Advanced Technology Series, Society of Petroleum Engineers: Richardson, TX, 1993; Vol. 1, No. 1, p 27.
75. Civan, F.; Knapp, R. M.; Ohen, H. A. *J. Pet. Sci. Eng.* **1989,** *3,* 65.
76. Geertsma, J. *SPEJ, Soc. Pet. Eng. J.* **1974,** *14,* 445.
77. Civan, F. *Proceedings of the Formation Damage Control Symposium;* Society of Petroleum Engineers: Richardson, TX, 1992; paper SPE 23787.
78. Payatakes, A. C.; Tien C.; Turian, R. M. *AIChE J.* **1974,** *20,* 900.
79. Mackie, R. I.; Homer, R. M. W.; Jarvis, R. J. *AIChE J.* **1987,** *33,* 1761.
80. Sharma, M. M. Ph.D. Dissertation, University of Southern California, 1985.
81. Sharma, M. M.; Yortsos, Y. C. *AIChE J.* **1987,** *33(10),* 1636.
82. Rege, S. D.; Fogler, H. S. *Chem. Eng. Sci.* **1987,** *42(7),* 1553.
83. Kirkpatrick, S. *Rev. Mod. Phys.* **1973,** *45(4),* 574.
84. Koplik, J. *J. Fluid Mech.* **1982,** *119,* 219.
85. Rege, S. D.; Fogler, H. S. *AIChE J.* **1988,** *34,* 1761.
86. Stein, P. C. D. Sc. Dissertation, Massachusetts Institute of Technology, Cambridge, MA, 1940.
87. Liu, X.; Civan, F. *Proceedings of the Production Operations Symposium;* Society of Petroleum Engineers: Richardson, TX, 1993; paper SPE 25429, p 231.
88. Lake, L. W. *Enhanced Oil Recovery;* Prentice Hall: Englewood Cliffs, NJ, 1989.
89. Aziz, K.; Settari, A. *Petroleum Reservoir Simulation;* Applied Science: London, 1979.
90. Scheuerman, R. F.; Bergersen, B. M. *J. Pet. Technol.* **1990,** *42,* 836–845.
91. Audibert, A.; Bailey, L.; Hall, P. L.; Keall, M.; Lecourtier, J. In *Physical Chemistry of Colloids and Interfaces in Oil Production;* Toulhoat, H.; Lecourtier, J. Eds.; Editions Technip: Paris, 1992; pp 203–209.
92. Clark, R. K.; Scheuerman, R. F.; Rath, H.; Van Laar, H. G. *J. Pet. Technol.* **1976,** *28,* 719–727.
93. Zaitoun, A.; Berton, N. *Proceedings of the Formation Damage Control Symposium;* Society of Petroleum Engineers: Richardson, TX, 1990; paper SPE 19416, pp 155–164.
94. Chakravorty, S. K.; Gateman, J. C.; Bohor, B. F.; Knutson, C. F. *J. Pet. Technol.* **1964,** *16,* 1107–1112.
95. Lauzon, R. V. *Oil & Gas J.* **1984,** July.
96. Hill D. G. Presented at the SPE Formation Damage Control Symposium, Lafayette, LA, March 24–25, 1982; paper SPE 10656.
97. Reed, M. G. *J. Pet. Technol.* **1972,** *34,* 860–864.
98. Vely, C. D. *J. Pet. Technol.* **1969,** *21,* 1111–1118.

Appendix A

Conservation Equation:

$$\frac{\partial(\sigma + \phi C)}{\partial t} + u\frac{\partial C}{\partial x} = 0$$

A linear relationship between porosity and deposited fines can be arranged.

$$\phi = \phi_0 - \frac{\sigma}{1 - \phi_d}$$

Conversion of coordinates:

$$\theta = t - \int_0^x \frac{\phi}{u}\,dx$$

Functional relationships:

$$C(x, t) = C'(x, \theta)$$
$$\sigma(x, t) = \sigma'(x, \theta)$$
$$\phi(x, t) = \phi'(x, \theta)$$

Partial derivatives, application of the chain rule:

$$\frac{\partial C}{\partial t} = \frac{\partial C'}{\partial \theta}\frac{\partial \theta}{\partial t} = \frac{\partial C'}{\partial \theta}\left(1 - \frac{1}{u}\int_0^x \frac{\partial \phi}{\partial t}dx\right)$$

$$\frac{\partial C}{\partial z} = \frac{\partial C'}{\partial z} + \frac{\partial C'}{\partial \theta}\frac{\partial \theta}{\partial x} = \frac{\partial C'}{\partial x} - \frac{\partial C'}{\partial \theta}\frac{1}{u}\int_0^x \frac{\partial \phi}{\partial x}\,dx = \frac{\partial C'}{\partial x} - \frac{\partial C'}{\partial \theta}\frac{\phi}{u}$$

Rearrangement of the conservation equation:

$$\frac{\partial(\sigma' + \phi'C')}{\partial \theta}\frac{\partial \theta}{\partial t} + u\left(\frac{\partial C'}{\partial x} + \frac{\partial C'}{\partial \theta}\frac{\partial \theta}{\partial x}\right) = 0$$

$$\frac{\partial(\sigma' + \phi'C')}{\partial \theta}\left(1 - \frac{\partial}{\partial t}\int_0^x \frac{\phi}{u}\,dx\right) + u\left(\frac{\partial C'}{\partial x} - \frac{\partial C'}{\partial \theta}\frac{\phi}{u}\right) = 0$$

$$\frac{\partial\sigma'}{\partial \theta} + u\frac{\partial C'}{\partial x} - \frac{\partial(\sigma' + \phi'C')}{\partial \theta}\frac{\partial}{\partial t}\int_0^x \frac{\phi}{u}\,dx + \frac{\partial\phi'C'}{\partial \theta} - \phi\frac{\partial C'}{\partial \theta} = 0$$

$$\frac{\partial\sigma'}{\partial \theta} + u\frac{\partial C'}{\partial x} - \frac{\partial(\sigma' + \phi'C')}{\partial \theta}\frac{\partial}{\partial t}\int_0^x \frac{\phi}{u}\,dx + C'\frac{\partial\phi'}{\partial \theta} = 0$$

$$\left(1 + \frac{C'}{1 - \phi_d}\right)\frac{\partial\sigma'}{\partial \theta} + u\frac{\partial C'}{\partial x} - \frac{\partial(\sigma' + \phi'C')}{\partial \theta}\frac{\partial}{\partial t}\int_0^x \frac{\phi}{u}\,dx = 0$$

Under the following assumptions,

$$1 + \frac{C'}{1 - \phi_d} \approx 1$$

$$\frac{\partial}{\partial t} \int_0^x \frac{\phi}{u} \, dx \approx 0$$

the classic filtration equation can be derived:

$$\frac{\partial \sigma'}{\partial \theta} + u \frac{\partial C'}{\partial x} = 0$$

RECEIVED for review July 28, 1994. ACCEPTED revised manuscript September 14, 1995.

Asphaltenes in Crude Oil and Bitumen: Structure and Dispersion

James G. Speight

Western Research Institute, 365 North 9th Street,
Laramie, WY 82070–3380

CRUDE PETROLEUM IS A MIXTURE OF COMPOUNDS boiling at different temperatures that can be separated into a variety of generic fractions by distillation and by fractionation (1). In fact, such methods provide a better sense of the overall composition of petroleum and the behavioral characteristics.

However, petroleum from different sources exhibits different characteristics, and the behavioral characteristics are often difficult to define with any degree of precision. As anticipated and inasmuch as there is a wide variation in the properties of petroleum, the proportions in which the different constituents occur will also vary widely (1–3). Thus, some crude oils have higher proportions of the lower boiling constituents, whereas others (such as bitumen, also referred to as natural asphalt) have higher proportions of the higher boiling constituents (often called the "asphaltic components" or "residuum"). It is these higher boiling constituents that often lead to problems during recovery and refining operations.

Petroleum can be considered to be a delicately balanced system insofar as the different fractions are compatible, provided there are no significant disturbances or changes made to the system. Such changes are (1) the alteration of the natural occurrences of the different fractions; (2) the chemical or physical alteration of the constituents as might occur during refining, especially changes that might be brought on by thermal processes; and (3) alteration of the polar group distribution as might occur during oxidation (i.e., asphalt manufacture) or the elimination of polar functions during processing. In addition, the sudden exposure of petroleum to air, as might occur during the initial stages of recovery operations, or the release of dissolved gases when a reservoir is first penetrated can also cause disturbances to the system.

All of the aforementioned occurrences disturb the petroleum system. However, when such disturbances occur, it is the higher molecular weight constituents that are most seriously affected. This can lead to incompatibility, which is variously referred to as precipitation, sludge formation, and the separation of coke precursors, depending on the circumstances. Thus, the dispersibility of the higher molecular weight constituents becomes an issue that needs to have attention. One of the ways by which this issue can be understood is to be aware of the chemical and physical character of the higher molecular weight constituents. By such means, the issue of dispersibility, and the attending issue of incompatibility, can be understood and even predicted.

The luxury of predictability is a valuable asset in understanding the behavior of petroleum during refining and recovery operations, and nowhere is it more helpful than in understanding the behavior of heavy oil and bitumen, which have greater proportions of the higher molecular weight constituents than conventional petroleum. Because of this (in part or in total), more problems exist in recovery and refining operations.

Understanding the nature of the higher molecular weight constituents is a first step in understanding dispersibility and incompatibility. In this chapter, the current understanding of the most complex (in terms of molecular weight and polarity) constituents of heavy oil and bitumen is reviewed. This fraction is often referred to as "asphaltenes" (Figure 1). Other complex entities also exist in the resin fraction but have received considerably less attention.

At this point, it is noteworthy that without the resin fraction, asphaltenes are generally nondispersible in the remainder of petroleum, thereby indicating that the resins are, under ambient conditions, a necessary constituent and that by their presence they prevent incompatibility (3, 4). This is only one of several factors that influence dispersibility or compatibility, and others will be noted, in turn, throughout the chapter.

Separation

Determination of the asphaltene fraction of petroleum has been investigated for most of this century (5–11), and therefore the art is not new. However, it is now generally accepted that asphaltenes are, by definition, a solubility class that is precipitated from petroleum, heavy oil, and bitumen by the addition of an excess of liquid hydrocarbon (11). The procedure not only dictates asphaltene yield but can also dictate the "quality" of the fraction (12–16). In fact, the very method of asphaltene separation is a prime example of the disturbance of the system by the addition of an external agent. Thus, during deasphalting, the dispersibility (or compatibility) of the asphaltenes in the system is changed. The

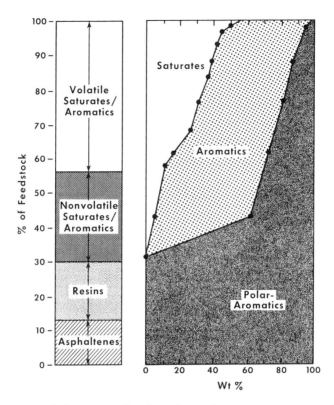

Figure 1.   Asphaltenes are also classed as polar aromatic compounds and are an extension of the petroleum continuum.

result is the separation (precipitation) of an asphaltene fraction. Various liquid solvents (or more correctly, nonsolvents) have been used for asphaltene separation. The most common liquids are low-boiling hydrocarbon liquids, with nonhydrocarbon liquids also receiving some, but much less, attention.

A recommended procedure for reproducibility is to ensure "stable" asphaltene yields from heavy oils. It is necessary to use >30 mL hydrocarbon/g feedstock; pentane or heptane, which are the preferable liquid hydrocarbons, although volatility constraints and stability of asphaltene-type are tending to favor the use of n-heptane; and 8- to 10-h contact time, which is the preferable period (*11*).

Hydrocarbon Solvents. Different feedstocks have different amounts of asphaltenes (Table I), which can influence properties. This is particularly true of different feedstocks from any one crude oil (Figure 2) in which the nonvolatile asphaltenes are concentrated in the residue or in the propane asphalt.

Table I.   Ranges for Asphaltene Occurrence
in Different Feedstocks

Substance	Asphaltenes	Resins	Oils
Petroleum	<0.1–12	3–22	67–97
Heavy oil	11–45	14–39	24–64
Residue[a]	11–29	29–39	?–49

NOTE: Values are ranges of composition in w/w%.
[a] Asphaltene occurrence in bitumen approximates that found in residue.

However, asphaltene separation is not as straightforward as was originally thought. For example, the use of "insufficient" amounts of the hydrocarbon will cause the resins, a fraction isolated at a later stage of the separation procedure by adsorption chromatography (Figure 3), to be separated with the asphaltene fraction either by inclusion within the asphaltene fraction or by adsorption onto the asphaltenes from the supernatant liquid (11, 14, 17). It is essential that a large excess of the hydrocarbon liquid be used to ensure yield and analytical data that are reproducible and within the limits of experimental error.

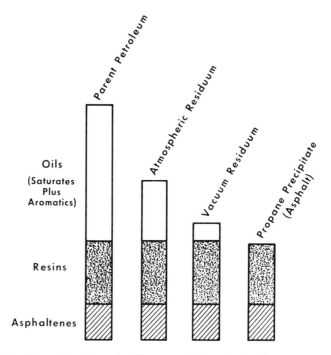

Figure 2.   Processing (e.g., distillation and deasphalting) concentrates asphaltenes into the nonvolatile and insoluble product.

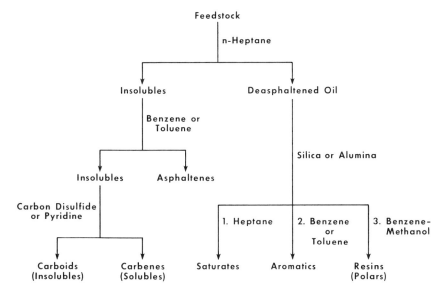

Figure 3. The nomenclature of petroleum fractions is a method of convenience and depends on the separation technique.

In summary, the definition of the asphaltenes, the resins, and the
nonvolatile oil fraction that occur in heavy oil and bitumen is an operational aid. Such separations cannot be based on any chemical or structural features of the constituent molecules (*11, 13, 18, 19*).

Nonhydrocarbon Solvents. Although an asphaltene fraction can
be removed from petroleum by using a wide variety of hydrocarbon
liquids (*14*), the use of nonhydrocarbon solvents as deasphalting media
and their influence on asphaltene dispersibility and compatibility has
also been investigated. Dispersibility of asphaltenes in petroleum is suggested to be conveniently related to the surface tension of the system
components (*8, 20, 21, 22, 23*). Obviously, asphaltene dispersion and
compatibility is complex and is dependent on several factors and varies
markedly with the character of the added liquid.

In summary, the separation of asphaltenes from petroleum, heavy
oil, and bitumen is not only a function of the amount of added diluent
but also a function of the type and composition of the diluent
(*14, 22, 24*).

Composition, Structure, and Functionality

Elemental Composition. The near constancy of the atomic hy-
drogen:carbon (H/C) ratio of asphaltenes (*1*) is surprising when the

number of possible molecular and elemental combinations involving the heteroelements are considered.

In contrast to the H/C ratio of asphaltenes, notable variations do occur in the proportions of the heteroelements, in particular in the proportions of oxygen and sulfur. Oxygen content varies from 0.3 to 4.9% and sulfur content varies from 0.3 to 10.3%. However, the nitrogen content of the asphaltenes has a somewhat lesser degree of variation (0.6–3.3% at the extremes). Exposing asphaltenes to atmospheric oxygen, however, can substantially alter the oxygen content, and exposing a crude oil to elemental sulfur, or even to sulfur-containing minerals, can result in excessive sulfur uptake. It is not unreasonable to suggest that oxygen and sulfur contents vary more markedly than does nitrogen content because of the potential for reaction between the petroleum material and oxygen or sulfur (and sulfur-containing minerals).

In addition, the use of different hydrocarbon liquids for the separation yields asphaltene fractions that are substantially different from each other (1, 25). For example, the H/C ratios of the heptane precipitate are lower than those of the pentane precipitate, indicating a higher aromaticity in the heptane precipitate. The N/C, O/C, and S/C atomic ratios are usually higher in the heptane precipitate, indicating higher proportions of the heteroelements in this material (26).

Fractional Composition.   The complexity of asphaltenes is indicated from studies using fractionation techniques (27–31).

Asphaltene fractionation using solid adsorbents has shown some success in defining asphaltenes as acidic and basic materials on the basis of separation using gel permeation chromatography (32), a transition metal oxide solid acid catalyst (33), silica gel chromatography (34), and ion exchange resins (31, 35). Analytical data also show that these fractions contain several heteroatoms per molecule and reinforce the concept of asphaltenes being complex mixtures of molecular sizes and various functional types (Figure 4) (18, 19).

Hydrocarbon Structures.   Early postulates of asphaltene structure centered around a variety of polymer structures based on aromatic systems (36, 37). More recent information has related to the structural parameters and carbon skeleton of petroleum fractions, and asphaltenes structures have been derived from spectroscopic studies of asphaltenes isolated from various petroleum and bitumen (38–46).

The data from these studies support the hypothesis that asphaltenes, viewed structurally, contain condensed polynuclear aromatic ring systems bearing alkyl side chains. These systems carry alkyl and alicyclic systems with heteroelements (i.e., nitrogen, oxygen, and sulfur) scattered

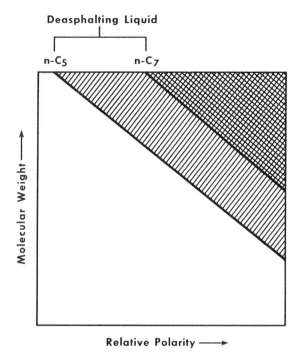

Figure 4. Asphaltenes separated from crude oil are a mixture of different molecular species that can be defined on the basis of polarity and molecular weight; the axes shown here are in arbitrary units but can be given in real numbers.

throughout in various, including heterocyclic, locations. With increasing molecular weight of the asphaltene fraction, both aromaticity and the proportion of heteroelements increase (*3, 47, 48*).

These findings led to the concept of large polynuclear aromatic systems, and efforts were made to describe the total structures of asphaltenes in accordance with magnetic resonance data and results of spectroscopic and analytical techniques (Figure 5) (*39, 41, 42, 49, 50*). However, the number of rings in such systems is open to question, apparently varying from as low as 6 to 15 or more.

The macromolecular structure of asphaltenes has also been subject to investigation insofar as the means by which the molecules can form a micelle is of importance to geochemists and to process chemists. X-ray analyses and molecular weight determination (*see* below) were the methods chosen to investigate the macromolecular structure of asphaltenes (*51*). The X-ray method is well documented in its use for carbon

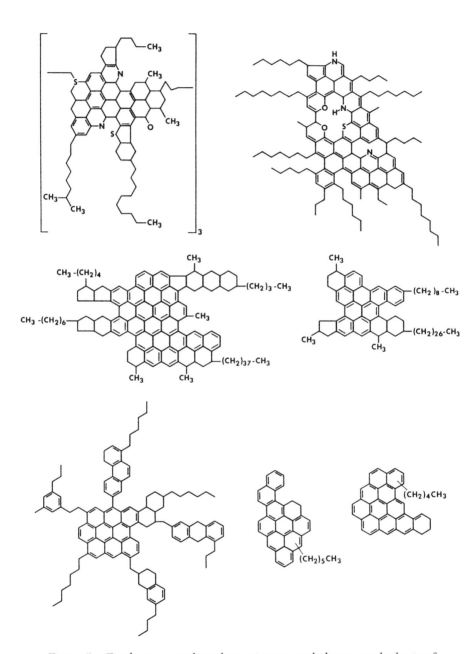

Figure 5.   Total structures have been given to asphaltenes on the basis of NMR investigations; these structures involve the use of large poynulcear aromatic systems, and heteroatoms are usually absent.

and graphite and yields information about the dimension of the unit cell such as interlamellar distance (c/2), layer diameter (L_a), height of unit cell (L_c), and number of lamellae (N_c) contributing to the micelle (Figure 6).

Heteroatom Structures. Studies on the disposition of nitrogen indicated the existence of nitrogen as various heterocyclic types (52–62). Primary, secondary, and tertiary noncyclic aromatic amines have not been detected in asphaltenes. There are also reports in which the organic nitrogen has been defined in terms of basic and nonbasic types (55). Evidence has been shown for the occurrence of carbazolic nitrogen (54, 57). Other studies have brought to light the occurrence of four-ring aromatic nitrogen species in petroleum (60).

Oxygen has been identified in carboxylic, phenolic, and ketonic locations (55, 56, 63–65) but is not usually regarded as being located primarily in heteroaromatic ring systems. In the context of polyhydroxy aromatic nuclei, it is of interest to note that pyrolysis of asphaltenes at 800 °C results in the formation of resorcinol (63), implying that such functions may indeed exist in the native material.

Sulfur occurs in systems such as benzothiophene, dibenzothiophene, and naphthenebenzothiophene (54, 55, 66–72). More highly condensed benzothiophene systems may also exist but are precluded from identification by low volatility. Investigations of the size distribution of the sulfur species indicate an increase in the size of the sulfur species from aromatics to resins to asphaltenes (73).

Metals (i.e., nickel and vanadium) also occur within the asphaltene fraction but are much more difficult to visualize as part of the asphaltene system. Nickel and vanadium occur as porphyrins (74–77), but whether these are an integral part of the asphaltene structure is not known. Some of the porphyrins can be isolated as a separate stream from petroleum (77).

Asphaltenes from different sources contain similar functionalities but with some variation in degree (1). This can be ascribed to the local and regional variations in the relative amounts of the various precursor molecules to variations in the maturation conditions.

Molecular Weight. Asphaltene molecular weights (vapor pressure osmometry) are dependent not only on the nature of the solvent but also on the solution temperature at which the determinations were performed (78). However, data from molecular weight determinations by the cryoscopic method (79) indicate that the absolute molecular weight of asphaltenes cannot be determined by any one method.

The data suggest that asphaltenes form molecular aggregates, even

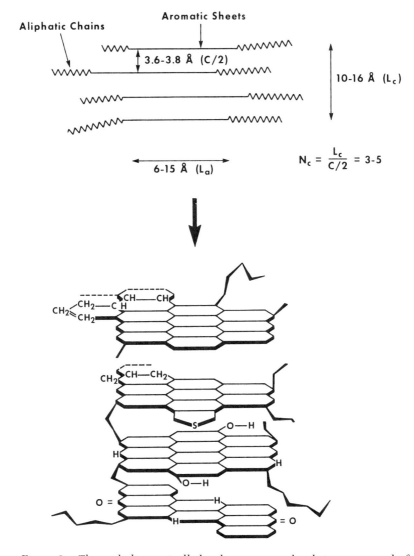

Figure 6. The asphaltene micelle has been proposed as being composed of a "stack" of asphaltene molecules associated either by σ bonds or through $\pi-\pi$ interactions.

in dilute solution (80), and this association is influenced by solvent polarity, asphaltene concentration, and the temperature of the determination (79, 81). In fact, it is strongly recommended that to negate the concentration effects and the temperature effects, the molecular weight determinations should be carried out at each of three different concentrations at three different temperatures. The data for each temperature are then extrapolated to zero concentration, and the zero concentration data at each temperature are then extrapolated to room temperature (82).

Thus, asphaltene molecular weights are truly variable. Data produced from the determinations by using highly polar solvents, however, indicate that the molecular weights, in solvents that prevent association, usually fall into the range of 1500–2500.

Molecular Models

Highly condensed polynuclear aromatic structures that have been used to explain the high yields of thermal coke from asphaltenes are generally not consistent with the behavior of asphaltenes during recovery or during processing and not with the natural product origins of petroleum. Many natural products contain small polynuclear aromatic systems, and large polynuclear aromatic systems are not usually found in nature in the types of source materials that were considered (83–86).

The concept of an asphaltene model that incorporates smaller polynuclear aromatic systems is more in keeping with the types of systems that occur in nature. Indeed, smaller polynuclear aromatic (and pseudoaromatic) systems are capable of producing high yields of thermal coke either because of the heteroatom content (87) or because of the presence of pendant alkyl moieties that have the capability of forming the internuclear cross-links (88) that can lead to coke. In this latter case, it is likely that the indigent alkyl chains can interact in this manner, or shorter alkyl chains, formed by thermolysis, can play the role of cross-linking agents.

If functionality is a requirement, polyhydroxy compounds are common in nature (86). Carbazole and amide types have also been recognized in natural product systems (86) as has nitrogen in many alkaloid systems (84).

Sulfur (as thiophene-type sulfur) is more difficult to identify, and its presence in the biological systems (other than the sulfur in the amino acids cysteine and methionine) is still speculative, although inorganic (mineral) sulfur has means by which it can be incorporated into petroleum as organic sulfur (89–91).

It is also noteworthy here that asphaltenes have been postulated to contain charged species within their structure (55), which, of course, may be analogous to the occurrence of zwitterion structures in proteins and peptides (92).

The manner in which these moieties occur within the asphaltene "molecule" must for the present remain largely speculative. Above all, asphaltenes are a solubility class and can be an accumulation of thousands of structural entities. However, caution is advised against combining a variety of identified products into one (albeit hypothetical) structure.

The danger in all of these studies is in attempting to link together fragmented molecules produced during, and identified as products of, thermal investigations into a structure that is believed to be real. Functionality and polynuclear aromatic ring systems are real, but combination of these parameters into one or even several structures can be misleading. However, no one method will guarantee the development of a suitable molecular model.

The effort must be multidimensional, taking into account all of the properties and characteristics of the material. Extreme caution is advised when average structures are used. For example, models that could better explain the asphaltene behavior should be selected from the various parts of the molecular weight–polarity diagram. The easiest would be to select the two extremes of this diagram and design highly polar and neutral species. This would be the amphoteric and substantially less polar (or near neutral) constituents of asphaltenes (18, 19, 93).

Theoretical models can be constructed from the data provided by a whole series of different investigations (25) but which focus on the incorporation of heteroatoms into the polynuclear aromatic systems (Figures 7 and 8). The incorporation of heteroatoms into such polynuclear aromatic systems reduces the volatiltity of the system, thereby allowing participation in coke formation. In addition, molecular interactions, such as hydrogen bonding and the formation of charge–transfer complexes, could also have a significant effect on volatility and, therefore, on coke yield.

Such models would be structurally compatible with the other constituents of petroleum and are able to represent the thermal chemistry and other process operations. It must also be recognized that although such models can have the large dimensions that have been proposed for asphaltenes, they can also be "molecular chameleons" insofar as they can vary in dimensions depending on the angle of rotation around an axis (Figure 9) or the freedom of rotation around one, or more, of the bonds.

Asphaltenes in Petroleum

The means by which the asphaltenes remain dispersed in the oil medium has been the subject of much speculation, not only by geochemists but also by process chemists and engineers.

Historical Aspects. An early hypothesis of the physical structure of petroleum (Figure 10) (94) suggested that asphaltenes are the centers of micelles formed by adsorption or even by absorption of part of the maltene fraction, that is, resin material, onto the surfaces or into the

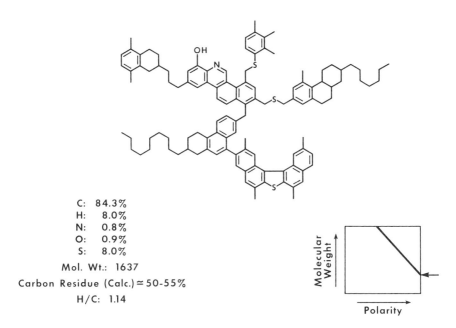

C: 84.3%
H: 8.0%
N: 0.8%
O: 0.9%
S: 8.0%
Mol. Wt.: 1637
Carbon Residue (Calc.) ≈ 50-55%
H/C: 1.14

Molecular Weight

Polarity

Figure 7.   Using a multidimensional approach, asphaltenes can be repre-
sented as more "open" natural product-type systems, such as this low mo-
lecular weight high polarity molecule.

interiors of the asphaltene particles. Thus, most of those substances with greater molecular weight and with the most pronounced aromatic nature are situated closest to the nucleus and are surrounded by lighter constituents of lesser aromatic nature. The transition of the intermicellular (dispersed or oil) phase is gradual and almost continuous.

Continued attention to this aspect of asphaltene chemistry has led to the postulate that asphaltene–asphaltene clusters form the micelle, although other options such as asphaltene–resin clusters are also possible.

The precise mechanism of asphaltene association has not been conclusively established, but hydrogen bonding (66, 95, 96) and the formation of charge–transfer complexes (66) have been cited as the causative mechanisms. Evidence exists that asphaltenes participate in such complexes (97, 98), but the exact chemical or physical manner in which they would form in petroleum is still open to discussion. Intermolecular hydrogen-bonding could also be involved in asphaltene association and may have a significant effect on observed molecular weights (95).

Solubility and Incompatibility.   Asphaltenes are, in fact, insoluble in the oil fraction (3, 4) and the likelihood that asphaltene dispersion is mainly attributable to the resins is possible.

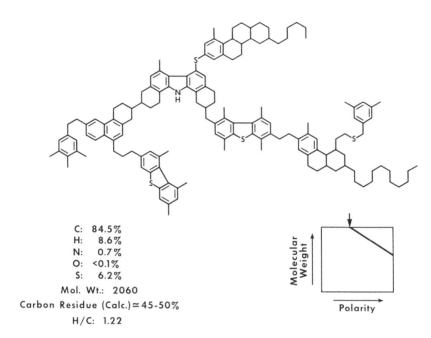

C: 84.5%
H: 8.6%
N: 0.7%
O: <0.1%
S: 6.2%
Mol. Wt.: 2060
Carbon Residue (Calc.) ≈ 45-50%
H/C: 1.22

Figure 8. A hypothetical model for a neutral–polar asphaltene might be as shown above.

Furthermore, the difficulty with which resins from one crude oil peptize (as in the case of a colloid, the terms peptization, dispersion, and solubilization are often used interchangeably to describe the means by which asphaltenes exist within petroleum) asphaltenes from a different crude oil and the instability of the "blend" indicate signifi-

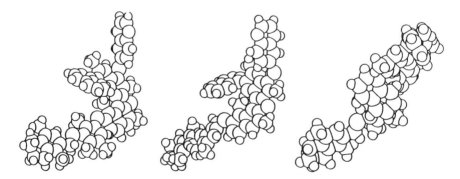

Figure 9. Molecular models of the low molecular weight high polarity asphaltene molecule show different three-dimensional aspects depending on bond rotation.

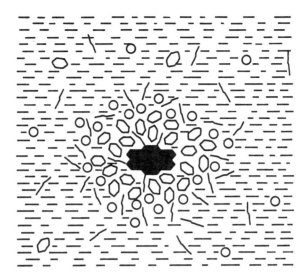

Figure 10. An early model of the physical structure of petroleum showed the asphaltenes dispersed and peptized by resin species.

cant structural differences among the asphaltenes and resins of various crude oils.

Colloid-Type Systems.   The evidence available in the literature appears to indicate that the hydrocarbon structures and some features, such as the various condensed ring systems, in different crude oils are similar. The variety of source materials involved in petroleum genesis, however, implies that, on an individual molecular scale, there may be substantial structural differences among the constituents of crude oil and bitumen.

A relatively high degree of aromaticity is generally prevalent in the asphaltenes and the resins. However, in some resins, the hydrocarbons show an increase in aliphatic material (e.g., more side chains) until, with considerable saturation, the oils, which contain numerous alkyl chains of varying length, are reached. The degree of aromaticity is important when the resins are desorbed. High aromaticity of the maltenes (i.e., that part of petroleum remaining after the asphaltenes have been removed, often referred to as the deasphalted or deasphaltened oil) indicates good solvency for the asphaltenes, and the solvent power of the maltenes is one of the most important factors in determining the physicochemical behavior of the petroleum colloid system. Therefore, petroleum is a complex system, with each fraction being dependent on other systems for complete mobility and solubility. It is presumed that the resins associate with the asphaltenes in the manner of an electron donor–acceptor.

The concept of hydrogen-bonding interactions as one of the means of association between the asphaltenes and resins has, however, led to a reconsideration of the assumed cluster as part of the micelle. Indeed, it appears that when resins and asphaltenes are present together, hydrogen bonding may be one of the mechanisms by which resin–asphaltene interactions are achieved. In some instances, it appears that resin–asphaltene interactions may be preferred over asphaltene–asphaltene interactions.

If the same intermolecular forces are projected to petroleum, it would perhaps not be surprising that asphaltenes exist not as larger agglomerations but as single entities that are peptized, and effectively dispersed, by the resins. In fact, it is quite likely that the micelle in petroleum consists of a single asphaltene molecule rather than as a stack of asphaltene molecules surrounded (peptized, dispersed) by resin molecules.

Asphaltene Dispersion and Process Chemistry and Physics

Without doubt, the most important asphaltene reactions, in the current context, are thermal reactions and the manner in which they relate to thermal and catalytic refining options. It is known that asphaltenes will produce significant quantities of distillable liquids in thermal processes, but the production of significant quantities of coke has always been an issue (1).

Understanding asphaltene chemistry from the behavior of model compounds is not as straightforward as it may appear. The individual reactions occurring in an extremely complex mixture, a multitude of secondary reactions, and the "interference" of the products from some of the constituents with those from other constituents of the mixture can make the thermal chemistry of asphaltenes extremely unpredictable (1).

As examples of the behavior of asphaltenes during refining, there are processes in which the dispersibility and compatibility or incompatibility of asphaltenes play an important role and which are worthy of consideration here. The first process is nonthermal and is the deasphalting process in which asphaltenes, and often resins, are discharged from the feedstock by the addition of hydrocarbon liquids. This is analogous to the laboratory separation procedure, with the exception that the process liquids are often the lower molecular weight hydrocarbons liquified under pressure. A similar situation occurs when asphaltenes are deposited on reservoir rock because of the increased solubility of hydrocarbon gases in the petroleum as reservoir pressure increases during maturation. In both cases, asphaltene dispersion is influenced by the addition of a component that decreases the ability of the surrounding medium to tolerate the "asphaltenes."

Another consideration of the deasphalting phenomenon is that whatever the micellar model, the surrounding "resinous sheath" is removed by the liquid and the asphaltenes, now denuded of this sheath, can no longer be dispersed in the liquid medium. Phase separation (precipitation) is the outcome.

Examples of thermal processes in which incompatibility might occur are the visbreaking process and the hydrocracking process. The visbreaking process is a means of reducing the viscosity of heavy feedstocks by "controlled" thermal decomposition. However, the process is often plagued by sediment ("sludge") formation that must be removed if the products are to meet fuel oil specifications.

There are several models that describe the thermal decomposition of asphaltenes (99–103) and are worthy of consideration. Using these models as a guide, it is reasonable to assume that during thermal processes, the asphaltene nuclear fragments become progressively more polar as the paraffinic fragments are stripped from the ring systems by scission of the bonds (preferentially) between the carbon atoms α and β to the aromatic rings (e.g., using the theoretical models, *see* Figures 11 and 12). The polynuclear aromatic systems denuded of the attendant hydrocarbon moieties are somewhat less soluble in the surrounding hydrocarbon medium than their "parent" systems, not only because of the solubilizing alkyl moieties but also because of the enrichment of the liquid medium in paraffinic constituents.

An analogy to the deasphalting process occurs, except the paraffinic material is a product of the thermal decomposition of the asphaltene molecules and is formed in situ rather than being added from an external source.

The visbreaker products might well contain polar species that have been denuded of some of the alkyl chains and that, on the basis of solubility, might be more rightly called carbenes and carboids (Figure 13). Moreover, an induction period may be required for phase separation or agglomeration to occur. Such products might, initially, be dispersible ("soluble") in the liquid phase, but after the induction period, cooling, or diffusion of the products, incompatibility (phase separation, sludge formation, agglomeration) occurs.

The presence of hydrogen changes the nature of the products, especially the coke yield by preventing the buildup of precursors that are incompatible in the liquid medium and form coke. But precisely how asphaltenes react with the catalysts is open to much more speculation.

Incompatibility is still possible when asphaltenes interact with catalysts, especially acidic support catalysts, through the functional groups, for example, the basic nitrogen species just as they interact with adsorbents. The possibility exists for interaction of the asphaltene with the catalyst through the agency of a single functional group in which the

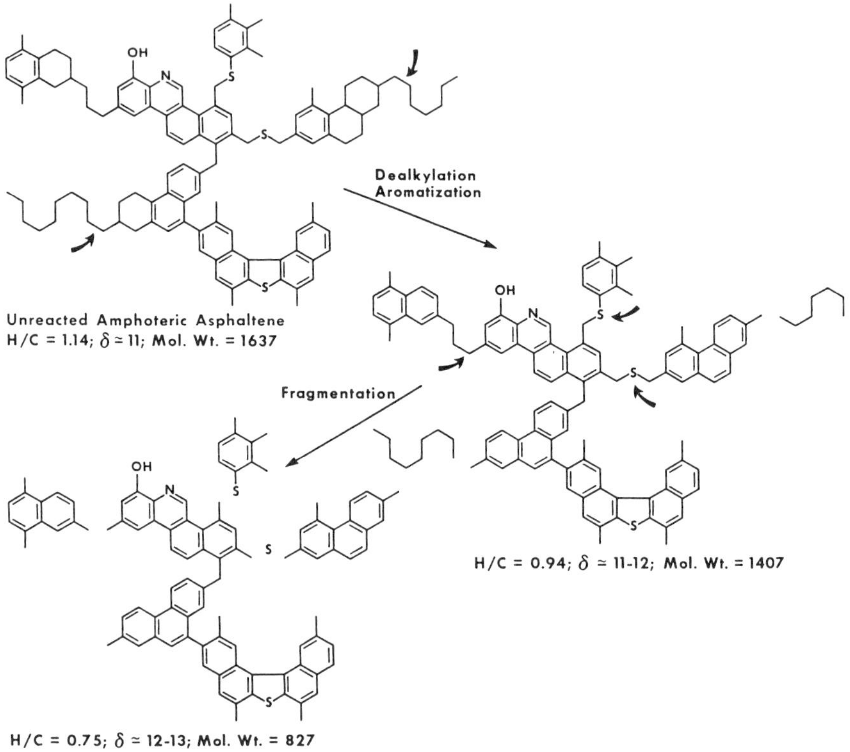

Figure 11. Molecular models can be used to show the possible changes during the thermal decomposition of asphaltenes; both chemical and physical changes can be illustrated.

remainder of the asphaltene molecule remains dispersed in the liquid phase. As a less desirable option, the asphaltene reacts with the catalyst at several points of contact, causing immediate loss of dispersibility (i.e., incompatibility) and deposition onto the catalyst surface.

Another area where loss of dispersibility might play a detrimental role, during processing or in the product, is in asphalt oxidation. The more polar species in a feedstock appear to oxidize first in an air-blowing operation, and after incorporation of oxygen to a limit, significant changes can occur in asphaltene molecular weight that is not so much because of oxidative degradation but to the incorporation of oxygen functions that interfere with the natural order of intramolecular structuring. The result is a poor grade asphalt where phase separation may already have occurred, or, should it occur at a later time, the result will be pavement failure because of a weakening of the necessary asphalt–aggregate interactions.

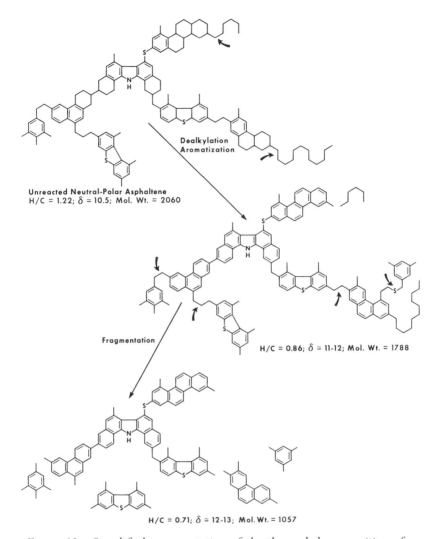

Figure 12. Simplified representation of the thermal decomposition of a neutral polar asphaltene.

A simple method of explaining the dispersibility and compatibility of asphaltenes in a variety of systems involves a comparison of the solubility parameter, δ, for petroleum fractions and for the solvents. Sufficient data exists to draw an approximate correlation between H:C atomic ratio and δ for hydrocarbons and the constituents of the lower boiling fractions of petroleum (25) and a parallel, or near parallel, line can be assumed that allows the solubility parameter of the asphaltenes and resins to be estimated (Figure 14).

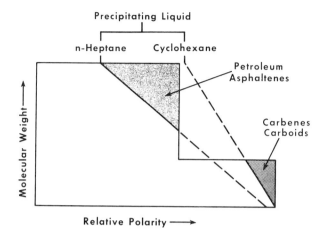

Figure 13. The molecular weight polarity diagram can also be used to illustrate the relative position of carbenes and carboids.

By this means, the solubility parameter of asphaltenes can be estimated to fall in the range of 9–12, which is also in keeping with the asphaltenes being composed of a mixture of different compound types and the accompanying variation in polarity. Removal of alkyl side chains from the asphaltenes will decrease the H/C ratio and increase the solubility parameter, thereby bringing about a concurrent decrease of the

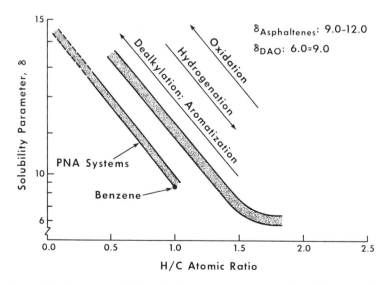

Figure 14. Estimates of the solubility parameter of crude oil fractions show that asphaltenes fall into the range of 9.0–12.0.

dispersibility and solubility of the asphaltene product in the hydrocarbon solvent.

In fact, on the molecular weight polarity-diagram for asphaltenes, carbenes and carboids can be shown as lower molecular weight, highly polar entities (Figure 13), in keeping with molecular fragmentation models (Figures 11 and 12).

Thus, crude oil may be considered to be a three-phase system (25). In this system, the two liquid phases are aromatics and saturates, whereas the asphaltenes exist as a soluble phase. As the liquid becomes progressively more paraffinic (i.e., less aromatic) or as the asphaltenes become more polar, the asphaltenes become progressively less dispersible. Finally, a point is reached where insolubility, or separation of the polar species from the liquid phase, occurs.

The initial fragmentation and aromatization of the asphaltenes yield isolatable products that are more aromatic than the original asphaltenes and have a slightly reduced molecular weight (82). By definition and character, these are carbenes. The next step will be the more complete fragmentation of the asphaltenes to produce the carboids that are, by definition, the immediate precursors to coke.

The use of a catalytic process for the conversion of high-asphaltene feedstocks adds another dimension to asphaltene science. Asphaltenes will interact with catalysts, especially acidic support catalysts through the basic nitrogen, just as they will interact with adsorbents. The means by which they do so is another issue, but evidence is available for interaction at a single functional group in which the remainder of the asphaltene molecule remains in the liquid phase. As a less desirable option, the asphaltene reacts with the catalyst at several points of contact, causing a more massive lay-down of coke on the catalyst surface (104).

Thus, asphaltene dispersibility and compatibility in the liquid medium is an important factor in process control (105) and needs to be addressed at some stage of the operation. The consequences of not doing so can be costly.

Conclusions

Petroleum is a delicately balanced system insofar as the different fractions are compatible, provided no significant disturbances or changes are made to the system. Such changes are the alteration of the natural occurrences of the different fractions; the chemical or physical alteration of the constituents as might occur during recovery and refining, especially changes that might be brought on by thermal processes; and alteration of the polar group distribution as might occur during oxidation (i.e., asphalt manufacture) or the elimination of polar functions during processing.

When such disturbances occur, it is the higher molecular weight constituents that are most seriously affected. This can lead to incompatibility (*103, 105*), which is variously referred to as loss of dispersibility, precipitation, and sludge formation. Thus, the dispersibility of the higher molecular weight constituents becomes an issue that needs attention. One way to understand this issue is to be aware of the chemical and physical character of the higher molecular weight constituents. By such means, the issue of dispersibility, and the attending issue of incompatibility, can be understood and even predicted.

References

1. Speight, J. G. *The Chemistry and Technology of Petroleum*, 2nd ed.; Marcel Dekker: New York, 1991.
2. Gruse, W. A.; Stevens, D. R. *The Chemical Technology of Petroleum*; McGraw-Hill: New York, 1960.
3. Koots, J. A.; Speight, J. G. *Fuel* **1975**, *54*, 179.
4. Swanson, J. M. *J. Phys. Chem.* **1942**, *46*, 141.
5. Marcusson, J. *Mitt. Materialpruf.* **1918**, *36*, 209.
6. Sachanen, A. *Petrol. Ztg.* **1925**, *21*, 1441.
7. Nellensteyn, F. J.; Roodenburg, N. M. *Chem. Ztg.* **1930**, *545*, 819.
8. Nellensteyn, F. J. *Chem. Weekbl.* **1931**, *28*, 313.
9. Hubbard, R. L.; Stanfield, K. E. *Anal. Chem.* **1948**, *20*, 460.
10. Eisma, E.; Krom, C. J. *J. Inst. Pet.* **1951**, *37*, 582.
11. Speight, J. G.; Long, R. B.; Trowbridge, T. D. *Fuel* **1984**, *63*, 616.
12. van Nes, K.; van Westen, H. A. *Aspects of the Constitution of Mineral Oils;* Elsevier: Amsterdam, Netherlands, 1951.
13. Girdler, R. B. *Proc. Assoc. Asphalt Paving Technol. Tech. Sess.* **1965**, *34*, 45.
14. Mitchell, D. L.; Speight, J. G. *Fuel* **1973**, *52*, 149.
15. Ali, L. H.; Al-Ghannan, K. A. *Fuel* **1981**, *60*, 1043.
16. Andersen, S. I.; Birdi, K. S. *Fuel Sci. Technol. Int.* **1990**, *8*, 593.
17. Ali, L. H.; Al-Ghannam, K. A.; Al-Rawi, J. M. *Iraqi J. Sci.* **1985**, *26*, 41.
18. Long, R. B. *Prepr. Am. Chem. Soc. Div. Pet. Chem.* **1979**, *24*, 891.
19. Long, R. B. In *Chemistry of Asphaltenes;* Bunger, J. W.; Li, N. C., Eds.; Advances in Chemistry 195; American Chemical Society: Washington, DC, 1981; pp 17–27.
20. Nellensteyn, F. J. In *The Science of Petroleum;* Dunstan, A. E.; Nash, A. W.; Brooks, B. T.; Tizzard, H. T., Eds.; Oxford University: Oxford, England, 1938; Vol. IV.
21. Pfeiffer, J. H. *The Properties of Asphaltic Bitumen;* Elsevier: Amsterdam, Netherlands, 1950.
22. Speight, J. G. Information Series No. 84; Alberta Research Council: Edmonton, Canada, 1979.
23. Monin, J. C.; Pelet, R. In *Advances in Organic Geochemistry;* Bjoroey, M., Ed.; John Wiley and Sons: New York, 1983.
24. Hotier, G.; Robin, M. *C. R. Seances Acad. Sci. Ser. 2* **1981**, *292*, 39.
25. Speight, J. G. In *Asphaltenes and Asphalts. Developments in Petroleum Science;* Yen, T. F.; Chilingarian, G. V., Eds.; Elsevier; Amsterdam, Netherlands, 1994; Vol. 40.
26. Speight, J. G.; Moschopedis, S. E. In *Chemistry of Asphaltenes;* Bunger, J. W.; Li, N. C., Eds.; American Chemical Society: Washington, DC, 1981; pp 1–15.

27. Bestougeff, M.; Darmois, R. *C. R. Hebd. Seances Acad. Sci.* **1947**, *224*, 1365.
28. Bestougeff, M.; Darmois, R. *C. R. Hebd. Seances Acad. Sci.* **1948**, *227*, 129.
29. Bestougeff, M. A.; Mouton, Y. *Bull. Liason Lab. Ponts Chausseus* Special Volume **1977**, 79.
30. Dron, R.; Bestougeff, M.; Voinovitch, I. A. *Rapp. Rech. LPC* **1978**, 75.
31. Speight, J. G. *Prepr. Am. Chem. Soc. Div. Pet. Chem.* **1986**, *31a*, 818.
32. Altgelt, K. H. *J. Appl. Polym. Sci.* **1965**, *9*, 3389.
33. Gould, K. A.; Grenoble, D. C.; Murrell, L. L.; Peters, W. J. M. U.S. Patent 4 422 929, 1983.
34. Selucky, M. L.; Kim, S. S.; Skinner, F.; Strausz, O. P. In *Chemistry of Asphaltenes;* Bunger, J. W.; Li, N. C., Eds.; Advances in Chemistry 195; American Chemical Society: Washington, DC, 1981; pp 83–118.
35. Francisco, M. A.; Speight, J. G. *Prepr. Am. Chem. Soc. Div. Fuel Chem.* **1984**, *29*, 36.
36. Hillman, E.; Barnett, B. *Proceedings of the 4th Annual Meeting of ASTM;* American Society for Testing and Materials: Philadelphia, PA, 1937; Vol. 37, p 558.
37. Murphy, B. *J. Inst. Pet.* **1945**, *31*, 475.
38. Yen, T. F.; Erdman, J. G. *Prepr. Am. Chem. Soc. Div. Pet. Chem.* **1962**, *7*, 99.
39. Yen, T. F. *Prepr. Am. Chem. Soc. Div. Pet. Chem.* **1972**, *17*, F102.
40. Haley, G. A. *Anal. Chem.* **1972**, *44*, 580.
41. Speight, J. G. *Appl. Spectrosc. Rev.* **1972**, *5*, 211.
42. Speight, J. G. *Appl. Spectrosc. Rev.* **1994**, *29*, 269.
43. Dickinson, E. M. *Fuel* **1980**, *59*, 290.
44. Bandurski, E. *Energy Sources* **1982**, *6*, 47.
45. Sadeghi, K. M.; Sadeghi, M. A.; Wu, W. H.; Yen, T. F. *Fuel* **1989**, *68*, 782.
46. Ali, L. H.; Al-Ghannam, K. A., Al-Rawi, J. M. *Fuel* **1990**, *69*, 519.
47. Yen, T. F. *Fuel* **1970**, *49*, 134.
48. Yen, T. F. *Prepr. Am. Chem. Soc. Div. Fuel Chem.* **1971**, *15*, 57.
49. Sawatsky, H.; Boyd, M. L.; Montgomery, D. S. *J. Inst. Pet.* **1967**, *53*, 162.
50. Witherspoon, P. A.; Winniford, R. S. In *Fundamental Aspects of Petroleum Geochemistry;* Nagy, B.; Colombo, U., Eds.; Elsevier; New York, 1967.
51. Yen, T. F.; Erdman, J. G.; Pollack, S. S. *Anal. Chem.* **1961**, *33*, 1587.
52. Helm, R. V.; Latham, D. R.; Ferrin, C. R.; Ball, J. S. *Chem. Eng. Data. Ser.* **1957**, *2*, 95.
53. Ball, J. S.; Latham, D. R.; Helm, R. V. *J. Chem. Eng. Data Ser.* **1959**, *4*, 167.
54. Clerc, R. J.; O'Neal, M. J. *Anal. Chem.* **1961**, *33*, 380.
55. Nicksic, S. W.; Jeffries-Harris, M. J. *J. Proc. Inst. Pet. London* **1968**, *54*, 107.
56. Moschopedis, S. E.; Speight, J. G. *Fuel* **1976**, *55*, 334.
57. Moschopedis, S. E.; Speight, J. G. *Prepr. Am. Chem. Soc. Div. Pet. Chem.* **1979**, *24*, 1007.
58. Moschopedis, S. E.; Hawkins, R. W.; Speight, J. G. *Fuel* **1981**, *60*, 397.
59. Guiochon, G. *Petroanalysis '81;* Crump, G. C., Ed.; John Wiley and Sons: Chicester, England, 1982.
60. Schmitter, J. M.; Ignatiadis, I.; Arpino, P. J. *Geochim. Cosmochim. Acta* **1983**, *47*, 1975.
61. Jacobson, J. M.; Gray, M. R. *Fuel* **1987**, *66*, 749.

62. Sinninghe Damste, J. S.; Eglinton, T. I.; de Leeuw, J. W. *Prepr. Am. Chem. Soc. Div. Fuel Chem.* **1991**, *36*, 710.
63. Ritchie, R. G. S.; Roche, R. S.; Steedman, W. *Fuel* **1979**, *58*, 523.
64. Ritchie, R. G. S.; Roche, R. S.; Steedman, W. *Chem. Ind.* **1979**, 25.
65. Speight, J. G.; Moschopedis, S. E. *Prepr. Am. Chem. Soc. Div. Pet. Chem.* **1981**, *26*, 907.
66. Yen, T. F. *Energy Sources* **1974**, *1*, 447.
67. Speight, J. G.; Pancirov, R. J. *Liq. Fuels Technol.* **1984**, *2*, 287.
68. Rose, K. D.; Francisco, M. A. *J. Am. Chem. Soc.* **1988**, *110*, 637.
69. Sinninghe Damste, J. S.; Eglinton, T. I.; de Leeuw, J. W.; Schenck, P. A. *Geochim. Cosmochim. Acta* **1989**, *53*, 873.
70. Sinninghe Damste, J. S.; Rijpstra, W. I. C.; Kock van-Dalen, A. C.; de Leeuw, J. W., Schenck, P. A. *Geochim. Cosmochim. Acta* **1989**, *53*, 1343.
71. George, G. N.; Gorbaty, M. L.; Kelemen, S. R. In *Geochemistry of Sulfur in Fossil Fuels*; Orr, W. L.; White, C. M., Eds.; ACS Symposium Series 429; American Chemical Society: Washington, DC, 1990; pp 220–230.
72. Strausz, O. P.; Lown, E. M.; Payzant, J. D. In *Geochemistry of Sulfur in Fossil Fuels*; Orr, W. L.; White, C. M., Eds.; ACS Symposium Series 429; American Chemical Society: Washington, DC, 1990; pp 366–396.
73. Hausler, D. W. *Spectrochim. Acta* **1985**, *40B*, 389.
74. Baker, E. W. In *Organic Geochemistry*; Eglinton, G.; Murphy, M. T. J., Eds.; Springer-Verlag: New York, 1969.
75. Yen, T. F. *The Role of Trace Metals in Petroleum*; Ann Arbor Science Publishers: Ann Arbor, MI, 1975.
76. *Metal Complexes in Fossil Fuels*; Filby, R. H.; Branthaver, J. F., Eds.; ACS Symposium Series 344; American Chemical Society: Washington, DC, 1987.
77. Branthaver, J. F. In *Fuel Science and Technology Handbook*; Speight, J. G., Ed.; Marcel Dekker: New York, 1990.
78. Moschopedis, S. E.; Fryer, J. F.; Speight, J. G. *Fuel* **1976**, *55*, 227.
79. Speight, J. G.; Moschopedis, S. E. *Fuel* **1977**, *56*, 344.
80. Winniford, R. S. *J. Inst. Pet.* **1963**, *9*, 215.
81. Speight, J. G.; Wernick, D. L.; Gould, K. A.; Overfield, R. E.; Rao, B. M. L.; Savage, D. W. *Rev. Inst. Fr. Pet.* **1985**, *40*, 51.
82. Speight, J. G. *Prepr. Am. Chem. Soc. Div. Pet. Chem.* **1987**, *32*, 413.
83. Fieser, L. F.; Fieser, M. *Natural Products Related to Phenanthrene*; Reinhold Publishing: New York, 1949.
84. Dalton, D. R. *The Alkaloids*; Marcel Dekker: New York, 1979.
85. Tyman, J. H. P. *Chem. Soc. Rev.* **1979**, *8*, 499.
86. Weiss, V.; Edwards, J. M. *The Biosynthesis of Aromatic Compounds*; John Wiley and Sons: New York, 1980.
87. Dussel, H. J.; Recca, A.; Kolb, J.; Hummel, D. O.; Stille, J. K. *J. Anal. Appl. Pyrolysis.* **1982**, *3*, 307.
88. NASA Technological Briefs; National Aeronautics and Space Administration: Washington, DC, 1994; Vol. 18, p 56.
89. Hobson, G. D.; Pohl, W. *Modern Petroleum Technology*; Applied Science Publishers: London, 1973.
90. Orr, W. L. *Prepr. Am. Chem. Soc. Div. Fuel Chem.* **1977**, *22*, 86.
91. Orr, W. L., Sinninghe Damste, J. S. In *Geochemistry of Sulfur in Fossil Fuels*; Orr, W. L.; White, C. M., Eds.; ACS Symposium Series 429; American Chemical Society: Washington, DC, 1990; pp 2–29.
92. Lehninger, A. L. *Biochemistry*; Worth Publishers: New York, 1970.

93. Speight, J. G.; Long, R. B. In *Atomic and Nuclear Methods in Fossil Energy Research*; Filby, R. H., Ed.; Plenum: New York, 1982; p 295.
94. Pfeiffer, J. P.; Saal, R. N. *Phys. Chem.* **1940**, *44*, 139.
95. Moschopedis, S. E.; Speight, J. G. *Proc. Assoc. Paving Technol.* **1976**, *45*, 78.
96. Acevedo, S.; Mendez, B.; Rojas, A.; Larisse, I.; Rivas, H. *Fuel* **1985**, *64*, 1741.
97. Penzes, S.; Speight, J. G. *Fuel* **1974**, *53*, 192.
98. Speight, J. G.; Penzes, S. *Chem. Ind.* **1978**, 729.
99. Magaril, R. Z.; Akensova, E. I. *Int. Chem. Eng.* **1968**, *8*, 727.
100. Magaril, R. Z.; Ramazaeva, L. F. *Izv. Vyssh. Uchebn. Zaved. Neft Gaz.* **1969**, *12*, 61.
101. Magaril, R. Z.; Ramazaeva, L. F.; Aksenova, E. I. *Khim. Tekhnol. Topl. Masel.* **1970**, *15*, 15.
102. Schucker, R. C.; Keweshan, C. F. *Prepr. Am. Chem. Soc. Div. Fuel Chem.* **1980**, *25*, 155.
103. Wiehe, I. A. *Prepr. Am. Chem. Soc. Div. Pet. Chem.* **1993**, *38*, 428.
104. Speight, J. G. In *Catalysis on the Energy Scene*; Kaliaguine, S.; Mahay, A., Eds.; Elsevier: Amsterdam, Netherlands, 1984.
105. Speight, J. G. *Proceedings of the 4th International Conference on the Stability and Handling of Liquid Fuels*; U.S. Department of Energy: Washington, DC, 1992; p 169.

RECEIVED for review July 5, 1994. ACCEPTED revised manuscript February 7, 1995.

Solids Production and Control in Petroleum Recovery

Blaine F. Hawkins, Ion Adamache, and Gerhard Leopold

Petroleum Recovery Institute, 100, 3512 33rd Street N.W., Calgary, Alberta T2L 2A6, Canada

Reservoir solids may be produced during exploitation of petroleum reservoirs. Solids management practices depend on the nature of the reservoir and the produced fluids. For conventional oil and gas reservoirs, operators normally have focused on prevention or control of solids influx into production wells. In Canada, exploitation of heavy oil reservoirs has been found to be more effective when solids are produced.

Solids production from conventional oil and gas reservoirs can have detrimental effects on well productivity and equipment. Therefore, the practice has been to attempt to prevent or control solids influx into the wells by using screens, gravel packs, or other means to restrict solids flow. If, when, where, and how to install solids control systems are governed by both cost and risk factors.

From a solids management point of view, production of heavy oil from unconsolidated formations poses different challenges than conventional oil and gas production. Production of large quantities of solids is critical for maintaining the productivity of wells in heavy oil reservoirs. This has necessitated the development of special pumping systems to transport fluids with high solids content.

Many reservoirs are now being exploited by horizontal wells, which have long horizontal contact with the reservoir. Solids control methods for horizontal wells have been adapted from those used in vertical wells. Special techniques and equipment have been developed to remove deposited solids from horizontal wells.

Specially designed equipment and techniques are required for the surface handling of high solids cut fluids. Separation, cleaning, and disposal of produced solids can be costly, especially if the solids must be oil-free to satisfy environmental regulations.

0065–2393/96/0251–0403$21.50/0

Background

Oilfield Operations.   Formation solids production and control are important aspects of oil and gas field operations. For those unfamiliar with the basic operations involved in oil well drilling, completion, and production, the following summary is provided. For a more detailed description of these operations, the reader is directed to Allen and Roberts (1) or Berger and Anderson (2).

The wells used for the exploitation of hydrocarbon reservoirs can be vertical, deviated, slanted, or horizontal. The well joins the downhole-producing zone with the surface facilities to bring oil or gas out in commercial quantities. A typical well-drilling system uses a bit at the end of a drill string through which a drilling fluid is circulated. The bit is actuated by rotating the drill string from the surface or by using a "downhole" motor or turbine hydraulically driven by the drilling fluid. The bit advances by simultaneous rotation and application of weight. The drilling fluid provides fluid head to control pressures that exist within the drilled formations. It also carries the drilling cuttings to surface and provides lubrication of the bit and drilling string. At various stages during drilling, casing is set and cemented in place to isolate the formations traversed by the well. Typically, a series of casings of decreasing diameter with well depth are installed, for example, surface casing, intermediate casing, production casing, and production liner. Various tubing strings may be installed inside the casing to permit production from different zones, installation of artificial lift systems (e.g., beam pumping, rotating pumps, gas lift), and transportation of the produced fluid to the surface.

The drilling fluid often contains solids, the purpose of which is to form a filter cake to restrict the drilling mud filtrate and solids from invading the productive formations. Even so, the drilling fluid can invade permeable petroleum formations, causing damage to the formation (permeability reduction in the vicinity of the wellbore) and decreasing well flow capacity. To provide a pathway for fluids to flow from the reservoir and to overcome the effect of the damaged zone, the production casing is perforated (3) across the desired interval using a "gun" with explosive charges. Perforation tunnels are created (Figure 1) through the damaged zone. The perforation tunnels may contain solid debris that must be removed before fluid can flow efficiently into the well. Often, the difference in pressure between the reservoir and the wellbore, the pressure drawdown, will dislodge and clean out the debris, permitting flow from the reservoir into the wellbore. Around the perforation tunnel is a region of crushed rock that has been compacted as a result of the perforation process (4–6). This zone is weak and subject to tensile failure, giving rise to production of solids (6). The produced solids can be transported to the surface as a suspension or they can accumulate in the wellbore, depending on factors such as flow velocity,

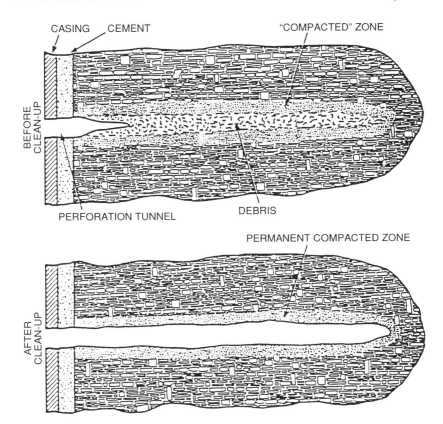

Figure 1. Diagram of a perforation cavity before and after cleanup, showing compacted zone. (Reproduced with permission from reference 5. Copyright 1991 Gulf Publishing Company.)

fluid density and viscosity, and density and concentration of the solid particles. Accumulation of solids, "sanding," can reduce or even stop hydrocarbon production (7, 8). When sanding occurs, it is necessary to conduct a remedial operation referred to as a workover. During a workover, the downhole equipment is removed and the solids are cleaned out of the well. If the problem continues, corrective measures may be applied to mitigate the influx of solids.

If the well productivity is less than its estimated potential, the well may be stimulated. Stimulation operations include techniques such as acidizing, surfactant treatment, and fracturing. In some cases, the need to stimulate the well may be avoided by underbalanced perforating, that is, the pressure in the wellbore is less than the formation pressure during perforating. In some cases, underbalanced drilling is applied to reduce formation damage.

Alternatively, the well may be completed "open hole," in which the productive formation is not cased or cemented. This type of completion is more likely to be done in consolidated reservoirs where the hole is stable.

The primary exploitation phase uses the energy (pressure) of the reservoir as the driving force for petroleum production. Over time, the reservoir pressure declines, and it may be necessary to implement secondary or tertiary recovery processes. Secondary processes involve injection of gas or water to provide pressure to drive the oil to producing wells. After primary and secondary recovery phases, more than 50% of the original oil in place often remains because of trapping phenomena, mainly due to capillary forces, or bypassing. Tertiary recovery processes, such as the injection of miscible solvents, surfactants, or steam, attempt to recover some of the remaining oil mainly by reducing capillary forces.

Horizontal Wells. Many reservoirs are now being exploited by using horizontal wells (9, 10) to enhance project economics. Horizontal wells differ from vertical wells in that from a certain depth, the "kick off" point, the well is curved gradually until it becomes more or less horizontal within the productive formation. The horizontal section of the well may extend hundreds of meters through the productive formation. The contact length of a horizontal well with the productive formation is often an order of magnitude or more greater than the contact length of a vertical well with the formation.

Horizontal wells have important advantages over vertical wells, primarily because of their longer contact length:

- Pressure gradient and fluid velocity around the wellbore can be an order of magnitude less.

- Petroleum production can be accelerated. The horizontal well production rate is normally two to five times larger (11) than vertical wells in the same formation, even though fluid velocity entering the well is lower.

- Water or gas coning problems can be mitigated.

- Better access to the reservoir can increase ultimate petroleum recovery by accessing zones of the reservoir that would otherwise be poorly produced.

Horizontal wells can be drilled with a larger well spacing than vertical wells and, like vertical wells, may be completed open or cased hole. Recent innovations such as multibranches and multilegs off of the main section increase the efficiency and productivity of horizontal wells (12, 13). However, new technology is still needed to overcome difficulties associated with the use of horizontal wells, such as effective clean-out of the long wellbores, once solids are deposited inside, and lifting higher fluid volumes, once multibranches or multilegs are in place.

Horizontal wells have some disadvantages. Economic risks are higher for horizontal wells because of their higher cost and greater potential for damage due to longer drilling time in the productive formation. They can cost several times more than a vertical well, depending on the drilling and completion techniques used. As more horizontal wells are drilled and completed in a given area, the incremental cost over a vertical well is reduced because of the experience gained. Typically, horizontal wells can drain only one layer or pay zone. However, "staircase" type wells, where the wellbore has horizontal sections drilled in more than one layer, or multileg wells with separate legs drilled into several layers from the same vertical casing, have been reported (*14*).

The path of the horizontal section is not strictly horizontal but has an undulating trajectory. Target deviations varying from ±2 to 3 m (i.e., level to within a window of 4–6 m diameter) are achieved more or less routinely, and target errors of as little as ±0.5 m (1-m window) may be achieved with some operational precautions and added expense. The undulating well profile can cause some difficulties in coning situations and complicate well control and cleaning operations. Should solids invade the horizontal wellbore, the particles may settle out in low spots or along the lower portion of the wellbore, forming a nonflowing layer of solids (Figure 2) (*15, 16*). The reader is directed to Chapter 4 for a discussion of solids transport in pipes.

Horizontal wells drilled in weakly consolidated sandstones tend to collapse, filling sections of the wellbore (*17, 18*). This can reduce or stop the fluid production from beyond the solids fill point. However, some formations that produce solids in vertical wells may not produce solids in horizontal wells because of the lower pressure drawdown and lower producing fluid velocities around the wellbore. During exploitation, other factors, such as declining reservoir pressure and increasing water production, can cause collapse of the horizontal hole, even in situations where the productive formations are stronger than typical unconsolidated or semiconsolidated sandstones. McLellan (*19*) has provided general guidelines for conducting stability assessments of horizontal wellbores. Proper completion technique is crucial (*20*).

Oilfield-Produced Solids. Typical reservoir rocks can be consolidated, partially consolidated (friable), or unconsolidated. The reservoir rock is composed of the load-bearing solids, "grains," and fine-grained solids, "fines," which are not part of the load-bearing structure. In consolidated rocks, the grains are held in place by cementing materials. In friable rocks, some weak cementing materials are present. In unconsolidated rocks, no cementing materials are present, but grains

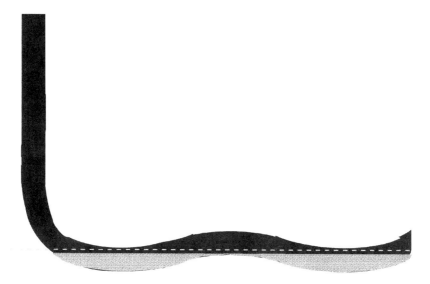

Figure 2.   Sand accumulation in horizontal wells. Complete plugging is possible when sinusoidal amplitude ≥ wellbore diameter. There is also the potential for development of longitudinal annular flow. (Reproduced with permission from reference 118. Copyright 1993 Nowsco Well Service Ltd.)

can be held together by capillary forces of the saturating fluids. The grains may consist of sands, cherts, or other mineral particles. Cements may consist of calcite, dolomite, silicate, clays, or other minerals.

Load-bearing solids tend to be larger, heavier particles. If dislodged, they can bridge across the perforations, preventing inflow of fluids, or they can be carried into wells where they can drop out, causing well plugging and necessitating clean-out operations. The term "solids control" refers to control of the load-bearing particles.

The nature of fines, together with the conditions under which they may be caused to migrate through the porous reservoir rock, is discussed in Chapter 7. Fines may be produced into the wellbore with the fluid. Being small particles of relatively low mass, they can be carried by produced fluids up the wellbore and to the surface. Because of their charged nature, fines can stabilize unwanted emulsions in surface equipment (21). Fines also can cause plugging of the reservoir pores, reducing fluid flow.

Other solids can form in wells. These include asphaltenes and paraffins (waxes), coke, hydrates, inorganic scales, and corrosion products. Consideration of these is beyond the scope of this chapter.

The reservoir solids are most conveniently classified according to their size. Because the particles are of irregular shape, classification is on the basis of "sieve opening" of sieve screens. Sieve opening is the width of the minimum square aperture through which the particle can pass. A number of standards have been developed for sieve screens (*see*, e.g., Appendix 4A of reference 1). The phi concept, in which ϕ is defined as

$$\phi = -\log_2 \text{ (size in mm)} = -\log_{10} \text{ (size in mm)}/\log_{10}2$$

provides a convenient means of evaluating particle size distributions (22). Particle sizes are also characterized qualitatively using the Wentworth class. These scales are compared in Table I. Fines typically would be particles that pass through sieve openings less than 0.05 mm.

Sand particles are often produced with fluids from wells that produce from loosely consolidated sandstone formations. Unconsolidated sandstone reservoirs with permeability of 0.5–8.0 darcy are most susceptible to sand production (6).

Impact of Solids Production. Solids production in vertical wells has been the cause of many costly operational problems, as follows:

- *Production losses.* The solids can bridge off in wells between the producing perforations and tubing inlet, reducing the rate of fluid production or shutting it off completely. If fluid flow velocities are not high enough to transport the solids to the surface, the solids can fill up the production liner or casing. Clean-out operations then may be required to remove solids from the well.

- *Equipment damage.* This can occur by erosion or plugging. A high velocity fluid carrying solid particles can erode parts of downhole or surface equipment such as casing, tubulars, wellheads, valves, piping, and fluid separator parts. Solids can accumulate in, plug, or damage components of surface production facilities such as pumps, filters, vessels, and tanks. A typical example is the premature failure of the downhole "sucker rod" pumps. The plunger–barrel couple can experience wear from solids entering between the two elements, the pump valves can wear, or the plunger can seize. Damaged equipment requires replacement, necessitating costly remediation operations. The production lost during equipment replacement increases the total cost. Oil spills may result from equipment failure, necessitating expensive cleanup activity.

- *Damage of downhole safety/well control components.* Solids can erode and destroy subsurface safety valves. Failure of these well safety devices can result in uncontrolled production situations with tremendous consequences, such as loss of life or large economic losses.

Table I.   U.S. Standard Sieves, Mesh Numbers, Phi
and Wentworth Class

Sieve Size Opening (mm)	U.S. Standard Sieve Mesh Number	Phi (ϕ)	Wentworth Class
			Boulder
256		−8	
			Cobble
64		−6	
			Pebble
4.0	5	−2	
3.36	6	−1.75	
2.83	7	−1.50	Granule
2.38	8	−1.25	
2.00	10	−1.00	
1.68	12	−0.75	
1.41	14	−0.50	Very Coarse Sand
1.19	16	−0.25	
1.00	18	0	
0.841	20	0.25	
0.707	25	0.50	Coarse Sand
0.595	30	0.75	
0.500	35	+1.00	
0.420	40	1.25	
0.354	45	1.50	Medium Sand
0.297	50	1.75	
0.250	60	+2.00	
0.210	70	2.25	
0.177	80	2.50	Fine Sand
0.149	100	2.75	
0.125	120	+3.00	
0.105	140	3.25	
0.088	170	3.50	Very Fine Sand
0.074	200	3.75	
0.0625	230	+4.00	
0.0526	270	4.25	
0.0442	325	4.50	Coarse Silt
0.0372	400	4.75	
		+5.00	0.031 mm
			Medium Silt
		+6	0.0156 mm
			Fine Silt
		+7	0.0078 mm
			Very Fine Silt
		+8	0.0039 mm
			Clay (everything smaller than 0.0039 mm)

NOTE: $\phi = -\log_2 \text{mm} = \dfrac{-\log_{10} \text{mm}}{\log_{10} 2}$ (mm = grain size in mm)

SOURCE: Reproduced with permission from reference 22. Copyright 1977 Colorado School of Mines.

- *Lifting problems.* Produced solids are denser than oil, water, or gas. In vertical lifting, solids slippage may occur, and the solid particles may sink relative to the entraining fluids. In this way the solids may concentrate in the vertical or inclined tubulars of the well, increasing the fluid head and reducing the fluid in-flow from the formation.

- *Casing collapse.* If large quantities of solids are produced, voids can form behind the casing, causing formation subsidence and transferring the overburden weight onto the casing, which can collapse.

- *Formation failure.* Excessive solids production can collapse relatively thin impermeable layers between a production zone and another zone. This can result in cross-flow between the zones or influx of unwanted gas or water.

- *Solids disposal.* This can be expensive, particularly if the solids are oil contaminated. On offshore locations, the cost could be large because of the need for additional treating capacity. Special facilities for cleaning and disposal of solids may be required because of the presence of oil, saltwater, and leachates.

Factors Affecting Solids Production and Control

Solids production usually results from mechanical failure of the reservoir rock. Factors affecting the nature and extent of the mechanical failure of the rock include (23)

- the natural rock strength
- the stresses to which the rock is subjected, whether naturally occurring or induced by the drilling operation
- the nature of the fluid that is produced into the wellbore
- the pressure drawdown between the reservoir and the wellbore
- the rate of fluid flow into the wellbore

Solids production is most common from geologically young sandstone reservoirs (i.e., Tertiary Age), which are usually located at relatively shallow depths. Solids production problems are less severe from older, deeper reservoir rocks that tend to be better consolidated. Solids production has been experienced worldwide in wells where oil or gas is produced from sandstone reservoirs and is often termed "sand production." Solids production is the preferred term because the produced solids typically consist of a mixture of grains, fines, and cement particles.

Solids production can be classified (24) as transient, continuous, or catastrophic. Transient solids production refers to situations where the concentration of produced solids decreases with time. This type of solids production is typical of well cleanup after completions or workovers and after water breakthrough. Continuous production occurs when some low level of solids production is tolerated. Typical levels for continuous solids production are 6–600 g/m^3 of fluid (24), a solids cut of less than 0.06%. Continuous production of solids with heavy oil from unconsolidated sandstone reservoirs can exceed 50% solids cut (25). Catastrophic solids production is caused by an event that allows a large slug of solids into a well, completely blocking the producing zone.

Rock Strength and Stress Factors.   The rock of a petroleum reservoir is subjected to stresses because of the weight of the rock above the reservoir (overburden). The stress field can be resolved into vertical and horizontal stresses in cartesian coordinates. In the general case, the stress experienced by an element of reservoir rock may be resolved into both normal and shear stresses. However, the coordinate axis system can be oriented such that the shear stresses are zero and the stress state of the rock is described by three orthogonal principal stresses, σ_1, σ_2, and σ_3 (Figure 3a).

When a well is drilled, the stress state of the rock around the wellbore is altered. The stresses around the wellbore are a function of position and, because of geometry, can be expressed in cylindrical coordinates (Figure 3b). The general solutions for the wellbore stresses can be quite complicated depending on the orientation of the well and the axes of

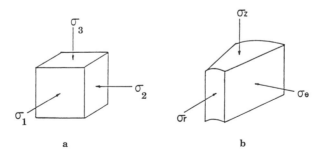

Figure 3.   Principal stresses to which a reservoir rock element is subjected (a) within the undisturbed reservoir and (b) in the wellbore region. (Reproduced with permission from reference 30. Copyright 1981 Society of Professional Well Log Analysts.)

principal stress (26–28). For the simple case of a well oriented along the vertical principal axis, the stresses at the wellbore wall (Figure 4) reduce to (28)

$$\sigma_z = \sigma_v - 2\mu(\sigma_H - \sigma_h) \cos 2\theta - p_{wb} \qquad \text{(axial stress)}$$

$$\sigma_\theta = \sigma_H + \sigma_h - 2(\sigma_H - \sigma_h) \cos 2\theta - p_{wb} \qquad \text{(tangential stress)}$$

$$\sigma_r = p_{wb} \qquad \text{(radial stress)}$$

where the principal stresses are the vertical stress, σ_v, the largest horizontal stress, σ_H, and the smallest horizontal stress, σ_h.

Also contributing to the overall stress experienced by the rock is the fluid pressure in the pores of the rock, p_f. This can relieve some of the compressive stresses by supporting the weight of the overburden. Variations of the above equations with corrections for fluid pressure have been used to estimate in situ stresses around the wellbore (29, 30).

When the stresses exceed the strength of the formation, failure of the rock can occur, leading to production of solids into the wellbore. Rocks can fail in tension when they are pulled apart, or under shear, when they are forced to deform because of compressive forces.

The Mohr circle technique is often used to define the stress envelope to which the rock is subjected (29–32). The radial and tangential stresses are plotted on the abscissa of a shear stress versus normal stress plot, and a circle of diameter $\sigma_r - \sigma_\theta$ is drawn through the points (Figure 5). A failure envelope is constructed as a straight line that intersects the ordinate at the initial shear strength of the rock, τ_i, with slope β. The angle β represents internal friction angle and is usually taken as 30° for unconsolidated sand. Any stress outside the envelope will result in rock failure. By plotting the static stress state and then redrawing the Mohr circle to intersect the failure envelope, the maximum safe drawdown pressure can be determined (Figure 6).

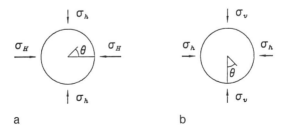

a   b

Figure 4.   Orientation of stresses around a borehole. (a) Vertical borehole in a formation with anisotropic horizontal stess. (b) Horizontal borehole in a formation with isotropic horizontal stress. (Reproduced with permission from reference 28. Copyright 1992 Elsevier.)

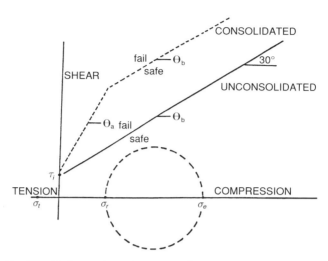

Figure 5.   The Mohr circle technique to determine stress regime. (Reproduced with permission from reference 30. Copyright 1981 Society of Professional Well Log Analysts.)

For unconsolidated sands, the failure and subsequent movement of the grains may lead to the formation of arches (Figure 7) that can stabilize the formation. Stability of the arches determines the degree to which solids production may occur from unconsolidated sands (27, 33–37). As long as stable arches are present, solids production may be relatively small. However, changes in operating conditions of the well (e.g., changing production rate) may result in collapse of the arches and solids production until new arches form (27).

Fluid Flow.   The major stress that tends to cause solids production during drilling or production operations usually results from fluid flow into or out of the reservoir. The fluid flow is proportional to the pressure difference between the reservoir and the wellbore. Whether the solids are water-wet or oil-wet may also affect their tendency to be produced. The drag exerted by the flowing fluid can separate individual grains and carry them into the wellbore. Drag forces increase with increasing pressure drawdown, fluid viscosity, and fluid flow rate. In horizontal wells, because of lower pressure gradient and fluid velocity around the wellbore, fluid drag forces on formation particles will be smaller, allowing higher solids-free fluid flow rates. Intergranular bonds and compressive stresses between the solid particles provide the major restraining forces against the fluid drag. Other restraining forces arise from intergranular friction, gravity forces, and capillary forces.

Solids production can be triggered by water breakthrough. This reduces the capillary forces between the grains, which then can be dis-

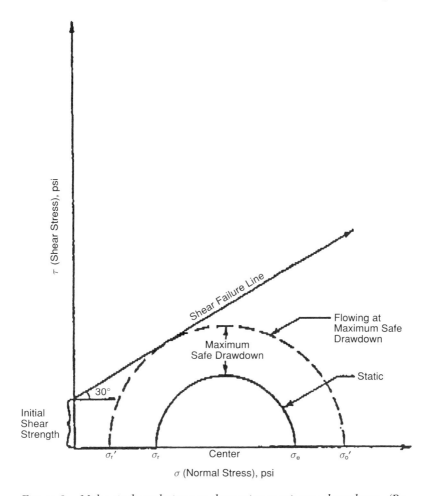

Figure 6.   Mohr circle technique to determine maximum drawdown. (Reproduced with permission from reference 32. Copyright 1989 Society of Petroleum Engineers.)

lodged and entrained in the fluid. The weakened, compacted zone around the perforation tunnel is especially susceptible to failure because of changes in drawdown and flow rate.

The compressive strength of a formation rock provides a measure of its solids-producing tendency. Formations with a compressive strength exceeding about 6900 kPa (1000 psi) will generally produce solid-free fluid, if good completion and production practices are observed. In moderately consolidated formations, the compressive strength is only about 690 kPa (100 psi) and solids production can be expected (*4*).

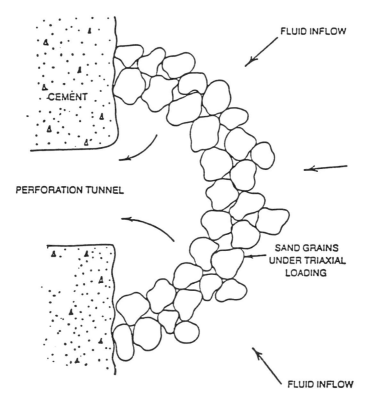

Figure 7.   Formation of sand arches stabilizes solids around the perforation. (Reproduced with permission from reference 5. Copyright 1991 Gulf Publishing Company.)

The influence of shear stress and pressure drawdown on solids production has been demonstrated in large-scale laboratory tests (38). Short bursts of solids, from perforation clean-up, and productivity improvement occurred after each increase in effective stress or drawdown.

Prediction of Solids Production.   In the development of conventional oil and gas reservoirs, solids management strategies are used to avoid solids production by proper production practices or to control or prevent solids production, if it is unavoidable. The use of solids control methods is usually based on experience with analogous offset wells in the area where a new well is to be drilled. Offset well data can be used effectively if the geological and reservoir characteristics and the well drilling, completion, and production procedures are similar. It is often difficult or expensive to recomplete wells if a solids production problem occurs after initial completion of the well. In some cases, productivity may be negatively impacted if solids control is retroactively installed in a well. For offshore wells in particular, it is more cost effective to implement solids control when the well is first drilled and completed. In other cases, developing new fields, for example, there may be no experience base from which solids production can be predicted (39). For this reason, operators have sought methods of predicting whether solids production is likely so that decisions regarding investment in solids control can be made based on some known risk factor.

During exploration and initial phases of development of a petroleum reservoir, indications of potential solids production problems can be obtained from such observations as

- missing portions of the cored interval or crumbly core, indicating partially consolidated or unconsolidated sands

- washouts, indicated by caliper logs or mechanical properties logs

- high drilling rate, which may indicate partially consolidated or unconsolidated sands

- solids production occurring during drill-stem tests or completion and production tests

- triggering solids production by increases in pressure drawdown or production rate or when water production begins

Prediction methods rely on correlations developed from field observations and experience, laboratory studies, or theoretical modeling. Methods based on field observations correlate solids production with some operating or reservoir parameter(s). This can be a single variable such as depth (39) or some combination of parameters. Tixier et al. (40) used the ratio of dynamic shear modulus to bulk compressibility, G/c_b, determined from mechanical property logs. Veeken et al. (24) used total drawdown pressure plotted versus sonic transit time to define a risk region (Figure 8). No solids production would be expected to occur below the risk region and no well likely could be produced without solids production above the risk region. More complicated correlations using multiple parameters have been applied (32). Multiparameter correlations require extensive data that often are not available. The correlations generally apply only to the geographical region or formation for which they were developed and thus have limited applicability to other areas.

Laboratory experiments allow investigation of solids production under controlled conditions. Hall and Harrisberger (33) and Bratli and Risnes (27) studied the stability of sand arches. More recently, Selby and Farouq Ali (36) investigated the effects of overburden pressure,

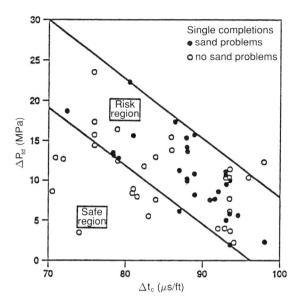

Figure 8.   Plot of drawdown vs. sonic transit time used to define a sand production "Risk region." (Reproduced with permission from reference 24. Copyright 1991 Society of Petroleum Engineers.)

flow rate, grain size and shape, and perforation size and shape on solids production. Radially loaded, thick-walled cylinders have been used in laboratory tests of compressive failure of sandstones (*24, 41, 42*). It has been observed that higher external loading is possible than would be expected from classic elastic–brittle failure theory. For example, Geertsma (*41*) provided data for a collection of sandstone samples from various locations around the world, showing that the ratio of actual failure pressure to yield pressure determined in thick-walled cylinder experiments varied from 1 to 16. For unconsolidated samples, solids production is dominated by flow rate and capillary forces, whereas boundary stress is a more important factor for friable and consolidated samples.

Theoretical methods of predicting solids production are often based on modeling the stability of the perforation cavity (*43–47*). These methods consider compressive failure, tensile failure, and erosion as mechanisms for failure of the solid matrix. The models thus need to consider the changing geometry of the perforation cavity over time. Perforation cavities can enlarge and join, leading to casing failure, or sloughing can occur in open hole completions.

Morita et al. (*45, 46*) used their models to predict transient and catastrophic solids production. They recommended avoiding rapid flow rate changes in low drawdown situations, keeping drawdown within a "critical cycle drawdown pressure." Stability is determined from its dependence on two parameters (Δp in the well and p of the formation). Morita et al. pointed out that solids problems are enhanced by water cut. This is not a new observation (48). However, Morita et al. explained the increase in solids cut at water breakthrough as due to the loss of capillary pressure holding the grains in place and an increase in drag forces after water breakthrough (this is because of the production practice of increasing total fluid production rate to maintain oil production rate after water breakthrough). Morita and Boyd (49) tested the Morita et al. models against field data. This necessitated the addition of "after water breakthrough" and "massive sand production" curves to the critical drawdown pressure diagram (Figure 9). Avoiding high drawdown in wells and decreasing the total flow after water breakthrough reduces the risk of solids production.

Weissenburger et al. (50) described an "engineering system" to predict solids production based on the Morita et al. model. It incorporates parameters such as oil type and bubble point pressure, to-

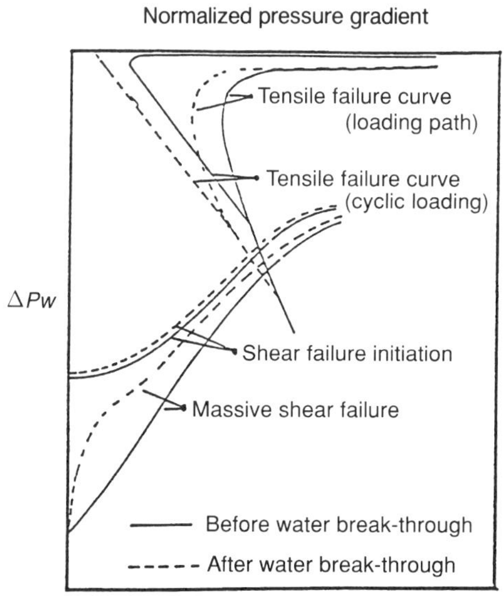

Figure 9.   Failure envelopes and critical maximum drawdown with influence of water breakthrough and massive shear failure. (Reproduced with permission from reference 49. Copyright 1991 Society of Petroleum Engineers.)

gether with geological aspects and rock properties. The "engineering system" has been used to evaluate alternative well-completion strategies (perforation size and phasing, operating limits to drawdown and flow rate, selective zone perforating, solids control method, and consolidation) based on geological description, log analysis, core testing, and simulation (51).

Veeken et al. (24) reviewed predictive models for solids production. Their assessment was that modeling of compressive failure is only qualitatively useful. This is because of the sensitivity of the results to the choice of yield envelope and failure criterion. Even so, this approach can be used to develop perforation strategy (density, phasing, and size), to select the stronger zones for perforation, and to provide guidelines for operation of a well (e.g., pressure drawdown, flow rate).

Modeling of tensile failure leads to the expression of a stability criterion in terms of the normalized drawdown pressure gradient, g_{pn}, at the cavity wall (24). For cylindrical geometry this is expressed as

$$g_{pn} \leq \sigma_\theta$$

However, the maximum tangential stress is limited by the unconfined compressive stress, σ_{ucs}, so that the criterion for stability can be expressed as:

$$g_{pn} \leq \sigma_{ucs}$$

Assuming uniform permeability and steady-state flow conditions, the stability criterion is expressed in terms of drawdown, Δp_{dd}, as

$$\Delta p_{dd} \leq \sigma_{ucs} \log (R_e/R)$$

where R is cavity radius and R_e is well drainage radius. Figure 10 (24) plots drawdown pressure against unconfined compressive strength. The line with slope of 0.5 provides a conservative prediction of solids production.

For the types of predictions described by Veeken et al. (24) and Morita et al. (45, 46) to be effective, formation strength, from actual measurements or log based correlations, is required.

Early modeling of wellbore stresses assumed that the rock behaved elastically (52). Bratli and Risnes (27) and Risnes et al. (53) included a plastic zone around the wellbore with variable permeability (Figure 11). In unconsolidated sands, the plastic zone is of the order of about 1 m radius; consolidated formations have a smaller plastic zone. For stress solutions with fluid flowing into an uncased wellbore (cylindrical geometry around the wellbore), a stability criterion that equates fluid flow parameters to rock strength parameters is found:

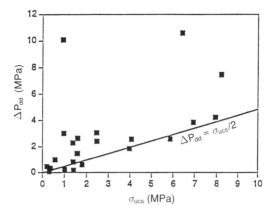

Figure 10. Use of plot of drawdown pressure against unconfined compressive strength to predict sand production. (Reproduced with permission from reference 24. Copyright 1991 Society of Petroleum Engineers.)

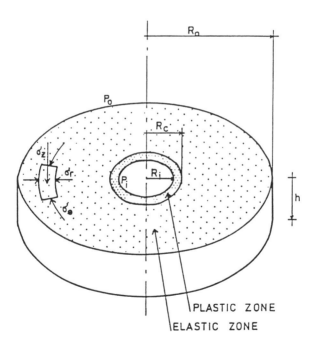

Figure 11. Prediction of solids production assuming a plastic zone around the wellbore. (Reproduced with permission from reference 53. Copyright 1982 Society of Petroleum Engineers.)

$$\frac{\mu q}{2\pi h k_c} = 2S_o \tan \alpha$$

This is equivalent to the criterion expressed by Veeken et al. (24), with the left side of the equation being pressure drawdown and the right side the unconfined compressive strength. In the case of homogeneous formation permeability, as flow rate increases, the radius of the plastic zone increases. When a certain critical flow rate is attained, the plastic zone grows through the whole formation, causing total formation collapse. In heterogeneous formations, with variable permeability in the plastic zone, thin inner concentric shells will collapse before total collapse.

In a cased and perforated well, only a slightly smaller plastic zone exists around the cased hole. In a perforated well in a poorly consolidated sand, the zone just around the wellbore is in a state of plastic stress, and greater flow rates can be attained without failure than will be possible in open hole situations. The production of solids will be governed by the stability of the sand arches behind the perforations. One of the important considerations in reducing solids production is to identify the poorly consolidated layers and avoid perforating them.

Vaziri (54) extended the Risnes model (53) by incorporating several features important to solution gas drive processes. Dissolved gas can come out of solution as the reservoir pressure is depleted below the oil "bubble point." Solution gas drive refers to oil production resulting from expansion of the gas phase. Vaziri assumed that liquid and gas form a single phase completely filling the pore space. Mechanical properties of the fluid (e.g., compressibility) vary with proportion of the gas phase and can be determined by application of Boyle's and Henry's laws. An expression for a fluid compressibility capacity, termed "fluid flexibility," of the following form is used:

$$\frac{1}{B_f} = \frac{(e_1 - S_1 e_1 + H S_1 e_1)^2}{e_0 (e_0 - S_0 e_0 + H S_0 e_0)(u_g^0 + p_a)} + \frac{S_1 e_1}{B_L e_0}$$

Vaziri also incorporated an expression for variation of permeability as a function of viscosity and changing porosity in the near wellbore region.

$$k = \frac{k_0 \, e_1^3}{\mu \, e_0^3} \frac{(1 + e_0)}{(1 + e_1)} S^3$$

The model couples the effects of stress and flow and accounts for varying permeability and compressibility of the pore fluid and nonlinear stress–strain behavior of the reservoir rock. The major factor responsible for fluid production is the compressibility attained by the pore fluid as a result of gas evolution. Transient flow rate is a function of the reservoir rock strength properties, the permeability, the fluid compressibility, and the pressure gradient near the well. Drilling the wellbore causes an enlarged cavity and a plastic zone. Instability around the wellbore occurs when a certain critical flow rate is exceeded. This flow rate is given as

$$q_c = \frac{4\pi Dkc' \tan \beta}{\gamma_f}$$

Santarelli et al. (42) developed a finite element simulator to model the near wellbore region of a perforated well. The risk of solids production, R, is defined as the ratio of the maximum effective stress around the perforation cavity (total stress less pore pressure), $\sigma_{max}{}^{*\prime}$, to the vertical in situ effective stress, σ_v'.

$$R = \frac{\sigma_{max}{}^{*\prime}}{\sigma_v'}$$

Santarelli et al. concluded that solids production risk is independent of perforation tunnel length and diameter. However, perforating in an underbalanced manner can cause perforation cavity enlargement, which will increase risk of solids production. Solids production risk also increases with perforation density because of mechanical interactions between perforations. In poorly cemented wells, there is an optimum perforation direction relative to the maximum horizontal stress that can reduce solids production risk to the same level as a perfectly cemented well.

Methods of Solids Control

Experience from many conventional oil and gas fields has shown that solids control should be installed before solids production occurs. It generally becomes more difficult to control further solids flow as the cumulative volume of produced solids increases. Consequently, in conventional oil and gas fields, initial solids control installations have proven to be consistently more successful than remedial treatments. Future enhancements will certainly improve the success of remedial treatments. The technology is continually advancing, and cost-effective through-tubing remedial methods are now available (55).

The completion of a well in an unconsolidated sandstone is more complicated than in a consolidated sandstone because of wellbore instability and the need to install solids control. Perforation geometry is an important consideration (3). At low fluid production rates, small quantities of solids may be produced, whereas at high fluid production rates, large quantities of solids may be carried in the production stream.

Solids production can be controlled by reducing fluid drag forces, by mechanically bridging the solids, or by increasing formation strength (1). Proper completion techniques and production practices can contribute to minimization of drag forces, whereas mechanical or chemical techniques address the bridging and formation strengthening approaches. The decision of whether to implement solids control is governed by both cost and risk factors. The initial cost of installation is matched with

- the risk that solids production will commence if no control is installed initially (with possible subsequent formation collapse)

- the cost of a subsequent remedial installation

- the risk that a remedial installation will fail or impair fluid production rates

Options for completing wells that provide physical restraint of solids movement include liners with holes or slots ("slotted liners"), prepacked screens, and gravel packing. These methods are based on sizing of the openings through which the fluids flow to promote formation of a stable particle bridge. The particle size distribution of the formation is used to select maximum screen-size opening or gravel size that will control the solids and permit an economic hydrocarbon fluid flow.

Formation consolidation using resins serves to increase the strength of the formation around the wellbore. This provides added resistance to fluid drag forces so that grains are not dislodged by the fluid at the desired production rates.

Slotted Liners or Prepacked Screens.   These devices are usually run on tubing and hung inside the casing opposite the solids-producing formation. This may be done in either an open or a cased hole. Slotted pipes, screens, and prepacked screens provide the lowest cost downhole filtering and are best suited to formations that are friable rather than completely unconsolidated. They are commonly used where reservoir permeability is greater than 1 μm^2 (1 darcy). Slotted liners and prepacked screens are used in only about 5% of solids-control completions (1). Opening sizes vary among these devices, with slotted liners having the largest openings and prepacked screens offering the finest filtering. Wire-wrapped screens have intermediate-sized openings.

Slotted liners or wire-wrapped screens can help prevent collapse of a horizontal hole, but they may be affected by plugging during placement if the horizontal section collapses while the screen is being run into the well. An alternative is to use prepacked liners or screens (56, 57). These can be screens to which gravel has been bonded or concentric screens or slotted liners packed with plastic consolidated gravel. Steel wool (58, 59) and permeable sintered steel (60) have been used in place of plastic consolidated gravel. Plugging can be minimized by using a solid-free filtered fluid system during placement. Removable screen coatings (acid or oil soluble or degraded by temperature) are sometimes used to prevent the assembly from plugging during placement in the well. The coating can then be removed by treatment with acid, oil, or solvent or the effect of bottom hole temperature before putting the well on production.

For applications in horizontal wells, screen elements are typically no longer than 6–10 m long so they can bend around the curved section of the well without difficulty. The screen–gravel wall thickness should be reasonably thin to provide a relatively large internal wellbore diameter and an adequate annular area for possible gravel packing. A cased hole completion, despite higher cost, can alleviate horizontal hole collapse during screen placement in the well. However, cementing of liners or casing in a weak sandstone formation can be problematic. Centralizers may become imbedded in the weak formation, and the displacement of drilling mud with spacer and preflush fluids ahead of the cement slurry may be inefficient, resulting in nonuniform placement of the cement slurry around the liner or casing. Cased hole completions allow selection of intervals for injection or production and more effective zone isolation. They also permit better handling of gas or water breakthrough in later stages of reservoir depletion. Prepacked liners have been used in horizontal wells in the Helder field (61) and in the Troll field with flow rates as high as 30,000 bbl/d (4770 m^3/d) (11).

Gravel Packing.   The technique of gravel packing was introduced in the water-well industry in the early 1900s, but it was not used by the petroleum industry until the 1930s. Being the oldest and most widely used solids control method, gravel packing accounts for approximately 75% of all solids control completions to date.

The normal gravel packing procedure is to place a screen in the casing or wellbore of an open hole completion opposite the production interval. Sized gravel is then pumped between the screen and the formation (Figure 12) using a viscous carrier fluid. Nonviscosified brine also may be used as a carrier fluid for the gravel (62, 63). The screen–slotted liner is located concentrically inside the layer of gravel to prevent gravel entry into the production tubing. The slots in the screen may be larger than for other methods and are generally only slightly smaller than the gravel. The formation grains bridge across the pores of the gravel pack, preventing formation solids from entering the well.

Gravel packs may be used in both open hole and the cased hole completions. In open hole gravel packs, there is no casing between the gravel pack and the formation. Open hole gravel pack completions pro-

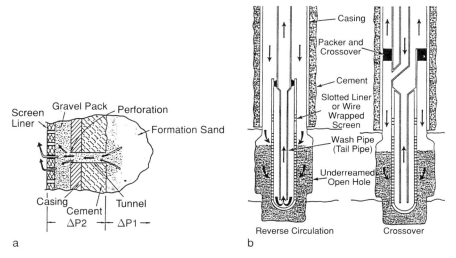

a   b

Figure 12.   (a) Schematic of a gravel pack completion. (Reproduced with permission from reference 138. Copyright 1982 Society of Petroleum Engineers–From Bell.) (b) Schematic of underreamed open hole gravel pack completion. (Reproduced with permission from reference 5. Copyright 1991 Gulf Publishing Company.)

vide high well productivity, but there are associated risks such as hole instability and inability to apply remedial techniques to exclude gas or water, which can break through to the well during reservoir depletion. The open hole gravel pack is the preferred mechanical control technique for high-productivity wells.

Gravel packing can give the highest well productivity in horizontal wells completed in weak formations compared with well completions using a slotted liner, wire-wrapped screen, or prepacked screen alone. However, there are difficulties in gravel packing a long undulating horizontal hole. These include

- the tendency of the hole to collapse during gravel packing operations
- the tendency of gravel to fall out of the carrying fluid in the horizontal wellbore annulus before the packing operation is completed
- reduction of the fluid velocity below the minimum necessary to transport the gravel, because of fluid losses into the formation
- the tendency of bow spring centralizers to embed into the formation rock, making the centralization of the screen in the undulating hole difficult

Because of these difficulties, only friable and consolidated sands that will not collapse before the gravel is put in place are amenable to gravel packing. However, during the life of the well, pressure decline and increased water production can cause collapse of the wellbore even in these relatively stronger sandstone formations.

The screens used may be either wire wrapped (for relatively shorter lengths) or prepacked (64, 65) and sealed with an acid or oil soluble material before installation. After installation, the sealing material is removed by treatment with acid or solvent. The gravel must be placed in the horizontal or highly deviated sections with an adequate gravel slurry, designed for the specific conditions of the application. Poor packing efficiency can result if the gravel drops out of the carrier fluid prematurely. Low-density gravel pack materials have been developed to avoid this difficulty (66). In very weak formations, a prepacked screen is recommended because it will provide solids control in zones not completely gravel packed.

Several operational aspects in gravel packing are important for the success of the application in horizontal wells (11). To reduce risk, the long horizontal sections can be gravel packed in several shorter intervals, without increasing operating time significantly. This technique has been applied in wells with open and cased hole completions with deviations up to 78° (11). It allows selective production, stimulation, or injection in different zones of the reservoir. The leak-off rate to the reservoir formation of the fluids (brine or intermediate viscosity gelled brine) used for carrying the gravel will determine the length of the zone that can be packed. Field and laboratory studies have shown that gravel is placed effectively by combining relatively low slurry viscosity and low gravel concentration with high pumping rate (65). Gravel can be placed with water if leak-off to the formation is controlled. Pump rate is selected based on borehole and screen geometry and fluid viscosity. If fluid leak-off is excessive, gravel concentration will increase, slurry velocity will decrease, and a bridge could form. Clearance is needed around the screen to allow normal gravel slurry transport. As an example, for gravel pack completions, the screen outside diameter should be about 7.5 cm less than the wellbore size. For screen or slotted liner completions without gravel packing, the screen should be about 3.75 cm less than wellbore size.

Zaleski (11) compared solids control alternatives in horizontal wells (Table II). Slotted liners are the most cost effective. Wire-wrapped screens can be two to three times more expensive than slotted liners for the same reservoir application. Gravel packing is adequate for unconsolidated reservoirs with high flow capacity where maximum performance is desired, as well as in formations with substantial clay content or fines. However, it can be 5–10 times more expensive than the other

Table II.   Comparison of Sand-Control Methods for Horizontal
and Highly Deviated Wells

Parameters Considered	Slotted Liner	Wire-Wrapped Screen	Prepacked Liner	Gravel Pack
Sand exclusion ability	4	3	2	1
Mechanical durability	3	4	2	1
Design flexibility	4	3	2	1
Long-term performance	4	3	2	1
Cost	1	2	3	4

NOTE: 1 is best and 4 is worst.
SOURCE: Reproduced with permission from reference 11. Copyright 1991 Society of Petroleum Engineers.

methods and is more suitable for larger high-flow capacity reservoirs. Open hole gravel packs are less expensive than cased hole gravel packs as a result of savings in casing, cementing, and perforating costs. However, cased hole completions provide more options for selective production and remedial operations.

Design of Screens and Gravel Packs.   Several authors describe design of screens and gravel packs in detail (4, 5, 67). The most important design consideration for screens and gravel packs is the proper sizing of liner and gravel pack openings relative to the size of the producing formation particles (5). This requires characterization of the formation solids. Sieve analysis provides particle size distribution (22). Other techniques that can contribute to formation solids characterization are scanning electron microscope analyses, X-ray diffraction, and thin-section analyses. These techniques provide evidence of the existence of grain cementation and the types and amounts of clay present. Some low porosity, low permeability sandstones often have clays and cementing materials that can be sensitive to completion, workover, or acid stimulation fluids. Coreflow tests using these fluids can identify potential problems, such as clay swelling, fines migration, plugging, scale precipitation, emulsion plugging, wettability reversal, or dissolution of cementing materials.

The solid samples can be obtained from full-size cores, sidewall cores, or perforation washing or bailing. Samples can also be separated from produced fluids. It is important to obtain representative samples. Sidewall cores typically are contaminated with the drilling fluid. Their smaller volume compared with a full-size core sample may restrict some laboratory analyses. A perforation washing or a bailed sample may be skewed toward large grain sizes. A produced sample may contain predominantly fines and be skewed toward small grain sizes. Sometimes, a "composite" sample may be prepared. This might be possible where solids production and sanding of the well occurs during production tests. The composite sample could consist of a mixture of produced samples collected before solids fill-up and bailed samples collected from different depths in the well after solids fill-up.

Figure 13 presents a particle size distribution curve obtained from a sieve analysis. In this figure, 10th percentile particle size, d_{10}, is defined as the point on the distribution scale where 10% by weight of the sample consists of larger particles and 90% by weight of the sample consists of smaller particles. Other significant points on the particle size distribution curve will be d_{40}, d_{50}, d_{70}, and d_{90}, which represent the particle sizes where the percent by weight of larger particles is 40%, 50%, 70%, and 90%, respectively. d_{50} represents the median diameter of the sample particles.

For formation solids characterization, a uniformity coefficient, C, given by the following expression, can be used (68):

$$C = \frac{d_{40}}{d_{90}}$$

If $C < 3$, the formation solids are uniform and relatively well sorted; for design purposes d_{10} can be used. If $C > 5$, the formation solids are nonuniform and relatively poorly sorted; for design purposes d_{40} can be used. If $C > 10$, the formation solids are very nonuniform; for design purposes d_{70} has been used (1).

The most often used design criterion (69) is to select median gravel size to be six times the median formation particle size, d_{50} (Figure 14). Others (70, 71) have recommended median gravel sizes ranging from 4 to 10 times d_{10}. Typical gravel sizes used for gravel packs and the average pore throat size of the gravel pack are given in Table III. Particle sizes for gravel pack materials are specified by U.S. mesh size range, for example, 20/40. Quality-control practices for selection of the gravel pack materials have been developed (72). Gravel pack design criteria based on gravel-formation particle ratio have been criticized (73) as an oversimplification of very complex phenomena related to solids transport (arch stability, hydrodynamic forces, filtration, and permeability reduction) that does not adequately consider the variation of formation particle size over the productive zone.

For formations with broad size distribution, Coberly (74) suggested that slot widths of slotted liners and wire-wrapped screens be two times d_{10}. For more uniform formations, slot size equal to the d_{10} has been recommended (75).

Resin Injection (Plastic Consolidation). Injection and solidification of resin binds the rock particles together, creating a stable matrix

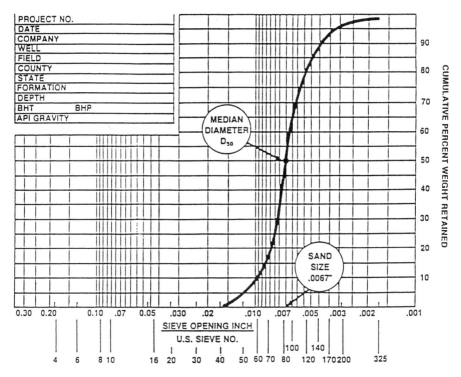

Figure 13. Typical particle size distribution. (Reproduced with permission from reference 131. Copyright 1989 Halliburton Energy Services.)

of permeable consolidated grains around the casing. The technique is applicable in small diameter casing and suitable for through-tubing application. Plastic consolidation methods can be used in abnormally high pressure wells and work effectively in formations with fine-grained sands difficult to control with a gravel packing system. The technique is also suitable for multiple producing zones. The quantity of injected resin is frequently a compromise between enhancing the consolidation strength and reducing the permeability.

In a typical treatment, a resin containing a catalyst is injected into the formation, generally through the perforations, to consolidate the sand grains in situ. The amount of catalyst determines the length of time it takes before the resin solidifies. The treatment fluid is followed by a postflush solution to maintain permeability of the formation in the vicinity of the wellbore. Phenolic, furan, or epoxy resins are commonly used. One disadvantage of resin consolidation solids control is that a notable reduction in permeability can result. Carlson et al. (6) reported that if a resin treatment is applied in an unconsolidated sandstone res-

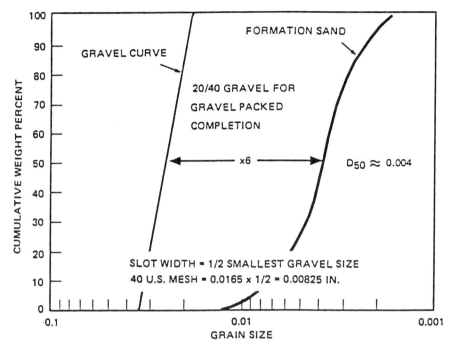

Figure 14. Example of a gravel pack design. (Reproduced with permission from reference 4. Copyright 1992 Society of Petroleum Engineers.)

Table III. Common Sizes of Gravel Packing Sand and Average Pore Throat Sizes

Gravel Size U.S. Mesh[a]	Median Grain Size[b]	Average Pore Throat[c]
12/20	1.257 (0.0495)	186 (0.0076)
16/20	0.889 (0.0350)	137 (0.0054)
20/40	0.635 (0.0250)	96 (0.0038)
30/50	0.445 (0.0175)	69 (0.0027)
40/60	0.340 (0.0134)	53 (0.0021)

[a] A minimum of 96% of the gravel sample should pass the coarse designated sieve and should be retained on the fine designate sieve. Not over 0.1% of the sample should be larger than the first sieve and not over 2% should be smaller than the last sieve (72).
[b] Values are in millimeters with inches in parentheses.
[c] Assumes normal particle size distribution, perfect spheres, and median grains = 6.5 times average pore throat. Values are in micrometers with inches in parentheses.

ervoir of 8 μm^2, the permeability may be reduced by 25% and the productivity by 10%.

To avoid some of the permeability reduction problems associated with injection of the resin mixture, Friedman et al. (76) used a catalyst that adsorbs onto the rock surface. Subsequent injection of the resin results in its polymerization in a thin layer around the rock grains, with minimum permeability reduction. This material is also suitable for use in thermal wells.

Thermal Well Completions.   An important recovery method for heavy oils from sandstone reservoirs involves the injection of steam into the reservoir to reduce the oil viscosity and stimulate oil flow. Cyclic steam stimulation is a typical application. It involves steam injection and subsequent production of oil and water through the same well after a "soak period." The process is repeated over several injection–production cycles. During this process, the reservoir rock surrounding the well is subjected to, and can be weakened by, thermal stresses. Consequently, the wells used for steam injection, "thermal wells," may have gravel pack completions. However, during steam injection, quartz gravel pack materials and formation cements can be dissolved by the high pH steam condensate (77, 78). The effectiveness of the gravel pack is thus reduced. The loss of silica cements can lead to the sandstone becoming unconsolidated and compound the problem.

Several materials have been proposed as alternatives to quartz gravel pack materials and have been tested for stability under steam injection conditions. Phenolic resin-coated gravel has been used in thermal wells (79). The coated grains agglomerate with temperature to form a consolidated pack that does not require a screen. Alumina-based materials have been reported to be stable in the high pH environment to over 300 °C (78), but others (80) found that dissolution of these materials can occur. Weaver and Knox (80) proposed an alternate material, "SRG," whose chemical formulation is not revealed. A steel wool filter has been used as an alternative to gravel packs in thermal wells (58, 59, 81, 82).

Duncan (83) reports high solids production during early production cycles of cyclic steam stimulation. Well completions designed to prevent inflow of silts and clays restricted fluid production. An operating strategy to produce the silts and clays by selecting screen size to control only the larger particles was adopted. Outside wire-wrapped screens on perforated casing were used. In this application, steam wells required solids control during the first two cycles, but screens were frequently removed after the second cycle.

Ammonium chloride and ammonium nitrate salts can be added to boiler feed water to reduce effluent pH and minimize silica dissolution (84).

Other Methods. Several other solids control methods have been used or proposed. An overbalanced resin surging method (85, 86) forces resin into perforations with an instantaneous release of overbalanced pressure. This method is used when a wellbore has existing perforations in an unconsolidated formation. Acid injection into the perforations converts the resin into a plastic that consolidates prepacked gravel or formation sand.

Hydraulic fracturing in conjunction with gravel packing (87–90) creates a short wide fracture that is packed with gravel (proppant) sized to prevent solids production. A high flow rate into the well can be maintained without solids production because fluid flows across the fracture face at rates below the critical solid production rate.

Perforation tunnel stability may be improved by orienting perforations in the direction of the principal horizontal stress (87).

Another technique to control solids inflow is to maintain a high level of grain-to-grain stress within the borehole wall. This technique was made possible by the practicality of long inflatable formation packers (5, 91).

Kantzas et al. (92) report laboratory studies that used a bacteria capable of precipitating calcium carbonate. Under appropriate conditions, the calcium carbonate can act as a cementing agent for unconsolidated sands. However, the process has drawbacks. In the laboratory tests, porosity was reduced by up to 50% and permeability was reduced by up to 90%.

Injection of sorted graded quartz sand into the formation zone adjacent to the perforations followed by placement of a screen with packer has been applied in Romania (93–95). The injected sand replaces the produced solids in the relaxed zone around the wellbore. In this way, the formation is repacked with a clean quartz sand, resulting in recompaction and restressing of the zone adjacent to the wellbore. This creates conditions comparable with those existing at the beginning of exploitation. The injection of the sorted quartz sand displaces some of the formation solids. The quartz sand, being uniformly sorted and clean, has higher porosity and permeability than the undisturbed formation. The velocity of fluids in the repacked zone is therefore lower, which reduces the entrainment of the formation particles. Also, the presence of the repacked zone impedes the collapse of upper zones, which could occur by producing large quantities of formation solids. Dislodging, sloughing, and collapse of interbedded shale layers (Figure 15), which would reduce the productivity of the well, is also prevented.

Hot air injection and cyclic in situ combustion have been used to consolidate sand (96). Oil in the near wellbore region undergoes cracking with the formation of coke, which serves to consolidate the sand grains.

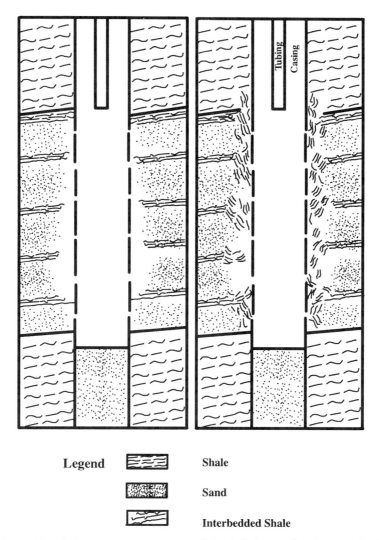

Legend   Shale

Sand

Interbedded Shale

Figure 15. Schematic representation of the dislodging, sloughing, and collapse of interbedded shale layers in the productive formation. (Reproduced with permission from reference 93. Copyright 1964.)

Cold Production of Heavy Oil from Unconsolidated Reservoirs

Efficient exploitation of some Canadian heavy oil reservoirs using vertical wells under primary conditions depends on producing the reservoir solids (25, 97–99). The reservoirs in question are generally unconsolidated and produce oil with viscosity in the range of 3000–20,000 mPa·s or more. Development of pumping systems that can effectively produce viscous oils containing large amounts of solids has contributed to the exploitation of these reservoirs.

Both oil production rates and overall oil recovery are much higher when the solids are produced into the wells and transported to the surface. Oil production rates have been reported (25, 98) to be more than 10 times the flow rate calculated using measured reservoir properties under pseudo-steady-state radial flow conditions ($10–20$ m^3/d compared with $1–2$ m^3/d). Initial sand cuts of 40–50% are not uncommon but decline over time and stabilize at lower values (2–4%). With low water cuts, solids cuts up to 60% can occur, and cumulative volume of solids produced from a single well can exceed 1000 m^3 (25). In conjunction with the increasing cumulative solids production, oil rates have been observed to increase over time (25). The carrying capacity of the viscous oil aids in producing the solids. With increasing water cut, the carrying capacity of the produced fluid drops and sanding of wells can occur. Solids control using gravel packs and screens often reduces oil production rate (98).

Operating practices that encourage the production of solids have been adopted. Perforation is carried out in an underbalanced condition using large diameter perforations. No solids control methods are implemented, and larger deeper sumps are drilled to provide maximum solids volume storage in the wells and minimize the need for cleanouts. After water breakthrough, total fluid rate is increased to maintain oil rate.

Even though the oils have a low gas-to-oil ratio, solution gas drive appears to be major factor (97, 100). Gas that comes out of solution is slow to coalesce because of the viscosity of the oil, and a "foamy oil" flow results. The nature of the foamy oil has been discussed (101), and it has been postulated that it is one of several factors contributing to the production of solids from these types of reservoirs.

Solids production from these heavy oil reservoirs was first discussed in some detail by Smith (97). Smith developed an analytical model to predict production, decline, recovery, pressure, and pressure-transient behavior, together with the large solids volume production and its effect on oil rate and well productivity. Smith's model incorporated time-dependent properties of the oil as a result of gas evolution and treated the unconsolidated reservoir sand as a soil in which cohesion relies only on the tension of the wetting phase. This is a similar, though simpler, approach than Vaziri's (54) finite element method. Smith developed a Darcy law formulation for compressible fluid flow

$$q_{sc} = 2\pi(kh/\mu_o)p_{sat}r\mathrm{d}p_{pD}/\mathrm{d}r$$

and a modified Horner pressure buildup relation for the compressible fluid situation

$$p_w^{\beta'+1}/p_{sat}^{\beta'} = p_i^{\beta'+1}/p_{sat}^{\beta'} - 921(\beta' + 1)(\mu/kh)q_{sc} \ln\left[(t_0 + \Delta t)/\Delta t\right]$$

Using the Horner relation, Smith demonstrated that the effective viscosity of the flowing fluid (foamy oil and solids) is much less than measured in the bulk single phase: 50–500 mPa·s compared with 1700–3500 mPa·s, for Lloydminster-type oils.

Smith evaluated field data and observed that lack of solids production resulted in low well productivity. One particular field test showed rapid tracer breakthrough in an offset well 30 m distant from the injection well. Smith concluded from this that the formation acts as "quicksand," just as the pressure is dropped below the current oil bubble point. Extensive solids production occurs at this time and will reoccur each time that a well is allowed to repressure and is pumped below its bubble point again. Smith's interpretation is based on the presence of loosely filled sand regions or channels with low fracture strengths. He postulated that the presence of gas as a finely dispersed bubble phase is a factor in the destabilization and flow of the solids. This reduces in situ stress and leaves the formation susceptible to fracture. Wong et al. (100) postulated that gas "ex-solution" creates a sharp pressure gradient front that destabilizes arches around the perforation cavity, allowing them to grow outward uniformly. Adamache (unpublished hypothesis, 1993) suggested that the foamy oil may be acting to transport the solids in a manner similar to froth flotation.

In the Celtic field (98), injectivity and minifrac tests confirm a low stress condition compared with initial conditions. Material balance considerations indicated that high levels of average gas saturation (>8%) were required to explain the oil recovery observed in the high production rate wells. Numerical simulation required an artificially high absolute permeability and high trapped gas saturation (35%) to history match the production satisfactorily. Results of independent laboratory studies of foamy oil flow confirm a high trapped gas saturation (101).

Production from the Celtic field was accompanied by high solids production, and oil production rate was a linear function of solids production rate (Figure 16). Proposed mechanisms include increased absolute permeability because of dilation of the reservoir sand, resulting from removal from the reservoir of substantial volumes of solids. McCaffery and Bowman (25) reported that during drilling through the Clearwater formation to lower intervals, circulation had been lost and both cement and lost circulation material appeared at offset wells some 100 m away. This suggests that the disturbed zone around the wellbore can extend significant distances in these reservoirs and corroborates the observations of the tracer test conducted by Smith (97). Tracer tests in the Clearwater formation (99) indicate a transit velocity between wells

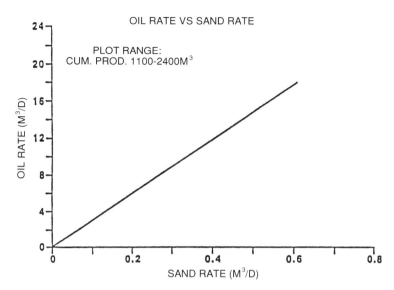

Figure 16.   Plot showing oil production rate as a function of sand production rate in the Celtic field. (Reproduced with permission from reference 98. Copyright 1992 Mobil Oil Canada.)

in excess of 7 m/min. This suggests that void conduits with little or no matrix material exist between the wells.

Proposed mechanisms of solids production from unconsolidated sand reservoirs have been discussed (102, 103). Dusseault and Santarelli (104) proposed a mechanism for massive solids production from poorly consolidated sandstones that was based on a general plastic yield of the reservoir brought about by a high pressure drawdown in the yielded region. The vertical stress that the reservoir experiences was also a contributing factor. Subsequently, Geilikman et al. (105–108) developed a model for continuous solids production from unconsolidated heavy oil reservoirs as a "yield front propagation." This is different from predictive models previously discussed (45, 46), which dealt with transient and catastrophic production but which did not discuss continuous production explicitly.

The Geilikman et al. model relates yield front velocity to the volumetric rate of solids production by mass balance. A front between yielded and intact zones moves radially away from the wellbore as solids production continues. Mass balance analysis allows determination of the position of the moving front. Fluid production enhancement is dependent on the instantaneous solids flux, as well as the radius of the yielded zone around the wellbore (i.e., the solids production history). Viscous resistance to oil flow is reduced when the granular matrix is also flowing. The volumetric fluid production, $Q_f(t)$, is given as

$$Q_f(t) = \frac{Q_0(t)\, \frac{k_y \phi_y}{k_i \phi_i} \ln \,(r_e/r_w) + \frac{q_s(t) \phi_y}{1 - \phi_y} \ln \,[R(t)/r_w]}{\ln \,[R(t)/r_w] + \frac{k_y \phi_y}{k_i \phi_i} \ln \,[r_e/R(t)]}$$

From this equation, enhancement of fluid production is a function of the current magnitude of solids production, $q_s(t)$, and the size of the yield zone, $R(t)$.

Geilikman et al. distinguished two regimes of fluid production enhancement. Short-term enhancement of the fluid production occurs in the earlier stages of production when solids flux is high and results from the reduction of viscous resistance to fluid flow because of simultaneous solid flow. The maximum short-term fluid production enhancement and maximum solids production rate do not coincide because both the instantaneous solids production rate and the position of the yield front determine the level of enhancement. These are opposing tendencies: solids production decreasing with time and yield zone radius increasing with time. Thus, a maximum in short-term enhancement will occur at some intermediate time. Long-term enhancement occurs in the later stages when the flux of solids production diminishes and results from the higher permeability of the enlarged yield zone around the wellbore.

The nature of the reservoir after such high solids production is still open to question. The possibilities include a radial relaxed zone around the well (105), wormholes or channels with loosely packed sand (97), or void conduits (99). Figure 17 depicts possible configurations of the reservoir as solids production proceeds.

Some unpublished evidence suggested that it is not necessary to produce the solids to have good oil productivity from reservoirs that are being exploited by horizontal wells. In early stages of production with horizontal wells, little solids production is observed. Slotted liners are often used to prevent collapse of the wellbore. Liners with slots or holes have been used in horizontal wells in the Pelican Lake area (12) and the Cactus Lake, North McLaren Pool (109). In the Cactus Lake project, solids production has not been observed nor have solids been found in any of the liners that have been reentered. Sametz (109) suggested that this is because the normal stress state has been preserved and that the drawdown, being distributed along the horizontal length of the well, is not sufficient to initiate solids production.

Pumping the High Solids Cut Fluids.   Although special pumping systems using plunger-type pumps have been used (110) or proposed (111, 112), the most effective pumping equipment for viscous high solids cut fluids has proven to be the progressive cavity pump (113) (Figure 18). Downhole parts rotate, eliminating rod fall problems that might

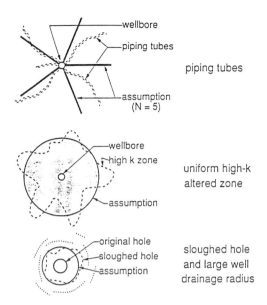

piping tubes

uniform high-k
altered zone

sloughed hole
and large well
drainage radius

Figure 17.   Schematic of radial and channel development of yielded zone in heavy oil unconsolidated reservoirs as a result of massive sand production. (Reproduced with permission from reference 102. Copyright 1993 Canadian Institute of Mining, Metallurgy, and Petroleum.)

occur with sucker rod beam pumping. Large diameter tubing is often used to reduce discharge pressure. Electric downhole progressive cavity pumps have been used in some horizontal wells.

The progressive cavity pump (PCP) uses a double helix stator and single helix rotor. A cavity exists between the rotor and stator that progresses up the pump barrel as the rotor is turned. The stator is usually moulded from a synthetic elastomer bonded to a steel tube and has a pitch length double that of the rotor. As the rotor turns, the fluid is forced to rise within the cavities as they progress from the bottom to the top of the pump (Figure 19). At 100% efficiency, an 18-stage PCP with 200 rpm operating speed and discharge volume of 200 ml per revolution can produce 56 m^3/d at a discharge pressure of 12,400 kPa.

The PCP is not a complete panacea for solids production from wells. Above 25% water cut, pumping problems because of "sanding-in" above the pump discharge have been observed (*114*). The pump suction intake can also be blocked, but this is a less common occurrence. Various solutions to the problem have been considered, including loading the annulus with produced oil (at the expense of increased bottom-hole pressure and some productivity impairment), cyclic loading of the tubing annulus, or reduction of hydraulic radius between pump rods and pro-

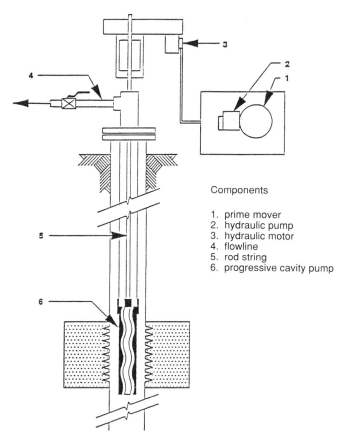

Components

1. prime mover
2. hydraulic pump
3. hydraulic motor
4. flowline
5. rod string
6. progressive cavity pump

Figure 18.   Progressive cavity pump system. (Reproduced with permission from reference 25. Copyright 1991 Amoco Canada Petroleum Ltd.)

duction tubing (increases pump discharge pressure causing premature pump failure).

Campbell (*114*) assessed settling rates of produced solids particles and determined a flow rate that would be able to keep the particles suspended and transport them to surface. A dual tubing string recirculation system was designed and is now used routinely. Produced water is recirculated down the well in one tubing string and back up with the produced fluids through a second tubing string. Fluid rates of more than 45 m^3/d are required when produced water cuts are in excess of 25%. This method has reduced lifting costs, well cleanout frequency, and pump wear.

Pumping with PCPs can be improved by "downhole emulsification" (*25, 115, 116*). Downhole emulsification uses surfactants to create oil-in-water emulsions of low viscosity. The benefits of downhole emulsi-

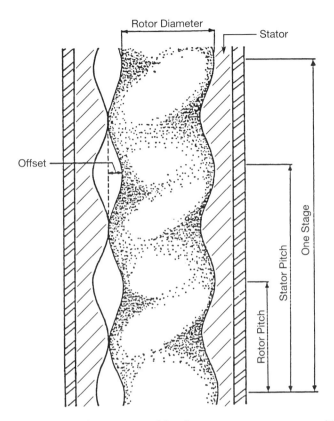

Figure 19.   Rotor and stator assembly of progressive cavity pump. (Reproduced with permission from reference 113. Copyright 1988 Canadian Institute of Mining, Metallurgy, and Petroleum.)

fication are related to reduced viscosity of the emulsion and include (1) the ability to use smaller equipment requiring lower torque, (2) lower pressures, (3) reduced power consumption, and (4) less pump wear. Sand cuts up to 16% have been produced with water-in-oil emulsions. A common practice is to create the emulsion on the surface using up to 3:1 produced crude-oil-to-water ratio. When applying this method of production, one needs to balance viscosity improvement (production increase, reduced equipment wear) with solids-carrying capacity.

One of the concerns regarding solids production during primary production from heavy oil reservoirs is the impact it may have on subsequent secondary or tertiary recovery. Chalaturnyk et al. (*103*) commented that subsequent thermal operations on wells that have produced a lot of solids will benefit from easier injection of steam. They thus maintain that cyclic steam processes and steam drive will be more manageable under such circumstances. However, both Chalaturnyk et al.

and Loughead and Saltuklaroglu (98) pointed out the difficulty of drilling in-fill wells because of lost circulation during drilling. Loughead and Saltuklaroglu suggested that the low stress state of the reservoir will not permit steam injection at pressures low enough to avoid fracturing.

Cleanout of Horizontal Wells.   We mentioned earlier solids dropout in horizontal wells because of low fluid velocity and low spots in the wellbore. These circumstances lead to the development of specialized wellbore cleanout technologies (117). A promising technique is the jet-pump–concentric coiled tubing system (118). Fluid is delivered through the inner portion of the coiled tubing through the jet pump. The deposited solids are fluidized by the jetting action of fluid, and the fluidized solids are taken up and transported to the surface through the outer section of the coiled tubing (Figure 20).

Sand Cleaning - Running In

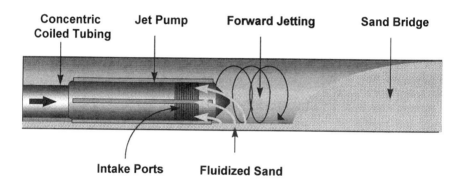

Sand Cleaning - Pulling Out

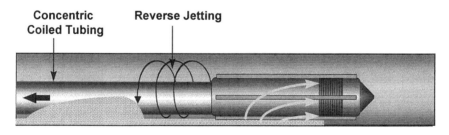

Figure 20. Jet-pump–concentric coiled tubing system for solids removal from horizontal wells. (Reproduced with permission from reference 118. Copyright 1993 Nowsco Well Service Ltd.)

Solids in Surface Production Facilities

Surface production facilities must be designed and operated to accommodate solids production. Produced solids collect in production equipment, separation equipment, and storage facilities in the regular routine of oilfield operations. The accumulated solids, which can total up to 20% of a vessel's volume during 1 year of operation, must eventually be removed before loss of equipment efficiency results. Before disposal of the removed materials, an effective process must be used to recover any crude oil bound to the solid material.

Storage Tanks.   Production storage tanks at centralized surface facilities, such as batteries and treatment plants, are often constructed with cone-shaped bottoms. Typically, such cone-shaped bottoms have a slope of 10–15°. (Steeper slopes, although more effective, require excessive portions of the overall tank volume.) These conical bottoms collect the solid material that settles over the course of daily operation. A ring of jets is frequently installed around the cone bottom to wash the accumulated solids into a "sand drain" at the bottom of the cone. The washing produces a slurry that is transferred through the drain to an underground storage tank. In the tank, the solids separate from the wash water through gravity settling and subsequently pumping the supernatant wash water out of the tank.

Production storage tanks located at individual wellsites generally do not have cone bottoms because draining is not a convenient procedure. Instead, the tank is cleaned on a regular basis by using a vacuum tank truck to remove the accumulated sediment from the tank floor. The vacuum truck is normally connected to a flow tee on a valve as close as possible to the bottom of the tank. High pressure water is fed through a "stinger" into the tank through the tee. The stinger is swept across the bottom of the tank and the fluidized solids are sucked into the vacuum truck.

A more traditional, but still common, method for removing the settled solids from a storage tank is to have field maintenance personnel enter the tank through an open manway at the bottom of the tank and dig out the accumulated solids using augers or shovels. It is wise to design and construct any tank that may be subject to solids accumulation with a manway for potential cleanout operations. Today, machines can eliminate the safety hazard to personnel entering storage tanks. Mechanized units are available that can be stationed at the manway door and scoop out the deposited solids.

Treating Vessels.   Horizontal heater treaters used in breaking produced oil–water emulsions use water jets and drains similar to those described for cone-bottom tanks to remove solids collected at the bottom of the vessel (*119*). Properly designed water jets (frequently called "sand jets") effectively fluidize the solids and flush the resulting slurry into the drain. In large vessels, where flow is exceptionally slow and laminar, the vessel may be divided by low baffles to create sections that can be cleaned individually.

Water-jet locations within the vessel must be carefully selected (*119*) to minimize erosion. The correct position and angle of the jet spray nozzles is important. Specifically, the following points have been emphasized (*120, 121*):

- In determining the clearing action of the jet, the most significant parameter is the liquid jet momentum.

- The jets are positioned approximately tangential to the vessel surface.

- Efficient solids removal is achieved if the nozzles are directed to sweep solids out of the vessel.

- The outer row of nozzles should be set fairly close to the vessel wall with a shallow impingement angle.

- Multiple tiers of spray nozzles are recommended for larger diameter vessels.

Nevertheless, caution must be taken in selecting an effective spray jet operating strategy and design. The violent action of the cleanout jets can develop turbulent flows sufficient to reform emulsions.

The free water knockout (FWKO) is a separation vessel used extensively in the oil industry to treat oilfield-produced emulsions. The FWKO separates the large volumes of free water associated with the produced fluid. Produced solids can collect in the FWKO. The use of water jets and drains is also an effective method for eliminating produced solids in the FWKO operation.

Flow Lines.   The removal of produced solids from flow lines connecting oilfield-treating equipment is dependent on the orientation of the specific length of flow line being considered. In vertical pipes, if the flow suddenly stops, the entrained particles simply settle to the bottom of the vertical section of the pipe. This settling results in plugging of the flow line. Consequently, vertical sections of production lines should be minimized. To accommodate any change in elevation, a gentle slope over a longer length of pipe would be advised.

To minimize the deposition of solids in horizontal lines, a steady flow rate with few interruptions is highly recommended. If the flow in a horizontal line is insufficient to keep the solids in suspension, they will settle and form a bed on the bottom of the pipe. As this bed thickens, the effective cross-sectional area of the line pipe is reduced, and the flow velocity is increased. Eventually, an equilibrium is established between the deposition and the reentrainment of solids at these elevated fluid velocities.

Solids Disposal Methods

The disposal of solids collected from heavy oil producing and treating operations has been an ongoing problem. In the past, the principal disposal method has been to spread the produced solids as an oil–produced solids–gravel mixture onto county roads. The use of the oily material in this way has certain advantages (*122*). The oily solids stabilize loose gravel or aggregate by filling the pore space. The oil not only controls dust but also, with time, binds the road surface. The sodium in the salt content also binds the road surface through the "hard pan" effect. The low level of volatile organics in heavy crude oil–solids mixtures has minimal environmental impact. Produced solids are generally not classed as a "dangerous oilfield waste" (*122, 123*).

Previously, municipalities and county boards allowed road spreading without cost to the oil companies (at times, even providing additional gravel). In effect, it had been a cheap paving and dust control method for these jurisdictions. Consequently, oil companies found it to be the least costly method of produced solids and "slop oil" disposal. However, there is concern with this method that heavy metals may leach into the surrounding groundwater.

In recent years, the increasing volumes of produced solids have made this option less feasible and practical. Cleaned solids are unsuitable for road surfacing material because of their low oil content.

Several solids disposal options can be considered as alternatives to road spreading: underground injection, landfill, land farming, and cleaning for resale. The principal constraints in using these methods are the large volumes of solids, inappropriate oil content (either too high or too low), high chlorides content, inconsistent quality of the mixture, environmental concerns (such as contamination of the surrounding groundwater by leachates), and the overall costs.

Costs of solids disposal by road spreading and underground injection have been reported (*124*) to be \$40/m^3 and \$36/m^3, respectively. A recent survey conducted by the authors indicated that current (1994) cost of solids treatment to remove oil and water is about \$50/m^3. Total disposal cost of the cleaned solids by landfill or land farming typically varies between \$110/m^3 and \$125/m^3, although costs as high as \$180/m^3 have been quoted.

Underground Injection. A cost-effective and environmentally safe method of solids disposal involves the injection of a slurry of solids

(20%) and water into an underground formation or cavern. A cavern suitable for underground solids disposal is created by drilling a well into a salt formation near to existing facilities. Salt is washed out of the formation using produced water from existing operations. The resultant saturated brine is then pumped into a water disposal well. Unlike natural gas storage caverns, where the cavern is developed to its final size before use as a storage facility, the development of a disposal cavern for solids will progress over the life of the project.

Disposal of produced solids by injection into a salt cavern (125) overcomes some environmental constraints. No preconditioning or further processing of the solids to remove oil and chlorides would be required, thus minimizing solids handling. However, solids injection into caverns does have the disadvantage that caverns are not conveniently located in all producing areas. Consequently, transportation costs to prospective cavern locations could be prohibitive for some producers.

The critical design parameters for the successful injection of solids are (124) as follows:

- injection pressure
- injection flow rate
- density of the solids/water slurry
- characteristics of the produced solids
- nature of the selected disposal formation

An underground formation for solids injection should be chosen as close as possible to the production area to minimize both operating and capital costs. The selected formation should also have the following characteristics:

- close proximity to existing facilities
- an uneconomic producing zone
- limited recompletion potential for production
- significant distance from offset production
- satisfactory isolation from producing zones

An alternative to injection of solids into a cavern is a technique termed slurry fracture injection (126–128). It involves mixing the solids into a slurry using fresh or produced water and injecting the slurry under high pressure into a suitable underground formation. At shallow depths, horizontal fractures are created that permit disposal of solids volumes of the order of tens of thousands of cubic meters. The carrying fluid bleeds off rapidly, leaving a body of solids entombed by the natural earth stresses. The shape and orientation of the injected body can be measured by analyzing surface displacement by tilt meter measurements (129).

Landfill. Industrial landfill sites have been used for "clean" solids disposal by oil producers. However, as with road disposal, the large volumes have also made this method less feasible. New locations have to be developed regularly, and extensive ongoing monitoring programs are required for existing sites. Transportation costs for the solids also increase as fill areas become more distant from producing areas.

Landfills are simply buried containers designed to hold solid wastes while at the same time preventing waste materials from having a negative environmental impact. Such precautions are necessary, especially if land reuse for alternate purposes is a key issue. Consequently, it is best to regard a landfill site as merely a temporary disposal facility. For example, if the contents of a landfill site present an environmental contamination problem, they may have to be either relocated or rendered environmentally neutral (i.e., having no residual biodegradable oil and no residual content of leachable material).

Solids that are environmentally inert can be placed into a simple pit without the need of a liner. A "wet" landfill is required if the solids contain any free liquid (i.e., liquid that can be readily drained or squeezed from the solids) or any leachable or putrescible materials. Oil, being biodegradable, is an example of a putrescible liquid; salt would be considered leachable. Wet landfills are located above the prevailing water table and must be isolated from it by either a synthetic or clay barrier. Once filled, wet landfills must be covered with the same low permeability material used for the barrier. A clay cap is especially good in preventing top soil from settling into the landfill material. However, a closed landfill may not allow biodegradation of any oil contaminant. Methane gas generated by the biodegradation of oil is often vented to atmosphere but could be collected for use as a fuel.

Guidelines for Landfill of Oily Wastes. Prevention of groundwater contamination is a primary objective in designing and establishing a landfill site. Regulatory approval for a landfill site will require that certain guidelines are met (*130*).

The following are generalized guidelines (*131*). Before proceeding to install any landfill site, it is mandatory that the appropriate regulatory agencies are consulted and proper approvals are in place:

1. *Site Restrictions.* The landfill (or treatment) site should be graded to prevent water accumulation or water runoff to water courses; located such that its bottom is well above the seasonal high of the water table; and located away from residential areas, wetlands, permafrost, critical wildlife habitats, drainage ways, flood plains, and, particularly, prime agricultural land.

2. **Operating Limitations.** The landfill (or treatment) site should be lined with a very low permeability clay or a synthetic liner to prevent migration of oil or leachates; covered to prevent any nuisance such as blowing sand; and designed so that any leachate, contaminated surface water, or groundwater can be contained and treated before release.

3. **Monitoring.** Groundwater and soil sample monitoring is required for the following parameters: before a landfill (or treatment) site is opened, during the site's operation, and for some time after its closure.

   a. groundwater
      - pH
      - major ions (e.g., Ca, Mg, Na, K, Cl, NO_3, SO_4)
      - heavy metals (e.g., Cd, Cr, Cu, Hg, Ni, Pb, Zn)
      - nitrogen (N)
      - TOC (total organic carbon)
      - TDS (total dissolved solids)
      - O&G (oil and grease)
      - COD (chemical oxygen demand)
      - electrical conductivity

   b. soil
      - classification (i.e., particle size determination)
      - pH
      - moisture content
      - major ions (e.g., Ca, Mg, Na, K, Cl, NO_3, SO_4)
      - heavy metals (e.g., Cd, Cr, Cu, Hg, Ni, Pb, Zn)
      - plant available nitrogen (N) and phosphorous (P) content
      - TOC (total organic carbon)
      - O&G (oil and grease)
      - electrical conductivity
      - cation ion exchange

Land Farming.   Land treatment is the most common method of oily solids disposal. The process is frequently referred to as "land farming," as the oily waste is spread thinly, worked into the soil, fertilized, and watered. Manure or straw is added to reduce erosion; lime addition reduces alkalinity. Commercially available strains of biodegrading bacteria may also be added to supplement the activity of naturally occurring strains. Because prevention of groundwater contamination is a serious concern, the same regulatory requirements applicable to landfill sites must be satisfied.

Practically all major oil-handling facilities incorporate plots of land for spreading oily wastes from spills and tanks and process vessel clean-outs. Careful design of the land farming site can restrict any potential contamination to the treatment site itself. Regular monitoring of the site and its immediate surroundings can reduce the possibility of groundwater contamination.

A variation to the land farming described above is to simply spread the oily solids and allow naturally occurring bacteria to biodegrade the material. Fresh water drilling mud is commonly disposed of in this manner. Free water is drained from the wet solids and any residual water is squeezed out. The resulting "paste" is spread thinly over the ground, not necessarily being worked into the soil. As in land farming, fertilizers or manure may be added to encourage biodegradation. The treatment area is used only once, not repeatedly. The effect on the soil is expected to be equivalent to a minor saltwater spill. The land is subsequently expected to recover the next year.

Cleaning for Resale or Disposal.   The solids, cleaned of oil and salt, may be marketed for construction or other industrial use. For example, a clean round-grained sand may be recycled as a frac or gravel pack material. Sand blasting and the manufacture of hard coatings, abrasives, and glass are other end uses for the cleaned solids. Frequently, the major obstacles to resale of the cleaned solids are the identification of prospective markets and the justification of the treatment and transportation costs to the client.

To remove the principal environmental contaminant, namely oil, three different techniques are commonly used: hot water wash (*132*), solvent extraction (*132*), and incineration and retorting. Simply storing the oily solids in a tank full of water will result in some separation of oil. For more effective removal of the oil, the solids are soaked in hot water. The hot water soaking is occasionally enhanced by circulating the water through jets installed in the bottom of the storage tank. Light hydrocarbons (i.e., solvents, diluents, diesel) can be used to dilute and soak heavier oils from the individual grains. The recovered liquid mixture is distilled and the solvent is recycled. Lighter solvents, such as propane, are used under pressurized conditions; the solvent is subsequently recovered through application of a vacuum. The third option, incineration and retorting, does not recover the oil. Oil can be removed by combustion in an incinerator; in retorting, the heat of combustion is used to recover the lighter oil fractions and the heavier components are coked and consumed as fuel. For both processes the solid is dried. Any residual salt content is recovered as either a leachate or as dust in the flue gas. To remove any salt still contained in the solids, the solids can be washed with fresh water. If fresh water is scarce, the solids may be washed in two stages: first with rinse water and then with fresh water. This fresh water subsequently becomes the rinse water for the next batch.

Toor (133) described an oil, water, and solids separator in which the solids are agitated in a solution of nonionic biodegradable detergent to emulsify the oil. Several such washing stages occur. The emulsion is separated from the solids in a hydrocyclone. The hydrocarbon content of the cleaned solids is of the order of 100 ppm.

Jamaluddin et al. (134) proposed flotation separation as a method to clean produced solids. In batch tests they successively reduced oil and salt content by more than 99%.

In offshore operations, it is desirable to discharge the produced solids overboard. Schlittler (135) described a batch process system that removed free oil from produced solids by means of a scrubber system fitted with rotating blades. The energy imparted by the blades promotes particle contact and shears the oil from the solid surfaces. The mixture of oil, solid, and water is separated by gravity in a wash tank. The cleaned solids are reported to contain less than 1 ppm oil, meeting or bettering standards for overboard discharge.

Regulatory.   Degradation of the soil in agricultural land, even on a temporary basis, may not be permitted. Various jurisdictions have established regulations or enacted legislation to prevent erosion and salt damage of fertile soil. Disposal of oily waste material onto roads has been approved by some jurisdictions (136) but not as a long-term disposal technique. Government regulatory authorities also established limits on oil and salt content for oil waste disposal (136, 137) and encourage petroleum operators to develop and use alternative techniques. In Alberta, for example, oily waste material and fluids may be applied to road surfaces if the following characterization criteria can be met and alternate practical technologies are not readily available:

1. No free water.

2. The hydrocarbon content must be $\geq 5\%$ as a calculated mix. The oil should be of a relatively high density (i.e., 900 kg/m^3 or greater).

3. A pH ≥ 6.

4. Total metals loading rate: Cd ≤ 1.6 kg/ha, Hg ≤ 1.0 kg/ha, Pb ≤ 200 kg/ha, Ni ≤ 100 kg/ha, V ≤ 150 kg/ha, B ≤ 20 kg/ha.

5. Other data required in the analysis but that are not part of the criteria are specific conductance and density of sample.

6. The material must not contain halogenated hydrocarbons, hazardous chemicals, refined or lube oils, or de-

terious substances such as filters, rags, vegetation, or other debris.

7. The maximum depth of application of material is limited to the smallest of the "Calculated Application Depths" determined for all parameters using the following formula:

$$d = 10,000L/[(D')(C)]$$

where d is the application depth in cm, L is the loading rate in kg/ha (*see* earlier), D' is the sample density in kg/m^3, and C is the concentration in mg/kg.

The following guidelines for the disposal of drilling mud for land spreading have been used (*131*):

- pH: neutral or $\geq 5.5 < $ pH $ < 8.5$
- electrical conductivity: <7.00 mmhos/cm at 25 °C
- Cl: 1000 mg/L (maximum)
- SO$_4$: 2000 mg/L (maximum)
- TDS (total dissolved solids): 4000 mg/L (maximum)

It should be noted that a regulatory agency may choose not to accept these guidelines when applied to solids disposal.

Summary

Solids production and control remain a high priority for the petroleum industry. In conventional oil and gas operations, control of produced solids is a primary focus. In recovery of heavy oil from unconsolidated reservoirs, production and handling of large quantities of produced solids present special challenges, both in the removal of solids from the wellbore and their subsequent disposal. The rapid expansion of horizontal well applications has led to some unique approaches to solids handling and is likely to lead to new technical advances as the horizontal drilling and completions technology continues to advance.

List of Symbols

B_f	bulk modulus of the fluid
B_L	liquid bulk modulus
C	uniformity coefficient used to characterize formation particle size distribution
c	uniformity coefficient
c'	cohesion intercept
c_b	bulk compressibility
C	concentration

d	application depth
D	thickness of the layer
D'	sample density
d_x	point on the particle size distribution scale that represents grain sizes in which $x\%$ cumulative by weight of the sample consists of larger particles
D_{50}	mean diameter (of a distribution)
e_0	initial void ratio
e_1	void ratio after a change in pressure
g_{pn}	normalized drawdown pressure gradient
G	dynamic shear modulus
h	height of producing layer, reservoir thickness
H	Henry's solubility constant
k	permeability
k_c	permeability in the plastic zone surrounding a well
k_i	permeability of the intact zone
k_o	oil permeability under fully saturated conditions
k_y	permeability of the yielded zone
L	loading rate
p	pressure
p_a	atmospheric pressure
p_f	fluid pressure in rock pores
p_{pD}	heavy-oil pseudopressure
p_i	initial pressure
p_{sat}	pressure at saturation
p_w	pressure at the well sandface
P_i	fluid pressure at inner boundary
P_{ld}	total drawdown (*see* Figure 8)
P_o	fluid pressure at outer boundary
q	fluid flow rate
q_c	critical flow rate
q_s	volumetric rate of solids production
q_{sc}	volumetric rate at standard conditions
Q_f	volumetric rate of fluid production
Q_o	fluid production rate with no solids production
r	radial position
r_e	radius of the reservoir
r_w	radius of the wellbore
R	cavity radius
R	risk of solids production
$R(t)$	radius of the yielded zone
R_c	radius of plastic zone
R_e	drainage radius
R_i	wellbore radius
R_o	outer boundary radius

S_o	inherent shear strength, cohesive strength
S_0	initial degree of saturation
S_1	degree of saturation after a change in pressure
t	time
t_c	transit time
t_0	initial time

Greek symbols

α	failure angle
β	internal friction angle
β'	compressibility constant
γ_f	fluid unit weight
Δ	difference
Δp_{dd}	drawdown pressure criterion
ΔP_{td}	total drawdown
Δt	time increment
θ	orientation of tangential stress relative to principal horizontal stress
μ	viscosity
μ_o	oil viscosity
ρ_w	not defined
σ_h	minimum principal horizontal stress
σ_H	maximum principal horizontal stress
σ_n	one of three orthogonal principal stresses ($n = 1, 2, 3$)
$\sigma_{max}*'$	maximum effective stress around the perforation cavity
σ_r	radial stress at wellbore wall (radial coordinates)
σ_{ucs}	unconfined compressive strength
σ_v	principal vertical stress
σ_v'	vertical in situ effective stress
σ_z	axial stress at wellbore wall (radial coordinates)
σ_θ	tangential stress at wellbore wall (radial coordinates)
τ_i	initial shear strength of rock
ϕ	negative logarithm to the base 2 of particle size in mm
ϕ_i	porosity of the intact zone
ϕ_y	porosity of the yielded zone

References

1. Allen, T. O.; Roberts, A. P. *Production Operations*, 2nd ed.; Oil & Gas Consultants International, Inc.: Tulsa, OK, 1982.
2. Berger, W. D.; Anderson, K. E. *Modern Petroleum—A Basic Primer of the Industry*, 3rd ed.; PennWell Publishing Company: Tulsa, OK, 1992.
3. Cosad, C. *Oilfield Rev.* **1992**, *4*, 54–69.
4. Penberthy, W. L. Jr.; Shaughnessy, C. M. *Sand Control*; SPE Series on Special Topics, Henry L. Doherty Series; Society of Petroleum Engineers: Richardson, TX; 1992; Vol. 1.

5. Suman, G. O. Jr.; Ellis, R. C.; Snyder, R. E. *Sand Control Handbook*; 2nd ed.; Gulf Publishing Company: Houston, TX; 1991.
6. Carlson, J.; Gurley, D.; King, G.; Price-Smith, C.; Waters, F. *Oilfield Rev.* **1992**, *4*, 41–53.
7. Muskat, M. In *Flow of Homogeneous Fluids*; McGraw Hill: New York, 1937; reprinted by International Human Resources Development Corporation: Boston, MA, 1982; p 439.
8. Muskat, M. In *Physical Principles of Oil Production*; McGraw Hill: New York, 1959; reprinted by International Human Resources Development Corporation: Boston, MA, 1981; p 258.
9. Butler, R. M. *Horizontal Wells for the Recovery of Oil, Gas and Bitumen*; Monograph No. 2; Petroleum Society of Canadian Institute of Mining, Metallurgy, and Petroleum: Calgary, Canada, 1994.
10. Joshi, S. D. *Horizontal Well Technology*; Pennwell: Tulsa, OK, 1991.
11. Zaleski, T. E., Jr. *J. Pet. Technol.* **1991**, *43*, 509–511.
12. Fontaine, T.; Hayes, L.; Reese, G. *Proceedings of the 9th Annual Heavy Oil and Oil Sands Technical Symposium*; Calgary, Canada, March 11, 1992.
13. Smith, R. C.; Hayes, L. A.; Wilkin, J. F. *Proceedings of the IADC/SPE Drilling Conference*; Dallas, TX, February 15–18, 1994; paper IADC/SPE 27436.
14. Cotton, D. *Proceedings of the SPE/CIM 3rd Annual One Day Conference on Horizontal Wells*; Calgary, Canada, November 15, 1993.
15. Gillies, R.; Shook, C.; Kristoff, B.; Parker, P. *Proceedings of the 11th Annual Heavy Oil and Oil Sands Technical Symposium*; Calgary, Canada, March 2, 1994.
16. Doan, Q.; Farouq Ali, S. M.; George, A. E. *Proceedings of the Canadian SPE/CIM/CANMET International Conference on Advances in Horizontal Well Applications*; Calgary, Canada, March 20–23, 1994; paper HWC94–31.
17. Sparlin, D. D.; Hagen, R. W. In *World Oil's Handbook of Horizontal Drilling and Completion Technology*; Gulf Publishing Company: Houston, TX, 1991; pp 59–63.
18. Sparlin, D. D.; Hagen, R. W. *Selection and Design of Sand Control Methods*; Short Course; Society of Petroleum Engineers: Richardson, TX, 1993.
19. McLellan, P. *Proceedings of the Canadian SPE/CIM/CANMET International Conference on Recent Advances in Horizontal Well Applications*; Calgary, Canada, March 20–23, 1994; paper HWC94–14.
20. Bell, S. *Pet. Eng. Int.* **1993**, *65*, 17–23.
21. Schramm, L. L. In *Emulsions: Fundamentals and Applications in the Petroleum Industry*; Schramm, L. L., Ed.; ACS Advances in Chemistry 231; American Chemical Society: Washington, DC, 1992; pp 1–49.
22. Hayes, J. R. In *Subsurface Geology*; LeRoy L. W.; LeRoy, D. O.; Rease, J. W., Eds.; Colorado School of Mines: Golden, CO, 1977; pp 61–74.
23. Anderson, R.; Coates, G.; Denoo, S.; Edwards D.; Risnes, R. *Tech. Rev.* **1986**, *34*, 29–32.
24. Veeken, C. A. M.; Davies, D. R.; Kenter, C. J.; Kooijman, A. P. *Proceedings of the SPE Annual Technical Conference and Exhibition*; Dallas, TX, October 6–9, 1991; paper SPE 22792.
25. McCaffery, W. J.; Bowman, R. D. *Proceedings of the Eighth Annual Heavy Oil and Oil Sands Technical Symposium*; Calgary, Canada, March 14, 1991.
26. Jaeger, J. C.; Cook, N. G. W. *Fundamentals of Rock Mechanics*, 3rd ed.; Chapman and Hall: London, 1979.

27. Bratli, R. K.; Risnes, R. *Soc. Pet. Eng. J.* **1981**, *21*, 236–248.
28. Fjaer, E.; Holt, R. M.; Horsrud, P.; Raaen, A. M.; Risnes, R. *Petroleum Related Rock Mechanics;* Elsevier: Amsterdam, Netherlands, 1992.
29. Pearson, R. M. In *Sand Control;* IHRDC Video Library for Exploration and Production Specialists, Module PE306 Workbook; International Human Resources Development Corporation: Boston, MA, 1988; pp 16–18.
30. Coates, G. R.; Denoo, S. A. *Transactions of the SPWLA 22nd Annual Logging Symposium;* Society of Professional Well Log Analysts: Houston, TX, 1981.
31. Barrow, D. C.; Lasseigne, C. A. *Proceedings of the SPE Annual Technical Conference and Exhibition;* Society of Petroleum Engineers: Richardson, TX, 1984; paper SPE 13087.
32. Ghalambor, A.; Hayatdavoudi, A.; Alcocer, C. F.; Koliba, R. J. *J. Pet. Technol.* **1989**, *41*, 1336–1343.
33. Hall, C. D. Jr.; Harrisberger, W. H. *J. Pet. Technol.* **1970**, *22*, 821–829.
34. Tippie, D. B.; Kohlhaas, C. A. *Proceedings of the SPE Annual Technical Meeting;* Society of Petroleum Engineers: Richardson, TX, 1973; paper SPE 4533.
35. Cleary, M. P.; Melvan, J. J.; Kohlhaas, C. A. *Proceedings of the SPE Annual Technical Conference and Exhibition;* Society of Petroleum Engineers: Richardson, TX, 1979; paper SPE 8426.
36. Selby, R. J.; Farouq Ali, S. M. *J. Can. Pet. Technol.* **1988**, *27(3)*, 55–63.
37. Yim, K.; Dusseault, M. B.; Zhang, L. *Proceedings of Eurock '94, Rock Mechanics in Petroleum Engineering;* A. A. Balkema: Rotterdam, Netherlands, 1994; pp 339–346.
38. Kooijman, A. P.; Halleck, P. M.; Veeken, C. A. M.; de Bree, P.; Kenter, C. J. *Proceedings of the SPE Annual Technical Conference and Exhibition;* Society of Petroleum Engineers: Richardson, TX, 1992; paper SPE 24798.
39. Lantz, J. R.; Ali, N. *J. Pet. Technol.* **1991**, *43*, 392–454.
40. Tixier, M. P.; Loveless, G. W.; Anderson, R. A. *J. Pet. Technol.* **1975**, *27*, 283–293.
41. Geertsma, J. *Soc. Pet. Eng. J.* **1985**, *25*, 848–856.
42. Santarelli, F. J.; Ouadfel, H.; Zundel, J. P. *Proceedings of the SPE Annual Technical Conference and Exhibition;* Society of Petroleum Engineers: Richardson, TX, 1991; paper SPE 22797.
43. Peden, J. M.; Yassin, A. A. M. *Proceedings of the SPE Annual Technical Conference and Exhibition;* Society of Petroleum Engineers: Richardson, TX, 1986; paper SPE 15406.
44. Tronvoll, J.; Morita, N.; Santarelli, F. J. *Proceedings of the SPE Annual Technical Conference and Exhibition;* Society of Petroleum Engineers: Richardson, TX, 1992; paper SPE 24799.
45. Morita, N.; Whitfill, D. L.; Massie, I.; Knudsen, T. W. *SPE Prod. Eng.* **1989**, *4*, 15–24.
46. Morita, N.; Whitfill, D. L.; Fedde, O. P.; Løvik, T. H. *SPE Prod. Eng.* **1989**, *4*, 25–33.
47. Morita, N. *SPE Drilling & Completion* **1994**, *9*, 227–235.
48. Lupu, P.; Adamache, I. *Pet. Gaze* **1954**, *4*, 139–146; 5, 199–211; 6, 233–242 (in Romanian).
49. Morita, N.; Boyd, P. A. *Proceedings of the SPE Annual Technical Conference and Exhibition;* Society of Petroleum Engineers: Richardson, TX, 1991; paper SPE 22739.
50. Weissenburger, K. W.; Morita, N.; Martin, A. J.; Whitfill, D. L. *Proceedings of the SPE Annual Technical Conference and Exhibition;* Society of Petroleum Engineers: Richardson, TX, 1987; paper SPE 16892.

51. Massie, I.; Nygaard, O.; Morita, N. *Proceedings of the SPE Annual Technical Conference and Exhibition;* Society of Petroleum Engineers: Richardson, TX, 1987; paper SPE 16893.
52. Paslay, P. R.; Cheatham, J. B. Jr. *Soc. Pet. Eng. J.* 1963, *3*, 85–94.
53. Risnes, R.; Bratli, R. K.; Horsrud, P. *Soc. Pet. Eng. J.* 1982, *22*, 883–898.
54. Vaziri, H. H. *Proceedings of the Petroleum Society of CIM Annual Technical Meeting;* Canadian Institute of Mining, Metallurgy, and Petroleum: Montreal, Canada, 1986; paper 86–37–75.
55. Restarick H. L., Jr.; Fowler, S. H., Jr.; Sedotal, W. P. *SPE Drilling & Completion* 1994, *9*, 236–243.
56. Escobar, J. A.; Turner, W. H. *Proceedings of the Latin American Petroleum Engineering Conference, II LAPEC;* Society of Petroleum Engineers: Richardson, TX, 1992; paper SPE 23642.
57. Restarick, H. L. *Proceedings of the SPE Middle East Oil Technical Conference and Exhibition;* Society of Petroleum Engineers: Richardson, TX, 1993; paper SPE 25547.
58. Toma, P.; Livesey, D. B.; Heidrick, T. R. *SPE Prod. Eng.* 1988, *3*, 249–257.
59. Toma, P.; Korpany, G.; King, R. W. *J. Can. Pet. Technol.* 1994, *30(4)*, 78–88.
60. Cole, R. C.; Arterbury, A. S.; Pearce, J. L. *Proceedings of the Tenth Symposium on Formation Damage Control;* Society of Petroleum Engineers: Richardson, TX, 1992; paper SPE 23770.
61. Murphy, P. J. *J. Pet. Technol.* 1990, *42*, 792–800.
62. Ledlow, L. B.; Johnson, M. H.; Richard, B. M.; Huval, T. J. *Proceedings of the SPE Annual Technical Conference and Exhibition;* Society of Petroleum Engineers: Richardson, TX, 1993; paper SPE 26543.
63. Johnson, M. H.; Montagna, J. N.; Richard, B. M. *Proceedings of the SPE International Symposium on Formation Damage Control;* Society of Petroleum Engineers: Richardson, TX, 1992; paper SPE 23774.
64. Harrison, D. J.; Johnson, M. H.; Richard, B. *Proceedings of the SPE California Regional Meeting;* Society of Petroleum Engineers: Richardson, TX, 1990; paper SPE 20027.
65. Ashton, J. P.; Liput, J.; Lemons, R.; Summerlin, J. *Proceedings of the SPE Annual Technical Conference and Exhibition;* Society of Petroleum Engineers: Richardson, TX, 1989; paper SPE 19718.
66. Price-Smith, C. *Offshore Inc. Oilmen (Int. Ed.)* 1993, *53*, 40–43.
67. *Sand Control;* Oil & Gas Consultants International, Inc.: Tulsa, OK, 1979.
68. Schwartz, D. H. *J. Pet. Technol.* 1969, *21*, 1193–98.
69. Saucier, R. J. *J. Pet. Technol.* 1974, *26*, 205–12.
70. Coberly, C. J.; Wagner, E. M. *Petroleum Technology;* AIME Technical Publication No. 960; American Institute of Mining, Metallurgical and Petroleum Engineers: New York, 1938; pp 1–20.
71. Tauch, G. H.; Corley, C. B. Jr. *Drilling Prod. Pract.*, 1958, *August*, 66–82.
72. *Recommended Practices for Testing Sand Used in Gravel Packing Operations*, 1st ed.; API Recommended Practice 58 (RP58); American Petroleum Institute: Dallas, TX, 1986.
73. Bouhroum, A.; Civan, F. *J. Can. Pet. Technol.* 1995, *34(1)*, 35–40.
74. Coberly, C. J. *Drilling Prod. Pract.* 1937, 189–201.
75. Rogers, E.B., Jr. *Oil & Gas J.* 1971.
76. Friedman, R. H.; Surles, B. W.; Kleke, D. E. *SPE Prod. Eng.* 1988, *3*, 167–168.

77. Reed, M. G. *J. Pet. Technol.* **1980**, *32*, 941–949.
78. Underdown, D. R.; Das, K. *J. Pet. Technol.* **1985**, *37*, 2006–2012.
79. Saunders, L. W.; McKinzie, H. L. *J. Pet. Technol.* **1981**, *33*, 221–228.
80. Weaver, J. D.; Knox, J. A. *J. Pet. Technol.* **1992**, *7*, 155–159.
81. Livesey, D. V.; Toma, P. U.S. Patent 4 434 054, 1984.
82. Harris, P.; Toma, P.; Rabeeh, S.; King, R. W. *J. Can. Pet. Technol.* **1991**, *30(4)*, 62–68.
83. Duncan, G. *J. Can. Pet. Technol.* **1984**, *23(2)*, 71–72.
84. Watkins, D. R.; Kalfayan, L. J.; Watanabe, D. J.; Holm, J. A. *J. Pet. Technol.* **1986**, *1*, 471–477.
85. Dees, J. M.; Handren, P. J. *J. Pet. Technol.* **1994**, *46*, 431–435.
86. Dees, J. M. *Proceedings of the SPE Annual Technical Conference and Exhibition;* Society of Petroleum Engineers: Richardson, TX, 1992; paper SPE 24841.
87. Abass, H. H.; Wilson, J. M.; Venditto, J. J.; Voss, R. E. *Proceedings of the SPE Production Operations Symposium;* Society of Petroleum Engineers: Richardson, TX, 1993; paper SPE 25494.
88. Hannah, R. R.; Pack, E. I.; Porter, D. A.; Black, J. W. *SPE Prod. & Facil.* **1994**, *9*, 262–266.
89. Sparlin, D. D.; Hagen, R. W. *World Oil* **1995**, *216*, 37–40.
90. Mathis, S. *Enhanced Prepacking Techniques;* Technical Services Report #414; Baker Hughes INTEQ: Houston, TX, 1994.
91. Suman, G. O., Jr. *Proceedings of the SPE Annual Fall Meeting;* Society of Petroleum Engineers: Richardson, TX, 1975; paper SPE 5717.
92. Kantzas, A.; Stehmeier, L.; Marentette, D. F.; Ferris, F. G.; Jha, K. N.; Mourits, F. M. *Proceedings of the Petroleum Society of CIM Annual Technical Meeting;* Canadian Institute of Mining, Metallurgy, and Petroleum: Montreal, Canada, 1992; paper CIM 92–46.
93. Adamache, I.; Ianculescu, E.; Pantea, I.; Birsan, P.; Circoana, V. *Pet. Gaze* **1964**, *15*, 542–549 (in Romanian).
94. Adamache, I.; Ghinea, M.; Pantea, I.; Stoican, V.; Birsan, P. *Sand Control During completion and Exploitation of Oil Wells;* Training Course for Specialization of Engineers in Oil and Gas Industry, Part A IIa; Centrul de Documentaire al Industriei Petrolui Si Chimiei: Bucharest, Romania, 1964; Vol. 3. (in Romanian).
95. Nicolescu, N. In *Well Workovers, Completions and Recompletions;* Editura Tehnica: Bucharest, Romania, 1981; pp 299–309.
96. Turta, A.; Socol, S.; Trasea, N.; Ilie, N; *Pet. Gaze* **1984**, *35*, 599–606.
97. Smith, G. E. *SPE Prod. Eng.* **1988**, *3*, 169–180.
98. Loughead, D. J.; Saltuklaroglu, M. *Proceedings of the Ninth Annual Heavy Oil and Oil Sands Technology Symposium;* Mobil Oil Canada: Calgary, Canada, March 11, 1992.
99. Squires, A. *Proceedings of the Tenth Annual Heavy Oil and Oil Sands Technical Symposium;* Calgary, Canada, March 9, 1993.
100. Wong, R. C. K.; Samieh, A. M.; Kuhlemeyer, R. L. *J. Can. Pet. Technol.* **1994**, *33*, 44–49.
101. Maini, B. B.; Sarma, H. K.; George, A. E. *J. Can. Pet. Technol.* **1993**, *32*, 50–54.
102. Dusseault, M. *J. Can. Pet. Technol.* **1993**, *32(9)*, 16–18.
103. Chalaturnyk, R. J.; Wagg, B. T.; Dusseault, M. B. *Proceedings of the SPE International Symposium on Formation Damage Control;* Society of Petroleum Engineers: Richardson, TX, 1992; paper SPE 23780.

104. Dusseault, M. B.; Santarelli, F. J. *Proceedings of the ISRM-SPE Symposium, Rock at Great Depth;* Maury, V.; Fourmaintraux, D., Eds.; A. A. Balkema: Rotterdam, Netherlands, 1989; Vol. 2, pp 789–797.
105. Geilikman, M. B.; Dusseault, M. B.; Dullien, F. A. L. *Proceedings of the Petroleum Society of CIM Annual Technical Conference;* Canadian Institute of Mining, Metallurgy, and Petroleum: Montreal, Canada, 1989; paper CIM 94–89.
106. Geilikman, M. B.; Dusseault, M. B.; Dullien, F. A. L. *Proceedings of the SPE International Symposium on Formation Damage Control;* Society of Petroleum Engineers: Richardson, TX, 1994; paper SPE 27343.
107. Geilikman, M. B.; Dusseault, M. B.; Dullien, F. A. L. *Proceedings of the SPE/DOE Ninth Symposium on Improved Oil Recovery;* Society of Petroleum Engineers: Richardson, TX, 1994; paper SPE/DOE 27797.
108. Geilikman, M. B.; Dusseault, M. B.; Dullien, F. A. L. Presented at the 4th European Conference on the Mathematics of Oil Recovery, Roros, Norway, June 7–10, 1994.
109. Sametz, P. D. *Proceedings of the Ninth Annual Heavy Oil and Oil Sands Technology Symposium;* Calgary, Canada, March 11, 1992.
110. Vonde, T. R. *J. Pet. Technol.* **1982,** *34,* 1951–57.
111. Ponder, M. U.S. Patent 4 921 407, 1990.
112. Sudol, T.; Ridley, R.; Nguyen, D. *Proceedings of the 5th Unitar Heavy Crude and Tar Sands International Conference;* Petroleos de Venezuela: Caracas, Venezuela, 1991.
113. Lea, J. F.; Anderson, P. D.; Anderson, D. G. *J. Can. Pet. Technol.* **1988,** *27,* 58–67.
114. Campbell, B. *Proceedings of the Ninth Annual Heavy Oil and Oil Sands Technical Symposium;* Calgary, Canada, March 11, 1992.
115. Browne, G.; Hass, G. H.; Sell, R. *Proceedings of the Ninth Heavy Oil and Oil Sands Technical Symposium;* Calgary, Canada, March 11, 1992.
116. Jamaluddin, A. K. M.; Nazarko, T. W. *Proceedings of the Petroleum Society of CIM 1993 Annual Technical Meeting;* Canadian Institute of Mining, Metallurgy, and Petroleum: Montreal, Canada, 1993; paper CIM 93–13.
117. Dedora, G. *Proceedings of the Eleventh Heavy Oil and Oil Sands Symposium;* Calgary, Canada, March 2, 1994.
118. Falk, K. *Proceedings of the Canadian SPE/CIM 3rd Annual One-Day Conference on Horizontal Wells;* Canadian Institute of Mining, Metallurgy, and Petroleum: Montreal, Canada, 1993.
119. Reid, D. *Proceedings of the Developments in Production Systems Conference;* IBC Technical Service Ltd.: London, 1993.
120. Skilbeck, F. *Revised Jet Wash Systems Designs for the Murchison and Hutton Platform Separators;* Conoco (UK) Ltd.: Aberdeen, UK, 1989.
121. Priestman, G. H.; Tippetts, J. R.; Dick, D. *The Design and Operation of Separator Sandwash Systems;* Elsevier: New York, 1991.
122. Cornwell, J. R. *Proceedings of the SPE/EPA Exploration and Production Environmental Conference;* Society of Petroleum Engineers: Richardson, TX, 1993; paper SPE 25930.
123. "Upstream Petroleum Waste Management Steering Committee Report to the ERCB on Recommended Oilfield Waste Management Requirements;" Informational Letter IL 93–8; Energy Resources Conservation Board: Calgary, Canada, September 2, 1993; and draft "Recommended Oilfield Waste Management Requirements;" Energy Resources Conservation Board, Calgary, Canada, August 24, 1993.

124. Cole, C. E.; Black, R. W. *Proceedings of the Seventh Annual Heavy Oil and Oil Sands Technical Symposium;* Calgary, Canada, March 14, 1990.

125. Lentz, R. W. *Proceedings of the Eighth Annual Heavy Oil and Oil Sands Technical Symposium;* Calgary, Canada, March 14, 1991.

126. Dusseault, M. B.; Bilak, R. *Proceedings of the Petroleum Conference of the South Saskatchewan Section;* Petroleum Society of Canadian Institute of Mining, Metallurgy, and Petroleum: Montreal, Canada, 1993.

127. Dusseault, M. B.; Bilak, R. A.; Bruno, M.; Rothenburg, L. *Proceedings of the Symposium on Scientific and Engineering Aspects of Deep Injection Disposal of Hazardous and Industrial Wastes;* Lawrence Berkeley Laboratory, Berkeley, CA, May 10–13, 1994.

128. Bruno, M. S.; Bilak, R. A.; Dusseault, M. B.; Rothenburg, L. *Proceedings of the SPE Western Regional Meeting;* Society of Petroleum Engineers: Richardson, TX, 1995; paper SPE 29646.

129. Rothenburg, L; Dusseault, M. B.; Bilak, R.; Bruno, M. *Proceedings of Eurock '94, Rock Mechanics in Petroleum Engineering;* A. A. Balkema: Rotterdam, Netherlands, 1994; pp 739–745.

130. *Guidelines for Industrial Landfills;* Alberta Environment: Edmonton, Canada, 1987.

131. Porteous, W. R. *Sand Production Management;* Petroleum Society of CIM Short Course; W.R. Porteous Engineering Ltd.: Calgary, Canada, 1989.

132. Wu, F. H. *Proceedings of the Eighth Annual Heavy Oil and Oil Sands Technical Symposium;* Calgary, Canada, March 14, 1991.

133. Toor, I. A. International Patent Application B09B 3/00; World Intellectual Property Organization; July 8, 1993.

134. Jamaluddin, A. K. M.; Kim, J. Y.; Gair, J. D. *J. Can. Pet. Technol.* **1994,** 33, 46–51.

135. Schlittler, W. J. *Proceedings of the 4th Annual Environmental, Safety and Health Conference and Exhibition;* Pennwell Conferences and Exhibitions: Houston, TX, 1993.

136. "Criteria for Disposal of Oily Waste to Roads;" General Bulletin GB 92–10; Energy Resources Conservation Board: Calgary, Canada, 1992.

137. "Storage Handling and Disposal of Oily Wastes;" Informational Letter IL 85–16, Energy Resources Conservation Board: Calgary, Canada, 1985.

138. Bell, W. T. Presented at the SPE International Petroleum Exhibition and Technical Symposium, Beijing, China, March 18–26, 1982; paper SPE 10033.

Received for review July 5, 1994. Accepted revised manuscript March 14, 1995.

NEAR-WELL AND OILWELL APPLICATIONS

Drilling Fluid Suspensions

Timothy G. J. Jones[1] and Trevor L. Hughes

Schlumberger Cambridge Research, High Cross Madingley Road, Cambridge CB3 0HG, United Kingdom

A review of the composition and properties of water- and oil-based drilling fluids is presented. The basic functions of drilling fluids are described. The composition of water- and oil-based drilling fluids is described, together with the functions of each component. An account of the rheology of drilling fluids is given: measurements (rig site and laboratory), rheological models, and the effect of drilling fluid composition and drilling fluid hydraulics (influence of rheology on frictional pressure losses and cuttings transport). The filtration (fluid loss) properties are also described with a comparison of static and dynamic filtration and an account of the effect of composition on filtration rates. A brief account is given of several problems associated with the management of drilling fluids: invasion and drilling induced formation damage, chemical aspects of wellbore stability in shales, solids control, environmental aspects of drilling fluid waste, and composition monitoring of drilling fluids.

The Functions of Drilling Fluids

THE PERFORMANCE OF A DRILLING FLUID is critical to the safe and efficient drilling of an oil well. Drilling fluids, or *muds* as they are commonly known, have evolved over most of this century from simple clay–water suspensions to the complex clay–polymer–electrolyte systems that are routinely used today. Allen (*1*) has given a brief history of rotary drilling and the use of drilling fluids. The annual worldwide expenditure on drilling fluids (excluding the previous Eastern Bloc countries) was estimated in 1990 to be $3.8–4.5 billion (*2*) with costs per well being in excess of $500,000 for the most demanding (deep, high temperature) wells.

[1] Current address: Schlumberger-Doll Research, Old Quarry Road, Ridgefield, CT 06877–4108.

0065–2393/96/0251–0463$32.50/0

Figure 1 shows a schematic of a wellbore during drilling operations. The drilling fluid is pumped down the drill pipe and out of the nozzles of the drill bit. The fluid returns to the surface in the annulus between the rotating drill pipe and the borehole wall (or casing in the upper sections) and is processed on the surface (removal of gas, drilled cuttings, and drilled fines). The drilling fluid is an open system that can lose and gain chemical components (suspended solids, dissolved ions, and organic components) in the borehole or in the processing on the surface. The analysis of cuttings and monitoring of the gas content in the returned drilling fluid are the realm of mud (or geological) logging (3), an operation that is distinct from drilling fluids engineering, which is the subject of this chapter.

Traditionally, the drilling fluid performs four basic functions and these functions are controlled by its chemical composition.

Hydrostatic Pressure.   During drilling, the column of rock originally occupying the hole is replaced by a column of drilling fluid. The

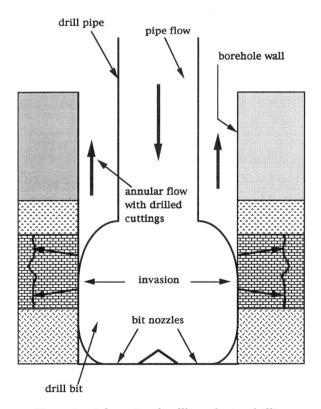

Figure 1.   Schematic of wellbore during drilling.

hydrostatic pressure exerted by the drilling fluid is determined by its density, which is controlled by the suspension of an insoluble mineral in the fluid. The commonest weighting agent is barite (or barytes, barium sulphate) and more than half of the world's annual production of 4–5 million tonnes is used as a weighting agent in drilling fluids (2). Other weighting agents that have been used are calcium carbonate and iron oxides.

The hydrostatic pressure exerted by the drilling fluid on the walls of the borehole must be carefully chosen. If the hydrostatic pressure is too large, then rock formations in contact with the fluid may fracture and large volumes of drilling fluid may be lost to the formation (lost circulation). If the hydrostatic pressure is too small, then the rock formation may fail by tensile failure. The effect of drilling fluid density on the mechanical stability of boreholes has been discussed by several authors (4, 5)

The hydrostatic pressure of the drilling fluid must also be sufficient to at least equal the pressure of the fluid (formation water, oil, or gas) in the permeable formations (sandstones, carbonates). If a formation that is penetrated contains a fluid whose hydrostatic pressure (pore pressure) is greater than that of the drilling fluid, then the fluid will enter the wellbore. The entry of the wellbore fluid into the well is termed a *kick* (6). If the wellbore fluid is gas, then its uncontrolled ascent to the surface constitutes a major danger; the gas will expand and increasingly displace the column of drilling fluid above it. The uncontrolled flow of reservoir fluids at the surface is termed a *blowout* (6). The density of the drilling fluid is the primary safety feature of the drilling fluid. Methods of evaluating formation pore pressures for drilling purposes have been described by Fertl (7).

Filtration (Fluid Loss) Control. The density of the drilling fluid is not, in general, able to be adjusted to ensure that its hydrostatic pressure is equal to that of the pore pressure of the drilled formations. It is common practice for the hydrostatic pressure of the drilling fluid to exceed the pore pressure, largely to ensure a safety margin against kicks. The drilling fluid will therefore tend to filter into (or invade) the permeable formations. Such invasion is not desirable, largely due to the impairment of producibility (permeability) of the near-wellbore region of the reservoir. The fluid loss of a drilling fluid is most commonly controlled by the use of a suspension of the clay bentonite (montmorillonite) that forms a filter cake of low permeability and limits the amount of filtrate lost to the formations. Water-soluble polymers are frequently added to reduce further the permeability of the filter cake.

Hole Cleaning. A major function of the drilling fluid is to clean the wellbore by carrying the drilled rock cuttings to the surface. The hole cleaning properties of a drilling fluid are determined by its rheology, which is controlled by its chemical composition. The rheology of drilling fluids is invariably designed to be non-Newtonian (8, 9), usually with a yield stress and a viscosity that decreases with increasing shear rate [i.e., the fluid is viscoplastic and usually thixotropic (8)]. Many concentrated solid–liquid suspensions, including bentonite–water suspensions, exhibit non-Newtonian rheology. High molecular weight water-soluble polymers are also used to viscosify drilling fluids and to produce non-Newtonian rheology.

The rheology of the drilling fluid is an important factor in determining the frictional pressures losses as it is pumped around the well (pressure drop in the drill pipe, across the bit nozzles, and in the annulus). The frictional pressures losses determine the pumping power required at the surface. An additional function of the circulating drilling fluid is to cool and lubricate the drill bit during drilling.

Wellbore Stability.   In addition to the problem of the purely mechanical stability of the wellbore related to the hydrostatic pressure of the drilling fluid, chemical–mechanical wellbore stability problems can result from interactions between the drilling fluid and certain formations. The mechanical properties of some drilled formations, such as salt and massive shale formations, can be severely weakened in the presence of drilling fluids. Salt sections, for example, can dissolve in water-based drilling fluids and can give rise to enlargement of the hole. The problem of drilling salt sections can be overcome by using salt-saturated water-based drilling fluids or oil-based drilling fluids.

A more pervasive problem is the maintenance of wellbore stability in shale formations [i.e., formations that have a high clay content, typically in excess of 50 wt% (10)]. In the presence of water, shales can take up water and swell and disperse or they can fracture. Problems associated with wellbore stability in shale sections are sticking of the drill pipe (usually termed stuck pipe), hole enlargement, and excessive generation of drilled solids.

Clearly, the formulation of drilling fluid composition must ensure that these four basic functions are met. The formulation is complex in view of the complexity of the fluid and the nature of its functions. An added complexity of drilling fluid composition is that it changes during the course of the drilling by the loss and addition of components. Fine drilled solids (e.g., dispersed clays) commonly contaminate the drilling fluid, together with ions from pore water influxes, whereas drilling fluid additives are removed or modified (e.g., polymer adsorption on cuttings, polymer degradation).

There is increasingly a fifth function that must be met, namely, to minimize the environmental impact of the drilling fluid and the drilled cuttings upon disposal. The increasing scrutiny of drilling waste and its environmental impact is having a marked effect on the design of drilling fluids, the use of particular additives and disposal methods.

The Composition of Drilling Fluids

Drilling fluids can be basically divided into water-based and oil-based fluids. A very large number of formulations have been used and advocated, particularly for water-based drilling fluids. A fairly complete listing of drilling fluid additives that are commercially available is published annually in the June edition of the journal *World Oil* (*11*).

Water-Based Drilling Fluids. Water-based drilling fluids consist essentially of either a freshwater- or seawater-based fluid with a weighting agent (usually barite), bentonite clay and several water-soluble polymers, together with added salts (commonly NaCl, KCl) and sodium or potassium hydroxide for pH control.

Solid Phases. The commonest weighting agent is barite, which, in its pure form, has a density of 4500 kg/m^3. The American Petroleum Institute (API) specification (*12*) for the density of barite is that it must be at least 4200 kg/m^3. Other API specifications are the soluble alkaline metal concentrations must not exceed 250 ppm and a maximum of only 3 wt% of the barite must have a particle size in excess of 75 μm with a maximum of 30 wt% having a particle size of less than 6 μm. Other suspended weighting agents are calcium carbonate and iron oxide [hematite (*12*)], although iron oxide is not often used due to its abrasiveness.

Bentonite is rock composed of the clay mineral montmorillonite, which is an expandable 2:1 aluminosilicate mineral (*13–15*). Figure 2 shows that the structure of the mineral consists of sheets or platelets that consist of a layer of octahedrally coordinated aluminum hydroxide sandwiched between two sheets of tetrahedrally coordinated silica (*13*). Although the thickness of the platelets is about 0.9 nm, the length can extend to about 1 μm. The bentonite platelets are therefore colloidal in size.

The colloid chemistry of clays, and bentonite in particular, is dominated by the presence of the fixed negative charge on the surface of the platelets and the nature of the exchange cation (usually sodium or calcium). The charge on the surface of the clay platelets arises from the isomorphous substitution of Si^{4+} by Al^{3+} or Fe^{3+} in the tetrahedral layer and the replacement of Al^{3+} by Fe^{2+} or Mg^{2+} in the octahedral layer. The greater part of the substitution (and hence charge) usually resides in the octahedral layer (*16*). The surface charge density of clay minerals is usually termed the cation exchange capacity (CEC) and expressed in milliequivalents (millimoles of univalent charge) per gram (or per 100 g)

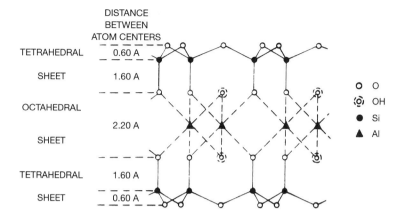

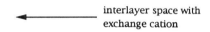

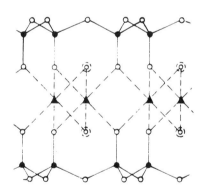

Figure 2.   Crystal structure of montmorillonite clay. (Reproduced with permission from reference 14. Copyright 1977 John Wiley and Sons.)

of dry clay. Bentonite has a CEC in the range 0.7–1.0 meq/g (*16*); the clays illite and kaolinite have CEC values in the ranges of 0.1–0.2 meq/g and 0.01–0.05 meq/g, respectively.

The crystallographic surface area of bentonite clay is about 750 m²/g (*17*), although the actual value that is obtained depends on the exchange cation. Quirk (*18*) has reported the solid–liquid surface area of bentonites with various alkali and alkaline earth exchange cations dispersed in aqueous solution. The sodium-exchanged form of bentonite has a measured surface area of 700 m²/g, which is close to the theoretical maximum. The calcium-exchanged form of bentonite has a solid–liquid surface area of $270 \ m^2/g$, indicating aggregation of the platelets.

The edges of clay platelets consist largely of Si–O and Al–O sites that behave as typical oxide surfaces in that their charge is determined by potential determining ions (i.e., H^+ and OH^- ions). In acid media, the edge charge will be positive due to proton adsorption, whereas in alkaline media the edge charge will be negatively charged (*19*).

The presence of the negative charges and compensating exchange cations on the surface of the clay platelets results in the electrostatic repulsion of the platelets in aqueous suspensions. In dilute electrolyte solutions, the clay platelets repel each other and disperse to form stable suspensions. The distribution of ions away from the surface of the clay platelets can be described by electrical double layer theory (*20*), *see also* Chapter 1 of this book; a schematic of the distribution of ions and the electrostatic potential Ψ between two clay platelets is shown in Figure 3. The extent of the electrostatic repulsion between the clay platelets can be modified by changing the ionic strength of the suspending electrolyte or the nature of the exchange cation. An increase

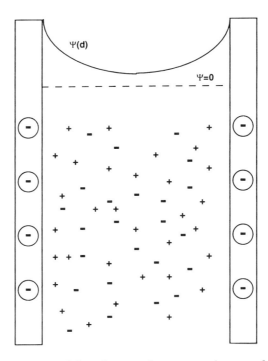

Figure 3.   Schematic of distribution of cations and anions between two charged clay platelets in an electrolyte solution. Variation of electrostatic potential Ψ with interplate separation d also shown.

in the ionic strength of the suspending electrolyte causes the zeta potential, the value of Ψ at some shear plane near the surface of the platelets, to decrease and thus to reduce the repulsion between them. The decreased electrostatic repulsion causes the platelets to aggregate or *flocculate*. Similarly, exchange cations characterised by either a low hydration energy (Cs^+, $(CH_3)_3N^+$) or high valency (Ca^{2+}, Al^{3+}) lower the zeta potential of the clay platelets and reduce the electrostatic repulsion between them. This aggregation or flocculation of the clay particles gives rise to the observed decrease in the solid–liquid interfacial area (*18*). Flocculation of the clay platelets leads to a loss in stability of the dispersion and flocs of clay platelets settle out of the suspension. *Gelation* is an extreme form of flocculation where all of the particles aggregate to form a single homogeneous network that exhibits some characteristics of a solid (e.g., elasticity).

Bentonite suspensions have frequently been used to test the applicability of electrical double layer theory. Lubetkin et al. (*21*), for example, have compacted clay suspensions of various ionic strengths to compare the measured variation of electrostatic repulsion (or swelling pressure) with interlayer spacing with that predicted by electrical double layer theory. The compaction of highly dispersed sodium- or lithium-exchanged bentonites is reasonably predicted by electrical double layer theory but the agreement is increasingly poor as the hydration energy of monovalent cations decreases and the ion exchange selectivity (ion binding) of the cation increases.

Water-Soluble Polymers.   A number of water-soluble polymers are used extensively to control the rheological, fluid loss and shale stabilising properties of water-based drilling fluids. List I lists commonly used polymers and Figure 4 shows the structure of several polymer molecules. Several recent reviews of water-soluble polymers in drilling fluids are available (*22–25*); McCormick et al. (*26*) have given a recent useful introduction to water-soluble polymers.

The most commonly used water-soluble polymers are the polysaccharides carboxymethyl cellulose (CMC) and xanthan gum (XC polymer). Reference 27 gives a comprehensive account of the manufacture and industrial uses of these polymers. Both of the polymers are polyelectrolytes and the sodium form is normally provided for use in drilling fluids. Low molecular weight CMC is commonly used to augment the fluid loss control of bentonite whereas higher molecular weight CMC is also used to viscosify drilling fluids. Xanthan gum is a biopolymer produced by the action of the bacterium *Xanthomonas campestris* on carbohydrates such as D-glucose. The polymer xanthan gum, which has been studied extensively for use in enhanced oil recovery (*28*), is commonly used to control the viscosity of water-based drilling fluids. The polysaccharide

List I. Water-Soluble Polymers Used in Drilling Fluids

- Polysaccharides
  —xanthan gum (XC)
  —carboxymethyl cellulose (CMC)
  —starch
  —guar gum
  —welan gum
- Other Natural Polymers
  —lignosulphonates
  —asphaltenes
- Synthetic Polymers
  —sodium polyacrylate
  —partially hydrolysed polyacrylamide (PHPA)
  —vinyl sulphonate–vinyl amide copolymer
  —polyamine cationic polymers
  —polyols

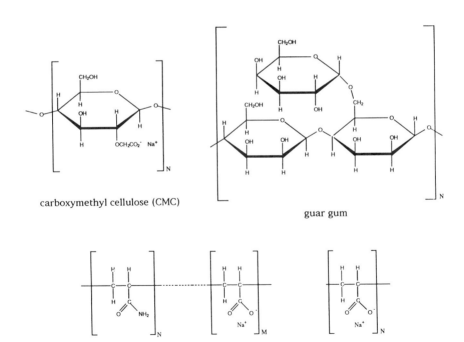

carboxymethyl cellulose (CMC)

guar gum

partially hydrolysed polyacrylamide (PHPA)   sodium polyacrylate

Figure 4.   Some common water-soluble polymers used in water-based drilling fluids.

welan gum, which is also a polyelectrolyte, has been used to viscosify water-based drilling fluids in the absence of bentonite (29). The neutral polysaccharide starch (27) has frequently been used to control the fluid loss of drilling fluids, despite its low solubility in water. The fluid loss action of starch is likely to be through a particle (or hydrogel) plugging mechanism.

The molecular weight of xanthan gum is typically of the order of $1-5 \times 10^6$ g/mole (30). The conformation of the xanthan gum molecule in aqueous solution is rodlike and shows little change in conformation on the addition of monovalent salts such as NaCl (31). The molecular weight of several CMC polymers has been recently determined to be in the range 50,000–275,000 g/mole (32). The degree of substitution (i.e., number of carboxylate groups per repeat unit) of oil field CMC is about 0.7, although so-called polyanionic cellulose appears to be CMC with a degree of substitution closer to 1.0 (32). CMC is not soluble in water when the degree of substitution is below about 0.4 (27). The CMC polymer molecule is also rigid but shows a larger change in conformation than xanthan gum as the ionic strength is increased; a random coil configuration is not reached, however, even at high ionic strengths ($I > 1.5$ M) (32).

A significant limitation in the use of polysaccharide polymers in water-based drilling fluids is their temperature stability. The polysaccharide polymers are prone to thermal-oxidative breakdown that reduces their molecular weight. Thomas (33) has observed that the upper temperature limit of starch is about 100 °C and 135 °C for CMC. Darley and Gray (25) have quoted comparable limiting temperatures. Other polysaccharide polymers are restricted to even lower temperatures. Guar gum, for example, is not used at temperatures above 70 °C (25), which limits its use to shallow wells.

Two other important classes of naturally occurring polymers are lignosulphonates and asphaltenes. Lignosulphonates (25, 34) have been very commonly used as dispersants in water-based drilling fluids (35), particularly with cross-linking of the lignosulphonates by chromium ions. The structure and molecular weight distribution of lignosulphonates, which are by-products of the pulp and paper industry, are not fully determined. Asphaltenes such as gilsonite (36) are frequently used in water-based drilling fluids for shale stabilization, although they are not soluble. The complex structure of asphaltenes has recently been discussed (37).

The temperature sensitivity of the polysaccharides, which is partly a consequence of the ether linkage in the backbone of the polymer (38), has lead to the increasing use of synthetic water-soluble polymers for high temperature drilling fluids. The use of synthetic polymers for fluid loss control in high temperature applications has been particularly im-

portant. Plank and Hamberger (*39*) have described a synthetic sulphonated polymer (complete structure not divulged) for fluid loss control
that is stable up to 200 °C. Other synthetic water-soluble polymers that
have been used for fluid loss control are vinylsulphonate–vinylamide
copolymers (*40*) and low molecular weight polyacrylates (*41*).

The most commonly used synthetic water-soluble polymers are the
polyacrylamides, largely in the form of the so-called partially hydrolyzed
polyacrylamides (PHPA) (*42–44*) that are polyacrylamide–polyacrylate
copolymers produced by hydrolyzing some fraction ($M/(M + N)$) of the
amide groups of polyacrylamide (*see* Figure 4). The PHPAs, which have
also been used extensively for enhanced oil recovery (*45*) and the control
of water mobility in oil production (*46*), are commonly used to prevent
the dispersion of shale cuttings during transport in the annulus. Typically
the weight average molecular weight of these polymers is in the range
$1–5 \times 10^6$ g/mole (*42, 43*). The PHPA polymers have been used to
clarify water and stabilize soils by flocculating solids (*47*).

Synthetic cationic polymers have been used in water-based drilling
fluids to control shale problems. Recent applications include the use of
low molecular weight polyamines (*48, 49*); some of these cationic polymers are also used for water treatment (*48*). The use of low molecular
weight polyols and glycols has recently been advocated with claims that
they act to control both fluid loss and shale hydration (*50*).

Oil-Based Drilling Fluids.   Oil-based drilling fluids fall into two
basic categories, namely, invert emulsion and oil (or non-emulsion) drilling fluids. The use of non-emulsion oil-based fluids has been limited,
largely due to the problem of obtaining sufficient viscosity or fluid loss
control (*51*). More recently, oil-soluble polymers have been used to give
viscosity and fluid loss control. An oil-based drilling fluid using calcium
carbonate and a resin for fluid loss control with an organophilic clay and
oil-soluble polymers for viscosity control has recently been formulated
(*52*).

The oil originally used in oil-based drilling fluids was either crude
or diesel oil. These oils have been largely replaced by refined mineral
oils with aromatic contents below about 0.25 wt% (*53*). Alternative oil
phases that have recently been introduced are poly(alphaolefins) (*54*)
and esters derived from vegetable oils (*55*). These and other synthetic
oils have been introduced in response to environmental pressures on
the disposal of waste oil.

The water-in-oil or invert emulsion oil-based drilling fluids provide
much of the required rheological and fluid loss properties. The emulsion
is stabilized by the use of emulsifiers that prevent the coalescence of
the droplets and formation of separated phases (*56, 57*). Figure 5 shows
the chemical structure of the two emulsifiers that are commonly used

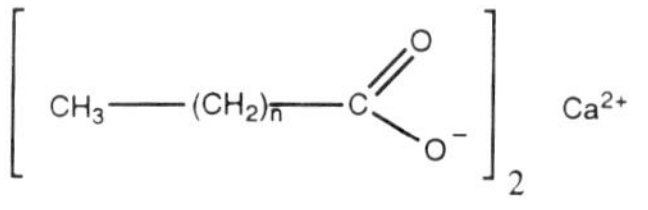

long-chain carboxylic acid (calcium soap)

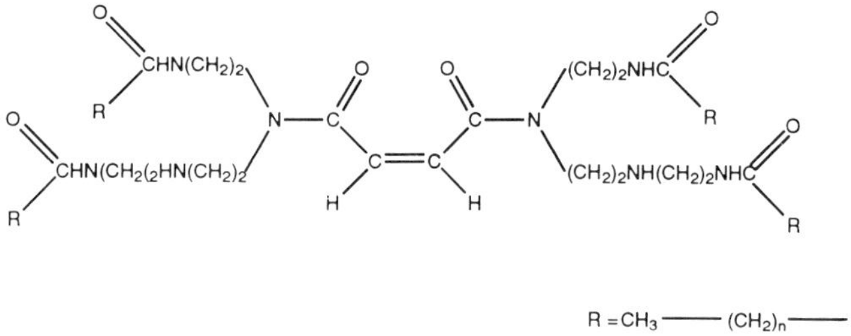

$R = CH_3 \text{———} (CH_2)_n \text{———}$

polyamide

Figure 5. Common emulsifiers used in invert emulsion oil-based drilling fluids.

in oil-based drilling fluids. The long-chain carboxylic acids are usually activated by the addition of calcium hydroxide to form calcium soaps that preferentially form water-in-oil emulsions (57).

It is well-known that emulsions can be stabilized by fine solid particles (58) in addition to chemical stabilizers (emulsifiers). A common addition to invert emulsion oil-based drilling fluids is a so-called organophilic clay, which is bentonite (or hectorite) clay saturated with an organic cation, most commonly a quarternary amine cation (59, 60). The organophilic clay disperses in oil and increases the viscosity, in a manner similar to the dispersion of sodium-exchanged (or, to a lesser extent, calcium-exchanged) bentonite in water. The organophilic clay is added to invert emulsion drilling fluids to increase the stability of the emulsion, to viscosify the oil phase, and to control fluid loss.

The invert emulsion drilling fluid is required to suspend both the weighting agent and the drilled solids (cuttings) generated during drilling. Barite is commonly used as the weighting agent and its hydrophilic surface is made hydrophobic by the emulsifiers in order to maintain the barite in the continuous oil phase. Calcium carbonate has been used as a weighting agent in invert emulsion drilling fluids as its surface is moderately hydrophobic. Excess emulsifier is required to ensure that the surface of the drilled cuttings is hydrophobic to maintain them predominantly in the oil phase.

Invert emulsion drilling fluids are commonly selected for their temperature stability and their ability to prevent the wellbore stability problems associated with the hydration of clays in shale formations. The thermodynamic activity a_w of the water in the aqueous (dispersed) phase is controlled by the addition of a salt (usually calcium chloride) to ensure that it is equal to or less than the activity of the water in the drilled shale formations. The emulsified layer around the water droplets is claimed to act as a semipermeable membrane that allows the transport of water into and out of the shale but not the transport of ions (*61*). When the activities (or, more strictly, the chemical potentials) of the water in the shale and invert emulsion are equal, then no net transport of water into or out of the shale occurs (i.e., the drilling fluid does not hydrate or dehydrate the shale). This equality of water activity has lead to the development of so-called balanced activity oil-based drilling fluids.

The Rheology of Drilling Fluids

Measurements and Rheological Models.   *Field Measurements.* The measurement of the rheology of drilling fluid samples at the rig site is made by two methods. The first method uses the Marsh funnel (Figure 6) to measure the time taken to empty a given volume (one U.S. quart or 946 mL) of drilling fluid. The time taken to empty the funnel is termed the Marsh funnel viscosity and typically ranges from 30–80 s;

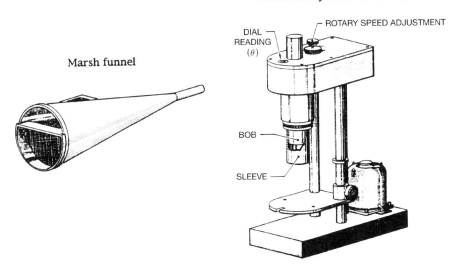

Figure 6.   Marsh funnel and concentric cylinder rheometer. (Reproduced with permission from reference 65. Copyright 1985 IHRDC.)

the Marsh funnel viscosity of water at room temperature is 26 s. The Marsh funnel viscosity (or any single-point measurement of apparent viscosity) cannot be used to obtain the rheological properties of non-Newtonian fluids such as drilling fluids. In addition, the Marsh funnel viscosity depends directly on the density of the drilling fluid. The Marsh funnel viscosity is regularly reported and gives a qualitative guide to changes in fluid rheology.

A more complete measurement of the rheology of a drilling fluid is obtained with a Fann or Baroid concentric cylinder rheometer (Figure 6). The outer cylinder rotates at a constant speed that produces laminar (couette) flow in the drilling fluid in the sample cup. The rotation of the drilling fluid produces a viscous drag on the inner cylinder that rotates until the torque generated by the rotation is balanced by the torque in the spring attached to the cylinder. The torque developed on the inner (stationary) cylinder is measured by its rotation θ (in degrees).

A number of Fann or Baroid rheometers have been used for the rig site measurement of fluid rheology (62, 63). The design of the rheometers is similar and the various viscometer types differ largely in the control of shear rate. Early models of the concentric cylinder rheometer were limited to two shear rate measurements made at rotation speeds of $r = 300$ rpm and $r = 600$ rpm. Later models have extended the number of rotational speeds (shear rates) at which the torque can be measured, enabling a more complete rheogram to be constructed.

The Fann rheometer has been calibrated to give direct readings of plastic viscosity (PV) and yield point (YP) as given by the simple Bingham plastic fluid model relating shear stress (τ) to shear rate ($\dot{\gamma}$):

$$\tau = \text{YP} + \text{PV}\dot{\gamma}. \tag{1}$$

The numerical value of the plastic viscosity PV in centipoise is given by the difference in the deflections at $r = 600$ rpm and $r = 300$ rpm

$$\text{PV} = \theta_{600} - \theta_{300} \tag{2}$$

and the numerical value of YP (in lb/100 sq ft) is given directly by

$$\text{YP} = \theta_{300} - (\theta_{600} - \theta_{300}) = \theta_{300} - \text{PV} \tag{3}$$

Figure 7 shows a schematic for the determination of PV and YP. The Fann reading θ_{300} at $r = 300$ rpm gives a direct reading of the viscosity of a Newtonian fluid in centipoise (1 cP = 1 mPa·s).

In addition to the plastic viscosity and yield point, the gel strength of the drilling fluid is measured after it has been at rest for a given period of time. The usual procedure (64) is to shear the sample and then allow the sample to be unsheared for 10 s or 10 min. After the prescribed

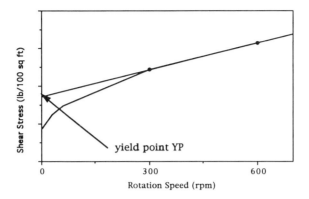

Figure 7.   Determination of yield point and plastic viscosity of drilling fluid using the Bingham plastic fluid model.

time the torque at which the inner cylinder of the rheometer just begins to slip is measured.

Rheological Models.   It is well known that the rheology of drilling fluids is often poorly described by the Bingham plastic model. The rheograms of drilling fluids frequently show curvature at low shear rate that is not predicted by equation 1. Figure 8, for example, shows a series of rheograms for simple bentonite–polymer (low molecular weight CMC) drilling fluids of varying polymer content (32). The rheograms do not appear to be linear even at the highest measured shear rates.

A number of rheological models have been used to describe the rheology of drilling fluids (and non-Newtonian fluids in general) (8, 9, 65–67). These models, it is stressed, have been obtained purely empirically by fitting τ as a function of $\dot{\gamma}$. A Newtonian fluid is characterized by

$$\tau = \eta\dot{\gamma} \tag{4}$$

where η is the Newtonian viscosity that is independent of shear rate. The most commonly used models to describe the rheology of drilling fluids are the Bingham plastic model

$$\tau = \tau_{o} + \eta_{p}\dot{\gamma} \tag{5}$$

where τ_{o} is the yield stress and η_{p} is the Bingham plastic viscosity and the power law model [also termed the Ostwald–de Waele model in the basic rheology literature (8)]

$$\tau = K\dot{\gamma}^{n} \tag{6}$$

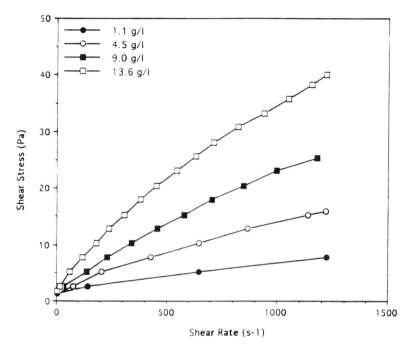

Figure 8. Typical rheograms of simple bentonite–polymer (CMC) drilling fluids. Drilling fluids contain 20.6 g/L bentonite and variable polymer concentrations (shown in g/L). (Reproduced with permission from reference 32. Copyright 1993 Society of Petroleum Engineers.)

where n is an exponent (often termed the flow behavior index) and K is a constant that is termed the consistency or consistency factor. Clearly when $n = 1$ Newtonian behavior is recovered and K is the Newtonian viscosity. Plastic fluids (or more strictly pseudoplastic fluids because the yield stress is zero) are characterized by $n < 1$. The dimensions of K $(Pa \cdot s^n)$ depend on the value n, which precludes it from being a fundamental material property of the fluid.

Several other models have been used that essentially incorporate a yield stress into the power law model. These models include the Herschel–Bulkley model

$$\tau = \tau_o + K\dot{\gamma}^n \qquad (7)$$

and the Robertson–Stiff model

$$\tau = K(C_o + \dot{\gamma})^n \qquad (8)$$

where C_o is a constant. Harris (8) has referred to the Herschel–Bulkley equation as a generalized Bingham equation. Graves and Collins (68) have modified the Bingham plastic model to account for curvature at low values of $\dot{\gamma}$:

$$\tau = (\tau_0' + \eta_p\dot{\gamma})(1 - \exp[-C\dot{\gamma})]) \tag{9}$$

where C is a constant. The Graves and Collins equation yields power law-like behavior as $\dot{\gamma} \rightarrow 0$ and the Bingham plastic model for large $\dot{\gamma}$. The parameter τ_0' is not a true yield stress but a measure of the displacement of the limiting high shear rate straight line from the origin.

A more specific model, where n has a fixed value, is given by the Casson equation:

$$\tau^{1/2} = k_0 + k_1\dot{\gamma}^{1/2} \tag{10}$$

where k_0^2 is the yield stress and k_1^2 is the high shear rate viscosity (32, 69).

The modified power law models (eqs 7 and 8) have previously been less useful for rig site use as the constants K and n have to be obtained by trial-and-error methods. The use of rig site computers has overcome these difficulties.

The shear stress τ is related to the measured deflection θ in the Fann rheometer by

$$\tau = 0.511\theta \tag{11}$$

where τ is measured in Pa and θ is measured in degrees. The choice of rheological model determines the relationship between the measured rotation speed r of the inner cylinder and the shear rate $\dot{\gamma}$. The shear rate $\dot{\gamma}_i$ of a Newtonian fluid on the wall of the inner cylinder is related to the rotation speed r (in rpm) of the inner cylinder by (65, 70)

$$\dot{\gamma}_i = \pi r\alpha^2/15(\alpha^2 - 1) \tag{12}$$

where α is the ratio of outer to inner cylinder radii. The Fann rheometer shown in Figure 6 is characterized by $\alpha = 1.068$ and equation 12 gives the familiar relationship $\dot{\gamma}_i = 1.700r$ (71). For a power law fluid, the equivalent form of equation 12 is

$$\dot{\gamma}_i = \pi r\alpha^{2/n}/15n(\alpha^{2/n} - 1) \tag{13}$$

which, of course, reduces to equation 12 when $n = 1$.

Laboratory Studies. Away from the rigors of the rig site measurement of rheology, a number of studies have been conducted to give a better understanding of drilling fluid rheology. In particular, use has been made of high temperature–high pressure (HT–HP) rheometers that enable the rheological properties of drilling fluids to be measured

under realistic down-hole conditions. Several representative studies are
now discussed.

The HT–HP rheology of invert emulsion oil-based drilling fluids has
attracted considerable attention in view of the marked pressure dependence of the viscosity (or consistency) and yield stress. Houwen and
Geehan (69) have measured the rheology of oil-based drilling fluids over
the temperature range 25–140 °C and at pressures from 1–1000 bar
(15–15,000 psi) using two Haake rheometers. The measured shear rate
of the two rheometers ranged from 0–2700 s^{-1} with an upper measurement of τ of 60 Pa. The HT–HP rheograms were analyzed using both
the Casson and Herschel–Bulkley equations. Figure 9 shows a series of
rheograms of invert-emulsion oil-based drilling fluids obtained at a pressure of 600 bar and over the temperature range 25–110 °C. The rheograms show curvature at low shear rate and suggest the presence of a
yield stress. Houwen and Geehan commented that the Casson equation
gave a consistently better fit than the Herschel–Bulkley equation to the
rheological data at high shear rate. A suggested cause of the increasingly
poor fit was that the Herschel–Bulkley equation predicted a continuous
decrease in the viscosity η_{HB} with increasing shear rate, namely,

$$\eta_{HB} = nK\dot{\gamma}^{n-1}; \qquad (n < 1) \qquad (14)$$

whereas the experimental data suggest that the high shear viscosity
tended to a constant (Newtonian) limit. The Casson model predicts a
constant high shear rate viscosity (k_1^2).

Figure 10 compares the pressure dependence of the rheological
parameters of the Casson model (high shear viscosity k_1^2), the Bingham

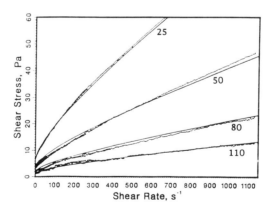

Figure 9.   Experimental and predicted rheograms for invert emulsion oil-
based drilling fluids. Rheograms collected at 600 bar hydrostatic pressure
and various temperatures (shown in degrees centigrade). (Reproduced with
permission from reference 69. Copyright 1986 Society of Petroleum Engineers.)

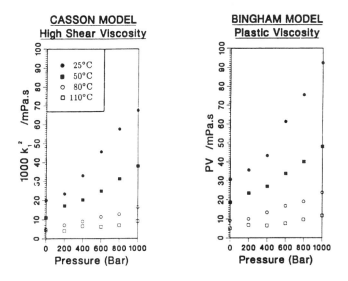

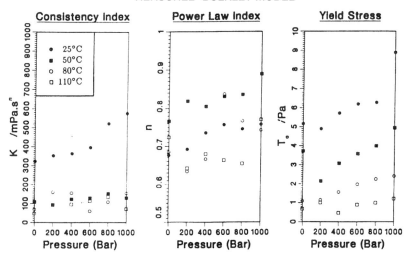

Figure 10.   Pressure dependence of parameters from various models of the rheology of invert emulsion oil-based drilling fluids at various temperatures. Casson high shear viscosity; Bingham plastic viscosity; consistency, power law exponent, and yield stress from Herschel–Bulkley model. (Reproduced with permission from reference 69. Copyright 1986 Society of Petroleum Engineers.)

plastic model (η_p) and the Herschel–Bulkley model $(K, n, \text{ and } \tau_o)$ at various temperatures. Figure 10 clearly shows the marked increase in viscosity (consistency) and yield stress with increasing hydrostatic pressure.

Houwen and Geehan (69) used an Arrhenius-like equation to model the temperature and pressure dependence of the Casson high shear viscosity $k_1{}^2$ and Casson yield stress $k_0{}^2$ by

$$k_i{}^2 = A_i \exp((E_i/RT) + (V_iP/RT)); \qquad i = 0, 1 \qquad (15)$$

where A_i is a pre-exponential factor, E_i is an activation energy, R is the gas constant, and V_i is a pressure coefficient that is the specific volume of the drilling fluid. Houwen and Geehan point out that the factor A_i is strictly dependent on temperature but over the temperature range studied no significant variation was observed. Figure 9 shows the goodness-of-fit of the models to the experimental data. Alderman et al. (72) have used a similar experimental and modeling approach for the HT–HP rheology of water-based drilling fluids. The HT–HP rheology was measured using a Haake HT–HP rheometer and a Bohlin HT–LP rheometer. The authors emphasized the problem of the time dependence of the rheology and the need to age (by shearing) the drilling fluids to eliminate hysteresis (thixotropy) in the rheograms. Figure 11 shows the removal of thixotropy in a simple unweighted bentonite water-based drilling fluid by shearing. The hysteresis was finally removed but only after shearing the drilling fluid for up to 30 min. The rheological measurements were therefore made on samples that represented drilling fluids after considerable circulation in the drill pipe and annulus.

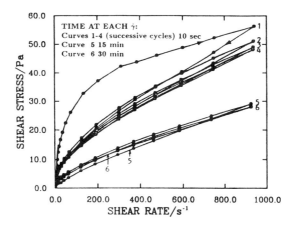

Figure 11. Rheograms of simple bentonite water-based drilling fluid (25 °C) showing the effects of thixotropy. (Reproduced with permission from reference 72. Copyright 1988 Society of Petroleum Engineers.)

Figure 12 shows the rheograms of a simple dispersed bentonite–
lignosulphonate drilling fluid as a function of temperature at a fixed
pressure of 400 bar. The form of the rheograms changes radically with
temperature, showing that while the viscosity decreases with increasing
temperature, the yield stress *increases*. The rheograms are well described
by the Herschel–Bulkley equation over the temperature and pressure
ranges studied. The Casson equation generally gives a less accurate fit
than the Herschel–Bulkley equation and the fit becomes less accurate
as the temperature increases. The Bingham equation gives a poor fit to
the experimental data at all temperatures.

Figure 13 shows the pressure dependence of the Herschel–Bulkley
parameters [n, K, τ_o, and η_{HB} (eq 14) at the highest shear rate] for a
weighted drilling fluid at 40 °C. In contrast to oil-based drilling fluids,
the rheology of water-based fluids shows little dependence on hydrostatic pressure.

Alderman et al. (72) modeled the temperature and pressure variation
of the high shear rate Herschel–Bulkley viscosity η_{HB} using

$$\eta_{HB} = A_w[1 + \delta\beta P] \exp((E_w + P(B_w T - C_w))/T). \qquad (16)$$

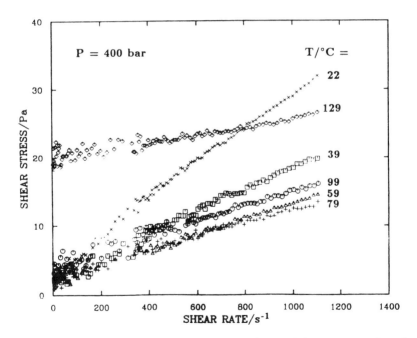

*Figure 12.   Rheograms of simple bentonite–lignosulphonate water-based
drilling fluids at constant hydrostatic pressure (400 bar) and various temperatures. (Reproduced with permission from reference 72. Copyright 1988
Society of Petroleum Engineers.)*

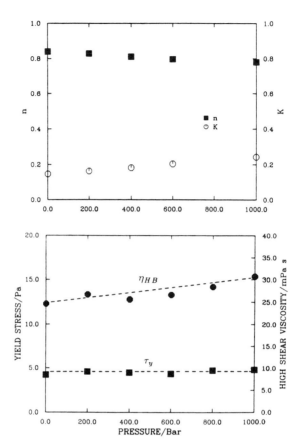

Figure 13. Pressure dependence of the rheological parameters from the Herschel–Bulkley model (n, K, τ_o, and a high shear rate viscosity η_{HB}) for a weighted water-based drilling fluid at 40 °C. (Reproduced with permission from reference 72. Copyright 1988 Society of Petroleum Engineers.)

The term $\delta\beta P$ represents the change in the volume fraction of solids: δ is an adjustable packing factor dependent on the volume fraction of solids and β is the isothermal compressibility of the drilling fluid. The parameter E_w is a constant that is essentially the activation energy E_a/R and B_w and C_w are constants that characterize the change in the viscosity of the aqueous phase with pressure. Alderman et al. found that the value of E_w was in the range 1.2–2.4×10^3 K, compared to 1.7×10^3 K for pure water. The corresponding range of activation energies is 10–20 kJ/mol for the drilling fluids and 14.1 kJ/mol for pure water.

Alderman et al. (72) described the Herschel–Bulkley yield stress τ_o by

$$\tau_0 = A_i \exp(-E_i/T); \qquad i = L, \, T < T^*; \quad i = H, \, T > T^* \qquad (17)$$

where $E_H \gg E_L$ and $E_L \approx 0$. The parameter T^* is a characteristic temperature below which τ_o shows little variation with temperature. Figure 14 shows the variation of η_{HB} and τ_o with T^{-1} at various pressures for a simple unweighted bentonite drilling fluid. The transition temperature T^* is about 57 °C, above which the bentonite drilling fluid shows marked gelation. The temperature T^* is likely to mark the sol–gel temperature for bentonite drilling fluids. Alderman et al. found that T^* was dependent on the surface chemistry of the bentonite and that T^* decreased as particle–particle attractions increased. The temperature T^* was found to increase on the addition of dispersants such as lignosulphonates.

The results of the study by Alderman et al. (72) are in generally good agreement with an earlier study of the rheology of water-based drilling fluids by Annis (73). Annis used a concentric cylinder rheometer that could work at temperatures up to 177 °C and pressures up to 67 bar (1000 psi). The yield stress of a bentonite suspension of concentration 43 g/L was observed to increase markedly above a temperature of about 93 °C; Annis observed that the rheology of bentonite suspensions became more non-Newtonian (more shear-thinning) as the temperature increased. Figure 15 shows rheograms of bentonite suspensions at various temperatures and in the presence of various simple additives. The addition of 1.4 g/L sodium hydroxide or sodium chloride to the bentonite

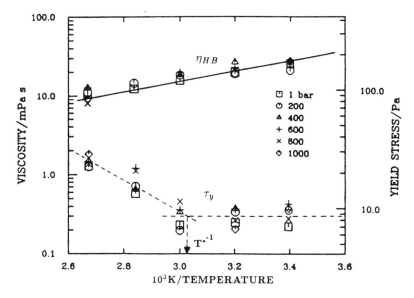

Figure 14. Temperature dependence of the Herschel–Bulkley high shear viscosity (η_{HB}) and yield stress (τ_o) of an unweighted water-based drilling fluid at various pressures. (Reproduced with permission from reference 72. Copyright 1988 Society of Petroleum Engineers.)

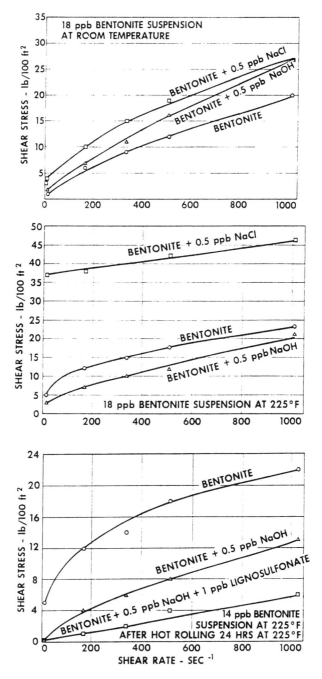

Figure 15.   Rheograms of bentonite–water suspensions at various temper-atures and in the presence of various additives (1 ppb = 2.85 g/L). (Re-produced with permission from reference 73. Copyright 1967 Society of Petroleum Engineers.)

suspension at room temperature had little effect on yield stress or plastic viscosity. At a temperature of 225 °F (107 °C) the addition of the sodium chloride resulted in a large increase in the yield stress of the suspension. In contrast the yield stresses of the neat bentonite slurry and a slurry with the addition of 1.4 g/L sodium hydroxide changed little with temperature. Addition of 2.9 g/L of a lignosulphonate thinner to the bentonite suspension with sodium hydroxide at 225 °F reduced the yield stress to almost zero.

Effect of Composition on Rheology. Relatively little systematic work has been published on the effects of drilling fluid composition on its rheology. The rheological properties of drilling fluids are primarily controlled by the concentration and molecular weight of water-soluble polymers (xanthan gum, high molecular weight CMC) and the concentration of suspended solids (bentonite with additional effects from the presence of the weighting agent and drilled solids) and the interactions (repulsive or attractive) between them.

Figure 16 shows the dependence of the Casson high shear viscosity (k_1^2) and Casson yield stress (k_0^2) on the concentration and molecular weight of the polymer CMC for simple bentonite–polymer drilling fluids of fixed bentonite content. Figure 17 shows the corresponding variation of k_0^2 and k_1^2 with bentonite concentration at fixed polymer content but variable molecular weight. The presence of the water-soluble polymers dominates the viscosity of the drilling fluids. For example, the bentonite suspension at a concentration of 70 g/L (Figure 17) has a viscosity k_1^2 of about 10 cp that is achieved by a solution of medium viscosity CMC at a concentration of 4.5 g/L. The viscosity of the drilling fluids increases as the molecular weight of the CMC polymer increases. Similarly the yield stress k_0^2 is increased by the presence of the polymers, although the low viscosity CMC has almost no effect on k_0^2.

Hughes et al. (*32*) have attempted to explain the rheology of CMC–bentonite drilling fluids by relating the high shear rate viscosity to the polymer concentration c_P and the intrinsic viscosity ($[\eta]$), which is a polymer size (or molecular weight) parameter. Hughes et al. found

$$\eta_R = (\eta - \eta_o)/\eta_o = k(c_P[\eta])^b \qquad (18)$$

where k and b are constants, η is the high shear viscosity of the drilling fluid, η_o is the high shear viscosity of the base fluid (i.e., the drilling fluid less the CMC polymer), and η_R is the reduced viscosity. A similar relationship had been found by Morris et al. (*74*) for polymer solutions at low shear rate. Figure 18 shows the plot of log (η_R) as a function of log ($c_P[\eta]$) for polymer–bentonite drilling fluids with CMC polymers of varying concentration and molecular weight, electrolyte concentration (*I*), and bentonite content. A single relationship is obtained for the poly-

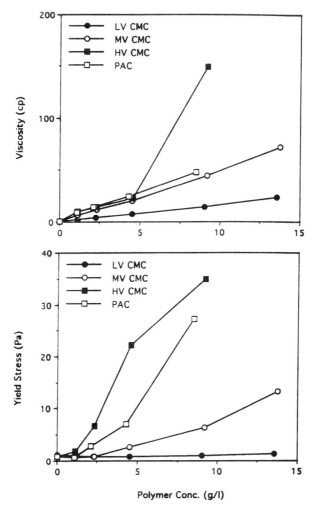

Figure 16.   Variation of Casson high shear viscosity ($k_1{}^2$) and yield stress ($k_0{}^2$) on polymer concentration for simple bentonite–polymer (CMC) drilling fluids. Bentonite content fixed at 20.6 g/L. (Reproduced with permission from reference 32. Copyright 1993 Society of Petroleum Engineers.)

mer–bentonite drilling fluids ($k = 6.92$, $b = 0.96$), although a different relationship is obtained for solutions of the polymers in the absence of bentonite (Figure 18) for reasons that are, as yet, not understood.

A decrease in the observed value of η may therefore occur by depletion of the polymer (decrease in c_P) or a decrease in $[\eta]$ which may be caused either by an increase in I (if the polymer is a polyelectrolyte) or a decrease in the average molecular weight (polymer degradation).

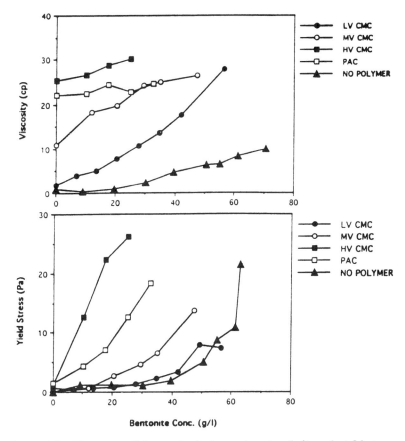

Figure 17. Variation of Casson high shear viscosity ($k_1{}^2$) and yield stress ($k_0{}^2$) on bentonite concentration for simple bentonite–polymer (CMC) drilling fluids. Polymer content fixed at 4.5 g/L. (Reproduced with permission from reference 32. Copyright 1993 Society of Petroleum Engineers.)

Heath and Tadros (75) have investigated the effects of pH and electrolyte content on the rheology of bentonite suspensions. Figure 19 shows the characteristic rheograms for bentonite suspensions (40 g/L) at various values of pH and ionic strength (sodium chloride concentration). Despite the curvature in the rheograms at low shear rate, Heath and Tadros used the Bingham plastic model to obtain the yield stress and the high shear rate viscosity. Figure 20 shows a summary of the variation of η_P and τ_0 with pH and ionic strength (I). The yield stress shows a clear minimum at pH = 7 for each value of I; τ_0 increases with increasing I. Alderman et al. (76) have also shown that τ_0 increases as the pH of a bentonite drilling fluid increases. Alderman et al. also showed

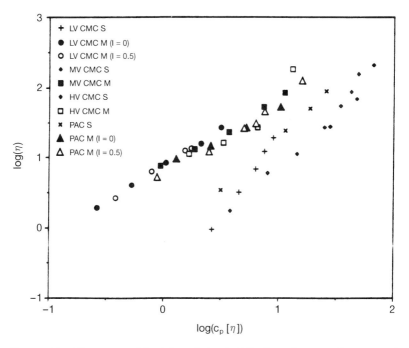

Figure 18.   Variation of log(η) with log(c_p[η]) for polymer solutions and bentonite–polymer drilling fluids (I = 0 and I = 0.5 molar). M = bentonite–polymer drilling fluid (bentonite content 20.6 g/L) and S = polymer solution. (Reproduced with permission from reference 32. Copyright 1993 Society of Petroleum Engineers.)

that τ_0 generally increased as the ratio of exchangeable sodium to calcium on the surface of the bentonite increased.

The high shear rate viscosity η_P shows less variation with pH or I (Figure 20). Heath and Tadros (75) attempted to explain the increase in τ_0 with increasing I and $|pH - 7|$ by the flocculation of bentonite clay and the formation of house-of-cards types of structures caused by edge-to-face attractive interactions (*see* reference 77 for a detailed discussion of flocculation and the formation of edge-to-edge and edge-to-face structures).

Heath and Tadros (75) showed that the addition of poly(vinyl alcohol) (PVA) to bentonite suspensions increased the yield stress and plastic viscosity for polymer concentrations up to about 5 g/L. At this PVA concentration only about 25% of the surface of the bentonite was covered with polymer. For higher polymer concentrations, the plastic viscosity and yield stress decreased with increasing polymer concentration. Heath and Tadros explain the increased yield stress and plastic viscosity by the flocculation of the bentonite by the PVA via a bridging mechanism.

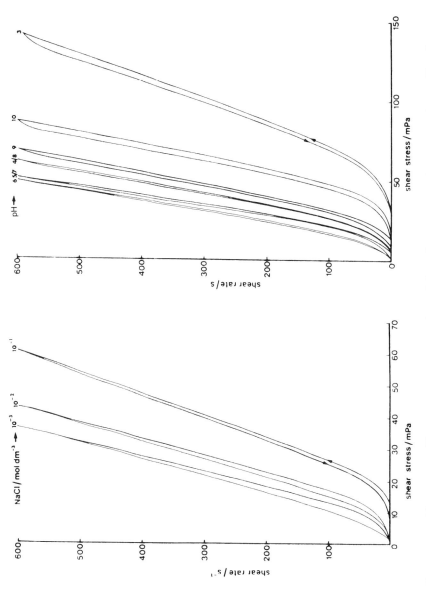

Figure 19. *Rheograms of bentonite suspensions (40 g/L) in the presence of various sodium chloride concentrations and values of pH. (Reproduced with permission from reference 75. Copyright 1983 Academic.)*

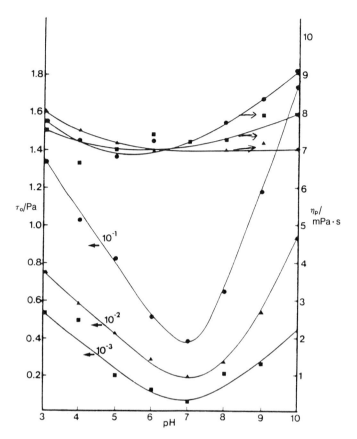

Figure 20.   Dependence of Bingham plastic viscosity (η_p) and yield stress (τ_o) of bentonite suspensions (40 g/L) on pH at various ionic strengths. (Reprinted with permission from reference 75. Copyright 1983 Academic.)

The rheology of drilling fluids may be modified by the modification of the interactions between the suspended particles. Polymeric dispersants (or thinners) such as lignosulphonates are frequently added to water-based drilling fluids to prevent flocculation of the solids (e.g., *see* Figure 15). Rabaioli et al. (78) have studied the adsorption of several polymeric dispersing agents on bentonite and their effect on the rheology of bentonite suspensions. Figure 21 shows the adsorption isotherms of the polymers sodium polyacrylate (weight average molecular weight 4000 g/mole), sodium lignosulphonate, ferrochrome lignosulphonate, and the synthetic copolymer polyacrylic ethyl sulphonic acid (PAA–AMES) on bentonite. The adsorption of the polyacrylic acid polymers was very low in comparison with the lignosulphonate polymers. The limiting amount of each of the polymers adsorbed on the bentonite (2–

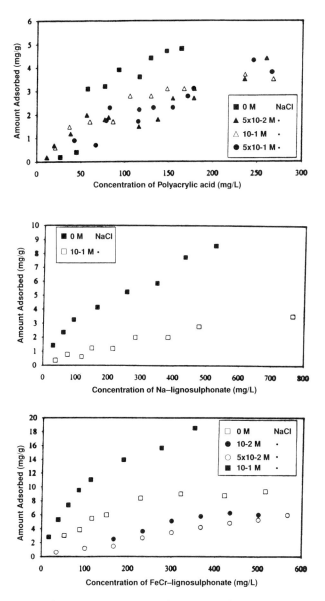

Figure 21. Adsorption isotherms of sodium polyacrylate, sodium ligno-sulphonate, ferrochrome lignosulphonate, and the copolymer PAA–AMES on bentonite at various ionic strengths. (Reproduced with permission from reference 78. Copyright 1993 Society of Petroleum Engineers.)

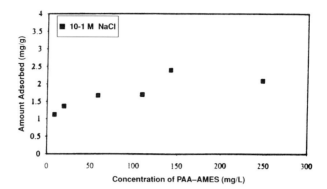

Figure 21.—Continued.

20 mg of polymer per gram of dry clay) was insufficient to form a mono-layer coverage. In contrast, Heath and Tadros (75) measured a limiting concentration of adsorbed (neutral) PVA of about 530 mg/g. Rabaioli et al. suggested preferential adsorption of the anionic polyelectrolytes on the edges of the platelets, although at pH = 10 the edges are likely to be negatively charged (19).

Despite the low levels of adsorption, the anionic polymers have a significant effect on the rheology of the bentonite suspensions. Figure 22 shows the dependence of the Bingham yield stress τ_o on bentonite concentration with and without added dispersants. Rabaioli et al. found the dependence of τ_o on the volume fraction of bentonite ϕ to be

$$\tau_o = A\phi^{4.8} \tag{19}$$

where A is a constant that is dependent on the presence of dispersant. The value of A for the bentonite–sodium polyacrylate suspension is smaller than for the bentonite–ferrochrome lignosulphonate suspension despite the higher levels of polymer adsorption in the latter suspension. The adsorption isotherms shown in Figure 21 cannot be compared on a molar basis as the molecular weight of the lignosulphonates are unknown.

The storage modulus G' obtained from low frequency (0.1–10 Hz) oscillatory rheological measurements (79) is related to ϕ (Figure 22) by

$$G' = B\phi^n \tag{20}$$

where B is a constant and the exponent n is a sensitive measure of par-ticle–particle interactions. When n is large (n = 10–20), G' is small and there are no attractive forces between the clay platelets. The flocculated bentonite suspensions are characterized by values of n of 2.5–3. The addition of the polymeric dispersants causes the value of n to increase

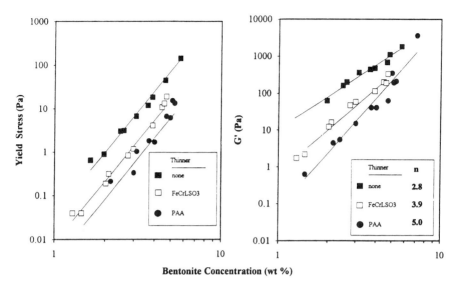

Figure 22. Dependence of Bingham yield stress and storage modulus (G') on bentonite content of bentonite suspensions in the presence of various additives. (Reproduced with permission from reference 78. Copyright 1993 Society of Petroleum Engineers.)

from a value of 2.8 for the bentonite suspension to $n = 3.9$ for the ferrochrome lignosulphonate–bentonite suspension and $n = 5.0$ for the sodium polyacrylate–bentonite suspension. The value of n is also increased by decreasing the ionic strength of bentonite suspensions (*78*).

Drilling Fluid Hydraulics. Drilling fluid hydraulics is concerned primarily with the fluid mechanics of large-scale flow of the drilling fluid in the drill pipe, nozzles, and annulus and the transport of cuttings in the annulus. Additional applications of drilling fluid hydraulics are displacement of the drilling fluid behind the casing by cement (*80*) and the rise velocity of gas bubbles that are formed in drilling fluids during gas kicks (*81*).

The objectives of a drilling hydraulics programme are to ensure good hole cleaning, to minimize frictional losses in the flow in the drill pipe and annulus and to minimize the erosion of open hole formations by the shear stress exerted on the borehole wall. Several texts have described drilling fluid (or non-Newtonian fluid) hydraulics in some detail (*82–85*). The bulk flow properties of the drilling fluid are calculated from the models $\tau = f(\dot{\gamma})$ used to fit the experimental rheological data.

Drill Pipe. Laminar Flow. Under normal oil well drilling conditions the flow of drilling fluid in the drill pipe and nozzles is turbulent.

A typical flow rate of 2 m³/min in the drill pipe (diameter = 0.1 m) gives a fluid velocity of 4.2 m/s, which causes fully turbulent flow in most drilling fluids. The calculation of the pressure drop down the length of the drill pipe is made somewhat complex by the flow being turbulent and the non-Newtonian rheology of the drilling fluid. These calculations are, however, important as about 30% of the total frictional pressure losses occur in the drill pipe.

The pressure drop ΔP down the length L of the pipe of radius R for a Newtonian fluid in laminar flow is given by

$$\Delta P = 8\eta L v_m/R^2 \tag{21}$$

where η is the Newtonian viscosity and v_m is the mean flow velocity (v_m = flow rate/cross-sectional area). The shear stress τ_w at the wall of the pipe is given by the general equation

$$\tau_w = R\Delta P/2L \tag{22}$$

which, for a Newtonian fluid (eq 21), gives

$$\tau_w = 4v_m\eta/R \tag{23}$$

and the corresponding expression for the wall shear rate is

$$\dot{\gamma}_w = 4v_m/R. \tag{24}$$

The analogous expressions for ΔP and $\dot{\gamma}_w$ for non-Newtonian fluids depend on the rheological model describing the fluid. The shear rate at the wall $\dot{\gamma}_w$ is given by the Rabinowitsch–Mooney equation (86, 87), which in its general form is independent of the rheology of the fluid:

$$\dot{\gamma}_w = 3v_m/R + (v_m/R)d \ln (4v_m/R)/d \ln (\tau_w) \tag{25}$$

where τ_w is given by equation 22. The derivative $d \ln (4v_m/R)/d \ln (\tau_w)$ is commonly denoted as $1/n'$ and the Rabinowitsch–Mooney equation becomes

$$\dot{\gamma}_w = (3n' + 1)v_m/n'R. \tag{26}$$

The Rabinowitsch–Mooney equation is important because it gives a general relationship between ΔP and flow rate (or v_m)

$$n' = d \ln (R\Delta P/2L)/d \ln (4v_m/R) \tag{27}$$

which is independent of the rheology of the fluid in laminar flow. The parameter n' gives a direct measure of the non-Newtonian behavior of the fluid. The condition $n' = 1$ describes a Newtonian fluid whereas $n' < 1$ describes a plastic (or pseudoplastic) fluid.

For a constant value of n' over a given range of shear rates, equation 27 integrates to give

$$R\Delta P/2L = K'(4v_m/R)^{n'} \tag{28}$$

and can be combined with equation 26 to give

$$\tau_w = K'(4n'/3n' + 1)^{n'}(\dot{\gamma}_w)^{n'} \tag{29}$$

which is the power law model (eq 6) for non-Newtonian fluids with $n = n'$ and $K = K'(4n'/3n' + 1)^{n'}$. It is stressed that the Rabinowitsch–Mooney equation gives a general description of the laminar flow of any fluid that is independent of its constitutive equation; the power law fluid model is obtained for the special case of constant n' over some range of $\dot{\gamma}$ where equation 27 can be directly integrated.

The expressions for ΔP (or τ_w) and $\dot{\gamma}_w$ depend on the particular choice of rheological model. For a power law fluid ($n = n'$ in eq 26)

$$\dot{\gamma}_w = (3n + 1)v_m/nR \tag{30}$$

and the wall shear stress τ_w is given by (from eq 6)

$$\tau_w = K(v_m(3n + 1)/nR)^n \tag{31}$$

which, with equation 22, yields

$$\Delta P = (2KL/R)(v_m(3n + 1)/Rn)^n \tag{32}$$

The development of the corresponding equations for non-Newtonian fluids with a yield stress (Bingham plastic, Casson, Herschel–Bulkley) is more complex because the part of the fluid for which $\tau < \tau_0$ will move as a rigid plug. Figure 23 shows a schematic of laminar and plug flow in a pipe. The inner core of radius R_p flows as a plug and the velocity profile is flat. The expressions for ΔP for fluids with a yield stress are not explicit and must be obtained iteratively. For example, the Bingham plastic model gives

$$24L\eta_p v_m/R^2\Delta P = 3 - 8L\tau_0/R\Delta P + (2L\tau_0/R\Delta P)^4 \tag{33}$$

which is the well-known Buckingham equation and it reduces to equation 22 when $\tau_0 = 0$. A common approximation (88) used to obtain simplified expressions for ΔP for fluids exhibiting a yield stress is to ignore the plug flow region. In the case of a Bingham plastic fluid, the term $(2L\tau_0/R\Delta P)^4$ in equation 33 is ignored to give

$$\Delta P = 8L\tau_0/3R + 8L\eta_p v_m/R^2. \tag{34}$$

LAMINAR FLOW

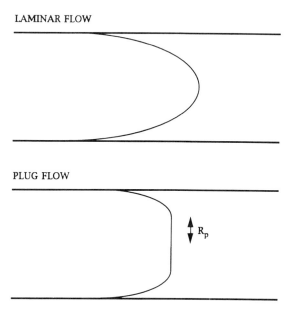

PLUG FLOW

R_p

Figure 23.   Schematic of laminar and plug flow in a pipe. Radius of central plug is R_p.

The corresponding approximations for the shear stress and shear rate at the wall are

$$\tau_w = 4\eta_p v_m/R + 4\tau_o/3 \tag{35}$$

and

$$\dot{\gamma}_w = 4v_m/R + \tau_o/3\eta_p. \tag{36}$$

Turbulent Flow.   The dimensionless Reynolds number Re of a New-tonian fluid flowing in a pipe of radius R can be defined by

$$Re = 2\rho v_m R/\eta \tag{37}$$

where ρ is the density of the fluid. The flow regime of a Newtonian fluid passes from laminar to turbulent when Re is approximately 2100. For the example of v_m = 2 m^3/min and R = 0.05 m with water (ρ = 1000 kg/m^3 and η = 0.001 Pa·s), Re = 3300, which is fully turbulent flow.

The definition of the Reynolds number for a non-Newtonian fluid is not uniquely defined as the viscosity is a function of the shear rate, that is, v_m. Several approaches have been used to obtain an expression for Re in non-Newtonian fluids. One approach is to define a mean viscosity η_m that is equal to the Newtonian viscosity giving the same frictional pressure drop ΔP_L in laminar flow

$$\eta_m = \Delta P_L R^2 / 8 L v_m \tag{38}$$

and the equivalent Reynolds number Re$'$ is defined by

$$\mathrm{Re}' = 2\rho v_m R / \eta_m = 16 \ \rho v_m{}^2 L / R \Delta P_L. \tag{39}$$

Metzner and Reed (89) and Dodge and Metzner (90) derived a generalized Reynolds number from the Rabinowitsch–Mooney equation

$$\mathrm{Re}'' = (2R)^{n'} \rho (v_m)^{2-n'} / K' 8^{n'-1} \tag{40}$$

where K' and n' are defined in equation 28. Equation 40 is readily applied to power law fluids. In each of these treatments turbulent flow is obtained when the equivalent Reynolds number is 2100.

In turbulent flow the frictional pressure drop down the drill pipe must be calculated from equations that have been determined empirically. The commonest method in drilling fluid hydraulics (91–93) is to use a friction factor f, the so-called Fanning friction factor, defined by the ratio of the wall shear stress τ_w to the kinetic energy per unit volume of the flowing fluid $0.5 \ \rho v_m{}^2$ (94)

$$f = \tau_w / 0.5 \ \rho v_m{}^2 = R \Delta P / L \rho v_m{}^2 \tag{41}$$

from which the pressure drop ΔP is given by

$$\Delta P = f L \rho v_m{}^2 / R \tag{42}$$

In laminar flow $f = 16/\mathrm{Re}$, whereas in turbulent flow the dependence of f on Re is a function of the specific rheological behavior of the fluid and roughness of the walls of the inside of the drill pipe (91). A number of functional relationships between f and Re have been proposed for turbulent flow. Ignoring the effects of the roughness of the surface of the drill pipe, f can be approximately related to Re by a generalized form of the well-known Blasius equation for Newtonian fluids (90, 95)

$$f = a/\mathrm{Re}^b \tag{43}$$

where a and b are constants that depend on n' (i.e., the rheology of the fluid). For Newtonian fluids the Blasius equation is

$$f = 0.078/\mathrm{Re}^{0.25} \tag{44}$$

and an approximate form of equation 42 is given by

$$\Delta P = 0.066 L \rho^{0.75} v_m{}^{1.75} \eta_e{}^{0.25} / R^{1.25} \tag{45}$$

where η_e is the effective viscosity of the fluid. Dodge and Metzner (90) derived the general relationship

$$f^{-0.5} = (4/n'^{0.75}) \log (f^{(1-0.5n')} \, \mathrm{Re}'') - 0.4/n'^{1.2} \qquad (46)$$

for any fluid with Re'' given by equation 40. Figure 24 shows the dependence of f on Re'' for various values of the exponent n'. Figure 24 shows that the onset of turbulent flow can occur over a range of Reynolds numbers that are determined by the value of n'. The data of Dodge and Metzner (90) show that the range of Re'' over which turbulence develops extends from $(3470-1370\ n')$ to $(4270-1370\ n')$.

The transition from laminar to turbulent flow in a non-Newtonian fluid depends on the rheological model used to describe it. The concepts of a critical friction factor or critical Reynolds number have been used to define the boundary. For a power law fluid the critical friction factor f_{cr} is given by (96)

$$f_{cr} = (1/(2 + n))^{(2+n)/(1+n)}(1 + 3n)^2/404n \qquad (47)$$

and, from $f = 16/\mathrm{Re}$, the critical Reynolds number Re^* is given by

$$\mathrm{Re}^* = 6464n/(1 + 3n)^2(1/(2 + n))^{(2+n)/(1+n)} \qquad (48)$$

which gives $\mathrm{Re}^* = 2100$ for a Newtonian fluid ($n = 1$). The value of Re^* increases as n decreases; for example, when $n = 0.5$, $\mathrm{Re}^* = 2381$.

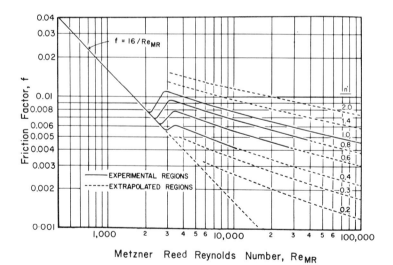

Figure 24.   Dependence of Fanning friction factor on Reynolds number for various values of exponent n' *for laminar and turbulent pipe flow. (Reproduced with permission from reference 90. Copyright 1959 American Institute of Chemical Engineers.)*

Hanks and Christiansen (97) have shown that equations 47 and 48 are not applicable to Bingham plastic fluids or those that exhibit a finite yield stress. Hanks (98) has developed the concept of the critical Reynolds number for Bingham plastics. From equation 33, the Bingham fluid Reynolds number can be defined by

$$\mathrm{Re_b} = ((1 - 4x/3 - x^4/3)/2x) \cdot \rho R^2 \tau_o / \eta_p^2 \tag{49}$$

where $x = \tau_o/\tau_w = 2\tau_o L/R\Delta P$. The Hedström number He is defined by

$$He = 4\rho R^2 \tau_o / \eta_p^2 \tag{50}$$

and equation 49 can be written as

$$\mathrm{Re_b} = ((1 - 4x/3 - x^4/3)/2x) \cdot He/8x \tag{51}$$

and the critical value of $\mathrm{Re_b}$ occurs when the critical value of x is given by (98)

$$He/16,800 = x_c/(1 - x_c)^3 \tag{52}$$

Figure 25 shows the dependence of the critical value of $\mathrm{Re_b}$ on He for a variety of Bingham plastics (98). The theoretical relationship, obtained

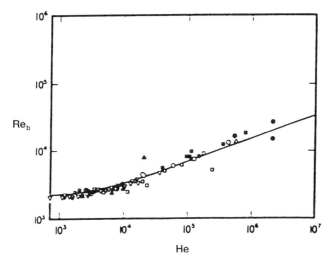

Figure 25.   Dependence of critical Bingham Reynolds number (Re_b) on Hedström number (He). (Reproduced with permission from reference 98. Copyright 1967 Society of Petroleum Engineers.)

from equations 51 and 52, gives an excellent fit to the experimental data.

For smooth pipes, the relationship between friction factor and Re_b is close to that found for Newtonian fluids (99). From equation 33, the friction factor for a Bingham plastic fluid is given by

$$f = (1/(1 - 4x/3 + x^4/3))8\eta_p/R\rho v_m \tag{53}$$

which approximates to

$$f = (1 + \tau_o R/3\eta_p v_m)8\eta_p/R\rho v_m \tag{54}$$

if the term representing plug flow ($x^4/3$) is ignored. The Bingham plastic Reynolds number is therefore approximated by

$$Re_b = 2R\rho v_m/((\eta_p(1 + \tau_o R/3\eta_p v_m)) \tag{55}$$

If the critical value of Re_b is assumed to be 2100, then the critical velocity v_{mc} of a Bingham fluid at which laminar flow becomes turbulent is

$$v_{mc} = 19(\tau_o/\rho)^{1/2} \tag{56}$$

for the condition $\tau_o R/3\eta_p v_m \gg 1$ that is usually obtained in drill pipe.

It is well known (100–102) that many of the water-soluble polymers used in drilling fluids, such as CMC, guar gum, and polyacrylamide, are effective drag reducers, (i.e., they reduce frictional pressure losses in turbulent flow). The value of the Reynolds number at the onset of turbulent flow may be increased to in excess of 10,000 with the addition of these drag reducing additives (103, 104). Figure 26, for example, shows the effect of the polymer CMC (concentration of 3 g/L) on the friction factor f. The origin of the drag reducing properties of these water-soluble polymers is not well understood, although it is related to their viscoelastic properties. Govier and Aziz (105) have reviewed the effects of viscoelastic additives on turbulent flow.

A comparison of several rheological models in predicting frictional pressure losses in laminar and turbulent flow has recently been made by Okafor and Evers (106). Comparison was made between the Bingham plastic, power law, and Robertson–Stiff rheological models. Figure 27 shows the fit of the 3 models to the experimental rheological data obtained for a bentonite drilling fluid ($\rho = 1066$ kg/m^3) characterized by a low yield stress. The Robertson–Stiff equation gives a better fit to the data than the Bingham plastic model but only slightly better than the power law model. The superior performance of the Robertson–Stiff model is, of course, a consequence of an additional fitting parameter.

Figures 28a and 28b compare the measured and calculated frictional pressure losses for laminar and turbulent flow of the drilling fluid through

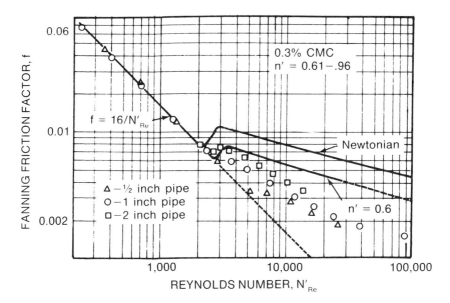

Figure 26.   Effect of the polymer CMC on the relationship between Fanning friction factor and Reynolds number in laminar and turbulent pipe flow. (Reproduced with permission from reference 90. Copyright 1959 American Institute of Chemical Engineers.)

a pipe using the power law and Bingham plastic rheological models. The power law model gives a reasonable prediction of the frictional pressure loss in laminar flow up to $v_m = 1.5$ m/s at which point the flow becomes turbulent. The power law model is less successful at predicting pressure losses in fully turbulent flow. The Bingham plastic model gives poorer predictions of frictional pressure in laminar flow but gives a good prediction of the onset of turbulent flow. Equation 55 gives a good prediction of the value of v_m at which the flow becomes turbulent. Figure 29 shows a summary of the calculations of frictional pressures in laminar and turbulent flows for power law and Bingham plastic fluids.

Annulus.   From the drill pipe, the drilling fluid is jetted out of the bit nozzles and into the annulus. The jets are designed to remove drilled cuttings away from the drill bit. The pressure drop across the nozzles accounts for about 60% of the total frictional pressures losses during the circulation of the drilling fluid. About 90% of the pressure drop across the nozzles is due to the turbulence at the outlet, with only about 10% due to the flow through the nozzles.

The return flow of the drilling fluid in the wellbore is considerably more complex than in the drill pipe as the flow is annular and the drilling fluid is rotated by the rotating drill pipe. In addition, the flow is fre-

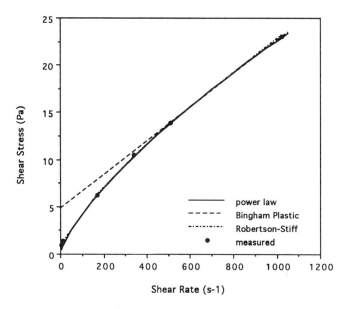

Figure 27. Comparison of fit of power law, Bingham plastic and Robertson–Stiff rheological models to experimental data from bentonite drilling fluid. (Data from reference 106.)

quently turbulent and in deviated wells the drill pipe is not in the center of the well (eccentric annulus). The pressure drop in the annulus accounts for about 10% of the total frictional pressure losses in normal drilling, although in slimhole drilling, where the annular gap is considerably smaller, the annular pressures losses can be up to 90% of the total (*107*).

The pressure drop ΔP down the length of a vertical annulus with laminar, non-rotational flow of a Newtonian fluid is given by (*108*)

$$\Delta P = 8\eta v_m L/G(R) \tag{57}$$

where

$$G(R) = R_B{}^2 + R_P{}^2 + (R_B{}^2 - (R_P{}^2 v)/\ln(R_P/R_B) \tag{58}$$

and R_B and R_P are the radii of the borehole and drill pipe, respectively.

Obtaining equivalent expressions for ΔP for non-Newtonian fluids is difficult and for most rheological models it is not possible to obtain analytical expressions. Good approximate solutions, however, can be obtained by replacing the narrow annulus geometry by a parallel plate model (*108, 109*), that is, annular flow is represented by slot flow. The

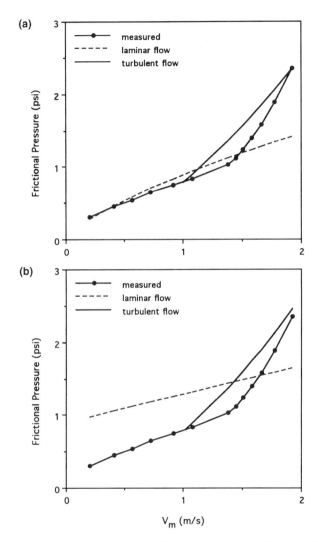

*Figure 28.   Comparison of calculated frictional pressures for flow of a dril-
ling fluid in a pipe using (a) power law and (b) Bingham plastic models.
(Data from reference 106.)*

plate width is $\pi(R_B + R_P)$ and the plate separation is $(R_B - R_P)$, giving
the required cross-sectional area of $\pi(R_B^2 - R_P^2)$. The approximate annular pressure drop for a Newtonian fluid is given by

$$\Delta P = 12\eta v_m L/(R_B - R_P)^2 \tag{59}$$

and the shear stress and shear rate at the borehole wall are given by

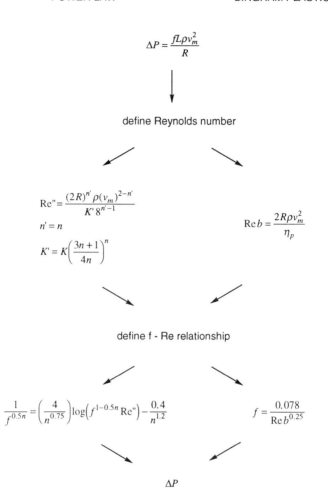

Figure 29. Calculation of frictional pressures in turbulent flow using the power law and Bingham plastic models.

$$\tau_w = 6v_m\eta/(R_B - R_P) \qquad (60)$$

and

$$\dot{\gamma}_w = 6v_m/(R_B - R_P). \qquad (61)$$

The annular pressure drop for a Bingham fluid is obtained from

$$12\eta_p v_m L/\Delta P(R_B{}^2 - R_P{}^2) = 1 - 3L\tau_o/\Delta P(R_B - R_P)$$
$$- 4(L\tau_o/\Delta P(R_B - R_P))^3 \qquad (62)$$

and the equivalent expression for a power law fluid is

$$6v_m(2KL/\Delta P(R_B - R_P))^{1/n}/(R_B - R_P) = 3n/(2n + 1). \qquad (63)$$

The slot flow approximation is recommended when $R_P/R_B \geq 0.2$–0.3, which is usually achieved in oil well drilling; the ratio R_P/R_B is in excess of about 0.8 for slimhole drilling (*107*).

The Reynolds number in annular flow is commonly defined by

$$\text{Re} = \rho v_m(R_B - R_P)/\eta_e \qquad (64)$$

or

$$\text{Re} = 2\rho v_m R_H/\eta_e \qquad (65)$$

where R_H is the hydraulic radius [$R_H = (R_B - R_P)/2$]. The onset of turbulence in annular flow also occurs at $\text{Re} = 2100$ and the frictional pressure drop can be calculated using equation 42 with R replaced by R_H. The limiting relationship $f = 24/\text{Re}$ is obtained for laminar flow in an annulus.

The frictional pressure gradient along the length of the annulus is in the same direction as the hydrostatic pressure gradient. The frictional pressure gradient can be combined with the hydrostatic pressure gradient to yield an effective density that is greater than the static fluid density. This combined density is termed the equivalent circulating density (ECD), which is given by

$$\text{ECD} = \rho + \Delta P/Zg \qquad (66)$$

where Z is vertical depth (taken as positive in the downward direction) and g is the acceleration due to gravity.

Marken et al. (*110*) have given a detailed account of the factors affecting annular pressure losses, including the effects of rotational flow in the annulus. It might be expected that rotational flow in the annulus would decrease the frictional pressure losses as the increased shear rate would lower the viscosity. However, Marken et al. observed an increase in frictional pressure due to the formation of Taylor vortices. McCann et al. (*107*) have observed increases in the frictional pressures in slimhole annuli due to annular rotation of the drilling fluid.

Cuttings Transport. *Vertical Wells.* During normal drilling the cuttings generated by the bit are transported to the surface in the annular flow of the drilling fluid. The ability of the drilling fluid to lift the cuttings back to the surface is dependent on a number of factors: drilling fluid rheology and density, annular geometry, and fluid velocity and cuttings size, shape, and density.

The volume fraction V_c of cuttings in the annulus is given by

$$V_c = \pi R_B^2 \cdot ROP \cdot (1 - \phi)/QR_t \qquad (67)$$

where ROP is the rate of penetration of the drilling, ϕ is the porosity of the drilled rock and Q is the flow rate of the drilling fluid. The parameter R_t, termed the transport ratio (111), is defined by

$$R_t = v_t/v_m \qquad (68)$$

where v_t is the transport velocity of the cuttings in the drilling fluid

$$v_t = v_m - v_s \qquad (69)$$

and v_s is the slip velocity of the cuttings in the drilling fluid. Clearly for hole cleaning the mud flow rate (v_m) must exceed the cuttings slip velocity (v_s). For a typical case, assuming for the sake of simplicity $v_s = 0$ ($R_t = 1$), with $Q = 2$ m³/min, ROP = 10 m/h, $\phi = 0.2$, and $2R_B =$ 12.25 in. (31.1 cm), the volume fraction of cuttings is $V_c = 0.005$ or 0.5%. For an average rock matrix density of 2700 kg/m³, the rate of cuttings production is about 1650 kg/h.

A cuttings Reynolds number Re_c can be defined by

$$Re_c = 2\rho v_s R_c/\eta_e \qquad (70)$$

where R_c is the radius of the cutting. Walker and Mayes (112) considered the flow regime to be turbulent for $Re_c > 100$.

The slip velocity v_s of a sphere in a Newtonian fluid of viscosity η is given by Stokes' well-known equation (113)

$$F = 6\pi R_s \eta v_s \qquad (71)$$

where F is the viscous drag and R_s is the radius of the sphere. Equation 71 is only valid for creeping flow (113), which is characterized by $Re_c < 0.1$. The terminal slip velocity of the cutting in the drilling fluid in a vertical well can be determined from equation 71 by replacing F by the weight of the particle

$$v_s = 2R_c^2(\rho_c - \rho)g/9\eta \qquad (72)$$

where ρ_c is the density of the cutting and g is the acceleration due to gravity. Clearly the greater the viscosity and density of the drilling fluid, the smaller the slip velocity.

To extend the prediction of v_s to flow regimes where $Re_c \gg 0.1$ it is necessary to use empirical correlations between Re_c and a friction factor, similar to the approach used for calculating frictional pressure losses in turbulent pipe flow. The cutting friction factor f_c is defined by

$$f_c = \tau_w/0.5\rho v_s^2 \qquad (73)$$

where, from equation 71,

$$\tau_w = F/\pi R_c^2 = 6\eta_e v_s/R_c \qquad (74)$$

giving

$$f_c = 12\eta_e/R_c\rho v_s \qquad (75)$$

The friction factor f_c (or drag coefficient (*114, 115*)) is related to the cutting Reynolds number by $f_c = 24/Re_c$ for $Re_c < 0.1$. Equations 72 and 75 combine to give a more general expression for the slip velocity

$$v_s = (8(\rho_c - \rho)gR_c/3\rho f_c)^{1/2} \qquad (76)$$

A number of experimentally determined relationships between f_c and Re_c have been published (*111, 114–117*). Figure 30 (from reference 114) shows a comparison between the published data of Walker and Mayes (*112*) and Moore (*116*); the combined data can be fitted by (*114*)

$$f_c = 1 + 40/Re_c \qquad (77)$$

over a wide range of Re_c.

The essential differences between the various treatments is an expression for η_e in Re_c (eq 70) in terms of the non-Newtonian rheology

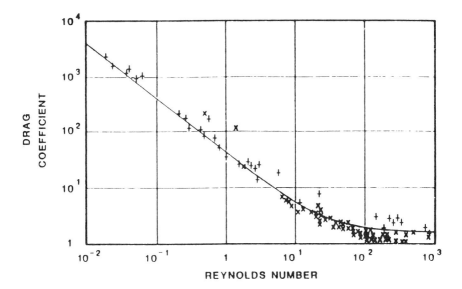

Figure 30. Dependence of cuttings drag coefficient (friction factor f_c) on cuttings Reynolds number. (Data from reference 112, plot symbol X; reference 116, plot symbol +; and reference 114.)

of the drilling fluid. Moore (*116*), for example, used the power law model to derive an expression for η_e for the drilling fluid in annular flow

$$\eta_e = K(2v_m(2n + 1)/n(R_B - R_P))^{n-1} \tag{78}$$

using

$$\dot{\gamma}_w = 2v_m(2n + 1)/n(R_B - R_P) \tag{79}$$

for the shear rate at the borehole wall. Sample and Bourgoyne (*117*) have discussed some of these early models in detail.

Gavignet and Wick (*115*) have given a treatment of the effect of the yield stress of a drilling fluid on cuttings settling velocity. Gavignet and Wick argued that in considering the drag forces on a settling cutting the effect of the yield stress on v_s must be separated from the more usual effect of viscosity (eq 75) because the cutting must overcome the yield stress to slip. In their treatment Gavignet and Wick related f_c to a modified Reynolds number D_Y, termed a dynamic parameter, defined by

$$D_Y = Re_{cy}/(1 + 7\pi Y/24) \tag{80}$$

where Re_{cy} is the cuttings Reynolds number obtained using the effective viscosity η_e derived from a polynomial used to fit the raw shear stress–shear rate rheological data and Y is the yield number defined by

$$Y = 2R_c\tau_0/v_s\eta_e \tag{81}$$

where τ_0 is the yield stress. Gavignet and Wick (*115*) found that both their experimental data and previously published data fell onto the $f_c - D_Y$ curve for Newtonian fluids. Note that an increase in τ_0 causes the slip velocity to decrease and R_t to increase via a decrease in D_Y and an increase in f_c.

Inclined and Horizontal Wells. The increase in the number of deviated and horizontal wells drilled over the last decade has focused attention on the problem of cuttings removal. The transport of cuttings in inclined boreholes changes markedly as a function of the angle of inclination and the treatment of cuttings transport developed for vertical wells no longer gives an adequate description. Tomren et al. (*118*) have conducted extensive laboratory experiments of cuttings transport in deviated wells as a function of fluid rheology, hole geometry, cuttings size, pipe rotation, and angle of inclination (χ). Figure 31 summarizes the main features of cuttings transport in near vertical ($\chi = 0–30°$), inclined ($\chi = 30–60°$), and steeply inclined wells ($\chi = 60–90°$). Above an angle of inclination of about 30°, a bed of cuttings will form on the lower side

$\chi = 0 - 30^{o}$

hole cleaning largely determined by
slip velocity of cuttings in drilling fluid;
hole cleaning with χ up to 30° not
significantly different from χ=0°.

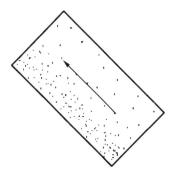

$\chi = 30 - 60^{o}$

bed of cuttings tends to form on lower
side of hole below critical velocity;
cuttings transport more efficient in laminar
flow for χ<45° and in turbulent flow for
χ>45°; at lower flow rates cuttings circulate
on lower wall; cuttings bed tends to slide
down hole when drilling fluid circulation
ceases.

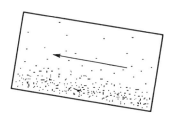

$\chi = 60 - 90^{o}$

formation of cuttings bed on low side of well;
critical velocity is in turbulent flow regime
and frequently at flow rates not attained in
normal drilling; in turbulent flow the
rheology of the drilling fluid has little effect
on cuttings transport; cuttings bed tends not
to slip down hole when drilling fluid
circulation ceases.

*Figure 31.   Summary of features of cuttings transport in inclined wells.
Arrows show direction of cuttings transport.*

of the well below some critical flow rate (or velocity) of the drilling
fluid.

Becker et al. (*119*) have studied the effects of drilling fluid rheology
and flow regime on cuttings transport in laboratory-scale inclined annuli.
For angles of inclination up to about 45°, cuttings transport was found
to be more effectively achieved with laminar flow; turbulent flow was
found to be more effective in steeply inclined and horizontal wells. In
turbulent flow the rheology of the drilling fluid had little influence on
cuttings transport. However, in the laminar flow regime, the effectiveness of hole cleaning exhibited a high degree of correlation with low

shear rate measurements such as the initial gel strength (64) and the shear stress at low shear rate (5 or 10 s^{-1}).

A number of models of cuttings transport in deviated wells have been developed in an attempt to predict this critical flow rate (120–123). A simple model of the cuttings bed by Gavignet and Sobey (120), for example, gives a reasonable prediction of the main features, including an estimate of the critical flow rate and the weak influence of drilling fluid rheology. Gavignet and Sobey predicted that for highly deviated wells ($\chi \geq 60°$), the critical flow rate would always occur when the drilling fluid was in turbulent flow. Luo et al. (122) have developed a physically based model to predict the critical flow rate of the drilling fluid to prevent (or remove) a stationary bed of drilled cuttings. The model is based on a regression analysis of the principal dimensionless groups associated with the dynamics of the cuttings on the borehole wall and the drilling fluid in the annulus. The two principal dimensionless groups identified by Luo et al. are

$$\Pi_1 = v_c^2 \rho / 2R_c(\rho_c - \rho)g \sin \chi \qquad (82)$$

and

$$\Pi_2 = 2R_c \rho v_c / \eta_a \qquad (83)$$

where v_c is the critical friction velocity of the drilling fluid on the borehole wall and is given by

$$v_c = (\tau_{wc}/\rho)^{1/2} \qquad (84)$$

where τ_{wc} is the critical shear stress at the borehole wall. The critical flow rate Q_c is obtained from v_c and the known annular geometry. Equation 82 is the ratio of the hydrodynamic and body forces on the cutting whereas equation 83 is equivalent to the cutting Reynolds number. The critical velocity is obtained from the regression

$$\Pi_1 = k\Pi_2^q \qquad (85)$$

where k and q are fitting constants. Equation 85 gives a generally good prediction of v_c for both laboratory flow loop and field data. The model has been used by Luo et al. (122) to isolate the effects of key variables on Q_c. For example, the critical flow rate Q_c varies approximately inversely with the density of the drilling fluid, due to the reduction in the buoyancy of the cuttings.

Filtration (Fluid Loss) Properties of Drilling Fluids

Static Filtration.   The static fluid loss of drilling fluids is a property that is regularly measured, both at the rig site during drilling operations and in the laboratory. The term static filtration is used to describe solid–liquid separation when the only movement of the drilling fluid is toward the filter medium. In contrast, the term dynamic filtration or cross-flow filtration is used to describe solid–liquid separation under conditions where the drilling fluid is also moving across the surface of the filter medium. The static fluid loss is measured in a standard API fluid loss cell (Figure 32) where a given volume of drilling fluid is filtered against a standard filter paper using a constant applied gas pressure of 100 psi (7 bars). The API fluid loss is the volume of filtrate collected after 30 min using a Whatman 50 (or equivalent) filter paper with a cross-sectional area of 45.8 cm^2 (*124, 125*).

Figure 33 shows the time dependence of the volume of static filtrate collected from a typical water-based drilling fluid. The initial volume collected at time $t = 0$ is termed the spurt loss and represents the uncontrolled flow of filtrate and fine solids into the filter medium; the spurt loss continues until a filter cake forms on the surface of the filter medium.

The functional form of the static filtration curve (Figure 33) may be readily derived if it is assumed that the flow of the filtrate through the filter cake is governed by Darcy's law (*126*). The flow rate of the filtrate dV_{fs}/dt is given by Darcy's equation

$$dV_{fs}/dt = k_c \Delta P_{fs} A / \eta_f h_c \tag{86}$$

where k_c is the permeability of the filter cake, A is the cross-sectional area of the filter medium, η_f is the filtrate viscosity and h_c is the cake thickness. The cake thickness h_c is a function of the cumulative filtrate volume V_{fs}. At some instant

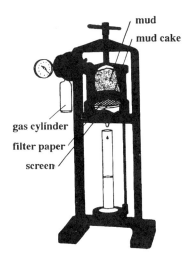

Figure 32.   API static fluid loss cell.

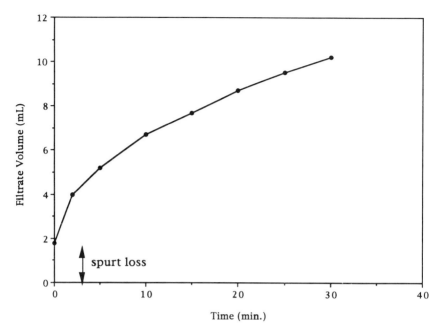

Figure 33. Time dependence of the cumulative filtrate volume from the static filtration of a bentonite–polymer drilling fluid.

$$dh_c = dV_c/A = (dV_m - dV_{fs})/A \qquad (87)$$

where dV_c is the volume of the filter cake formed from a volume dV_m of drilling fluid. A parameter ξ can be defined by

$$\xi = dV_{fs}/dV_m \qquad (88)$$

which is the volume of filtrate per unit volume of filtered drilling fluid. Equation 87 can be integrated to give

$$h_c = V_{fs}(\xi^{-1} - 1)/A \qquad (89)$$

assuming that ξ is a constant that does not change during the course of the filtration. Substitution of equation 89 into Darcy's equation (equation 86) and integration gives

$$V_{fs}^2 = 2k_c \Delta P_{fs} A^2 t/\eta_f (\xi^{-1} - 1) \qquad (90)$$

which describes the well-known dependence of the cumulative filtrate V_{fs} volume on $\sqrt{t}$ (*127–130*). Note the derivation of equation 90 assumes that k_c and ξ^{-1} are constant during the course of the filtration. Equation 90 has been expressed as (*129, 130*)

$$V_{fs}/A = St^{1/2} \qquad (91)$$

where S is the desorptivity, a term used in soil physics, given by

$$S^2 = 2k_c\Delta P_{fs}/\eta_f(\xi^{-1} - 1). \qquad (92)$$

The static filtration flow rate is given by

$$dV_{fs}/Adt = 0.5St^{-1/2} \qquad (93)$$

which decreases monotonically with time.

Equation 90 predicts that V_{fs} varies with $\sqrt{\Delta P_{fs}}$. In practice V_{fs} is a much weaker function of ΔP_{fs} (127–130), which is a consequence of the compressibility of the filter cake. Sherwood et al. (130) found that for bentonite filter cakes

$$S = 49.4\Delta P_{fs}{}^{0.16} \qquad (94)$$

(for ΔP_{fs} in bars and S in $\mu m/s^{1/2}$) where the exponent of 0.16 is typical for compactible filter cakes.

The derivation of equation 90 requires that the parameter ξ is a constant for the cake as a whole. The compressibility of the filter cake, however, gives rise to a non-uniform liquid: solid ratio through the filter cake. Figure 34, from Sherwood et al. (130), shows the variation of the void fraction e_c, defined by

$$e_c = 1-\phi/\phi \qquad (95)$$

at distance X from the filter medium. The profile at $t = 30$ min shows the composition gradient in the filter cake that would typically be obtained in a static API filtration cell; e_c varies from a value of about 10

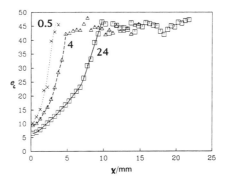

Figure 34.   Variation of void fraction e_c in static filter cake as a function of distance X from filter medium. Filtration times are shown in hours. (Reproduced from reference 130. Copyright 1991 American Chemical Society.)

at the filter medium ($X = 0$) to an average value of about 45 in the bulk drilling fluid. Horsfield et al. (131) have obtained similar moisture profiles using proton NMR imaging. Excellent agreement was observed between the profiles obtained from NMR imaging and destructive sectioning of the cake.

The compressibility of the cake is represented by the dependence of e_c on ΔP_{sf}. Sherwood et al. (130) fitted experimental data by

$$e_c = 6.0 \Delta P_{fs}^{-0.52} \tag{96}$$

over a filtration pressure range of 0–70 bars. Over the pressure range of the API filtration test, e_c is linear in $\ln \Delta P_{fs}$

$$e_c = 6.29 - 2.63 \ln \Delta P_{fs}. \tag{97}$$

Relationships of the type shown in equation 97 have been obtained for compacted clays (132) and soils (133) and are predicted by simple electrical double layer theory (21).

It is well known that the compressibility and permeability of clay filter cakes (and clay compacts in general) are a strong function of clay composition (particularly exchange cation) and saturating solution. Figure 35, for example, shows the dependence of filter cake permeability k_c as a function of e_c for various ratios of the exchange cations potassium and sodium (134). The permeability was observed to scale as

$$k_c \propto e_c^m \tag{98}$$

where m $\approx$ 2 for K/Na = 0 and $m \approx$ 4 for K/Na = 2.3.

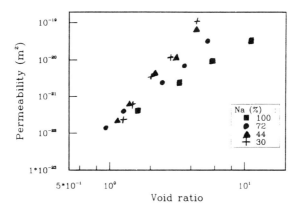

Figure 35.   Dependence of filter cake permeability k_c on void fraction for various ratios Na/(Na + K) of exchange cations. (Reproduced with permission from reference 134. Copyright 1991 Mineral Society.)

Dynamic Filtration.   Dynamic (or cross-flow) filtration is a considerably more complex separation process than static filtration as the
rate of filtration is a strong function of the flow of the drilling fluid. A
number of studies (*see* reference 128 and the references cited therein)
have demonstrated the various factors that control dynamic filtration rates.

The elaborate nature of the equipment needed to study realistic
dynamic filtration on representative rock samples has largely precluded
routine measurements of the dynamic fluid loss of drilling fluid samples
at the rig site during drilling operations. Figure 36, for example, shows
the equipment used by Fordham and co-workers (*129, 135*) to measure
dynamic filtration rates of water-based drilling fluids. Similarly large
and complex equipment has been used recently by Jiao and Sharma

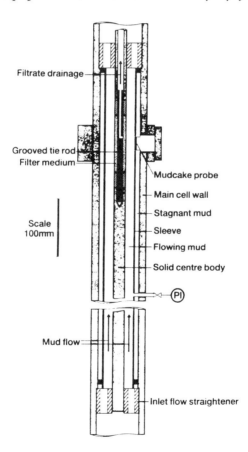

*Figure 36.   Schematic of annular cell for measurement of dynamic fluid
loss of drilling fluids. (Reproduced with permission from reference 135.
Copyright 1989 Advance Publications Ltd.)*

(*136*) and Warren, Smith, and Ravi (*137*). Recently, Chenevert et al. (*138*) have devised a dynamic filtration test cell that uses the geometry of a static filtration cell with a rotating cone to provide a well-defined shear rate on the surface of the cake.

Figure 37 shows a schematic of the dynamic filtration of a drilling fluid. The drilling fluid is filtered across the filter medium at the pressure difference ΔP_{fd} while being subjected to flow (assumed laminar in the following discussion), which gives rise to a shear stress τ_c at the cake–fluid interface where the shear rate is $\dot{\gamma}_c$.

Figure 38 compares the time dependence of the cumulative filtrate volume (V_{fd} or V_{fs}) from a water-based drilling fluid collected under conditions of static and dynamic filtration (*135*). Similar behavior has been observed with the dynamic filtration of invert emulsion oil-based drilling fluids (*136*). It is clear that the cumulative filtrate volume increases as $\dot{\gamma}_c$ increases. Typically, during the first few hours of filtration, the static and dynamic filtration rates are equal and show no sensitivity to shear rate. Fordham and co-workers (*129, 135*) have termed this curved region of the dynamic filtration curve quasi-static and have demonstrated that equations 90 or 91 describe the dynamic fluid loss behavior. A consequence of the insensitivity of dynamic fluid loss to changes in $\dot{\gamma}_c$ during quasi-static filtration is that S is independent of $\dot{\gamma}_c$, a result that Fordham and Ladva (*135*) have demonstrated for simple bentonite water-based drilling fluids (Figure 39).

At longer times, the cumulative filtrate volume V_{fd} shows a linear dependence on time, indicating that the filter cake has reached a constant thickness. The limiting (constant) dynamic filtration rate is a marked function of $\dot{\gamma}_c$ and τ_c. The limiting dynamic filtration rate Q_{fd} (= dV_{fd}/Adt as $t \rightarrow \infty$) has been scaled by $\dot{\gamma}_c$ and τ_c by (*135*)

$$Q_{fd} = Q_o(\dot{\gamma}_c/\Gamma)^{0.69} \tag{99}$$

$$Q_{fd} = Q_o(\tau_c/T)^{1.18} \tag{100}$$

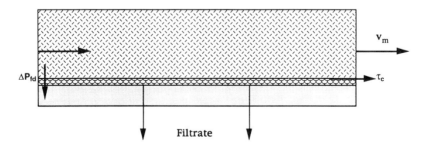

Figure 37.   Schematic of dynamic filtration of drilling fluid against a filter medium.

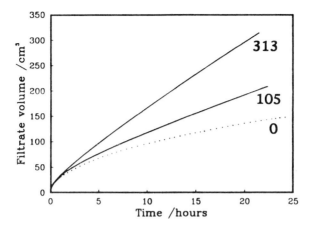

Figure 38. Time dependence of cumulative filtrate volume from a water-based drilling fluid under various conditions of shear. Shear rates are shown in s^{-1}. (Reproduced with permission from reference 135. Copyright 1989 Advance Publications Ltd.)

where Γ and T are arbitrary scaling factors and Q_o is a constant. Figure 40 shows the scaling of Q_{fd} by $\dot{\gamma}_c$ and τ_c. Equations 99 and 100 show that τ scaled as $\dot{\gamma}^{0.59}$, which is a smaller exponent than the average value of 0.75 obtained for the exponent in the Herschel–Bulkley model of the drilling fluid's rheology. However, Fordham and Ladva point out that the scatter in the $Q_{fd} - \tau_c$ plot is such that an exponent of unity in

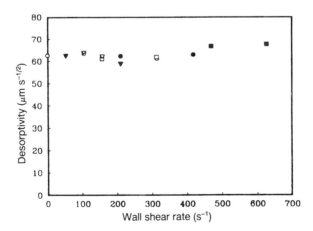

Figure 39. Dependence of desorptivity S on shear rate τ_c. (Reproduced with permission from reference 135. Copyright 1989 Advance Publications Ltd.)

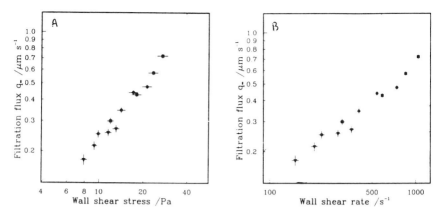

Figure 40.   Scaling of limiting rate of dynamic filtration with (A) shear stress τ_c and (B) shear rate $\dot{\gamma}_c$. (Reproduced with permission from reference 135. Copyright 1989 Advance Publications Ltd.)

equation 100 would also give a good fit to the data where τ would scale as $\dot{\gamma}^{0.69}$.

Fordham and co-workers (*129, 135*) have defined a critical time, t_c, where the filtration passes from quasi-static to dynamic. The critical time scales approximately as

$$t_c = \tau_o (\Gamma/\dot{\gamma}_c)^{1.32} \tag{101}$$

where the exponent is approximately twice that obtained in equation 99.

Figure 41 shows the dependence of S and Q_{fd} on ΔP_{fd} for simple bentonite water-based drilling fluids (*134*). The desorptivity S shows only a weak dependence on filtration pressure for $\Delta P_{fd} < 0.5$ MPa whereas Q_{fd} shows no dependence on ΔP_{fd} over the pressure range 0–20 MPa.

Bezemer and Havenaar (*139*) observed that Q_{fd} was directly proportional to $\dot{\gamma}_c$, which is a stronger dependence than found by Fordham and co-workers (eq 99). Figure 42a shows the dependence of Q_{fd} on $\dot{\gamma}_c$ for two fluid types (muds 2 and 3 in Bezemer and Havenaar's Table 2). A least-squares fit to the data shows some scatter but an approximately linear relationship is observed. Bezemer and Havenaar also found that h_c was inversely proportional to $\dot{\gamma}_c$ for a given drilling fluid. Figure 42b shows the variation of $1/h_c$ with $\dot{\gamma}_c$ for Bezemer and Havenaar's muds 2 and 3. Mud 2 shows a reasonable fit to $h_c \dot{\gamma}_c$ = constant but the data for mud 3 show considerable scatter. The dependence of $1/h_c$ on $\dot{\gamma}_c$ for Fordham and Ladva's simple bentonite drilling fluids is also shown in Figure 42b. Fordham and Ladva (*135*) observed that the determination

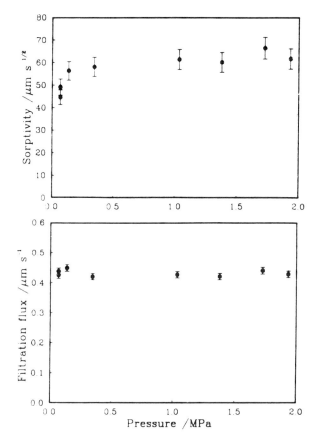

Figure 41.   Dependence of desorptivity (S) and limiting rate of dynamic filtration (Q_{fd}) on pressure (ΔP_{fd}). (Reproduced with permission from reference 135. Copyright 1989 Advance Publications Ltd.)

of h_c can be subject to considerable errors; they estimated the tolerance of the measurements of h_c to have been ±0.5 mm.

Fordham and Ladva (*140*) have argued the existence of a critical filtration rate below which there is no further growth of the filter cake (i.e., h_c is constant). The critical filtration rate is the limiting filtration rate Q_{fd}, which is achieved at constant cake thickness. Fordham and Ladva further argued that if a filter medium exhibited a Darcy flow rate below Q_{fd}, then no filter cake would form under dynamic conditions. This hypothesis was tested using a sandstone filter of low permeability [Ohio Sandstone, average permeability 0.067 millidarcies (0.067 $\times 10^{-3}$ μm^2)] and controlling the dynamic filtration rate by varying ΔP_{fd}. Figure 43 shows the cumulative filtrate volume as a function of time for various values of ΔP_{fd}. At low values of ΔP_{fd} (≤1.5 MPa) no filter cake

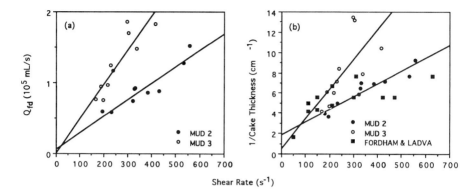

Figure 42. Dependence of (a) limiting dynamic filtration rate (Q_{fd}) and (b) inverse of filter cake thickness ($1/h_c$) on shear rate of drilling fluid. (Muds 2 and 3 from reference 139 with Fordham and Ladva's data from reference 135.) Lines show least squares fit to data from muds 2 and 3.

formed and the filtrate flux dV_{fd}/dt increased linearly with ΔP_{fd}; the variation of dV_{fd}/dt with ΔP_{fd} was controlled by the permeability of the filter medium. For $\Delta P_{fd} \geq 1.5$ MPa, a filter cake was formed and dV_{fd}/dt was controlled by the permeability of the filter cake and insensitive to further increases in ΔP_{fd}. For comparison, Figure 43 also shows the formation of a filter cake on a filter medium of significantly higher permeability [Portland Limestone, average permeability 730 millidarcies ($0.73 \ \mu m^2$)]. After an initial spurt loss that was significantly higher than that experienced in the Ohio sandstone samples, the limiting filtrate flux into the Portland Limestone was approximately equal to that obtained from the Ohio sandstone samples when a filter cake was formed (i.e., when the critical filtration flux was exceeded). Figure 43 shows that when a filter cake is formed, the limiting filtration rate is determined by the permeability of the filter cake and not by the permeability of the filter medium.

Despite the importance of determining the dynamic fluid loss of drilling fluids, there are few theoretical treatments of the dynamic filtration process. In particular, it is not possible to predict dynamic filtration rates from static filtration rates and a knowledge of $\dot{\gamma}_c$ or τ_c from the fluid rheology and v_m. Outmans (*127*) has given a theoretical treatment of both static and dynamic filtration. Outmans identified three phases of dynamic filtration: (1) build up of filter cake with declining filtration rate; (2) constant filter cake thickness but gradually decreasing filtration rate due to compaction; (3) constant filtration rate. A constant value of h_c is reached when the shear stress τ_c equals the resistance to

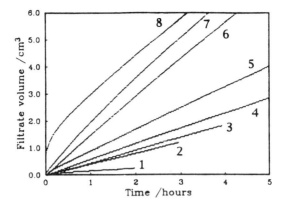

Figure 43.   Time dependence of cumulative filtrate volume from dynamic filtration into media of high and low permeability. Filtration into Ohio sandstone at various ΔP_fd (MPa): 1. 0.14; 2. 0.55; 3. 0.60; 4. 0.97; 5. 1.45; 6. 1.93; 7. 1.93. Filtration into Portland limestone at 8. 1.93 MPa. (Reproduced with permission from reference 140. Copyright 1992 Academic.)

erode the filter cake, $f_c P_s$, where P_s is the cohesive matrix stress and f_c is a friction factor. The limiting dynamic filtration rate derived by Outmans is

$$Q_{fd} = k_c (\tau_c/f_c)^\varsigma / \eta_f \varsigma \delta \qquad (102)$$

where ς is a constant that is equivalent to the exponent in equation 94 and is in the range 0.1–0.15, and δ is the thickness of a surface layer of the filter cake that is subject to shear. Note that Q_{fd} scales as $\tau_c^{0.15}$ in Outmans' theory whereas Fordham and co-workers (*129, 135*) have found a scaling of $\tau_c^{1.2}$; these two scalings can only be reconciled if δ in equation 102 scales approximately as τ_c^{-1}.

The dynamic filtration theory of Outmans (*127*) requires experimental terms such as particle–particle stresses, particle friction factors, and thickness of a shear zone within the filter cake that would be difficult to determine. However, the qualitative picture of dynamic filtration presented by Outmans, namely, irreversible adhesion of solid particles up to a certain thickness that is determined by the shear stress (or shear rate) at the surface of the cake, accords with the experiments of Fordham and co-workers (*129, 135*). Once a filter cake has formed under dynamic conditions, it is difficult to remove it by subsequent changes in $\dot{\gamma}_c$ or v_m. Figure 44 shows the effect of changes in the flow rate on cumulative filtrate volume. The limiting filtration rate obtained when the initial flow rate of the drilling fluid was 1.8 m³/h remained unaltered when the flow rate of the drilling fluid was increased to 7.0 m³/h in a step-wise manner. The increased shear stress on the filter cake was insufficient to reduce h_c and increase the filtration rate. Further confirmation of the erosion resistance of filter cakes derived from water-based drilling fluids has come from the measurements of Sherwood et al. (130) who measured the yield stress τ_{yc} of filter cakes formed from water-based drilling fluids and found a relationship of the form

$$\tau_{yc} = B_c \phi^{3.0} \qquad (103)$$

where B_c is a constant (≈ 30 bars). It was assumed that the filter cake would erode when $\tau_c > \tau_{yc}$, thus giving rise to an increased filtration rate. No significant increases in filtration rate were predicted for

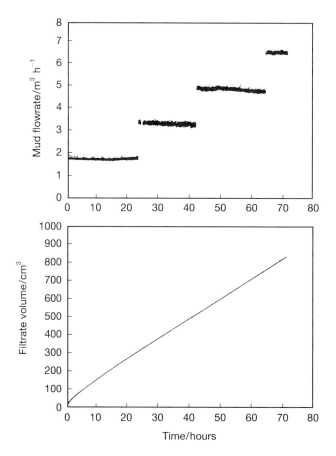

Figure 44.   Effect of changes in flow rate of drilling fluid on rate of dynamic filtration. (Reproduced with permission from reference 135. Copyright 1989 Advance Publications Ltd.)

$\tau_c \leq 1000$ Pa, which is more than an order of magnitude larger than the values of τ_c obtained in the experiments of Fordham and Ladva (*135*). Bentonite filter cakes formed from the dynamic filtration of water-based drilling fluids show considerable resistance to erosion. The filter cakes formed from the dynamic filtration of invert emulsion oil-based drilling fluids have been found to be less shear resistant than those from water-based drilling fluids (*136*).

Effect of Composition on Fluid Loss.   There have been many studies and tests of the effects of various additives (usually polymeric) on the static fluid loss of drilling fluids. There have been many fewer studies on the effect of these additives on dynamic filtration rates and very few that compare their effect on both dynamic and static rates.

Hughes et al. (*32*) have studied the effect of bentonite clay content and CMC content on the static fluid loss of simple bentonite–polymer drilling fluids. Figure 45 shows the dependence of the API static fluid loss (after 30 min) on bentonite content with no polymer and in the presence of polymer. The polymer-free API fluid loss V_{30} (in mL) is related to bentonite concentration C_{ben} (in g/L) by

$$V_{30} = 198C_{\text{ben}}^{-0.69} \tag{104}$$

or, from equation 91 with $t = 1800$ s,

$$S = 1.02C_{\text{ben}}^{-0.69} \tag{105}$$

with S in mm/s$^{1/2}$. The weight-average molecular weight M_w of the CMC polymers was in the range 57,000–270,000 g/mol and the value of M_w had no significant effect on S. However, a concentration of 4.5 g/L of each of the CMC polymers had a marked effect on S. The variation of S with bentonite content in the presence of 4.5 g/L of each of the polymers can be approximated by

$$S = 0.003C_{\text{ben}}^{-0.46} \tag{106}$$

Figure 46 shows the dependence of V_{30} on CMC polymer content at a fixed bentonite content. At $I = 0$, where the CMC polymers are fully extended, there is no effect of M_w on V_{30}, although only 1 g/L of any of the CMC polymers causes V_{30} to decrease by a factor of 2 over the neat bentonite suspension. An effect of M_w and/or charge density of the polymer (degree of substitution) is observed for $I = 0.5$ molar.

Burchill et al. (*141*) have measured V_{30} as a function of polymer concentration and ionic strength for a number of synthetic polymers, including poly(vinyl alcohol) (**PVA**), poly(acrylic acid) (**PAA**) and sodium polyacrylate. Figure 47 shows a summary of results for several polymers.

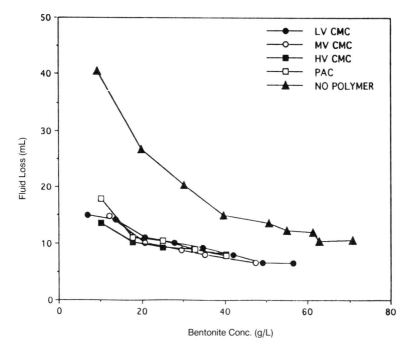

Figure 45.   Dependence of API static fluid loss of bentonite–polymer (CMC) drilling fluids on bentonite concentration at constant polymer content (4.5 g/L). (Reproduced with permission from reference 32. Copyright 1993 Society of Petroleum Engineers.)

The polyacrylate polymers and a derivative of a vinyl acetate: maleic anhydride copolymer cause V_{30} to decrease monotonically with increasing polymer concentration, similar to the CMC polymers (Figure 46). The polymers PVA and poly(vinyl pyridinium) (PVP) hydrochloride markedly increased V_{30} at low concentration; at concentrations above 1 g of polymer per gram of added bentonite PVA functions as a static fluid loss additive. The maximum in the API fluid loss at low PVA concentrations approximately coincides with the maximum in the yield stress and plastic viscosity found by Heath and Tadros (75). The increased static fluid loss is consistent with Heath and Tadros's conclusion that bentonite is flocculated by low concentrations of PVA. The concentration of PVA required to decrease V_{30} below that of the neat bentonite suspension is significantly larger than the concentration of CMC, where effective static fluid loss control can be achieved at polymer: bentonite weight ratios of about 0.1 g/g. More effective fluid loss control has been achieved with other synthetic polymers such as poly(vinyl sulphonate)–poly(vinyl amide) copolymer (40) and other sulphonated polymers (39).

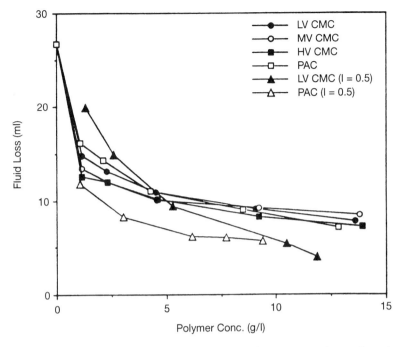

Figure 46.   Dependence of API static fluid loss of bentonite–polymer (CMC) drilling fluids on polymer concentration at constant bentonite content (20.6 g/L). (Reproduced with permission from reference 32. Copyright 1993 Society of Petroleum Engineers.)

It has long been known that fluid loss additives can be very different in their effects on the static and dynamic filtration rates of both oil- and water-based drilling fluids. The static fluid loss of most invert emulsion oil-based drilling fluids is usually low. However, dynamic fluid loss rates from oil-based drilling fluids are shear sensitive and the dynamic fluid loss rates exceed those of most water-based drilling fluids under comparable conditions. This effect had been observed by Ferguson and Klotz (*142*) and more recently by Jiao and Sharma (*136*). Figure 48, from Ferguson and Klotz (*142*), compares the rates of static and dynamic fluid loss from an invert emulsion oil-based drilling fluid. The very low rate of static filtration gives way to a high dynamic fluid loss that increases markedly with increasing v_m (and hence $\dot{\gamma}_c$). The addition of asphaltenes to invert emulsion drilling fluids has been demonstrated to reduce the spurt loss but they have little effect on the limiting rates of dynamic filtration (*136*).

A comparison of the effect of well-known fluid loss additives on the rates of static and dynamic filtration was made by Kreuger some 30

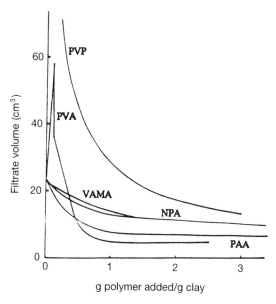

*Figure 47.   Dependence of API static fluid loss on bentonite–polymer sus-
pensions on polymer type and concentration at constant bentonite content
(10 g/L). PVA: polyvinyl alcohol; PAA: polyacrylic acid; NPA: sodium polyacrylate; PVP: polyvinyl pyridinium hydrochloride; VAMA: vinyl acetate
maleic anhydride copolymer. (Reproduced with permission from reference
141. Copyright 1983 Mineral Society.)*

years ago (*143*). Figure 49 shows the effect on static and dynamic filtration rates on the addition of five fluid loss additives, including quebracho, which is a tannin extracted from trees (*25*). The lignosulphonate
and quebracho additives had little effect on static fluid loss but dramatically increased the rate of dynamic filtration. In contrast, the polyacrylate additive reduced the static filtration rate but had little effect on the
dynamic rate. The fluid loss additives CMC and starch showed the ability
to decrease modestly both static and dynamic filtration rates. With all
five additives there is an indication that the rate of dynamic fluid loss
begins to increase as the concentration of the additive continues to
increase.

Plank and Gossen (*144*) have reported on the temperature dependence of API static filtrate volumes of simple drilling fluids using various
polymeric additives. Figure 50 shows the dependence of filtrate volume
on temperatures for a base bentonite fluid and with added starch, polyanionic cellulose, and a synthetic sulphonated polymer (*39*). Starch is
an effective fluid loss additive up to about 100 °C whereas the polyanionic cellulose begins to lose its effectiveness at about 140 °C; these

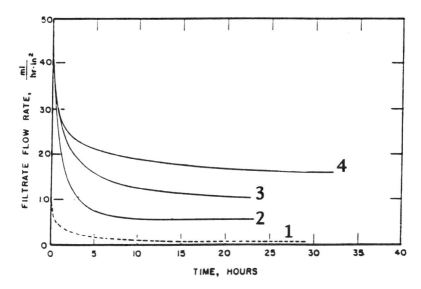

Figure 48. Comparison of rates of filtration of an invert emulsion oil-based drilling fluid under static and various dynamic conditions. 1. $v_m = 0$ (extrapolated static filtration rate); 2. $v_m = 0.6$ m/s; 3. $v_m = 1.5$ m/s; 4. $v_m = 2.9$ m/s. (Reproduced with permission from reference 142. Copyright 1954 American Institute of Mechanical Engineers.)

temperature limits agree with those reported by Thomas (33). The synthetic sulphonated polymer maintains effective fluid loss control to temperatures in excess of 170 °C.

Structure of Filter Cakes. A number of studies have sought to relate the microstructure of bentonite (and other clay) filter cakes to their mode of formation (static or dynamic conditions) and the presence of various fluid loss additives and other suspended solids. One of the earliest studies was made by Hartmann et al. (145) who compared the structure of filter cakes grown under static and dynamic conditions from drilling fluids of varying compositions. Figures 51a and 51b compare the scanning electron microscope (SEM) micrographs of samples of freeze-dried filter cake obtained from the static filtration of a simple bentonite drilling fluid formulated in water and a calcium chloride solution (0.18 M calcium concentration), respectively. The SEM micrograph of the bentonite filter cake formed in water shows a characteristic honeycomb structure; similar structures have been observed by Plank and Gossen (144) and recently by Longeron et al. (146). The addition of calcium chloride to the drilling fluid was observed by Hartmann et al. to result in a less ordered cake structure with a larger average pore

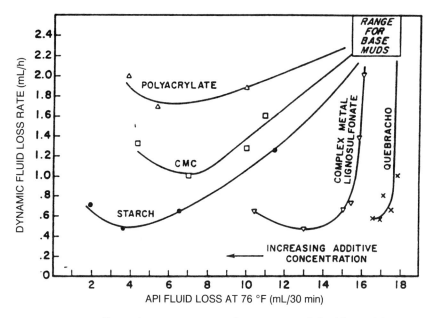

Figure 49.   Effects of various common dispersants and fluid loss additives on the static and dynamic fluid loss of a water-based drilling fluid. (Reproduced with permission from reference 143. Copyright 1963 Society of Petroleum Engineers.)

size and hence with higher permeability (API static fluid loss increased from 10 to 55 mL). Plank and Gossen (*144*), however, did not observe any loss to the honeycomb structure on addition of comparable concentrations of NaCl, CaCl$_2$, and MgCl$_2$ to the drilling fluid. The structure of filter cakes formed by the dynamic filtration of drilling fluids with a high gel strength was not observed by Hartmann et al. (*145*) to be significantly different from those formed by static filtration. However, when the gel strength was low, the dynamic filter cakes were observed to be oriented by both the flow and the fabric of the underlying rock (*145*).

Plank and Gossen (*144*) have studied the effects of various polymeric fluid loss additives on the structure of filter cakes formed under static conditions. Figures 52a and 52b show **SEM** micrographs of filter cakes formed from drilling fluids containing starch and a synthetic sulphonated fluid loss additive, respectively. The starch causes numerous bridges to form across the walls of the pores, often with branching. The synthetic sulphonated polymer gave rise to small rods (~4 µm in length) attached to the pore walls of the cake and projecting into the pore space (Figure 52b). In contrast, a low molecular weight polyanionic cellulose fluid loss additive ($M_w \sim$ 90,000 g/mole) formed very few bridges. In the

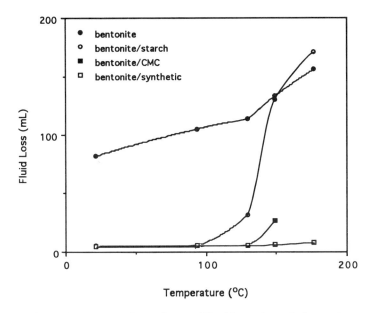

Figure 50.   Temperature dependence of fluid loss of simple bentonite water-based drilling fluids with varying polymeric fluid loss additives. (Data from reference 144.)

presence of divalent cations (Ca^{2+}, Mg^{2+}) aggregates of polyanionic cellulose could be observed.

Mechanisms of Fluid Loss Control.   The mechanisms of fluid loss control are not known in detail. The action of additives ranges from straightforward pore bridging–blocking such as bentonite, starch, and asphaltenes to more complex effects with CMC and PAA where the polymeric additives may adsorb on bentonite clay platelets and prevent flocculation by steric and/or electrostatic stabilization.

One mechanism for the mode of action of polymeric fluid loss additives that is frequently cited is the increase in filtrate viscosity. Such a mechanism is, in general, not tenable. A comparison of Figures 16, 17, 45, and 46 shows that although the weight-average molecular weight for the various CMC polymers has a marked effect on both the viscosity of the drilling fluid and its filtrate, there is no significant effect on static fluid loss.

It has proved difficult to unravel the mechanism(s) whereby polymers such as CMC can reduce the rates of static fluid loss. Heinle et al. (*147*) suggest that the reduction in static fluid loss at high ionic strength is

(a)

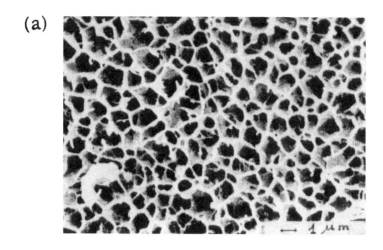

(b)

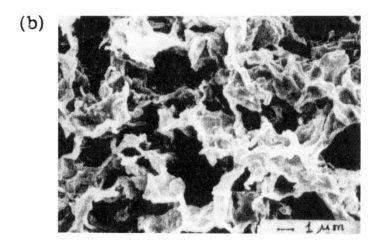

Figure 51.   SEM micrographs of filter cake formed under static conditions from simple bentonite drilling fluid formulated with (a) water and (b) 0.18 M CaCl₂. (Reproduced with permission from reference 145. Copyright 1988 Society of Petroleum Engineers.)

achieved by adsorption of CMC on bentonite. However Heinle et al. demonstrate that there is little adsorption of CMC on bentonite as $I \rightarrow$ 0. Rabaioli et al. (78) also found that low molecular weight dispersants, which frequently act as fluid loss additives, exhibit little adsorption on bentonite. If the polymeric fluid loss additives are functioning with little adsorption, then a mechanism such as depletion stabilization (148) must be sought. The effects of shear on depletion stabilization may markedly differentiate additives in static and dynamic regimes.

(a)

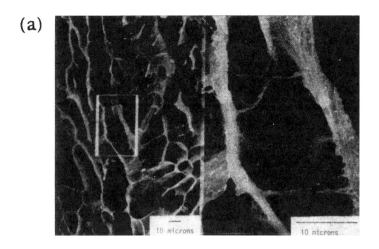

(b)

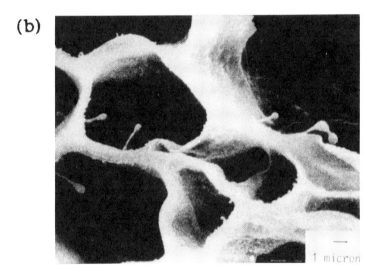

Figure 52. SEM micrographs of filter cake formed under static conditions from simple bentonite drilling fluids containing the fluid loss additives (a) starch and (b) synthetic sulphonated polymer. (Reproduced with permission from reference 144. Copyright 1991 Society of Petroleum Engineers.)

Drilling Fluids: Some Practical Problems

Invasion and Drilling Related Formation Damage. The influx of particulates and filtrate into the near wellbore region of permeable formations during static or dynamic filtration has a number of consequences, two of the most important being displacement of wellbore fluids (invasion) and a reduction in formation permeability. The impact of invasion and drilling related formation damage on the evaluation of, and production from, hydrocarbon reservoirs is briefly discussed.

It is well known that the spurt loss of drilling fluid into permeable formations is associated with the invasion of solid particles (so-called deep-bed filtration) to form an internal filter cake (149). Large-scale laboratory simulations of static and dynamic filtration at downhole conditions have shown that the spurt loss can last from 30 s to 5 min (137). Jiao and Sharma (150) have measured changes in permeability of core samples that have been subjected to invasion from various water-based drilling fluids under conditions of dynamic filtration. Figure 53 shows the time dependence of the measured permeability k_a of the core samples, normalized to their initial value k_b when saturated with a 0.51 M NaCl solution, during the dynamic filtration of three water-based drilling fluids and the injection of filtrate. The permeability was determined from the pressure drop at various locations along the length of the core. The measurement at port 1 includes the permeability of the filter cake and consequently the ratio $k_a/k_b \ll 1$.

Two mechanisms are causing the reduction in permeability of the core samples, namely, the invasion of bentonite particles from the drilling fluid and the mobilization of clay mineral fines in the core samples (i.e., the swelling and dispersion of the clay minerals in the core samples when the 0.51 M NaCl solution is replaced by filtrate of low ionic strength). The injection of the low ionic strength filtrate in the absence of any drilling fluid solids caused k_a/k_b to decrease to about 0.3. The dynamic filtration of the 20 g/L bentonite suspension resulted in a reduction in k_a by both invaded particles and fines migration; the 40 g/L bentonite suspension caused a reduction in k_a by fines migration only. The suspension of 40 g/L bentonite in a 0.34 M NaCl solution caused less reduction in k_a and the initial value was more nearly approached as the direction of flow in the core was reversed. Jiao and Sharma (150) suggested that the NaCl solution had caused the bentonite to flocculate, resulting in the bentonite exhibiting a larger effective particle size and a lower tendency to invade the core sample. Table I shows a summary of the drilling fluids used in the various invasion experiments and compares the cumulative filtrate volume collected after 600 min.

Figure 54 shows the dependence of k_a/k_b for the core samples invaded during the dynamic filtration of bentonite–electrolyte drilling fluids in the presence of various dispersants and fluid loss additives. The highly dispersed bentonite–lignosulphonate drilling fluid gave rise to low values of k_a/k_b throughout the core sample indicating permeability reduction by both fines mobilization and invasion of the dispersed bentonite particles. The addition of 0.34 moles of NaCl, which partly flocculated the bentonite particles, largely eliminated the reduction of k_a

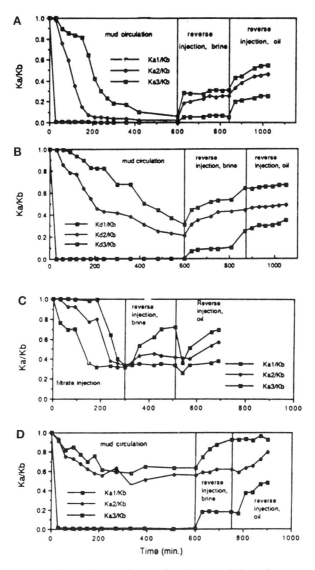

Figure 53. Time dependence of normalized permeability of water-saturated sandstone samples during dynamic filtration of various water-based drilling fluids and injection of filtrate. (A) 20 g/L bentonite suspension; (B) 40 g/L bentonite suspension; (C) injection of fresh-water filtrate; (D) 40 g/L bentonite in 0.51 M NaCl. (Reproduced with permission from reference 150. Copyright 1992 Society of Petroleum Engineers.)

Table I.   Summary of Dynamic Fluid Loss in Sandstone Cores

Series	Drilling Fluid	Cumulative Filtrate Vol. (mL) After 600 mins
1 (Figure 53)	20 g/L bentonite	47
	40 g/L bentonite	20
	40 g/L bentonite; 0.51 M NaCl	48
2 (Figure 54)	40 g/L bentonite; 30 g/l LS[a]	35
	40 g/L bentonite; 0.34 M NaCl; 30 g/L LS	20
	40 g/L bentonite; 0.34 M NaCl; 3 g/L CMC	48
	40 g/L bentonite; 0.34 M NaCl; 10 g/L Starch	40
3 (Figure 55)	40 g/L bentonite; 0.34 M NaCl	25
	40 g/L bentonite; 0.34 M NaCl; 3 g/L CMC	23
	40 g/L bentonite; 0.34 M NaCl; 30 g/L LS	9.5

[a] LS is lignosulphonate
SOURCE: Data are taken from reference 150.

in the deeper sections of the core. The presence of the CMC and starch fluid loss additives caused some reduction in k_a by particle invasion that was largely removed by back flushing.

The invasion of drilling fluid filtrate and subsequent damage in formations containing hydrocarbon is of particular importance. Figure 55 shows the effect of the invasion of filtrates from various water-based drilling fluids on the permeability to oil of core samples saturated with oil and water at irreducible water saturation (150). The oil permeability decreased during the invasion of the water-based filtrate, although as the water saturation increased, some decrease in the permeability of oil was expected due to the effect of saturation on relative permeability (151). The dynamic filtration rates of the bentonite–NaCl and bentonite–NaCl–lignosulphonate drilling fluids into the core samples were approximately halved by the presence of the oil. In contrast, the rate of dynamic filtration from the bentonite–NaCl–CMC drilling fluid was not affected by the presence of the oil in the core. Jiao and Sharma (136) similarly found a lower rate of invasion of filtrate from an invert emulsion oil-based drilling fluid in a core sample fully saturated with water than the corresponding rate in a core saturated with oil at irreducible water saturation.

The invasion of drilling fluid filtrate into permeable zones that contain hydrocarbon adds significantly to the complexity of evaluating the

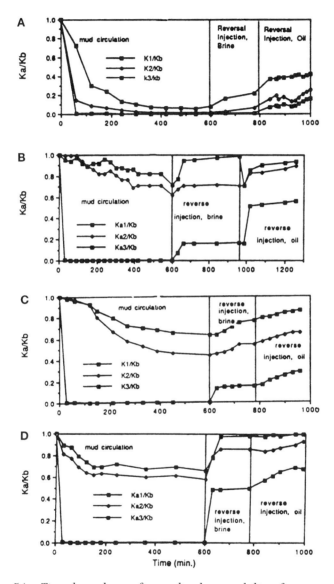

Figure 54.   Time dependence of normalized permeability of water-saturated sandstone samples during dynamic filtration of various water-based drilling fluids. (A) 40 g/L bentonite + 30 g/L lignosulphonate; (B) 40 g/L bentonite + 0.34 M NaCl + 30 g/L lignosulphonate; (C) 40 g/L bentonite + 3 g/L CMC + 0.34 M NaCl; (D) 40 g/L bentonite + 10 g/L starch + 0.34 M NaCl. (Reproduced with permission from reference 150. Copyright 1992 Society of Petroleum Engineers.)

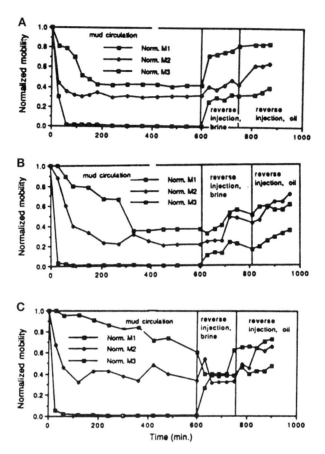

Figure 55.   Time dependence of normalized permeability of oil-saturated sandstone samples during dynamic filtration of various water-based drilling fluids. (A) 40 g/L bentonite + 0.34 M NaCl; (B) 40 g/L bentonite + 3 g/L CMC + 0.34 M NaCl; (C) 40 g/L bentonite + 30 g/L lignosulphonate + 0.34 M NaCl. (Reproduced with permission from reference 150. Copyright 1992 Society of Petroleum Engineers.)

hydrocarbon saturations of these zones using electric logs (152, 153). The estimate of the true formation resistivity before any invasion has taken place requires an invasion diameter to be selected. The invasion front is usually assumed to be sharp and resulting from pistonlike displacement of borehole fluids by the invading filtrate. Lane (154) has recently presented simulations of filtrate invasion into hydrocarbon-bearing formations.

   The spontaneous potential (SP) log (155) is an example of a borehole measurement that arises as a direct consequence of the invasion of filtrate into a permeable zone (Figure 56). The SP has two electrochemical

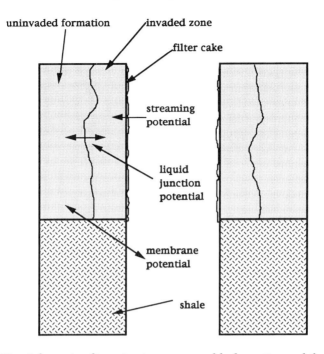

Figure 56. Schematic of invasion into a permeable formation and the components of the spontaneous potential (SP).

components: a liquid junction (diffusion) potential in the permeable formation that arises from the difference in ionic concentrations between the formation water and the filtrate and a membrane (Nernst) potential that arises from the difference in ionic concentrations between the formation water and the filtrate (*155*) across the shale. A third electrokinetic contribution to the SP is the streaming potential that arises from the flow of filtrate through the filter cake (*155, 156*). Wyllie (*156*) found that the streaming potential E_s was related to the static filtration pressure ΔP_{fs} by

$$E_s = c\Delta P_{fs}{}^y \tag{107}$$

where c and y are constants. The exponent y was determined to be in the range 0.57 to 0.90 for a range of drilling fluids. Wyllie observed that the exponent y was significantly larger than the exponent relating the rate of static filtration to ΔP_{fs} (eq 94).

Chemical Aspects of Wellbore Stability. A considerable (and continuing) effort has been made to develop water- and oil-based drilling fluids that will drill through the massive shale sections, which commonly overlay hydrocarbon reservoirs, with minimal problems. Common problems associated with drilling shale sections are excessive production of drilled solids, hole enlargement, the sticking of the drill pipe, and even hole collapse.

The basic problem of the interaction of drilling fluids with shale formations is an imbalance in the chemical potential of the water in the drilling fluid and in the shale. During the compaction of the shale, water is expelled and the clay–water ratio increases (*see* eq 96, for example). The presence of the exchange cations associated with the surface of the clay causes the water activity in the shale a_w^{sh} to decrease as the water content decreases. The chemical potential μ_w^{sh} of the water in the shale is given by

$$\mu_w^{sh} = \mu_w^\circ + RT \ln a_w^{sh} + (P^{sh} - P^\circ)V_m \qquad (108)$$

where μ_w° is the chemical potential of water in a defined standard state at pressure P°, P^{sh} is the pressure of the water in the shale, and V_m is the partial molar volume of water in the shale. The corresponding expression for the chemical potential of water in a drilling fluid is

$$\mu_w^f = \mu_w^\circ + RT \ln a_w^f + (P - P^\circ)V_m. \qquad (109)$$

The condition for hydration of the shale is $\mu_w^f > \mu_w^{sh}$, which gives (assuming the values of μ_w° and V_m are equal in the drilling fluid and the shale)

$$RT \ln a_w^f/V_m + P > RT \ln a_w^{sh}/V_m + P^{sh}. \qquad (110)$$

The imbalance in the chemical potential of the water in the shale and drilling fluid results in a tendency for water to enter the shale. Equation 110 is applicable to both water- and oil-based drilling fluids. When $\mu_w^f = \mu_w^{sh}$ equation 110 gives the well-known expression for the swelling pressure $(P^{sh} - P)$ between the shale and the drilling fluid. The permeability of shales is very low and the rate of filtration into the shale will be below the critical filtration rate (*140*) and no filter cake will form on the surface of the shale.

The response of the shale to the gradient in the chemical potential of water is a fundamental chemical–mechanical property of the shale that is largely determined by its composition. Several classifications of shale response to drilling fluids have been made [e.g., O'Brien and Chenevert (*157*)]. The shales that contain a high concentration of montmorillonite clay and that are geologically of recent age (frequently of Tertiary age) have little cementation and their response to the imbalance in the chemical potential of water is swelling (the uptake of water and increase in a_w^{sh}) and dispersion ($a_w^{sh} \rightarrow a_w^f$). These are termed mud-making or *gumbo shales* and can result in initial borehole closure followed by hole enlargement after erosion of the weak shale by the shear stress exerted by the flowing drilling fluid. The dispersion of the shale results in a significant increase in the fraction of colloidal solids in the drilling fluid and its viscosity and yield stress are both increased. Sherwood (*158*) has recently incorporated the thermodynamics of swelling into a Biot poroelastic model of a swelling shale.

The other extreme of behavior is represented by shales that are highly cemented and unable to swell (a_w^{sh} constant) and the response to the imbalance of chemical potential is an increase in P^{sh}. The shales will tend to exhibit tensile failure by fracturing and large angular fragments may detach from the borehole wall. These shales, which are usually geologically older, contain mostly illite and kaolinite clays. Intermediate behavior of shale hydration has been observed (*157*).

The imbalance in the chemical potential of water between the drilling fluid and the shale can be reduced by the addition of electrolyte to the drilling fluid. The addition of soluble calcium salts (e.g., $CaCl_2$) would be an effective way to reduce a_w^f, but the presence of high concentrations of calcium would severely flocculate the bentonite clay and would precipitate the water-soluble polymers. Potassium salts, particularly KCl, have been used most frequently to reduce a_w^f, with concentrations up to 2 molar being commonplace. Although sodium salts are more effective than potassium salts at reducing a_w^f at the same molar concentration, sodium salts are generally less effective than potassium salts at reducing shale hydration. The reason is that the presence of the potassium both lowers a_w^f and raises a_w^{sh} by exchange with the sodium and calcium cations that most commonly saturate the shale. The increase in a_w^{sh} by replacing sodium and calcium ions by potassium ions at constant water content is evident from the water vapor adsorption isotherms of homoionic clays (*see* references 159 and 160, for example).

The approach outlined previously is also applicable to the aqueous phase in an invert emulsion oil-based drilling fluid. The chemical potential of the water in the aqueous (dispersed) phase is usually controlled by the concentration of calcium chloride. The transport of water between the shale and the aqueous phase of the invert emulsion is less complex than with water-based drilling fluids, because with the emulsions there is no cation exchange between the ions in the fluid and in the shale. The thin emulsified layer surrounding the water droplets is postulated to act as a semipermeable membrane that allows only the passage of water (*61*).

Solids Control.   The circulating drilling fluid is an open system to which solids are added both at the surface (bentonite and barite to control properties) and in the borehole from drilled formations. The drilled cuttings are removed from the drilling fluid on the surface by a vibrating sieve called a *shale shaker*. The sieve mesh size is typically 75 μm, which is sufficiently large to ensure that barite particles (mean particle size 30–40 μm) are not removed from the drilling fluid. The finer drilled solids (fines) will pass through the sieve and if no further processing is done, then the drilled fines will be pumped back down the wellbore and remain part of the drilling fluid. The build up of drilled fines (particularly clays from swelling–dispersing shale sections) will lead to an increase in the viscosity of the drilling fluid. The use of chemical dispersants such as lignosulphonates can reduce the viscosity and the tendency of the solids to gel at high temperatures, but eventually the drilling fluid must be diluted and excess fluid discarded. A further problem in the build up of drilled solids is that they form poor quality filter cakes and give rise to higher rates of fluid loss.

A more effective strategy for dealing with drilled solids is to remove them continuously from the circulating drilling fluid by the use of solids control equipment. Figure 57 shows a schematic of the solids control equipment (*161*) such as shale shaker, degasser (to remove entrained gas derived from drilled formations; *see* reference 3), and centrifuges. Additional solids control equipment not shown in Figure 57 is a hydrocyclone (*162*).

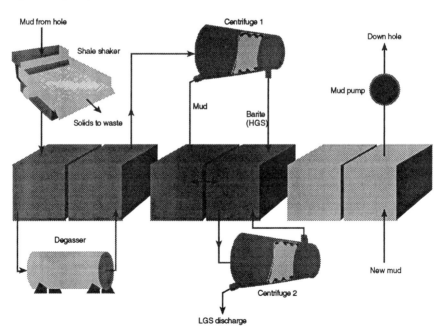

Figure 57.   Schematic of solids control equipment on a drilling rig. (Reproduced with permission from reference 161. Copyright 1994 Schlumberger.)

Figure 58, from Froment et al. (*163*), shows the range of particle sizes of solids in drilling fluids and the regions of particle sizes over which the various items of the solids control equipment are operating. It is clear that the particle sizes of drilled solids extends over a wide range, from clay particles (<1 μm) to large cuttings (> 1 mm).

The rheological properties of the drilling fluid have a marked influence on the performance of solids control equipment. Froment et al. (*163*) have pointed out that an increase in the viscosity of the drilling fluid will decrease the flow rate capacity of the shale shaker and will increase the minimum particle size of the solids in the separated stream from a hydrocyclone that is returned to the circulating drilling fluid. For example, Figure 59 shows the particle size distribution of the solids in the under flow from a hydrocyclone. The density and viscosity of the drilling fluid are observed to have a marked effect on the separation characteristics of the hydrocyclone.

Environmental Aspects of Drilling Waste Disposal.   One of the most important developments in drilling fluids over the past decade has been the increased scrutiny by government agencies and environmental pressure groups of drilling fluid waste (*164, 165*). Of the total waste generated by the oil industry in exploration and production op-

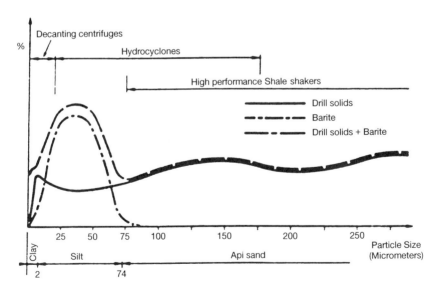

Figure 58.   Particle size distribution of API grade barite and drilled solids in typical drilling fluids. The ranges of particle sizes of solids removed by various items of solids control equipment are shown. (Reproduced with permission from reference 163. Copyright 1986 Society of Petroleum Engineers.)

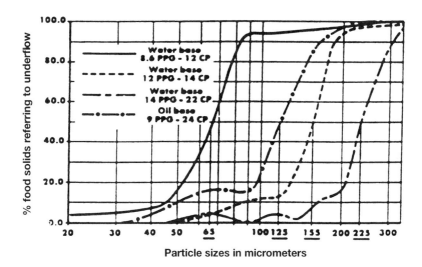

Figure 59.   Effects of density and viscosity of water- and oil-based drilling fluids on particle size distribution of solids separated by a hydrocyclone. (Reproduced with permission from reference 163. Copyright 1986 Society of Petroleum Engineers.)

erations, only about 2% is drilling waste; some 98% of the waste is formation water that is coproduced with the hydrocarbon (*166*). Hudgins (*167*) has given a recent account of chemical usage in offshore drilling operations in the North Sea. In 1989, 446 exploration and development wells were drilled consuming 146,656 tonnes of chemicals, of which an estimated 57% were discharged. Of the total usage of drilling fluid chemicals the breakdown was: 120,794 tonnes of barite and bentonite; 17,609 tonnes of inorganics (NaCl, KCl, Ca(OH)$_2$, NaOH, etc.); 8253 tonnes of other chemicals (polymers, surfactants, emulsifiers, etc.). The ratio of the usage of the inorganic and other chemicals used in water-based and oil-based fluids was 3:1.

The disposal of drilling fluid waste has become a major factor in the use of chemicals in water-based drilling fluids and in the use of oil-based drilling fluids because the dumping of waste from conventional (i.e., non-synthetic) oil-based fluids (drilling fluid and cuttings contaminated with oil from the fluid) is increasingly prohibited.

Drilling fluid products that are to be disposed of from offshore platforms and landfill sites must be tested for their effects on animal and plant life. Components that have been identified as potentially toxic and hazardous to the environment are oil (diesel or mineral), heavy metals (Cr, Hg, Zn, Pb, etc.), and chlorides.

A number of waste disposal techniques have been recently investigated and several are now briefly discussed. One of the largest disposal

problems is waste from oil-based drilling fluids. The continued use of
oil-based drilling fluids, with their attendant advantages, is dependent
on finding a solution to the problem of waste disposal.

An increasingly common method for the offshore disposal of oil-
based drilling waste is the injection of spent drilling fluid and ground
cuttings into non-producing formations that have been fractured for the
purpose. Crawford and Lescarboura (*168*) have studied the problem of
the injection of oil-based waste produced by a platform in the Norwegian
sector of the North Sea into subsea formations. Crawford and Lescarboura (*168*) estimated that the drilling of the required 56 development
wells would produce 67,500 tonnes of cuttings, giving a volume of about
37,500 m³. The cuttings were to be ground and mixed with seawater at
a water: cuttings volume ratio of 3:1, giving a total injection volume of
150,000 m³. The composition of the slurry (by volume) would be 80%
water, 6% oil, and 14% solids.

Rheological and fluid loss tests were performed on the slurry. The
experimental rheogram of the slurry was obtained with a Fann rheometer
(Figure 6) and fitted using the Herschel–Bulkley equation (eq 7). Figure
60 shows a plot of ln $(\tau - \tau_o)$ as a function of ln $\dot{\gamma}$ at 3 °C and 38 °C
(representing surface and downhole temperatures). The values of K and
n were estimated to be 0.38 Pa · s^n and 0.15 at 3 °C and 0.47 Pa · s^n and
0.09 at 38 °C. The flow of the slurry in a pipe was studied to measure
frictional pressure losses encountered during the pumping into the fractured formation. Figure 61 shows the frictional pressure losses across a

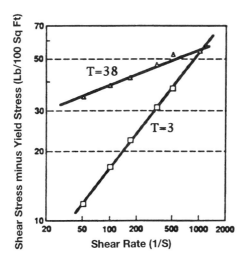

*Figure 60.   Rheograms of cuttings contaminated with oil-based drilling fluid
slurries in seawater at 3 °C and 38 °C. (Reproduced with permission from
reference 168. Copyright 1993 Society of Petroleum Engineers.)*

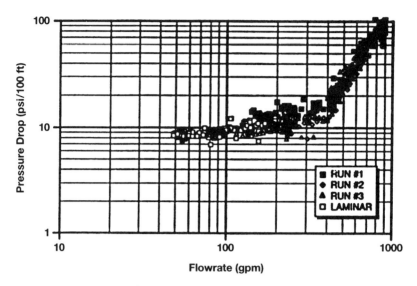

Figure 61.   Measured frictional pressure losses of cuttings–seawater slurry during pipe flow as a function of flow rate. (Reproduced with permission from reference 168. Copyright 1993 Society of Petroleum Engineers.)

3 in. (7.6 cm) diameter pipe of length 40 ft (12.2 m). At a flow rate of 1.5 m³/min, the flow became turbulent and the frictional pressure losses increased rapidly with increasing flow rate. Crawford and Lescarboura (*168*) attempted to predict the pressure drop-flow rate behavior using a recent algorithm developed by Reed and Pilehvari (*169*). The frictional pressure drop in laminar flow was well predicted but the model underestimated the pressure drop obtained in turbulent flow.

The cuttings slurry was to be used to fracture the formation into which it was to be injected. A critical aspect of the hydraulic fracturing of permeable rock formations is the prevention of pressure decline in the fracture by fluid loss into the formation (*170*). The values of the desorptivity S obtained from the static fluid loss of the cuttings–seawater slurry were in the range 390–1950 μm/s$^{1/2}$, which were very much larger than the values found for oil- and water-based drilling fluids. Rather than add fluid loss additives to the slurry (and add to the disposal cost), Crawford and Lescarbara (*168*) argued that the slurry should be used to fracture a low permeability (shale) formation and to use the high fluid loss to stop the propagation of the fracture when it reached a permeable formation. Sirevåg and Bale (*171*) arrived at a similar conclusion; they obtained values of S = 157 μm/s$^{1/2}$ for their cuttings–seawater slurries.

Other methods of disposing of cuttings from oil-based drilling fluids have been investigated and include cuttings cleaning (using solvents, surfactant solutions, and supercritical fluids) and incineration (combustion of oil on cuttings). Oil-based drilling fluid waste from onshore drilling operations has been disposed of by landfilling, solidification, mixing with flyash for road making, and land farming. The land farming method (172) involves the spreading of oil-based drilling fluid waste on farm land to degrade the oil over a period of time.

A radically different approach to the disposal of water-based drilling fluid waste is to use it as the mix water for the cement system that will bond the casing to the borehole wall (173, 174). The cement uses blast furnace slag in place of conventional Portland cement. The blast furnace slag is activated by Na_2CO_3-NaOH and the fluid loss additives used in the drilling fluid can be used to control the fluid loss of the cementing slurry. A cementing slurry of density 1820 kg/m^3 was reported to consist of 135 kg of blast furnace slag, 6.3 kg of Na_2CO_3, 2.3 kg of NaOH, 1.8 kg of lignosulphonate, and 0.107 m^3 of water-based drilling fluid (ρ = 1340 kg/m^3) (173).

Monitoring of Drilling Fluids. During the course of drilling operations, while the drilling fluid is being circulated, its composition is subject to change. The changes in composition arise from the loss and/or alteration of components (ion exchange, adsorption of organics on drilled cuttings, polymer degradation, etc.) and the uptake of material from the wellbore (dispersed drilled solids, influx of ions from formation water). The composition of the drilling fluid may therefore move away from the specified formulation, which may have been developed after considerable effort and cost.

The complexity of both water- and oil-based drilling fluids has precluded the routine monitoring of chemical composition. The current methods of monitoring the chemical composition at the rig site during drilling, as described by the API Recommended Practice (175), are limited to the measurement of the concentrations of a few ions in the filtrate of water-based drilling fluids (Cl^-, Ca^{2+} + Mg^{2+}, K^+) and the pH of the filtrate. The concentrations of the remaining major ions (Na^+, SO_4^{2-}, CO_3^{2-}), the suspended solids, and the organics are not monitored routinely during drilling operations. A total solids concentration can be estimated from a combination of the density of the drilling fluid and the solids content, obtained by retorting the drilling fluid (176). The state of the drilling fluid and the need for treatment (e.g., addition of drilling fluid additives) are generally determined by the routine measurement of fluid rheology and fluid loss.

Hughes and Jones and co-workers (177–179) have developed an ion chromatography technique for the analysis of filtrates obtained from water-based drilling fluids. The use of ion chromatography enables a full analysis of the ionic concentration of all the major cations and anions in the filtrate; the technique can also be adopted to determine the con-

centration of heavy metals in filtrates (*177*). The regular analysis of filtrates obtained from drilling fluid samples from the inlet and outlet
streams enables the change in filtrate composition due to interaction
with the wellbore and drilled cuttings to be evaluated. Figure 62, for
example, shows the changes in the calcium and sulphate ion concentrations as a function of drilled depth during the drilling of an evaporite
sequence. The presence of the anhydrite ($CaSO_4$) in the drilled formations caused the drilling fluid to become saturated with respect to
calcium and sulphate ions. The presence of the high calcium ion concentration has a deleterious effect on the bentonite and water-soluble
polymers in the drilling fluid. The high calcium concentration manifested
itself most clearly by the high values of static (API) fluid loss. The drilling
fluid was treated by addition of $NaHCO_3$ and $NaOH$ to precipitate the
calcium ions. Figure 62 shows that the calcium ion concentration decreased more rapidly than the sulphate ion concentration. The decrease
in the calcium ion concentration was accompanied by a decrease in the
static fluid loss.

Houwen et al. (*180*) have recently demonstrated the application of
a rig site X-ray fluorescence (XRF) spectrometer to determine the concentration of barite in both water- and oil-based drilling fluids. The XRF
measurement enables the total solids content of the drilling fluid to be
broken down into the so-called high gravity solids (barite) and low-
gravity solids (bentonite and drilled solids). Houwen et al. (*180*) have
shown that the current routine technique of estimating the barite con-

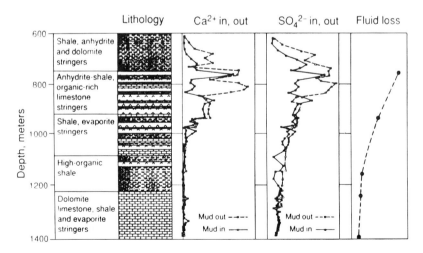

*Figure 62. Changes in the calcium and sulphate concentration of a water-
based drilling fluid during drilling of a complex evaporite sequence. Scales:
Ca^{2+} 0–16 mM; SO_4^{2-} 0–40 mM; API (static) fluid loss 10–20 mL. (Reproduced with permission from reference 178. Copyright 1992.)*

centration using a retort and fluid density is subject to gross errors. Figure 63 shows a comparison of the relative concentrations of high- and low-gravity solids determined by XRF and the retort–density method in fluid samples obtained from a well. The comparison shows that the retort–density method consistently overestimated the barite content of the drilling fluid, whereas the XRF measurement showed a significantly lower barite content and higher concentration of drilled solids.

A comprehensive analysis of the solids and organics content of water-based drilling fluids has been demonstrated by Hughes and Jones and co-workers (*181*) using Fourier transform infrared (FTIR) spectrometry with a multivariate calibration model. Figure 64 shows a typical example of a diffuse reflectance spectrum of the solid and polymer residue of a dried sample of water-based drilling fluid. The characteristic absorbance bands of barite and bentonite are clearly observed, together with a complex of C–H stretching bands due to the presence of organic components. Figure 65 shows the prediction of the concentration of the components

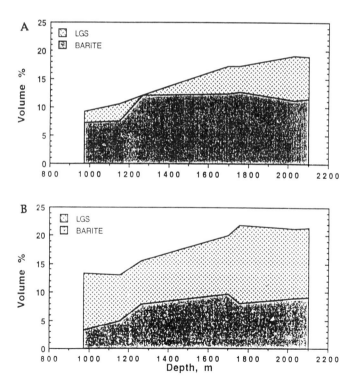

Figure 63. Comparison of the depth dependence of concentration of barite and low-gravity solids (LGS) in a circulating drilling fluid determined by (A) retort; (B) XRF. (Reproduced with permission from reference 180. Copyright 1993 Society of Petroleum Engineers.)

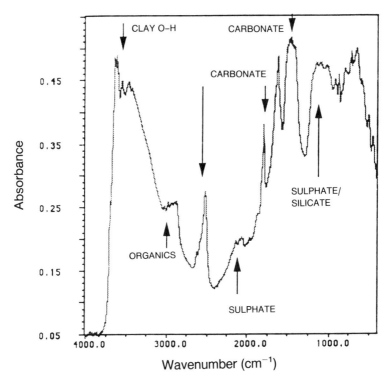

Figure 64.   Typical diffuse reflectance infrared spectrum of a dried water-based drilling fluid. (Reproduced with permission from reference 181. Copyright 1991 Society of Petroleum Engineers.)

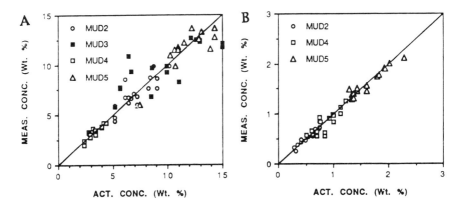

Figure 65.   Prediction of (a) bentonite and (b) CMC content in various types of dried water-based drilling fluids. (Reproduced with permission from reference 181. Copyright 1991 Society of Petroleum Engineers.)

bentonite and CMC in dried test fluid samples of accurately known concentration. The two components are typically predicted to within a relative accuracy of ±10%. Jones et al. (*182*) have recently demonstrated that the concentrations of components in both water- and oil-based drilling fluids may be quantified in the liquid state using attenuated total reflectance (ATR)–FTIR spectrometry. Figure 66 shows the FTIR–ATR spectrum of an invert emulsion oil-based drilling fluid. The presence of the continuous oil phase, discrete aqueous phase, and suspended solids (organophilic clay, barite, and calcium carbonate) can be readily observed. Quantitative analysis of components in both oil- and water-based drilling fluids has been made (*182*). The absorbance of the characteristic Si–O stretching band in the ATR spectra of freshly prepared bentonite suspensions increases systematically with time (*182*). This increase appears to be due to the decrease in the particle size of the bentonite that arises from the hydration and dispersion of the bentonite clay in water.

Murch et al. (*183*) have recently described the application of the on-line measurement of the physical properties of drilling fluids that may eliminate the need for laborious manual rig site testing. The fluids monitoring system (FMP) constructed by Murch et al. gave continuous

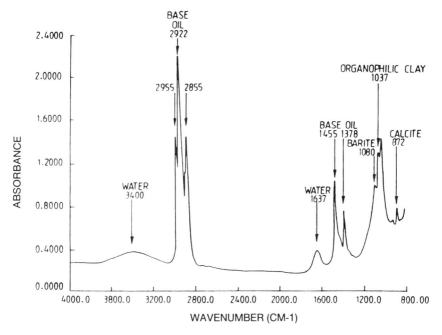

Figure 66.   Typical attenuated total reflectance infrared spectrum of an invert emulsion oil-based drilling fluid. (Reproduced with permission from reference 182. Copyright 1994.)

measurements of electrical conductivity, temperature, pH, density, and rheology. The rheology of the flowing drilling fluid was determined by using the pressure drop across three pipe rheometers of differing diameter (and hence v_m; see eq 33, for example). Figure 67 (from reference 161) shows a comparison of the plastic viscosity PV obtained from a standard Fann viscometer (Figure 6) and the pipe rheometer on the FMP. The measurement of the Bingham plastic viscosity obtained by the two methods shows good agreement.

The combination of physical and chemical measurements previously described has the capability of giving a full description of the performance and composition of the drilling fluid. These measurements form the basis of a process-control approach to drilling fluids and the ability to correct for changes away from the optimal performance of the fluid by controlled additions of chemicals or changes in the solids control procedure. Such an approach should minimize the use of drilling fluid additives and the dilution–disposal of drilling fluids and should reduce the problems of waste disposal.

Conclusions

The areas of science and technology that are spanned by drilling fluids are considerable: rheology, fluid mechanics, colloid science, polymer

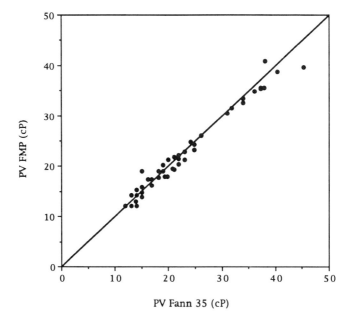

Figure 67.   Comparison of Bingham plastic viscosity (in cP) obtained from a Fann concentric viscometer and an on-line pipe rheometer. (Data from reference 161.)

chemistry, rock mechanics, analytical chemistry, process control, and increasingly, environmental monitoring and toxicology. There is a considerable and continuing research and development activity in drilling fluids technology, much of which is driven by the need to reduce drilling costs (minimize drilling fluid-related problems) and to produce environmentally acceptable drilling fluids.

The current environmental scrutiny on drilling fluid waste is set to continue, and increasing constraints are likely be placed on the disposal of drilling fluid additives. The continuing use of oil-based drilling fluids is dependent on environmentally acceptable methods of waste disposal. If these methods cannot be found, then water-based drilling fluids that routinely can operate at high temperatures ($>140\ °C$) and can match the shale stabilizing performance of oil-based drilling fluids must be sought.

Developments in drilling practices are also placing increasing demands on drilling fluids. The increasing need to drill extended horizontal wells to maximize oil recovery is requiring the development of drilling fluids that can clean the hole, can maintain stable wellbores in highly deviated holes, and can minimize formation damage.

Acknowledgments

The authors thank Geoff Maitland, Thomas Geehan, Otto Houwen, John Sherwood, Chris Hall, Edmund Fordham, Gerry Meeten, Phil Fletcher, Patrick Tomkins, and Sarah Pelham for many discussions on drilling fluids over many years. The authors would also like to thank Peter Bern (British Petroleum) for advice and discussion of his work.

List of Symbols

a, A	constants (eq 19)
a_w	thermodynamic activity of water in aqueous phase
a_w^f	thermodynamic activity of water in drilling fluid
a_w^{sh}	thermodynamic activity of water in shale
A	cross-sectional area (m^2)
A_i, A_w	preexponential factor in Arrhenius equations (Pa or Pa · s)
b, B	constants
B_w	constant in equation 16 (Pa^{-1})
c, C	constants
C	constant in Graves and Collins' rheological model (eq 9)
C_{ben}	concentration of bentonite (g/L)
C_o	yield stress constant, Robertson–Stiff model (s^{-1})
C_P	polymer concentration (g/L)
C_w	constant in equation 16 (K/Pa)

d	distance between clay platelets (m)
D_y	dynamic parameter (modified Reynolds number)
ΔP_L	frictional pressure drop in laminar flow (Pa)
e_c	void fraction of filter cake
ECD	equivalent circulating density (kg/m^3)
E_H, E_i, E_L, E_w	activation energy (expressed in J/mol or K when divided by the gas constant)
E_s	streaming potential (volts)
f	Fanning friction factor
f_c	critical friction factor
f_{cr}	is labeled here as critical friction factor, but is on the list as cutting friction factor
F	viscous drag force (N)
g	acceleration due to gravity (m/s^2)
G'	storage modulus (Pa)
h_c	thickness of filter cake (m)
H_e	Hedström number
I	ionic strength (molar)
k	constant
k_a	permeability of flow damaged core sample (m^2)
k_b	permeability of undamaged core sample (m^2)
k_c	permeability of filter cake (m^2)
$k_0{}^2$	Casson yield stress (Pa)
$k_1{}^2$	Casson high shear rate viscosity (Pa·s)
K	consistency of power law fluid (Pa·s^n)
K'	constant (Pa·s^n)
L	length (m)
m	constant
M_w	molecular weight (g/mol)
n	flow consistency index of power law fluid
n	constant, measure of particle–particle interaction
n'	d ln (τ_w)/d ln $(4v_m/R)$
P	pressure (Pa)
P_{fd}	pressure drop across filter cake in dynamic filtration (Pa)
P_{fs}	pressure-drop across filter cake in static filtration (Pa)
P^o	reference hydrostatic pressure (Pa)
P^{sh}	pressure of water in shale (Pa)
PV	plastic viscosity (Pa·s)
Q	flow rate of drilling fluid (m^3/s)
Q_c	critical flow rate of drilling fluid below which a bed of cuttings will form in a deviated well (m^3/s)
Q_{fd}	limiting dynamic filtration rate (m^3/s)
Q_o	constant (m^3/s)
r	rotation speed of concentric cylinder viscometer (rpm)

R	gas constant (8.3143 J/K/mol)
R	inner radius of drill pipe (m)
R_B	radius of borehole (m)
R_c	cuttings radius (m)
Re	Reynolds number
Re*	critical Reynolds number
Re'	equivalent Reynolds number
Re''	generalized (Dodge–Metzner) Reynolds number
Re_b	Bingham fluid Reynolds number
Re_c	cutting Reynolds number
Re_{cy}	cuttings Reynolds number from effective viscosity derived from a polynomial
R_H	hydraulic radius $(R_B - R_P)/2$ (m)
ROP	rate of penetration of drilling (m/h)
R	radius of drill pipe (m)
R_p	radius of unsheared plug in pipe during plug flow (m)
R_s	radius of sphere (m)
R_t	cuttings transport ratio
S	desorptivity $(m/s^{1/2})$
t	time (s)
t_c	critical time (s)
T	arbitrary scaling factor
T	absolute temperature (K)
$T*$	characteristic temperature (K)
v_c	critical friction velocity of fluid at borehole wall (m/s)
v_m	mean velocity (m/s)
v_{mc}	critical mean velocity of drilling fluid at onset of turbulence (m/s)
v_s	slip velocity of cutting in drilling fluid (m/s)
v_t	transport velocity of cutting in drilling fluid (m/s)
V_c	volume fraction of cuttings in drilling fluid
V_{fd}	cumulative volume of filtrate from dynamic filtration (m^3)
V_{fs}	cumulative volume of filtrate from static filtration (m^3)
V_i	pressure coefficient (specific volume of drilling fluid) (m^3/mol)
V_m	partial molar volume of water (m^3/mol)
V_{30}	API fluid loss (mL)
x	τ_o/τ_w
x_c	critical value of x
X	distance from filter medium (m)
Y	yield number
YP	yield point (Pa)
Z	vertical depth (m)

Greek

α	ratio of radii of outer to inner cylinders in concentric cylinder viscometer
β	isothermal compressibility (Pa^{-1})
$\dot{\gamma}$	shear rate (s^{-1})
$\dot{\gamma}_c$	shear rate at interface between drilling fluid and filter cake (s^{-1})
$\dot{\gamma}_i$	shear rate on inner cylinder of concentric cylinder viscometer (s^{-1})
$\dot{\gamma}_w$	wall shear rate (s^{-1})
Γ	arbitrary scaling factor
δ	adjustable packing factor
η	Newtonian viscosity ($Pa \cdot s$)
η_a	apparent viscosity ($Pa \cdot s$)
η_e	effective viscosity ($Pa \cdot s$)
η_f	filtrate viscosity ($Pa \cdot s$)
η_{HB}	Herschel–Bulkley viscosity ($Pa \cdot s$)
η_o	viscosity of solvent ($Pa \cdot s$)
η_p	Bingham plastic viscosity ($Pa \cdot s$)
η_R	reduced viscosity
$[\eta]$	intrinsic viscosity of polymer in solution (L/g)
$\theta_{300}, \theta_{600}$	deflection of torque spring on Fann rheometer at 300 and 600 rpm (degrees)
Θ	deflection of torque spring on Fann rheometer (degrees)
μ_w^f	chemical potential of water in drilling fluid (J/K/mol)
μ_w^o	chemical potential of water in a reference state (J/K/mol)
μ_w^{sh}	chemical potential of water in shale (J/K/mol)
ξ	constant
ξ	differential change in filtrate volume with respect to mud volume
Π_1	dimensionless group
Π_2	dimensionless group
ρ	fluid density (kg/m^3)
ρ_c	cuttings density (kg/m^3)
τ	shear stress (Pa)
τ_c	shear stress at cake–fluid interface
τ_o	yield stress (Pa)
τ_w	wall shear stress (Pa)
τ_{wc}	critical wall shear stress (Pa)
τ_{yc}	yield stress of filter cake (Pa)
ϕ	volume fraction of solids in drilling fluid or filter cake; porosity of rock formation

χ   angle of inclination of wellbore (degrees)
Ψ   electrostatic potential (volts)

References

1. Allen, J. H. In *Drilling and Drilling Fluids*; Chiligarian, G. V.; Vorabtr, P., Eds.; Elsevier: Amsterdam, Netherlands, 1981; pp 1–16.
2. Offshore Business *Oilfield Mud and Cementing*; Smith Rea Energy Analysts: Canterbury, England, 1990; No. 33.
3. *Mud Logging: Principles and Interpretation*; Whittaker, A., Ed.; International Human Resources Development Corp.: Boston, MA, 1985.
4. Bruce, S. Presented at the IADC/SPE Drilling Conference, Houston, TX, February/March 1990; paper SPE 19942.
5. Tan, C. P.; Willoughby, D. R. Presented at the 68th Annual Technical Conference of the Society of Petroleum Engineers, Houston, TX, October 1993; paper SPE 26325.
6. Moore, P. L. *Drilling Practices Manual*; The Petroleum Publishing Co.: Tulsa, OK, 1974; pp 298–309.
7. Fertl, W. H. *Abnormal Formation Pressures*; Elsevier: Amsterdam, Netherlands, 1976.
8. Harris, J. *Rheology and Non-Newtonian Flow*; Longman: London, 1977; pp 1–22.
9. Nguyen, Q. D.; Boger, D. V. *Annu. Rev. Fluid Mech.* **1992,** *24,* 47–88.
10. Shaw, D. B.; Weaver, C. E. *J. Sediment. Petrol.* **1965,** *35,* 213–222.
11. *World Oil* **1994,** *215(6),* 51–90.
12. American Petroleum Institute *Specification for Drilling-Fluid Materials*; 15th ed.; API Spec. 13A; American Petroleum Institute: Washington, DC, 1993.
13. Grim, R. E. *Clay Mineralogy,* 2nd ed.; McGraw-Hill: New York, 1968; pp 77–92.
14. Van Olphen, H. *An Introduction to Clay Colloid Chemistry,* 2nd ed.; Wiley: New York, 1977; pp 57–82.
15. MacEwan, D. M. C.; Wilson, M. J. In *Crystal Structure of Clay Minerals and Their X-ray Identification*; Brindley, G. W.; Brown, G., Eds.; Mineralogical Society: London, 1981; pp 197–214.
16. Schultz, L. G. *Clays Clay Miner.* **1969,** *17,* 114–149.
17. Van Olphen, H. *An Introduction to Clay Colloid Chemistry,* 2nd ed.; Wiley: New York, 1977; p 254.
18. Quirk, J. P. *Isr. J. Chem.* **1968,** *6,* 213–234.
19. James, R. O.; Park, G. A. In *Surface and Colloid Science*; Matijevic, E., Ed.; Plenum: New York, 1982; Vol. 12, pp 119–216.
20. Van Olphen, H. *An Introduction to Clay Colloid Chemistry,* 2nd ed.; Wiley: New York, 1977; pp 29–44.
21. Lubetkin, S. D.; Middleton, S. R.; Ottewill, R. H. *Philos. Trans. R. Soc. London Ser. A* **1984,** *311,* 353.
22. Carico, R. D.; Bagshaw, F. R. Presented at SPE–AIME Production Technology Symposium, Hobbs, NM, October 1978; paper SPE 7747.
23. Chatterji, J.; Borchardt, J. K., *J. Pet. Technol.* **1981,** *33,* 2042–2056.
24. Estes, J. C. In *Water-Soluble Polymers: Beauty with Performance*; Glass, J. E., Ed., Advances in Chemistry 213; American Chemical Society: Washington, DC, 1986; pp 155–170.

25. Darley, H. C. H.; Gray, G. R. *Composition and Properties of Drilling and Completion Fluids*, 5th ed.; Gulf Publishing Company, Houston, TX, 1988; pp 542–617.
26. McCormick, C. L.; Bock, J.; Schulz, D. N. In *Encyclopedia of Polymer Science and Engineering*, 2nd ed.; Wiley: New York, 1989; Vol. 17 pp 730–784.
27. *Handbook of Industrial Gums and Resins*; Davidson, R. L., Ed.; McGraw-Hill: New York, 1980.
28. Littmann, W. *Polymer Flooding*; Elsevier: Amsterdam, Netherlands, 1988; pp 30–32.
29. Beihoffer, T. W.; Dorrough, D. S.; Deem, C. K.; Schmidt, D. D.; Bray, R. P. *Oil & Gas J.* 1992, 90(11), 47–52.
30. Hunt, J. A.; Young, T. S.; Willhite, G. P. Presented at the SPE/DOE 5th Symposium on Enhanced Oil Recovery, Tulsa, OK, April 1986; paper SPE 14949.
31. Tinland, B.; Rinaudo, M. *Macromolecules* 1989, 22, 1863–1868.
32. Hughes, T. L.; Jones, T. G. J.; Houwen, O. H. *SPE Drill. Complet.* 1993, 8, 157–164.
33. Thomas, D. C. *Soc. Pet. Eng. J.* 1982, 22, 171–180.
34. Park, L. S. *SPE Drill. Eng.* 1988, 3, 311–314.
35. Lauzon, R. V.; Short, J. S. Presented at the 54th Annual Technical Conference of the Society of Petroleum Engineers, Las Vegas, NV, September 1979; paper SPE 8225.
36. Davis, N.; Tooman, C. E. *SPE Drill. Eng.* 1989, 4, 47–56.
37. Tissot, B. P.; Welte, D. H. *Petroleum Formation and Occurrence*, 2nd ed.; Springer-Verlag: Berlin, Germany, 1984; pp 403–408; Strausz, O. P.; Mojelsky, T. W.; Lowm, E. M. *Fuel* 1992, 71, 1355–1363.
38. Tyssee, D. A.; Vetter, O. J. *J. Pet. Technol.* 1981, 21, 721–730.
39. Plank, J. P.; Hamberger, J. V. Presented at the SPE European Petroleum Conference, London, October 1988; paper SPE 18372.
40. Hille, M. Presented at the International Symposium on Oilfield Geothermal Chemistry, Phoenix, AZ, April 1985; paper SPE 13558.
41. Carney, L. L.; Guven, N.; McGrew, G. T. Presented at the California Regional Meeting of the Society of Petroleum Engineers, San Francisco, CA, 1982; paper SPE 10736.
42. Clark, R. K.; Scheuerman, R. F.; Rath, H.; Van Laar, H. G. *J. Pet. Technol.* 1976, 28, 719.
43. Clark, R. K. In *Water-Soluble Polymers: Beauty with Performance*; Glass, J. E., Ed.; Advances in Chemistry 213; American Chemical Society: Washington, DC, 1986; pp 171–181.
44. Bol, G. M. Presented at the IADC/SPE Drilling Conference, Dallas, TX, February 1986; paper SPE 14802.
45. Shupe, R. D. *J. Pet. Technol.* 1981, 21, 1513–1529.
46. Lockhart, T. P. *SPE Adv. Technol. Ser.*, 1994, 2(2), 199–205.
47. Theng, B. K. G. *Formation and Properties of Clay–Polymer Complexes*; Elsevier: Amsterdam, Netherlands, 1979; pp 95–108.
48. Beihoffer, T. W.; Dorrough, D. S.; Schmidt, D. D. Presented at the IADC/SPE Drilling Conference, Houston, TX, February 1990; paper SPE 19953.
49. Retz, R. H.; Friedheim, J.; Lee, L. J.; Welch, O. O. Presented at the Offshore Europe Conference, Aberdeen, Scotland, September 1991; paper SPE 23064.

50. Reid, P. I.; Elliot, G. P.; Minton, R. C.; Chambers, B. D.; Burt, B. A. Presented at the SPE/EPA Exploration Production Environment Conference, San Antonio, TX, March 1993; paper SPE 25989.
51. Vorabutr, P.; Chilingarian, G. V. In *Drilling and Drilling Fluids;* Chilingarian, G. V.; Vorabutr, P., Eds.; Elsevier, Amsterdam, Netherlands, 1981; pp 365–398.
52. Carter, T. S.; Faul, G. F. Presented at the 67th Annual Technical Conference of the Society of Petroleum Engineers, Washington, DC, October 1992; paper SPE 24590.
53. Jacques, D. F.; Newman, H. E.; Turnbull, W. B. Presented at the IADC/ SPE Drilling Conference, New Orleans, LA, February 1992; paper SPE 23881.
54. Friedheim, J. E.; Pantermuehl, R. M. Presented at the SPE/IADC Drilling Conference, Amsterdam, Netherlands, February 1993; paper SPE 25753.
55. Kenny, P. *Oil & Gas J.* 1993, *91(45)*, 88–91; Carlson, T.; Hemphill, T. Presented at the SPE International Petroleum Conference and Exhibition, Vera Cruz, Mexico, October 1994; paper SPE 28739.
56. Gopal, E. S. R. In *Emulsion Science;* Sherman, P., Ed.; Academic: London, 1968; pp 1–75.
57. Schramm, L. L. In *Emulsions: Fundamentals and Applications in the Petroleum Industry;* Schramm, L. L., Ed.; Advances in Chemistry 231; American Chemical Society: Washington, DC, 1992; pp 1–49.
58. Kitchener, J. A.; Mussellwhite, P. R. In *Emulsion Science;* Sherman, P., Ed.; Academic: London, 1968; pp 77–130.
59. Van Olphen, H. *An Introduction to Clay Colloid Chemistry,* 2nd ed.; Wiley: New York, 1977; pp 171–173, 183–185.
60. Salisbury, D. P.; Walker, N. S. *Spectroscopy* 1986, *1*, 44–47.
61. Chenevert, M. E. *J. Pet. Technol.* 1970, *22*, 309–1316.
62. Tschirley, N. K. In *Drilling and Drilling Fluids;* Chiligarian, G. V.; Vorabtr, P., Eds.; Elsevier, Amsterdam, Netherlands, 1981; pp 125–128.
63. Darley, H. C. H.; Gray, G. R. *Composition and Properties of Drilling and Completion Fluids,* 5th ed.; Gulf Publishing Company: Houston, TX, 1988; pp 97–102.
64. Tschirley, N. K. In *Drilling and Drilling Fluids;* Chiligarian, G. V.; Vorabtr, P., Eds.; Elsevier, Amsterdam, Netherlands, 1981; pp 128–129.
65. *Theory and Application of Drilling Fluid Hydraulics;* Whittaker, A., Ed.; International Human Resources Development Corp.: Boston, MA, 1985; pp 24–38.
66. Pal, R.; Yan, Y.; Masliyah, J. In *Emulsions: Fundamentals and Applications in the Petroleum Industry;* Schramm, L. L., Ed.; Advances in Chemistry 231; American Chemical Society: Washington, DC, 1992; pp 131–170.
67. Whorlow, R. W. *Rheological Techniques,* 2nd ed.; Ellis Horwood: Chichester, England, 1992; pp 8–18.
68. Collins, R. E.; Graves, W. G. Unsolicited SPE paper, June 1978; paper SPE 7654.
69. Houwen, O. H.; Geehan, T. Presented at the 61st Annual Technical Conference of the Society of Petroleum Engineers, New Orleans, LA, October 1986; paper SPE 15416.
70. Wilkinson, W. L. *Non-Newtonian Fluids: Fluid Mechanics, Mixing and Heat Transfer;* Pergamon: London, 1960; pp 21–26, 130–134.
71. Darley, H. C. H.; Gray, G. R. *Composition and Properties of Drilling and Completion Fluids,* 5th ed.; Gulf Publishing Company: Houston, TX, 1988; pp 188–201.

72. Alderman, N. J.; Gavignet, A.; Guillot, D.; Maitland, G. C. Presented at the 63rd Annual Technical Conference of the Society of Petroleum Engineers, Houston, TX, October 1988; paper SPE 18035.
73. Annis, M. R. *J. Pet. Technol.* **1967,** *19,* 1074–1080.
74. Morris, E. R.; Cutler, A. N.; Ross-Murphy, S. B.; Rees, D. A.; Price, J. *Carbohydr. Polym.* **1981,** *1,* 5–21.
75. Heath, D.; Tadros, T. F. *J. Colloid Interface Sci.* **1983,** *93,* 307–319.
76. Alderman, N. J.; Ram Babu, D.; Hughes, T. L.; Maitland, G. C. *Spec. Chem. Prod. Mark. Appl.* **1989,** *9,* 314–316, 318, 326.
77. Van Olphen, H. *An Introduction to Clay Colloid Chemistry,* 2nd ed.; Wiley: New York, 1977; pp 95–98.
78. Rabaioli, M. R.; Miano, F.; Lockhart, T. P.; Burrafuto, G. Presented at the SPE International Symposium on Oilfield Chemistry, New Orleans, LA, March 1993; paper SPE 25179.
79. Marin, G. In *Rheological Measurement;* Collyer, A. A.; Clegg, D. W., Eds.; Elsevier: London, 1988; pp 297–343.
80. Guillot, D.; Hendriks, H.; Callet, F.; Vidick, B. In *Well Cementing;* Nelson, E. B., Ed; Schlumberger Educational Services: Houston, TX, 1990; pp 5.1–5.37.
81. Johnson, A. B.; White, D. B. *Int. J. Multiphase Flow* **1993,** *19,* 921–941.
82. Wilkinson, W. L. *Non-Newtonian Fluids: Fluid Mechanics, Mixing and Heat Transfer;* Pergamon: London, 1960; pp 50–95.
83. Govier, G. W.; Aziz, K. *The Flow of Complex Mixtures in Pipes;* Kreiger: Huntington, NY, 1977; pp 182–266.
84. *Theory and Application of Drilling Fluid Hydraulics;* Whittaker, A., Ed.; International Human Resources Development Corp.: Boston, MA, 1985.
85. Bourgoyne, A. T.; Millheim, K. K.; Chenevert, M. E.; Young, F. S. *Applied Drilling Engineering;* Society of Petroleum Engineers: Richardson, TX, 1991; pp 113–189.
86. Wilkinson, W. L. *Non-Newtonian Fluids: Fluid Mechanics, Mixing and Heat Transfer;* Pergamon: London, 1960; pp 30–33.
87. Govier, G. W.; Aziz, K. *The Flow of Complex Mixtures in Pipes;* Kreiger Publishing Company: Huntington, NY, 1977; pp 183–185.
88. *Theory and Application of Drilling Fluid Hydraulics;* Whittaker, A., Ed.; IHRDC: Boston, MA, 1985; pp 66–67.
89. Metzner, A. B.; Reed, J. C. *AIChEJ* **1955,** *1,* 434–440.
90. Dodge, D. W.; Metzner, A. B. *AIChEJ* **1959,** *5,* 189–204.
91. *Theory and Application of Drilling Fluid Hydraulics;* Whittaker, A., Ed.; IHRDC: Boston, MA, 1985; pp 70–75.
92. Bourgoyne, A. T.; Millheim, K. K.; Chenevert, M. E.; Young, F. S. *Applied Drilling Engineering;* Society of Petroleum Engineers: Richardson, TX, 1991; pp 144–149.
93. Bobok, E. *Fluid Mechanics for Petroleum Engineers;* Elsevier: Amsterdam, Netherlands, 1993; pp 286–336.
94. Bird, R. B.; Stewart, W. E.; Lightfoot, E. N. *Transport Phenomena;* Wiley: New York, 1960; pp 181–183.
95. Govier, G. W.; Aziz, K. *The Flow of Complex Mixtures in Pipes;* Kreiger: Huntington, NY, 1977; p 152.
96. Ryan, N. W.; Johnson, M. M. *AIChEJ* **1959,** *5,* 433–435.
97. Hanks, R. W.; Christiansen, E. B. *AIChEJ* **1962,** *8,* 467–471.
98. Hanks, R. W. *AIChEJ* **1963,** *9,* 306–309; Hanks, R. W.; Pratt, D. R. *Soc. Pet. Eng. J.* **1967,** *7,* 342–346.

99. Govier, G. W.; Aziz, K. *The Flow of Complex Mixtures in Pipes;* Kreiger: Huntington, NY, 1977; pp 227–228.

100. Savins, J. G. *Soc. Pet. Eng. J.* **1964,** *4,* 203–214.

101. Lumley, J. L. *Annu. Rev. Fluid Mech.* **1969,** *1,* 367–384.

102. Morgan, S. E.; McCormick, C. L. *Prog. Polym. Sci.* **1990,** *15,* 103–145.

103. Metzner, A. B.; Park, M. G. *J. Fluid Mech.* **1964,** *20,* 291–303.

104. Seyer, F. A.; Metzner, A. B. *Can. J. Chem. Eng.* **1969,** *47,* 525–529.

105. Govier, G. W.; Aziz, K. *The Flow of Complex Mixtures in Pipes;* Kreiger: Huntington, NY, 1977; pp 237–248.

106. Okafor, M. N.; Evers, J. F. Presented at the Western Regional Meeting SPE, Bakersfield, CA, March/April 1992; paper SPE 24086.

107. McCann, R. C.; Quigley, M. S.; Zamora, M.; Slater, K. S. *SPE Drill. Complet.* **1995,** *10,* 96–103.

108. Bourgoyne, A. T.; Millheim, K. K.; Chenevert, M. E.; Young, F. S. *Applied Drilling Engineering;* Society of Petroleum Engineers: Richardson, TX, 1991; pp 149–150.

109. *Theory and Application of Drilling Fluid Hydraulics;* Whittaker, A., Ed.; International Human Resources Development Corp.: Boston, MA, 1985; pp 91–109.

110. Marken, C. D.; He, X.; Arild, S. Presented at the 67th Annual Technical Conference of the Society of Petroleum Engineers, Washington, DC, October 1992; paper SPE 24598.

111. Sample, K. J.; Bourgoyne, A. T. Presented at the 53rd Annual Technical Conference of the Society of Petroleum Engineers, Houston, TX, October 1978; paper SPE 7497.

112. Walker, R. E.; Mayes, T. M. *J. Pet. Technol.* **1975,** *27,* 893–900.

113. Bird, R. B.; Stewart, W. E.; Lightfoot, E. N. *Transport Phenomena;* Wiley: New York, 1960; pp 56–60.

114. *Theory and Application of Drilling Fluid Hydraulics;* Whittaker, A., Ed.; International Human Resources Development Corp.: Boston, MA, 1985; pp 119–128.

115. Gavignet, A. A.; Wick, C. J. *SPE Drill. Eng.* **1987,** *2,* 309–315.

116. Moore, P. L. *Drilling Practices;* The Petroleum Publishing Company: Tulsa, OK, 1974; pp 228–240.

117. Sample, K. J.; Bourgoyne, A. T. Presented at the 52nd Annual Technical Conference of the Society of Petroleum Engineers, Denver, CO., October 1977; paper SPE 6645.

118. Tomren, P. H.; Iyoho, A. W.; Azar, J. J. *SPE Drill. Eng.* **1986,** *1,* 43–56.

119. Becker, T. E.; Azar, J. J.; Okrajni, S. S. *SPE Drill. Eng.* **1991,** *6,* 16–24.

120. Gavignet, A. A.; Sobey, I. J. *J. Pet. Technol.* **1989,** *41,* 916–921.

121. Peden, J. M.; Ford, J. T.; Oyeneyin, M. B. Presented at Europec 90, The Hague, Amsterdam, October 1990; paper SPE 20925; Gao, E.; Ford, J.; Oyeneyin, M. B.; Larrcia, M.; Peden, J. M.; Parker, D. Presented at SPE Computer Conference, New Orleans, LA, July 1993; paper SPE 26217.

122. Luo, Y. Ph.D. Thesis, Heriot-Watt University, UK, 1988; Luo, Y.; Bern, P. A.; Chambers, B. D. Presented at the IADC/SPE Drilling Conference, New Orleans, LA, February 1992; paper SPE 23884.

123. Larsen, T. I.; Pileharvi, A. A.; Azar, J. J. Presented at the SPE Rocky Mountain Region/Low Permeability Reservoirs Symposium, Denver, CO, April 1993; paper SPE 25872.

124. Tschirley, N. K. In *Drilling and Drilling Fluids;* Chiligarian, G. V.; Vorabtr, P., Eds.; Elsevier, Amsterdam, Netherlands, 1981; pp 131–136.

125. Darley, H. C. H.; Gray, G. R. *Composition and Properties of Drilling and Completion Fluids*, 5th ed.; Gulf Publishing Company: Houston, TX, 1988; pp 102–103.
126. Scheidegger, A. E. *The Physics of Flow Through Porous Media*; University of Toronto: Toronto, Canada, 1957; pp 54–69.
127. Outmans, H. D. *Soc. Pet. Eng. J.* **1963**, *3*, 236–244.
128. Darley, H. C. H.; Gray, G. R. *Composition and Properties of Drilling and Completion Fluids*, 5th ed.; Gulf Publishing Company: Houston, TX, 1988; pp 282–320.
129. Fordham, E. J.; Ladva, H. K. J.; Hall, C.; Baret, J-F.; Sherwood, J. D. Presented at the 63rd Annual Technical Conference of the Society of Petroleum Engineers, Houston, TX, October 1988; paper SPE 18038.
130. Sherwood, J. D.; Meeten, G. H.; Farrow, C. A.; Alderman, N. J. *J. Chem. Soc. Faraday Trans.* **1991**, *87*, 611–618.
131. Horsfield, M. A.; Fordham, E. J.; Hall, C.; Hall, L. D. *J. Magn. Reson.* **1989**, *81*, 593–596.
132. Chilingar, G. V.; Knight, L. *Bull. Am. Assoc. Pet. Geol.* **1960**, *44*, 101–106.
133. Nagaraj, T. S.; Srinivasa Murthy, B. R. *Géotechnique* **1986**, *36*, 27–32.
134. Denis, J. H.; Keall, M. J.; Hall, P. L.; Meeten, G. H. *Clay Miner.* **1991**, *26*, 255–268.
135. Fordham, E. J.; Ladva, H. K. J. *Physicochem. Hydrodyn.* **1989**, *11*, 411–439.
136. Jiao, D.; Sharma, M. M. *SPE Drill. Complet.* **1993**, *3*, 165–169.
137. Warren, B. K.; Smith, T. R.; Ravi, K. M. Presented at the Western Regional Meeting of the Society of Petroleum Engineers, Anchorage, AL, May 1993; paper SPE 26069.
138. Chenevert, M. E.; Al-Abri, S.; Jin, L. *Oil & Gas J.* **1994**, *92*, 62–66.
139. Bezemer, C.; Havenaar, I. *Soc. Pet. Eng. J.* **1966**, *6*, 292–298.
140. Fordham, E. J.; Ladva, H. J. K. *J. Colloid Interface Sci.* **1992**, *148*, 29–34.
141. Burchill, S.; Hall, P. L.; Harrison, R.; Hayes, M. H. B.; Langford, J. I.; Livingstone, W. R.; Smedley, R. J.; Ross, D. K.; Tuck, J. J. *Clay Miner.* **1983**, *18*, 373–397.
142. Ferguson, C. K.; Klotz, J. A. *Trans. AIME* **1954**, *201*, 29–42.
143. Kreuger, R. F. *J. Pet. Technol.* **1963**, *15*, 90–98.
144. Plank, J. P.; Gossen, F. A. *SPE Drill. Eng.* **1991**, *6*, 203–208.
145. Hartmann, A.; Özeler, M.; Marx, C.; Neumann, H.-J. *SPE Drill. Eng.* **1988**, *3*, 395–402.
146. Longeron, D.; Argillier, J-F.; Audibert, A. Presented at the European Formation Damage Conference, The Hague, Netherlands, May 1995; paper SPE 30089.
147. Heinle, S. A.; Shah, S.; Glass, J. E. In *Water-Soluble Polymers: Beauty with Performance*; Glass, J. E., Ed.; Advances in Chemistry 213; American Chemical Society: Washington, DC, 1986; pp 183–196.
148. Napper, D. H. *Polymeric Stabilisation of Colloidal Dispersions*; Academic: London, 1983; pp 379–413.
149. Darley, H. C. H.; Gray, G. R. *Composition and Properties of Drilling and Completion Fluids*, 5th ed.; Gulf Publishing Company: Houston, TX, 1988; pp 491–522.
150. Jiao, D.; Sharma, M. M. Presented at the SPE International Symposium on Formation Damage Control, Lafayette, LA, February 1992; paper SPE 23823.
151. Dake, L. P. *Fundamentals of Reservoir Engineering*; Elsevier: Amsterdam, Netherlands, 1978; pp 121–124.

152. Serra, O. *Fundamentals of Well-Log Interpretation. 1. The Acquisition of Logging Data;* Elsevier: Amsterdam, Netherlands, 1984; pp 54–56.

153. Allen, D.; Auzerais, F.; Dussan, E.; Goode, P.; Ramakrishnan, T. S.; Schwarz, L.; Wilkinson, D.; Fordham, E. J.; Hammond, P. S.; Williams, R. *Oilfield Rev.* **1991**, *3(3)*, 10–23.

154. Lane, H. S. Presented at the SPWLA 34th Annual Logging Symposium, Calgary, Canada, June 1993; paper D.

155. Serra, O. *Fundamentals of Well-Log Interpretation. 1. The Acquisition of Logging Data;* Elsevier: Amsterdam, Netherlands, 1984; pp 77–88.

156. Wyllie, M. R. J. *Trans. AIME* **1951**, *192*, 1–16.

157. O'Brien, D. E.; Chenevert, M. E. *J. Pet. Technol.* **1973**, *25*, 1089–1100.

158. Sherwood, J. D. *Proc. Roy. Soc. London Ser. A.* **1993**, *440*, 365–377.

159. Calvet, R. *An. Agron.* **1973**, *24*, 77–94.

160. Keren, R.; Shainberg, I. *Clays Clay Miner.* **1979**, *27*, 145–150.

161. Bloys, J. B.; Davis, N.; Smolen, B.; Bailey, L.; Houwen, O. H.; Reid, P. I.; Sherwood, J. D.; Fraser, L.; Hodder, M. *Oilfield Rev.* **1994**, *6(2)*, 33–43.

162. Bourgoyne, A. T.; Millheim, K. K.; Chenevert, M. E.; Young, F. S. *Applied Drilling Engineering;* Society of Petroleum Engineers: Richardson, TX, 1991; pp 57–60, 67–72.

163. Froment, T. D.; Rodt, G. M.; Houwen, O. H.; Titreville, B. Presented at the IADC/SPE Drilling Conference, Dallas, TX, February 1986; paper SPE 14753.

164. Engelhardt, F. R.; Ray, J. P.; Gillam, A. H. *Drilling Wastes;* Elsevier: London, 1989.

165. Derkies, D. L.; Souders, S. H. Presented at the SPE/EPA Exploration Production Environmental Conference, San Antonio, TX, March 1993; paper SPE 25934.

166. Kabrick, R. M.; Rogers, L. A. Presented at the 67th Annual Technical Conference of the Society of Petroleum Engineers, Washington, DC, October 1992; paper SPE 24564.

167. Hudgins, C. M. *J. Pet. Technol.* **1994**, *46*, 67–74.

168. Crawford, H. R.; Lescarboura, J. A. Presented at the 68th Annual Technical Conference of the Society of Petroleum Engineers, Houston, TX, October 1993; paper SPE 26382.

169. Reed, T. D.; Pilehvari, A. A. Presented at the Production Operations Symposium, Oklahoma City, OK, March 1993; paper SPE 25456.

170. Ben-Naceur, K. In *Reservoir Stimulation*, 2nd ed.; Economides, M. J.; Nolte, K. G., Eds.; Prentice Hall: Englewood Cliffs, NJ, 1989; pp 3.16–3.22.

171. Sirevåg, G.; Bale, A. Presented at the SPE/IADC Drilling Conference, Amsterdam, Netherlands, February 1993; paper SPE 25758.

172. Wilton, B. S.; Bloys, J. B.; Watts, R. D.; Hipp, B. W. Presented at the 67th Annual Technical Conference of the Society of Petroleum Engineers, Washington, DC, October 1992; paper SPE 24566.

173. Nahm, J. J.; Javanmardi, K.; Cowan, K. M.; Hale, A. H. Presented at the SPE/EPA Exploration Production Conference, San Antonio, TX, March 1993; paper SPE 25988.

174. Schlemmer, R. P.; Branam, N. E.; Edwards, T. M.; Valenziano, R. C. Presented at the 68th Annual Technical Conference of the Society of Petroleum Engineers, Houston, TX, October 1993; paper SPE 26324.

175. *Recommended Practice Standard Procedure for Field Testing Water-Based Drilling Fluids,* 1st ed.; API Recommended Practice 13B–1 (RP 13B–1); American Petroleum Institute: Washington, DC, 1990.

176. *Recommended Practice Standard Procedure for Field Testing Oil-Based Drilling Fluids,* 2nd ed.; API Recommended Practice 13B-2 (RP 13B-2); American Petroleum Institute: Washington, DC, 1991.
177. Hughes, T. L.; Jones, T. G. J.; Geehan, T. *SPE Drill. Complet.* **1995,** *10,* 255-264.
178. Hall, C.; Fletcher, P.; Hughes, T. L.; Jones, T. G. J.; Maitland, G. C.; Geehan, T. *Oil and Gas Technology in a Wider Europe: Proceedings of the 4th EC Symposium;* Petroleum Science and Technology Institute: Aberdeen, 1992; pp 511-525.
179. Hughes, T. L.; Jones, T. G. J.; Geehan, T.; Tomkins, P. G. Presented at the SPE/IADC Drilling Conference, Amsterdam, Netherlands, February 1993; paper SPE 25703.
180. Houwen, O. H.; Gilmour, A.; Sanders, M. W.; Anderson, D. R.; Prouvost, L. P.; White, D. B. Presented at the SPE/IADC Drilling Conference, Amsterdam, Netherlands, February 1993; paper SPE 25704.
181. Hughes, T. L.; Jones, T. G. J.; Tomkins, P. G.; Gilmour, A.; Houwen, O. H.; Sanders, M. W. Presented at the 1st International Conference on Health Safety Environment, The Hague, Netherlands, November 1991; paper SPE 23361.
182. Jones, T. G. J.; Hughes, T. L.; Tomkins, P. G. U.S. Patent 5 306 909, 1994.
183. Murch, D. K.; White, D. B.; Prouvost, L. P.; Michel, G. L.; Ford, D. H. Presented at the IADC/SPE Drilling Conference, Dallas, TX, February 1994; paper SPE 27447.

NOTE: All papers with SPE numbers are available through the Society of Petroleum Engineers, P.O. Box 833836, Richardson, TX 75083-3836.

RECEIVED for review July 28, 1994. ACCEPTED revised manuscript October 12, 1995.

Suspensions in Hydraulic Fracturing

Subhash N. Shah

School of Petroleum and Geological Engineering, The University
of Oklahoma, Norman, OK 73019-0628

*Suspensions or slurries are widely used in well stimulation and
hydraulic fracturing processes to enhance the production of oil and
gas from the underground hydrocarbon-bearing formation. The
success of these processes depends significantly upon having a
thorough understanding of the behavior of suspensions used.
Therefore, the characterization of suspensions under realistic conditions, for their rheological and hydraulic properties, is very important. This chapter deals with the state-of-the-art hydraulic fracturing suspension technology. Specifically it deals with various types
of suspensions used in well stimulation and fracturing processes,
their rheological characterization and hydraulic properties, behavior of suspensions in horizontal wells, review of proppant settling velocity and proppant transport in the fracture, and presently
available measurement techniques for suspensions and their merits.
Future industry needs for better understanding of the complex behavior of suspensions are also addressed.*

THE HYDRAULIC FRACTURING TECHNIQUE for well stimulation has been
used successfully since 1946 as a means of increasing oil and gas production. Since its inception, the science of hydraulic fracturing technology has advanced considerably. In the early 1950s, the pioneer
operating and service companies contributed in the cooperative development of well stimulation technique. In the 1970s, the U.S. government
provided funds to develop tight gas sands and unconventional energy
resources such as coal bed methane. The most significant advance of
the technology was made in the 1980s when emphasis was placed on
stimulating medium to high permeability reservoirs. The hydraulic fracturing technology has been successfully applied in sand control when
producing soft and unconsolidated formations. Treatment optimization
and improving economics must become the focus of the industry as it

0065–2393/96/0251–0565$15.75/0
© 1996 American Chemical Society

moves into the future. In the past decade, exciting progress has been made in the development of new fluids and proppant, accurate downhole tools, and highly sophisticated interpretation techniques for monitoring hydraulic fracturing treatments (1–4).

In the application of hydraulic stimulation process, a viscous fluid (generally a non-Newtonian fluid) is pumped down into the well under high pressure to initiate and extend induced fractures. After this stage, another viscous-fluid stage containing proppant (solid particles; we use "proppant" and "particle" interchangeably) is injected down into the wellbore at high rate and pressure to maintain the fracture geometry created by the previous clean viscous-fluid stage. The other function of the proppant-laden fluid (suspension or slurry; we use "proppant-laden fluid," "suspension," and "slurry" interchangeably) stage is to keep the created fracture open after pumping has ceased. Fluids containing proppant account for 20–80% of the total volume of a fracturing treatment (5). Therefore, it is imperative to understand the behavior of these suspensions.

Fracture geometry and extension during treatment depend largely on the rheological characteristics of the clean viscous fluid and suspensions or slurries prepared with viscous carrier fluids. Particle settling and distribution in the fracture also are affected significantly by suspension properties.

We discuss the suspensions used in well stimulation and hydraulic fracturing processes. The following sections pertain to various types of suspensions used in well stimulation and fracturing processes, their rheological characterization and hydraulic properties, behavior of suspensions in horizontal wells, a state-of-the-art review of proppant settling velocity and proppant transport in the fracture, presently available measurement techniques for suspensions and their merits, and, finally, a summary and conclusions on the use of suspensions in well stimulation. Future industry needs for better understanding of the complex behavior of suspensions are also mentioned in this section.

Types of Suspensions

A fracturing fluid is used primarily to wedge open and extend a fracture hydraulically and to transport and distribute the proppant along the fracture. Success of a fracturing treatment, among other things, highly depends on selection of the proper fracturing fluid. Selection must be made based on known fluid and reservoir properties. Fluid properties strongly govern fracture propagation and the proppant distribution and placement. Today, many different types of fracturing fluids are available for reservoirs, ranging from shallow low temperature formations to those that are deep and hot. Comprehensive details and their design information are outside the scope of this chapter. For the treatment of a specific formation, one must consult published engineering and laboratory test data available from various sources, such as service companies.

Fracturing fluids are classified as Newtonian, polymer solutions, cross-linked polymer solutions, emulsions, micellar solutions, foams, and gelled-organic liquids in solution with a liquefied gas (*4, 6–8*). As mentioned previously, proppant or solids are often used to hold the fracture open after the fracturing treatment has been completed. The solid particles become lodged in the fracture, keeping it open and providing a channel for the oil and gas to flow more easily from the formation to the wellbore. Because of its low cost, silica sand is by far the most commonly used proppant today. Many manufactured materials, such as sintered bauxite (aluminum oxide), alumina, cordierite, mullite, silicon carbide, and some ceramic oxides, are also used (*9–12*). Coated proppants, such as resin-coated sands, are also being used (*13*). The manufactured materials and coated proppants are significantly stronger than sand and are used in deep formations where high in situ stresses crush sand or where large proppant permeability is needed.

Suspensions or slurries used for well stimulation are prepared by adding a known quantity of proppant to the carrier fluid of choice to obtain a desired proppant concentration. The proppant concentration is usually referred to as the amount of solids added in a gallon of carrier fluid. A 20–40 U.S. mesh size is the most commonly used proppant for hydraulic fracturing. However, other particle sizes such as 8–12, 8–16, 10–20, 12–20, and 40–60 mesh are also used.

The carrier fluids are non-Newtonian in nature and provide high viscosity and excellent drag-reducing characteristics. These fluids include polymer solutions, cross-linked polymer solutions, emulsions, and foam fluids.

Many types of water-soluble polymers such as guar gum, hydroxypropyl guar (HPG), carboxymethyl hydroxypropyl guar, and hydroxyethyl cellulose can be added to water to make it more viscous to improve proppant transport properties. For stability at higher temperatures, metal ion cross-linkers such as borate, Ti(IV), Zr(IV), and Al(III) are also used. For water-sensitive formations, oil-based fluids such as gelled kerosene, diesel, distillates, and many crude oils are used. For better fluid-loss behavior, polymer emulsions are also sometimes used. Polymer oil-in-water emulsions normally consist of 60–70% liquid hydrocarbon as an internal phase and 30–40% gelled water as the external phase. Because of their high solubility, alcohol-based fracturing fluids are used for treating gas-producing formations. Foams are primarily used in low pressure reservoirs to improve cleanup and minimize formation damage. Suspensions or slurries are prepared with all of these carrier fluids. Thus, it is very important to understand the behavior of these suspensions.

Rheological Characterization

In the early days of fracturing treatments, proppant concentrations were 1–6 lb/gal, but now concentrations of 14 and 16 lb/gal are not uncommon. The effect of proppant on fluid viscosity (under laminar flow in the fracture) at low solid concentrations is not significant but is at higher solid concentrations. With recent advances in placement of high proppant concentrations and the desire to predict fracture geometry and extension accurately, it has become important to include the effect of proppant concentration on the viscosity of fracturing fluids in the currently available fracture design simulators.

Rheological characterization of neat fluid (i.e., fluid without proppant) is relatively well understood. However, the rheological characterization of fracturing suspensions or slurries is not well known. The primary reason for not determining the rheological properties of slurries is the difficulty of particle settling at low shear rates while making measurements with a rotational-type viscometer, a pipe viscometer, or a slot-flow device. The proppant must be kept in uniform suspension to obtain meaningful data.

Particle shape, size, and density; polymer gelling-agent concentration; solids concentration; test temperature; and fracture shear rate affect viscosity increase that results from the addition of a solid in the fracturing fluid.

Gardner and Eikerts (14) used a large closed-loop pipe viscometer to evaluate the effect of 20–40 mesh sand on the viscosity of aluminate-cross-linked 60-lb carboxymethyl hydroxyethylcellulose/1000 gal gel. They concluded that the effect of sand on gel viscosity was a larger increase in slurry viscosity in the fracture than predicted from Newtonian slurry models.

Ely et al. (15) studied the effect of sand concentration on 40 lb HPG/1000 gal fluid using a Brookfield viscometer and reported an increase in viscosity caused by the addition of sand. For 50 lb/1000 gal cross-linked HPG fluid containing 2, 4, and 6 lb/gal sand at 190°F, the power law flow behavior index, n, decreased whereas the consistency index, K, increased with increasing sand concentration.

For fracturing suspensions prepared with only linear HPG, Nolte (16) recommended the following modified Landel et al.'s (17) equation for Newtonian suspensions.

$$\mu_r = [1 - (\phi/\phi_{max})]^{-2.5n} \qquad (1)$$

where μ_r is the relative viscosity and is the ratio of the suspension viscosity to the carrier fluid viscosity at the same shear rate, ϕ is the volume fraction solids and ϕ_{max} is the maximum packing concentration. For uniform spherical particles, ϕ_{max}, the maximum packing concentration varies between 0.74 (body-centered cubic) and 0.52 (simple cubic lattice). Nolte simply modified the power of -2.5 in Landel et al.'s equation to $-2.5n$. The modified Landel et al.'s equation 1 is independent of shear rate.

Results (*16*) of tests conducted in a specially designed cup and bob with 20, 30, and 40 lb HPG/1000 gal fluids containing 18–25 mesh neutrally buoyant beads at four volume fractions of particles were reported. It was claimed that for all polymer and particle loadings, rheograms from 0 to 150 s^{-1} essentially produced the same n value as the corresponding fluid without particles. The K values, however, were higher because of the particles. The results are compared with Landel et al.'s equation 1 and with the following equation of Frankel and Acrivos (*18*).

$$\mu_r = 9/8 \left[\frac{(\phi/\phi_{max})^{1/3}}{1 - (\phi/\phi_{max})^{1/3}} \right] \tag{2}$$

Landel et al.'s equation appears to be reasonably good with less viscous suspensions, whereas the Frankel and Acrivos equation seems to fit the data with more viscous slurries and at higher particle concentrations. It is assumed that all of Nolte's data are taken at room temperature.

Keck et al. (*19*) studied the effect of proppant on the effective viscosity of non-Newtonian fluids and presented the following modified Eiler's (*20*) expression that includes the effect of shear rate, temperature, gel concentration, and proppant concentration.

$$\mu_r = \left[1 + \{0.75(e^{1.5n} - 1)e^{-(1-n)\dot{\gamma}/1000}\} \frac{1.25\phi}{1 - 1.5\phi} \right]^2 \tag{3}$$

Tests were conducted with a rotational viscometer with 30, 40, and 60 lb HPG/1000 gal fluids containing neutrally buoyant 60–100 mesh styrene divinylbenzene beads at concentrations up to 12 lb/gal and temperatures up to 65.5 °C. Data were gathered only at three shear rates: 5, 170, and 1000 s^{-1}. Their modified Eiler's equation was based on correlating relative viscosity as a function of clean fluid n values, solids concentration, and fracture shear rate. The gel concentration and temperature effects were incorporated into n. Figure 1 depicts the effect of polymer concentration on the relative viscosity of suspension at a shear rate of 170 s^{-1} and 23.9 °C. It can be seen that the lower polymer concentration has the greater viscosity ratio than the higher concentrations and that the difference between these increases with volume fraction solids.

By using field and service company data, Jennings (*21*) developed the following two simple empirical equations, one for non-cross-linked

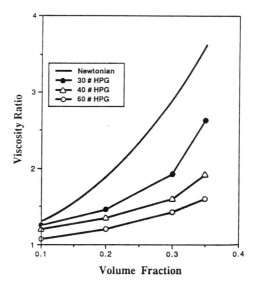

Figure 1. Experimental viscosity ratio vs. volume fraction: polymer concentration effects. (Reproduced with permission from reference 19. Copyright 1992 Society of Petroleum Engineers.)

fluids (base gels) and one for cross-linked gels, to estimate relative viscosity. These correlations are a function of sand concentration only and are independent of gelling agent concentration, temperature, and fracture shear rate. Non-cross-linked fluids are defined as follows:

$$\mu_{am} = \mu_a (0.21c + 1) \tag{4}$$

where c is the sand concentration in lb/gal. Cross-linked fluids are defined as follows:

$$\mu_{am} = \mu_a(0.075c + 1) \tag{5}$$

Shah (22, 23) investigated the rheological behavior of moderately concentrated (up to 35 vol%) suspensions prepared with non-Newtonian carrier fluids. Experimental data on the Poiseuille slit flow of suspensions of sand in HPG solutions were presented. Flow data were gathered by pumping various non-Newtonian slurries into an 8-ft-tall vertical apparatus, with care taken to avoid settling difficulties.

The polymer solutions or base gels and suspensions exhibited pseudoplastic non-Newtonian behavior, and they were characterized by the following Ostwald–de Waele or power law fluid model.

$$\tau = K(\dot{\gamma})^n \tag{6}$$

or

$$\mu_a = (\tau/\dot{\gamma}) = K(\dot{\gamma})^{n-1} \tag{7}$$

Figure 2 presents an example of a logarithmic plot of the apparent viscosity versus slot shear rate data of 60 lb HPG/1000 gal fluid with and without proppant at 60 °C. It can be seen that for the shear rate range studied, a linear relationship between apparent viscosity and shear rate on a logarithmic scale for all fluids with and without sand is evident. This means that the rheological behavior of these fluids, HPG solutions as well as slurries, can be adequately described by the power law expression given by equation 6. Further, a significant increase in apparent viscosity of HPG solution is seen because of the addition of sand. The viscosity increase is more pronounced at lower shear rates than at higher shear rates.

The n_s and K_s values are plotted as a function of volume fraction solids in Figure 3 for fluids tested at 60 °C. It can be seen in this figure for these fluids, as the volume fraction solids increases, the n_s values decrease whereas the K_s values increase. This means the degree of the non-Newtonian character of the base gels increases by adding sand. The assumption that the n_s values of sand-laden fluids remain the same as the carrier fluid n by previous investigators does not seem to be valid.

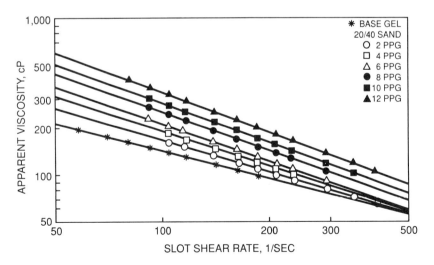

Figure 2.   Viscosity vs. slot shear rate, 60-lb HPG slurries at 60 °C. (Reproduced with permission from reference 22. Coypright 1993 Society of Petroleum Engineers.)

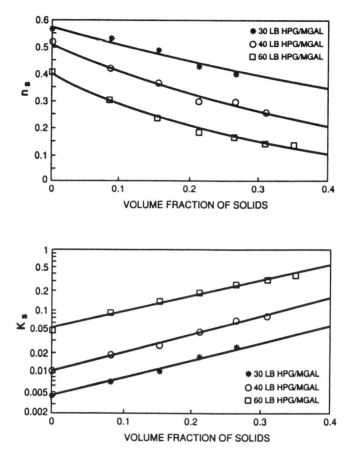

Figure 3.   K_s and n_s vs. volume fraction of solids at 60 °C. (Reproduced with permission from reference 22. Copyright 1993 Society of Petroleum Engineers.)

The dramatic effect on n_s and K_s is seen by increasing solids content. The n_s values, however, seem to level off or reach asymptotic values at higher solids concentration.

Regression analysis was performed on n_s and K_s data of Figure 3 by using the equations of the following form:

$$n_s = a_n \cdot (\exp) b_n \phi \qquad (8)$$

$$K_s = a_k \cdot (\exp) b_k \phi \qquad (9)$$

where a_n, b_n, a_k, and b_k are constants. The equations 8 and 9 reduce to the HPG solution for zero volume fraction solids. The variations of constants a_n and a_k and exponents b_n and b_k in equations 8 and 9 as a function

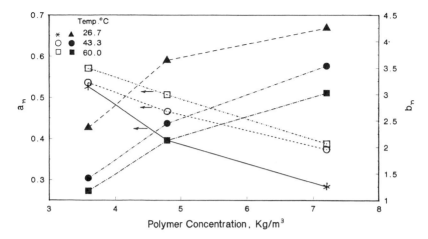

Figure 4.   Parameters a_n and b_n vs. polymer concentration. (Reproduced with permission from reference 23. Copyright 1993 Society of Petroleum Engineers.)

of polymer concentration and temperature are presented in Figures 4 and 5.

Figure 6 depicts the effect of volume fraction of solids on the relative viscosity for 40 lb HPG/1000 gal fluid at 100 s^{-1} and at test temperatures of 26.7, 43.3, and 60 °C (23). A Newtonian relative viscosity–concentration curve based on Thomas (24) and Maron and Pierce (25) correlations are shown for comparison.

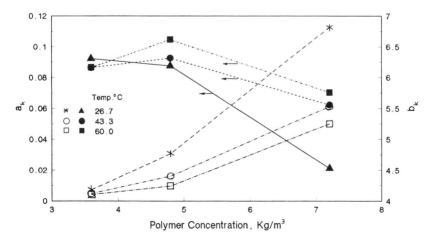

Figure 5.   Parameters a_k and b_k vs. polymer concentration. (Reproduced with permission from reference 23. Copyright 1993 Society of Petroleum Engineers.)

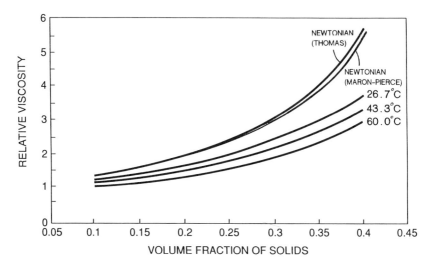

Figure 6.   Effect of sand concentration on relative viscosity. HPG 4.8 kg/ m³; shear rate = 100 s⁻¹. (Reproduced with permission from reference 23. Copyright 1993 Society of Petroleum Engineers.)

It can be seen from Figure 6 that the relative viscosity increases gradually at lower sand concentrations but increases very rapidly at higher sand concentrations. At low particle concentrations, the hydro-dynamic interaction is a dominating factor, whereas at higher particle concentrations, interparticle interaction becomes a dominating factor. Furthermore, in Figure 6, relative viscosities of all non-Newtonian slurries at all temperatures are substantially lower than those predicted by either Thomas equation or Maron–Pierce equation for Newtonian slurries. Large errors may be committed by not using proper equations for non-Newtonian slurries.

The effects of polymer concentration, shear rate, and temperature were also investigated. As the polymer concentration increases (power law exponent n decreases, that is, the viscosity of the carrier fluid or suspending liquid increases), the relative viscosity of the pseudoplastic slurry decreases. It is important to know that the more viscous the fluid (lower n value), its relative viscosity will deviate further from the Newtonian predictions. Relative viscosity decreases as the shear rate increases. Again, the effect is more pronounced at higher solid concentration than at low solid concentration. Temperature also has a dramatic effect on the relative viscosity of slurries. This effect is partly due to the reduction in carrier fluid viscosity because of thermal effects. The relative viscosity increases as the temperature increases. Overall, Shah's experimental results agree more closely with Keck's than with any other reported study.

It should be remembered that unlike Nolte's proposed equation, Shah's results show definite shear rate-dependent relative viscosities. Also, power law exponent, n_s, of slurry is not the same as the carrier fluid n as assumed by Nolte. Prud'homme (26) has also reported decreasing n values with increasing volume fraction solids.

Hydraulic Properties

Hydraulic properties, that is, friction-loss calculations of proppant-laden fluids or slurries, are very important not only in the design of any hydraulic fracturing treatment but also in real-time monitoring of fracturing treatments. Recent advances (27, 28) in real-time fracture analysis have necessitated an accurate knowledge of bottomhole treating pressure (BHTP). To estimate BHTP, an accurate prediction of friction pressures of fluids in the flow conduit is required. It is possible to obtain the BHTP from the surface pressure with the following equation:

$$p_{wt} = p_{wh} + p_h - p_f \qquad (10)$$

The wellhead pressure, p_{wh}, can be measured with good accuracy. The hydrostatic pressure, p_h, can be calculated accurately with recent advances in radioactive densimeters. Methods for predicting friction pressures of clean fracturing fluids have been well established. The friction pressures of gelled fluids through tubular goods, p_f, are more difficult to predict accurately when they are cross-linked or contain proppants.

Several studies have been reported to determine friction losses in turbulent flow of slurries. Hannah et al. (29) presented an approach in which they compared expressions for the friction pressure of the slurry and clean fluid. In their analysis, they assumed Blasius'(30) turbulent Fanning friction factor versus Reynolds number equation for Newtonian fluids. The following expression for estimating slurry friction pressure knowing the clean fluid friction pressure is proposed.

$$p_t = p_{fl}(\mu_r)^{0.2}(\rho_r)^{0.8} \qquad (11)$$

where μ_r is the relative viscosity (ratio of slurry apparent viscosity to fluid apparent viscosity) and ρ_r is the relative density (ratio of slurry density to fluid density). As is evident from the data shown in Figure 7, despite some data scatter, the correlation seems to work reasonably well for HPG fluids and for sand concentration up to 7.5 lb/gal.

Shah and Lee (31) presented an approach for predicting friction pressures of fracturing slurries that is based on an analytical method and uses nondimensional quantities. From flow data for various slurries in multiple pipes, generalized correlations were developed and presented that incorporated variables such as particle size, particle density, particle

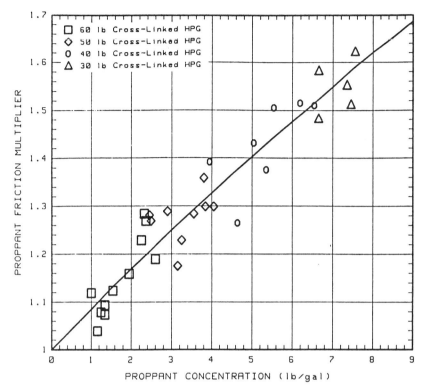

Figure 7.   Correlation for proppant friction correction factor—5.5 in. csg, 1.9-in. tubing annulus. (Reproduced with permission from reference 29. Copyright 1983 Society of Petroleum Engineers.)

concentration, fluid density, flow rate, polymer gel concentration, and pipe size.

Following Molerus and Wellmann's (32) analytical approach, the friction pressure of slurry, p_t, is expressed as the sum of friction pressure of clean fluid, p_{fl}, and an additional friction pressure, p_p, caused by particle present in the fluid. The experimental data are correlated with the expression of the form

$$(v_s/v) = v_{s\infty} + s_1(N_{Frp})^{-s_2} \qquad (12)$$

where v is the average fluid velocity, v_s is the mean slip velocity between particle and fluid, $v_{s\infty}$ is an infinite slip velocity parameter, N_{Frp} is particle Froude number, and s_1 and s_2 are empirical parameters determined from experimental data.

The clean fluid friction pressure can be estimated with either the recently published correlations and method described (33) or the aid of

other similar methods. The additional friction pressure resulting from
the presence of proppant can be estimated with the aid of the diagram
presented in Figure 8. From this figure, knowing N_{Frp} and $N_{Fr}*$ supplies
the value of (v_s/v). With this value, the dimensionless pressure drop, p_d,
is calculated as follows:

$$p_d = \frac{(v_s/v)^2}{1 - (v_s/v)} \tag{13}$$

The additional pressure drop, p_p, is then calculated with the following
equation:

$$p_p = p_d \cdot \frac{c_v(\rho_p - \rho_f)Lg}{(v_t/v)^2} \tag{14}$$

where c_v is the fractional volumetric concentration of proppants and is
defined as

$$c_v = \frac{(\rho_s - \rho_f)}{(\rho_p - \rho_f)} \tag{15}$$

The friction pressure of slurry, p_t, is then simply a summation of p_{fl}
and p_p.

   The percent friction pressure increase over base gel was found to
be a strong function of sand concentration and flow rate. At a constant

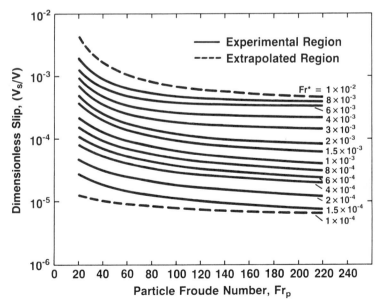

*Figure 8.   Master curve. (Reproduced with permission from reference 31.
Copyright 1986 Society of Petroleum Engineers.)*

flow rate, the percent friction pressure correction increases as sand concentration increases. At a given sand concentration, however, the correction decreases as rate increases. At the same flow rate, percent correction for more viscous slurry is lower than that for less viscous slurry. Furthermore, slurry shows higher percent correction when it is pumped down the annulus than tubing. Also, as particle size increases, the percent friction pressure correction increases significantly. At lower proppant concentrations, the effect of particle density is minimal, but at higher proppant concentrations, this effect may be important.

Figure 9 shows a plot of pressure drop versus flow rate for fluids containing 40 lb HPG/1000 gal with and without sand in a 2⅞-in field-sized tubing. An excellent agreement is seen between measured pressures and laboratory predicted pressures.

Figure 10 depicts a comparison between predicted percent friction pressure increase over base gel and actual field data acquired during a fracturing treatment. A reasonably good agreement between the two is seen.

Based on Bowen's (34) procedure, Lord and McGowen (35) developed a statistical based correlation of drag ratio versus fluid velocity.

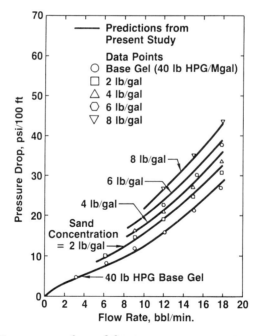

Figure 9.   Comparison of actual friction pressures at various flow rates for 40 lb HPG/1000 gal and sand-laden 40 lb HPG/1000 gal fluids in 2⅞-in. tubing. (Reproduced with permission from reference 31. Copyright 1986 Society of Petroleum Engineers.)

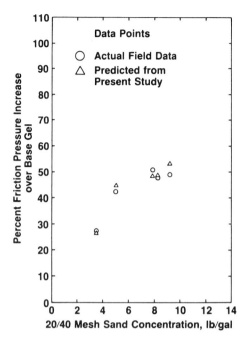

Figure 10.   Comparison of field data with predictions. (Reproduced with permission from reference 31. Copyright 1986 Society of Petroleum Engineers.)

The laboratory data of Shah and Lee (*31*) with various HPG solutions and slurries were used for this purpose. Their initial correlation given by equation 16 below was shown to considerably overpredict friction loss of dense slurries during injections down long vertical tubing strings.

$$\ln (1/\sigma) = 2.1505 - 8.024/v - 0.2365 \, G/V$$
$$- 0.1639 \ln G - 0.05266p \cdot e^{1/G} \qquad (16)$$

where σ is the drag ratio and is defined as the ratio of frictional pressure drop of gel or slurry and water, v is the average fluid velocity in ft/s, G is the HPG gel concentration in lb/1000 gal, and p is the proppant concentration in lb/gal.

The authors speculated that some heterogeneous flow phenomena, such as particle migration to the centerline, could possibly be responsible for the observation of lower than expected friction loss values during the treatment.

Figure 11 depicts a comparison of laboratory data to correlation predictions of drag ratio versus velocity. It can be seen that the correlation developed is diameter invariant. The laboratory correlation was

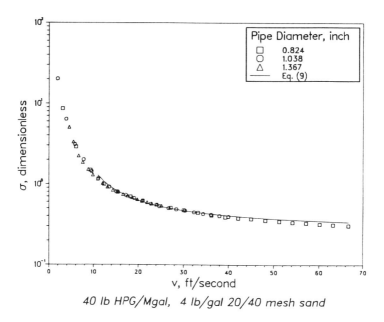

40 lb HPG/Mgal, 4 lb/gal 20/40 mesh sand

Figure 11. Comparison of laboratory data to correlation predictions of drag ratio vs. velocity. (Reproduced with permission from reference 35. Copyright 1986 Society of Petroleum Engineers.)

later "calibrated" with actual fracturing slurry friction loss values obtained from several tubing treatments. This correlation is given by

$$\ln(1/\sigma) = 2.38 - 8.024\, v - 0.2365\, G/V \qquad (17)$$
$$- 0.1639 \ln G - 0.028\, p \cdot e^{1/G}$$

The field calibrated correlation (equation 17) provided improved plots of net pressure versus time for analysis (Figure 12).

Keck et al. (*19*) proposed the following Dodge and Metzner (*36*) type equation for the clean fluids:

$$(1/\sqrt{f}) = 14.9\, n^{-1.6} d^{0.13} \log(N_{\text{Res}}\sqrt{f}) - 53.6 n^{-1.9} d^{0.27} \qquad (18)$$

where d is the diameter in feet and n is the power law flow behavior index. N_{Res} is the solvent Reynolds number. It was found that the results were insensitive to the power law consistency index, K. On a logarithmic scale, for a given pipe size the plot of $1/\sqrt{f}$ versus $(N_{\text{Res}}\sqrt{f})$ is a straight line (Figure 13). This equation was developed from the yard tests conducted with linear HPG solutions in 2- to 4-inch diameter pipes. It should be noted that for $n = 1$, equation 18 above does not reduce to a Newtonian equation of Dodge and Metzner. Furthermore, like Dodge and Metzner's Newtonian equation, equation 18 is not dimensionless.

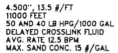

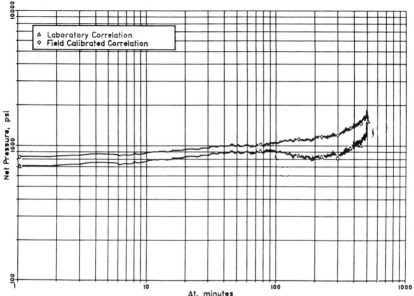

Figure 12. Field case comparison of laboratory and field calibrated correlation predictions of net pressure vs. time. (Reproduced with permission from reference 35. Copyright 1986 Society of Petroleum Engineers.)

For estimating friction pressure of slurries, Keck et al. (*19*) recommend to multiply the clean fluid friction pressure estimated by equation 18 above by the following correction factor:

$$M = \mu_r^{0.55} \cdot \rho_r^{0.45} \tag{19}$$

where μ_r and ρ_r are fluid relative viscosity and relative density.

This equation was derived by following procedure similar to that given by Hannah et al. (*29*) but using the Virk (*37*) equation. It is claimed that the equation is valid for turbulent flows of viscoelastic fluids and is applicable to most of the available field data.

The relative viscosity in equation 19 is obtained from the following modified Eiler equation:

$$\mu_r = \left[1 + \{0.75(e^{1.5n} - 1)e^{-(1-n)\dot{\gamma}/1000}\} \frac{1.25\phi}{1 - 1.5\phi} \right]^2 \tag{20}$$

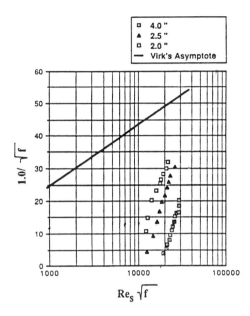

Figure 13.   Prandtl–Von Karman Plot for 40 lb yard test results in turbulent flow. (Reproduced with permission from reference 19. Copyright 1992 Society of Petroleum Engineers.)

For friction loss of slurries Cooper et al. (38) presented the friction correction factor shown in Figure 14. The authors claim that it is based on a theoretical method of predicting friction pressures of non-Newtonian fluids. The method assumes vertical flow, no hold-up effect, and pseudohomogeneous characteristics. However, the details of the method are not presented in open literature. It is shown in their article that the modeled effect on friction pressures is consistent with a field study conducted by Swanson and Meeken (39).

Tan and McGowen (40) derived an empirical correlation for predicting tubular friction pressure for CO_2-energized fluids from both field and laboratory data. Results of several case histories showed significant improvement in predicting friction pressure for CO_2-energized fluids as compared with existing correlations being used in the oil industry. They expanded Lord and McGowen's (35) correlation to account for the effect of CO_2 quality (10–70%) on the friction pressure and used the experimental procedure developed by Tan and Shah (41) to obtain the friction pressure correction factor for time and shear-dependent fracturing fluids normally used in conjunction with CO_2. Lord and McGowen's correlation was modified to include a friction pressure correction factor term for the liquid phase and quality term for CO_2. Thus, following empirical equation for CO_2 foam slurries is suggested:

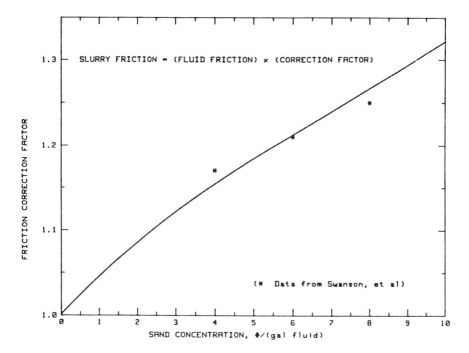

Figure 14.   Correction factors for effect of sand on slurry friction pressure as used by the computer's model. (Reproduced with permission from reference 38. Copyright 1983 Society of Petroleum Engineers.)

$$\ln (1/\sigma) = 1/\{2.38 - 8.024\, v - 0.2365 \cdot G/V - 0.1639 \cdot \ln (G)\}^{**\,(-1)}$$
$$- p_1 \cdot p \cdot e^{(1/G)} + \ln (M) - C_1 Q^{c_2} \tag{21}$$

For treatment with linear gel, equation 21 reduces to the original Lord and McGowen's correlation, that is, equation 17. The authors recommended the constant internal phase (CIP) approach originally proposed by Harris et al. (42) to minimize friction pressure increase during proppant stages. The CIP involves maintaining a constant liquid rate and decreasing gas flow rate when proppant is added. The authors claim that the application of CIP has greatly improved the control of foam fracturing treatments, especially with CO_2 treatments down small tubular string such as $2\frac{3}{8}$ or $2\frac{7}{8}$ in and has allowed the placement of much higher proppant concentrations into the formation with less friction loss in tubulars.

By using Fanning friction factor, generalized Reynolds number, and CIP approach, Kenyon (43) analyzed Tan and McGowen (40) field data for CO_2 foam fluids and developed the following correlation:

$$p_f/L = f\,(\mathrm{Re}'_\phi)(2\rho_{\mathrm{mix}}v^2/d) \tag{22}$$

where f and Re'_ϕ are the Fanning friction factor and Reynolds number. The μ_r is relative viscosity and is evaluated from

$$\mu_r = (1 - \phi)^{-1.7} \tag{23}$$

with

$$\phi = \phi_{CO_2} + 0.5\phi_{\mathrm{sand}} \tag{24}$$

$$\left(\mathrm{Re}'_\phi = \left(\rho_{\mathrm{mix}}d^n v^{2-n}\right)/\left(K8^n\left(\frac{3n + 1}{4n}\right)^n \mu_r\right)\right) \tag{25}$$

Figure 15 shows the friction factor correlation using the modified internal phase volume concept. According to the author, the proposed correlation appears promising, but considerable scatter is observed if one attempts comparison with time-dependent field data or CO_2 quality greater than 75%. Somewhat unstable friction pressure is noted at high CO_2 qualities. The author speculates this is due to some loss of internal phase or loss of external phase drag reduction at high CO_2 qualities.

Behavior of Suspensions in Horizontal Wells

Horizontal well drilling has increased throughout the industry in recent years and is being evaluated as an economical means to improve hydrocarbon production from some reservoirs (44–47). Perforation and fracture are the two major concerns for proppant transport when a conven-

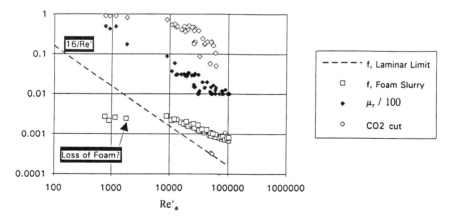

Figure 15.   Fanning friction factor with internal phase method. (Reproduced with permission from reference 43. Copyright 1993 Society of Petroleum Engineers.)

tional vertical wellbore is fractured. Proppant transport through the wellbore to the fracture initiation points is a major consideration when a horizontal well is fractured, regardless of whether the entry points are treated individually or simultaneously. When a single entry point is treated, a low fluid velocity associated with treatment through the casing may contribute to proppant deposition along the wellbore. When multiple entry points are simultaneously fractured, fluid and proppant must be distributed at fracturing rates to several points along the wellbore; therefore, surface injection rates and their contribution to proppant transport diminish very rapidly as treating fluid is diverted to individual fractures down the wellbore. An inadvertent shutdown during pumping may also deposit proppant that must later be resuspended either to continue injection or to remove it from the wellbore.

To design a successful hydraulic fracturing treatment for horizontal wells, accurate information on the transport properties of slurry in horizontal pipe is required. One must know the critical deposition and resuspension velocities of various fluids in horizontal pipe flow.

Extensive literature available from the slurry pipeline industry can be used to help planners predict the critical deposition velocities of solids suspended in Newtonian fluids such as water while being transported through horizontal pipes. Figure 16 shows typical wall shear stress versus nominal shear rate responses for horizontal homogeneous and heterogeneous flow systems. At high nominal shear rate values, heterogeneous fluid response (curve A) tends to parallel homogeneous or carrier fluid response (curve B). As the heterogeneous slurry velocity decreases, the solids concentration gradient increases until either a stationary or slowly moving particle bed appears along the pipe bottom.

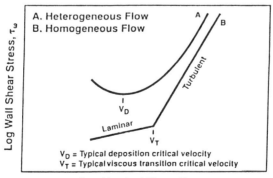

Figure 16.   Wall shear stress vs. nominal shear rate for homogeneous and heterogeneous slurries. (Reproduced with permission from reference 50. Copyright 1991 Society of Petroleum Engineers.)

The slurry velocity at which a particle bed forms is defined as critical deposition velocity, V_D, and represents the lower pump rate limit for minimum particle settling. A further decrease in slurry velocity leads to increased friction loss, as indicated by a characteristic hook upward of curve A, and may also lead to pipe plugging. After shutdown, if flow rate over the settled solids is gradually increased, a response similar to curve A of Figure 16 is once again obtained. With increasing nominal shear rate, wall shear stress decreases until a minimum is reached and then increases rapidly thereafter. The fluid velocity that corresponds to this minimum stress value is the critical resuspension velocity, V_s.

Oroskar and Turian (48) developed a critical deposition velocity correlation based on balancing energy required to suspend particles with energy dissipated by an appropriate fraction, F, of turbulent eddies present in the flow. They found F to be usually very close to unity (>0.95) and therefore its inclusion, especially when raised to a fractional power, has essentially no influence on correlation predictions. Their equation appears in the following form:

$$\frac{V_D}{\sqrt{gd_p(F_s - 1)}} = 1.85\, c^{0.1536}(1 - c)^{0.3564}$$
$$\times (dp/d)^{-0.378} \cdot N_{Rep}'^{0.09} \cdot F^{0.30} \tag{26}$$

The data were extracted from a number of experimental investigations reported in the literature for the development of the previous equation. The Oroskar–Turian correlation and others appearing in the literature were all developed to describe critical deposition velocity of Newtonian carrier fluids with various solid types, sizes, and concentrations.

Shah and Lord (49, 50) generalized the Oroskar–Turian correlation to increase its capability to correlate critical velocity measurements with non-Newtonian carrier fluids. The parameter F was eliminated from the generalization because of its insignificant contribution to the correlation results and because it would be undefined for the laminar flows associated with many of the non-Newtonian fluid measurements. The generalized form of equation 26, which can be applied to either critical deposition or resuspension velocity, is as follows:

$$\frac{V_D}{\sqrt{gd_p(F_s - 1)}} = Y c^{0.1536}(1 - c)^{0.3564} \times (d_p/d)^{-w} \cdot N_{Re}'^Z \cdot F^{0.30} \tag{27}$$

where coefficient Y and exponents w and z are adjustable constants that can be evaluated by regression analysis for particular critical velocity data sets. Apparent viscosity, μ_a, is substituted into the modified Reynolds number expression to N_{Re}' further generalize equation 26 to non-Newtonian fluids. The rheological properties and model parameters of fluids tested are listed in Table I.

Table I.   Rheological Properties and Model Parameters
of Fluids Tested at 26.7 °C

Fluid	n	K$(Pa \cdot s^n)$	y	w	z
1.2 kg/m³ HPG	0.941	0.00188	3.4260	0.6696	−0.1481
2.4 kg/m³ HPG	0.719	0.0546	2.2573	0.4691	0.0294
4.8 kg m³ HPG	0.486	0.6942	0.1743	0.4830	0.5276
7.2 kg/m³ HPG	0.377	2.7195	0.2360	0.4513	0.4274
4.8 kg/m³ cross-linked HPG	0.446	0.9719	0.0639	0.3786	0.7858
7.2 kg/m³ cross-linked HPG	0.304	4.3091	0.0613	0.4356	0.6476
Polymer emulsion	0.516	2.9733	0.2075	−0.067	1.0297

SOURCE: Reproduced with permission from reference 23. Copyright 1993 American Institute of Chemical Engineers.

Figure 17 shows critical deposition velocities for various fluids containing 6 lb/gal 20–40 mesh sand in various field-size tubular goods. Note that critical velocities are very dependent on pipe size. Much greater velocities are required for the larger pipes to minimize particle settling. Furthermore, the critical deposition velocities of all fracturing slurries are substantially lower than those of sand–water slurries, and the critical deposition velocities of cross-linked slurries are substantially lower than those of base fluid slurries. Slightly higher critical resuspension velocities than critical depositional velocities for all fluids were reported.

Higher critical deposition and resuspension velocities are also required for proppant that is denser than sand. For less viscous fluids,

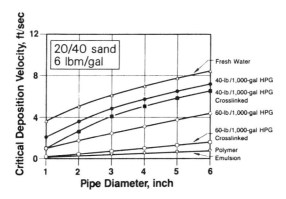

Figure 17.   Effect of pipe diameter on critical deposition velocity (based on model predictions). (Reproduced with permission from reference 49. Copyright 1990 Society of Petroleum Engineers.)

both velocities increase slightly with increasing proppant concentration; however, they are independent of proppant concentration for more viscous fluids.

Proppant Transport

One of the most important factors in the effectiveness of the hydraulic fracturing treatment is the ability to predict the settling velocity of proppant under fracture conditions. The transport of proppant and the final distribution of proppant in the fracture highly depends on the accurate estimation of settling velocity of proppant. The length of the propped fracture, the conductivity of the propped fracture, and height of the propped fracture are consequently affected by the settling velocity.

In the last decade, several investigators have addressed the subject of particle motion in fracturing fluids. Experiments to determine particle settling velocities under dynamic conditions involved large vertical fracture flow models (51–54), flow loops with single particles suspended in a vertical column (55), and rotating concentric cylinder devices in which particle fall rates could be observed (56–60). A detailed discussion regarding correlations developed from the data obtained from these apparatuses is beyond the scope of this chapter. A comprehensive list of available correlations has been presented by Daneshy (61). The following are, however, some of the correlations presently available that one can use to estimate single particle settling velocity in fracturing fluids: Novotny (60)

$$C_D = \frac{24}{N_{Rep}} \qquad N_{Rep} < 0.1 \qquad (28)$$

Shah (55)

$$\sqrt{C_D^{2-n} \cdot N'_{Rep}{}^2} = A N'_{Rep}{}^B + c \qquad 0.01 < N'_{Rep} < 100 \qquad (29)$$

where N'_{Rep} is the particle Reynolds number based on the apparent viscosity; Meyers (62)

$$C_D = \left[\frac{24}{N'_{Rep}{}^{1.1}} \left(\frac{24}{N'_{Rep}{}^{0.9}} x^2 + 7.5 \right) \right]^{1/2} \qquad N'_{Rep} < 500 \qquad (30)$$

Clark and Guler (58) and Clark (59)

$$V_e = a \cdot v_t \qquad (31)$$

where a is the correlation coefficient that depends on the type of gel and cross-linker; Acharya (63)

$$V_t = \left[\frac{(\rho_p - \rho_f)gd_p{}^{n+1}}{18 \, Kf_1(n)}\right]^{1/n} \qquad N'_{Rep} < 2 \qquad (32)$$

where

$$f_1 = 3^{3(n-1)/2}\left[\frac{33n^5 - 63n^4 - 11n^3 + 97n^2 + 16n}{4n^2(n + 1)(n + 2)(2n + 1)}\right] \qquad (33)$$

and

$$V_t = \frac{3\rho_f}{4(\rho_p - \rho_f)gd_p}\left[\frac{24f_1(n)}{N'_{Rep}} + \frac{f_2(n)}{N'_{Rep}f_3(n)}\right]^{-1/2} \qquad 2 < N'_{Rep} < 500 \quad (34)$$

$$f_2(n) = 10.5n + 3.5 \qquad (35)$$

$$f_3(n) = 0.32n + 0.13 \qquad (36)$$

and Roodhart (54)

$$V_t = \frac{gd_p{}^2(\rho_p - \rho_f)}{18}\left[\frac{1}{\mu_0} + \frac{1}{K(\dot{\gamma}_z)^{n-1}}\right] \qquad (37)$$

As mentioned earlier, during the hydraulic fracturing process, proppant slurry is pumped. After pumping has ceased, the particles settle in fracturing gels in a multiparticle environment. Proppant settling in slurries is a very complex phenomenon compared with single-particle settling in a fluid. Very little literature exists that explains the complex hindered-settling effects of particles in non-Newtonian fracturing gels. Without adequate theory for description of this phenomenon, most fracturing design simulators today use correlations for single particle settling velocities in fracturing gels, and from the theory of Newtonian suspensions, correct these velocities to account for the hindered settling effects (e.g., *see* reference 60).

Recent studies tried to answer some of the questions related to proppant settling in fluids in a multiparticle environment.

Kirkby and Rockefeller (64) conducted static sedimentation experiments to evaluate the settling behavior of various nonflowing slurries with the differential pressure technique for measuring proppant concentration and settling velocity in concentrated slurries. They concluded from their study that a proppant clustering phenomenon occurs in fracturing fluids at typical proppant concentrations, resulting in average static slurry settling velocities that are considerably greater than those of single particles (Figure 18).

Dunand and Soucemarianadin (65) used viscometric measurements to study sedimentation of single particles and suspensions in static HPG fracturing fluids. They showed that the viscosity of these fluids can be

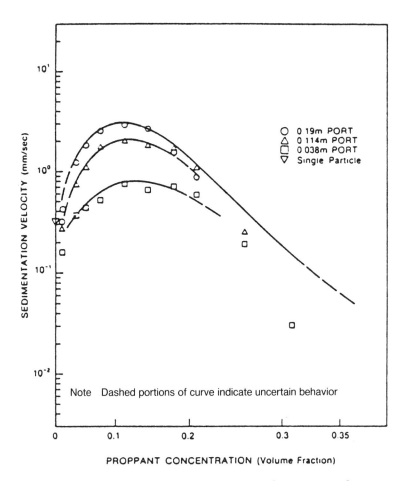

Figure 18.   Slurry settling velocity of 40–45-mesh Ottawa sand in 0.48% by weight hydroxypropylguar gum. (Reproduced with permission from reference 64. Copyright 1985 Society of Petroleum Engineers.)

calculated by the Ellis model and that theoretical predictions and experimentally measured velocities of single particles agree well. Using gamma ray attenuation measurements, Dunand and Soucemarianadin studied the settling behavior of concentrated proppant suspensions. They claimed that in fluids with identical single particle settling velocities, the average settling rate of a concentrated suspension in a static non-Newtonian fluid is two to three times higher than in a corresponding Newtonian fluid.

McMechan and Shah (66) conducted large-scale testing of the settling behavior of proppants in fracturing fluids with a slot configuration to model realistically the conditions observed in a hydraulic fracture. The test apparatus consisted of a 0.5 × 8 in rectangular slot 14.5 ft high, faced with Plexiglas and equipped with pressure taps at 1-ft intervals. This allowed both the qualitative visual observations and quantitative density measurements for calculation of proppant concentrations and settling velocities. Non-cross-linked as well as cross-linked slurries were evaluated. Clustering was found to occur in linear gels at sand concentrations below 10 lb/gal, which resulted in a much higher settling velocity than the single particle settling velocity. At sand concentrations above 10 lb/gal, hindered settling effects were found to be dominant. The hindered settling behavior agreed with a more general non-Newtonian power law form of the Barnea–Mizrahi (67) equation. Figures 19 and 20 depict sand concentration versus position in the test cell for various fluids tested after 10 and 30 min, respectively.

Clearly and Fonseca (68) reported that the mechanism of convective settling has profound implications for the design and execution of hydraulic fracturing treatments. They argued that current fracturing design simulators do not consider the two most important phenomena occurring during fracture flow: convective downward motion of heavier proppant-laden stages and encapsulation of viscous fluids by low viscosity fluids. The efforts reported by the authors are aimed at quantifying the phenomenology involved and incorporating the results into a fracture simulator, developed for the Gas Research Institute.

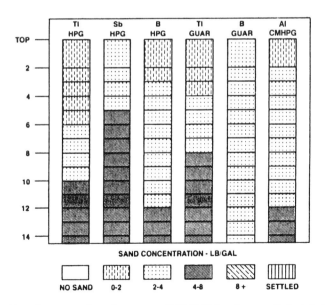

Figure 19.   Proppant distribution in 40 lb/1000-gal cross-linked fluid with 4 lb/gal 20–40-mesh sand (time = 10 min). (Reproduced with permission from reference 66. Copyright 1991 Society of Petroleum Engineers.)

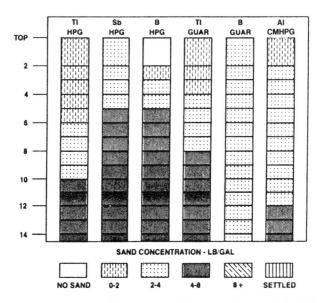

Figure 20.   Proppant distribution in 40 lb/1000-gal cross-linked fluid with 4 lb/gal 20–40-mesh sand (time = 30 min). (Reproduced with permission from reference 66. Copyright 1991 Society of Petroleum Engineers.)

To gain insight into the single particle settling rates and proppant transport of fracturing fluids, Conway et al. (69) constructed a slot model that contained two 4 ft high × 8 ft long × $^5/_{16}$ in. wide Plexiglas sections. In this study, in particular, the investigators concentrated on trying to understand qualitatively the effects of slot entrance on the observed particle transport characteristics. The quantification of the factors responsible for the particle segregation and static slurry settling rate in non-Newtonian fluids, however, was not possible. Figure 21 shows the effect of perforations on sand distribution in 1.5% polyacrylamide. With only the center perforation open, the largest variance in sand concentration is seen from top to bottom and the maximum sand concentration is about 80% higher than initially entered the slot. With two perforations open, the average sand concentration is not that much different, and with three perforations open the sand concentration across the upper two-thirds of the slot is about that which is entering the slot. This study emphasizes the importance of perforation placement in the flow of suspensions through slot.

Measurement Techniques

As previously described, the measurement of slurries, either for their rheological characterization or for their hydraulic properties, is not an

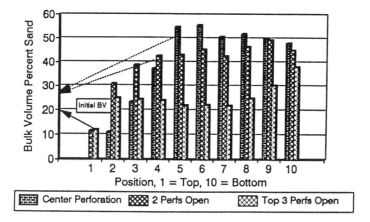

Figure 21. Effect of perforations on sand distribution in 1.5% polyacrylamide. (Reproduced with permission from reference 69. Copyright 1992 Society of Petroleum Engineers.)

easy task. Many problems are encountered during these measurements, and one must overcome these problems to obtain quality data. Numerous techniques have been attempted by several investigators. The available devices have their relative advantages and disadvantages.

Measurements with concentric cylinder viscometers are relatively easy and require small volumes of test fluid, but particle settling and fluid segregation are problems associated with them. Furthermore, most commercial viscometers have very small clearance between the two cylinders and thus are not designed to test fracturing slurries. Rotational viscometers with wider gap are recommended for slurry measurements. Measurements with some slurries at higher rotational speed may produce erroneous data because of centrifugal forces (segregation) in the test gap itself. Generally, attempts are made to provide some stirring of the fluid just before making the measurements, but this approach is not convenient or practical.

The wiped-disk viscometer was recently introduced as a means to provide slurry rheology data. In this device, because the fluid sample has to be captured and held for a period of time, some segregation of flow occurs that influences the rheology of fluid to be investigated. De Kruijf et al. (70) introduced a helical screw viscometer for determining the rheology of proppant-laden linear gels. The authors claim it is highly suitable for these types of fluids because it has a large measuring gap compared with the proppant particle diameter, and the fluid is continuously circulated so that the proppant particles cannot settle to the bottom of the rheometer as it normally happens with a traditional rotational viscometer.

Researchers at the University of Oklahoma (71) also investigated the rheology and proppant transport characteristics of fracturing fluids in high pressure (1200 psi), high temperature (107 °C), 10 ft long × 7 ft high fracture flow simulator. The slot width can be dynamically adjusted from 0 to 1.25 in., which provides data over a wide range of shear rate. One wall of the flow channel consists of 12 separate moveable steel platens, which are 28 in. square and are laid out in a 3 × 4 matrix. The opposite wall is a continuous 6-in.-thick steel slab. Each moveable platens is covered by a 1-in.-thick replaceable facing. The facing simulates rock surfaces and can be made with desired permeabilities. The simulator is very well instrumented. Laser Doppler velocimeters are used to measure point velocities as well as velocity profiles across the gap (72). A system of transmitting fiber-optic bundles mounted in the facings on one side of the apparatus, combined with receiving fibers on the other side of the apparatus, provides a "vision system" for monitoring proppant concentration in the apparatus and for monitoring settled bed formation (73, 74).

While testing slurries, either in a tube or slot, the flow rate has to be sufficiently high to keep the particles suspended and thus to produce a homogeneous slurry. This is not always possible if one is interested in gathering rheological data at low shear rates. These devices require larger fluid volume and are relatively more time consuming. Tube or pipe viscometers are ideal for determining hydraulic properties of slurries because the data are gathered at higher rates in turbulent flow. Because of high rates, there is a very little risk of particle settling.

Conclusions

The state-of-the-art hydraulic fracturing slurry technology is not as far advanced as it should be. Significant progress has been made in the last decade toward measuring rheological and hydraulic properties of fracturing slurries. Advances have also been made in characterization methods and testing procedures. However, increased efforts are needed to achieve thorough understanding of true behavior of slurries under field conditions. Better instruments need to be designed to provide more accurate slurry property data. Influence of multiparticle environment on settling is not well understood. Effects of fluid elasticity and viscosity on particle settling are also not known thoroughly. The effects of fluid elasticity and particle migration on the hydraulic properties of slurry in highly turbulent flow conditions in pipe flow should also be investigated further.

List of Symbols

a	correlation coefficient
c	sand concentration
c_1, c_2	coefficients
c_v	fractional volumetric concentration of proppants
C_D	drag coefficient
d	pipe diameter
d_p	particle diameter
f	Fanning friction factor
$f_1, f_2, f_3,$	functions
F	fraction of eddies with velocities exceeding hindered settling velocity
F_s	ratio of particle to fluid densities
g	acceleration of gravity
G	gelling agent concentration (lb/1000 gal)
K	consistency index, power law model constant
K_s	consistency index for slurry
L	pipe length
M	correction factor
n	flow behavior index, power law model constant
n_s	flow behavior index for slurry
N_{Frp}	particle Froude number
$N_{Fr}*$	terminal Froude number
N_{Re}	Reynolds number
N'_{Re}	modified Reynolds number
N_{Rep}	proppant Reynolds number based on effective viscosity
N'_{Rep}	particle Reynolds number based on apparent viscosity
N_{Res}	solvent Reynolds number
p	proppant concentration (lb/gal)
p_1	coefficient
p_d	dimensionless pressure drop
p_f	friction pressure
p_{fl}	friction pressure of clean or neat fluid
p_h	hydrostatic pressure
p_p	additional pressure drop caused by proppant
p_t	total pressure drop of proppant-laden fluid
p_{wh}	wellhead pressure
p_{wt}	BHTP
Q	flow rate
Re'_ϕ	Reynolds number

s_1, s_2	constants
v	average fluid velocity
V_D	critical depositional velocity
v_e	experimental settling velocity
v_s	mean slip velocity
V_s	critical resuspension velocity
$v_{s\infty}$	infinite slip velocity
v_t	particle terminal settling velocity
w	exponent
Y	coefficient
z	exponent
$\dot{\gamma}$	shear rate
$\dot{\gamma}_I$	shear rate imposed by the fluid motion
$\dot{\gamma}_z$	shear rate at the location with the lowest effective viscosity
μ	viscosity
μ_a	apparent viscosity of base fluid
μ_o	effective zero shear viscosity
μ_a^*	apparent viscosity based on effective shear rate
μ_{am}	apparent viscosity of proppant-laden fluid
μ_r	relative viscosity
ρ_{mix}	mixture density
ρ_f	fluid density
ρ_p	particle density
ρ_r	relative density
ρ_s	slurry density
σ	drag ratio
τ	shear stress
τ_w	wall shear stress
ϕ	volume fraction of solids
ϕ_{max}	maximum packing concentration

References

1. Hassebroek, W. E.; Waters, A. B. *J. Pet. Technol.* **1964,** *July,* 760.
2. Abou-Sayed, A. S. *Pet. Eng. Int. Suppl.* **1993,** *October.*
3. Von Flatern, R. *Pet. Eng. Int.* **1993,** *October,* 26.
4. Veatch, R. W. *J. Pet. Technol.* **1983,** *April,* 677; **1983,** *May,* 853.
5. Constien, V. G. In *Reservoir Stimulation,* 2nd ed.; Prentice Hall: Englewood Cliffs, NJ, 1989; Chapter 5.
6. Shah, S. N.; Lord, D. L.; Tan, H. C. Presented at the 1992 SPE International Meeting on Petroleum Engineering, Beijing, China, March 24–27, 1992; paper SPE 22391.
7. Buechley, T. C.; Lord, D. L. *Oil & Gas J.* **1973** *September 17,* 84.
8. Roodhart, L. P. Presented at the 1987 SPE/DOE Low Permeability Reservoirs Symposium, Denver, CO, May 18–19, 1987; paper SPE 16413.

9. Cutler, R. A.; Jones, A. H.; Swanson, S. R.; Carroll, H. B., Jr. Presented at the 1981 SPE/DOE Low Permeability Gas Reservoirs Symposium, Denver, CO, May 27–29, 1981; paper SPE 9869.

10. Neal, E. A.; Parmley, J. L.; Colpays, P. J. Presented at the 1977 SPE Annual Technical Conference and Exhibition, Denver, CO, October 9–12, 1977; paper SPE 6816.

11. Callanan, M. J.; McDaniel, R. R.; Lewis, P. E. Presented at the 1983 SPE/DOE Low Permeability Gas Reservoirs Symposium, Denver, CO, March 13–16, 1983; paper SPE 11633.

12. Cutler, R. A.; Enniss, D. O.; Jones, A. H. Presented at the 1983 SPE/DOE Low Permeability Gas Reservoirs Symposium, Denver, CO, March 13–16, 1983; paper SPE 11634.

13. Sinclair, A. R.; Graham, J. W. Presented at the 1978 ASME Energy Technical Conference, Houston, TX, November 5–9, 1978.

14. Gardner, D. C.; Eikerts, J. V. Presented at the 1982 SPE Annual Technical Conference and Exhibition, New Orleans, LA, September 26–29, 1982; paper SPE 11066.

15. Ely, J. W.; Haskett, S. A.; Holditch, S. A. *J. Pet. Technol.* **1989**, *November,* 1194.

16. Nolte, K. G. Presented at the 1988 SPE Eastern Regional Meeting, Charleston, WV, November 1–4, 1988; paper SPE 18537.

17. Landel, R. F.; Moser, B. G.; Bauman, A. J. *Fourth International Congress on Rheology;* Brown University Proceedings, 1963, Part 2; Interscience Publishers: New York, 1965; p 663.

18. Frankel, N. A.; Acrivos, A. *Chem. Eng. Sci.* **1967**, *22*, 847–853.

19. Keck, R. G.; Nehmer, W. L.; Strumolo, G. S. *SPE Prod. Eng.* **1992**, *February,* 21.

20. Eilers, V. H. *Kolloid-Z* **1941**, 97, 313.

21. Jennings, A. R., Jr. Presented at the 1990 SPE Annual Technical Conference and Exhibition, New Orleans, LA, September 23–26, 1990; paper SPE 20641.

22. Shah, S. N. *SPE Prod. & Facil.* **1993**, *May,* 123–130.

23. Shah, S. N. *AIChE J.* **1993**, *39(2)*, 207–214.

24. Thomas, D. G. *J. Colloid Sci.* **1965**, *20*, 267–277.

25. Maron, S. H.; Pierce, P. E. *J. Colloid Sci.* **1956**, *11*, 80–95.

26. Prud'homme, R. K. "Progress Report;" Gas Research Institute: Chicago, IL, May 25, 1989.

27. Nolte, K. G.; Smith, M. B. *J. Pet. Technol.* **1981**, *September,* 1767–1775.

28. Erdle, J. C.; et al. Presented at the 1981 SPE Annual Technical Conference and Exhibition, San Antonio, TX, October 4–7, 1981; paper SPE 10310.

29. Hannah, R. R.; Harrington, L. J.; Lance, L. C. Presented at the 1983 SPE Annual Technical Conference and Exhibition, San Francisco, CA, October 5–8, 1983; paper SPE 12062.

30. Blasius, H. *Mitt. Forsch.* **1913**, *1*, 131.

31. Shah, S. N.; Lee, Y. N. *SPE Prod. Eng.* **1986**, *November,* 437–445.

32. Molerus, O.; Wellmann, P. *Chem. Eng. Sci.* **1981**, *10*, 1623–1632.

33. Shah, S. N. *Oil & Gas J.* **1984**, *January 16,* 92–98.

34. Bowen, R. L., Jr. *Chem. Eng.* **1961**, *July 24,* 143.

35. Lord, D. L.; McGowen, J. M. Presented at the 1986 SPE Annual Technical Conference and Exhibition, New Orleans, LA, October 5–8, 1986; paper SPE 15367.

36. Dodge, D. W.; Metzner, A. B. *AIChE J.* **1959**, *5(2)*, 189–204.

37. Virk, P. S. *AIChE J.* **1975,** *21(4),* 625–656.
38. Cooper, G. D.; Nelson, S. G.; Schopper, M. D. Presented at the 1983 SPE Annual Technical Conference and Exhibition, San Francisco, CA, October 5–8, 1983; paper SPE 12063.
39. Swanson, G. S.; Meeken, R. B. Presented at the 1981 SPE California Regional Meeting, Bakersfield, CA, March 25–26, 1981; paper SPE 9935.
40. Tan, H. C.; McGowen, J. M. Presented at the 1991 SPE Rocky Mountain Regional Meeting and Low Permeability Reservoirs Symposium, Denver, CO, April 15–17, 1991; paper SPE 21856.
41. Tan, H. C.; Shah, S. N. Presented at the 1991 International Symposium on Oil Field Chemistry, Anaheim, CA, February 20–22, 1991; paper SPE 21034.
42. Harris, P. C.; Klebenow, D. E.; Kundert, D. P. Presented at the 1988 SPE Rocky Mountain Regional Meeting, Casper, WY, May 16–18, 1988; paper SPE 17532.
43. Kenyon, D. E. Presented at the 1993 SPE Gas Technology Symposium, Calgary, Canada, June 28–30, 1993; paper SPE 26152.
44. Petzet, G. A. *Oil & Gas J.* **1988,** *April 11,* 15–17.
45. Bosio, J.; Reiss, L. H. *Oil & Gas J.* **1988,** *March 21,* 71.
46. Tolle, G.; Dellinger, T. *Oil & Gas J.* **1986,** *May 26,* 80–86.
47. Littleton, J. *Pet. Eng. Int.* **1986,** *April,* 43–46.
48. Oroskar, A. R.; Turian, R. M. *AIChE J.* **1980,** *26(4),* 550–558.
49. Shah, S. N.; Lord, D. L. *SPEDE* **1990,** *September,* 225–232.
50. Shah, S. N.; Lord, D. L. *AIChE J.* **1991,** *37(6),* 863–870.
51. Schols, R. S.; Visser, W. Presented at the 1974 SPE European Spring Meeting, Amsterdam, Netherlands, May 29–30, 1974; paper SPE 4834.
52. Sievert, J. A.; Wahl, H. A.; Clark, P. E.; Harkin, M. W. Presented at the 1981 SPE/DOE Low Permeability Symposium, Denver, CO, May 27–29, 1981, paper SPE/DOE 9865.
53. Medlin, W. L.; Sexton, J. H.; Zumwalt, G. L. Presented at the 1985 SPE Annual Technical Conference and Exhibition, Las Vegas, NV, September 22–25, 1985; paper SPE 14469.
54. Roodhart, L. P. Presented at the 1985 SPE/DOE Low Permeability Gas Reservoirs Symposium, Denver, CO, May 19–22, 1985; paper SPE 13905.
55. Shah, S. N. *SPEJ, Soc. Pet. Eng. J.* **1982,** 164–170; *Trans. AIME* 273, Part 2.
56. Hannah, R. R.; Harrington, L. J. *J. Pet. Technol.* **1981,** *May,* 909–913.
57. Harrington, L. J.; Hannah, R. R.; Williams, D. Presented at the SPE Annual Technical Conference and Exhibition, Las Vegas, NV, September 23–26, 1979; paper SPE 8342.
58. Clark, P. E.; Guler, N. Presented at the 1983 SPE/DOE Low Permeability Gas Reservoirs Symposium, Denver, CO, March 14–16, 1983; paper SPE 11636.
59. Clark, P. E.; Quadir, J. A. Presented at the 1981 SPE/DOE Low Permeability Gas Reservoirs Symposium, Denver, CO, May 27–29, 1981; paper SPE 9866.
60. Novotny, E. J. Presented at the SPE Annual Technical Conference and Exhibition, Denver, CO, October 9–12, 1977; paper SPE 6813.
61. Daneshy, A. A. In *Recent Advances in Hydraulic Fracturing;* Society of Petroleum Engineers: Richardson, TX, 1989; Chapter 10, pp 210–222.
62. Meyers, B. R. *Oil & Gas J.* **1986,** *May 26,* 71–77.
63. Acharya, A. *SPE Pet. Eng.* **1986,** *March,* 104–110.

64. Kirkby, L. L.; Rockefeller, H. A. Presented at the 1985 SPE/DOE Low Permeability Gas Reservoirs Symposium, Denver, CO, May 19–22, 1985; paper SPE 13906.
65. Dunand, A.; Soucemarianadin, A. Presented at the 1985 SPE Annual Technical Conference and Exhibition, Las Vegas, NV, September 22–25, 1985; paper SPE 14259.
66. McMechan, D. E.; Shah, S. N. *SPE Pet. Eng.* **1991,** *August,* 305–312.
67. Barnea, E.; Mizrahi, J. *Chem. Eng. J.* **1973,** *5,* 171–189.
68. Cleary, M. P.; Fonseca, A., Jr. Presented at the 1992 SPE Annual Technical Conference and Exhibition, Washington, DC, October 4–7, 1992; paper SPE 24825.
69. Conway, M.; Penny, G.; Slabaugh, B. *An Investigation of the Rheology and Proppant Transport of Common Fracturing Fluids;* Rheology Consortium Report; Stim-Lab, Inc.: Duncan, OK, 1992.
70. de Kruijf, A.; Davies, D. R.; Fokker, P. A. Presented at the 1994 SPE Formation Damage Symposium, Lafayette, LA, February 8–10, 1994; paper SPE 27398.
71. Rein, R. G., Jr.; Lord, D. L.; Shah, S. N. Presented at the SPE Annual Technical Conference and Exhibition, Houston, TX, October 3–6, 1993; paper SPE 26524.
72. Mears, R. B.; Sluss, J. J., Jr.; Fagan, J. E. Presented at the 1993 SPE Annual Technical Conference and Exhibition, Houston, TX, October 3–6, 1993; paper SPE 26619.
73. *Fracturing Fluid Characterization Facility, Annual Report, Aug. 1991–Dec. 1992;* The University of Oklahoma: Norman, OK, December 1992.
74. In *FOCUS—Tight Gas Sands;* Gas Research Institute: Chicago, IL, December 1991; pp 1–27.

NOTE: All papers with SPE numbers are available through the Society of Petroleum Engineers, P.O. Box 833836, Richardson, TX 75083–3836.

RECEIVED for review July 5, 1994. ACCEPTED revised manuscript January 12, 1995.

Principles and Applications of Cement Slurries

D. Guillot and J. F. Baret

Etudes et Productions, Schlumberger—Division Dowell, 26, rue de la Cavee, BP202, 92142 Clamart Cedex, France

The basic principles of oil well cementing are first described together with the main critical engineering constraints encountered during a primary cementing operation, that is, wellbore control, mud displacement, fluid loss control, and gas migration. Then cement slurry properties that are relevant to the process are reviewed and the procedures used to measure these properties are discussed. Particular attention is given to rheological measurements that can be affected by wall slip when using coaxial cylinder or pipe flow viscometers. This is followed by an overview of the additives that are used to obtained the required cement slurry or set cement properties, like weighting agents/extenders, retarders, dispersants, fluid loss agents, and antisettling agents. Some typical mechanisms of action of these additives are briefly discussed.

Functions of Cementing Slurries

Introduction to Oil Well Cementing.   Applications of cement slurries to oil well drilling are several. The main operation, called primary cementing, consists of placing such fluids in the annular space between the borehole and a cylindrical tube—the casing—which is used to line the borehole at several stages during drilling (Figure 1). The cement, once set, supports the casing and seals off the annulus. This fulfills the main objectives of the operation, that is,

- isolate the well bore from formation fluids
- prevent migration of formation fluids in the annulus behind the casing
- protect the casing against corrosion by formation fluids

0065–2393/96/0251–0601$15.75/0

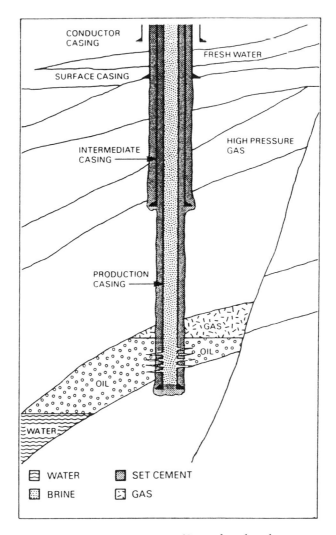

Figure 1. Primary casing cementing. (Reproduced with permission from reference 1. Copyright 1984 Dowell Schlumberger.)

To supplement a faulty primary cement job, cement slurries are also circulated behind the casing between two sets of perforations at the bottom and at the top of the desired interval. This is called remedial cementing.

Two other types of operations involve cement slurries. They are used to plug back a well from a deeper to a shallower depth for a number of reasons: abandonment, whip stocking, lost circulation, or shutting off water. Squeeze cementing consists of forcing a column of cement slurry under pressure against the formation into channels behind casing or into perforation tunnels. The objective here consists of plugging or sealing off fissures or perforations. A typical primary cementing operation will now be described in more detail.

Description of a Typical Primary Cementing Operation. At the very beginning of a primary cementing operation, the upper part of the borehole is cased, whereas the bottom part that has just been drilled is not. The borehole is full of drilling mud. The density of this fluid is such that the well bore is stable and that the formation fluids are contained, that is, at any depth the hydrostatic pressure due to the drilling mud is higher than the pressure of the formation fluids and smaller than the fracturing pressures of the intersected formations. These constraints also apply to the static and dynamic pressures during the overall cementing operation until the cement is set.

As explained earlier, the objective of a primary cementing operation is to line the borehole from the bottom depth up to at least a few hundred feet into the upper cased section. To accomplish this task, the casing, possibly equipped with centralizers, is first run into the borehole. At the end of this process, both the casing itself and the annulus between the bore hole and the casing are full of drilling mud.

The next step consists of replacing the mud in the annulus with a cement slurry. To this effect, a fluid called a spacer—to separate the drilling mud from the cement slurry—followed by the cement slurry is pumped into the casing. These fluids are then displaced from the casing into the annulus, usually with the drilling mud, until the cement slurry has reached its required position in the annulus (Figure 2). The end of the cementing job is triggered by a pressure increase due to a mechanical plug that lands on a special equipment at the bottom of the casing. This device, called the top plug, is inserted into the casing between the cement slurry and the displacement fluid.

To prevent fluid intermixing in the casing, other mechanical plugs called bottom plugs can be placed in between other fluids. The bottom plugs are equipped with a rupture disk that breaks as they land at the bottom of the casing.

To make a primary cementing operation successful, several conditions have to be satisfied. They are described in the following section.

Engineering Constraints During Cementing Operations. The engineering constraints related to a primary cementing operation are numerous. One can differentiate those that are linked to the placement process itself from those that are linked to the sealing material that is used, that is, cement slurries.

Constraints Related to the Placement Process. As mentioned earlier, during the operation, at any depth the hydrostatic and dynamic

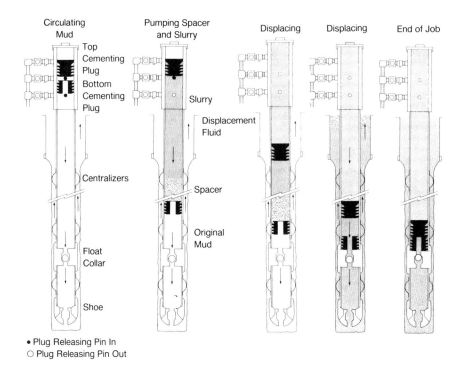

Figure 2. Principles of cement placement for a primary cementing job. (Reproduced with permission from reference 1. Copyright 1984 Dowell Schlumberger.)

(hydrostatic + friction pressure) pressures should be maintained in between two boundaries: a lower and an upper limit (Figure 3). This constraint has two mains consequences. First, cement slurries must be designed with densities varying in a wide range, typically from less than 1000 (8.34) to more than 2400 kg/m^3 (20 lb/gal). Second, the apparent viscosity of cement slurries should not be too high; the same criterion applies to mixture of cement slurries with other fluids it may intermix with during the placement, for example, the spacer and the drilling mud. (An absolute upper limit to the apparent viscosity is difficult to define as it is operation dependent. In practice though, the viscosity at infinite shear rate rarely exceeds 500 cp).

The second set of constraints is related to the displacement of the drilling mud from the annulus. This is a fairly difficult technical issue, that is, the displacement of a non-Newtonian fluid by a series of non-Newtonian fluids—most of them exhibiting a yield stress—in an eccentric annulus (the casing is rarely perfectly centered in the hole). At each interface, drilling mud–spacer and spacer–cement slurry, the process is governed by buoyancy forces and viscous forces.

The role of buoyancy forces is obvious. Assuming viscous forces are

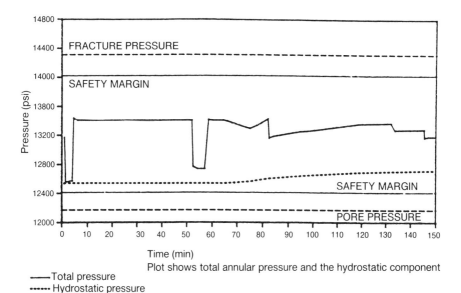

Figure 3. Well safety: the placement pressure at a given depth should remain between pore and fracture pressure during all the duration of the job.

negligible, if the displacing fluid has a higher density than the fluid it displaces, which is most of the time the case, the latter will "float" on top of the former, leading to a flat interface in the annulus. Therefore, the larger the density differences between the fluids involved the better the quality of the displacement.

The role of viscous forces is more controversial. Several techniques are applied, all of them requiring a more or less stringent control of the rheology of the displacing fluids (spacer and cement slurry). The governing parameter is the flow rate that should be as high as possible regardless of the flow regime. The only implication is that the dynamic pressure during placement should be smaller than the fracture pressure of the formations that defines an upper limit to the apparent viscosity of the cement slurry (and of course to the flow rate). The flow regime of the spacer (possibly of the cement but in practice this almost never the case) should be turbulent in the annulus during a minimum amount of time. This technique has little implication on the rheology of the cement slurry as in most circumstances it is viscous enough when compared with the spacer for the interface to be flat. It is recognized as giving extremely good results in particular when water, possibly mixed with mud dispersants, is used as the spacer. When the flow regime is laminar, the apparent viscosity of any displacing fluid should be higher than that of the fluid it displaces either at an average shear-rate representative of the flow conditions or in the whole shear-rate range encountered. The latter condition implies for example that a cement slurry exhibits a yield stress at least higher than that of the drilling mud it will displace.

A third constraint is related to the potential cross flow situation in the annulus across permeable formations. The main flow path is of course along the well bore axis. But because, during the operation, the pressure in the annulus is always higher than the pore pressure of the formation fluids, the annular fluids are likely to leak off into the formation pores. This phenomenon should be minimized as the objective is obviously not to "loose" these fluids into the formations. Another possible consequence is the damage that the "lost" fluid may cause to the permeability of hydrocarbon bearing formations thus making the production of formation fluids more difficult.

On top of these constraints, the density, rheology, and thermal properties of cement slurries play a role in another simultaneous process that takes place during the placement in the annulus, that is, the thermal exchanges that are going to define the temperature history seen by the different fluids elements. The displacement and the thermal exchanges consist of a fully coupled problem that is never treated as such because this would be too complicated to simulate in a reasonable amount of time. More of the results of the simulation would be jeopardized by the lack of information on the temperature dependence of the different fluid rheologies as discussed later. The rheology of the cement slurries is not designed to control the thermal exchanges. Their rheologies are designed rather to optimize mud displacement, and the resulting temperature history is possibly taken into account in some testing procedures, like thickening time tests.

Constraints Related to the Nature of Cement Slurries. Some of them have already been addressed, like the constraints on density and rheology. Because cement slurries are suspensions of solid particles in a water-base fluid, there are several other constraints that should be added to the previous ones. The slurry should remain homogeneous during the whole placement process and even after placement until it is set. This means, for example, that the solid particles, the density of which differs from that of the base fluid, should not settle under gravity forces neither under static nor under dynamic conditions. The potential cross-flow process that was described previously is actually a cross-filtration process. In general, the wellbore fluids involved in drilling and cementing operations are suspensions of solid particles that are too big to penetrate into the formation pores. Because these fluids are overbalanced with respect to the formation fluids, a filtration cake is formed at the annular fluid–formation interface. So cement slurries are placed into an annulus that usually have a drilling mud cake preestablished across permeable formations. Because this mud cake is difficult to remove during cement placement, the cement cake develops on top of a mud cake as schematized in Figure 4 (2). During cement placement, the axial annular flow tends to erode the cake and a steady filtration velocity is reached (3). The effect of fluid loss is mainly an increase in solid content of cement the slurry, with its consequences on slurry density and rheology (4). After placement, a static filtration process occurs, that is, the only flow is the filtration one, perpendicular to the bore hole surface, and the cake grows. If the cake is growing too much, it may bridge the annulus and prevent hydrostatic pressure to be transmitted downhole jeopardizing the control of the well (Figure 5). Finally, there are constraints that are related to the nature of the solid particles in suspension, that is, portland cement.

Fluid loss rate is given by Darcy's law:

$$Q = K_f/\mu \cdot \Delta P/L \cdot A$$

where Q is the filtration rate, K_f the permeability, μ the viscosity, $\Delta P/L$ the pressure gradient across the cake, and A the area.

In dynamic mode, the cake is not growing, the volume of fluid lost, v_f, is proportional to time:

$$v_f = k/\mu \cdot \Delta P/L \cdot A \cdot t$$

In static mode, the cake thickness, L, is increasing with time proportionally to the volume of fluid lost:

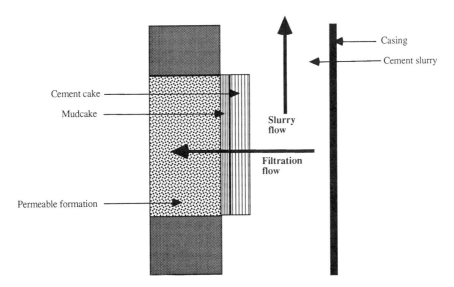

Figure 4. Schematic of filtration cakes.

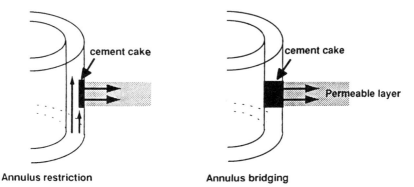

Annulus restriction   Annulus bridging

*Figure 5.   Cake growth may stop the cement flow during placement or pre-
vent hydrostatic pressure transmission in the static regime. (Reproduced
with permission from reference 4. Copyright 1984 Society of Petroleum Engineers.)*

$$A \cdot L \propto v_f$$

Therefore,

$$V \propto A \, (2 \, k/\mu \cdot \Delta P \cdot t)^{1/2}$$

The volume of fluid loss, and the cake size too, varies with the square
root of time.

The first one has to do with the setting of the cement slurry that is
accelerated by an increase of the temperature (and also of the pressure
to a lesser extent). It is obviously important that the cement slurry remains
"liquid" during the placement period. But once in place it must set within
a reasonable amount of time for the next operations to be performed as
soon as possible (further drilling, well completion, and so on).

The second one is related to the inability of the cementing material
to transmit pressure during the liquid to solid transition. After placement,
once static, the cement slurry ceases to transmit the full hydrostatic
pressure because it builds up gel strength as a structure is establishing
its matrix. Once self-supporting, the pore pressure of this structure becomes controlled by its shrinkage (volume reduction due to the ongoing
chemical reactions) and by its permeability. This leads to two typical
situations. First, across a permeable formation, the formation fluid in-
vades the porosity of the cement to compensate for chemical shrinkage,
and the cement pore pressure is governed by the pore pressure of the
formation. Second, across an impermeable zone, the cement pore pressure decreases down to the local vapor pressure. This downhole pressure
reduction in the annulus starts as soon as the cement slurry is static and
continues as it is setting. At one point in time, it may lead to the migration

of formation fluids in the annulus toward zones of lower pressure, that is, generally speaking, upwards (Figure 6), if the gel strength and the permeability of the building structure permit. This phenomenon can be catastrophic when the fluid is compressible (gas) as it expands and possibly forms channels while migrating.

The Required Properties of the Cementing Material and Their Measurements

Density. A cement slurry is designed at a density such that the well is under control during the whole process, that is, the static and dynamic pressures remain in between two limits across the entire open hole section during the whole operation.

In the laboratory, the density of a cement slurry is usually derived from the measured mass or volume of the different ingredients and from their respective densities. When a measurement is performed, it is usually done with pressurized equipment like a pressurized mud balance (Appendix C of reference 6) as cement slurries always contain a small quantity of air bubbles (these are trapped into the slurry and difficult to eliminate because of the fluid yield stress). In the field, measuring slurry density is a key issue as it is much more cumbersome if not impossible (for continuous mix operations) to measure the mass or volume of the different ingredients. Different types of equipments are used like pressurized mud balances, radioactive densitometers, or vibrating tube devices. Once the slurry is mixed at the right density, it is of utmost importance to make sure it is stable.

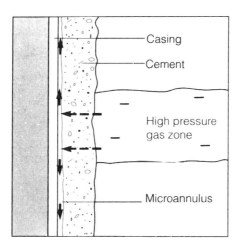

Figure 6. Cement matrix gas channeling. (Reproduced with permission from reference 5. Copyright 1991 Schlumberger.)

Stability.   Because cement slurries are water-based suspensions of particles in which density differs from water density, these particles are likely to settle (or sometimes float) either under static conditions or under dynamic conditions. This of course is not acceptable as it may have dramatic consequences on well bore control (Figure 7) and on various aspects of the process as cement slurry properties depend on the liquid-to-solid ratio of the suspension (thickening time, rheology, and so on). The stability of cement slurries is usually tested under static conditions. Two kind of tests are being performed. The objective of the free water test is to determine if the interstitial water will tend to separate from the bulk of the slurry. To this effect the slurry is poured in a graduated cylinder and left static. After 2 hours, the volume of supernatant fluid at the top of the slurry column is recorded as the slurry free water (Appendix M of reference 6 and Section 7 of reference 7). The settling test evaluates the possible sedimentation of solid particles below the supernatant fluid. This is done by measuring the density profile of the same cement column once it is set. Both tests can be run under realistic conditions of temperature and pressure. There are two key issues in the interpretation of these tests: the length scale, that is, how to extrapolate the results of a laboratory test performed on a 240-mm cement column to a several hundred meter one in the field, and the effect of the wellbore inclination on these phenomenon.

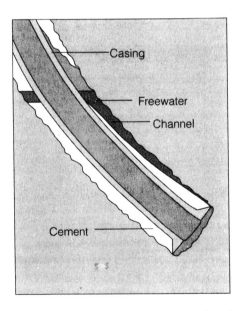

Figure 7.   Free water channel due to settling. (Reproduced with permission from reference 5. Copyright 1991 Schlumberger.)

Another aspect of cement slurry stability is the stability under dynamic conditions. Dynamic conditions are usually more severe than static ones because cement slurries are shear-thinning. This is a problem in the laboratory as the solid particles may settle while the fluid is being sheared (thickening time, rheology), and also in the field especially if the well is deviated from vertical. But there is currently no standard test in the industry to evaluate the stability of cement slurries under dynamic conditions.

Consistency: Thickening Time Measurements.   During a cementing operation, cement slurry rheology must remain low enough to be placed in the desired downhole location. But once in place it must set within a reasonable amount of time. Therefore, from a placement point of view, it is important to predict when this transition from a liquid state to a solid state will occur. This is why more than 50 years ago the industry developed laboratory equipment to investigate the time that a cement slurry will remain pumpable under simulated wellbore conditions of temperature and pressure. These procedures and equipment have been standardized. The test is referred to as the thickening time test. It is the first basic design test that almost every cement slurry undergoes before a cementing operation. The specifications and operational procedures are described in Section 9 of API Spec. 10A (7) and in Appendix E of API Spec. 10 (6).

The cement slurry that is contained in a cylindrical cup is sheared by the relative movement of a paddle (either the cup or the paddle can be rotated). The torque necessary to maintain the paddle in a fixed position or to rotate it at a specified rotational speed (150 rpm) is recorded as a function of time while the slurry is submitted to a given temperature and pressure history. The torque reading, after proper calibration, is converted to a consistency expressed in Bearden units (Bc). A simple conversion of this unit to a common viscosity unit cannot be done with non-Newtonian fluids.

Figure 8 represents an ideal thickening time curve. The slurry consistency remains fairly low (below 30 Bc) up to the point of departure at which time it increases sharply. Simultaneously, the temperature of the slurry increases, which indicates that the slurry starts to set. The test is stopped when a consistency of 100 Bc is reached. For such ideal cases, the pumping time, whether defined as the point of departure or as the time to reach whatever arbitrary consistency, is not controversial. But all thickening time curves are not as ideal as the one shown in Figure 8. The rate of increase of the consistency after the point of departure can be much slower, which makes the pumping time more difficult to define. This may or may not be related to the fact that an increase in

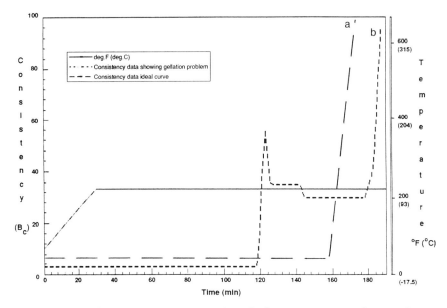

Figure 8. Thickening time curves. (a) Ideal curve corresponding to the beginning of cement setting and (b) increase in cement slurry consistency due to gellation before setting.

the slurry consistency can be due to the gellation of the fluid (this question will be addressed below) rather than to the setting of the cement.

From a placement point of view, the analysis of the thickening time curve raises two questions. Up to which consistency can a given cement slurry be considered as "pumpable"? Generally speaking, it is not possible to answer such a question without measuring the true rheology of the fluid. The objective of the thickening time test is just to detect the transition from a "liquid" to a "solid" or possibly to a gelled state under realistic downhole conditions (Unfortunately, rheological measurements on cement slurries are not always performed under realistic conditions of temperature and pressure—*see* the following section). Hence, different rules of thumb are used in the industry to define the pumping time (point of departure, 30 Bc as in Section 9 of reference 7). Is the thickening time curve sensitive to the shear history experienced by the fluid? The answer is unfortunately yes, but only to a certain extent. This effect has to be compared with the effect of two other key parameters, that is, temperature and pressure. A complete discussion about these effects is outside the scope of this chapter. From a rheological point of view, this sensitivity to shear history would imply that the test should be run under a more controlled shear field. This is what Purvis et al. (8) attempted to do by replacing the standard paddle by an helical screw.

Rheology: Flow Curves. The rheological properties of cement slurries are important to many aspects of a cementing operation. They are obviously a key in the calculation of friction pressures, but they also play a role in the efficiency of the displacement process as well as on the thermal exchanges occurring during the placement.

Field Practice. The rheology of cement slurries is most of the time measured in the laboratory using a concentric cylinder viscometer.

Measurements at atmospheric pressure. The basic features of the equipment and the testing procedure are described in Appendix H of reference 6. The method assumes that the fluid is time-independent. The preheated fluid, contained in a large thermocontrolled cup, is sheared between an outer cylinder, the rotor, and an inner cylinder, the bob. The outer cylinder is rotated at a constant velocity (maximum velocity is 600 rpm), whereas the torque exerted by the fluid on the inner cylinder is being measured. The characteristics of the geometry are

$$R_1 = 0.679 \text{ in. } (17.25 \text{ mm})$$
$$R_2 = 0.725 \text{ in. } (18.42 \text{ mm})$$
$$\text{Gap} = 0.046 \text{ in. } (1.17 \text{ mm})$$
$$l = 1.500 \text{ in. } (38.0 \text{ mm})$$

The test consists of shearing the fluid at the highest rotational speed (300 rpm) for 1 min, before recording the corresponding torque. The rotational speed is then decreased step by step until a complete stop, and the corresponding readings are recorded after 20 s of rotation at each rotational velocity. Because the test is performed at atmospheric pressure, the maximum testing temperature is around 85 °C (185 °F).

High temperature and pressure measurements. Equipment has been developed to overcome the temperature limitation of the standard basic equipment. The geometry can be either similar or much bigger than the geometry of the atmospheric equipments. The main difference is that pressure can be applied on the tested fluid that allow measurements to be performed up to typically 250 °C and 130 MPa. Most of them are also equipped with a recirculating device that ensures the slurry remains homogeneous during the much longer shearing times required by high temperature and pressure testing (slurry stability under shear is addressed below).

Data analysis. The torque–rotational velocity data are usually fitted to a two parameter model, either the power law or the Bingham plastic model, and sometimes to a three-parameter model like the Herschel–Bulkley model.

$$\tau = k \times \dot{\gamma}^n$$
$$\tau = \tau_y + \mu_p \times \dot{\gamma}$$
$$\tau = \tau_y + k \times \dot{\gamma}^n$$

Figure 9 gives an example of two cement slurries that can be reasonably well described by such models. But the behavior of cement slurries is not always as simple as these two examples; several problems can flaw the results.

Problems Encountered in the Rheological Characterization of Cement Slurries. *Stability.* As discussed earlier, stability is a key aspect in the design of cement slurries. A cement slurry may be perfectly stable under static conditions and not stable in a coaxial cylinder viscometer because of the combined action of gravity and centrifugal forces. Sedimentation in cement paste during the rheological test may lead to large errors in the measurements as shown by Bhatty and Banfill (9). But the role of centrifugal forces is far from being negligible at a rotational speed of 600 rpm. Both phenomenon may explain why the reproducibility of viscometric measurements is drastically improved when reducing the maximum rotational speed from 600 to 300 rpm as discussed by Beirute *(10)*. A hysteresis in the torque readings measured while increasing the rotational speed first and then while decreasing it may be interpreted as an effect of particle segregation (this could also be due to another time dependent effect: thixotropy or rheopexy).

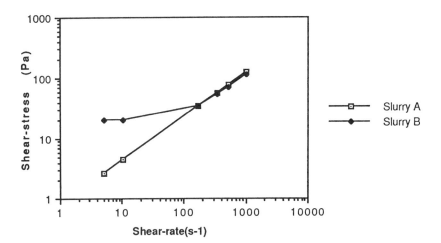

Figure 9. Rheological measurements performed with a coaxial cylinder viscometer. (A) Slurry showing a power law behavior and (B) slurry showing a Bingham plastic behavior.

Slippage. The viscometric measurements performed on cement slurries with coaxial cylinder viscometers were shown to be sensitive to the surface roughness of the cylinders by several authors (*11*). Although the exact cause for such a behavior is not fully understood, it can be explained by a depleted layer in particle concentration close to the smooth cylindrical surfaces. Mannheimer (*12, 13*) has done a thorough analysis of this phenomenon using experimental data generated with smooth surface cylinders. Assuming it can be accounted for by a slip velocity, V_s, of the tested fluid at the cylinder walls, he was able to calculate how V_s is affected by shear stress from experiments performed with two different geometries (V_s is assumed to depend only on shear stress, and the radius ratio of the cylinders has to be close enough to 1).

Figure 10 shows the results obtained on different cement formulations. Up to three different flow patterns can be observed. First, below a given critical shear stress, which is cement formulation dependent, the flow is entirely governed by slip at the wall. This is what Mannheimer calls the 100% slip situation in which no meaningful rheological information can be derived. As shear stress increases, the flow behavior is partially governed by wall slip partially by shear within the fluid itself. In this region, experimental data must corrected for wall slip (Figure

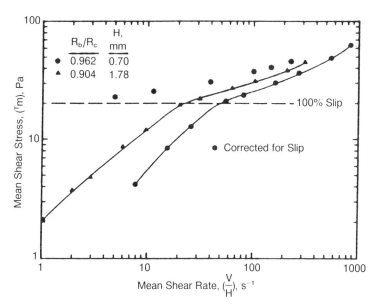

Figure 10. Rheological measurements performed with a coaxial cylinder viscometer and different geometries. The fact that the curves are not superimposed implies that the measurements are affected by wall slip. (Reproduced with permission from reference 13. Copyright 1991 E.&F.N. Spon.)

11). At even higher shear stresses, wall slip can possibly disappear and the flow is entirely governed by shear within the tested material itself.

One consequence of wall slip is that a cement slurry may appear to flow below its yield stress. In other words, the yield point cannot be obtained by extrapolation of the flow curve to zero shear rate. When fitting the raw data to a given rheological model, it is of course necessary to eliminate those that are affected by wall slip. Because rheological measurements are usually performed with a single geometry, in theory this is not feasible. But the experience shows that data in the 100% slip region can be detected. They show up on a loglog plot as a separate linear curve at low shear rates (looking like a Newtonian plateau) as shown on Figure 12. This being said, with a single geometry, it is not possible to determine by how much the remaining data is affected by wall slip.

Pipe Flow. The rheological properties of cement slurries have also been characterized using pipe viscometers (*13, 15–18*). The experimental results show different flow patterns. In some circumstances, relatively large pipe diameters and high shear rates, the flow curves, and shear-stress at the wall versus Newtonian shear rate at the wall are independent of pipe diameter (*16, 17*). The reverse situation is observed when pipe diameters are relatively small, of the order of 5 mm (*13, 15*). The diameter dependency of the flow curves can be explained, as discussed

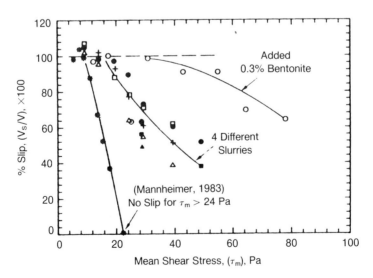

Figure 11. Ratio of the slip velocity V_s *to the fluid mean velocity* V *as a function of the measured shear stress. (Reproduced with permission from reference 13. Copyright 1991 E.&F.N. Spon.)*

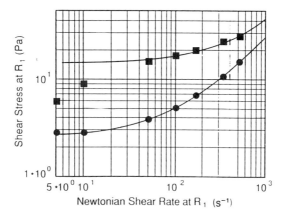

Figure 12.   Coaxial cylinder viscometer measurements. Key: ● slurry is not affected by wall slip; ■ slurry affected by wall slip at 5 and 10 s⁻¹. (Reproduced with permission from reference 14. Copyright 1990 Schlumberger Educational Services.)

above, by the presence of a high water content slip layer that forms at the surface of the cylinders. A similar approach as for coaxial flow can be used to account for this phenomenon. Therefore, wall slip can be accounted for by performing flow experiments in the same shear stress range with pipes having different diameters but the same relative surface roughness as shown by Mannheimer (*13*).

Comparison Between Different Viscometers.   To validate their rheological measurements, several authors have tried to compare the results obtained using coaxial cylinder and pipe viscometers. Their findings are not necessarily in agreement. Bannister (*15*) was able to predict the frictional pressure drops of a cement slurry in a 1.815-in. ID pipe from pipe viscometer data corrected for wall slip. Mannheimer, who tried to reconcile coaxial cylinder and pipe viscometer data, both of them being corrected for wall slip was successful with one cement slurry formulation, but the approach failed with another one (*13*). Denis et al. (*16*) showed good agreement between coaxial cylinder and pipe viscometer data above a critical shear rate—or shear stress—that is pipe diameter dependent.

However, Shah et al. (*18*) proposed an empirical correlation between the rheological properties of a large number of cement slurries as measured with the two types equipments (modified Fann35, and pipe nominal IDs of $\frac{1}{2}$, $\frac{3}{4}$, 1, and $1\frac{1}{4}$ in.). These slurries are assumed to behave as Bingham plastic fluids and

$$\mu_p \text{ (pipe)} = 0.962 \times [\mu_p \text{ (coaxial)}]^{0.9815}$$

$$\tau_y \text{ (pipe)} = 1.273 \times [\tau_y \text{ (coaxial)}] - 1.612$$

where plastic viscosities are in millepascal seconds and yield stresses in pound force per 100 ft² (the constants in the yield stress equation have been corrected because in the analysis of their pipe and coaxial flow data Shah et al. (*18*) used the Newtonian shear rate instead of the non-Newtonian one).

Rheology: Gel Strength Measurements.   Measuring the gel strength of cement slurries is important. This property will affect, for example, the start-up pressure after a temporary shutdown or the void filling properties of a given cement. Its a key to the understanding of the ability of a standing cement column to transmit pressure.

Field Practice.   The "dynamic" test described previously is often followed by gel strength measurements. For this purpose, the rotor is maintained static during the desired rest period (10 s, 1 min, 10 min). Then rotation is resumed at a very low speed (3 rpm). The torque increases until the gel breaks and the gel strength is determined from the maximum torque recorded. This procedure is affected by several drawbacks. First, with standard field equipment, the stiffness of the torque-measuring device is not high enough, and the inertia of the measuring device is far from being negligible. These effects were shown by Spears et al. (*19*) to underestimate the gel strength of drilling muds. But even if these problems were solved, the measurement would quite often be affected by slippage at the wall as explained above. For these reasons, the measurements performed according to this procedure are usually not reliable, sometimes even on a relative basis, that is, when comparing two cement slurries.

Stirring Paddle Devices.   In an attempt to overcome these problems as well as the temperature limitation (85 °C) of standard oil field equipment, other devices, similar to the equipment used to measure the thickening time, have been developed (*20*). The cement slurry contained in a cylindrical cup is usually stirred with a paddle under pressure and temperature. This allows the simulation of the shear history encountered by the fluid during placement. Then, the rotational speed is reduced to a very low value—typically 0.003 rpm—and the torque on the paddle is measured as a function of time. The main advantages of such a technique is that measurements are performed under realistic conditions of pressure, temperature, and shear history. On the other hand, the analysis of the data is not straightforward as the stress distribution in these devices is not known, and it is not clear whether or not measurements are affected by wall slip layers.

Vane Method.   Another technique was proposed by Duzy and Boger (*21*) to allow a better characterization of the stress field. It consists of replacing the two cylinders of a coaxial cylinder viscometer by a vane. With such a flow geometry, the yield surface is within the fluid itself rather than at the interface between the fluid and the cylinders, which means that measurements are not affected by wall slip. This technique also gives access to an estimate of the shear modulus and of the yield strain. Such measurements can be performed under realistic conditions of pressure and temperature as shown by Keating and Hannant (22).

Fluid Loss.   The ability of a cement slurry to prevent its dehydration through the formation of a filter cake after placement in the annulus is evaluated using a standard testing procedure (Appendix F of reference 6). The filtrate loss of around 300 cm^3 of slurry is measured at 1000 psi differential pressure across a standard filtration medium (325 mesh, 45 μm, screen supported by 60 mesh, 250 μm, screen) in a 2.13-in. diameter cylindrical cell. The duration of the test is 30 min. If all of the filtrate passes through the screen in less than 30 min, the filtrate volume is extrapolated to 30 min, assuming it varies as the square root of time.

Additives Used to Achieve the Required Properties of Cement Slurries

Density.   A suspension of portland cement particles in water can only be used in a narrow density range. Acceptance tests for oil well cements are run at a specified density within this narrow range. For example, class G cements are mixed at 44% water by weight of cement (BWOC) that, for an average cement specific gravity of 3.15, gives a solid volume fraction of 42% and a density of 1.89 g/cm^3 or 15.8 lb/gal. As density decreases below or increases above this "optimum" value, such a suspension is less and less stable (sedimentation) or more and more viscous. Therefore, it is necessary to add other material(s) to the suspension to adjust its density in a wider range.

Weighting agents. At the high end of the spectrum, cement slurry densities are carefully designed by adding dense materials. To increase the slurry density, the most commonly used material is hematite (Fe$_2$O$_3$). Its density is about 5 g/cm^3, it is sieved at 100 μm (*see* Figure 13), and it has no significant chemical interaction with cement. Barite (BaSO$_4$, density 4.33 g/cm^3) is sometimes used as a weighting agent for cement even though it provides less stability (sedimentation) to the slurry, but it is readily available on all drilling rigs because it is a major component for drilling fluids. Episodically, more exotic minerals can be used, for instance, manganese oxide or ilmenite (FeTiO$_3$).

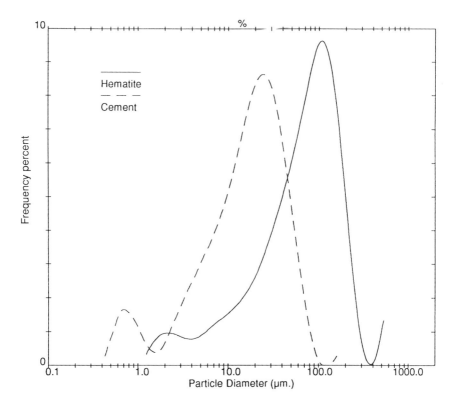

Figure 13.   Particle size distribution of a cement and a conventionally used hematite.

Extenders. At the low end of the spectrum, lightweight slurries are obtained either by adding more water or by blending light solids with the cement. In the first case, additives are required to prevent sedimentation in such a low solid content suspension. The most common of these additives is bentonite, which is well-known for its gel-forming capability inducing non-Newtonian behavior and providing antisettling properties to the suspensions even at low solid volume fraction. This property of bentonite is used in drilling fluids to carry the cuttings outside the hole (23). In cement slurries, bentonite allows the addition of 5.3% additional mix water by weight of cement for each 1% of bentonite.

Other solids are used as "extenders" of cement slurries because, with a high surface area and a strong interaction with water, they immobilize a significant fraction of the water. This hydration water is structured and thus contributes to the rheological behavior of the suspension as part of the solid phase. Materials of this type are diatomaceous earth, fly ash, and pozzolan.[1] Pozzolans are the most important group of cement extenders because they are not just inert fillers but chemically react with calcium hydroxide and thus contribute to the final strength of the system. One generic weakness of low solid volume fraction slurries being the low compressive strength they develop, pozzolanic addition is often a significant improvement.

Sodium silicates are also used to provide the fluid with a yield stress large enough to hold the particles at high water content. The mechanism is completely different from that of bentonite platelets that, having opposite charges on the faces and on the edges, gel the fluid by forming card house structures. Here, sodium silicate reacts with lime or calcium chloride to form a calcium silicate gel. It is this gel that provides the yield stress required to hold the particles.

Another approach to decrease slurry weight can be to add light solids like coal, asphaltite minerals, hollow glass, or ceramic microspheres. Finally, foams can be prepared using air or nitrogen and a cement slurry as the continuous phase to densities as low as 0.8 g/cm^3 (*24*).

Stability.   It has been mentioned already that the amount of water that can be added to a cement slurry is limited by the requirement for the suspension to remain stable (Figure 7). That is, a cement slurry placed in the well annulus should display a limited density gradient from top to bottom. The American Petroleum Institute (API) has defined a limit of 1.4% of "free water" acceptable for a cement slurry column standing 2 h in a 250-mL glass cylinder (*7*). Free water, the appearance of a particle free layer of liquid on top of the rest of the suspension, is one aspect of the more complex process of sedimentation. Depending on the level of flocculation of the slurry, (Figure 14), sedimentation may occur either in a Stokesian mode, where each particle sediments individually, only the hydrodynamic interactions of the other particles interfering to modify the settling velocity given by Stokes' law, or, if physicochemical interactions are present, the particle network sediments collectively, it subsides, liberating some free water on top of the suspension (*25*).

By letting the cement set in a cylinder and then slicing the hard column, it is possible to obtain the density gradient formed at the onset of the cement setting (*26*). The parameter controlling this sedimentation can be described in the Bingham fluid terminology as a yield stress (*see* next section). Indeed, ideally a particle will not fall if the yield stress of the fluid times some linear function of its area is greater than its buoyant

---

[1]ASTM designation of pozzolans is as follows: "A siliceous or a siliceous and aluminous material, which in itself possesses little or no value, but will, in finely divided form and in presence of moisture, chemically react with calcium hydroxide at ordinary temperatures to form compounds possessing cementitious properties."

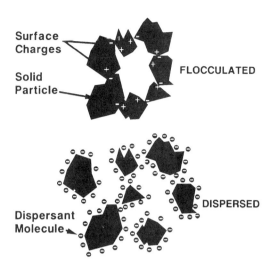

Figure 14.   Cement particle floc dispersed with polyanions.

weight (27). [The settling velocity of an isolated sphere in a Bingham fluid depends on a dimensionless group:

$$Y = \frac{3\tau_y}{r(\rho_s - \rho_f)g} = \frac{4\pi r^2 \tau_y}{4/3\, \pi r^3\, (\rho_s - \rho_f)g}$$

which represents the ratio of a force due to the fluid yield stress to the external force due to gravity. The particle will remain static when this ratio is smaller than 0.286 (27). A fluid of a given yield stress holding a 10-μm hematite particle will hold a 18.6-μm cement particle, and a 24.2-μm silica particle.] The suspension is stable if the gel is strong enough to hold the particles. It will be seen below that slurry yield stress is essentially adjusted with the level of dispersant. At high concentration of dispersant, cement particles are repulsive and individual (Figure 15). Consequently, they settle. At low dispersant concentrations, cement particles flocculate in an infinite network that is stable. This description of stable cement slurry as a gelled solid is an oversimplification because it is not an absolute immobility of the particles that is sought but a slow enough process to make it negligible in the time frame corresponding to the set of the cement. Thus, what matters is the viscosity at very low shear rates that, corresponding to the velocity of a cement particle falling in the gravity field (on the order of 10^{-2} s^{-1}), will be chosen relatively high. However, to be easily pumpable, the slurry should display a low viscosity at high shear rates (usual pumping rate corresponds to a shear rate on the order of 10^2 s^{-1}). This implies that cement slurries have to

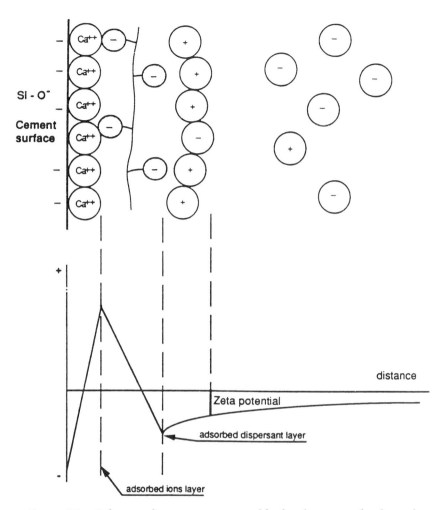

Figure 15.   Calcium adsorption is responsible for the positively charged patches on cement grains. A dispersant oligomer is shown adsorbed to the calcium layer.

be shear thinning (*see* next section). Biopolymers like polysaccharides can be added to impart such a behavior to cement slurries.

In the case where other minerals have been added to the formulation, hematite to densify or silica for strength retrogression, sedimentation is compounded with a segregation process. Here, the density gradient created is not just due to a variation in particle concentration but also to the fact that bigger and heavier particles tend to fall faster. For example, the concentration of hematite particles or of coarse silica particles (sand) can be higher at the bottom of a set sample than at the top, whereas the trend can be the opposite for the concentration of silica particles of size similar to cement.

Rheology.   When cement powder is mixed with water, the rheology of the suspension thus obtained is well described by the Bingham model for viscoplastic fluids, that is, the shear stresses measured in a viscometer for various values of the shear rate fit reasonably well a straight line going through an ordinate at the origin called "yield stress". The microscopic explanation of the yield stress can be found in particle interactions that induce the formation of macroscopic flocs. Indeed, cement particles are polymineralic materials, essentially calcium silicates and calcium aluminates (Figure 16), which lose part of their hydroxyls (the interstitial water pH rises to about 13) and calcium ions that readsorb preferentially on some minerals, transforming the particle surface into a negatively/positively charged patchwork. Linking complementary patches, a network develops throughout the suspension, providing the gel strength rheologically measured as a yield stress. The development of the network is time-dependent, inducing thixotropic behavior.

Dispersants are polyanionic polymers, mostly poly(naphthalene sulfonate) (PNS) and polymelamine sulfonate, which adsorb on the positively charged zones of the particle surfaces and suppress the interactions required to build the gel structure. Thus, the slurry is dispersed (Figure 17). With large quantities of dispersant, it is possible to obtain a New-

CLINKER GRAIN STRUCTURE

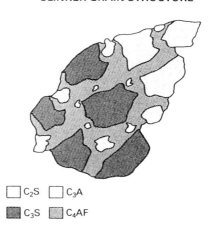

☐ C_2S ☐ C_3A

▨ C_3S ▨ C_4AF

Figure 16.   A polymineral cement grain. Beside gypsum, which is added during grinding of the clinker, cement is constituted of four main components: tricalcium silicate (C_3S), dicalcium silicate (C_2S), tricalcium aluminate (C_3A), and tetracalcium aluminoferrite (C_4AF). (Reproduced with permission from reference 1. Copyright 1984 Dowell Schlumberger.)

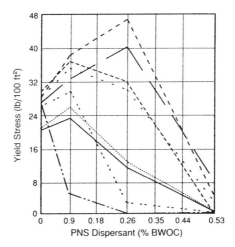

Figure 17. Yield stress as a function of dispersant concentration for different cements. The yield stress begins to increase with dispersant concentration and then decreases. This is due to the originally smaller number of negatively charged sites that increases with the dispersant to match the number of positively charged sites (yield stress maximum). The decreasing parts of the curves correspond to the adsorption of polyanions on the positive sites. (Reproduced with permission from reference 28. Copyright 1986 Elsevier.)

tonian fluid, that is, with no yield stress, but then, as noted before, the slurry is not stable anymore. A highly dispersed slurry permits sedimentation (Figure 18).

In most practical cases, oil field cement slurries contain several water-soluble organic additives. Therefore, cement interstitial liquid is an aqueous solution that is likely not to behave as a Newtonian fluid. Specifically, if the organic additives are long-chain polymers, the interstitial fluid will display a pseudoplastic behavior, as described, for instance, by the power law model. In turn, the slurry will display a yield pseudoplastic behavior as described for example by the Herschel–Bulkley model (*see* previous sections).

Fluid Loss. Cement slurry is in contact with the huge geological formation area of the hole that constitutes the outer surface of the annulus. This surface is often permeable, and because the slurry is at higher pressure than the formation fluids, a permeation flow occurs radially from the cement. Fluid loss control is a performance required for most of the cement slurries; it is achieved by reducing the permeability of the cake formed or by increasing the viscosity of the interstitial liquid. Hydrosoluble polymers of any significant molecular weight can be used to viscosify water, and hydoxyethyl cellulose is often used with cement, at least at low temperatures (<100 °C), because it does not have strong

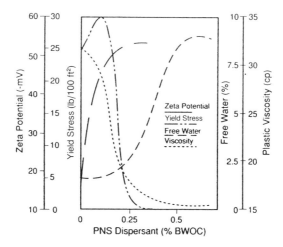

Figure 18.   The yield stress is reflecting changes in zeta-potential. When enough negative charges, brought by the dispersant, are adsorbed onto the particle surfaces the slurry deflocculates and the yield stress is zero. When its value falls below a threshold (1 or 2 lbf/100 ft²), the slurry becomes unstable, and some free water develops. (Reproduced with permission from reference 29. Copyright 1986 Schlumberger.)

interaction with cement particles. At higher temperatures, synthetic terpolymers [mostly based on 2-acrylamide-2-methyl propane sulfonic acid (AMPS)] are used as fluid loss control agents. The viscosification of the water induces also a viscosification of the slurry that may be a serious drawback in many situations. In these cases, particulate materials like latices are used to reduce cake permeability without increasing the viscosity.

Fluid-to-Solid Transition: The Cement Setting.   Once in place in the annulus, the cement slurry remains almost unchanged during a lapse of time, the dormant or induction period (Figure 19), which depends on the temperature, the cement brand, and the concentration of retarder or accelerator (Figure 20). The length of this dormant period is an element of any slurry design that is carefully adjusted because any early setting of the cement may lead to the loss of the well, whereas an excessive waiting time may unduly increase rig costs. Generally, the well temperature is such (most of the bottom-hole-circulating-temperatures range between 50 and 150 °C) that the cement needs to be retarded. Several classes of chemicals are used as retarders: in the lower temperature range lignosulfonates are convenient and at higher temperature (above 100 °C) hydroxycarboxylic acids such as glucoheptonic acid are generally preferred. As shown in Figure 21, the cement setting retardation varies largely with temperature.

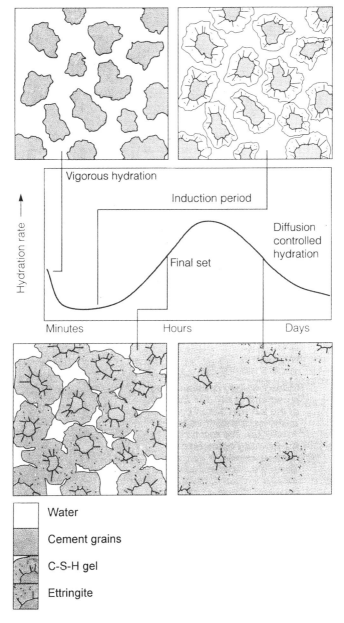

Figure 19.   Cement hydration process. Calcium silicate hydrates to form C–S–H, a quasi-amorphous gel of composition close to $C_3S_2H_3$. The excessive rate of hydration of the aluminate phase is controlled by gypsum through the formation of a calcium trisulfoaluminate hydrate, ettringite. (Reproduced with permission from reference 5. Copyright 1991 Elsevier.)

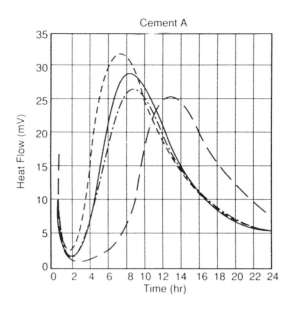

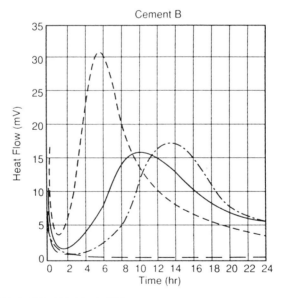

Figure 20.   Calorimetric behavior of two cements with different additives showing the exothermic effect of cement hydration. (Reproduced with permission from reference 14. Copyright 1990 Schlumberger Educational Services.)

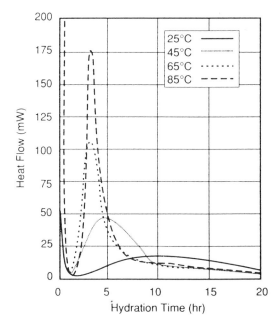

Figure 21. At higher temperature, cement hydration proceeds sooner and quicker. (Reproduced with permission from reference 14. Copyright 1990 Schlumberger Educational Services.)

The nomenclature in cement chemistry is the following: C is CaO, A is Al_2O_3, S is SiO_2, F is Fe_2O_3, H is H_2O. Therefore, C_3S is Ca_3SiO_5 (tricalcium silicate), C_2S is Ca_2SiO_4 (dicalcium silicate), C_3A is $Ca_3Al_2O_6$ (tricalcium aluminate), C_4AF is $Ca_4Al_2Fe_2O_{10}$ (tetracalcium aluminoferrite), and C–S–H is $xCaO \cdot ySiO_2 \cdot zH_2O$ (variable composition). This pressure decline is compounded with another effect known as "chemical shrinkage." Cement hydration is a chemical reaction in which anhydrous minerals are reacting with water to form hydrates. Hydrates have an overall lower volume than the sum of the initial volumes of water plus anhydrous minerals. Hence, as hydration proceeds, the total volume of the sample decreases (Figure 22). However, this shrinkage is concomitant with the strength development of the cement that therefore prevents in most cases any significant movement of the outside boundaries of the sample. The few percent chemical shrinkage results in the creation of only a few tenths of a percent of bulk shrinkage, the rest of it providing some extra porosity. In Figure 23, it is shown that if the sample is enclosed in a sealed container, this process induces a drastic pressure reduction.

Cement is made of four main phases: two silicates, C_3S and C_2S, and two aluminates, C_3A and C_4AF. The silicate phases comprise more than

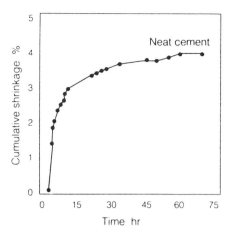

Figure 22.   Cement chemical shrinkage. (Reproduced with permission from reference 5. Copyright 1991 Schlumberger.)

70% of the total material, and C_3S is the principal constituent (Table I). The aluminate phases are the most reactive, and to prevent an early set of these phases, 3–5% of gypsum is added to the cement clinker to induce the precipitation of ettringite needle-shaped crystallites onto the aluminate surfaces. When all the gypsum is consumed, the hydration resumes and this is the end of the dormant period. Simultaneously, silicates have been through a dissolution–precipitation process: ions, mostly Ca^{2+} and OH^-, are released in the water and precipitate as calcium silicate hydrate, the C–S–H gel, and calcium hydroxide (portlandite) (30). The formation of these amorphous solids and crystals modify

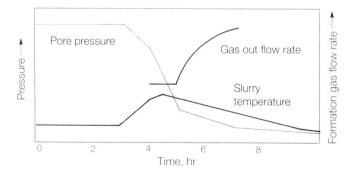

Figure 23.   Cement pore pressure declines due to gellation (self-supporting column) and chemical shrinkage. It occurs simultaneously with the maximum of the hydration exotherm, and if gas is present, it is at this stage gas will penetrate into the cement. (Reproduced with permission from reference 5. Copyright 1991 Schlumberger.)

Table I. Analysis of Three Oil-Well Cements by Inductive Coupled Plasma

Compound	Sample 1	Sample 2	Sample 3
CaO	66.3	65.7	63.2
SiO_2	22.3	20.8	19.1
Fe_2O_3	5.7	6.2	5.3
Al_2O_3	3.4	3.6	4.1
SO_3	1.9	2.2	2.3
MgO	1.3	0.8	2.2
Na_2O	0.1	0.1	0.2
K_2O	0.6	0.5	0.8
Insoluble Residue	1.1	0.1	0.1
Loss Ignition	1.1	1.3	0.5

NOTE: all values are weight percent.

the rheological properties of the slurry. Before solidification, cement slurries go through a gellation phase, and yield stress increases significantly as hydration goes on. For a critical value of the yield stress, the slurry is becoming self-supporting, that is, the gel is strong enough to hold the weight of the cement. This has a major consequence on the overall equilibrium in the well because the subtle balance of the hydrostatic pressure that has been carefully designed in between pore and fracture pressures is disrupted and the fluid column hydrostatic falls below the pore pressure.

Figure 6 schematizes how any overpressured fluid in the formation may flow into the well, gas being particularly hazardous. Of course, cement is in place to hinder gas or formation fluid progression, but when the solid structure is in process of consolidation it is not yet very strong and still very porous. Special systems have been developed to reduce this transition time in which the cement is neither fluid nor solid and to reduce the permeability of the gelled slurry; they are mainly based on the use of latex.

Set Cement. At the end of the transition phase, cement begins to develop strength; commonly, except for lightweight cements where the amount of water may prevent the rapid development of the solid structure, systems used in the oil field have 24 h compressive strength of the order of 20 MPa. Often, specifications are given by local authorities on the type of compressive strength cement should display at surface after 24 h; generally, it is about 4 MPa. This strength is required to hold the casing in place under reasonable stresses.

The outer face of the cement sheath placed in the annulus will be in contact with all the fluids of the geological formations during the whole life of the well. Some of these fluids are very aggressive toward cement. Brines containing sulfate or carbonic acid may have a deleterious effect on most of the cements. In these types of environments, special formulations have been developed that tend to reduce the permeability of the set cement to a minimum. Moreover, oil well cements have been classified by the API and a high sulfate-resistant type has been defined with a C_3A content lower than 3% (Table II). Indeed, this phase is the responsibility of the expansion and the resulting cracking occurring in cement where sulfate induces ettringite formation after the matrix has solidified.

List of Symbols

A	filtration area
g	gravitational acceleration
H	annular gap of a coaxial cylinder viscometer
k	consistency index of a power law fluid, or of an Herschel–Bulkley fluid
k_f	cake permeability
L	cement filter cake length
l	viscometer length
n	power law index of a power law fluid, or of an Herschel–Bulkley fluid
Q	filtration rate
r	radius of a spherical particle
R_1	coaxial cylinder viscometer inner radius
R_2	coaxial cylinder viscometer outer radius
t	time
v_f	fluid loss volume
V	fluid mean velocity
V_s	slip velocity
Y	dimensionless settling number
$\dot{\gamma}$	shear rate
$\Delta P/\Delta L$	pressure gradient
μ	water viscosity
μ_p	plastic viscosity of a Bingham plastic fluid
ρ_f	density of a fluid
ρ_s	density of a solid particle
τ	shear stress
τ_m	mean annular shear stress in a coaxial cylinder viscometer
τ_y	yield stress of a Bingham plastic fluid

Table II. Chemical Requirements for API Oil-Well Cements by Cement Class

Cement	A	B	C	D, E, F	G	H
Ordinary Grade (O)						
Magnesium oxide (MgO), maximum, percent	6.0	NA	6.0	NA	NA	NA
Sulfur trioxide (SO_3), maximum, percent	3.5[a]	NA	4.5	NA	NA	NA
Loss on ignition, maximum, percent	3.0	NA	3.0	NA	NA	NA
Insoluble residue, maximum, percent	0.75	NA	0.75	NA	NA	NA
Tricalcium aluminate (C_3A), maximum, percent	NR	NA	15	NA	NA	NA
Moderate Sulfate-Resistant Grade (MSR)						
Magnesium oxide (MgO), maximum, percent	NA	6.0	6.0	6.0	6.0	6.0
Sulfur trioxide (SO_3), maximum, percent	NA	3.0	3.5	3.0	3.0	3.0
Loss on ignition, maximum, percent	NA	3.0	3.0	3.0	3.0	3.0
Insoluble residue, maximum, percent	NA	0.75	0.75	0.75	0.75	0.75
Tricalcium silicate (C_3S), maximum, percent	NA	NR	NR	NR	58[b]	58[b]
minimum, percent	NA	NR	NR	NR	48[b]	48[b]
Tricalcium aluminate (C_3A), maximum percent[b]	NA	8	8	8	8	8
Total alkali content expressed as sodium oxide (Na_2O) equivalent, maximum, percent	NA	NR	NR	NR	0.75[c]	0.75[c]
High Sulfate-Resistant Grade (HSR)						
Magnesium oxide (MgO), maximum, percent	NA	6.0	6.0	6.0	6.0	6.0
Sulfur trioxide (SO_3), maximum, percent	NA	3.0	3.5	3.0	3.0	3.0
Loss on ignition, maximum, percent	NA	3.0	3.0	3.0	3.0	3.0
Insoluble residue, maximum, percent	NA	0.75	0.75	0.75	0.75	0.75
Tricalcium silicate (C_3S) maximum, percent	NA	NR	NR	NR	65[b]	65[b]
minimum, percent	NA	NR	NR	NR	48[b]	48[b]
Tricalcium aluminate (C_3A), maximum, percent	NA	3[c]	3[c]	3[b]	3[b]	3[b]
Tetracalcium aluminoferrite (C_4AF) plus twice the tricalcium aluminate (C_3A), maximum, percent	NA	24[c]	24[c]	24[b]	24[b]	24[b]
Total alkali content expressed as sodium oxide (Na_2O) equivalent, maximum, percent	NA	NR	NR	NR	0.75[c]	0.75[c]

NOTES: Methods covering the chemical analyses of hydraulic cements are described in ASTM C114: *Standard Methods for Chemical Analysis of Hydraulic Cement.* NR is no requirement and NA is not available.

[a] When the tricalcium aluminate content (expressed as C_3A) of the cement is 8% or less, the maximum SO_3 content shall be 3%.

[b] The expressing of chemical limitations by means of calculated assumed compounds does not necessarily mean that the oxides are actually or entirely present as such compounds. When the ratio of the percentages of Al_2O_3 to Fe_2O_3 is 0.64 or less, the C_3A content is zero. When the Al_2O_3 to Fe_2O_3 ratio is greater than 0.64, the compounds shall be calculated as follows: C_3A is $(2.65 \times \% \ Al_2O_3) - (1.69 \times \% \ Fe_2O_3)$ and C_4AF is $3.04 \times \% \ Fe_2O_3$. $C_3S = (4.07 \times \% \ CaO) - (7.60 \times \% \ SiO_2) - (6.72 \times \% \ Al_2O_3) - (1.43 \times \% \ Fe_2O_3)(2.85 \times \% \ SO_3)$. When the ratio of Al_2O_3 to Fe_2O_3 is less than 0.64, the C_3S shall be calculated as follows: C_3S is $(4.07 \times \% \ CaO) - (7.60 \times \% \ SiO_2) - (4.48 \times \% \ A_2O_3) - (2.86 \times \% \ Fe_2C_3) - (2.85 \times \% \ SO_3)$.

[c] The sodium oxide equivalent (expressed as Na_2O equivalent) shall be calculated by the formula: Na_2O equivalent = $(0.658 \times \% \ K_2O) + (\% \ Na_2O)$.

SOURCE: Produced with permission from reference 6. Copyright 1995 American Petroleum Institute.

References

1. Dowell Schlumberger *Cementing Technology;* Nova Communication Ltd.: London, 1984.
2. Baret, J. F. Presented at the International Meeting on Petroleum Engineering of the Society of Petroleum Engineers, Tianjin, China, November 1–4, 1988; paper SPE 17630.
3. Fordham, E. J.; Ladva, H. K. J.; Hall, C.; Baret, J. F.; Sherwood, J. Presented at the Annual Technical Conference and Exhibition of the Society of Petroleum Engineers, Houston, TX, October 2–5, 1988; paper SPE 18038.
4. Daccord, G.; Baret, J. F. *SPE Drill. Completion* **1994,** *9,* 133–138.
5. *Oilfield Review;* Elsevier Science Publishers B.V.: Amsterdam, The Netherlands, 1991; p 2.
6. *Specification for Materials and Testing for Well Cements,* 5th ed.; API Specification 10; American Petroleum Institute: Washington, DC, July 1, 1990.
7. *Specification for Cements and Materials for Well Cementing,* 22nd ed.; API Specification 10A; American Petroleum Institute: Washington, DC, January 1, 1995.
8. Purvis, D. L.; Mueller, D. T.; Dawson, J. C.; Bray, W. S. Presented at the 68th Annual Technical Conference and Exhibition of the Society of Petroleum Engineers, Houston, TX, October 3–6, 1993; paper SPE 26576.
9. Bhatty, J. I.; Banfill, P. F. G. *Cem. Concr. Res.* **1982,** *12,* 69–78.
10. Beirute, R. M. *Oil Gas J.* **1986,** *Sept.* 22, 36–38.
11. Dimond, C. R.; Tattersall, G. H. In *Hydraulic Cement Pastes: Their Structure and Properties;* Maxwell-Cook, P. V., Ed.; Publication 15.171; Cement and Concrete Association, Viewpoint Publication: Sheffield, England, 1976; pp 118–133.
12. Mannheimer, R. J. *Oil Gas J.* **1983,** *81(49),* 144–147.
13. Mannheimer, R. J. *Rheology of Fresh Cement and Concrete;* E.&F.N. Spon: London, 1991; pp 147–158.
14. *Well Cementing;* Nelson, E. B., Ed.; Elsevier Science B.V.: Amsterdam, The Netherlands, 1990.
15. Bannister, C. E. Presented at the 55th Annual Fall Technical Conference and Exhibition of the Society of Petroleum Engineers and the American Institute of Mining, Metallurgical and Petroleum Engineers, Dallas, TX, September 21–24, 1980; paper SPE 9284.
16. Denis, J. H.; Guillot, D. J. Presented at the Drilling Conference of the Society of Petroleum Engineers and of the International Association of Drilling Contractors, New Orleans, LA, March 15–18, 1987; paper SPE/IADC 16137.
17. Guillot, D. J.; Denis, J. H. Presented at the European Petroleum Conference of the Society of Petroleum Engineers, London, October 16–19, 1988; paper SPE 18377.
18. Shah, S. N.; Sutton, D. L. *SPE Prod. Eng.* **1990,** *5(4),* 415–424.
19. Spears, R. C. A.; Holme, K. R.; Tung, M. A. *Rheol. Acta* **1987,** *26,* 447–452.
20. Sabins, F. L.; Tinsley, J. M.; Sutton, D. L. Presented at the 55th Annual Fall Technical Conference and Exhibition of the Society of Petroleum Engineers and the American Institute of Mining, Metallurgical and Petroleum Engineers, Dallas, TX, September 21–24, 1980; paper SPE 9285.
21. Duzy, N. Q.; Boger, D. V. *J. Rheol.* **1983,** *27,* 321–349.
22. Keating, J.; Hannant, D. J. *Rheology of Fresh Cement and Concrete;* E.&F.N. Spon: London, 1991, pp 137–146.

23. Bourgoyne, A. T.; Chenevert, M. E.; Millheim, K. K.; Young, F. S. In *Applied Drilling Engineering;* Evers, J. F.; Pye, D. S., Eds.; SPE Textbook Series; Society of Petroleum Engineers: Richardson, TX, 1991; Vol. 2, p 41.

24. de Rozières, J.; Ferrière, R. F. Presented at the Drilling Conference of the International Association of Drilling Contractors and of the Society of Petroleum Engineers, Houston, TX, February 27–March 2, 1990; paper IADC/ SPE 19935.

25. Stiles, D. A.; Baret, J. F. Presented at the Rocky Mountain Regional/Low Permeability Reservoirs Symposium of the Society of Petroleum Engineers, Denver, CO, April 12–14, 1993; paper SPE 25866.

26. Greaves, C.; Hibbert, A. *Oil Gas J.* 1990, *Feb. 20,* 35–40.

27. Beris, A. N.; Tsamopoulos, J. A.; Armstrong, R. C.; Brown, R. A. *J. Fluid Mech.* 1985, *158,* 219–244.

28. Michaux, M.; Défossé, C. *Cem. Concr. Res.* 1986, *16,* 23–30.

29. Michaux, M.; Oberste-Padtberg, R.; Défossé, C. *Cem. Concr. Res.* 1986, *16,* 921–930.

30. Taylor, H. F. W. *Cement Chemistry;* Academic: London, 1990; pp 199–242.

RECEIVED for review July 28, 1994. ACCEPTED revised manuscript June 1, 1995.

Suspensions in Surface Operations

13

Suspensions in the Hot Water Flotation Process for Canadian Oil Sands

Robert C. Shaw[1], Laurier L. Schramm[*,2], and Jan Czarnecki[1]

[1]Edmonton Research Centre, Syncrude Canada Ltd., 9421 17th Avenue, Edmonton, Alberta T6N 1H4, Canada
[2]Petroleum Recovery Institute 100, 3512 33rd Street N.W., Calgary, Alberta T2L 2A6, Canada

Suspensions are created and must be processed during the application of the hot water flotation process to Canada's Athabasca oil sands, a large-scale commercial application of mined oil sands technology. These suspensions are more than just two-phase dispersions, being comprised of not only solids and water but also dispersed oil and gas. As such, they form interesting petroleum industry suspensions. A review of the hot water flotation process is presented with an emphasis on the occurrence, nature, and properties of suspensions.

Oil Sands

Oil sands are unconsolidated sandstone deposits of a very heavy hydrocarbon: bitumen. Bitumen is chemically similar to conventional oil but has comparatively high density (low gravity per American Petroleum Institute standards) and high viscosity. Based on United Nations Institute for Training and Research discussions aimed at establishing definitions for heavy crude oil and oil (tar) sands (*1–3*), bitumen can be placed in the context of other crude oils as shown in Table I.

Oil sand deposits are present in many locations around the world and appear to be similar in many respects (*4–6*), occurring along the rim of major sedimentary basins, mainly in either fluviatile or deltaic environments containing sands of high porosity and permeability. Reviews are available for most locations worldwide (*4, 6–13*). Although superficially similar, not all deposits are the same when subjected to a detailed examination. There are significant variations in composition

* Corresponding author.

Table I.   Hydrocarbon Definitions

Hydrocarbon	Viscosity Range (mPa·s at deposit temperature)	Density Range (kg/m³ at 15.6 °C)
Crude oils	<10,000	<934
Heavy crude oil	<10,000	934–1000
Extra heavy crude	<10,000	>1000
Bitumen/tar	>10,000	>1000

and structures, both of which influence amenability to hot water flotation.

Numerous estimates of the oil capacity of world oil sands have been given. One estimate of 2100 billion barrels of oil is almost as much as the world's total discovered medium and light gravity oils in place (5). Of the estimated total, about 91% is contained in the Canadian and Venezuelan deposits. Of the four Canadian deposits, the largest, Athabasca deposit, forms the world's largest self-contained accumulation of hydrocarbons: it is at least four times the size of the largest conventional oil field (Ghawar in Saudi Arabia) (5). Of the Athabasca's estimated 600 billion barrels of bitumen, perhaps 60 billion barrels could be recovered by surface mining oil sand followed by a beneficiation process.

Development of the Hot Water Separation–Flotation Process

Accounts of early exploration and examination of the Athabasca deposit can be found elsewhere (6, 14–16); only an overview is given here. Preliminary geological descriptions of the deposit were made in the early 1800s (15) followed by attempts to separate the bitumen from the sands. One method was to collect the material oozing from exposed faces along river beds and boil it to help settle out sand and remove lighter components to yield a tar. In Europe at around the same time, tar was being obtained from oil sands by treating crushed oil sand with boiling water and skimming off the oil that floated to the surface (17). The first hot water extraction plants were built in Alberta in the 1920s. Clark's hot water process, patented in 1928 (18), involved mixing together oil sand, electrolyte (e.g., sodium silicate), and hot water and keeping the thick slurry at 75–100 °C. The slurry was then diluted with more hot water in a separating bath so the solids could sink and be removed. Water-containing bitumen rose to the surface and was skimmed off. Further electrolyte (e.g., calcium chloride) was added to clarify the water by aiding the removal of silicate, silt, and clay. In Clark's process, the main aim was to remove solids. At this stage of development, little or no specific attention to flotation aspects was given.

In the 1930s and 1940s, many "commercial" or large pilot operations were underway, which resulted in feasible large-scale commercial
development of oil sands (*19*). In the 1960s, Great Canadian Oil Sands
(later, Suncor Inc.) built a commercial plant to use the hot water flotation
process to produce bitumen at a level of 45,000 barrels per day (increased to 50,000–60,000 barrels per day in the 1970s). In the 1970s,
a consortium of companies, known as Syncrude Canada Ltd., completed
construction of another commercial oil sands plant using the hot water
flotation process to produce bitumen at a level of 125,000 barrels per
day. In these large operations, the oil sands are first mined, and the
bitumen is extracted by a hot water flotation process. The extracted
bitumen is subsequently upgraded by refinery processes to produce
synthetic crude oil.

Nature and Variability of Oil Sands

In oil sand processing, the general principles of mineral flotation apply,
yet oil sand composition and structure, and their variations, have a great
impact on the way the flotation must be operated. General descriptions
of the geology of the Canadian deposits can be found in several books
(*15, 20–23*). Considerable effort has gone into describing the geological
aspects of oil sand deposits, including subdivisions into depositional environments. This approach is based on the principle that, for example,
rivers deposit different sands in a different geometry from lakes or
oceans. Within the environments are further subdivisions called facies,
the scale of subdivision being governed by the need to categorize the
major features of a deposit (*24*). Some summary treatments have been
given by Carrigy, Mossop, Stewart, and MacCallum and others (*21, 22,
25–30*).

According to Carrigy and Kramers (*21*), all of the oil sand deposits
in Alberta occur in sandstones of early Cretaceous (ca. 110 million years)
age. Sediments were brought in to the deposit area from both uplift in
the Rocky Mountain area (providing sediments from the west that had
previously been under the sea) and from the Canadian Shield that was
being eroded (providing sediments from the east). In this era, most of
the sediments were deposited in a fluvial environment, although some
occurred in a marine environment in northern Alberta. In later eras,
the Boreal Sea had moved southward, causing the deposition of glau-
conitic sandstones. Increased tectonism in western and southwestern
Alberta then caused an influx of sediments (mainly of volcanic origin)
into central Alberta, forcing a retreat of the Boreal Sea to the north and
northwest. This caused a return to the continental-type deposition in
some areas, but in other areas marine conditions prevailed for a longer
time. Eventually, the Boreal Sea from the north and the Gulfian Sea

from the southwest merged to form a continuous sea over most of Alberta. In this environment, the marine upper Cretaceous sediments were deposited. The result of these kinds of processes is that the oil sands occur as a mixture of kinds of sediments, overlain by varying thickness of non-oil-bearing formations and that the Alberta deposits each have their own depositional characteristics.

The Athabasca deposit is the largest of the Canadian oil sand deposits (21, 31) and is the only one shallow enough to be amenable to surface mining techniques. Here, the bitumen is contained mostly in the McMurray formation, which lies over limestones and under marine shales. The McMurray formation is a drainage basin that filled in with sediments, and at different times the sea alternately flooded (estuarine deposits) and then receded (fluvial deposits) so that a number of distinct depositions can be discerned (22, 29, 31, 32). The bulk of the sediments appear to be the result of estuarine phases, with increasing marine invasion at later dates. Such sequences, each layering and disrupting with fluvial or marine movements, led finally to a system of sediments with great diversity. Accordingly, the oil-bearing sands in this deposit have great variability in their compositions and properties.

Origin of the Oil.   Some variability occurs in oil content among the sedimentary facies, which raises the question of the origin of the oil. Although controversial (6, 22, 33–37), most workers agree that the oil, in some mobile form, was able to migrate into the sediments and then alter in place because of some combination of the following (6, 22, 23, 35):

- lighter components may have evaporated or migrated away from the bulk oil

- they also may have dissolved somewhat into the pore water

- in shallow depths oxidation may have occurred (increasing the average molecular mass and viscosity of the oil)

- anaerobic bacteria may have degraded the lighter hydrocarbons

Any combination of these factors would have resulted in a residuum of the heavier components, hence bitumen. There is a connection between the origin of the oil and the structure of oil sand. If it is believed that the sand grains in the oil sand are covered by a film of water, then one might suspect that oil migration and generation took place when the sands were wet (e.g., not trapped in dry sand beaches or dunes). In such a case, one could anticipate that simply filling in the voids between solid particles should result in a simple relationship between oil content (grade) and particle size distribution.

Composition and Properties of Oil Sands. The oil sands resemble conventional oil deposits in many respects, but there are important differences (6, 36, 38, 39). Athabasca oil sand consists mainly of quartz sand. The mineral fraction is predominantly quartz grains, with smaller amounts of feldspar grains, mica flakes, and clays (32, 36, 38, 40). The clays in this deposit are predominantly kaolinite and illite with some chlorite. Some tables of mineral and bitumen compositions of Athabasca oil sands are given by Camp (39). In general, the oil-bearing sands are fine to very-fine grained (62.5–250 μm). Figure 1 shows the three major categories.

The oil-bearing sands have fairly high porosities (25–35%) compared with the 5–20% porosity of most petroleum reservoirs (32). The high porosity is achieved by having a low occurrence of mineral cements. According to Mossop (36), "It is because the sediment is not consolidated, with no cementing agent to indurate the material and give it strength, that the deposits are called oil sands, not sandstones." The permeability of the non-oil-bearing sands is high, whereas in oil-bearing sands the bitumen (being immobile in the deposit conditions) impedes the flow of fluids through the matrix. The ranges of porosity and permeabilities encountered in the Athabasca mining area are 30–35% and up to 5000 md (millidarcy), respectively (30). The permeability of undisturbed oil-bearing sand can be as low as about 50 md or less, whereas that for oil sand that has been mined and dumped is on the order of

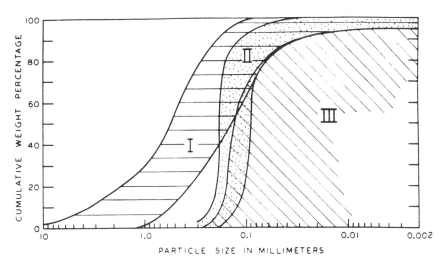

Figure 1. Distribution of grain sizes in Athabasca oil sand according to Carrigy in terms of (I) coarse-grained sands, (II) fine-grained sands, and (III) very-fine-grained sands and silts. (Reproduced with permission from reference 38. Copyright 1967 Elsevier.)

several darcys (41). The extent to which the voids in the grain structure are filled with fluids is represented by saturation. The highest oil content Athabasca oil sands have oil saturations of about 18% w/w (36% v/v) and water saturations of about 2% w/w (4% v/v) (36).

The depositional environments, porosities, permeablilties, and bitumen saturations are related. Where sediment transport and deposition were originally slow, relatively large amounts of silt and clay deposited. When oil migrated into such areas of low porosity and permeability, little was retained. Regions in which there were originally strong currents deposited larger grains and little fine-grained material. When oil migrated into these environments of high porosity and permeability, relatively large amounts of bitumen were trapped. Accordingly (26, 36), the best ore bodies are those located along deep river or estuarine channels. The strong influence of clay content on oil saturation has been emphasized by Carrigy (42), who surmised that the ability of clays to absorb large amounts of water reduced permeability to oil, causing lower oil contents. Although various conventions exist for describing saturations (e.g., reference 31), for oil sands amenable to mining and hot water flotation an appropriate set of definitions are as follows: rich oil sand containing 12% and higher bitumen (usually 12–14%), average grade 10–11%, and lean grade 6–9%. Lower than 6% is usually not considered to be of "ore-grade" quality.

Bitumen Viscosity and Density. The viscosity of Athabasca bitumen in-place is so high, about 1,000,000 mPa·s at reservoir temperature, that the bitumen in oil sand is practically immobile. This makes the bitumen difficult to displace in attempts at in situ recovery (36, 43), but it gives oil sand enough material strength to be mineable. Once mined, and when beneficiating oil sand, bitumen viscosity has some important influences. Several studies of the rheological properties of Athabasca bitumen have been reported (44–46). For practical purposes, Athabasca bitumen can be considered (44, 45) to be Newtonian in character (except at quite low temperatures and very low shear rates) in contrast to that in other world locations. For example, bitumen in the Utah deposits of the United States is markedly non-Newtonian (47), more viscous, and therefore requires the addition of diluent during the conditioning step (48–50). The temperature dependence of Athabasca bitumen viscosity is shown in Figure 2.

Density is of fundamental importance to any flotation process. Ward and Clark (46) and Camp (39) reported densities and temperature dependence for Athabasca bitumens, an example of which is shown in Figure 3. From this figure, the variation is comparatively slight. Many authors point out that the normal hot water process temperature of 80 °C (176 °F) is approximately the temperature in which the bitumen

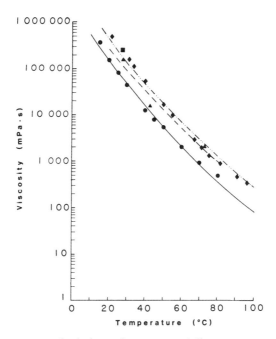

Figure 2. Viscosity of Athabasca bitumens at different temperatures, based on three different reported studies: Schramm and Kwak (45) (●), Dealy (44) (▲ and ■), and Jacobs (92) (♦). (Reproduced with permission from reference 45. Copyright 1988 Canadian Institute of Mining, Metallurgy, and Petroleum.)

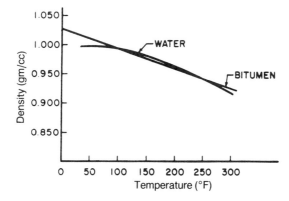

Figure 3. Densities of water and an Athabasca bitumen at different temperatures according to Camp. (Reproduced from reference 39. Copyright 1976 Cameron Engineers, Inc.)

density is smallest relative to water and imply that it is by this density difference that the bitumen flotation takes place, but in fact the situation is more complex in a practical flotation circuit, as is discussed in a later section.

Arrangement of the Phases.   The arrangement of phases in oil sands is largely determined by their volume fractions: quartz sand forms the bulk of the material with either the bitumen (in rich oil sands) or the water (in lean oil sands) forming the continuous phase. Of course, bicontinuous structures are possible as well. It may not seem obvious that the origin of the oil would strongly influence the arrangement of the phases, but wetting behavior influences both oil saturation and the location of fine solids. It is commonly stated that the Athabasca oil sands are water-wet, as illustrated in the oil sand structures of Figure 4 (39)

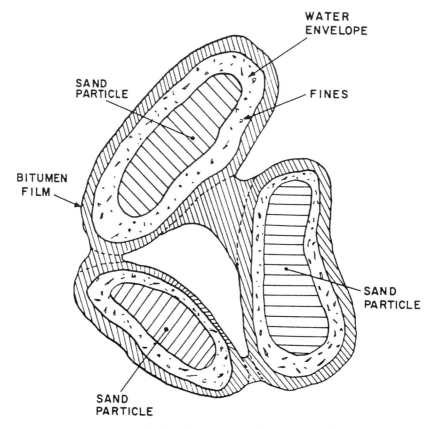

Figure 4.   An early model for the structure of Athabasca oil sand according to Camp. (Reproduced with permission from reference 93. Copyright 1963 Alberta Research Council.)

and Figure 5 (*51*). Examination of the references cited in support of this conclusion turns up a surprising lack of evidence. Originally, Clark (*52*) advanced the postulate that "The water is probably present as a film around the sand particles. The oil surrounds the moist sand grains as an envelope." Clark reasoned that water is known to wet quartz and silicate minerals in preference to oils (*see* reference 53), and his microscopic observations (*52*) showed that when oil sand samples were freshly cut clean (wet), quartz particles were exposed as the oil envelopes were ruptured and sheared away. Ball (*54, 55*) observed that dry heating did not cleanly separate oil, whereas boiling in water did, and inferred that in dry heating the water film evaporates, leaving oil to attach to the solids. A more recent NMR study (*56*) found evidence for bulk water between sand grains but also water bound directly to clay mineral surfaces. All of these results are consistent with those from a theoretical proposal by Takamura (*51*), that connate water exists as pendular rings around sand grain contact points and as roughly 10-nm-thick films on sand grain surfaces. A summary illustration is shown in Figure 5. In the same work (*51*), microscopic observation of displacement of water-wet sand grains by an immersion oil did indeed show evidence for pendular rings and thin films. Yet, just as for Clark's early observations, these experiments provide better support for an explanation of the hot water process displacement of oil than they do for a model of oil sand structure. Also, a contrary point of view has been advanced by Zajic et al. (*57*) based on freeze-fracture transmission electron microscopy, which showed that the water phase was present as droplets emulsified in the bitumen, in the space between the solids, and apparently not on the

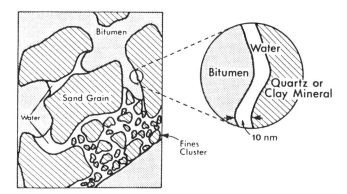

Figure 5. A refined model for the structure of Athabasca oil sand according to Takamura. (Reproduced with permission from reference 51. Copyright 1982 Canadian Society for Chemical Engineering.)

sand or clay surfaces (the micrograph resolution was claimed to be ±0.01 μm).

If the oil migrated into a water or mineral environment, then a correlation should be found between the non-solid void space and bitumen saturation, modified to account for the films of water thought to be held by sand grains and clay particles. Carrigy (42) related grain size distribution for a number of Athabasca oil sands to the variation in oil content as shown in Figure 6; oil sands containing progressively more clay-sized (<2 μm) materials have lower oil contents.

In summary, there is a general consensus that, for the most part, the mineral grains in Athabasca oil sand are water-wet and that most of the bitumen is not in direct contact with the mineral phase but instead separated by at least a thin film of water. There remains some reason to believe that a fraction of the solids are, however, oil-wetted. The separation of most of the oil from solids by a water film is widely held to be the characteristic difference between Athabasca oil sand and oil sand from other oil sand deposits in the world (e.g., California, New Mexico, or Utah) that are thought to consist of oil-wet solids. These "oil-wet" oil sands are considered to be more difficult to beneficiate using hot

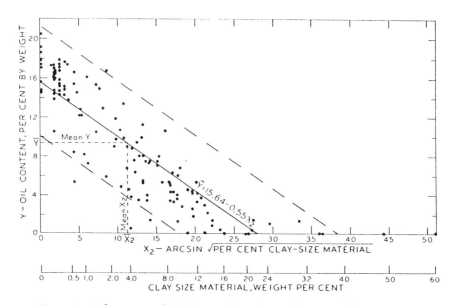

Figure 6.   Oil contents of Athabasca oil sands as a function of the percentage of clay-sized particles (<2 μm), according to Carrigy. (Reproduced with permission from reference 42. Copyright 1962 Society for Sedimentary Geology.)

water flotation because of the difficulty in dislodging bitumen from an oil-wet surface (*50, 58*).

Commercial Hot Water Flotation Process

We outline the basic steps in the commercial process (Figure 7) and their place in an integrated oil sands-synthetic crude oil production process. Although based on Syncrude Canada's operation, much of the process is similar to that of Suncor (e.g., reference 59). Additional details are available in the technical and patent literature (*6, 13, 39, 59–65*).

Before oil sand is mined, some 30 m of overburden material must be removed (an illustration of the vertical section is shown in Figure 8). The mining of the oil sand (ore) body, which is about 60 m thick, is accomplished by large draglines that dig the oil sand from an open pit. Typically, the mining operation must remove 0.5 tonne of overburden and mine 2 tonne of oil sand of about 10% bitumen content to yield 1 barrel of oil after extraction. Obviously, as the grade of oil sand decreases, additional tonnes must be mined and processed to yield the same amount of oil. Therefore, a commercial operation has an economic grade limit dictated by the trade-off between the mining and processing costs versus the value of the oil.

The mined oil sand is free-casted onto windrows from which bucketwheel reclaimers load the oil sand onto conveyors that carry it to surge pile and dump pockets. (In the Suncor operation, the oil sand is mined directly by bucketwheel excavators and conveyed to a surgebin.) From the surge pile and dump pockets, a complex arrangement of feeding devices and conveyors are used to deliver oil sand to tumblers at a uniform feed rate.

Figure 8 shows an example of a section of core hole from the Mildred Lake area of the Athabasca deposit. In mining, a dragline digs from the bottom up, collecting oil sand and clay (lens) bands to various degrees. A certain degree of mixing occurs during the casting onto windrows, reclaiming, transferring via conveyors, and dumping into surge piles. Despite this mixing, the delivered oil sand is neither homogeneous nor uniform. Because oil sands with different nature and composition are associated with different conditions for optimal separation and flotation, bitumen process control strategies are very important.

The oil sand is next conditioned by slurrying with water in rotating horizontal drums, called tumblers. Heat and shear are used to overcome the forces holding oil sand lumps together. In this ablation process, successive layers of each lump are warmed and sheared off until everything is dispersed. Besides stirring to maintain a state of suspension, a

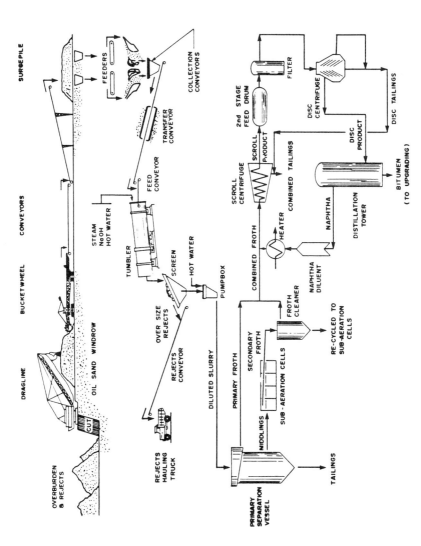

Figure 7. Diagram of a commercial oil sands mining and hot water flotation process. (Reproduced with permission from reference 45. Copyright 1988 Canadian Institute of Mining, Metallurgy, and Petroleum.)

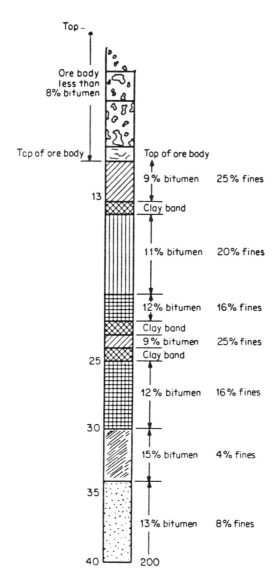

Figure 8. Cross-section of a core hole from the Mildred Lake area of the Athabasca oil sands deposit. (Reproduced with permission from reference 59. Copyright 1984 McGraw-Hill.)

number of other things must happen in the conditioning step. The bitumen has to be separated from the solids (which make up about 70% w/w of the slurry) and prepared for separation from the aqueous phase. Steam is added to raise the (exit) temperature to 80 °C. Air is not sparged in directly but becomes worked in to aerate the bitumen by inclusion of about 30% v/v gas (39). Sodium hydroxide is added to raise the solution pH. The amounts of the reagents added are typically in the proportion oil sand:water:NaOH(20%) = 1:0.2:0.0012 by mass. An appreciable time is required to achieve good distribution of the bitumen, minerals, and reagents and to allow chemical and surface reactions to occur. Within 5 min or so, a quasi-steady state is reached, probably not full thermodynamic equilibrium.

The slurry is discharged from the tumblers onto vibrating screens and washed with hot spray water to remove oversized solids and undigested oil sand lumps. This process may also provide additional air entrainment and hence further aeration of bitumen. Additional hot water is added to the slurry, which is then pumped to primary separation vessels. The rejected solids (about 5% of the original oil sand) are conveyed out of the plant for disposal.

The diluted (flooded) slurry contains about 7% aerated bitumen droplets, 43% water, and 50% suspended solids. The aerated bitumen droplets have the lowest density and rise (float) to the surface of the primary separation vessel, a large vessel with a cylindrical upper section and a conical lower section. The vessel is maintained in a quiescent condition to facilitate this flotation, as well as the settling of coarse solids to the bottom. The slurry is retained here for about 45 min. Because the process is continuous, the presence of fine minerals (e.g., clays) makes this vessel susceptible to solids buildup that can increase the viscosity (39, 66, 67). To maximize the flotation and sedimentation processes, the middlings region viscosity and density are kept low by adjusting the flood water addition and middlings removal rates. Mechanical rakes at the bottom of the vessel keep the coarse rapidly settling solids moving toward the bottom from which they are withdrawn as a concentrated suspension (primary tailings).

The smaller suspended solids that do not settle rapidly and the smaller and poorly aerated bitumen droplets that do not float rapidly are all drawn off in a slurry from the middle of the vessel (middlings). The bitumen droplets in middlings have either too little air content or too small diameters for rapid enough flotation. The middlings stream and primary tailings stream contain enough bitumen that they are combined and pumped to a special tailings oil recovery (TOR) flotation circuit (68). The TOR circuit is not shown in Figure 7. The middlings from this TOR vessel are then pumped to a scavenging (secondary) flotation circuit for additional bitumen flotation. Here, conventional flotation cells, using vigorous agitation and air sparging, are used to cause further bitumen aeration and flotation. This results in a secondary froth that is much more contaminated with water and solids than is the primary froth and typically contains 15% bitumen, 20% solids, and 65% water. Meanwhile, the TOR froth is recycled into the flooded slurry that is fed into the primary separation vessels. The TOR tailings are combined with the tailings from the scavenging circuits. Other variations of this process are also practiced.

Most of the bitumen is recovered as a primary froth that is skimmed from the top of the vessel into a launder extending around the upper lip of the separation vessel. Primary froth typically consists of 60% bitumen, 10% solids, and 30% water. It also contains air and so is compressible and more viscous than bitumen (45). To make it easier to pump, it is deaerated in towers by causing it to cascade through shed decks, flowing against the upward flow of steam. The froth from the secondary flotation is "cleaned" in stirred thickeners to remove some of the water and solids and then deaerated. Because of its higher quality, primary froth does not need the difficult cleaning and is more highly valued than secondary froth. Thus, much emphasis is directed at optimizing the primary froth yield in the process (e.g., reference 63). The primary and secondary froths, once deaerated, are combined into a single feed for a froth treatment process. This deaerated froth contains about 60% oil, 30% water, and 10% solids, which is essentially the same as primary froth because secondary froth accounts for only about 5% of the total bitumen production. A review of froth structure and properties is given in reference 69.

A froth treatment process is used to remove water and fine solids from the combined froth (6, 13, 39, 59, 61). The froth is first diluted with heated naphtha in about 1:1 volume ratio and then centrifuged in scroll centrifuges (at about 350 × g) to remove coarse solids (greater than 44 μm). The diluted froth is next filtered and pumped to disc-nozzle centrifuges where higher g forces (about 2500 × g) are used to remove essentially all of the remaining solids and most of the water. After stripping off the naphtha, the bitumen is upgraded into synthetic crude oil.

The tailings from the primary and secondary flotation processes are combined and transported to a tailings pond, as are the aqueous tailings from the froth treatment process. Some of the supernatant water from the pond can be recycled into the process. The remainder of the tailings undergo a slow consolidation into sludge. Numerous attempts to accelerate the process have been tried and (usually) are found to be either ineffective or too expensive (59). A review of the tailings and sludge problems is given in Chapter 14. The magnitude of the tailings problem can be appreciated if it is remembered that the predicted ultimate area of the tailings pond from a Syncrude-sized and type operation is over 20 km^2.

Elements of the Physical-Chemical Basis of the Process

The physical-chemical basis of the process is not yet fully understood. The commercial extraction and flotation plants are controlled using empirical relations involving oil sand grade and fine solids content information (60, 62). Sanford (60) found several important correlations, first between the <44 μm fine solids size fraction and the <5 μm fine solids size fraction as shown in Figure 9. This correlates with the bitumen content in oil sand as shown in Figure 6 and also correlates with the amount of process aid (sodium hydroxide) addition required for optimal hot water flotation process efficiency as shown in Figure 10. Taking Figures 6, 9, and 10 together leads to the main method of commercial process control: the bitumen content of oil sand feed entering the plant is determined on-line by infrared reflectance and used to estimate the level of fine solids in the feed and hence to the level of process aid addition required. Despite optimizing for each quality of feed, Figure 11 shows how oil recoveries become progressively poorer with decreasing grade of oil sand (70). For grades of below 10%, bitumen content recoveries of less than 90% are obtained, whereas at the same time lower energy efficiencies in the process are obtained. Improved empirical correlations are continually being discovered for these and other anom-

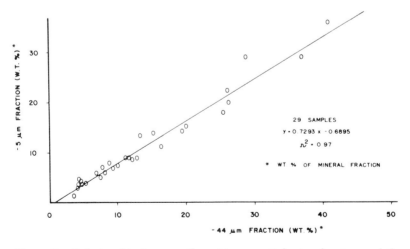

Figure 9.   Relationship between the <44 μm particle size fraction and the <5 μm particle size fraction in Athabasca oil sands determined by Sanford. (Reproduced with permission from reference 60. Copyright 1983 Canadian Society for Chemical Engineering.)

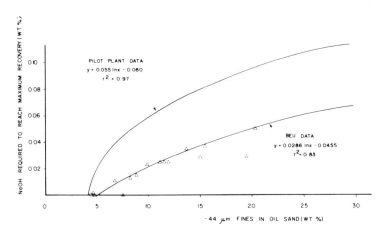

Figure 10. Relationship between the <44 μm particle size fraction in Athabasca oil sands and the amount of NaOH addition required to optimize the hot water flotation process, as determined by Sanford. (Reproduced with permission from reference 60. Copyright 1983 Canadian Society for Chemical Engineering.)

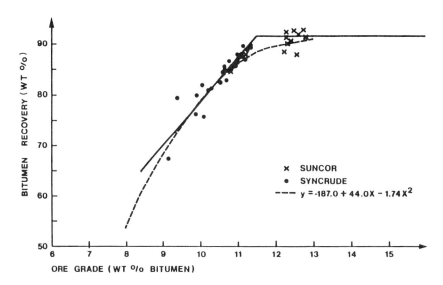

Figure 11. Bitumen recovery versus oil sand grade based on mean commercial operating plant data to 1980, according to Houlihan. (Reproduced with permission from reference 70. Copyright 1982 United Nations Institute for Training and Research.)

alous oil sands (71). Improved mechanistic information could be used to develop improved process aids, process controls, and even alternate processes.

Many subprocesses are required to carry out an efficient separation and flotation. Figure 12 shows some of the elementary processes. Although the real phenomena may not be entirely subdividable in this way or take place in exactly the order assigned, it will be apparent that the tumbler and primary flotation vessel combine quite a few simultaneous or nearly simultaneous elementary process steps, and this makes the interaction of process variables difficult to predict. In consequence, much hot water flotation process optimization research involves test processing in a laboratory- or pilot-scale apparatus. The small-scale observations are used to describe the flotation behavior of the oil sand and infer what will happen at the full-scale plant level. Many different laboratory- and pilot-scale investigations into oil sand processing have been conducted over the past 60 years or so. A practical standard hot water flotation process batch extraction unit and test procedure has evolved in which small (0.5 kg) samples of homogenized oil sand are processed. A detailed description is given elsewhere (72), part of which follows:

> The extraction cell is jacketed to allow for constant temperature during operation, the height is sufficient to provide a quiescent zone, and the cell is square to facilitate slurrying and subsequent agitation without the need for baffles. Air is added through the impellor shaft, which allows good control even at low addition rates and avoids problems of plugging by fine solids .... The air introduction feature permits aeration of the slurry during mixing and allows one to simulate the secondary recovery step ... subsequent to primary recovery. After extraction is complete, tailings can easily be removed through a valve on the bottom of the cell.... The impellor speed is controlled by a variable-speed fractional horsepower motor.

Figures 13 and 14 show the unit and illustrate some of the steps and variables involved. This test is reproducible and sensitive enough to be useful for evaluating new process aids (chemicals), process variables, and determining the processibility of different oil sand samples.

An example of a continuous pilot-scale experimental extraction circuit has been established by Syncrude and has, to some degree, been described in the literature (61, 69, 72, 73). In this particular unit, larger amounts, 2000–3000 kg/h, of oil sand are processed continuously. An illustration of the circuit is given in Figure 15. It is a scaled-down version of the continuous commercial process, although the addition of sophisticated measuring sensors and computer control allow more careful

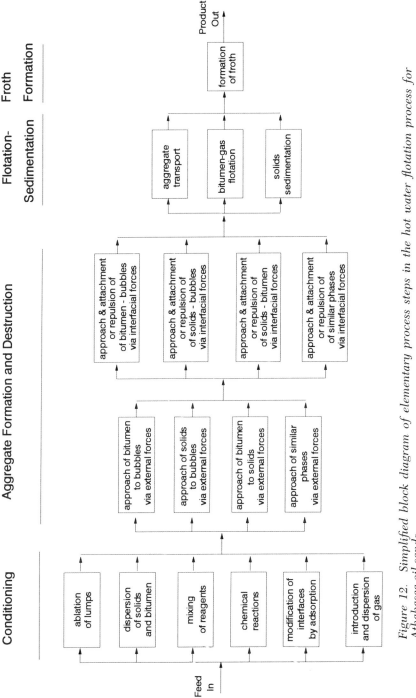

Figure 12. Simplified block diagram of elementary process steps in the hot water flotation process for Athabasca oil sands.

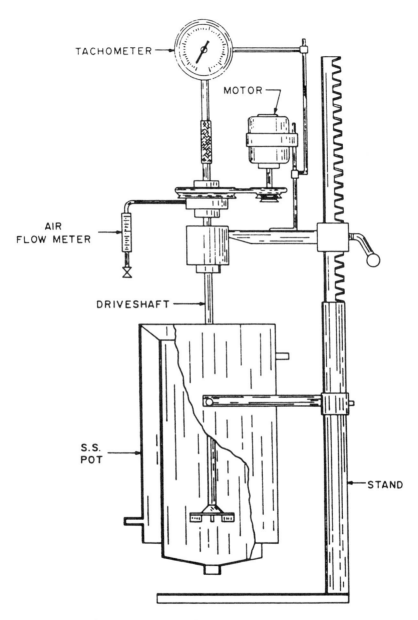

Figure 13.   Schematic diagram of the laboratory standard batch extraction unit apparatus, after Sanford and Seyer. (Reproduced with permission from reference 72. Copyright 1979 Canadian Institute of Mining, Metallurgy, and Petroleum.)

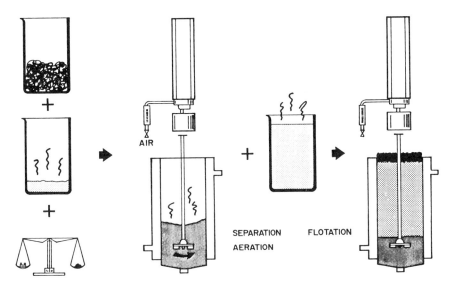

Figure 14.   Illustration of the steps involved in conducting a batch hot water flotation process test, after Schramm and Smith. (Reproduced with permission from reference 82. Copyright 1989 Alberta Oil Sands Technology and Research Authority.)

monitoring and mass-balancing than is possible in the full-scale commercial process. The smaller circuit is thus better suited to research studies. Sanford (60) showed that results from the batch and pilot processes described above can be correlated. As shown in Figure 10, the batch test results establish trends that translate directly to the pilot scale. Absolute process recoveries are translated only with difficulty due to unavoidable differences incorporated into the processing in batch mode at such a small scale. The larger pilot process, being continuous and closely modeled after the commercial process, yields results that compare quite well with that process.

Concept of Processibility.   Although there are many variables, process efficiency is more sensitive to some variables than to others (62, 74–76). Early studies led to the identification of base (NaOH) addition level as the preferred process variable, and since then much work has been aimed at determining how much base is needed. It was at first thought that the process must be operated at generally "alkaline pH" (18, 52, 77). Further research involved study of an increasing number of oil sands, which led to the discovery that the process could be controlled to achieve good bitumen separation and flotation efficiency by maintaining a constant pH. This was specified at different values; for example, Bowman (78) recommended the middlings layer pH be kept

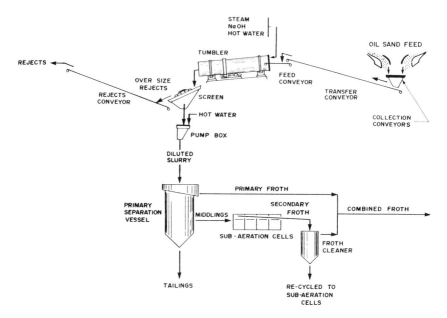

Figure 15. Diagram of the Syncrude 2270 kg/h continuous experimental extraction circuit (pilot plant). (Reproduced with permission from reference 66. Copyright 1989 Canadian Institute of Mining, Metallurgy, and Petroleum.)

in the region 7–8.5, whereas Innes and Fear (62) and Floyd et al. (79) recommended the pH range 8.0–8.5. It was eventually discovered by Sanford (80) that pH was not the important parameter but rather NaOH addition level and that it should be regulated in response to fines level in the feed (Figure 10). Figure 16 shows processibility curves for four oil sands of differing composition. The term processibility refers to the primary bitumen (oil) recovery versus process aid (NaOH addition) relationship for a given oil sand and means, in essence, the NaOH addition level required to achieve maximum primary oil recovery. This forms a partial means for categorizing oil sands.

Conditioning.   The hot water flotation process for oil sands is a separation process in which the objective is to separate bitumen from mineral particles by exploiting differences in their surface properties. The slurry conditioning process involves many process elements as illustrated in Figure 12. Given that ablation and mixing, mass and heat transfer, and chemical reactions are accommodated, the conditioning step involves separating bitumen from the sand and mineral particles.

Disengagement of bitumen from solids will be favored if their respective surfaces can be made more hydrophilic because a lowering of

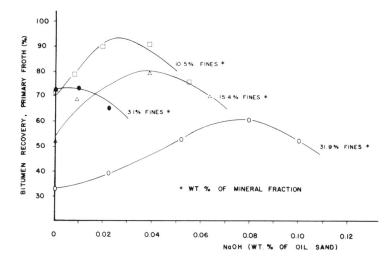

Figure 16.   Processibility curves (laboratory batch extraction) for four oil sands of different composition. (Reproduced with permission from reference 60. Copyright 1983 Canadian Society for Chemical Engineering.)

free energy will accompany the separation. The practical separation is enhanced by a combination of the effects of mechanical shear and disjoining pressure. Adopting the water-wet model for Athabasca oil sand, we know that a thin aqueous film already separates the bitumen from the sand (Figure 5), so this preexisting separation has to be enhanced.

The need to attain alkaline conditions in the oil sand slurry has already been emphasized. The main role of the base is to produce (saponify) natural surfactants from the bitumen (60, 72), and the action of these anionic surfactants has been studied in some detail (81–85). Under suitable alkaline conditions, both ionization of functional groups at the bitumen surface (51, 86) and adsorption of the anionic natural surfactant molecules at the bitumen–aqueous interface (81, 82, 85) occur. Figure 17 shows how addition of NaOH in the process increases the electrophoretic mobilities of dispersed bitumen droplets and fine solid particles. The same trends have since been independently confirmed by Hupka and Miller (87) and Drelich et al. (88). The ionization of surface groups and adsorption of charged surfactants cause increased electrostatic repulsion that increases the disjoining pressure in the aqueous film separating the bitumen and solids. Figure 18 illustrates the effects of increasing (NaOH addition) versus decreasing (CaCl$_2$ addition) this disjoining pressure in the film. Increased disjoining pressure together with the applied mechanical and thermal energy causes the separation of bitumen from solids. At this stage, the bitumen has been separated, but also the fine solids have been dispersed.

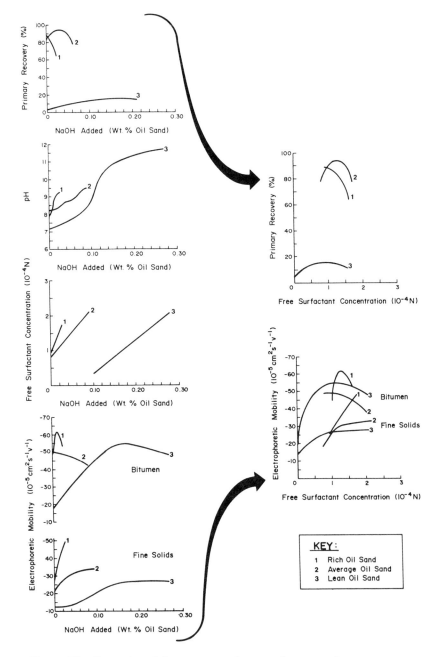

Figure 17.   Examples of the connection between hot water flotation process efficiency and measured chemical and physical properties. (Reproduced with permission from reference 85. Copyright 1987 Canadian Society for Chemical Engineering.)

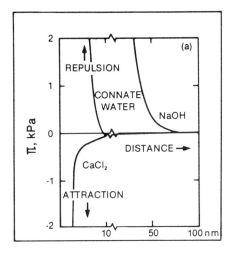
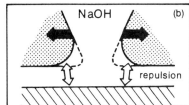
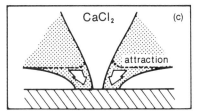

Figure 18. Illustration of the effects of increasing (NaOH addition) or decreasing (CaCl₂ addition) the thin aqueous film disjoining pressure (π) between bitumen and sand. (Reproduced with permission from reference 94. Copyright 1983 Canadian Institute of Mining, Metallurgy, and Petroleum.)

Additionally, bitumen–air attachment has to occur. How this happens is not entirely clear. The process conditions that most favor bitumen–solids separation, that is, a high degree of electrostatic repulsion due to charged surfactant molecules at the interfaces, also will tend to act in opposition to gas–bitumen attachment because the gas bubbles also acquire a surface charge of the same sign (Schramm, L. L.; Smith, R. G., Syncrude Canada Ltd., Edmonton, Alberta, Canada, unpublished results). However, in mineral flotation in which there is gas–solid attachment without filming, such electrostatic repulsion is not as important a factor as are inertia effects when the particles and bubbles are larger than, say, 10–40 μm in diameter. It is possible for bitumen droplets to attach to gas bubbles and form bubble droplet pairs or aggregates, as in mineral flotation, and Houlihan (Houlihan, R. N., Syncrude Canada Ltd., Edmonton, Alberta, Canada, unpublished results) found that for low alkali addition levels or reduced temperature conditions, bitumen droplets will attach to air bubbles as discrete particles. However, it appears that for the most part a balance of interfacial tensions in the system favors filming of the bitumen around the gas bubbles. Both laboratory studies (Schramm, L. L.; Smith, R. G., Syncrude Canada Ltd., Edmonton, Alberta, Canada, unpublished results) and Houlihan's continuous extraction circuit studies (Houlihan, R. N., Syncrude Canada Ltd., Edmonton, Alberta, Canada, unpublished results) indicate that under normal processing conditions, the bitumen will preferentially encapsulate air bubbles. Figures 19b and 20 show the spontaneous filming of bitumen

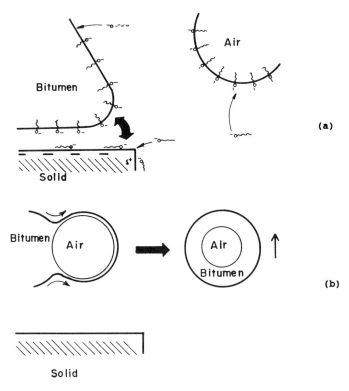

Figure 19.   Illustration of two of the steps in the hot water flotation process: (a) the separation of bitumen from solids showing the adsorption of naturally produced surfactants and (b) the attachment and filming of bitumen around gas bubbles. (Reproduced from reference 69. Copyright 1994 American Chemical Society.)

around a gas bubble brought into contact with the solution–bitumen interface. Similar observations have been made independently by Miller and colleagues (58, 88, 89). These aerated bitumen globules are what float upward in the flotation vessels to form froth. Gas-bubble–mineral-particle attachment also occurs for the fraction of mineral particles that are not hydrophilic, so that some bubble–particle aggregates also form.

Flotation.   After screening, washing, and dilution (flooding), the suspension pumped to primary flotation vessels comprises bitumen droplets, aerated bitumen globules, dispersed coarse and fine solids, and probably bubble–particle aggregates. This suspension is fed into a point somewhat higher than the center of the vessel as shown, for a continuous extraction circuit vessel, in Figure 21. The aim of the process here is to selectively separate the dispersed aerated bitumen droplets

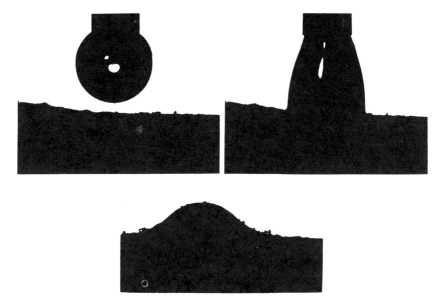

Figure 20.   Photographic sequence in which, viewing the images clockwise, an air bubble on the tip of a capillary, is pushed down through an alkaline solution at 80 °C until it just touches a layer of bitumen coated onto a silica surface. The bitumen spontaneously spreads over the surface of the bubble, causing it to detach from the capillary and become engulfed. Note the obvious presence of solid particles in the bitumen on the surface in the lower photo. (Photomicrographs by L. L. Schramm, reproduced from reference 69. Copyright 1994 American Chemical Society.)

or globules from the aqueous suspension of solid particles (which will form the tailings). Considering the structure of the bitumen globules, high shear rates are avoided to prevent destruction of the bitumen envelope and premature release of the gas. In continuous operation, the flotation vessel is considered to contain three layers: the froth layer at the top, a coarse solids layer (from which the primary tailings are withdrawn) at the bottom, and a middlings layer in the center. The middlings layer is the largest.

The bitumen droplet or globule diameters and size distribution are related to slurry conditioning variables such as the nature of the oil sand and the chemicals used to process the ore. Houlihan (Houlihan, R. N., Syncrude Canada Ltd., Edmonton, Alberta, Canada, unpublished results) made extensive direct photomicrographic studies of rising bitumen droplet diameters and rise velocities in the primary separation vessel of a continuous pilot-scale extraction circuit. He observed that droplet size distributions for a given feed are relatively narrow and vary somewhat for different grades of oil sand being processed. The rising bitumen

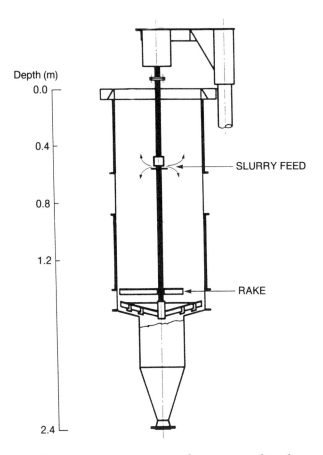

Figure 21. Illustration of the primary flotation vessel in the continuous experimental extraction circuit (pilot plant) of Figure 15. (Adapted with permission from reference 66. Copyright 1989 Canadian Institute of Mining, Metallurgy, and Petroleum.)

globules were for the most part spherical, consisting of a thin film of oil enveloping an air bubble, and under good processing conditions had diameters of about 280 ± 140 µm (total range from 125 to 400). The thickness of the bitumen film has been estimated to be about 30 µm, so that a mean overall droplet size of 280 µm would imply gas bubble sizes (in the interior of the globules) of about 216 ± 140 µm. Bitumen droplet or globule size was found to be invariant with depth in the separation vessel, implying that bitumen droplets do not coalesce as they rise through the middlings phase (Houlihan, R.N., Syncrude Canada Ltd., Edmonton, Alberta, Canada, unpublished results).

The bitumen globules must rise through the suspension of dispersed fine particles and settling coarse particles. As a starting point for predicting the velocity of rising bitumen droplets, one can apply the terminal Stokes velocity for spherical droplets rising in laminar flow, which is given by v_t as

$$v_t = -[(\rho_s - \rho)gD_p{}^2]/(18\eta) \tag{1}$$

where ρ_s and ρ are the droplet and medium densities, respectively, g is the gravitational constant, D_p is the droplet diameter, η is the medium viscosity, and rising droplets are taken to have a positive velocity. Use of equation 1 involves assuming slow-moving spherical droplets in a dilute dispersion and neglecting the influence of any colloidal forces between dispersed species. Even allowing for the inadequacies of equation 1 in this application, for practical purposes the rise velocity will still be proportional to the density difference and inversely proportional to the medium viscosity.

If the flotation medium is taken to be the aqueous clay suspension less the bitumen droplets, then the medium viscosity can be estimated (*90*) as

$$\eta = (\eta_B X)/[10^{1.82(1-X)}] \tag{2}$$

where η_B is the measured apparent bulk viscosity and X is the volume fraction of the medium. In the middlings layer, the bitumen content is on the order of volume fraction 0.04, so $\rho \approx \rho_B$; also $X \approx 0.96$, and from equation 2,

$$\eta \approx 0.81\eta_B \tag{3}$$

Based on the bitumen film thickness and average globule diameter estimates given earlier, one might estimate the density of the bitumen globules to be about 0.5×10^3 kg·m^{-3}; however, because of the additional entrained solids content, the density of the bitumen globules are actually typically on the order of 0.90×10^3 kg·m^{-3} under process conditions.

Equation 1 shows that although the density difference between bitumen globules and the particle suspension drives the flotation, the suspension viscosity can inhibit efficient flotation by reducing the rise velocity. The middlings zone suspension consists of fine solid particles, mostly kaolinite and illite clays, that can be strongly interacting depending on pH and clay concentration, as shown in Figure 22 for a model system obtained from process tailings. (Here, high shear viscosities are shown in order to be comparable with the high shear in situ process viscosity measurements shown later in Figure 23.) The significantly higher non-Newtonian viscosities and yield stresses associated with lower pH (pH 6) compared with higher pH (pH 8) can be explained in terms of edge-face clay particle interactions (*66*).

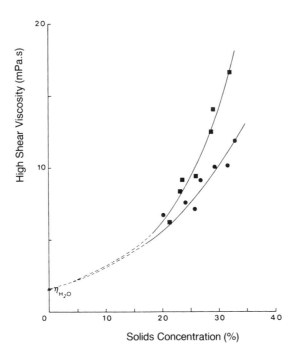

Figure 22. The variation in high shear viscosity with solid (clay) concentration for model system (tailings) suspensions at pH 6 (■) and 8 (●). (Reproduced with permission from reference 66. Copyright 1989 Canadian Institute of Mining, Metallurgy, and Petroleum.)

The enhanced particle interactions at lower pH values have an influence on flotation performance. Figure 23 shows the variation of in situ suspension viscosity with depth in the continuous extraction circuit flotation vessel of Figure 21, for two different NaOH process aid addition levels. The density variation throughout most of the vessel (excluding the froth) was slight. It can be seen that for the zero NaOH addition level condition, a zone of much increased suspension viscosity, by more than an order of magnitude, occurs just below the slurry feed plate. The solution pH was about 6. The figure shows that under the increased NaOH addition level condition, yielding a solution pH of about 8, only quite low suspension viscosities are measured, except at the bottom of the froth layer.

Using the measured in situ viscosities and suspension densities determined from discrete samples (Figure 24), bitumen globule rise velocities have been estimated (66) from equations 1–3. In Figure 25, Stokes rise velocities are calculated for 0.26 mm diameter globules as a function of depth in the vessel. For bitumen globules at a depth of 0.8 m in the vessel (just below the slurry feed plate), the zero NaOH addition

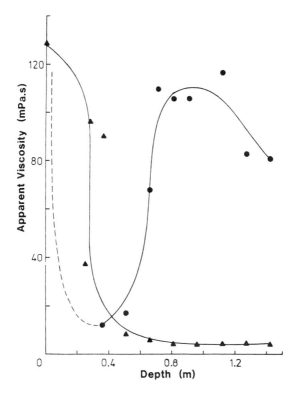

Figure 23. The variation of in situ suspension viscosity with depth in the primary flotation vessel of Figure 21, for two different process aid addition levels: zero NaOH addition (●) and 0.025 mass% relative to the oil sand (▲). The high viscosities at shallow depths correspond to the froth layer. (Reproduced with permission from reference 66. Copyright 1989 Canadian Institute of Mining, Metallurgy, and Petroleum.)

condition causes a suspension viscosity high enough that about 100 min would be required to reach the froth layer from which they can be recovered (66). However, in this continuous flow vessel, the residence time is about 45 min; so before the globules could rise to the froth layer, they would instead be swept out with the middlings stream. The increased viscosity in the central region of the vessel can also promote fine solids buildup. This can cause a spiraling increase in solids concentration and viscosity such that the middlings region viscosity becomes so high that the vessel is inoperable (62, 67). For the same bitumen globules, the increased NaOH addition causes a suspension viscosity low enough that only 5 min would be required to reach the froth layer (66). This is well within the residence time of 45 min; therefore, the globules can rise to the froth layer and be recovered.

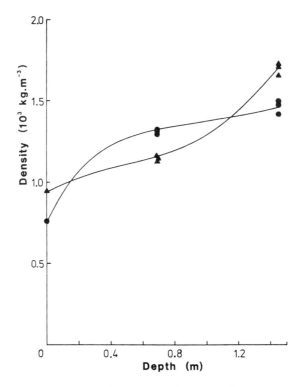

Figure 24. The variation of suspension density with depth in the primary flotation vessel of Figure 21, for two process aid addition levels corresponding to Figure 23: zero NaOH addition (●) and 0.025 mass% relative to the oil sand (▲). (Reproduced with permission from reference 66. Copyright 1989 Canadian Institute of Mining, Metallurgy, and Petroleum.)

A final comment on the relationship between viscosity and density in this system should be made. It has often been claimed that a simple relationship exists between suspension viscosity and density in these process suspensions (e.g., reference 91). This has been applied to the calculation, rather than measurement, of process suspension viscosities based on easily measured process densities. The results of process measurements as shown in Figures 23 and 24 show that the actual relationship is more complex and depends on other process variables as well (67).

At the surface of the primary separation vessels, successfully floated bitumen globules form a froth layer. Under normal vessel loading conditions, the rate of bitumen droplet coalescence is slow relative to the rate at which bitumen droplets collect at the interface; a moderate depth of froth is needed to allow water drainage to take place, with release of nonfloated particles that have been, unavoidably, entrained to some

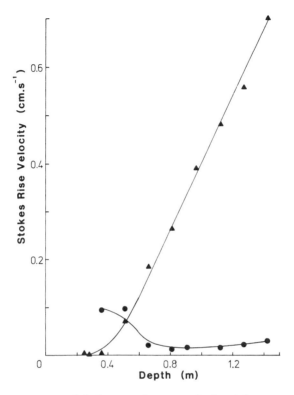

Figure 25. *Bitumen globule rise velocities calculated from measured viscosity and density for 0.26 mm diameter globules as a function of depth in the primary flotation vessel. (Reproduced with permission from reference 66. Copyright 1989 Canadian Institute of Mining, Metallurgy, and Petroleum.)*

extent between the bubbles (*see* reference 69). The froth layer can be removed mechanically from the surface by scraping over the "lip" of the cell, where it collapses and flows away in a launder; the basic separation is accomplished.

At the bottom of the primary separation vessels, successfully sedimented tailings are withdrawn, combined with other tailings, and transported to a tailings pond. Tailings ponds form other examples of important petroleum industry suspensions and are discussed in Chapter 14.

Summary

We have shown that to apply the hot water flotation process to Athabasca oil sands, one must deal with suspensions of various kinds. The initial oil sands slurry contains rocks and particles from which bitumen must be separated. The actual flotation medium contains a suspension of particles whose viscosity must be sufficiently low to allow aerated bitumen globules to float (cream) at a practical rate. Although the hot water process is typically viewed as a problem in oil–water separation, emulsification, and aeration, it is apparent that particle suspensions are created as well and that the nature and properties of the particle suspensions can have a significant influence on process efficiency.

List of Symbols

v_t terminal Stokes velocity for spherical droplets rising in laminar flow
ρ_s droplet density
ρ medium density
g gravitational constant
D_p droplet diameter
η medium viscosity (total suspension less bitumen globules)
η_B apparent bulk viscosity
X volume fraction of the medium (total suspension less bitumen globules)

Acknowledgment

We thank Syncrude Canada Ltd. and the Petroleum Recovery Institute for permission to publish this work.

References

1. Martinez, A. R. In *The Future of Heavy Crude and Tar Sands;* Meyer, R. F.; Wynn, J. C; Olson, J. C., Eds.; United Nations Institute for Training and Research: New York, 1982; pp ixvii–ixviii.
2. Danyluk, M.; Galbraith, B.; Omana, R. In *The Future of Heavy Crude and Tar Sands;* Meyer, R. F.; Wynn, J. C.; Olson, J. C., Eds.; United Nations Institute for Training and Research: New York, 1982; pp 3–6.
3. Khayan, M. In *The Future of Heavy Crude and Tar Sands;* Meyer, R. F.; Wynn, J. C.; Olson, J. C., Eds.; United Nations Institute for Training and Research: New York, 1982; pp 7–11.
4. Walters, E. J. In *Oil Sands Fuel of the Future;* Hills, L. V., Ed.; Memoir 3, Canadian Society of Petroleum Geologists: Calgary, Canada, 1974; pp 240–263.
5. Demaison, G. J. In *The Oil Sands of Canada–Venezuela 1977;* Redford, D. A.; Winestock, A. G., Eds.; CIM Special Volume 17; Canadian Institute of Mining, Metallurgy and Petroleum: Calgary, Canada, 1977; pp 9–16.
6. Ruhl, W. *Tar (Extra Heavy) Sands and Oil Shales;* Ferdinand Enke: Stuttgart, 1982.
7. *Major Tar Sand and Heavy Oil Deposits of the United States;* Interstate Oil Compact Commission: Oklahoma City, OK, 1984.
8. Gutierrez, F. J. In *The Future of Heavy Crude Oils and Tar Sands;* Meyer, B. F.; Steele, C. T., Eds.; McGraw-Hill: New York, 1981; pp 107–117.

9. Rosing, K.E. In *The Future of Heavy Crude Oils and Tar Sands;* Meyer, B. F.; Steele, C. T., Eds.; McGraw-Hill: New York, 1981; pp 124–133.
10. Khalimov, E. M.; Muslimov, R. Kh.; Yudin, G. T. In *The Future of Heavy Crude Oils and Tar Sands;* Meyer, B. F.; Steele, C. T., Eds.; McGraw-Hill: New York, 1981; pp 134–138.
11. Valera, R. In *The Future of Heavy Crude Oils and Tar Sands;* Meyer, B. F.; Steele, C. T., Eds.; McGraw-Hill: New York, 1981; pp 254–263.
12. Phizackerly, P. H.; Scott, L. O. In *Bitumens, Asphalts and Tar Sands;* Chilingarian, G. V.; Yen, T. F., Eds.; Elsevier: Amsterdam, Netherlands, 1978; pp 57–92.
13. Towson, D. In *Kirk–Othmer Encyclopedia of Chemical Technology,* 3rd ed.; Wiley: New York, 1983; Vol. 22, pp 601–627.
14. Carrigy, M. A. In *Guide to the Athabasca Oil Sands Area;* Carrigy, M. A.; Kramers, J. W., Eds.; Alberta Research Council: Edmonton, Canada, 1974; pp 173–185.
15. Fitzgerald, J. J. *Black Gold with Grit;* Gray's Publishing Co.: Sidney, British Columbia, Canada, 1978.
16. Spragins, F. K. In *Bitumens, Asphalts and Tar Sands;* Chilingarian, G. V.; Yen, T. F., Eds.; Elsevier: Amsterdam, Netherlands, 1978; pp 93–122.
17. Ells, S. C. "Bituminous Sands of Northern Alberta Occurrence and Economic Possibilities"; Report No. 632; Mines Branch, Canada Department of Mines: Ottawa, Canada, 1926.
18. Clark, K. A. Can. Patent 289,058, 1929.
19. Blair, S. M. "Report on the Alberta Bituminous Sands;" King's Printer: Ottawa, Canada, 1951.
20. *Oil Sands Fuel of the Future;* Hills, L. V., Ed.; Memoir 3; Canadian Society of Petroleum Geologists: Calgary, Canada, 1974.
21. *Guide to the Athabasca Oil Sands Area;* Carrigy, M. A.; Kramers, J. W., Eds.; Information Series 65; Research Council of Alberta: Edmonton, Canada, 1973.
22. Stewart, G. A.; MacCallum, G. T. *Athabasca Oil Sands Guide Book;* Canadian Society of Petroleum Geologists: Calgary, Canada, 1978.
23. *Bitumens, Asphalts and Tar Sands;* Chilingarian, G. V.; Yen, T. F., Eds.; Elsevier: Amsterdam, Netherlands, 1978.
24. Walker, R. G. In *Facies Models;* Walker, R. G., Ed.; Reprint Series 1; Geoscience Canada: Edmonton, Canada, 1981; pp 1–7.
25. Carrigy, M. A. *Geology of the McMurray Formation; Part III, General Geology of the McMurray Area;* Memoir 1; Research Council of Alberta: Edmonton, Canada, 1959.
26. Mossop, G. D. In *Facts and Principles of World Petroleum Occurrence;* Miall, A. D., Ed.; Memoir 6; Canadian Society of Petroleum Geologists: Calgary, Canada, 1980; pp 609–632.
27. Mossop, G. D.; Flach, P. D. *Sedimentology* **1983,** *30,* 493.
28. Stewart, G. A. In *K.A. Clark Volume;* Carrigy, M. A., Ed.; Alberta Research Council: Edmonton, Canada, 1963; pp 15–27.
29. Carrigy, M. A. *Sediment. Geol.* **1967,** *1,* 327.
30. Jardine, D. In *Oil Sands Fuel of the Future;* Hills, L. V., Ed.; Memoir 3; Canadian Society of Petroleum Geologists: Calgary, Canada, 1974; pp 50–67.
31. Pow, J. R.; Fairbanks, G. H.; Zamora, W. J. In *K.A. Clark Volume;* Carrigy, M. A., Ed.; Alberta Research Council: Edmonton, Canada, 1963; pp 1–14.

32. O'Donnell, N. D.; Jodrey, J. M. *Proceedings of the SME of AIME Meeting;* Society of Mining Engineers of the American Institute of Mining, Metallurgical and Petroleum Engineers: New York, 1984; preprint 84–421.

33. *Bitumens, Asphalts and Tar Sands;* Chilingarian, G. V.; Yen, T. F., Eds.; Elsevier: Amsterdam, Netherlands, 1978; pp 1–15.

34. Silverman, S. R. In *Bitumens, Asphalts and Tar Sands;* Chilingarian, G. V.; Yen, T. F., Eds.; Elsevier: Amsterdam, Netherlands, 1978; pp 17–25.

35. Kendall, G. H. *Proceedings of the Geological and Mineralogical Associations of Canada;* Department of Geology, University of Alberta, Edmonton, Canada, 1976.

36. Mossop, G. D. *Science (Washington, D.C.)* **1980,** *207,* 145.

37. Vigrass, L. W. *Am. Assoc. Pet. Geol. Bull.* **1968,** *52,* 1984.

38. Carrigy, M. A. In *Proceedings of the 7th World Petroleum Congress;* Elsevier: Amsterdam, Netherlands, 1967; Vol. 3, pp 573–581.

39. Camp, F. W. *Tar Sands of Alberta, Canada,* 3rd ed.; Cameron Engineers Inc.: Denver, CO, 1976.

40. Carrigy, M. A. "Criteria for Differentiating the McMurray and Clearwater Formations in the Athabasca Oil Sands"; Bulletin 14; Research Council of Alberta: Edmonton, Canada, 1963.

41. Clark, K. A. *Trans. Can. Inst. Min. Metall.* **1960,** *63,* 151.

42. Carrigy, M. A. *J. Sediment. Petrol.* **1962,** *32,* 312.

43. Farouq Ali, S. M. In *Oil Sands Fuel of the Future;* Hills, L. V., Ed.; Memoir 3; Canadian Society of Petroleum Geologists: Calgary, Canada, 1974; pp 199–211.

44. Dealy, J. M. *Can. J. Chem. Eng.* **1979,** *57,* 677.

45. Schramm, L. L.; Kwak, J. C. T. *J. Can. Pet. Technol.* **1988,** *27,* 26.

46. Ward, S. H.; Clark, K. A. "Determination of the Viscosities and Specific Gravities of the Oils in Samples of Athabaska Bituminous Sand"; Report No. 57; Research Council of Alberta: Edmonton, Canada, 1950.

47. Christensen, R. J.; Lindberg, W. R.; Dorrence, S. M. *Fuel* **1984,** *63,* 1312.

48. Misra, M.; Miller, J. D. *Mining Eng.* **1980,** *32,* 302.

49. Hupka, J.; Miller, J. D.; Cortez, A. *Min. Eng.* **1983,** *35,* 1635.

50. Yang, Y. J.; Bukks, K.; Miller, J. D. *Energy Process. Can.* **1989,** *82,* 14.

51. Takamura, K. *Can. J. Chem. Eng.* **1982,** *60,* 538.

52. Clark, K. A. *Trans. Can. Inst. Min. Metall.* **1944,** *47,* 257.

53. Bartell, F. E.; Miller, F. L. *Ind. Eng. Chem.* **1932,** *24,* 335.

54. Ball, M. W. *Trans. Can. Inst. Min. Metall.* **1941,** *44,* 58.

55. Ball, M. W. *Bull. Am. Assoc. Pet. Geol.* **1935,** *19,* 153.

56. Sobol, W. T.; Schreiner, L. J.; Miljkovic, L.; Marcondes-Helene, M. E.; Reeves, L. W.; Pintar, M. M. *Fuel* **1985,** *64,* 583.

57. Zajic, J. E.; Cooper, D. G.; Marshall, J. A.; Gerson, D. F. *Fuel* **1981,** *60,* 619.

58. Miller, J. D.; Misra, M. *Fuel Proc. Technol.* **1982,** *6,* 27.

59. Erskine, H. L. In *Handbook of Synfuels Technology;* Meyers, R. A., Ed.; McGraw-Hill: New York, 1984; pp 5–79.

60. Sanford, E. C. *Can. J. Chem. Eng.* **1983,** *61,* 554.

61. Schutte, R.; Ashworth, R. W. In *Ullmans Encyklopadie der Technischen Chemie;* Verlag Chemie GmbH: Weinheim, Germany, 1979; Vol. 17.

62. Innes, E. D.; Fear, J. V. D. *Proceedings of the 7th World Petroleum Congress;* Elsevier: Amsterdam, Netherlands, 1967; Vol. 3, pp 633–650.

63. Schramm, L. L.; Smith, R. G. Can. Patent 1,188,644, 1985; U.S. Patent 4,462,892, 1984.

64. Perrini, E. M. *Oil from Shale and Tar Sands;* Noyes Data Corp.: Park Ridge, NJ, 1975.
65. Ranney, M. W. *Oil Shale and Tar Sands Technology;* Noyes Data Corp.: Park Ridge, NJ, 1979.
66. Schramm, L. L. *J. Can. Pet. Technol.* **1989,** *28,* 73–80.
67. Schramm, L. L. Can. Patent 1,232,854, 1988; U.S. Patent 4,637,417, 1987.
68. Adam, D. G. *Proceedings of Advances in Petroleum Recovery and Upgrading Technology;* Alberta Oil Sands Technology and Research Authority: Edmonton, Canada, 1985.
69. Shaw, R. C.; Czarnecki, J.; Schramm, L. L.; Axelson, D. In *Foams: Fundamentals and Applications in the Petroleum Industry;* Schramm, L. L., Ed.; Advances in Chemistry 242; American Chemical Society: Washington, DC, 1994; pp 423–459.
70. Houlihan, R. In *The Future of Heavy Crude and Tar Sands;* Meyer, R. F.; Wynn, J. C.; Olson, J. C., Eds.; United Nations Institute for Training and Research: New York, 1982; pp 1076–1086.
71. Smith, R. G.; Schramm, L. L. *Fuel Proc. Technol.* **1989,** *23,* 215.
72. Sanford, E. C.; Seyer, F. A. *CIM Bull.* **1979,** *72,* 164.
73. Cymbalisty, L. M. *Proceedings of the 30th Canadian Chemical Engineering Conference;* Canadian Society for Chemical Engineering: Edmonton, Canada, 1980.
74. Seitzer, W. H. Presented at the Proceedings ACS Division of Petroleum Chemistry, San Francisco, CA, April 1968.
75. Malmberg, E. W.; Bean, R. M. Presented at the Proceedings ACS Division of Petroleum Chemistry, San Francisco, CA, April 1968.
76. Malmberg, E. W.; Bean, R. M. Presented at the Proceedings ACS Division of Petroleum Chemistry, San Francisco, CA, April 1968.
77. Clark, K. A.; Pasternack, D. S. *Ind. Eng. Chem.* **1932,** *24,* 1410.
78. Bowman, C. W. U.S. Patent 3,623,971, 1971.
79. Floyd, P. H.; Schenk, R. C.; Erskine, H. L.; Fear, J. V. D. Canadian Patent 841,581, 1970; U.S. Patent 3,401,110, 1968.
80. Sanford, E. U.S. Patent 4,201,656, 1980.
81. Schramm, L. L.; Smith, R. G. *Colloids Surf.* **1985,** *14,* 67.
82. Schramm, L. L.; Smith, R. G. *AOSTRA J. Res.* **1989,** *5,* 87.
83. Schramm, L. L.; Smith, R. G.; Stone, J. A. *AOSTRA J. Res.* **1984,** *1,* 5.
84. Schramm, L. L.; Smith, R. G.; Stone, J. A. *AOSTRA J. Res.* **1985,** *1,* 147.
85. Schramm, L. L.; Smith, R. G. *Can. J. Chem. Eng.* **1987,** *65,* 799.
86. Takamura, K.; Chow, R. S. *Colloids Surf.* **1985,** *15,* 35.
87. Hupka, J.; Miller, J. D. *Int. J. Min. Process.* **1991,** *31,* 217.
88. Drelich, J.; Hupka, J.; Miller, J. D.; Hanson, F. V. *AOSTRA J. Res.* **1992,** *8,* 139.
89. Misra, M.; Aguilar, R.; Miller, J. D. *Sep. Sci. Technol.* **1981,** *16,* 1523.
90. Foust, A. S.; Wenzel, L. A.; Clump, C. W.; Maus, L.; Andersen, L. B. *Principles of Unit Operations;* Wiley: New York, 1960; pp 449–453.
91. Graybill, J. B.; White, C. N.; Loveland, J. W. Can. Patent 889,823, 1972; U.S. Patent 3,530,042, 1970.
92. Jacobs, F. A. M.Sc. Thesis, University of Calgary, Calgary, Canada, 1978.
93. Cottrell, J. H. In *K.A. Clark Volume;* Carrigy, M. A., Ed.; Alberta Research Council: Edmonton, Canada, 1963; pp 193–206.
94. Takamura, K.; Chow, R. S. *J. Can. Pet. Technol.* **1983,** *22,* 22.

RECEIVED for review September 6, 1994. ACCEPTED revised manuscript January 17, 1995.

Nature and Fate of Oil Sands Fine Tailings

Randy J. Mikula[1], Kim L. Kasperski[1], Robert D. Burns[2], and Mike D. MacKinnon[3]

[1] Fuel Processing Laboratory, Western Research Centre, CANMET, P.O. Bag 1280, Devon, Alberta T0C 1E0, Canada
[2] Suncor Oil Sands Group, P.O. Box 4001, Fort McMurray, Alberta T9H 3E3, Canada
[3] Syncrude Canada Ltd., P.O. Box 5790, Station L, Edmonton, Alberta T6C 4G3, Canada

The chemical and physical properties of clay suspensions produced during oil production from oil sands are described. With a composition of approximately 70 wt% water (with some unrecovered bitumen) and 30 wt% solids (>90% less than 44 μm in size), these clay suspensions consolidate very slowly. Clay aggregate or floc morphology has been shown to be a function of the water chemistry and can be manipulated to produce a tailings suspension that is easier to consolidate and dewater. Commercial oil sands processing has been going on in northeastern Alberta since 1967, and in that time approximately 250 million m³ of this difficult to dewater clay suspension has been produced. The reclamation options for this material (mature fine tailings) on a commercial scale are also outlined.

CONTINUED DEVELOPMENT OF THE LARGE OIL SANDS DEPOSITS found in northeastern Alberta is critical in meeting Canada's present and future energy requirements. At present, two commercial oil sands processing plants are in operation in the Athabasca oil sands deposit (Figure 1). Great Canadian Oil Sands (now Suncor Oil Sands Group) started production in 1967, whereas Syncrude Canada Ltd. started in 1978.

The combined production of synthetic crude oil (SCO) by Syncrude and Suncor is more than 250,000 barrels (about 40,000 m³) per day, which represents over 15% of Canada's crude oil requirements. Syncrude Canada Ltd. processes bitumen and produces SCO at about three

0065–2393/96/0251–0677$18.75/0

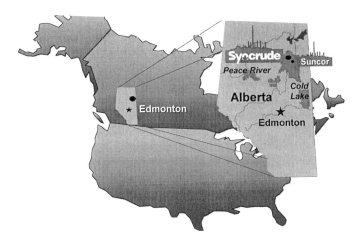

*Figure 1.   Map showing the location of the Athabasca oil sands deposit and
the Syncrude and Suncor extraction operations.*

times the rate of Suncor (Table I). Since commercial operations began
in 1967, a total of over 1 billion barrels of SCO have been produced
from the processing of more than 2 billion tonnes of oil sand (Figure 2).
A schematic illustrating the relationship between mining, extraction,
and tailings in the oil sands operations is shown in Figure 3.

Oil production from the oil sands more closely resembles a mining
operation than a conventional oil development. The oil sands are mined
from open pit mines at rates of about 450,000 tonnes per day (Table I).
At both plants, bitumen is separated from the ore using the Clark hot

Table I.   Average Amount of Materials Handled Daily at the Syncrude
and Suncor Oil Sands Operations in 1993

Materials	Syncrude	Suncor
Oil sand processed (t)	350,000	110,000
Average grade of OS (wt % bitumen)	10.8	12.2
Bitumen extracted $(m^3)^a$	35,000	12,000
Synthetic crude oil produced $(m^3)^a$	30,000	10,000
Rejects from extraction (t)	25,000	6,000
Total tailings (t)	570,000	295,000
Tailings sand (>22 μm) (t)	230,000	75,000
Tailings fines (<22 μm) (t)	40,000	10,000
Tailings water (m^3)	300,000	210,000
Raw water import (Athabasca River) (m^3)	90,000	28,000
Recycle water from settling basins (m^3)	250,000	185,000

[a] To convert to barrels: 1 t = 6.3 barrels.

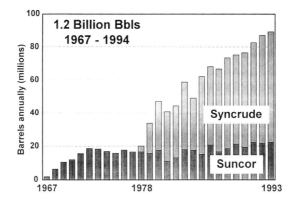

Figure 2. Syncrude and Suncor oil production history.

water extraction (CHWE) process, which involves dispersion of the oil sands with water, steam, mechanical conditioning, and caustic (NaOH) with the subsequent separation of bitumen from the sand solids. Unlike conventional oil reservoirs that typically produce less than 30% of the hydrocarbon in place, the oil sands extraction process is between 90 and 95% efficient. The extracted bitumen is upgraded to SCO by coking and hydrocracking processes at the site, and the SCO is then pipelined to market.

Extraction and upgrading lead to the production of large volumes of liquid and solid waste materials that must be reclaimed (Table I). For each barrel of synthetic crude produced, about 2 tonnes of ore must be processed, with a resulting waste product of about 1.8 tonnes of solid tailings and about 2 m^3 of wastewater. When the removal of overburden and rejects is considered, the development of oil sands is very much a problem of material handling and waste management. The nature and quality of waste materials produced require the development of acceptable reclamation options.

A significant byproduct of bitumen extraction is the tailings stream, which contains water, mineral components, and unrecovered bitumen. The minerals consist of a coarse (>22 μm) sand component and a fines (<22 μm) fraction made up of silts and clays (predominantly kaolinite and illite).

The dispersive nature of the oil sands extraction process does not affect the settling properties of the coarse solids. These solids segregate from the fines and are used for building the dikes that surround the settling ponds that contain the process recycle water. The fines that are not retained with the beach solids run off into the tailings pond as a thin slurry and consolidate very slowly; full consolidation of untreated tailings is estimated to take thousands of years. Large quantities of this suspension

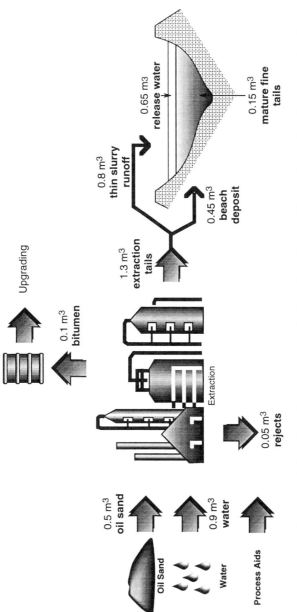

Figure 3. Schematic of the oil sands operation, including the volume of fine tails produced compared with other process stream volumes.

of fine clay minerals [so-called mature fine tailings (MFT)] have built up under the recycle water layer contained in the settling ponds. A photograph of this MFT material is shown in Figure 4. The MFT inventory at the Syncrude and Suncor sites has reached some 250 million m^3 (Figure 5).

Understanding the nature and properties of this waste component is necessary to ensure full utilization of the massive oil sands resource. Although reviews of MFT properties and management are available (*1*–*7*), the aim of this chapter is to present recent advances in understanding the source, properties, and treatment of oil sands fine tailings. It is only recently that MFT has replaced sludge as the descriptor for this difficult to handle tailings material. As a result, most of the literature before 1991 refers to sludge, whereas most of the literature after 1991 uses the term MFT.

Origin of Fine Tailings

The properties of the accumulating MFT are dominated by the characteristics of the fines fraction. Figure 6 is a generalized schematic of the oil sands extraction process illustrating the sources of fine tailings, coarse rejects, primary separation vessel tailings, and flotation cell (froth cleaner or secondary separation) tailings. A fourth source of tailings is the fines separated from the combined froths in the froth treatment plant. The extraction tails and froth treatment plant tails are pumped to tailings ponds as separate streams. The former are enriched in coarse solids, whereas the latter are enriched in fine solids. Furthermore, as noted earlier, a fraction of the fine tailings is retained in the beach sands and does not contribute to the buildup of MFT. Figure 7 illustrates the mining and extraction processes and tailings basins. The tailings slurry from extraction is pumped to the settling basin where it segregates, with the coarser material dropping in the dikes and beaches and the fines dominating the suspension. As shown in Figure 8, the thin fine tailings slurry initially settles rapidly to a certain density, after which the settling slows dramatically. The fines settle over a period of days to about 20% solids and then consolidate over a period of 2–5 years to approximately 30% solids. Further consolidation is very slow.

The slow settling and consolidation after the suspension, reaching about 30% solids, have been attributed to the presence of unrecovered bitumen (Figure 9) (*5*) or to the presence of fine clays and amorphous materials, which either may hold amounts of water disproportionate to their concentration or may form an ordered floc structure at 30 wt% solids (*4*). One explanation emphasizes the possible role of soluble organic surfactants that modify the clay surfaces, the effect of strongly bound organic material on the minerals, asphaltenic components from

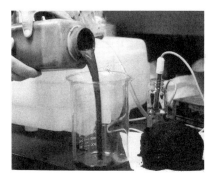

Figure 4.   Photograph of typical MFT showing the mud-like nature of this suspension.

the oil sand, and polysaccharides from biological activity (*8–17*). Another explanation for MFT stability concerns the identification of very small (<100 nm) clay minerals that can be extracted from MFT (*18–20*).

Although disagreement exists as to whether the organic or mineral model best describes the intractable nature of MFT, studies since 1991 have favored the mineral model of MFT stability (*18, 21–26*). This is largely based on the identification of mineral interactions that can be modeled with ultrafine clay components and appropriate dispersing water chemistry (*18, 20, 21*). However, with any system of colloidal particles, surface phenomena dominate, and even trace amounts of organic material associated with the minerals may have important effects. In any case, the fundamental floc structure or interparticle aggregation ultimately determines the bulk tailings behavior.

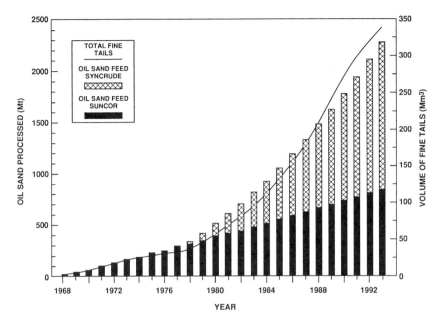

Figure 5. *Cumulative mass of oil sand feed processed and fine tails pro-duced by Syncrude and Suncor.*

Tailings Properties

Tailings Mineralogy. Tailings minerals consist of sand, clays, amorphous oxides, and trace metals. The sand is 97.5–99% SiO_2, 0.5–0.9% Al_2O_3, and 0.1–0.9% Fe (*4, 27, 28*). The oil sands, and hence the clay minerals found in the fine tails suspension, come from the McMurray Formation. The majority of clays in this formation are kaolinite and illite with traces of smectites, chlorite, vermiculite, and mixed-layer clays (*5, 29*). The upper McMurray Formation has a larger amount of smectites, whereas the lower McMurray Formation has larger amounts of vermic-ulite and mixed-layer clays. However, in both areas, kaolinite and illite are still the predominant clay minerals (*5*).

A breakdown of the mineral compositions of a typical oil sand, over-burden clay layers in the McMurray Formation, and the MFT are given in Figure 10. The figure shows that kaolinite and illite dominate the clay fraction in all of these materials, with smectites generally found only in the overburden and in intercalated clay lenses in the oil sands. The amounts and types of clay minerals found in tailings ponds vary considerably, but assays of tailings pond MFT show mostly kaolinite and illite (*30–32*). Montmorillonite has been reported in the Suncor MFT

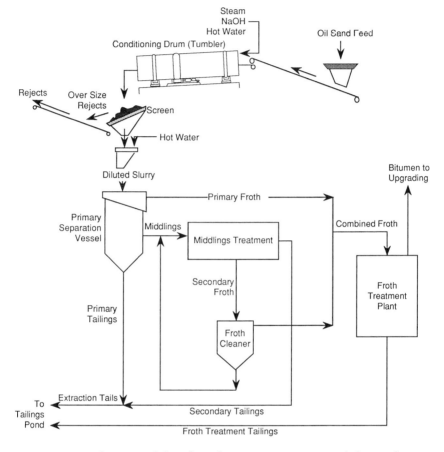

Figure 6.   Schematic of the oil sands extraction process and the resulting tailings suspensions.

suspension (8, 30), and trace amounts of a swelling clay have also been detected in Syncrude MFT samples. Analysis of both Suncor and Syncrude fine tails showed about 1–5 wt% iron (2) (amount varying with depth) and trace amounts of aluminum and silicon oxides (4, 32). Further work is required to determine the mineralogical form of these materials (i.e., amorphous or crystalline). If they are present as amorphous material, they could play an important role in the water-holding behavior of MFT.

Heavy metals detected in tailings include Ti, Zr, Fe, V, Mg, Mn, Al, Pb, Zn, Nb, and Mo (33–35). The most important are Ti and Zr, as these have commercial value and are present in quantities that might warrant their recovery from the tailings (34, 36).

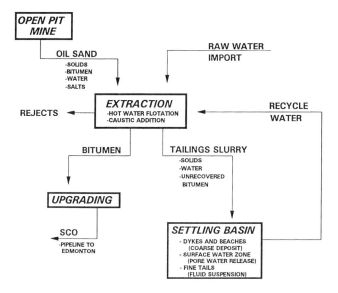

CROSS SECTION OF OIL SANDS TAILINGS SETTLING BASIN

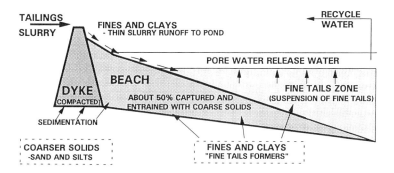

Figure 7.   Schematic of the fate of the oil sands tailings suspensions.

MacKinnon and Sethi (2) prepared a detailed multielemental analysis of the composition of fine tails solids, and they indicate similar elemental distributions in both Syncrude and Suncor tailings ponds.

Composition of MFT Pore Water.   Water is the major component of MFT, and its composition and the way in which it moves through the fine tails structure (hydraulic conductivity, permeability) are important in determining how fine tails develop and ultimately will impact on fine tails production. This impact could be determined by either the inorganic or organic composition of the aqueous phase.

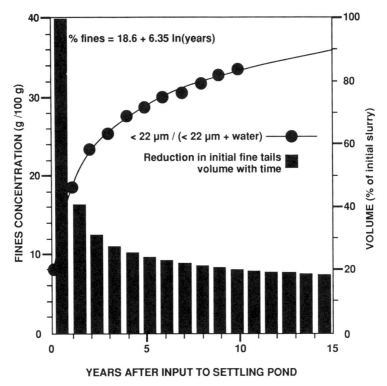

Figure 8. Graphs showing the change in fines concentration (<22 μm material) as tailings volume is reduced with time in the tailings suspension settling pond.

The main sources of inorganic ions found in tailings water are (1) oil sand connate water (water in the spaces between the sand grains), (2) makeup water taken from the Athabasca River for use in plant processes, and (3) chemicals added during hot water extraction and other processes. These sources contribute ions that accumulate in the pond water that is recycled to oil sands processing.

Most of the added inorganics come from the oil sand itself. The highest grade of oil sand contains 2–3 wt% (10–15% of pore volume) connate water; as ore grade decreases, connate water content increases (37). Connate water in oil sand mined at Syncrude has a composition equivalent to 30–100 mM (1.7–5.8 g L^{-1}) NaCl, with small amounts of K^+, Ca^{2+}, Mg^{2+}, and SO_4^{2-} (30, 37). Connate water in Suncor oil sand has a lower salinity (30). Several other sources of inorganics are available. River water contributes to the ionic content of tailings (38). Up to 0.06

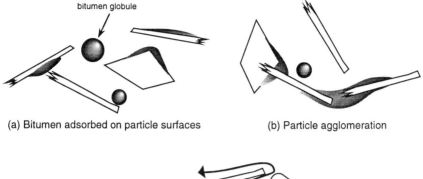

(a) Bitumen adsorbed on particle surfaces   (b) Particle agglomeration

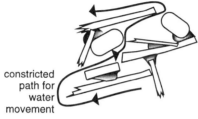

(c) Reduction of sludge permeability

Figure 9.   Organic model for slow consolidation of the tailings suspension. Modifications to the concept as shown take into account that concentrations of organic surfactants bind or gel the solids into a stable floc or aggregate structure that is slow to dewater or consolidate. (Reproduced with permission from reference 1. Copyright 1991 Alberta Oil Sands Technology and Research Authority.)

kg Na^+ is added as NaOH per tonne of oil sand at Syncrude (3); smaller amounts are added at Suncor. Other plant processes such as boiler-feed water treatment (which uses $Al_2(SO_4)_3$ and H_2SO_4) and bitumen upgrading (which produces sour water), add SO_4^{2-} to the tailings water (30). The ionic contribution from these process aids is still minor compared with that from the oil sand itself.

Bicarbonate (HCO_3^-) concentration increases in the tailings pond water because of its presence in the oil sands feed as well as from the adsorption of CO_2 during aeration in the conditioning stage in the extraction plant and, to a lesser extent, from absorption into the relatively high pH tailings pond water (30).

The composition of the aqueous phase in the tailings pond is not static as all process water and liquid waste is stored in the tailings ponds and therefore the ions accumulate. An ion such as Cl^- is considered to be conserved because its concentration is unaffected by chemical or physical processes (2). However, reactions that occur that could affect the concentration of other ions in the aqueous phase include ion ex-

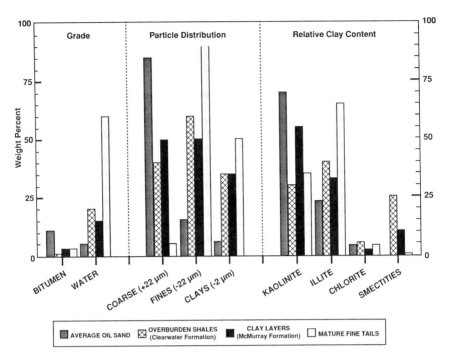

Figure 10. Weight percent bitumen and water (grade); weight percent coarse, fine, and clay-size particles (particle distribution); and weight percent clay (clay content), for the oil sand, overburden, clay layers, and the MFT suspension (27).

change between dissolved cations and those adsorbed on clay surfaces (2, 22, 30) and reactions responding to equilibria between solids and their dissolved ions (3, 39). It is also possible that the dissociation of solid-associated organic complexes (39) and biological interactions (3) may contribute to the aqueous inorganic makeup of the tailings pond water. These equilibria define the MFT water chemistry and therefore affect the clay interaction and floc structure that determines the bulk properties and water-holding capacity of the fine tails suspension.

In Table II, the compositions of surface-zone recycle water and fine tails pore water from Syncrude and Suncor are compared (2). Over time, the recycle process water is becoming more brackish (especially at Syncrude). Concentration of the major ions in the tailings ponds as a function of depth is shown in Figure 11. The concentrations of major ions in tailings water have increased with time: for example, at Syncrude, from 1979 to 1993, there was a 200–300% increase in conserved ions such as Na^+ and Cl^-. At Suncor, absolute changes were smaller (2, 3, 35, 40). However, no evidence exists of trace metal buildup over time (2).

Table II. Composition of Surface Zone Recycle Water and Fine Tails Pore Water from the Syncrude and Suncor Settling Basins

Variables	Syncrude Mildred Lake Settling Pond[a]		Suncor Tailings Ponds[b]	
	Surface Zone	Fine Tails	Surface Zone	Fine Tails
General				
pH (units)	7.8–8.0	8.1–8.5	7.7–8.1	8.0–8.4
Conductivity ($\mu S/cm$)	2500	1300–1600	1100–1200	1400–1600
Mineral solids (wt%)	0.03	10–60	0.5–2.0	10–60
Hydrocarbon content (wt%)	<0.001	0.5–4	<0.01	1–8
Dissolved organic carbon (mg C/L)	50–60	55–65	55–70	60–80
Ammonia (mg N/L)	5–7	3–5	5–8	4–15
Major ions (mg/L)				
Cations				
Na^+	600	450–500	350–400	350–450
K^+	10	9–12	10	10–25
Mg^{2+}	4	2–4	4	10–25
Ca^{2+}	7	3–6	7	4–20
Anions				
F^-	4.0	7.0	3	5
Cl^-	300	140–160	40–50	20–40
SO_4^{2-}	210	2–10	80–100	10–70
HCO_3^-	850–950	950–1050	800–900	800–1400
Trace elements (mg/L)				
Al	0.1–0.2	0.5–2.50	1–3	1–10
Si	2–3	4–8	5–10	5–20
Fe	0.10	0.2–0.5	0.1–0.5	0.2–5.0
B	3	3	3	3–4

[a] Results from 1993 survey.
[b] Results from 1992 survey.

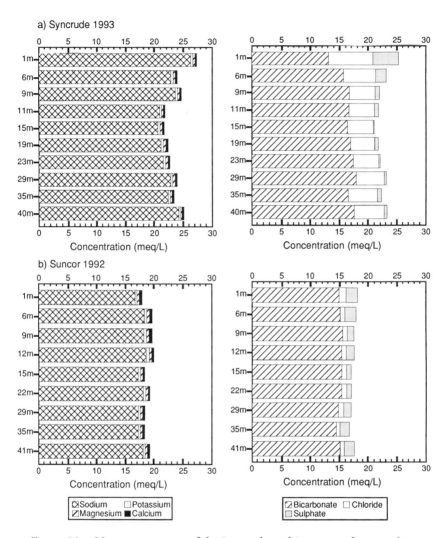

Figure 11.   Major ion content of the Syncrude and Suncor settling ponds as a function of depth.

In addition to the inorganic components, there is a fraction of water-soluble organics released during the hot water extraction process. Over the years, the dissolved organic content has not changed at Suncor or Syncrude. Most of this organic material is polar, including organic acids (carboxylic and sulphonic surfactants). Although there is a considerable hydrocarbon phase in MFT (unrecovered bitumen), the amount actually dissolved is small (Table III).

Table III.   **Fractionation of Organic Matter in Samples Collected from the Syncrude Settling Basin**

| Sample Source | Organic Carbon Content (mg C/L) | Acid Extract[a] (ppm) | Base/Neutral Extract[b] (ppm) | | | |
| | | | Total | Alumina Column Fractions | | |
				Fraction 1	Fraction 2	Fraction 3
Surface water zone (recycle water)						
Total (includes SPM)	100	96	4.8	0.460	<0.20	1.2
Solids free	50–70	94	3.3	<0.20	<0.20	<0.20
Fine tails zone (below water/ fine tails interface)						
Total (includes SPM and Bitumen) (mg/kg of dry weight)	>25,000	1,000–10,000	25,000–40,000	10,000–15,000	8,000–10,000	8,000–15,000
Pore water (mg/L)	60–80	60	4.2	<0.20	<0.20	<0.20

[a] Acid (pH = 2) partition with dichloromethane Soxhlet extraction of sample.

[b] Base/neutral partition with dichloromethane extract of sample. Base/neutral extract separated on an alumina column into three fractions based on polarity. NB. Volatile components less than C_7 not included.

Settling and Consolidation Behavior of the Fine Tailings Suspension

Figures 12a and 12b summarize the solids and particle size distributions in both the Suncor and Syncrude tailings ponds. Although there is a sharp change in density at the interface between the recycle water and the fine tails, the gradual densification or settling of the mineral sus-

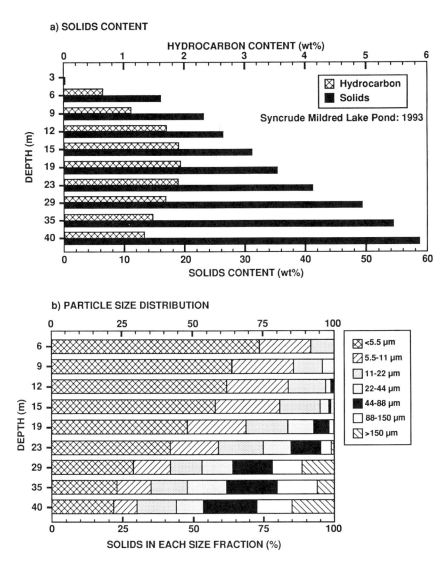

Figure 12. Solids and hydrocarbons content and particle size distribution as a function of depth in the Syncrude and Suncor settling ponds.

pension with depth in the two ponds is apparent. The higher solids content toward the bottom of the ponds is not indicative of consolidation of the fines but rather of coarse sand that has settled through the fine tailings. As Figure 13 shows, the Si:Al ratio for both Syncrude and Suncor increases significantly as quartz becomes the dominant mineral (due to sand at the bottom of the ponds).

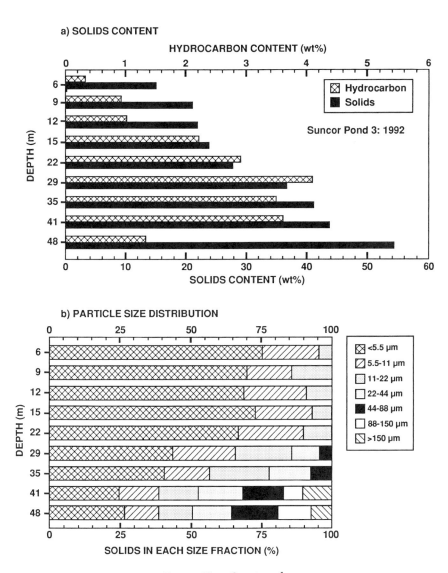

Figure 12.—Continued.

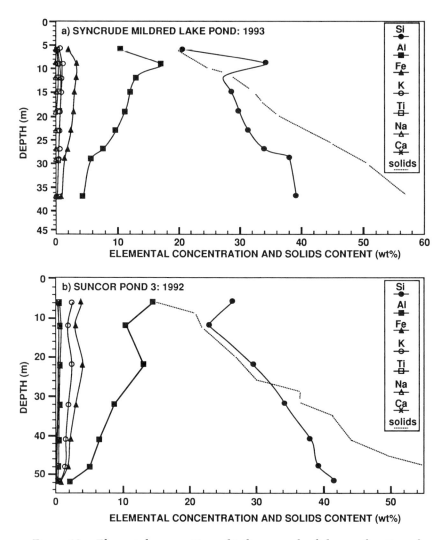

Figure 13.   Elemental composition of tailings pond solids as a function of depth (3).

The temperature profiles in Figure 14 clearly indicate that the Syncrude and Suncor ponds have higher temperatures in the winter months than one might expect, especially the Suncor ponds. The reasons for this are not clear.

Settling of these fine tails suspensions (fluid fine tails that are not trapped or entrained in the beach sand) and the ultimate formation of MFT in the settling ponds can be divided into three stages: Stokes law sedimentation, during which the particles sediment according to their

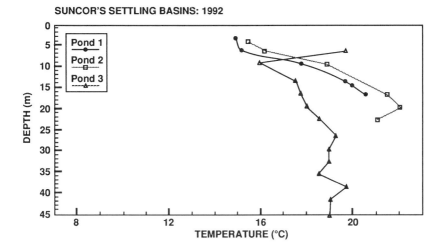

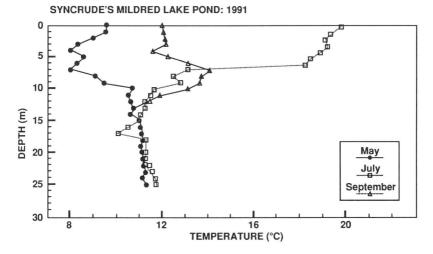

Figure 14.   Temperature profile of the Syncrude and Suncor tailings pond as a function of depth (Reproduced with permission from reference 3. Copyright 1989 Alberta Oil Sands Technology and Research Authority.)

size and relative density; hindered settling, during which the particles are in close enough proximity to influence each other and the solids can be considered to be settling as a solid mass; and consolidation, during which a structure is set up in the suspension so that an effective inter-particulate stress develops between the tailings particles (*41, 42*). Consolidation is controlled by the permeability and compressibility of the system and is induced by the weight of the particles (*43*). As a result of the consolidation process, excess pore pressure is set up in the suspension (i.e., a pressure in excess of that calculated from the hydrostatic head), which gradually dissipates as the fine tails consolidate, eventually reaching a value of zero in the fully consolidated system. In the present tailings system, this is expected to take centuries.

As shown in Figure 8 (*41*), the tailings solids sediment rapidly over a period of months up to approximately 20% solids (Stokes law sedimentation up to 5% solids, hindered sedimentation after that). The fine tails then consolidate over the next 5 years (or 3–5 years for some ponds; reference 44) to approximately 30% solids, at which point consolidation slows dramatically. Scott and Dusseault (*40*) postulated that this drop in the consolidation rate is due to the high concentration of unrecovered bitumen in the MFT (which leads to a hydraulic conductivity 0.1–0.01 of that of equivalent material with no bitumen) and the low effective density of the MFT solids (*5, 41, 45–48*). As discussed earlier, the interaction of mineral components has also been postulated to explain the lack of consolidation. This implies a floc structure determined by the clay mineral interactions that are, in turn, a function of the water chemistry. Syncrude MFT (fines content > 30 wt%) have a low permeability of $1–0.01$ nm s^{-1} (*46, 47*). The dewatering behavior is completely dominated by this low permeability and the long drainage paths that result when the MFT are stored in deep ponds (*49*).

The rheology of the fine tailings sample has been investigated extensively (*43, 49–52*). MFT exhibit both shear thinning and thixotropy. Gel strength or yield point has also been investigated and found to have a distinct time dependence (*51–53*). This time dependence is indirect evidence of the formation of a floc structure or the establishment of distinct interparticle orientations. After shearing of the MFT sample, yield points were considerably smaller than those for the same MFT sample that had been allowed to sit undisturbed for a period of time. This effect is shown in Figures 15–17, where the thixotropy and the yield point increase with time after shearing. The shearing (to disturb the floc structure) serves to start the experiments with a suspension with a consistent degree of dispersion. The thixotropy, or shear thinning behavior, is seen in the hysteresis in the shear rate versus shear stress plots.

To the extent that this behavior is due to the time dependence of floc or structure formation in the suspension, it should be possible to directly observe the floc formation using microscopy. Investigations using cryogenic electron microscopy showed a definite structure of the clay mineral flocs or aggregates that is also sensitive to vibration or disturbance of the sample, analogous to the time dependence of the yield points or gel strength (*24, 25*). Figure 18 shows a micrograph of a relatively undisturbed MFT suspension with a large sand grain supported by the clay structure. The figure also shows a higher magnification view of what might be termed a "card house-like" structure, although

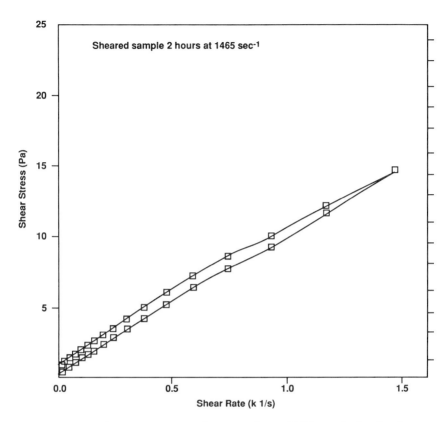

Figure 15. Shear stress versus shear rate for an MFT sample that has been sheared to break up the floc or aggregate structure. Very little yield stress is evident in this sample.

the lower magnification photograph shows what appears to be a predominantly edge to edge clay aggregate structure. The possibility of this structure being a function of the sample preparation cannot be discounted, although similar structures were observed in unfrozen samples using confocal microscopy (Figure 19). This shows that cryogenic electron microscopy is useful for probing the aggregate or floc structure in these suspensions. An undisturbed MFT sample is shown in Figure 18, whereas Figures 20 and 21 show the loss of structure with varying degrees of sample agitation. This is analogous to the changes in the bulk rheology shown in Figures 15–17. Whether this structure is a classical card house orientation of clay platelets, as described by earlier authors (5, 41, 50), or a predominantly edge to edge clay structure is not an important distinction (53, 54). The findings that bulk behavior is related to the observed floc morphology and that the floc structure can be altered

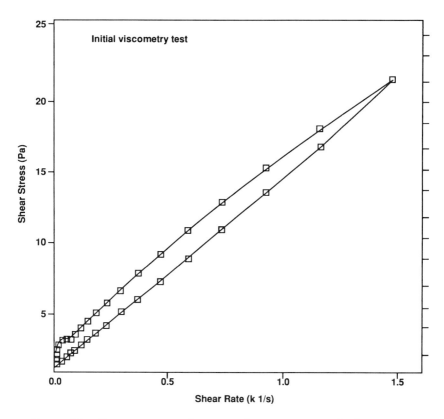

Figure 16. Shear stress vs. shear rate for an MFT sample that was not initially sheared. A significant yield stress is evident relative to the sheared sample in Figure 15. The thixotropy, or shear thinning (the area enclosed by the hysteresis curve), is also greater than for the sample that was sheared. The shear stress on the return curve (lower curve) is always lower after shearing (the upper part of the curve).

by changing water chemistry or by mechanical agitation are important elements in understanding MFT.

Settling of MFT was found to be enhanced under certain conditions, thought to be related to low-frequency vibration and subsequent collapse of the clay floc structure (55). Figure 22 illustrates this behavior in which settling was significantly enhanced during specific times of the year when it appeared that vibrations were causing further settling.

The results of another experiment offering indirect evidence that the clay particles form a particular floc structure are shown in Figure 23. Centrifugation of tailings fines (7 wt% solids) at moderate speed (230 × g) lead to significantly more settling than occurred during centrifugation of fines that have been allowed to settle to the point where

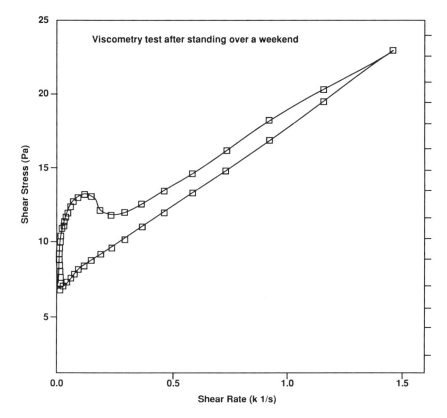

Figure 17. Shear stress vs. shear rate for an MFT sample that was allowed to sit for 72 h. A significant increase in yield point and thixotropy is noted compared with Figures 15 and 16.

the clay flocs or aggregates have formed some structural strength (20 wt% solids). At lower centrifugation speeds (115 × g), the fines and settled fines behave similarly (Figure 24) (as with 1 × g settling), indicating that a certain energy input is required to collapse the clay aggregate or floc structure.

The observed microscopic structure or particular interparticle associations were found to correlate with bicarbonate ion concentration, although, as mentioned earlier, small concentrations of organic surfactants released by the caustic extraction process (that would also correlate with bicarbonate concentration) cannot be ruled out. Figure 25 shows MFT created without caustic [the other six lease owners (OSLO) process]. No structure comparable with that observed in the MFT from the Clark extraction process is evident in this sample despite its similar particle size distribution and weight percent solids. Kaolinite clays in deionized

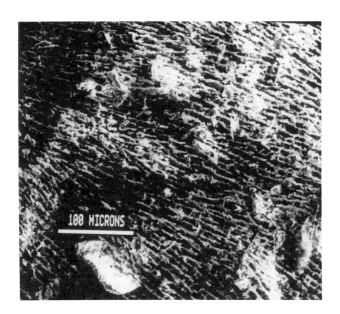

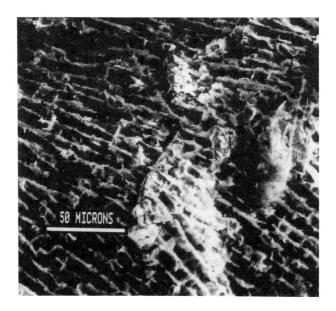

Figure 18.   Electron micrograph of a cryogenically prepared MFT sample. A distinct ordering of the clays can be noted as well as a large sand grain that is supported in the clay structure. The lower photograph shows the structure at higher magnification where the edge to edge arrangement emphasized in the upper figure is not as pronounced.

Figure 19. Confocal micrograph of an MFT sample that was unfrozen and observed using the confocal optical technique. The orientation or structuring of the clay component is identical to that observed with the cryogenic sample preparation for the electron microscope (Figure 18). Scale bar, 100 μms. Photograph courtesy of V. A. Munoz.

water (Figure 26) also show no evidence of the floc structure shown in Figure 18. Addition of bicarbonate to the OSLO MFT (Figure 27, top) produced a floc or aggregate structure similar to that in the Clark MFT. Addition of acid and lime (Figure 27, bottom) to Clark MFT, which removes bicarbonate produces a structure similar to the OSLO MFT (Figure 25).

The importance of water chemistry in determining the properties of the MFT suspension was investigated using several techniques. Figures 28 and 29 show that the amount of water released during centrifugation and the specific resistance to filtration (SRF) both have a distinct pH dependence (56, 57). Figure 30 shows the same pH dependence for the electrosonic amplification (ESA) signal and the elastic modulus (G'). The ESA is related to the electrophoretic mobility and the charge on the particle surface. The higher the absolute value of the charge, the more dispersed the suspension. Similarly, G' is also related to the degree of dispersion in the suspension. Higher G' values correspond to less disperse suspensions because G' is a measure of the elastic response of the system

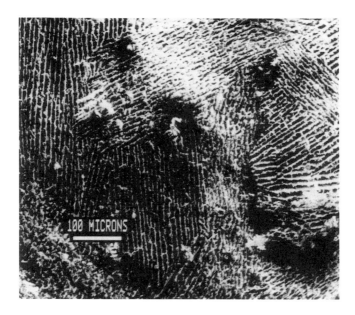

Figure 20.   Electron micrograph of an MFT sample that was vibrated to destroy the floc structure and allowed to sit for only a short time before being rapidly frozen. This image shows that the long range structure in Figure 18 has been disturbed.

to a shear stress. The degree of dispersion of the suspension is demonstrated by the minimum amount of water released (Figure 28) during low-speed centrifugation and the maximum resistance to filtration that correspond closely to the natural pH of the MFT suspension. Understanding the relationship between the suspension water chemistry, floc structure, and bulk MFT behavior is important in developing fine tailings treatment options.

MFT Management

Several approaches are being examined for handling MFT: reduction of the quantities of material produced, reduction of current inventory, and secure storage. The buildup of MFT in the recycle water ponds (Figure 5) is a significant reclamation challenge. Several reclamation options for fine tails have been developed, but the range of options is limited by the cost of handling the huge amounts of material involved, as well as technological feasibility and environmental constraints.

The search for solutions to any processing or reclamation problem such as the formation of MFT almost always involves one of two approaches: treating the tailings after they are formed or preventing for-

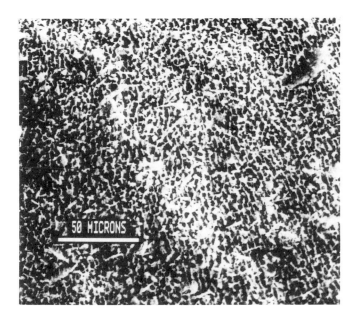

Figure 21. Electron micrograph of an MFT sample that was vibrated and then frozen immediately. The floc structure observed in Figures 18 and 20 is now almost completely destroyed. This series of electron micrographs illustrates the changes in floc or aggregate structure that can be modified with input of mechanical energy. These changes in floc structure are analogous to the series of sheared samples investigated using rheology (Figures 15–17).

mation in the first place. The proven efficiency of the existing CHWE flotation process over a wide range of ore grades and operating conditions makes it difficult to replace. This is despite certain drawbacks such as the dispersive nature of the process that contributes to the formation of the MFT suspension. Areas currently under investigation follow.

Selective Mining. One way to reduce the amount of MFT produced is to use selective mining to minimize the mining of clay bands and lenses with the oil sands. This reduces the amount of clays dispersed in the extraction process. Suncor has implemented selective mining by moving from dragline mining to truck-and-shovel mining.

Extraction Process Changes. A variety of extraction processes has claimed to achieve comparable bitumen recoveries from oil sands with less dispersion of the fine minerals. Two processes have been developed to the level of significant pilot demonstrations. One is the OSLO process, which involves flotation of the bitumen using mineral flotation

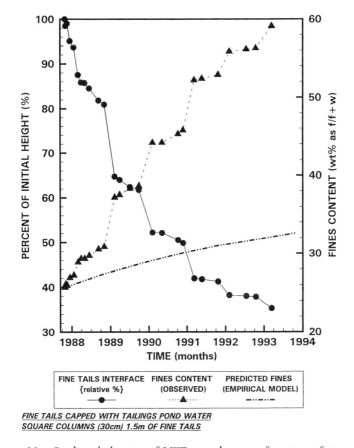

Figure 22.   Settling behavior of MFT samples as a function of time. The enhanced settling observed in the winter months is thought to be due to vibrations from nearby truck traffic. The samples are kept in a heated building that might be subjected to more vibration when the outside ground is frozen. The dashed line shows the fines content that would be predicted from the empirical model based on the data in Figure 8.

methods and reagents (58). On a pilot scale, tailings from the OSLO process were reclaimed after only a few years by surcharging the fines with coarse sand from the primary extraction. The absence of NaOH and the resulting lower pH (relative to the conventional hot water process) may lead to significantly less dispersion, a favorable floc structure, and greater gel strength in the resulting fines. These fine tailings are capable of supporting a sand surcharge and subsequently consolidating because of the applied stress.

Another process (demonstrated by Bitmin Resources) also attempts to minimizes fines dispersion in the extraction operation by minimizing

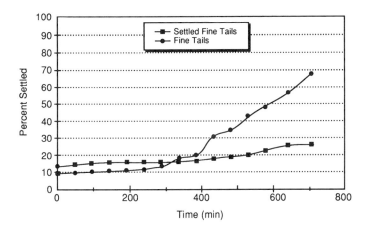

Figure 23. Centrifugation (230 × g, 1200 rpm) of a settled fine tailings sample compared with the fine tailings. The floc structure that is set up during settling at 1 × g (to approximately 15 wt% solids) inhibits further settling even under higher g forces. The fine tailings that have not settled to form a floc structure (approximately 7 wt% solids) are able to settle past this (metastable) point under 230 × g centrifugation.

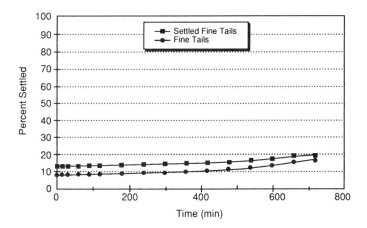

Figure 24. With even lower speed centrifugation (115 × g, 600 rpm) of a settled fine tailings sample compared with the unsettled fine tailings, the settling behavior for the two samples is similar because the floc structure is able to form at 600 rpm (115 × g) in a similar manner to settling at 1 × g. This gives some indication of the strength of the particle aggregates and the conditions under which they will form.

Figure 25.   MFT from the OSLO extraction process that shows no evidence of the structure found in the Clark hot water process MFT.

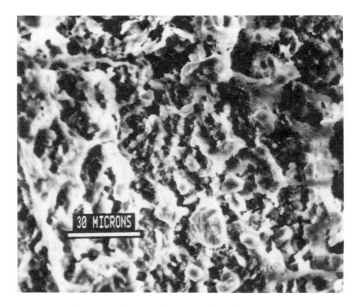

Figure 26.   Kaolinite in deionized water, also showing no evidence of preferred orientation or floc structure as was found in the Clark MFT.

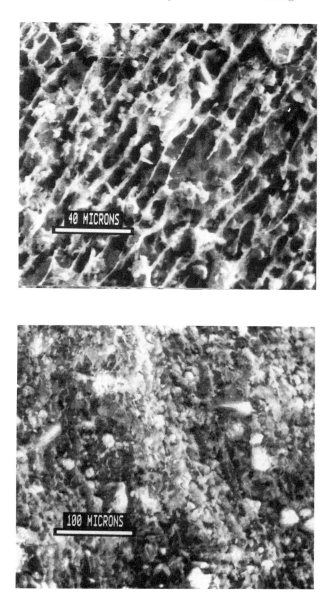

Figure 27. The OSLO MFT with added bicarbonate showing structuring that is similar to that in the Clark MFT. The lower photograph shows the Clark or CHWE MFT after treatment with acid and lime to reduce bicarbonate content and adjust the pH. There is very little evidence of structuring in this chemically treated sample. This change in the microscopic floc or aggregate structure between the Clark and OSLO MFT samples from these two extraction processes illustrates the importance of water chemistry in determining tailings behavior.

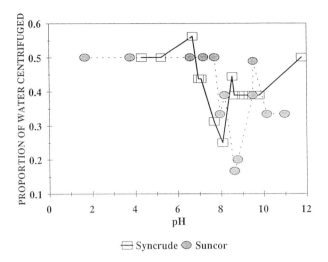

Figure 28. Water released during low-speed centrifugation of MFT as a function of pH. The minimum amount of water released is near pH 8.5 that corresponds approximately to the buffered pH of the bicarbonate system.

mechanical energy input and maintaining a near neutral pH (59). In this case, the tailings are flocculated, combined, filtered, and reclaimed as solid tailings. Figure 31 shows the dewatered tailings stream and subsequent tailings handling.

Production of a Geotechnically Stable Material. The existing commercial operations have produced significant quantities of MFT that will eventually be reclaimed regardless of future process changes. A variety of potential reclamation processes has been investigated with varying degrees of success. Mixing MFT with the smectitic clay overburden was successfully demonstrated but will be an option only on leases where sufficient amounts of suitable overburden exist (60, 61). Other methods include flocculation and filtration of the tailings (62), oil agglomeration processes (63), natural evaporation (64), hydrocyclone densification (65), spiking the total tailings to entrain fines with the coarse sand to minimize the amount of fines that end up as MFT (66), and others (67–84).

Enhanced Dewatering. Two of the most successful reclamation options include the use of freeze–thaw densification and the creation of nonsegregating mixtures (66, 67). Freeze–thaw processes use the cold winter season to freeze the suspension, concentrating the particles and releasing water from the suspension during thawing. Several investigators have investigated the factors that determine freeze–thaw be-

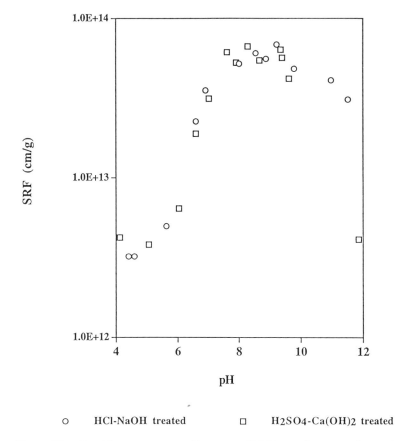

o   HCl-NaOH treated   □   H2SO4-Ca(OH)2 treated

Figure 29.   Specific resistance to filtration of MFT as a function of pH. The greatest resistance to filtration is also near pH 8.5, analogous to the behavior shown in Figures 27 and 28, illustrating the importance of water chemistry in determining MFT behavior.

havior, including freezing rate and chemical modifications to enhance water release (67, 68). The freeze–thaw process extracts water normally tied up by the clay–water suspension and, in the process of growing a crystalline water structure, forces individual clay particles together. Figure 32 shows the typical reticular ice structure and the individual soil peds or clumps that are formed as a result of the freezing process. The gaps that remain after formation of the ice structure provide an excellent path for water to drain upward. The freezing process removes water from the suspension in the process of growing ice crystals, mechanically forcing clay flocs or aggregates together. The resulting increase in ion concentration also provides an additional chemical driving force for dewatering.

SUNCOR TAILINGS

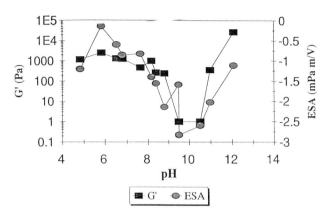

Figure 30.   The ESA and G′ of mature fine tailngs as a function of pH. The maximum in the absolute value of the ESA indicates a maximum of electrical dispersion force between the particles in suspension. The minimum in G′ also indicates a relatively disperse suspension.

Freeze–Thaw Processing.   Freeze–thaw processing of oil sands fine tails has been demonstrated to be effective in reducing volumes of the clay suspension on a large scale. Field tests carried out from 1992 to 1994 demonstrated the effectiveness of freeze–thaw dewatering, as well as the feasibility of reclaiming a dry landscape incorporating significant quantities of fine tailings.

Suncor's predecessor, Great Canadian Oil Sands, investigated the freeze–thaw process with dilute suspensions of fine tailings (10 wt% solids) in the early 1970s. Laboratory studies showed that volume reduction was feasible, but the process was not economically attractive. Dewatering only advanced to the 30–40 wt% solids range, and this was found to occur naturally through settling. Work by Johnson and Syncrude Canada Ltd. revived interest in the process when laboratory and small field tests were carried out from 1985 to 1987. This work showed that it was possible to achieve a solid product after several freeze–thaw cycles (67).

Sego and colleagues (68–72) set up a laboratory that continues to investigate the freeze–thaw process. Recent work shows that the amount of thaw strain volume reduction in the thawed material, after free water is decanted, is related to the extraction process chemistry that produced the fine tailings and subsequent chemical treatment of the fine tailings.

Suncor processed 100,000 cubic yards (over 75,000 m^3) of oil sands fine tailings through single freeze–thaw cycles in the winters of 1992 and 1993 and 1993 and 1994. Approximately 75,000 cubic yards

Figure 31. Tailings handling in the Bitmin process in which the fine tailings are flocculated, combined with the coarse tailings, and filtered. The material is clearly very stable. Photograph courtesy of Bitmin Resources.

Figure 32.   Photograph of the ice structure that is formed during freezing of MFT. The ice lenses and subsequent soil ped structures provide efficient drainage paths for the release of water.

(57,000 m³) of acid-and lime-treated fine tailings (400 ppm sulphuric acid, 400 ppm lime) and 25,000 cubic yards (19,000 m³) of untreated pond 1 fine tailings were processed each winter. The second winter's test was done to verify the results of the first winter.

Field tests show that fine tailings suspensions containing 30 wt% mineral could be frozen up to depths of 3 m and, upon thawing and draining, would achieve solids contents of 75 wt%. Generally, results of field tests have been better than the results of similar tests carried out in the laboratory (Figure 33). The explanation appears to be that the thawing process is significantly slower in the field than in the laboratory (months vs. 24 h). The freezing and thawing fine tailings spend significant periods near the freezing point, and water is free to move about and produce larger ice crystals.

The field tests have investigated several methods of applying the freeze–thaw process: (1) serial application of thin layers of fine tailings on top of previously frozen material, (2) application of thin layers on top of the frozen surface of a fine tailings reservoir, and (3) annual in-

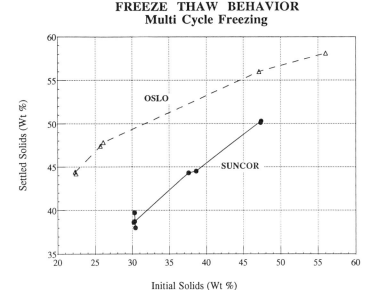

Figure 33. The increase in weight percent solids as a function of freezing cycle for various MFT samples (D. Sego, unpublished data).

cremental freezing of a fine tailings reservoir to freeze and dewater the top layer of fine tailings, which settles after dewatering, exposing fresh material to the surface for freezing the following year.

Experiments in the field investigated only acid- and lime-treated fine tailings and untreated material. Laboratory studies show that acid-only treatment is effective in increasing thaw strain, as is lime treatment by itself. It has also been shown that treatment with $CaSO_4$ produces similar results. Sego and colleagues reported that single-cycle thaw strain is related to the exchangable sodium ratio of tailings suspensions. The field tests demonstrated that it was feasible to freeze up to 2.5 m of fine tails and then to thaw and dewater the tailings. In this case, bottom drainage conditions were ideal as the 60 × 100-m test cells were constructed on sand tailings with improvement dykes. The 2.5 m of frozen fine tailings yielded 0.75 m of dewatered clay after one season. The second year of field tests involved improvements in field application techniques but were done on top of the previous year's test clay deposit to verify the process results in the absence of bottom drainage and determine the effectiveness of consolidation as a result of additional loading.

Based on these field tests, practical application of the freeze–thaw process on a large scale is believed to be feasible. To effectively use the infrastructure and impoundment areas necessary for commercial application, freeze–thaw processing of fine tailings requires long time frames, about 10–20 years. Under these circumstances, freeze–thaw drying farms of several square kilometers could process up to 100 million m³ of fine tailings at costs between $0.10 and $0.40 per m³, depending on the cost of chemicals and whether the farm was integrated into the mining operation. Furthermore, analysis of release water from freeze–thaw tests of acid- and lime-treated fine tailings show it to be nontoxic. Potentially, it could be released to the environment after further treatment.

Nonsegregating Tailings.   Nonsegregating tailings (NST) are the product of chemical treatment to flocculate the coarse sand component of the tailings with the fine clay minerals. The tailings do not segregate into coarse and fine materials, and the association of the coarse solids imparts an internal stress that significantly enhances the consolidation of the nonsegregating material. Nonsegregating mixtures of sand tailings and fine tailings brings into play the force of gravity (the weight of sand particles that are effectively suspended) in the suspension of fines and water to force the water from between the clay flocs or aggregates.

The nonsegregating mixture has a higher permeability than a clay suspension of a similar void ratio, resulting in more rapid dewatering. The addition of the sand provides self-weight force transferred from one sand grain to another and to the clay aggregates within the sand structure. NST are used in the Florida phosphate industry in which montmorillonite clay is the dominant fine mineral, although unlike oil sands tailings, no chemical additives are required. Oil sands tailings and fine tailings mixtures require a chemical additive to produce nonsegregating behavior. They also require minimum concentrations of clay, as the clay is responsible for adhering to and supporting the sand particles. Scott et al. (73), in a study of Syncrude tailings, found that the addition of lime would produce nonsegregating behavior (74). NST contain between 50 and 60 wt% solids when deposited. They undergo a period of low-stress sedimentation and initial consolidation, in which most of the free water is released as the mixture settles to 70-plus wt% solids (Figure 34). At this point, an effective stress starts to develop and continues to increase as additional loading is applied in the form of added layers of NST (75–79). Scott et al. (72) developed a nomogram that defines the fines, water, and sand proportions that make a nonsegregating mixture. Chemical treatment can enhance this behavior by moving the NST region to higher fines:sand ratios. Figure 35 shows the nomogram that defines NST behavior, whereas Figure 36 illustrates the principle of NST water release by the stress that is applied to the MFT by the self-weight of the coarse tailings that are suspended in the fines aggregate structure.

Implementation of NST mixtures would eliminate fine tailings production from the present Clark hot water process bitumen extraction

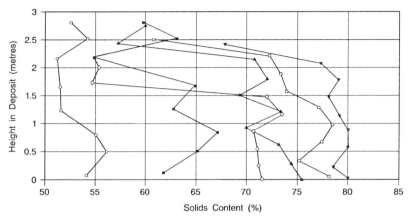

Figure 34. The increase in weight percent solids with time for one of the Suncor field test cells. The increasing solids content (approaching that of a stable soil) with constant fines ratio indicates that the suspension is non-segregating.

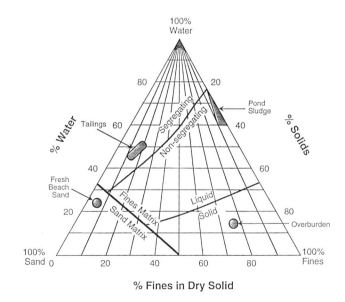

Figure 35. Nomogram of the fines, water, and sand tailings mixtures, defining the fines and sand content that results in nonsegregating mixtures (18, 35, 79, 80).

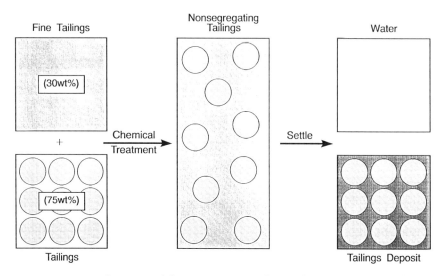

Figure 36. *Schematic of the NST process: chemical treatment of the fine tailings such that the floc or aggregate structure can support the coarse tailings, followed by the consolidation of the combined tailings to release water, and create a more stable tailings deposit.*

plants. In 1993, Suncor conducted a field test that involved forming a nonsegregating mixture from extraction plant tailings that were hydrocycloned to produce a thickened sand mixture containing 75 wt% solids. This stream was mixed with approximately an equal volume of fine tailings containing 30 wt% solids, and the mixture was pumped through a pipeline. Sulphuric acid and lime were injected into the pipeline in turn. Over the course of the field test, five trenches, 60 m long, 15 m wide, and 4.5 m deep, were filled with the nonsegregating mixture. Layers 0.45 m thick were placed approximately every 1–2 weeks and allowed to settle before the next layer was placed. Approximately one third of the volume was released as relatively clean water during the first week of settling. This release water was not acutely toxic and could probably be released safely after further treatment. This field test demonstrated that the NST process was practical on a large scale and that the hydraulic pipeline properties of the NST are superior to those of ordinary tailings.

Fine Tails Capping.   Another reclamation option being studied is the capping of mature fine tails with a water layer, in which a lake ecosystem develops (75–77). In this reclamation option, the fine tails material remains as a fluid. As a result, it must be deposited in geotechnically and hydrogeologically secure areas. Utilization of mined out pits provides the opportunity for below grade storage areas that will minimize the potential for release of the contained fine tails. The fine tails will be capped with a layer of water of sufficient depth to prevent mixing of the layers and to allow the formation of a stable self-sustaining lake ecosystem that is isolated from direct contact with the local environment. A significant research effort has been expended to assess the feasibity of this approach (78). At present, research supports the fine tails capping approach as being technically and environmentally acceptable for the long-term safe reclamation of fine tails.

As discussed earlier, the mature fine tails material has relatively high yield strength and thixotropic or gel-like character (51, 53). It has relatively high viscosity under low shear rates. At very low shear rates it is shear thickening, and it is only at higher shear rates that it is shear thinning. These properties make its disposal as a fluid an operational alternative to the other reclamation methods. Geotechnically, fine tails are weak, not stackable, and unable to support any loading. Direct loading of this material with a solid capping layer is not likely without first dewatering or mixing with other materials as described previously. In a fine tails capping scenario, mature or aged fine tails (>35% solids by weight) would be transferred from an active settling basin to the mined-out pit (77). The fine tails would then be capped with a layer of water, which would develop into a lake similar to those in the surrounding boreal forest region. A littoral zone around the edge of the lake would be enhanced to reduce erosion and increase use by waterfowl. Although slow, further dewatering of the fine tails zone would continue, and over a period of centuries, it is predicted to consolidate to a nonfluid material.

During the initial period of its capping, the potential for redisturbance of the reclaimed fine tails solids by wind or thermal mixing across the fine tails to water interface must be considered. Such an event could lead to high suspended clay loadings into the lake environment that would be detrimental to biota in the water zone. By proper design of the physical aspects of these capped systems and given the rheological properties of the fine tails, such events can be minimized (79). A water capped fine tails lake system is expected to become a stable self-sustaining ecosystem with acceptable levels of biological diversity and productivity. In addition, such a system will provide a safe long-term storage area for the fine tails and will isolate any biologically detrimental components in the fine tails from impact on the biota of the region.

Ultimately, some combination of methods might be used for fine tailings reclamation. Athough freeze–thaw, nonsegregating mixtures, and water capping are all being tested successfully at significant pilot scales, they are not necessarily competing technologies. For instance, freeze–thaw and nonsegregating mixture methods may be used to minimize the amount of material that will ultimately be water capped, especially in cases where tailings properties prevent complete dewatering

to a material with a significant strength. Furthermore, complete reclamation will probably involve a variety of options, depending on the
particular mining plan, overburden properties, and amount of MFT.
The freeze–thaw, NST, and water capping methods are discussed in
detail in references 68 and 73–75.

Summary

Tailings properties and tailings mitigation processes have been under
investigation for almost 30 years. In the last 5 years, a concerted research
effort and new mine and lease development options have accelerated
progress in understanding the composition and structure of oil sands
fine tailings and in the development of treatment options. Much of this
recent research effort has been under the auspices of the Fine Tailings
Fundamentals Consortium, involving industry, universities, and government.

   It is understood that the fine clay minerals that make up the MFT
suspension have a particular floc or aggregate structure due to the mineralogy and water chemistry of the tailings. The stable nature of these
suspensions is a distinct function of the pH of the system and the dispersive nature of the Clark hot water process. The dispersion of the
suspension is enhanced by pH (near 8.5) and the mechanical energy
used to disperse the suspension of bitumen, sand, clays, and water in
the extraction process. Ideally, future oil sands development will address
the process issues that contribute to the dispersion of fines and creation
of stable clay suspensions. However, the current operations have a significant lifetime ahead and an increasingly important role in maintaining
Canada's continued energy self-sufficiency. Because the properties of
the MFT suspensions are defined by the surface properties of the fine
clays and the chemistry of interstitial water, fundamental studies are
likely to be important in the ultimate understanding of the stable nature
of the MFT.

   Research supported through the Fine Tailings Fundamentals Consortium has contributed significantly to the establishment of options for
mitigation of the MFT produced in the hot water process. The amount
of MFT material to be handled will likely ultimately exceed 1 billion
m^3. Any reclamation scheme will be costly, even just in terms of moving
the material to treatment and disposal areas. Treatment and reclamation
of existing tailings will most likely involve an integrated approach by
using a combination of fines reduction, entrapment, dewatering, and
storage that will give an economical and environmentally acceptable
solution. The fundamental understanding of the fine tailings properties
developed over the years has helped to establish the utility of these
mitigation methods. Although the new technologies identified for the

treatment and disposal of the MFT represent a significant research and development effort, considerable developmental work remains to be done to support implementation of the treatment options that will ultimately be commercialized.

Acknowledgments

R. Mikula and K. Kasperski acknowledge the support of the Panel for Energy Research and Development. The members of the Fine Tails Fundamentals Consortium include Syncrude, Suncor, Natural Resources Canada (CANMET), the National Research Council, the Alberta Research Council, Alberta Department of Energy, Environment Canada, OSLO (a consortium of the Other Six Lease Owners), and AOSTRA (Alberta Oil Sands Technology and Research Authority). We thank C. W. Angle, J. Kan, V. A. Munoz, D. Scott, D. Sego, W. Strand, R. Zrobok, and Y. Xu for helpful discussions and for contributing some of the figures. We also thank Syncrude and Suncor for permission to release previously unpublished information in this chapter.

References

1. Kasperski, K. L., *AOSTRA J. Res.* **1991**, 8.
2. MacKinnon, M.; Sethi, A. *Proceedings of the Fine Tailings Fundamentals Symposium Oil Sands, Our Petroleum Future;* Edmonton, April 1993; F2.
3. MacKinnon, M. D. *AOSTRA J. Res.* **1989**, 5, 109.
4. Yong, R. N.; Sethi, A. J. *J. Can. Pet. Technol.* **1978**, *17*, 76.
5. Scott, J. D.; Dusseault, M. B.; Carrier, W. D. III. *Appl. Clay Sci.* **1985**, *1*, 207.
6. Camp, F. W. *Can. J. Chem. Eng.* **1977**, 55, 581–590.
7. Hepler, L. G.; Smith, R. G. *The Alberta Oil Sands: Industrial Procedures for Extraction and Some Recent Fundamental Research;* Alberta Oil Sands Technology and Research Authority: Edmonton, Canada, 1994.
8. Kessick, M. A. *CIM Bull.* **1978**, *71*, 80.
9. Moschopedis, S. E.; Fryer, J. F.; Speight, J. G. *Fuel* **1977**, *56*, 109.
10. Hocking, M. B.; Lee, G. W. *Fuel* **1977**, *56*, 325.
11. Levine, S.; Sanford, E. In *Proceedings of the 30th Canadian Chemical Engineering Conference, Edmonton, Alberta, Oil Sands, Coal and Energy Modelling,* Wanke, S. E., Ed.; CSChE Publications: 1980, Vol. IV, p 1112.
12. Speight, J. G.; Moschopedis, S. E. *Fuel Proc. Technol.* **1977/78**, *1*, 261.
13. Speight, J. G.; Moschopedis, S. E. *Surface and Interfacial Phenomena Related to the Hot Water Processing of Athabasca Oil Sands;* Alberta Research Council Information Series 86; Alberta Research Council: Edmonton, Canada, 1980.
14. Kessick, M. A. *J. Can. Pet. Technol.* **1979**, *18*, 49.
15. Ignasiak, T. M.; Zhang, Q.; Kratochvil, B.; Maitra, C.; Montgomery, D. S.; Strausz, O. P. *AOSTRA J. Res.* **1985**, *2*, 21.
16. Sengupta, S.; Hall, E. S.; Tollefson, E. L. In *The Fourth UNITAR/UNDP International Conference on Heavy Crude and Tar Sands Proceedings;* Meyer, R. F.; Wiggins, E. J., Eds.; Alberta Oil Sands Technology and Research Authority: Edmonton, Canada, 1989; Vol. 5, p 45.

17. Kotlyar, L. S.; Sparks, B. D.; Kodama, H. *AOSTRA J. Res.* **1990**, *6*, 41.
18. Sparks, B. D.; Kotlyar, L. S.; Majid, A. A. *The Petroleum Society of CIM and AOSTRA 1991 Technical Conference;* Banff, Canada, April 21–24, 1991; p 117–1.
19. Sethi, A. J. *Proceedings of the Petroleum Society of the CIM Sludge Symposium;* Alberta Oil Sands Technology and Research Authority: Edmonton, Canada, 1991.
20. Yong, R. N.; Sethi, A. J. Can. Patent 1190498, 1985.
21. Mikula, R. J.; Angle, C. W.; Zrobok, R.; Kan, J.; Xu, Y. *Factors That Determine Oil Sands Sludge Properties;* CANMET Report 93–40 (CF); Natural Resources Canada: Devon, Canada, 1993.
22. Angle, C. W.; Zrobok, R.; Hamza, H. A. *The Petroleum Society of CIM and AOSTRA 1991 Technical Conference;* Banff, Canada, April 21–24, 1991; p 125–1.
23. Srinivasan, N. S.; Spitzer, J. J.; Hepler, L. G. *J. Can. Pet. Technol.* **1982**, *21*, 29–32.
24. Mikula, R. J.; Munoz, V. A.; Lam, W. W.; Payette, C. *Proceedings of the Fine Tailings Fundamentals Symposium Oil Sands, Our Petroleum Future;* Edmonton, Canada, April 1993; F4.
25. Mikula, R. J.; Payette, C.; Munoz, V. A.; Lam, W. W. *The Petroleum Society of CIM and AOSTRA 1991 Technical Conference;* Banff, Canada, April 21–24, 1991; p 120–1.
26. Hardy Associates Ltd. "Athabasca Oil Sands Tailings Disposal Beyond Surface Mineable Limits," Alberta Environment Report, 1979.
27. Longstaffe, F. J. "Clay Mineral Characteristics of Three Different Types of Oil-Sand," Alberta Oil Sands Technology and Research Authority Agreement #295 (360), Final Report, 1983.
28. Roberts, J. O. L.; Yong, R. N.; Erskine, H. L. In *Proceedings of the Applied Oilsands Geoscience Conference;* Edmonton, Canada, June 11–13, 1980.
29. Lane, S. J.; Khan, F. Z.; Tonelli, F. A. "Dry Tailings Disposal from Oil Sands Mining," Report to Environmental Protection Service, Environment Canada and Department of Supply and Services Canada, 1984.
30. Kessick, M. A. *Int. J. Miner. Process.* **1980**, *6*, 277.
31. Sethi, A. J. *The Petroleum Society of CIM and AOSTRA 1991 Technical Conference;* Banff, Canada, April 21–24, 1991; p 126–1.
32. Majid, A.; Ripmeester, J. A.; Davidson, D. W. *Organic Matter Absorbed to Heavy Metal Minerals in Oil Sands;* Report No. C1095–82S; National Research Council of Canada: Ottawa, Canada, 1982.
33. Kramers, J. W.; Brown, R. A. S. *CIM Bull.* **1976**, *69*, 92.
34. MacKinnon, M. D.; Retallack, J. T. *Preliminary Characterization and Detoxification of Tailings Pond Water at the Syncrude Canada Ltd. Oil Sands Plant, in Land and Water Issues Related to Energy Development, 4th Conference;* Rand, P. J., Ed.; Ann Arbor Science: Ann Arbor, MI, 1982; p 185.
35. Trevoy, L. W.; Schutte, R.; Goforth, R. R. *CIM Bull.* **1978**, *71*, 175.
36. Takamura, K. *Can. J. Chem. Eng.* **1982**, *60*, 538.
37. Kupchanko, E. E. *Preliminary Compatibility Studies of GCOS Sand Tailings Pond Water with Athabasca River;* Environmental Health Services Division Report; Alberta Department of Health: Canada, 1968.
38. Jonasson, R. G.; Schutte, R.; Danielson, L. J.; Zhou, Z. *The Petroleum Society of CIM and AOSTRA 1991 Technical Conference;* Banff, Canada, April 21–24, 1991; p 122–1.
39. Burchfield, T. E.; Hepler, L. G. *Fuel* **1979**, *58*, 745.

40. Scott, J. D.; Dusseault, M. B. In *Proceedings of the 33rd Canadian Geotechnical Conference;* Canadian Geotechnical Society: Canada, 1980.
41. Yong, R. N.; Elmonayeri, D. *Can. Geotech. J.* **1984,** *21,* 644.
42. Pollock, G. W. M.Sc. Thesis, Department of Civil Engineering, University of Alberta, Edmonton, Canada, 1988.
43. Yong, R. N.; Sheeran, D. E.; Sethi, A. J.; Erskine, H. L. *Proceedings of the 6th Symposium on Engineering Applications of Mechanics/33rd Annual Technical Meeting of the Petroleum Society of the CIM;* Calgary, Canada, June 6–9, 1982.
44. Dusseault, M. B.; Scott, J. D. *J. Part. Sci. & Technol.* **1983,** *1,* 295.
45. Devenny, D. W. *Oil Sand Tailings Integrated Planning to Provide Long Term Stabilization;* Canadian Land Reclamation Association: Canada, 1977.
46. Suthaker, N. N.; Scott, J. D. *Proceedings of the 47th Canadian Geotechnical Engineering Conference;* Halifax, Canada, 1994.
47. Devenny, D. *Proceedings of the Fine Tailings Fundamentals Symposium Oil Sands, Our Petroleum Future;* Edmonton, Canada, April 1993; F13.
48. Schutte, R.; Czarnecki, J. A.; Liu, J. K. *The Petroleum Society of the CIM and AOSTRA, 1991 Technical Conference Proceedings;* Banff, Canada, April 1991; p 119.
49. Sheeran, D.; Burns, R.; Gaston, L. *The Petroleum Society of CIM and AOSTRA 1991 Technical Conference;* Banff, Canada, April 21–24, 1991; Paper 91–116.
50. Schramm, L. L. *J. Can. Pet. Technol.* **1989,** *28,* 73.
51. Danielson, L. J.; MacKinnon, M. D. *AOSTRA J. Res.* **1990,** *6,* 99.
52. Zrobok, R.; Angle, C. W. *The Petroleum Society of CIM and AOSTRA 1991 Technical Conference;* Banff, Canada, April 21–24, 1991; p 129–1.
53. Zrobok, R.; Kan, J.; Angle, C. W.; Hamza, H. A. *Proceedings of the Fine Tailings Fundamentals Symposium Oil Sands, Our Petroleum Future;* Edmonton, Canada, April 1993; F9.
54. Babchin, A.; Reitman, V; Rispler, K. *The Petroleum Society of CIM and AOSTRA 1991 Technical Conference;* Banff, Canada, April 21–24, 1991; p 127–1.
55. MacKinnon, M. D. "Effects of Vibration on the Rate of Fine Tails Densification, Year 2;" Researcher Report to the Fine Tails Fundamentals Consortium, September 1991.
56. Kan, J; Zrobok, R. "Changes of Ion Concentrations with Time in pH-Adjusted Oil Sands Tailings;" CANMET Report 93–38 (CF); Natural Resources Canada: Devon, Canada, 1993.
57. Xu, Y; Mikula, R. J. "The Filterability and Surface Tension of Mature Fine Tailings: The Effect of pH and Calcium Ions;" CANMET Report 93–37; Natural Resources Canada: Devon, Canada, 1993.
58. Sury, K. N.; Paul, R.; Derenewski, T. M.; Schulz, D. G. *The New OSLO Process, Oil Sands, Our Petroleum Future Conference Proceedings;* Edmonton, Canada, April 1993.
59. Strand, W. Bitmin Resources, personal communication.
60. Dusseault, M. B.; Ash, P. O.; Scott, J. D. *Proceedings of the 6th International Conference Expansive Soils;* 1987.
61. Lord, E. R. F.; Maciejewski, W.; Cymerman, G.; Lahaie, R. *Proceedings of the Fine Tailings Fundamentals Symposium Oil Sands, Our Petroleum Future;* Edmonton, Canada, April 1993; F22.
62. Hamshar, J. A.; Gregoli, A. A.; Rimmer, D. P.; Yildirim, E. *Proceedings of the Fine Tailings Fundamentals Symposium Oil Sands, Our Petroleum Future;* Edmonton, Canada, April 1993; F14.

63. Majid, A.; Sparks, B. D. *Proceedings of the Fine Tailings Fundamentals Symposium Oil Sands, Our Petroleum Future;* Edmonton, Canada, April 1993; F15.
64. Burns, R.; Cuddy, G.; Lahaie, R. *Proceedings of the Fine Tailings Fundamentals Symposium Oil Sands, Our Petroleum Future;* Edmonton, Canada, April 1993; F16.
65. Shaw, W.; Beck, J.; Livingstone, W. *Proceedings of the Fine Tailings Fundamentals Symposium Oil Sands, Our Petroleum Future;* Edmonton, Canada, April 1993; F19.
66. Cuddy, G.; Mimura, W.; Lahaie, R. *Proceedings of the Fine Tailings Fundamentals Symposium Oil Sands, Our Petroleum Future;* Edmonton, Canada, April 1993; F21.
67. Johnson, R. L.; Bork, P.; Allen, E. A. D.; James, W. H.; Koverny, L. "Oil Sands Sludge Dewatering by Freeze Thaw and Evapotransperation;" Report # RRTAC 93-8 (271); Alberta Conservation and Reclamation Council: Canada, 1993.
68. Sego, D. C.; Dawson, R.; Burns, R.; Lowe, L.; Derenewski, T.; Johnson, R. *Proceedings of the Fine Tailings Fundamentals Symposium Oil Sands, Our Petroleum Future;* Edmonton, Canada, April 1993; F17.
69. Sego, D. C.; Dawson, R. F.; Derenewski, T.; Burns, B. *Proceedings of the 7th International Cold Regions Engineering Specialty Conference;* Edmonton, Canada, March 7–9, 1994.
70. Dawson, R. F.; Sego, D. C. *Proceedings of the 46th Canadian Geotechnical Conference;* Saskatoon, Canada, April 4–7, 1993.
71. Sego, D. C.; Dawson, R. F.; Derenewski, T.; Burns, B. *Proceedings of the 1993 Eastern Oil Shale Symposium;* Lexington, MA, November 16–19, 1993.
72. Sego, D. C.; Proskin, S. A.; Burns, R. *Proceedings of the 46th Canadian Geotechnical Conference;* Saskatoon, Canada, April 4–7, 1993.
73. Scott, D.; Liu, Y.; Caughill, D. L. *Proceedings of the Fine Tailings Fundamentals Symposium Oil Sands, Our Petroleum Future;* Edmonton, Canada, April 1993; F18.
74. Derenewski, T.; Mimura, W. *Proceedings of the Fine Tailings Fundamentals Symposium Oil Sands, Our Petroleum Future;* Edmonton, Canada, April 1993; F20.
75. Gulley, J.; MacKinnon, M. D. *Proceedings of the Fine Tailings Fundamentals Symposium Oil Sands, Our Petroleum Future;* Edmonton, Canada, April 1993; F23.
76. MacKinnon, M. D.; Boerger, H. *Assessment of a Wet Landscape Option for Disposal of Fine Tails Sludge from Oil Sands Processing;* Petroleum Society of the Canadian Institute of Mining, Metallurgy, and Petroleum: Calgary, Canada, 1991; Vol. 3, pp 124-1–124-15.
77. Boerger, H.; MacKinnon, M. D.; Van Meer, T.; Verbeek, A. In *Proceedings of the Second International Conference on Environmental Issues Management of Waste in Energy and Minerals Production;* Singhal, R. K.; Mehrotra, A. K.; Fytas, K.; Collins, J. L., Eds.; 1992; p 1249.
78. Nelson, L. R.; Gulley, J. R.; MacKinnon, M. *Proceedings of UNITAR Conference: Heavy Crude and Tar Sands—Fueling for a Clean and Safe Environment;* Houston, TX, February 1995.
79. Lawrence, G. A.; Ward, P. R. B.; MacKinnon, M. D. *Can. J. Civil Eng.* **1991,** *18,* 1047.
80. Liu, Y.; Caughill, D. L.; Scott, J. D.; Burns, R. *The 47th Canadian Geotechnical Conference;* Halifax, Canada, 1994.

81. Caughill, D. L.; Scott, J. D.; Liu, Y.; Burns, R.; Shaw, W. H. *The 47th Canadian Geotechnical Conference;* Halifax, Canada, 1994.
82. Morgenstern, N. R.; Scott, J. D. Presented at the ASCE Specialty Conference, New Orleans, LA, February 1995.
83. Scott, J. D.; Cymerman, G. J. *Proceedings of a Symposium Held in Conjunction with the ASCE Convention;* San Francisco, CA, October 1984.
84. Dusseault, M. B.; Scott, J. D.; Ash, P. O. *Proceedings of the Fourth UNITAR/ UNDP International Conference on Heavy Crude and Tar Sands;* Meyer, R. F.; Wiggins, E. J.; Eds.; 1989; Vol. 1, p 161.

RECEIVED for review July 5, 1994. ACCEPTED revised manuscript March 2, 1995.

GLOSSARY AND INDEXES

Glossary

Suspension Terminology

Laurier L. Schramm

Petroleum Recovery Institute 100, 3512 33rd Street N.W., Calgary, Alberta
T2L 2A6, Canada

*This glossary provides brief explanations for important terms in
the study of suspensions that occur in the petroleum industry,
whether such studies involve fundamental principles, experimental
investigations, or industrial applications. Even when restricted to
petroleum industry applications, the suspension field encompasses
aspects of many disciplines and comprises a large body of terminology. This selection of frequently used terms includes scientific
terms relating to the principles underlying suspension stability and
properties. Cross-references for the more important synonyms and
abbreviations are also included.*

MANY KINDS OF SUSPENSIONS OCCUR in the petroleum industry; some
are useful and some are unwanted, some are carefully formulated, and
some result from a separate process and must be dealt with as received.
A number of scientific and engineering disciplines are involved in their
study, treatment, and use. Because each field brings with it some of its
own specialized terminology, many terms can come to be associated
with suspensions and their properties. This glossary encompasses the
more important terms used in the chapters of this book. Some basic
knowledge of physical chemistry, geology, and chemical engineering is
assumed; otherwise, I tried to include as many important suspension
terms as possible and included cross-references for the more important
synonyms and abbreviations. The glossary also contains nonsuspension
terms from the broader fields of colloid science and petroleum science.

For terms drawn from fundamental colloid and interface science, I
placed much reliance on the recommendations of the IUPAC Commission on Colloid and Surface Chemistry (*1*). Many of the terms included
here were drawn from *The Language of Colloid and Interface Science*

0065–2393/96/0251–0727$21.25/0
© 1996 American Chemical Society

(2) to which the reader is referred for broader coverage. Several other sources were particularly helpful (3–6).

Ablation   The reduction of particles into smaller sizes due to erosion by other particles or the surrounding fluid. May also refer to the size reduction of liquid droplets due to erosion, as in the processing of an oil sand slurry in which the oil (bitumen) is very viscous.

Absolute Viscosity   Viscosity measured by using a standard method, with the results traceable to fundamental units. Absolute viscosities are distinguished from relative measurements made with instruments that measure viscous drag in a fluid without known or uniform applied shear rates. *See* Viscosity.

Absorbance   In optics, a characteristic of a substance whose light absorption is being measured. The Beer–Lambert law gives the ratio of transmitted (I) to incident (I_0) light as $\log(I/I_0) = alc$, where a is the absorptivity, l is the optical path length, and c is the concentration of species in the optical path. The logarithmic term is called the absorbance.

Absorption   The increase in quantity (transfer) of one material into another or of material from one phase into another phase. Absorption may also denote the process of material accumulating inside another.

Activated Carbon   Carbonaceous material (such as coal) that has been treated, or activated, to increase the internal porosity and surface area. This treatment enhances its sorptive properties. Activated carbon is used for the removal of organic materials in water and wastewater treatment processes. Also termed activated charcoal.

Adhesion   The attachment of one phase to another. *See* Work of Adhesion.

Adhesional Wetting   The process of wetting when a surface (usually solid), previously in contact with gas, becomes wetted by liquid. This term is sometimes used to describe wetting that includes the formation of an adhesional bond between the liquid and the phase it is wetting. *See also* Immersional Wetting, Spreading Wetting, Wetting.

Adsorbate   A substance that becomes adsorbed at the interface or into the interfacial layer of another material, or adsorbent. *See* Adsorption.

Adsorbent   The substrate material onto which a substance is adsorbed. *See* Adsorption.

Adsorption   The increase in quantity of a component at an interface or in an interfacial layer. In most usage it is positive, but it can be negative (depletion); in this sense, negative adsorption is a different process from desorption. Adsorption may also denote the process of components accumulating at an interface.

Adsorption Isotherm   The mathematical or experimental relationship between the equilibrium quantity of a material adsorbed and the composition of the bulk phase at constant temperature. The adsorption isobar is the analogous relationship for constant pressure, and the adsorption isostere is the analogous relationship for constant volume. *See also* Brunauer–Emmett–Teller Isotherm, Langmuir Isotherm.

Advancing Contact Angle   The dynamic contact angle that is measured when one phase is advancing, or increasing its area of contact, along an interface while in contact with a third immiscible phase. It is essential to state through which phase the contact angle is measured. *See also* Contact Angle.

Aerosol   Colloidal dispersions of liquids or solids in a gas. Distinctions may be made among aerosols of liquid droplets (e.g., fog, cloud, drizzle, mist, rain, spray) and aerosols of solid particles (e.g., fume and dust).

Agitator   Mechanical mechanisms that mix and recirculate colloidal dispersions within vessels. The mechanisms may include propellors, paddles, turbines, or shaking devices.

Agglomeration   The aggregation of particles, droplets, or bubbles in a dispersion. This term is sometimes used to indicate a combination of aggregation and coalescence processes. *See* Spherical Agglomeration.

Agglutination   *See* Aggregation.

Aggregate   A group of species, usually droplets, bubbles, particles or molecules, that are held together in some way. A micelle can be considered to be an aggregate of surfactant molecules or ions.

Aggregation   The process of forming a group of droplets, bubbles, particles, or molecules that are held together in some way. This process is sometimes referred to interchangeably as coagulation or flocculation, although in some usage these refer to aggregation at the primary and secondary minimum, respectively. The synonym agglutination has also been used. The reverse process is termed deflocculation or peptization. *See also* Electrocoagulation, Polymer Bridging, Primary Minimum, Sweep Flocculation. For suspensions, coagulation usually refers to the formation of compact aggregates, whereas flocculation refers to the formation of a loose network of particles, often due to polymer bridging. In montmorillonite clay suspensions, coagulation refers to dense aggregates produced by face–face oriented particle associations, whereas flocculation refers to loose aggregates produced by edge–face or edge–edge oriented particle associations. *See also* reference 7.

Aging   The properties of many colloidal systems may change with time in storage. Aging in suspensions usually refers to aggregation, that is, coagulation or flocculation. It is also used to describe the process of recrystallization, in which larger crystals grow at the expense of smaller ones, that is, Ostwald ripening.

Agitator Ball Mill   A machine for the comminution, or size reduction, of minerals or other materials. Such machines crush the input material by wet grinding in a cylindrical rotating bin containing grinding balls. Colloidal size particles can be produced with these mills.

Amott–Harvey Test   *See* Amott Test.

Amott Test   A measure of wettability based on a comparison of the amounts of water or oil imbibed into a porous medium spontaneously and by forced displacement. Amott test results are expressed as a displacement-by-oil (δ_o) ratio and a displacement-by-water ratio (δ_w). In the Amott–Harvey test, a core is prepared at irreducible water saturation and then an Amott test is run. The Amott–Harvey relative displacement (wettability) index is then calculated as $\delta_w - \delta_o$, with values ranging from -1.0 for complete oil-wetting to 1.0 for complete water-wetting. *See also* reference 8, Wettability, Wettability Index.

Amott Wettability Index   *See* Amott Test.

Andreason Pipet   A graduated cylinder having provision for withdrawing subsamples from the bottom. Used to study sedimentation in the determination of particle sizes. *See* Stokes' Law.

Anion-Exchange Capacity   The capacity for a substrate to adsorb anionic species while simultaneously desorbing (exchanging) an equivalent charge quantity of other anionic species. Example: this property is sometimes used to characterize clay minerals that often have very large cation-exchange capacities but may also have significant anion-exchange capacities. *See also* Ion Exchange.

Anisokinetic Sampling   *See* Isokinetic Sampling.

Anisotropic   A material that exhibits a physical property, such as light transmission, differently in different directions.

API Gravity   A measure of the relative density (specific gravity) of petroleum liquids. The API gravity, in degrees, is given by $°API = (141.5/\text{relative density}) - 131.5$, where the relative density at temperature t ($°C$) = (density at t)/(density of water at 15.6 $°C$).

Apparent Viscosity   Viscosity determined for a non-Newtonian fluid without reference to a particular shear rate for which it applies. Such viscosities are usually determined by a method strictly applicable to Newtonian fluids only.

Asphalt   A naturally occurring hydrocarbon that is a solid at reservoir temperatures. An asphalt residue may also be prepared from heavy (asphaltic) crude oils or bitumen, from which lower boiling fractions have been removed.

Asphaltene   A high molecular mass, polyaromatic component of some crude oils that also has appreciable quantities of sulfur, nitrogen, oxygen, and metal contents. In practical work, asphaltenes are usually defined operationally by using a standardized separation scheme. One such scheme defines asphaltenes as those components of a crude oil or bitumen that are soluble in toluene but insoluble in n-pentane.

Atterberg Limits   A group of (originally) seven limits of soil consistency, or relative ease with which material can be deformed or made to flow. The only Atterberg limits that are still in common use are the liquid limit, plastic limit, and plasticity number. *See* references 9 and 10.

Attractive Potential Energy   *See* Gibbs Energy of Attraction.

Attrition   The reduction of particle size by friction and wear. *See also* Ablation, Comminution.

Backscattering   *See* Light Scattering.

Basic Sediment and Water   That portion of solids and aqueous solution in an emulsion that separates out on standing or is separated by centrifuging in a standardized test method. Basic sediment may contain emulsified oil as well. Also referred to as BS&W, Bottom Settlings and Water, and Bottom Solids and Water.

Batch Mixer   A type of processing equipment in which the entire amount of material to be used is put into the mixer and mixed for a definite period of time, with multiple recirculation of material through the mixing zone, in contrast to what happens in a continuous mixer. After the mixing period, the whole amount of material is removed from the mixer.

Bed Knives   The stationary cutting blades in a cutting mill machine for comminution.

Beer–Lambert Law   *See* Absorbance.

Beneficiation   In mineral processing, any process that results in a product having an improved desired mineral content. Example: froth flotation.

BET Isotherm   *See* Brunauer–Emmett–Teller Isotherm.

Bingham Plastic Fluid   *See* Plastic Fluid.

Bitumen   A naturally occurring viscous hydrocarbon having a viscosity greater than 10,000 mPa·s at ambient deposit temperature and a density greater than 1000 kg/m^3 at 15.6 °C. *See* references 11–13. In addition to high molecular mass hydrocarbons, bitumen contains appreciable quantities of sulfur, nitrogen, oxygen, and heavy metals.

Boltzmann Equation   A fundamental equation giving the local concentrations of ions in terms of the local electric potential in an electric double layer. *See also* Poisson–Boltzmann Equation.

Bottle Test   In water treatment, a standard test method in which either the coagulant dosage is varied or the solution pH is varied for a given coagulant dosage to optimize the coagulation of solids. Frequently termed jar test. This term has different meanings in foams and in emulsions; *see* reference 2.

Bottom Settlings and Water   *See* Basic Sediment and Water.

Bottom Solids and Water   *See* Basic Sediment and Water.

Bright–Field Illumination   A kind of illumination for microscopy in which the illumination of a specimen is arranged so that transmitted light remains in the optical path of the microscope and is used to form the magnified image. This is different from the arrangement in Dark–Field Illumination.

Brownian Motion   Random fluctuations in the density of molecules at any location in a liquid, due to thermal energy, cause other molecules and small dispersed particles to move along random pathways. The random particle motions are termed Brownian motion and are most noticeable for particles smaller than a few micrometers in diameter.

Brunauer–Emmett–Teller Isotherm   (BET Isotherm) An adsorption isotherm equation that accounts for the possibility of multilayer adsorption and different enthalpy of adsorption between the first and subsequent layers. Five "types" of adsorption isotherm are usually distinguished. These are denoted by roman numerals and refer to different characteristic shapes. *See* Adsorption Isotherm.

BS&W   *See* Basic Sediment and Water.

Capillarity   A general term referring either to the general subject of or to the various phenomena attributable to the forces of surface or interfacial tension. The Young–Laplace equation is sometimes referred to as the equation of capillarity.

Capillary   A tube having a very small internal diameter. Originally, the term referred to cylindrical tubes whose internal diameters were of similar dimension to hairs.

Capillary Flow   Liquid flow in response to a difference in pressures across curved interfaces. *See also* Capillary Pressure.

Capillary Forces   The interfacial forces acting among oil, water, and solid in a capillary or in a porous medium. These determine the pressure difference (capillary pressure) across an oil–water interface in the capillary or in a pore. Capillary forces are largely responsible for oil entrapment under typical petroleum reservoir conditions.

Capillary Number   (N_c) A dimensionless ratio of viscous to capillary forces. One form gives N_c as Darcy velocity times viscosity of displacing phase divided by interfacial tension. It is used to provide a measure of the magnitude of forces that trap residual oil in a porous medium.

Capillary Pressure   The pressure difference across an interface between two phases. When the interface is contained in a capillary, it is sometimes referred to as the suction pressure. In petroleum reservoirs, it is the local pressure difference across the oil–water interface in a pore contained in a porous medium. One of the liquids usually preferentially wets the solid, and therefore the capillary pressure is normally taken as the pressure in the nonwetting fluid minus that in the wetting fluid.

Capillary Viscometer   An instrument used for the measurement of viscosity in which the rate of flow through a capillary under constant applied pressure difference is determined. This method is most suited to the determination of Newtonian viscosities. There are various designs, among which are the Ostwald and Ubbelohde types.

Carbon Black   Carbon particles of very small size; hence, large specific surface area capable of acting as an effective adsorbent for some substances.

Cation-Exchange Capacity   The capacity for a substrate to adsorb hydrated cationic species while simultaneously desorbing (exchanging) an equivalent charge quantity of other cationic species. Example: this property is used to characterize clay minerals that may have very large cation-exchange capacities and also significant anion-exchange capacities. *See also* Ion Exchange.

CCC   *See* Critical Coagulation Concentration.

CCT   *See* Critical Coagulation Temperature.

Centrifuge   An apparatus in which an applied centrifugal force is used to achieve a phase separation by sedimentation or creaming. For centrifuges operating at very high relative centrifugal forces (so-called *g* forces), the terms supercentrifuge (approximately tens of thousands RCF or *g*s) or ultracentrifuge (approximately hundreds of thousands RCF or *g*s) are used. *See also* Relative Centrifugal Force.

Centrifuge Test   *See* Bottle Test.

CFC   Critical flocculation concentration. *See* Critical Coagulation Concentration.

CFT   Critical flocculation temperature. *See* Critical Coagulation Temperature.

Charge Density   In colloidal systems, the quantity of charge at an interface expressed per unit area.

Charge Reversal   The process wherein a charged substance is caused to take on a new charge of the opposite sign. Such a change can be brought about by any of oxidation, reduction, dissociation, ion exchange, or adsorption. Example: the adsorption of cationic polymer molecules onto negatively charged clay particles can exceed the requirements for charge neutralization and thus cause charge reversal.

Chemical Adsorption   *See* Chemisorption.

Chemisorption   (Chemical adsorption) The adsorption forces are of the same kind as those involved in the formation of chemical bonds. The term chemisorption is used to distinguish chemical adsorption from physical adsorption, or physisorption, in which the forces involved are of the London–van der Waals type. Some guidelines for distinguishing between chemisorption and physisorption are given by IUPAC in reference 1.

Classifier   A machine used to separate particles of specified size ranges. Wet classifiers include settling tanks, centrifuges, hydrocyclones, and vibrating screens. Dry classifiers, also termed air classifiers, use gravity or centrifugal settling in gas streams. *See also* reference 14.

Classifier Mill   A kind of mechanical impact mill or jet mill for size reduction (comminution) that also incorporates a particle classifier.

Clays   Clay minerals refers to the aluminosilicate minerals having two- or three-layer crystal structure. These minerals typically exhibit high specific surface area, significant surface charge density (cation-exchange capacity), and low hydraulic conductivity. Examples: montmorillonite, kaolinite, illite. "Clays" is sometimes used to distinguish particles having sizes of less than about 2–4 μm, depending on the size classification system used. In this sense the term includes any suitably fine-grained solid, including nonclay minerals.

CMC   *See* Critical Micelle Concentration.

Coadsorption   The adsorption of more than one species simultaneously.

Coagulation   *See* Aggregation.

Coagulum   The dense aggregates formed in coagulation are referred to, after separation, as coagulum. *See* Aggregation.

Coalescence   The merging of two or more dispersed species into a single one. Coalescence reduces the total number of dispersed species and also the total interfacial area between phases. In emulsions and foams, coalescence can lead to the separation of a macrophase, in which case the emulsion or foam is said to break. The coalescence of solid particles is termed sintering.

Co-ions   In systems containing large ionic species (colloidal ions, membrane surfaces, and so on), co-ions are those that, compared with the large ions, have low molecular mass and the same charge sign. For example, in a suspension of negatively charged clay particles containing dissolved sodium chloride, the chloride ions are co-ions and the sodium ions are counterions. *See also* Counterions.

Collector   A surfactant used in froth flotation to adsorb onto solid particles, make them hydrophobic, and thus facilitate their attachment to gas bubbles. *See also* Froth Flotation.

Colloid Mill   A high-shear mixing device used to prepare colloidal dispersions of particles or droplets by size reduction (comminution). Also termed dispersion mill.

Colloid Osmotic Pressure   When a colloidal system is separated from its equilibrium liquid by a semipermeable membrane, not permeable to the colloidal species, the colloid osmotic pressure is the pressure difference required to prevent transfer of the dissolved noncolloidal species. Also referred to as the Donnan pressure. The reduced osmotic pressure is the colloid osmotic pressure divided by the concentration of the colloidal species. *See also* Osmotic Pressure.

Colloid Stability   Colloid stability is high if a specified process that causes the colloid to become a macrophase, such as aggregation, does not proceed at a significant rate. Colloid stability is different from thermodynamic stability (*see* reference 1). The term colloid stability must be used with reference to a specific and clearly defined process, for example, a colloidally metastable emulsion may signify a system in which the droplets do not participate in aggregation, coalescence, or creaming at a significant rate. *See also* Kinetic Stability, Thermodynamic Stability.

Colloid Titration   A method for the determination of charge, and the zero point of charge, of colloidal species. The colloid is subjected to a potentiometric titration with acid or base to determine the amounts of acid or base needed to establish equilibrium with various pH values. By titrating the colloid in different, known concentrations of indifferent electrolyte, the point of zero charge can be determined as the pH for which all the isotherms intersect. *See also* Point of Zero Charge.

Colloidal   A state of subdivision in which the particles, droplets, or bubbles dispersed in another phase have at least one dimension between about 1 and 1000 nm. *See also* Colloidal Dispersion.

Colloidal Dispersion   A system in which colloidal species are dispersed in a continuous phase of different composition or state.

Colloidal Processing   In ceramics, a variation of slip-casting in which a stabilized colloidal dispersion of particles is poured into a mold for sintering.

Comminution   The reduction of particles, or other dispersed species, into smaller sizes. Examples of comminution machines include agitator ball mills, colloid mills, cutting mills, disk mills, jet mills, mechanical impact mills, ring-roller mills, and roll crushers. *See also* reference 14, Ablation.

Compaction   *See* Subsidence.

Concentric Cylinder Rheometer   *See* Rheometer.

Condensation Methods   Used for preparing colloidal dispersions in which either precipitation from solution or chemical reaction is used to create colloidal species. The colloidal species are built up by deposition on nuclei that may be of the same or different chemical species. If the nuclei are of the same chemical species, the process is referred to as homogeneous nucleation; if the nuclei are of different chemical species, the process is referred to as heterogeneous nucleation. *See also* Dispersion Methods.

Cone–Cone Rheometer   *See* Rheometer.

Cone–Plate Rheometer   *See* Rheometer.

Confocal Microscopy   A microscopic technique used to produce three-dimensional images of specimens that actually have considerable depth. A series of shallow depth-of-field image slices through a thick specimen are obtained. The three-dimensional image is then obtained by reconstruction so that no out-of-focus elements contribute to the final image.

Consistency   An empirical or qualitative term referring to the relative ease with which a material can be deformed or made to flow. It is a reflection of the cohesive and adhesive forces in a mixture or dispersion. *See also* Atterberg Limits.

Contact Angle   When two immiscible fluids (e.g., liquid–gas or oil–water) are both in contact with a solid, the angle formed between the solid surface and the tangent to the fluid–fluid interface intersecting the three-phase contact point. It is essential to state through which phase the contact angle is measured. By convention, if one of the fluids is water, then the contact angle is measured through the water phase; otherwise, the contact angle is usually measured through the most dense phase. Distinctions may be made among advancing, receding, or equilibrium contact angles. Contact angles are important in areas such as liquid wetting, imbibition, and drainage.

Contact-Angle Hysteresis   A phenomenon manifested by differing values of advancing and receding contact angles in the same three-phase contact system. Both may differ from the equilibrium contact angle. *See also* Contact Angle.

Continuous Phase   In a colloidal dispersion, the phase in which another phase of particles, droplets, or bubbles is dispersed. Sometimes referred to as the external phase. Continuous phase is the opposite of dispersed phase. *See also* Dispersed Phase.

Couette Flow   The flow of liquid in the annulus between two concentric cylinders that rotate at different speeds. In the Couette rheometer, one cylinder rotates and torque is measured at the other. *See also* Rheometer.

Couette Rheometer   *See* Couette Flow.

Coulter Counter Technique   A particle- or droplet-sizing technique in which the flow of dispersed species in a capillary, between charged electrodes, causes changes in conductivity that are interpreted in terms of the sizes of the species. Coulter is the brand name for the automated counter. *See also* Sensing Zone Technique.

Counterions   In systems containing large ionic species (colloidal ions, membrane surfaces, and so on), counterions are those that, compared with the large ions, have low molecular mass and opposite charge sign. For example, clay particles are usually negatively charged and are naturally associated with exchangeable counterions such as sodium and calcium. In the early literature, the term Gegenion was used to mean counterion. *See also* Co-ions.

Critical Coagulation Concentration   (CCC) The electrolyte concentration that marks the onset of coagulation of dispersed species. The

CCC is very system-specific, although the variation in CCC with electrolyte composition has been empirically generalized. *See also* Schulze–Hardy Rule.

Critical Coagulation Temperature   (CCT) The minimum temperature to which a dispersion must be raised to induce coagulation. *See also* Critical Coagulation Concentration.

Critical Flocculation Concentration   *See* Critical Coagulation Concentration.

Critical Flocculation Temperature   *See* Critical Coagulation Temperature.

Critical Micelle Concentration   (CMC) The surfactant concentration above which molecular aggregates, termed micelles, begin to form. In practice, a narrow range of surfactant concentrations represents the transition from a solution in which only single unassociated surfactant molecules (monomers) are present to a solution containing micelles. Useful tabulations are given in references 15 and 16. *See also* Micelle.

Critical Surface Tension of Wetting   The minimum, or transition, surface tension of a liquid for which it will no longer exhibit complete wetting of a solid. This value is usually taken to be characteristic of a given solid and is sometimes used as an estimate of the solid's surface tension. For a given solid, it is typically determined by plotting the cosine of contact angles between the solid of interest and a series of liquids versus the surface tensions of those liquids (a Zisman plot). The surface tension extrapolated to zero contact angle is the critical surface tension of wetting of the solid. *See also* Hydrophobic Index.

Critical Wetting Surface Tension   *See* Critical Surface Tension of Wetting.

Crude Oil   A naturally occurring hydrocarbon produced from an underground reservoir. In the petroleum field, distinctions drawn among light, heavy, extra-heavy, and bituminous crude oils are made. *See also* references 11–13, Asphalt, Bitumen, Extra-Heavy Crude Oil, Heavy Crude Oil, Light Crude Oil, Oil.

Cryogenic SEM   *See* Freeze-Fracture Method, Electron Microscopy.

Cryogenic TEM   *See* Freeze-Fracture Method, Electron Microscopy.

Cutting Mill   A machine for the comminution, or size reduction, of materials. Such machines use a rotating shaft on which is mounted a series of cutting knives that interleave with a series of separately mounted stationary knives. Cutting mills can reduce materials to particles on the order of 100 μm.

Darcy's Law   *See* Permeability.

Dark-Field Illumination   A kind of illumination for microscopy in which the illumination of a specimen is arranged so that transmitted light falls out of the optical path of the microscope and only light scattered by a dispersed phase is observed. It is used to detect the presence of dispersed species that are smaller than the resolving power of the microscope. Sometimes termed dark-ground illumination. A microscope using this principle is referred to as an ultramicroscope. Example: commonly used in particle microelectrophoresis. Compare with Bright-Field Illumination.

Dark-Field Microscope   *See* Dark-Field Illumination, Ultramicroscope.

Deborah Number   In rheology, the dimensionless ratio of relaxation time for a process to the time of observation of that process.

Debye Forces   Attractive forces between molecules due to dipole-induced dipole interaction. *See also* Dispersion Forces.

Debye–Hückel Parameter   *See* Debye Length.

Debye–Hückel Theory   A description of the behavior of electrolyte solutions in which ions are treated as point charges and their distribution is described in terms of a competition between electrical forces and thermal motion. *See also* DLVO Theory.

Debye Length   A parameter in the Debye–Hückel theory of electrolyte solutions, κ^{-1}. The Debye length is also used in the DLVO theory, where it is referred to as the electric double-layer thickness and represents the distance over which the potential falls to about one-third of the value of the surface potential. *See* Electric Double-Layer Thickness.

Debye Parameter   *See* Debye Length.

Deflocculation   The reverse of aggregation (or flocculation or coagulation). Peptization means the same thing.

Depletion Flocculation   The flocculation of dispersed species induced by the interaction of adsorbed polymer chains. *See* Sensitization.

Depletion Stabilization   The stabilization of dispersed species induced by the interaction (steric stabilization) of adsorbed polymer chains. Also called steric stabilization. *See* Protection.

Depressant   Any agent that may be used in froth flotation to selectively reduce the effectiveness of collectors for certain mineral components. *See also* Froth Flotation.

Desorption The process by which the amount of adsorbed material becomes reduced. That is, the converse of adsorption. Desorption is a different process from negative adsorption. *See also* Adsorption.

Dialysate *See* Dialysis.

Dialysis A separation process in which a colloidal dispersion is separated from a noncolloidal solution by a semipermeable membrane, that is, a membrane that is permeable to all species except the colloidal-sized ones. Osmotic pressure difference across the membrane drives the separation. The solution containing the colloidal species is referred to as the retentate or dialysis residue. The solution that is free of colloidal species is referred to as the dialysate or permeate; at equilibrium (no osmotic pressure difference), this solution is referred to as the equilibrium dialysate. *See also* Ultrafiltration.

Dialysis Residue *See* Dialysis.

Dielectric Constant *See* Permittivity.

Differential Viscosity The rate of change of shear stress with respect to shear rate taken at a specific shear rate ($\eta_D = d\,\tau/d\,\dot{\gamma}$).

Diffuse Double Layer *See* Diffuse Layer, Electric Double Layer.

Diffuse Layer The Gouy layer, in an electric double layer.

Diffusiophoresis The movement of a colloidal species in response to the concentration gradient of another dissolved, noncolloidal, solute.

Dilatant A non-Newtonian fluid for which viscosity increases as the shear rate increases. The process is termed shear-thickening.

Discontinuous Phase *See* Dispersed Phase.

Discreteness of Charge Charged colloidal species usually obtain their charge from a collection of discrete charge groups present at their surfaces. This discreteness of charge is, however, frequently approximated as a uniform surface charge distribution in descriptions of colloidal phenomena (e.g., DLVO theory). *See also* Esin–Markov Effect.

Disk Attrition Mill *See* Disk Mill.

Disk Mill A machine for the comminution, or size reduction, of wood products or other material. Such machines crush the input material between two grinding plates mounted on rotating disks. Also termed disk attrition mill.

Dispersant Any species that may be used to aid in the formation of a colloidal dispersion. Often a surfactant.

Disperse   *See* Dispersion.

Dispersed Phase   In a colloidal dispersion, the phase that is distributed in the form of particles, droplets, or bubbles into a second immiscible phase that is continuous. Also referred to as the disperse, discontinuous, or internal phase. *See also* Continuous Phase.

Disperse Phase   *See* Dispersed Phase.

Dispersing Agent   *See* Dispersant.

Dispersion   (Colloids) A system in which finely divided droplets, particles, or bubbles are distributed in another phase. As it is usually used, dispersion implies a distribution without dissolution. A suspension is an example of a colloidal dispersion; *see also* Colloidal.

(Fluid-Flow Phenomena)   The mixing of one fluid in another immiscible fluid by convection and molecular diffusion during flow through capillary spaces or porous media.

(Groundwater Contamination)   The mixing of a contaminant with a noncontaminant phase. The mixing is due to the distribution of flow paths, tortuosity of flow paths, and molecular diffusion.

Dispersion Forces   Interaction forces between any two bodies of finite mass. Sometimes called van der Waals forces, they include the Keesom orientation forces between dipoles, Debye induction forces between dipoles and induced dipoles, and London (van der Waals) forces between two induced dipoles. Also referred to as Lifshitz–van der Waals forces.

Dispersion Methods   The class of mechanical methods used for preparing colloidal dispersions in which particles or droplets are progressively subdivided. *See also* Condensation Methods.

Dispersion Mill   *See* Colloid Mill.

Dissolved-Gas Flotation   *See* Froth Flotation.

DLVO Theory   An acronym for a theory of the stability of colloidal dispersions developed independently by B. Derjaguin and L. D. Landau in one research group and by E. J. W. Verwey and J. Th. G. Overbeek in another. *See* Chapter 1. The theory was developed to predict the stability against aggregation of electrostatically charged particles in a dispersion.

Donnan Pressure   *See* Colloid Osmotic Pressure.

Dorn Potential   *See* Sedimentation Potential.

Double Layer   *See* Electric Double Layer.

Drag   The force due to friction experienced by a moving dispersed species.

Dynamic Interfacial Tension   *See* Equilibrium Surface Tension.

Dynamic Surface Tension   *See* Equilibrium Surface Tension.

EDL   *See* Electric Double Layer.

Elastic Scattering   *See* Light Scattering.

Electric Double Layer   (EDL) An idealized description of the distribution of free charges in the neighborhood of an interface. Typically, the surface of a charged species is viewed as having a fixed charge of one sign (one layer), whereas oppositely charged ions are distributed diffusely in the adjacent liquid (the second layer). The second layer may be considered to be made up of a relatively more strongly bound Stern layer in close proximity to the surface and a relatively more diffuse layer (Gouy layer, or Gouy–Chapman layer) at greater distance.

Electric Double-Layer Thickness   A measure of the decrease of potential with distance in the diffuse part of an electric double layer. It is the distance over which the potential falls to $1/e$, or about one-third, of the value of the surface or Stern layer potential, depending on the model used. Also termed the Debye length.

Electrocoagulation   Coagulation induced by exposing a dispersion to an alternating electric field gradient between two sacrificial metal electrodes. Electrocoagulation is apparently due to a combination of the alternating electric field and the adsorption on dispersed particles, or droplets, of ions solubilized from the electrodes. *See also* reference 17.

Electrocratic   A dispersion stabilized principally by electrostatic repulsion.

Electrocrystallization   A kind of electrodeposition in which ions from solution become deposited on or into an electrode surface and then participate in crystallization, the building up of old crystals, or the growing of new ones at the electrode surface.

Electrodecantation   A separation process in which a colloidal dispersion is separated from a noncolloidal solution by an applied electric field together with the force of gravity. Also called electrophoresis convection.

Electrodeposition   The deposition of dissolved or dispersed species on an electrode under the influence of an electric field.

Electrodialysate   *See* Electrodialysis.

Electrodialysis   Separation process similar to dialysis and ultrafiltration in which a colloidal dispersion is separated from a noncolloidal solution by a semipermeable membrane, that is, a membrane that is permeable to all species except the colloidal-sized ones. Here, an applied electric field (rather than osmotic pressure or an applied pressure) across the membrane drives the separation. As in dialysis and ultrafiltration, the solution containing the colloidal species is referred to as the retentate or dialysis residue. However, the solution that is free of colloidal species is referred to as electrodialysate rather than dialysate because the composition is usually different from that produced by dialysis. *See also* Dialysis, Ultrafiltration.

Electroendosmosis   *See* Electroosmosis.

Electroformed Sieve   *See* Particle Size Classification.

Electrokinetic   A general adjective referring to the relative motions of charged species in an electric field. The motions may be either of charged dispersed species or of the continuous phase, and the electric field may be either an externally applied field or created by the motions of the dispersed or continuous phases. Electrokinetic measurements are usually aimed at determining zeta potentials.

Electrokinetic Potential   *See* Zeta Potential.

Electron Microscopy   Three types: transmission electron microscopy (TEM), scanning electron microscopy (SEM), and scanning transmission electron microscopy (STEM). TEM is analogous to transmitted-light microscopy but uses an electron beam rather than light and uses magnetic lenses to produce a magnified image on a fluorescent screen. In SEM, a surface is scanned by a focused electron beam and the intensity of secondary electrons is measured and used to form an image on a cathode-ray tube. In STEM, a surface is scanned by a very narrow electron beam that is transmitted through the sample. The intensities in the formed image are related to the atomic numbers of atoms scanned in the sample.

Electroosmosis   The motion of liquid through a porous medium caused by an imposed electric field. The term replaces the older terms electrosmosis and electroendosmosis. The liquid moves with an electroosmotic velocity that depends on the electric surface potential in the stationary solid and on the electric field gradient. The electroosmotic volume flow is the volume flow rate through the porous plug and is usually expressed per unit electric field strength. The electroosmotic pressure is the pressure difference across the porous plug that is required to just stop electroosmotic flow.

Electroosmotic Pressure   *See* Electroosmosis.

Electroosmotic Velocity   *See* Electroosmosis.

Electrophoresis   The motion of colloidal species caused by an imposed electric field. The term replaces the older term cataphoresis. The species move with an electrophoretic velocity that depends on their electric charge and the electric field gradient. The electrophoretic mobility is the electrophoretic velocity per unit electric field gradient and is used to characterize specific systems. An older synonym, no longer in use, is kataphoresis. The term microelectrophoresis is sometimes used to indicate electrophoretic motion of a collection of particles on a small scale. Previously, microelectrophoresis was used to describe the measurement techniques in which electrophoretic mobilities are determined by observation through a microscope. The recommended term for these latter techniques is now microscopic electrophoresis (*see* reference 1).

Electrophoresis Convection   *See* Electrodecantation.

Electrophoretic Mobility   *See* Electrophoresis.

Electrophoretic Velocity   *See* Electrophoresis.

Electrorheological Fluid   A dispersion of microscopic particles suspended in a low permittivity, low conductivity liquid such that the dispersion's flow properties change in the presence of electric fields. Also termed ER fluids; the electrorheological effect is also termed the Winslow effect. When an electric field is applied to an ER fluid, the polarizable particles become electric dipoles and can align to form chains and more complex structures; as a result the fluid becomes more viscous and may form a gel. Example: cornstarch dispersed in corn oil and subjected to an electric field gradient of about 10,000 V/cm. *See also* reference 18.

Electrosteric Stabilization   The stabilization of a dispersed species by a combination of electrostatic and steric repulsions. An example is the stabilization of suspended solids by adsorbed polyelectrolyte molecules.

Electroviscous Effect   Any influence of electric double layer(s) on the flow properties of a fluid. The primary electroviscous effect refers to an increase in apparent viscosity when a dispersion of charged colloidal species is sheared. The secondary electroviscous effect refers to the increase in viscosity of a dispersion of charged colloidal species that is caused by their mutual electrostatic repulsion (overlapping of electric double layers). An example of the tertiary electroviscous effect would be for polyelectrolytes in solution where changes in polyelectrolyte molecule conformations and their associated effect on solution apparent viscosity occur.

Elutriation   The separation of smaller lighter particles from larger heavier particles because of the flow of surrounding fluid that tends to "carry" the lighter particles.

Emulsion   A dispersion of droplets of one liquid in another immiscible liquid in which the droplets are of colloidal or near-colloidal sizes. The term emulsion may also be used to refer to colloidal dispersions of liquid crystals in a liquid. Emulsions were previously referred to as emulsoids, meaning emulsion colloids.

Energy of Adhesion   *See* Work of Adhesion.

Energy of Cohesion   *See* Work of Cohesion.

Energy of Immersional Wetting   *See* Work of Immersional Wetting.

Energy of Separation   *See* Work of Separation.

Engulfment   The process in which a particle dispersed in one phase is overtaken by an advancing interface and surrounded by a second phase. Example: when a freezing front (the interface between a solid and its freezing liquid phase) overtakes a particle, the particle will either be pushed along by the front or else it will be engulfed by the front, depending on its interfacial tensions with the solid and with the liquid.

Enhanced Oil Recovery   The third phase of crude oil production, in which chemical, miscible or immiscible gas, or thermal methods are applied to restore production from a depleted reservoir. Also known as tertiary oil recovery. *See* Primary Oil Recovery, Secondary Oil Recovery.

Enmeshment   *See* Sweep Flocculation.

Enthalpy Stabilization   *See* Steric Stabilization.

Entropy Stabilization   *See* Steric Stabilization.

Equation of Capillarity   *See* Young–Laplace Equation.

Equilibrium Contact Angle   The contact angle that is measured when all contacting phases are in equilibrium with each other. The term arises because either or both of the advancing or receding contact angles may differ from the equilibrium value. It is essential to state which interfaces are used to define the contact angle. *See also* Contact Angle.

Equilibrium Dialysate   *See* Dialysis.

Equilibrium Surface Tension   Surface or interfacial tensions may change dynamically as a function of the age of the surface or interface. Thus, the dynamic (preequilibrium) tensions are distinguished from the limiting, or equilibrium, tensions.

Equivalent Spherical Diameter   The diameter of a sedimenting species determined from Stokes' law assuming a spherical shape. Also referred to as the Stokes diameter or (divided by a factor of 2) the settling radius.

ER Fluid   *See* Electrorheological Fluid.

Esin Markov Effect   The change in zero point of charge of a species that occurs when the electrolyte can become specifically adsorbed. In the presence of indifferent electrolytes, the zero point of charge is a constant.

External Phase   *See* Continuous Phase.

External Surface   When a porous medium can be described as consisting of discrete particles, the outer surface of the particles is termed the external surface. *See also* Internal Surface.

Extinction Coefficient   *See* Absorbance.

Extra-Heavy Crude Oil   A naturally occurring hydrocarbon having a viscosity less than 10,000 mPa·s at ambient deposit temperature and a density greater than 1000 kg/m^3 at 15.6 °C. *See* references 11–13.

Feret's Diameter   A statistical particle diameter; the length of a line drawn parallel to a chosen direction and taken between parallel planes drawn at the extremities on either side of the particle. This diameter is thus the maximum projection of the particle onto any plane parallel to the chosen direction. The value obtained depends on the particle orientation; thus, these measurments have significance only when a large enough number of measurements are averaged together. *See also* Martin's Diameter.

Film   Any layer of material that covers a surface and is thin enough to not be significantly influenced by gravitational forces. *See also* Monolayer Adsorption.

Film Flotation Technique   *See* Hydrophobic Index.

Film Water   In soil science, the film of water that remains, surrounding soil particles, after drainage. This layer may range from several to hundreds of molecules thick and comprises water of hydration plus water trapped by capillary forces.

Filter Ripening   In water filtration, the process in which deposition of an initial layer of particles causes the filter surface to take on a nature more similar to the particles to be removed. This process enhances the filtering (hence, removal) of the particles.

Filtration   The process of removing particles or large molecules from a fluid phase by passing the fluid through some medium that will not permit passage of the particles or large molecules. The filtration medium may comprise woven fabric or metal fibers, porous media, or other materials. In water treatment, filtration refers to sand or mixed-bed granular filters that are used to remove colloidal and larger sized particles.

Floc *See* Flocculation.

Flocculation A kind of aggregation. The products of the flocculation process are referred to as flocs or floccules. *See* Aggregation.

Flocculation Value *See* Critical Coagulation Concentration.

Floccule *See* Flocculation.

Flotation *See* Froth Flotation, Sedimentation.

Fluorescence Microscopy Light microscopy in which ultraviolet light is used to induce fluorescence in a specimen. The fluorescent light is then used to form the magnified image in either transmitted- or reflected-light modes (the ultraviolet light is filtered out at this stage).

Fly Knives The rotating cutting blades in a cutting mill machine for comminution.

Foam A dispersion of gas bubbles in a liquid, in which at least one dimension falls within the colloidal size range. Thus, a foam typically contains either very small bubble sizes or, more commonly, quite large gas bubbles separated by thin liquid films. The thin liquid films are called lamellae (or laminae). Sometimes distinctions are drawn as follows. Concentrated foams, in which liquid films are thinner than the bubble sizes and the gas bubbles are polyhedral, are termed polyederschaum. Low-concentration foams, in which the liquid films have thicknesses on the same scale or larger than the bubble sizes and the bubbles are approximately spherical, are termed gas emulsions, gas dispersions, or kugelschaum. *See also* Froth, Solid Foam.

Forward Scattering *See* Light Scattering.

Fractal A structure that has an irregular shape under all scales of measurement. The fractal dimension of a species is the exponent D to which a characteristic dimension must be raised to obtain proportionality with the overall size of the species. Fractal dimensions are used for species having a dimensionality of between 2 and 3, such as many particle aggregates.

Fractal Dimension *See* Fractal.

Free Energy A measure of the balance of energetic and entropic forces in a system. For systems maintained at constant pressure, the free energy is referred to as the Gibbs free energy (now frequently termed Gibbs energy); $G = H - TS$, where H is the enthalpy, T is temperature, and S is entropy. For systems maintained at constant volume, the free energy is referred to as the Helmholtz free energy (Helmholtz energy); $A = E - TS$, where $E = H - PV$ (P is pressure and V is volume).

Free Molecules   In polymer- or surfactant-containing systems, the molecules of polymer or surfactant dissolved in solution, that is, those that are not adsorbed or precipitated. For surfactant solutions, free surfactant includes those molecules present in micelles.

Free Surfactant   *See* Free Molecules.

Freeze-Fracture Method   A sample preparation technique used in electron microscopy in which specimens are quickly frozen in a cryogen and then cleaved to expose interior surfaces. In some techniques, the sample is then observed directly in an electron microscope equipped with a cryogenic stage; in other cases, the cleaved sample is coated with a metal coating to produce a replica, which is observed in the electron microscope. *See also* Electron Microscopy.

Friction Factor   In the rheology of a dispersion, the friction factor relates to the dissipation of energy due to friction at the surfaces of the dispersed species (i.e., due to drag).

Froth   A type of foam in which solid particles are also dispersed in the liquid (in addition to the gas bubbles). The solid particles may even be the stabilizing agent. The term froth is sometimes used to refer simply to a concentrated foam, but this usage is not preferred.

Froth Flotation   A separation process using flotation, in which particulate matter becomes attached to gas (foam) bubbles. The flotation process produces a product layer of concentrated particles in foam termed froth. Variations include dissolved-gas flotation, in which gas is dissolved in water that is added to a colloidal dispersion. As microbubbles come out of solution, they attach to and float the colloidal species. *See also* Scavenging Flotation.

Gegenion   *See* Counterions.

Gel   A suspension or polymer solution that behaves as an elastic solid or semisolid rather than a liquid. A dried-out gel is termed a xerogel. Examples: gels of gelatin solutions or of clay suspensions. *See also* reference 19.

Generalized Plastic Fluid   A fluid characterized by both of the following: the existence of a finite shear stress that must be applied before flow begins (yield stress) and pseudoplastic flow at higher shear stresses. *See also* Plastic Fluid.

g Forces   *See* Relative Centrifugal Force.

Gibbs Energy of Attraction When two dispersed-phase species approach, they may attract each other as a result of such forces as the London–van der Waals forces. The Gibbs energy of attraction may be thought of as the difference between Gibbs attractive energies of the system at a specified separation distance and at infinite separation. Although IUPAC (*1*) has discouraged the use of the synonyms potential energy of attraction and attractive potential energy, they are still in common usage. *See also* Gibbs Energy of Interaction, Gibbs Energy of Repulsion.

Gibbs Energy of Interaction When two dispersed-phase species approach, they experience repulsive and attractive forces such as electrostatic repulsion and van der Waals attraction. The Gibbs energy of interaction may be thought of as the difference between Gibbs energies of the system at a specified separation distance and at infinite separation. An example of the dependence of Gibbs energy of interaction and distance of separation is that calculated from DLVO theory. Although IUPAC (*1*) has discouraged the use of the synonyms potential energy of interaction and total potential energy of interaction, they are still in common usage. *See also* DLVO Theory, Gibbs Energy of Attraction, Gibbs Energy of Repulsion, Primary Minimum.

Gibbs Energy of Repulsion When two dispersed-phase species approach, they may repel each other as a result of such forces as electrostatic repulsion. The Gibbs energy of repulsion may be thought of as the difference between Gibbs repulsive energies of the system at a specified separation distance and at infinite separation. Although IUPAC (*1*) has discouraged the use of the synonyms potential energy of repulsion and repulsive potential energy, they are still in common usage. *See also* Gibbs Energy of Attraction, Gibbs Energy of Interaction.

Gibbs Free Energy Now frequently termed Gibbs energy. *See* Free Energy.

Gouy–Chapman Layer *See* Electric Double Layer.

Gouy–Chapman Theory A description of the electric double layer in a colloidal dispersion in which one layer of charge is assumed to exist as a uniform charge distribution over a surface and the counterions are treated as point charges distributed throughout the continuous dielectric phase.

Gouy Layer *See* Electric Double Layer.

Granule Sometimes used to describe particles having sizes greater than about 2000 μm, depending on the classification system used. Also called gravel.

Gravel   *See* Granule.

g   *See* Relative Centrifugal Force.

Hamaker Constant   In the description of the London–van der Waals attractive energy between two dispersed bodies, such as particles. The Hamaker constant is a proportionality constant characteristic of the internal atomic packing and polarizability of the particles. Also termed the van der Waals–Hamaker constant. *See also* Dispersion Forces.

Hammer Mill   A device for reducing the particle size of a solid, for example, a pigment, that uses centrifugal force to drive the solid between rotating "hammers" and a stationary ring-shaped "anvil."

Head Group   The lyophilic functional group in a surfactant molecule. In aqueous systems the polar group of a surfactant. *See also* Surfactant, Surfactant Tail.

Heavy Crude Oil   A naturally occurring hydrocarbon having a viscosity less than 10,000 mPa·s at ambient deposit temperature and a density between 934 and 1000 kg/m^3 at 15.6 °C. *See* references 11–13.

Helmholtz Condenser   *See* Helmholtz Double Layer.

Helmholtz Double Layer   A simplistic description of the electric double layer as a condenser (the Helmholtz condenser) in which the condenser plate separation distance is the Debye length. The Helmholtz layer is divided into an inner Helmholtz plane (IHP) of adsorbed, dehydrated ions immediately next to a surface, and an outer Helmholtz plane (OHP) at the center of a next layer of hydrated, adsorbed ions just inside the imaginary boundary where the diffuse double layer begins. That is, both Helmholtz planes are within the Stern layer.

Helmholtz Energy   Helmholtz free energy. *See* Free Energy.

Helmholtz Free Energy   Now frequently termed Helmholtz energy. *See* Free Energy.

Helmholtz Plane   *See* Helmholtz Double Layer.

Helmholtz–Smoluchowski Equation   *See* Smoluchowski Equation.

Hemimicelle   An aggregate of adsorbed surfactant molecules that may form beyond monolayer coverage, the enhanced adsorption being due to hydrophobic interactions between surfactant tails. Hemimicelles (half-micelles) have been considered to have the form of surface aggregates or of a second adsorption layer with reversed orientation, somewhat like a bimolecular film. In bilayer surfactant adsorption, the term admicelles has also been used (20).

Henry Equation   A relation expressing the proportionality between electrophoretic mobility and zeta potential for different values of the Debye length and size of the species. *See also* Electrophoresis, Hückel Equation, Smoluchowski Equation.

Heterocoagulation   The coagulation of dispersed species of different types or having different states of surface electric charge.

Heterodisperse   A colloidal dispersion in which the dispersed species (droplets, particles, and so on) do not all have the same size. Subcategories are paucidisperse (few sizes) and polydisperse (many sizes). *See also* Monodisperse.

Heterovalent   Species having different valencies, as opposed to homovalent for the same valency. In ion exchange, it means that the adsorbing and desorbing species have different charges.

Hofmeister Series   *See* Lyotropic Series.

Homogenizer   Any machine for preparing colloidal systems by dispersion. Examples: colloid mill, blender, ultrasonic probe.

Homovalent   Species having the same valencies, as opposed to heterovalent for different valencies. In ion exchange, it means that the adsorbing and desorbing species have the same charge.

Hückel Equation   A relation expressing the proportionality between electrophoretic mobility and zeta potential for the limiting case of a species that can be considered small and with a thick electric double layer. *See also* Electrophoresis, Henry Equation, Smoluchowski Equation.

Hydrophilic   A qualitative term referring to the water-preferring nature of a species (atom, molecule, droplet, particle, and so on). For suspensions, hydrophilic usually means that the particles prefer the aqueous phase over the oil phase. In this example, hydrophilic has the same meaning as oleophobic, but such is not always the case.

Hydrophobic   A qualitative term referring to the water-avoiding nature of a species (atom, molecule, droplet, particle, and so on). For suspensions hydrophobic usually means that the particles prefer the oil phase over the aqueous phase. In this example, hydrophobic has the same meaning as oleophilic, but such is not always the case. A functional group of a molecule that is not very water soluble is referred to as a hydrophobe.

Hydrophobic Effect   The partitioning of a substance from an aqueous phase into (or onto) another phase due to its hydrophobicity. Often characterized by an octanol–water partitioning coefficient.

Hydrophobic Index   An empirical measure of the relative wetting preference of very small solid particles. In one test method, solid particles of narrow size range are placed on the surfaces of a number of samples of water containing increasing concentrations of alcohol (thus providing a range of solvent surface tensions). The percentage alcohol solution at which the particles just begin to become hydrophilic and sink is taken as the hydrophobic index. The corresponding solvent surface-tension value is taken as the critical surface tension of wetting. The technique is also referred to as the film-flotation technique (*21*) or sink–float method. *See also* Critical Surface Tension of Wetting.

Hyperfiltration   *See* Ultrafiltration.

IHP   Inner Helmholtz plane, *see* Helmholtz Double Layer.

Imbibition   The displacement of a nonwetting phase by a wetting phase in a porous medium or a gel; the reverse of drainage.

Immersional Wetting   The process of wetting when a solid (or liquid) that is initially in contact with gas becomes completely covered by an immiscible liquid phase. *See also* Adhesional Wetting, Spreading Wetting, Wetting.

Inclined-Plate Settling   *See* Lamella Settling.

Indifferent Electrolyte   An electrolyte whose ions have no significant effect on the electric potential of a surface or interface as opposed to potential-determining ions that have a direct influence on surface charge. This distinction is most valid for low electrolyte concentrations. Example: for the AgI surface in water, $NaNO_3$ would be an indifferent electrolyte, but both Ag^+ and I^- would be potential-determining ions.

Induced Gas Flotation   *See* Froth Flotation.

Induction Forces   Debye forces. *See* Dispersion Forces.

Inelastic Scattering   *See* Light Scattering.

Inherent Viscosity   In solutions and colloidal dispersions, the natural logarithm of the relative viscosity, all divided by the solute or dispersed-phase concentration. $\eta_{Inh} = C^{-1} \ln(\eta/\eta_o)$. In the limit of vanishing concentration, it reduces to the intrinsic viscosity. Also termed the logarithmic viscosity number.

Ink Bottle Pore   A description of one kind of shape of dead-end pore in a porous medium in which a narrow throat is connected to a larger pore body. *See also* Porous Medium.

Inner Helmholtz Plane   (IHP) *See* Helmholtz Double Layer.

Inner Potential In the diffuse electric double layer extending outward from a charged interface, the electrical potential at the boundary between the Stern and the diffuse layer is termed the inner electrical potential. Synonyms include the Stern layer potential or Stern potential. *See also* Electric Double Layer, Zeta Potential.

Interface The boundary between two immiscible phases, sometimes including a thin layer at the boundary within which the properties of one bulk phase change over to become the properties of the other bulk phase. An interfacial layer of finite specified thickness may be defined. When one of the phases is a gas, the term surface is frequently used.

Interfacial Film A thin layer of material positioned between two immiscible phases, usually liquids, whose composition is different from either of the bulk phases.

Interfacial Layer The layer at an interface that contains adsorbed species. Also termed the surface layer.

Interfacial Potential *See* Surface Potential.

Interfacial Tension *See* Surface Tension.

Interferometry An experimental technique in which a beam of light is reflected from a film. Light reflected from the front and back surfaces of the film travels different distances and produces interference phenomena, a study of which allows calculation of the film thickness.

Intermediate Pore An older term, now replaced by mesopore.

Intermicellar Liquid An older term for the contiuous (external) phase in micellar dispersions. *See also* Continuous Phase, Micelle.

Internal Phase *See* Dispersed Phase.

Internal Surface In porous media the surface contained in pores and throats that are in communication with the outside space. Media having internal porosity also have internal surface area that may be available for sorption reactions. *See also* Activated Carbon.

Intrinsic Viscosity The specific viscosity divided by the dispersed-phase concentration in the limits of both the dispersed-phase concentration approaching infinite dilution, and of shear rate approaching zero ($[\eta] = \lim_{c \to 0} \lim_{\dot{\gamma} \to 0} \eta_{SP}/C$). Also termed limiting viscosity number.

Inverse Micelle A micelle that is formed in a nonaqueous medium, thus having the surfactants' hydrophilic groups oriented inward away from the surrounding medium.

Ion Exchange   A special kind of adsorption in which the adsorption of an ionic species is accompanied by the simultaneous desorption of an equivalent charge quantity of other ionic species. Ion exchange is commonly used for removing hardness and other metal ions in water treatment. The ion-exchange media can be arranged to provide a specific selectivity.

Ionic Strength   A measure of electrolyte concentration given by $I = 1/2 \sum c_i z_i^2$, where c_i are the concentrations, in moles per liter, of the individual ions, i, and z_i are the ion charge numbers.

Irridescent Layers   *See* Schiller Layers.

Isodisperse   *See* Monodisperse.

Isoelectric   A particle or ionic macromolecule that exhibits no electrophoretic or electroosmotic motion.

Isoelectric Point   The solution pH or condition for which the electrokinetic or zeta potential is zero. Under this condition, a colloidal system will exhibit no electrophoretic or electroosmotic motions. *See also* Point of Zero Charge.

Isokinetic Sampling   Collecting samples of a flowing dispersion using a method in which the sampling velocity (in the sampling probe) is equal to the upstream local velocity. If these velocities are not the same (anisokinetic sampling), then fluid streamlines ahead of the probe will be distorted; collection of particles or droplets will be influenced by their inertia, which varies with particle size, and sampling will not be representative.

Isotherm   The mathematical representation of a phenomenon occurring at constant temperature. *See also* Adsorption Isotherm.

Jar Test   For water treatment, *see* Bottle Test.

Jet Impingement   A dispersion technique in which a jet of liquid is directed at a surface or at a jet of another liquid.

Jet Mill   A machine for the comminution, or size reduction, of mineral or other particles. Such machines accelerate feed particles in a jet and cause size reduction by promoting interparticle and particle–wall collisions at high speed. Very small-sized particles can be produced with these mills. Also termed jet pulverizers.

Jet Pulverizer   *See* Jet Mill.

Kataphoresis   *See* Electrophoresis.

Keesom Forces   *See* Dispersion Forces.

Kinematic Viscosity Absolute viscosity of a fluid divided by the density.

Kinetic Stability Although most colloidal systems are metastable or unstable with respect to the separate bulk phases, they may have an appreciable kinetic stability. That is, the state of dispersion may exist for an appreciable length of time. Colloidal species can come together in very different ways; therefore, kinetic stability can have different meanings. A colloidal dispersion can be kinetically stable with respect to coalescence but unstable with respect to aggregation, or a system could be kinetically stable with respect to aggregation but unstable with respect to sedimentation. It is crucial that stability be understood in terms of a clearly defined process. *See also* Colloid Stability, Thermodynamic Stability.

Köhler Illumination In microscopy, the illumination provided through a condenser lens system adjusted to produce optimum brightness with uniform illumination of a sample.

Lamella Settling A process for phase separation based on density differences. A commercial lamella settler for suspensions comprises a stack of parallel plates spaced apart from each other and inclined from the horizontal. The space between each set of plates forms a separate settling zone. The feed is pumped into these spaces, at a point near the longitudinal middle of the plates. The less dense phases rise to the underside of the upper plates and flow to the tops of those plates. Meanwhile, the more dense phases settle down to the upperside of the lower plates and flow to the bottoms of those plates. Product is collected at the tops of the plate stack, and tailings are collected at the bottom of the plate stack. Such an inclined lamella-settling process is much more efficient than vertical gravity separation. Also termed inclined plate settling or inclined tube settling.

Laminar Flow A condition of flow in which all elements of a fluid passing a certain point follow the same path or streamline; there is no turbulence. Also referred to as streamline flow.

Langmuir Isotherm An adsorption isotherm equation that assumes monolayer adsorption and constant enthalpy of adsorption. The amount adsorbed per mass of adsorbent is proportional to equilibrium solute concentration at low concentrations and exhibits a plateau or limiting adsorption at high concentrations. *See* Adsorption Isotherm.

Laplace Flow *See* Capillary Flow.

Latex A dispersion (suspension or emulsion) of polymer in water. Latex rubber, a solid, is produced either by coagulating natural latex or by synthetic means through emulsion polymerization. Example: latex paint is a latex containing pigments and filling additives.

Lens   In physics, any piece of material or device that concentrates or disperses an incident beam of light, sound, electrons, or other radiation. In colloid science, a nonspreading droplet of liquid at an interface is said to form a lens. The lens is thick enough for its shape to be significantly influenced by gravitational forces. In geology, a specific geological layer resembling a convex lens. Example: clay mineral lens.

Lifshitz–van der Waals Forces   *See* Dispersion Forces.

Light Crude Oil   A naturally occurring hydrocarbon having a viscosity less than 10,000 mPa·s at ambient deposit temperature and a density less than 934 kg/m^3 at 15.6 °C. *See* references 11–13.

Light Scattering   Light will be scattered (deflected) by local variations in refractive index caused by the presence of dispersed species depending on their size (*1*). In elastic scattering, no wavelength shift occurs; in inelastic scattering, wavelength shifts occur because of molecular transitions; and in quasielastic scattering, wavelength shifts and line broadening occur because of time-dependent processes. In light scattering, the scattering plane contains both the incident light beam and the line that connects the center of the scattering system to the point of observation. The scattering angle lies in this plane and is measured clockwise, viewing into the incident beam. By this measure, forward scattering is at a scattering angle of zero and backscattering is at a scattering angle of 180°. *See also* Mie Scattering, Nephelometry, Rayleigh Scattering, Tyndall Scattering.

Limiting Sedimentation Coefficient   *See* Sedimentation.

Limiting Viscosity Number   *See* Intrinsic Viscosity.

Lipophilic   The (usually fatty) organic-liquid-preferring nature of a species. Depending on the circumstances may also be a synonym for oleophilic.

Lipophobic   The (usually fatty) organic-liquid-avoiding nature of a species. Depending on the circumstances may also be a synonym for oleophobic.

Liquid Limit   The minimum water content for which a small sample of soil or similar material will barely flow in a standardized test method (*9, 10*). Also termed the upper plastic limit. *See also* Atterberg Limits, Plastic Limit, Plasticity Number.

Logarithmic Viscosity Number   *See* Inherent Viscosity.

London Forces   *See* Dispersion Forces.

Lower Plastic Limit   *See* Plastic Limit.

Lyocratic   A dispersion stabilized principally by solvation forces. Example: the stability of aqueous biocolloid systems can be explained in terms of hydration and steric stabilization. *See also* Electrocratic.

Lyophilic   General term referring to the continuous-medium- (or solvent)-preferring nature of a species. *See* Hydrophilic.

Lyophilic Colloid   An older term used to refer to single-phase colloidal dispersions. Examples: polymer and micellar solutions. Other synonyms no longer in use: semicolloid or half-colloid.

Lyophobic   General term referring to the continuous-medium- (or solvent)-avoiding nature of a species. *See* Hydrophobic.

Lyophobic Colloid   An older term used to refer to two-phase colloidal dispersions. Examples: suspensions, foams, emulsions.

Lyotropic Series   A series and order of ions indicating, in decreasing order, their effectiveness in influencing the behavior of colloidal dispersions. Also termed Hofmeister series. Example: the following series shows the effect of different species on coagulating power. Cations: Cs^+ > Rb^+ > K^+ > Na^+ > Li^+. Anions: CNS^- > I^- > Br^- > Cl^- > F^- > NO_3^- > ClO_4^-.

Macroion   A charged colloidal species whose electric charge is attributable to the presence at the surface of ionic functionalities.

Macromolecule   A large molecule composed of many simple units bonded together. Macromolecules may be naturally occurring, such as humic substances, or synthetic, such as many polymers.

Macropore   *See* Pore.

Magnetophoretic Mobility   The mobility of a paramagnetic or ferromagnetic particle moving under the influence of an external magnetic field. The magnetophoretic mobility equals the particle velocity, relative to the medium, divided by the magnetic-field gradient at the location of the particle (22). This is analogous to the definition of electrophoretic mobility.

Martin's Diameter   A statistical particle diameter; the length of a line drawn parallel to a chosen direction such that it bisects the area of a particle. The value obtained depends on the particle orientation, and so these measurments have significance only when a large enough number of measurements are averaged together. *See also* Feret's Diameter.

Mass-Area Mean Diameter   An average particle diameter calculated from measurement of average particle area.

Mature Fine Tailings   (MFT) The bottom layer of high solids concentration particles in a mature tailings pond. Mature fine tailings refers to the sediment layer for which all practical particle settling and dewatering has already occurred and in which the solids concentration remains practically constant. In oil sands process tailings, mature fine tailings is what was formerly termed tailings sludge. *See* Chapter 14.

Mechanical Impact Mill   A machine for the comminution, or size reduction, of mineral or other particles. Such machines pulverize feed particles (typically about 10 mm initially) by causing them to strike a surface at high speed. Very small-sized particles can be produced with these mills.

Mechanical Syneresis   Any process in which syneresis is enhanced by mechanical means. *See also* Syneresis.

Mercury Porosimetry   *See* Porosimeter.

Mesopore   *See* Pore.

Metastable   *See* Thermodynamic Stability.

MFT   *See* Mature Fine Tailings.

Micelle   An aggregate of surfactant molecules or ions in solution. Such aggregates form spontaneously at sufficiently high surfactant concentration, above the critical micelle concentration. The micelles typically contain from tens to hundreds of molecules and are of colloidal dimensions. If more than one kind of surfactant forms the micelles, they are referred to as mixed micelles. If a micelle becomes larger than usual as a result of either the incorporation of solubilized molecules or the formation of a mixed micelle, then the term swollen micelle is applied. *See also* Critical Micelle Concentration, Inverse Micelle.

Microelectrophoresis   *See* Electrophoresis.

Micronizing   The process by which a solid is reduced to particle sizes of less than about 100 μm, using any type of particle size reduction equipment. Examples: micronized talc, micronized pigment.

Micropore   *See* Pore.

Microscopic Electrophoresis   *See* Electrophoresis.

Microscopy   Light microscopy involves the use of light rays and lenses to observe magnified images of objects. The magnified image may be formed from transmitted light for transparent materials or from reflected (incident) light for opaque materials. In each case, different illuminating modes are used and the light used may be visible, infrared, or ultraviolet. *See also* Bright-Field Illumination, Dark-Field Illumination, Köhler Illumination. Different viewing modes may be used, such as polarizing, fluorescence, phase contrast, and interference contrast. A derived technique is confocal microscopy. An analogous technique, electron microscopy, involves the use of electrons rather than light. *See also* Confocal Microscopy, Electron Microscopy.

Mie Scattering   Light will be scattered (deflected) by local variations in refractive index caused by the presence of dispersed species, depending on their size. The scattering of light by species whose size is much less than the wavelength of the incident light is referred to as Rayleigh scattering, and it is termed Mie scattering if the species' size is comparable with that of the incident light. Also termed Lorenz–Mie scattering. *See also* Light Scattering, Rayleigh Scattering.

Monodisperse   A colloidal dispersion in which all the dispersed species (droplets, particles, and so on) have the same size. Otherwise, the system is heterodisperse (paucidisperse or polydisperse).

Monolayer Adsorption   Adsorption in which a first or only layer of molecules becomes adsorbed at an interface. In monolayer adsorption, all of the adsorbed molecules will be in contact with the surface of the adsorbent. The adsorbed layer is termed a monolayer or monomolecular film.

Monolayer Capacity   In chemisorption, the amount of adsorbate needed to satisfy all available adsorption sites. For physisorption, the amount of adsorbate needed to cover the surface of the adsorbent with a complete monolayer.

Monomolecular Layer   *See* Monolayer Adsorption.

Motionless Mixer   *See* Static Mixer.

Moving Boundary Electrophoresis   An indirect electrophoresis technique for particles too small to be viewed. This principle is used in the Tiselius apparatus. Here a colloidal dispersion is placed in the bottom of a U-tube, the upper arms of which are filled with a less dense liquid that both provides the boundaries and makes the connections to the electrodes. Under an applied electric field the motions of the ascending and descending boundaries are measured.

Multilayer Adsorption   Adsorption in which the adsorption space contains more than a single layer of molecules; therefore, not all adsorbed molecules will be in contact with the surface of the adsorbent. *See also* Brunauer–Emmett–Teller Isotherm, Monolayer Adsorption.

Naphtha   A petroleum fraction that is operationally defined in terms of the distillation process by which it is separated. A given naphtha is thus defined by a specific range of boiling points of its components.

Negative Adsorption   *See* Adsorption.

Negative Tactoids   *See* Tactoid.

Nephelometry   The study of the light-scattering properties of dispersions. In general, a nephelometer is an instrument capable of measuring light scattering by dispersions at various angles. *See also* Light Scattering, Turbidity, Tyndall Scattering.

Newtonian Flow   Fluid flow that obeys Newton's law of viscosity. Non-Newtonian fluids may exhibit Newtonian flow in certain shear-rate or shear-stress regimes. *See also* Newtonian Fluid.

Newtonian Fluid   A fluid or dispersion whose rheological behavior is described by Newton's law of viscosity. Here shear stress is set proportional to shear rate. The proportionality constant is the coefficient of viscosity, or simply, viscosity. The viscosity of a Newtonian fluid is a constant for all shear rates.

Non-Newtonian Flow   Fluid flow that does not obey Newton's law of viscosity. Non-Newtonian fluids may exhibit non-Newtonian flow only in certain shear-rate or shear-stress regimes. A number of categories of non-Newtonian flow are distinguished, including dilatant, pseudoplastic, thixotropic, rheopectic, and rheomalaxic. *See also* Newtonian Fluid.

Non-Newtonian Fluid   A fluid whose viscosity varies with applied shear rate (flow rate). *See* Newtonian Fluid.

Nonwetting   *See* Wetting.

Nuclei   As a solute becomes insoluble, the formation of a new phase has its origin in the formation of clusters of solute molecules, termed germs, that increase in size to form small crystals or particles, termed nuclei. One means of preparing colloidal dispersions involves precipitation from solution onto nuclei, which may be of the same or different chemical species. *See also* Condensation Methods.

Oden's Balance   An apparatus for determining sedimentation rates in which a balance pan is immersed in a sedimentation column and is used to intercept and accumulate sedimenting particles, whose mass can be determined as a function of time.

OHP   Outer Helmholtz plane. *See* Helmholtz Double Layer.

Oil   Liquid petroleum (sometimes including dissolved gas) that is produced from a well. In this sense oil is equivalent to crude oil. The term oil is, however, frequently more broadly used and may include, for example, synthetic hydrocarbon liquids, bitumen from oil (tar) sands, fractions obtained from crude oil, and liquid fats (e.g., triglycerides). *See also* Crude Oil.

Oleophilic   The oil-preferring nature of a species. A synonym for lipophilic. *See also* Hydrophobic.

Oleophobic   The oil-avoiding nature of a species. A synonym for lipophobic. *See also* Hydrophilic.

Orientation Forces   Keesom forces. *See* Dispersion Forces.

Orthokinetic Aggregation   The process of aggregation induced by hydrodynamic motions such as stirring, sedimentation, or convection. Orthokinetic aggregation is distinguished from perikinetic aggregation, the latter being caused by Brownian motions.

Osmometer   An instrument for determining the osmotic pressure exerted by solvent molecules diffusing through a semipermeable membrane in contact with a solution or hydrophilic colloidal dispersion. *See also* Colloid Osmotic Pressure, Osmotic Pressure.

Osmosis   The process in which solvent will flow through a semipermeable membrane (permeable to solvent but not to solute) from a solution of lower dissolved solute activity (concentration) to that of higher activity (concentration). *See also* Osmotic Pressure.

Osmotic Pressure   The pressure difference required to prevent osmosis, or the spontaneous movement of solvent across a semipermeable membrane separating pure solvent from a solution of dissolved species. *See also* Colloid Osmotic Pressure.

Ostwald Ripening   The process by which larger droplets or particles grow in size in preference to smaller droplets or particles because of their different chemical potentials. *See also* Aging.

Ostwald Viscometer   *See* Capillary Viscometer.

Outer Helmholtz Plane   (OHP) *See* Helmholtz Double Layer.

Pallmann Effect   *See* Suspension Effect.

Particle Size Classification   The separation or determination of particles into different size ranges. A number of classification systems are used, some of which correspond to physical means of separations such as by sieves.

Paucidisperse   A colloidal dispersion in which the dispersed species (droplets, particles, and so on) have a few different sizes. Paucidisperse is a category of heterodisperse systems. *See also* Monodisperse.

Peptization   The dispersion of an aggregated (coagulated or flocculated) system. Deflocculation means the same thing.

Peptizing Ions   *See* Potential-Determining Ions.

Percolation   A condition in a dispersed system in which a property such as conductivity increases strongly at a critical concentration, termed the percolation threshold, as a result of the formation of continuous conducting paths, termed infinite clusters.

Percolation Threshold   *See* Percolation.

Perikinetic Aggregation   The process of aggregation when induced by Brownian motions. Perikinetic aggregation is distinguished from orthokinetic aggregation, the latter being caused by hydrodynamic motions such as sedimentation or convection.

Permeability   A measure of the ease with which a fluid can flow (fluid conductivity) through a porous medium. Permeability is defined by Darcy's law. For linear, horizontal, isothermal flow permeability is the constant of proportionality between flow rate times viscosity and the product of cross-sectional area of the medium and pressure gradient along the medium.

Permeate   *See* Dialysis.

Permittivity   A measure of the ability of a medium to affect an applied electric field. It is the constant of proportionality between the force acting between two point charges and the product of the two charges divided by the square of the distance separating them. The relative permittivity of a material equals the permittivity of the material multiplied by that of vacuum. An older term for relative permittivity is dielectric constant.

Petroleum   Refers to any hydrocarbon or hydrocarbon mixture, usually liquid, but sometimes solid or gaseous.

Photomicrograph   A photographic image formed by a microscope. The resulting photographic image is much larger than the original object being photographed. This is not the same as a microphotograph.

Photozone Counter   A particle- or droplet-sizing technique, analogous to the electrical sensing-zone methods, that relies on visible light absorption in sample introduced into a small chamber. The particles or droplets must be greater than the wavelength of the light used, to minimize scattering (>1 μm). *See* Sensing Zone Technique.

Physical Adsorption *See* Chemisorption.

Physisorption *See* Chemisorption.

Pickering Emulsion An emulsion stabilized by fine particles. The particles form a close-packed structure at the oil–water interface, with significant mechanical strength, which provides a barrier to coalescence.

Pigment Insoluble material that is finely divided, micronized (for example), and uniformly dispersed in a formulated system for the purpose of coloring it or making it opaque. Examples: TiO_2 in soap bars and paints; iron oxides in eye makeup and paints.

Pigment Grind Pigment particles dispersed in a liquid, such as castor oil. *See also* Roll Mill.

Plastic Flow The deformation or flow of a solid under the influence of an applied shear stress.

Plastic Fluid A fluid characterized by both of the following: the existence of a finite shear stress that must be applied before flow begins (yield stress) and Newtonian flow at higher shear stresses. May be referred to as Bingham plastic. *See also* Generalized Plastic Fluid.

Plastic Limit The minimum water content for which a small sample of soil or similar material will barely deform or crumble in a standardized test method (9, 10). Also termed the lower plastic limit. *See also* Atterberg Limits, Liquid Limit, Plasticity Number.

Plasticity Index *See* Plasticity Number.

Plasticity Number The difference between the liquid limit and the plasticity limit of a soil or similar material (9, 10). Also termed the plasticity index. *See also* Atterberg Limits, Liquid Limit, Plastic Limit.

Point of Zero Charge The condition, usually the solution pH, at which a particle or interface is electrically neutral. This is not always the same as the isoelectric point, which refers to zero charge at the shear plane that exists a small distance away from the interface. *See also* Colloid Titration.

Poiseuille's Law Poiseuille flow is the steady flow of incompressible fluid parallel to the axis of a circular pipe or capillary. Poiseuille's law is an expression for the flow rate of a liquid in such tubes. It forms the basis for the measurement of viscosities by capillary viscometry.

Poiseuille Flow *See* Poiseuille's Law.

Poisson–Boltzmann Equation   A fundamental equation describing the distribution of electric potential around a charged species or surface. The local variation in electric-field strength at any distance from the surface is given by the Poisson equation, and the local concentration of ions corresponding to the electric-field strength at each position in an electric double layer is given by the Boltzmann equation. The Poisson–Boltzmann equation can be combined with Debye–Hückel theory to yield a simplified, and much used, relation between potential and distance into the diffuse double layer.

Poisson Equation   A fundamental equation describing the reduction in electric-field strength that occurs with increasing distance away from a charged species in a dielectric medium. In electric double-layer theory, the effects of the various ion charges are averaged into layers by assuming charge distribution to be a continuous function of distance away from a charged surface. The Poisson equation gives the relationship between the volume charge density at a point in solution and the potential at that same point. This equation can be combined with the Boltzmann equation and Debye–Hückel theory to yield a simplified, and much used, relation between potential and distance into the diffuse double layer.

Polar Group   *See* Head Group.

Polar Substance   A substance having different, usually opposite, characteristics at two locations within it. Example: a permanent dipole. Increasing polarity generally increases solubility in water.

Polyampholyte   *See* Polyelectrolyte.

Polydisperse   A colloidal dispersion in which the dispersed species (droplets, particles, and so on) have a wide range of sizes. Polydisperse is a category of heterodisperse systems. *See also* Monodisperse.

Polyelectrolyte   A kind of colloidal electrolyte comprising a macromolecule that, when dissolved, dissociates to yield a polyionic parent macromolecule and its corresponding counterions. Also termed a polyion, polycation, or polyanion. Similarly, a polyelectrolyte may be referred to in certain circumstances as a polyacid, polybase, polysalt, or polyampholyte (1). Example: carboxymethylcellulose.

Polyion   *See* Polyelectrolyte.

Polymer   A molecule that is made up of many repeating units, or groups, of atoms. Sometimes termed homopolymer to distinguish from copolymers such as block copolymers.

Polymer Bridging   A mechansim of aggregation or flocculation in which long-chain polymers adsorb onto particle surfaces leaving loops and ends extending out into solution. If these loops and ends contact and adsorb onto another particle, then a so-called bridge is formed. *See* reference 23. *See also* Aggregation.

Polymer Colloid   A dispersion of colloidal size polymer particles in a nonsolvent medium. Example: submicroscopic latex spheres, prepared either by emulsion polymerization or by seeded emulsion polymerization, are used for a veriety of calibration purposes in colloid science. *See* reference 24.

Polysalt   *See* Polyelectrolyte.

Pore   In porous media, the interconnecting channels forming a continuous passage through the medium are made up of pores, or openings, which may be of different sizes. Macropores have diameters greater than about 50 nm. Mesopores have diameters of between about 2.0 and 50 nm. Micropores have diameters of less than about 2.0 nm.

Porosimeter   An instrument for the determination of pore size distribution by measuring the pressure needed to force liquid into a porous medium and applying the Young–Laplace equation. If the surface tension and contact angle appropriate to the injected liquid are known, pore dimensions can be calculated. A common liquid for this purpose is mercury; hence, the term mercury porosimetry.

Porosity   The ratio of the volume of all void spaces to total volume in a porous medium. In geology primary porosity refers to initial, or unweathered, media, and secondary porosity refers to that associated with weathered media.

Porous Medium   A solid containing voids or pore spaces. Normally such pores are quite small compared with the size of the solid and well-distributed throughout the solid. In geologic formations, porosity may be associated with unconsolidated (uncemented) materials, such as sand, or a consolidated material, such as sandstone.

Potential-Determining Ions   Ions whose equilibrium between two phases, frequently between an aqueous solution and a surface or interface, determines the difference in electrical potential between the phases, or at the surface. Example: for the AgI surface in water both Ag^+ and I^- would be potential-determining ions. If such ions are responsible for the stabilization of a colloidal dispersion, they are referred to as peptizing ions. *See also* Indifferent Electrolyte.

Potential Energy of Interaction   *See* Gibbs Energy of Interaction.

Power-Law Fluid   A fluid or dispersion whose rheological behavior is reasonably well described by the power-law equation. Here shear stress is set proportional to the shear rate raised to an exponent n, where n is the power-law index. The fluid is pseudoplastic for $n < 1$, Newtonian for $n = 1$, and dilatant for $n > 1$.

Primary Electroviscous Effect   *See* Electroviscous Effect.

Primary Minimum   In a plot of Gibbs energy of interaction versus separation distance, two minima may occur. The minimum occurring at the shortest distance of separation is referred to as the primary minimum and that occurring at larger separation distance is termed the secondary minimum.

Primary Oil Recovery   The first phase of crude oil production, in which oil flows naturally to the wellbore. *See also* Enhanced Oil Recovery, Secondary Oil Recovery.

Proppants   Solid particles in a suspension that is injected into a petroleum reservoir to maintain open fractures that have been artificially induced. The proppant-filled fractures then remain permeable to the flow of fluids. The suspension used for this purpose is sometimes termed proppant-laden fluid. *See also* Chapter 11.

Proppant-Laden Fluid   *See* Proppants.

Protected Lyophobic Colloids   *See* Sensitization.

Protection   The process in which a material adsorbs onto particle surfaces and thereby makes a suspension less sensitive to aggregation. *See also* Sensitization.

Protective Colloid   A colloidal species that adsorbs onto and acts to "protect" the stability of another colloidal system. The term refers specifically to the protecting colloid and only indirectly to the protected colloid. Example: when a lyophilic colloid such as gelatin acts to protect another colloid in a dispersion by conferring steric stabilization. *See also* Protection.

PSD   Particle size distribution.

Pseudoplastic   A non-Newtonian fluid whose viscosity decreases as the applied shear rate increases, a process that is also termed shear-thinning. Pseudoplastic behavior may occur in the absence of a yield stress and also after the yield stress in a system has been exceeded (i.e., once flow begins).

Pulp   In mineral processing, a slurry of crushed ore dispersed in water.

Rayleigh Ratio   In light scattering, the ratio of intenstites of incident to scattered light at some specified distance.

Rayleigh Scattering   Light will be scattered (deflected) by local variations in refractive index caused by the presence of dispersed species and depending on their size. The scattering of light by species whose size is much less than the wavelength of the incident light is referred to as Rayleigh scattering, and it is termed Mie scattering if the species' size is comparable with that of the incident light. An example of Rayleigh scattering is that due to molecules in the atmosphere that scatter blue light from the sun's white-light illumination and cause the sky to appear blue while the sun appears orange–yellow. *See also* Light Scattering, Mie Scattering.

RCF   *See* Relative Centrifugal Force.

Receding Contact Angle   The dynamic contact angle that is measured when one phase is receding, or reducing its area of contact, along an interface while in contact with a third immiscible phase. It is essential to state which interfaces are used to define the contact angle. *See also* Contact Angle.

Reduced Osmotic Pressure   *See* Colloid Osmotic Pressure.

Reduced Sedimentation Coefficient   *See* Sedimentation.

Reduced Viscosity   For solutions or colloidal dispersions, the specific increase in viscosity divided by the solute or dispersed-phase concentration, respectively ($\eta_{Red} = \eta_{SP}/C$). Also termed the viscosity number.

Refracted Light   Light that has changed direction by passing from one medium through another in which its wave velocity is different.

Reflected Light   Light that strikes a surface and is redirected.

Relative Centrifugal Force   (RCF) When a centrifuge is used to enhance sedimentation or creaming, the centrifugal force is equal to mass times the square of the angular velocity times the distance of the dispersed species from the axis of rotation. The square of the angular velocity times the distance of the dispersed species from the axis of rotation, when divided by the gravitational constant, g, yields the relative centrifugal force or RCF. RCF is not strictly a force but rather the proportionality constant. It is substituted for g in Stokes' law to yield an expression for centrifuges and is used to compare the relative sedimentation forces achievable in different centrifuges. Because RCF is expressed in multiples of g, it is also termed g force or simply gs.

Relative Permittivity   *See* Permittivity.

Relative Viscosity   In solutions and colloidal dispersions, the viscosity of the solution or dispersion divided by the viscosity of the solvent or continuous phase, respectively ($\eta_{Rel} = \eta/\eta_o$). Also termed the viscosity ratio.

Relative Viscosity Increment   *See* Specific Increase in Viscosity.

Relaxation Time   The time required for the value of a changing property to be reduced to $1/e$ of its initial value.

Repeptization   Peptization, usually by dilution, of a once-stable dispersion that was aggregated (coagulated or flocculated) by the addition of electrolyte.

Repulsive Potential Energy   *See* Gibbs Energy of Repulsion.

Resistazone Counter   *See* Sensing Zone Technique.

Retarded van der Waals Constant   *See* Hamaker Constant.

Retentate   *See* Dialysis, Electrodialysis, Ultrafiltration.

Reverse Osmosis   *See* Ultrafiltration.

Reversible Sol–Gel Transformation   *See* Thixotropic.

Rheology   Strictly, the science of deformation and flow of matter. Rheological descriptions usually refer to the property of viscosity and departures from Newton's law of viscosity. *See also* Rheometer.

Rheomalaxis   A special case of time-dependent rheological behavior in which shear-rate changes cause irreversible changes in viscosity. The change can be negative, as when structural linkages are broken, or positive, as when structural elements become entangled (like work-hardening).

Rheometer   Any instrument designed for the measurement of non-Newtonian and Newtonian viscosities. The principal class of rheometer consists of the rotational instruments in which shear stresses are measured while a test fluid is sheared between rotating cylinders, plates, or cones. Examples of rotational rheometers: concentric cylinder, cone–cone, cone–plate, double-cone–plate, plate–plate, and disc. *See* reference 25.

Rheopexy   Dilatant flow that is time-dependent. At constant applied shear rate, viscosity increases, and in a flow curve, hysteresis occurs (but opposite to the thixotropic case).

Ring–Roller Mill   A machine for the comminution, or size reduction, of minerals. Such machines crush the input material between a stationary ring and vertical rollers revolving inside the ring. Particle sizes of as low as about 30 μm can be produced.

Roll Crusher   A machine for the comminution, or size reduction, of mineral lumps or stones. Such machines crush the input material between a plate and revolving roller or between more than one roller.

Roll Mill   A device for imparting shear to a dispersion for the purpose of reducing the particle size of the dispersed material. Somewhat similar to a roll crusher. Examples of dispersions processed over a roll mill are pigment grinds (pigments dispersed in a fluid such as castor oil) and soap formulations (where the dispersed material includes fragrance oil droplets and pigments).

Rotational Rheometer   *See* Rheometer.

Roughness Factor   The factor by which the surface area of a nonporous solid is greater than that calculated from the macroscopic dimensions of the surface.

Rule of Schulze and Hardy   *See* Schulze–Hardy Rule.

Sand   A term used to distinguish particles having different sizes in the range between about 50–63 μm and about 2000 μm, and with several subcategories, very fine, fine, medium, coarse, and very coarse defined according to the operational scale adopted. *See* references 26 and 27.

Scanning Electron Microscopy   (SEM) *See* Electron Microscopy.

Scanning Transmission Electron Microscopy   (STEM) *See* Electron Microscopy.

Scavenging Flotation   A flotation separation process, in which particles or droplets become attached to gas bubbles that are injected (sparged) into the flotation medium. Also termed induced gas flotation. Example: the froth flotation of bitumen. *See also* Froth Flotation.

Schiller Layers   The layers of particles that may be formed during sedimentation such that the distances between layers are on the order of the wavelength of light, leading to iridescent, or Schiller layers.

Schlieren Optics   An optical arrangement designed to allow detection of density gradients occurring in fluid flow. Typically, a narrow beam of light is collimated by one lens and focused on a knife-edge by a second lens. A density gradient in a fluid, between the lenses, causes a diffraction pattern to appear beyond the knife-edge.

Schulze–Hardy Rule   An empirical rule summarizing the general tendency of the critical coagulation concentration (CCC) of a suspension, an emulsion, or other dispersion to vary inversely with the sixth power of the counterion charge number of added electrolyte. *See also* Critical Coagulation Concentration.

Secondary Minimum   *See* Primary Minimum.

Secondary Oil Recovery   The second phase of crude oil production, in which water or an immiscible gas are injected to restore production from a depleted reservoir. *See also* Enhanced Oil Recovery, Primary Oil Recovery.

Sediment   The process of sedimentation in a dilute dispersion generally produces a discernable, more concentrated dispersion that is termed the sediment and has a volume termed the sediment volume.

Sedimentation   The settling of suspended particles or droplets due to gravity or an applied centrifugal field. The rate of this settling is the sedimentation rate (or velocity). The sedimentation rate divided by acceleration is termed the sedimentation coefficient. The sedimentation coefficient extrapolated to zero concentration of sedimenting species is termed the limiting sedimentation coefficient. The sedimentation coefficient reduced to standard temperature and solvent is termed the reduced sedimentation coefficient. If extrapolated to zero concentration of sedimenting species, it is termed the reduced limiting sedimentation coefficient. Negative sedimentation is also called flotation. Flotation in which droplets rise upward is also called creaming. Flotation in which particulate matter becomes attached to gas bubbles is also referred to as froth flotation. *See also* Froth Flotation, Subsidence, Svedberg.

Sedimentation Coefficient   *See* Sedimentation.

Sedimentation Equilibrium   The state of a colloidal system in which sedimentation and diffusion are in equilibrium.

Sedimentation Field Strength   *See* Sedimentation Potential.

Sedimentation Potential   The potential difference at zero current caused by the sedimentation of dispersed species. This mechanism of potential difference generation is known as the Dorn effect; accordingly, the sedimentation potential is sometimes referred to as the Dorn potential. The sedimentation may occur under gravitational or centrifugal fields. The potential difference per unit length in a sedimentation potential cell is the sedimentation field strength.

Sediment Volume   *See* Sediment.

SEM   Scanning electron microscopy. *See* Electron Microscopy.

Sensing Zone Technique   A general term used to refer to any of the particle- or droplet-sizing techniques that rely on (usually) conductivity or capacitance changes in sample introduced between charged electrodes. Also termed resistazone counter. An example is the Coulter counter. Sensing zone technique is also used with reference to similar techniques that use light absorption or scattering instead of electrical properties, *see* Photozone Counter.

Sensitization   The process in which small amounts of added hydrophilic colloidal material make a hydrophobic colloid more sensitive to coagulation by electrolyte. Higher additions of the same material usually make the dispersion less sensitive to coagulation, and this is termed protective action or protection. The protected colloidally stable dispersions that result in the latter case are termed protected lyophobic colloids.

Settling Radius   *See* Equivalent Spherical Diameter.

SFA   *See* Surface Force Apparatus.

Shear   The rate of deformation of a fluid when subjected to a mechanical shearing stress. In simple fluid shear, successive layers of fluid move relative to each other such that the displacement of any one layer is proportional to its distance from a reference layer. The relative displacement of any two layers divided by their distance of separation from each other is termed the shear or the shear strain. The rate of change with time of the shear is termed the shear rate or the strain rate.

Shear Plane   Any species undergoing electrokinetic motion moves with a certain immobile part of the electric double layer that is commonly assumed to be distinguished from the mobile part by a sharp plane, the shear plane. The shear plane is also termed the surface of shear. The zeta potential is the potential at the shear plane. *See also* Zeta Potential.

Shear Rate   *See* Shear.

Shear Stress   A certain applied force per unit area is needed to produce deformation in a fluid. For a plane area around some point in the fluid and in the limit of decreasing area the component of deforming force per unit area that acts parallel to the plane is the shear stress.

Shear Thickening   The increasing viscosity of a non-Newtonian fluid as the applied shear rate increases. *See also* Dilatant.

Shear Thinning   The decreasing viscosity of a non-Newtonian fluid as the applied shear rate increases. *See also* Pseudoplastic.

Sieve   *See* Particle Size Classification.

Silt   A term used to distinguish particles having sizes of greater than about 2–4 μm and less than about 50–63 μm, depending on the operational scale adopted. *See also* Clays, Sand.

Sink–Float Method   *See* Hydrophobic Index.

Sintering   The coalescence or merging of two or more solid particles into a single particle.

Slip-Casting   In ceramics, the process in which a slurry of dispersed particles is poured into a mold, the liquid removed, and the particles sintered to form the final product.

Sludge   *See* Mature Fine Tailings.

Slurry Quality   In suspensions, the concentration of solid particles.

Smoluchowski Equation   In electrophoresis, a relation expressing the proportionality between electrophoretic mobility and zeta potential for the limiting case of a species that can be considered to be large and have a thin electric double layer. Also termed Helmholtz–Smoluchowski Equation. *See also* Electrophoresis, Henry Equation, Hückel Equation.

Soil   Naturally occurring unconsolidated material, whether mineral or organic, that is on the earth's surface and is capable of supporting plant growth.

Sol   A colloidal dispersion. In some usage the term sol is used to distinguish dispersions in which the dispersed-phase species are of very small size so that the dispersion appears transparent.

Solid Aerosol   *See* Aerosol.

Solid Emulsion   A colloidal dispersion of a liquid in a solid. Examples: opal, pearl.

Solid Foam   A colloidal dispersion of a gas in a solid. Example: polystyrene foam.

Solid Suspension   A colloidal dispersion of a solid in another solid. Example: ruby-stained glass, a dispersion of gold particles in glass.

Solubilization   The process by which the solubility of a solute is increased by the presence of another solute. Micellar solubilization refers to the incorporation of a solute (solubilizate) into or on micelles of another solute to thereby increase the solubility of the first solute.

Solvophoresis   A variant of diffusiophoresis. If a particle is immersed into a mixed solvent in which a concentration gradient exists, the particle will tend to move in the direction of increasing concentration of the solvent component that is preferentially adsorbed onto its surface (28).

Sorbate   A substance that becomes sorbed into an interface or another material or both. *See also* Sorption.

Sorbent   The substrate into which or onto which a substance is sorbed or both. *See also* Sorption.

Sorption   A term used in a general sense to refer to either or both of the processes of adsorption and absorption.

Sorptive   Material that is present in one or both of the bulk phases bounding an interface and capable of becoming sorbed.

Specific Increase in Viscosity   The relative viscosity minus unity. Also referred to as specific viscosity or relative viscosity increment.

Specific Surface Area   *See* Surface Area.

Specific Viscosity   *See* Specific Increase in Viscosity.

Spherical Agglomeration   The process of separating particles from their suspension by selective wetting and agglomeration with a second immiscible liquid. The second liquid preferentially wets the particles and causes particle adhesion by capillary liquid bridges (29). The process of agglomeration thus includes both aggregation and coalescence. *See also* Agglomeration.

Spreading   The tendency of a liquid to flow and form a film coating an interface, usually a solid or immiscible liquid surface, in an attempt to minimize interfacial free energy. Such a liquid forms a zero contact angle as measured through itself.

Spreading Pressure   *See* Surface Pressure.

Spreading Wetting   The process of wetting in which a liquid, already in contact with a solid (or second, immiscible liquid) surface, spreads over the solid surface, thereby increasing the interfacial area of contact between them. *See also* Adhesional Wetting, Immersional Wetting, Wetting.

Stability   *See* Colloid Stability, Thermodynamic Stability.

Standard Sedimentation Coefficient   A synonym for reduced sedimentation coefficient. *See* Sedimentation.

Static Mixer   A device for mixing components in a solution or dispersion without moving mechanical elements. Stationary flow guiding elements are built into a device, frequently a section of pipe, and induce mixing and dispersion by repeatedly dividing and recombining partial streams of the flowing material. Also termed motionless mixers.

Static Surface Tension   A synonym for the equilibrium surface tension or interfacial tension. *See* Surface Tension.

STEM   Scanning transmission electron microscopy. *See* Electron Microscopy.

Steric Stabilization   The stabilization of dispersed species induced by the interaction (steric stabilization) of adsorbed polymer chains. Example: adsorbed proteins stabilize the emulsified oil (fat) droplets in milk by steric stabilization. Also termed depletion stabilization. *See also* Protection.

Stern Layer   The layer of ions in an electric double layer that, hydrated or not, lie adjacent to the surface (adsorbed ions). The rest of the electric double layer is often distinguished as the diffuse part, where assumptions such as that treating the ions as point charges can more reasonably be made. *See also* Electric Double Layer, Helmholtz Double Layer.

Stern Layer Potential   *See* Inner Potential.

Stern Potential   *See* Inner Potential.

Stokes Diameter   *See* Equivalent Spherical Diameter.

Stokes' Law   A relation giving the terminal settling velocity of a sphere as $2r^2\Delta\rho g/(9\eta)$, where r is the sphere radius, $\Delta\rho$ is the density difference between the phases, g is the gravitational constant, and η is the external-phase viscosity.

Strain, Strain Rate   *See* Shear Rate.

Streaming Potential   The potential difference at zero current created when liquid is made to flow through a porous medium.

Streamline Flow   *See* Laminar Flow.

Stress   *See* Shear Stress.

Subsidence   The process of sedimentation in which the settling of suspended particles results in a dense compaction, or coagulation, of particles in which liquid is squeezed out. Geologically, significant compaction of clay layers due to lowering of the water table (dewatering).

Substrate   A material that provides a surface or interface at which adsorption or other phenomena take place.

Suction Pressure   *See* Capillary Pressure.

Supercentrifuge   *See* Centrifuge.

Surface   *See* Interface.

Surface-Active Agent   *See* Surfactant.

Surface Area   The area of a surface or interface, especially that between a dispersed and a continuous phase. The specific surface area is the total surface area divided by the mass of the appropriate phase.

Surface Charge   The fixed charge that is attached to, or part of, a colloidal species' surface and forms one layer in an electric double layer. There is thus a surface-charge density associated with the surface. *See also* Electric Double Layer.

Surface Conductivity   The excess conductivity, relative to the bulk solution, in a surface or interfacial layer per unit length. Also termed the surface excess conductivity.

Surface Coverage   The ratio of the amount of adsorbed material to the monolayer capacity. The definition is the same for either of monolayer and multilayer adsorption.

Surface Force Apparatus   (SFA) An apparatus used to obtain information about the nature and range of interparticle forces (the interparticle forces are not directly determined). In one widely used design, thin sheets of mica are cleaved and glued to cylindrical glass forms that are then placed in cross-cylinder geometry with the mica sheets facing each other. The mica sheets may be prepared with coatings that may be of interest. An electrolyte of surfactant solution is placed between the cylinders and the upper cylinder lowered toward the lower under a known load. An equilibrium separation distance (film thickness) will be attained that is measured by light interferometry. Measurements are made for varying loads, varying solution compositions, in different liquids, and with or without surface coatings. *See* reference 30.

Surface Layer   *See* Interfacial Layer.

Surface of Shear   *See* Shear Plane.

Surface Potential   The potential at the interface bounding two phases, that is, the difference in outer (Volta) potentials between the two phases.

Surface Pressure   Actually an analog of pressure; the force per unit length exerted on a real or imaginary barrier separating an area of liquid or solid that is covered by a spreading substance from a clean area on the same liquid or solid. Also referred to as spreading pressure.

Surface Tension   The contracting force per unit length around the perimeter of a surface is usually referred to as surface tension if the surface separates gas from liquid or solid phases and interfacial tension if the surface separates two nongaseous phases. Although not strictly defined the same way, surface tension can be expressed in units of energy per unit surface area. For practical purposes, surface tension is frequently taken to reflect the change in surface free energy per unit increase in surface area.

Surfactant   Any substance that lowers the surface or interfacial tension of the medium in which it is dissolved. Soaps (fatty acid salts containing at least eight carbon atoms) are surfactants. Detergents are surfactants, or surfactant mixtures, whose solutions have cleaning properties. Also referred to as surface-active agents or tensides. In some usage surfactants are defined as molecules capable of associating to form micelles.

Surfactant Tail   The lyophobic portion of a surfactant molecule. It is commonly a hydrocarbon chain containing eight or more carbon atoms. *See also* Head Group.

Suspended Sediment   The insoluble particulate matter in natural water bodies such as rivers, lakes, and oceans.

Suspension   A system of solid particles dispersed in a liquid. Suspensions were previously referred to as suspensoids, meaning suspension colloids. Aside from the obvious definition of a colloidal suspension, a number of operational definitions are common in industry, such as any dispersed matter that can be removed by a 0.45 μm nominal pore size filter.

Suspension Effect   A finite potential, the Donnan potential, may exist between a suspension and its equilibrium solution. Also referred to as the Pallmann or Wiegner effect.

Svedberg   A unit of the sedimentation coefficient equal to 10^{-13} s.

Swamping Electrolyte   An excess of indifferent electrolyte that severely compresses electric double layers and minimizes the influence of electric charges borne by large molecules or dispersed colloidal species.

Sweep Flocculation   A mechansim of aggregation or flocculation in which particles are enmeshed by a coagulant matrix. The particles are aggregated not due to charge neutralization but rather to enmeshment. Example: the rapid precipitation of a metal hydroxide from supersaturated solution where the settling fluffy hydroxide particles trap and enmesh other suspended particles. *See* reference 23.

Swelling   The increase in volume associated with the uptake of liquid or gas by a solid or a gel.

Swelling Pressure   The pressure difference between a swelling material and the bulk of fluid being imbibed that is needed to prevent additional swelling. *See also* Swelling.

Swollen Micelle   *See* Micelle.

Syneresis   The spontaneous shrinking of a colloidal dispersion due to the release and exudation of some liquid. Frequently occurs in gels and foams but also occurs in flocculated suspensions. Mechanical syneresis refers to enhancing syneresis by the application of mechanical forces.

Tactoid In clay suspensions, the thin sheet-like or plate-like particles may aggregate to form stacks of particles in face-to-face orientation, which are termed tactoids. In coacervation processes tactoid has another meaning. *See* reference 2.

TDS Total dissolved solids.

TEM Transmission electron microscopy. *See* Electron Microscopy.

Tenside *See* Surfactant.

Tensiometer Any instrument used to measure surface and interfacial tension.

Tertiary Oil Recovery *See* Enhanced Oil Recovery.

Thermodynamic Stability In colloid science, the terms thermodynamically stable and metastable mean that a system is in a state of equilibrium corresponding to a local minimum of free energy (*1*). If several states of energy are accessible, the lowest is referred to as the stable state and the others are referred to as metastable states; unstable states are not at a local minimum. Most colloidal systems are metastable or unstable with respect to the separate bulk phases. *See also* Colloid Stability, Kinetic Stability.

Thickness of the Electric Double Layer *See* Electric Double-Layer Thickness.

Thixotropic Pseudoplastic flow that is time-dependent. At constant applied shear rate, viscosity gradually decreases, and in a flow curve hysteresis occurs. That is, after a given shear rate is applied and then reduced, it takes some time for the original dispersed species' alignments to be restored. Thixotropy in gels is sometimes termed reversible sol–gel transformation.

Tiselius Apparatus An apparatus for the determination of electrophoretic mobilities. *See* Moving Boundary Electrophoresis.

Tortuosity In porosimetry evaluations, experimental data tend to be interpreted in terms of a model in which the porous medium is taken to comprise a bundle of cylindrical pores having radius *r*. If the Young–Laplace equation is then applied to the data, an effective value of *r* can be calculated, even though this model ignores the real distribution of irregular channels. The calculated *r* value is sometimes considered to represent the radius of an equivalent cylinder or, alternatively, is termed the tortuosity.

Total Potential Energy of Interaction   *See* Gibbs Energy of Interaction.

Transitional Pore   An older term now replaced by mesopore. *See* Pore.

Transmission Electron Microscopy   (TEM) *See* Electron Microscopy.

Turbidimetry   *See* Turbidity.

Turbidity   The property of dispersions that causes a reduction in the transparency of the continuous phase due to light scattering and absorption. Turbidity is a function of the size and concentration of the dispersed species. The turbidity coefficient is simply the extinction coefficient in the Beer–Lambert equation for absorbance when light scattering rather than absorbance proper is being studied (hence turbidimetry). *See also* Nephelometry.

Turbulent Flow   A condition of flow in which all components of a fluid passing a certain point do not follow the same path. Turbulent flow refers to flow that is not laminar, or streamline.

Tyndall Beam   *See* Tyndall Scattering.

Tyndall Effect   When a sufficiently dilute dispersion of small particles or droplets is viewed directly against an illuminating light source, it will appear to be completely transparent. In contrast, when the same dispersion is viewed from the side (at a right angle to the illuminating beam) and against a dark background, the dispersion appears somewhat turbid and blue-white in color. The scattered light is due to Tyndall scattering and the optical effect is referred to as the Tyndall effect. *See* Tyndall Scattering.

Tyndall Scattering   A process that produces a colored beam of light scattered by uniform dispersion of particles whose size approaches the wavelength of the incident light. The scattered light is referred to as a Tyndall beam. The wavelength of scattered light varies with the angle of observation (Tyndall spectra), and this feature allows particle size to be calculated.

Ubbelohde Viscometer   *See* Capillary Viscometer.

Ultracentrifuge   *See* Centrifuge.

Ultrafiltrate   *See* Ultrafiltration.

Ultrafiltration   A separation process somewhat like dialysis in which a colloidal dispersion is separated from a noncolloidal solution by a semipermeable membrane, that is, a membrane that is permeable to all species except the colloidal-sized ones. Here an applied pressure (rather than osmotic pressure) across the membrane drives the separation. As in dialysis the solution containing the colloidal species is referred to as the retentate or dialysis residue. However, the solution that is free of colloidal species is referred to as ultrafiltrate rather than dialysate, because the composition is usually different from that produced by dialysis. Also referred to as hyperfiltration or reverse osmosis. *See also* Dialysis.

Ultramicroscope An optical microscope that uses dark-field illumination to make visible extremely small (submicrometer-sized) particles or droplets. Also termed Dark-Field Microscope. *See also* Dark-Field Illumination.

Ultrasonic Dispersion The use of ultrasound waves to achieve or aid in the dispersion of particles or droplets.

Upper Plastic Limit *See* Liquid Limit.

U.S. Bureau of Mines Wettability Test *See* Wettability Index.

van der Waals Adsorption An older term now replaced by physical adsorption or physisorption. *See also* Chemisorption.

van der Waals Forces *See* Dispersion Forces.

van der Waals–Hamaker Constant *See* Hamaker Constant.

Velocity Gradient A parameter that indicates the intensity of mixing. It is a function of the power input, the reactor volume, and the fluid viscosity. Higher velocity gradients are used in coagulation where the goal is to disperse the coagulant to the particle surfaces. Lower velocity gradients are used in flocculation where the goal is particle collisions and aggregation, and higher gradients would break up flocs.

Viscoelastic A liquid (or solid) with both viscous and elastic properties. A viscoelastic liquid will deform and flow under the influence of an applied shear stress, but when the stress is removed the liquid will slowly recover from some of the deformation.

Viscoelastometer An instrument for studying viscoelastic fluids. Viscoelastometers may be used to apply a constant shear stress so that the resulting deformation can be determined (creep curve) or to apply a sudden deformation and determine the stress needed to maintain the deformation (stress relaxation).

Viscoelectric Constant A reflection of the increase in viscosity of a liquid due to the presence of an electric field. It is given by the increase in viscosity divided by the viscosity in the absence of an electric field and divided also by the square of the electric-field gradient.

Viscometer Any instrument used in the measurement of viscosity. In most cases the term is applied to instruments capable of measuring only

Newtonian viscosity and not capable of non-Newtonian measurements. *See* Rheometer.

Viscosity   A measure of the resistance of a liquid to flow. It is properly the coefficient of viscosity and expresses the proportionality between shear stress and shear rate in Newton's law of viscosity. Many equations have been used to predict the viscosities of colloidal dispersions; *see* reference 2.

Viscosity Number   *See* Reduced Viscosity.

Viscosity Ratio   *See* Relative Viscosity.

Washburn Equation   An equation describing the extent of displacement of one fluid by another in a capillary tube or cylindrical pore in a porous medium. If h is the depth of penetration of invading fluid and dh/dt is the rate of penetration, then $dh/dt = \gamma r \cos \theta/(4\eta h)$, where γ is the interfacial tension, r is the capillary radius, θ is the contact angle, and η is the viscosity of the invading fluid. It is used in the evaluation of porosimetry data and may be used to provide information about contact angles, capillary radii, and pore radii, depending on the experiments conducted.

Wet   *See* Wettability.

Wet Oil   An oil containing free water or emulsified water.

Wettability   A qualitative term referring to the water- or oil-preferring nature of surfaces, such as mineral surfaces. Wettability may be determined by direct measurement of contact angles or inferred from measurements of fluid imbibition or relative permeabilities. Several conventions for describing wettability values exist. *See also* Amott Test, Contact Angle, Wettability Index, Wetting.

Wettability Index   A measure of wettability based on the U.S. Bureau of Mines wettability test in which the wettability index (W) is determined as the logarithm of the ratio of areas under the capillary pressure curves for both increasing and decreasing saturation of the wetting phase. Complete oil-wetting occurs for $W = -\infty$ (in practice about -1.5), and complete water-wetting occurs for $W = \infty$ (in practice about 1.0). Another wettability index is derived from the Amott–Harvey test. *See also* reference 8, Amott Test, Wettability.

Wetting   A general term referring to one or more of the following specific kinds of wetting: adhesional wetting, spreading wetting, and immersional wetting. Frequently used to denote that the contact angle between a liquid and a solid is essentially zero and there is spontaneous spreading of the liquid over the solid. Nonwetting, on the other hand, is frequently used to denote the case where the contact angle is greater than 90° so that the liquid rolls up into droplets. *See also* Contact Angle, Wettability.

Wetting Tension   The work done on a system during the process of immersional wetting, expressed per unit area of the phase being immersionally wetted. *See also* Immersional Wetting.

Wiegner Effect   *See* Suspension Effect.

Work of Adhesion   The energy of attraction between molecules in a phase. Defined as the work per unit area done on a system when two phases meeting at an interface of unit area are separated reversibly to form unit areas of new interfaces of each with a third phase.

Work of Cohesion   The work per unit area done on a system when a body of a phase is separated reversibly to form two bodies of the phase, each forming unit areas of new interfaces with a third phase.

Work of Immersional Wetting   *See* Wetting Tension.

Work of Separation   Synonym for the work of adhesion.

X-ray Diffraction   (XRD) A technique in which the scattering of X-rays by a crystal lattice is measured and used to determine the crystal's structure.

XRD   *See* X-ray Diffraction.

Yield Stress   For some fluids, the shear rate (flow) remains at zero until a threshold shear stress is reached; this is termed the yield stress. Beyond the yield stress flow begins. Also termed the yield value.

Yield Value   *See* Yield Stress.

Young's Equation   A fundamental relationship giving the balance of forces at a point of three-phase contact. For a gas–liquid–solid system, Young's equation is $\gamma_{SL} + \gamma_{LG} \cos \theta = \gamma_{SG}$, where γ_{SL}, γ_{LG}, and γ_{SG} are interfacial tensions between solid–liquid, liquid–gas, and solid–gas, respectively, and θ is the contact angle of the liquid with the solid, measured through the liquid.

Young–Laplace Equation   The fundamental relationship giving the pressure difference across a curved interface in terms of the surface or interfacial tension and the principal radii of curvature. In the special case of a spherical interface, the pressure difference is equal to twice the surface (or interfacial) tension divided by the radius of curvature. Also referred to as the equation of capillarity.

Zeolites   A class of aluminosilicate minerals having large cavities in their crystal structures. These allow ion exchange of large ions and also can permit the size-selective passage of organic molecules. They are used as ion exchangers, molecular sieves, and catalysts.

Zero Point of Charge   *See* Point of Zero Charge.

Zeta Potential   Strictly called the electrokinetic potential, the zeta potential refers to the potential drop across the mobile part of the electric double layer. Any species undergoing electrokinetic motion, such as electrophoresis, moves with a certain immobile part of the electric double layer that is assumed to be distinguished from the mobile part by a distinct plane, the shear plane. The zeta potential is the potential at that plane and is calculated from measured electrokinetic mobilities (e.g., electrophoretic mobility) or potentials (e.g., sedimentation potential) by using one of a number of available theories.

Zimm Plot   A graph used in the determination of root-mean-square end-to-end distances of dispersed species based on light-scattering data.

Zisman Plot   *See* Critical Surface Tension of Wetting.

References

1. *Manual of Symbols and Terminology for Physicochemical Quantities and Units, Appendix II;* Prepared by IUPAC Commission on Colloid and Surface Chemistry; Butterworths: London, 1972. *See also* the additions in *Pure Appl. Chem.* **1983**, *55*, 931–941; *Pure Appl. Chem.* **1985**, *57*, 603–619.
2. Schramm, L. L. *The Language of Colloid and Interface Science;* American Chemical Society: Washington, DC, 1993.
3. Williams, H. R.; Meyers, C. J. *Oil and Gas Terms,* 6th ed.; Matthew Bender: New York, 1984.
4. *A Dictionary of Petroleum Terms,* 2nd ed.; Petroleum Extension Service, University of Texas at Austin: Austin, TX, 1979.
5. *The Illustrated Petroleum Reference Dictionary,* 2nd ed.; Langenkamp, R. D., Ed.; PennWell Books: Tulsa, OK, 1982.
6. *McGraw-Hill Dictionary of Scientific and Technical Terms,* 3rd ed.; Parker, S. P., Ed.; McGraw-Hill: New York, 1984.
7. van Olphen, H. *An Introduction to Clay Colloid Chemistry,* 2nd ed.; Wiley-Interscience: New York, 1977.
8. Anderson, W. G. *J. Pet. Technol.* **1986**, *38(12)*, 1246–1262.
9. *Glossary of Soil Science Terms;* Soil Science Society of America: Madison, WI, 1987.
10. *Glossary of Terms in Soil Science;* Research Branch, Agriculture Canada: Ottawa, Canada; revised 1976.
11. Martinez, A. R. In *The Future of Heavy Crude and Tar Sands;* Meyer, R. F.; Wynn, J. C.; Olson, J. C., Eds.; United Nations Institute for Training and Research: New York, 1982; pp ixvii–ixviii.
12. Danyluk, M.; Galbraith, B.; Omana, R. In *The Future of Heavy Crude and Tar Sands;* Meyer, R. F.; Wynn, J. C.; Olson, J. C., Eds.; United Nations Institute for Training and Research: New York, 1982; pp 3–6.

13. Khayan, M. In *The Future of Heavy Crude and Tar Sands;* Meyer, R. F.; Wynn, J. C.; Olson, J. C., Eds.; United Nations Institute for Training and Research: New York, 1982; pp 7–11.

14. Kukla, R. J. *Chem. Eng. Prog.* **1991,** *87,* 23–35.

15. Myers, D. *Surfactant Science and Technology;* VCH: New York, 1988.

16. Rosen, M. J. *Surfactants and Interfacial Phenomena;* Wiley: New York, 1978.

17. Donini, J. C.; Angle, C. W.; Hassan, T. A.; Kasperski, K. L.; Kan, J.; Kar, K. L.; Thind, S. S. In *Emerging Separation Technologies for Metals and Fuels;* Lakshmanan, V. I.; Bautista, R. G.; Somasundaran, P., Eds.; The Minerals, Metals & Materials Society: Warrendale, PA, 1993; pp 409–424.

18. Gast, A. P.; Zukoski, C. F. *Adv. Colloid Interface Sci.* **1989,** *30,* 153–202.

19. Almdal, K.; Dyre, J.; Hvidt, S.; Kramer, O. *Polym. Gels Networks* **1993,** *1,* 5–17.

20. Harwell, J. H.; Hoskins, J. C.; Schechter, R. S.; Wade, W. H. *Langmuir* **1985,** *1(2),* 251–262.

21. Fuerstenau, D. W.; Williams, M. C. *Colloids Surf.* **1987,** *22,* 87–91.

22. Schramm, L. L.; Clark, B. W. *Colloids Surf.* **1983,** *7,* 135–146.

23. Halverson, F.; Panzer, H. P. In *Kirk–Othmer Encyclopedia of Chemical Technology,* 3rd ed.; Wiley: New York, 1980; pp 489–523.

24. van den Hul, H. J.; Vanderhoff, J. W. In *Polymer Colloids;* Fitch, R. M., Ed.; Plenum: New York, 1971; pp 1–27.

25. Whorlow, R. W. *Rheological Techniques;* Ellis Horwood: Chichester, England, 1980.

26. Scholle, P. A. *Constituents, Textures, Cements, and Porosities of Sandstones and Associated Rocks;* Memoir 28; American Assocciation of Petroleum Geologists: Tulsa, OK, 1979.

27. Blatt, H.; Middleton, G.; Murray, R. *Origin of Sedimentary Rocks,* 2nd ed.; Prentice-Hall: Englewood Cliffs, NJ, 1980.

28. Kosmulski, M.; Matijević, E. *J. Colloid Interface Sci.* **1992,** *150(1),* 291–294.

29. Capes, C. E.; McIlhinney, A. E.; Sirianni, A. F. In *Agglomeration 77;* American Institute of Mining, Metallurgical, and Petroleum Engineers: New York, NY, 1977; pp 910–930.

30. Luckham, P. F.; de Costello, B. A. *Adv. Colloid Interface Sci.* **1993,** *44,* 183–240.

RECEIVED for review July 5, 1994. ACCEPTED revised manuscript February 3, 1995.

Author Index

Affiliation Index

Subject Index

Bestsellers from ACS Books

The ACS Style Guide: A Manual for Authors and Editors
Edited by Janet S. Dodd
264 pp; clothbound ISBN 0–8412–0917–0; paperback ISBN 0–8412–0943–X

The Basics of Technical Communicating
By B. Edward Cain
ACS Professional Reference Book; 198 pp;
clothbound ISBN 0–8412–1451–4; paperback ISBN 0–8412–1452–2

Chemical Activities (student and teacher editions)
By Christie L. Borgford and Lee R. Summerlin
330 pp; spiralbound ISBN 0–8412–1417–4; teacher ed. ISBN 0–8412–1416–6

Chemical Demonstrations: A Sourcebook for Teachers,
Volumes 1 and 2, Second Edition
Volume 1 by Lee R. Summerlin and James L. Ealy, Jr.;
Vol. 1, 198 pp; spiralbound ISBN 0–8412–1481–6;
Volume 2 by Lee R. Summerlin, Christie L. Borgford, and Julie B. Ealy
Vol. 2, 234 pp; spiralbound ISBN 0–8412–1535–9

Chemistry and Crime: From Sherlock Holmes to Today's Courtroom
Edited by Samuel M. Gerber
135 pp; clothbound ISBN 0–8412–0784–4; paperback ISBN 0–8412–0785–2

Writing the Laboratory Notebook
By Howard M. Kanare
145 pp; clothbound ISBN 0–8412–0906–5; paperback ISBN 0–8412–0933–2

Developing a Chemical Hygiene Plan
By Jay A. Young, Warren K. Kingsley, and George H. Wahl, Jr.
paperback ISBN 0–8412–1876–5

Introduction to Microwave Sample Preparation: Theory and Practice
Edited by H. M. Kingston and Lois B. Jassie
263 pp; clothbound ISBN 0–8412–1450–6

Principles of Environmental Sampling
Edited by Lawrence H. Keith
ACS Professional Reference Book; 458 pp;
clothbound ISBN 0–8412–1173–6; paperback ISBN 0–8412–1437–9

Biotechnology and Materials Science: Chemistry for the Future
Edited by Mary L. Good (Jacqueline K. Barton, Associate Editor)
135 pp; clothbound ISBN 0–8412–1472–7; paperback ISBN 0–8412–1473–5

For further information and a free catalog of ACS books, contact:
American Chemical Society
Distribution Office, Department 225
1155 16th Street, NW, Washington, DC 20036
Telephone 800–227–5558

Highlights from ACS Books

Good Laboratory Practice Standards: Applications for Field and Laboratory Studies
Edited by Willa Y. Garner, Maureen S. Barge, and James P. Ussary
ACS Professional Reference Book; 572 pp; clothbound ISBN 0–8412–2192–8

Silent Spring Revisited
Edited by Gino J. Marco, Robert M. Hollingworth, and William Durham
214 pp; clothbound ISBN 0–8412–0980–4; paperback ISBN 0–8412–0981–2

The Microkinetics of Heterogeneous Catalysis
By James A. Dumesic, Dale F. Rudd, Luis M. Aparicio, James E. Rekoske,
and Andrés A. Treviño
ACS Professional Reference Book; 316 pp; clothbound ISBN 0–8412–2214–2

Helping Your Child Learn Science
By Nancy Paulu with Margery Martin; Illustrated by Margaret Scott
58 pp; paperback ISBN 0–8412–2626–1

Handbook of Chemical Property Estimation Methods
By Warren J. Lyman, William F. Reehl, and David H. Rosenblatt
960 pp; clothbound ISBN 0–8412–1761–0

Understanding Chemical Patents: A Guide for the Inventor
By John T. Maynard and Howard M. Peters
184 pp; clothbound ISBN 0–8412–1997–4; paperback ISBN 0–8412–1998–2

Spectroscopy of Polymers
By Jack L. Koenig
ACS Professional Reference Book; 328 pp;
clothbound ISBN 0–8412–1904–4; paperback ISBN 0–8412–1924–9

Harnessing Biotechnology for the 21st Century
Edited by Michael R. Ladisch and Arindam Bose
Conference Proceedings Series; 612 pp;
clothbound ISBN 0–8412–2477–3

From Caveman to Chemist: Circumstances and Achievements
By Hugh W. Salzberg
300 pp; clothbound ISBN 0–8412–1786–6; paperback ISBN 0–8412–1787–4

The Green Flame: Surviving Government Secrecy
By Andrew Dequasie
300 pp; clothbound ISBN 0–8412–1857–9

For further information and a free catalog of ACS books, contact:
American Chemical Society
Distribution Office, Department 225
1155 16th Street, NW, Washington, DC 20036
Telephone 800–227–5558